Scanning Electron Microscopy and X-Ray Microanalysis

Joseph I. Goldstein
Dale E. Newbury
Joseph R. Michael
Nicholas W.M. Ritchie
John Henry J. Scott
David C. Joy

Scanning Electron Microscopy and X-Ray Microanalysis

Fourth Edition

 Springer

Joseph I. Goldstein
University of Massachusetts
Amherst, MA, USA

Joseph R. Michael
Sandia National Laboratories
Albuquerque, NM, USA

John Henry J. Scott
National Institute of Standards and Technology
Gaithersburg, MD, USA

Dale E. Newbury
National Institute of Standards and Technology
Gaithersburg, MD, USA

Nicholas W.M. Ritchie
National Institute of Standards and Technology
Gaithersburg, MD, USA

David C. Joy
University of Tennessee
Knoxville, TN, USA

ISBN 978-1-4939-6674-5   ISBN 978-1-4939-6676-9   (eBook)
https://doi.org/10.1007/978-1-4939-6676-9

Library of Congress Control Number: 2017943045

Printed on acid-free paper

This Springer imprint is published by Springer Nature
The registered company is Springer Science+Business Media LLC
The registered company address is: 233 Spring Street, New York, NY 10013, U.S.A.

Preface

This is not your father's, your mother's, or your grandparent's Scanning Electron Microscopy and X-Ray Microanalysis (SEMXM). But that is not to say that there is no continuity or to deny a family resemblance. SEMXM4 is the fourth in the series of textbooks with this title, and continues a tradition that extends back to the "zero-th edition" in 1975 published under the title, "Practical Scanning Electron Microscopy" (Plenum Press, New York). However, the latest edition differs sharply from its predecessors, which attempted an encyclopedic approach to the subject by providing extensive details on how the SEM and its associated devices actually work, for example, electron sources, lenses, electron detectors, X-ray spectrometers, and so on.

In constructing this new edition, the authors have chosen a different approach. Modern SEMs and the associated X-ray spectrometry and crystallography measurement functions operate under such extensive computer control and automation that it is actually difficult for the microscopist-microanalyst to interact with the instrument except within carefully prescribed boundaries. Much of the flexibility of parameter selection that early instruments provided has now been lost, as instrumental operation functions have been folded into software control. Thus, electron sources are merely turned "on," with the computer control optimizing the operation, or for the thermally assisted field emission gun, the electron source may be permanently "on." The user can certainly adjust the lenses to focus the image, but this focusing action often involves complex interactions of two or more lenses, which formerly would have required individual adjustment. Moreover, the nature of the SEM field has fundamentally changed. What was once a very specialized instrument system that required a high level of training and knowledge on the part of the user has become much more of a routine tool. The SEM is now simply one of a considerable suite of instruments that can be employed to solve problems in the physical and biological sciences, in engineering, in technology, in manufacturing and quality control, in failure analysis, in forensic science, and other fields.

The authors also recognize the profound changes that have occurred in the manner in which people obtain information. The units of SEMXM4, whether referred to as chapters or modules, are meant to be relatively self-contained. Our hope is that a reader seeking specific information can select a topic from the list and obtain a good understanding of the topic from that module alone. While each topic is supported by information in other modules, we acknowledge the likelihood that not all users of SEMXM4 will "read it all." This approach inevitably leads to a degree of overlap and repetition since similar information may appear in two or more places, and this is entirely intentional.

In recognition of these fundamental changes, the authors have chosen to modify SEMXM4 extensively to provide a guide on the actual use of the instrument without overwhelming the reader with the burden of details on the physics of the operation of the instrument and its various attachments. Our guiding principle is that the microscopist-microanalyst must understand which parameters can be adjusted and what is an effective strategy to select those parameters to solve a particular problem. The modern SEM is an extraordinarily flexible tool, capable of operating over a wide range of electron optical parameters and producing images from electron detectors with different signal characteristics. Those users who restrict themselves to a single set of operating parameters may be able to solve certain problems, but they may never know what they are missing by not exploring the range of parameter space available to them. SEMXM4 seeks to provide sufficient understanding of the technique for a user to become a competent and efficient problem solver. That is not to say that there are only a few things to learn. To help the reader to approach the considerable body of knowledge needed to operate at a high degree of competency, a new feature of SEMXM-4 is the summary checklist provided for each of the major areas of operation: SEM imaging, elemental X-ray microanalysis, and backscatter-diffraction crystallography.

Readers familiar with earlier editions of SEMXM will notice the absence of the extensive material previously provided on specimen preparation. Proper specimen preparation is a critical step in solving most problems, but with the vast range of applications to materials of diverse character, the topic of specimen preparation itself has become the subject of entire books, often devoted to just one specialized area.

Throughout their history, the authors of the SEMXM textbooks have been closely associated as lecturers with the Lehigh University Summer Microscopy School. The opportunity to teach and interact with each year's students has provided a very useful experience in understanding the community of users of the technique and its evolution over time. We hope that these interactions have improved our written presentation of the subject as a benefit to newcomers as well as established practitioners.

Finally, the author team sadly notes the passing in 2015 of Professor Joseph I. Goldstein (University of Massachusetts, Amherst) who was the "founding father" of the Lehigh University Summer Microscopy School in 1970, and who organized and contributed so extensively to the microscopy courses and to the SEMXM textbooks throughout the ensuing 45 years. Joe provided the stimulus to the production of SEMXM4 with his indefatigable spirit, and his technical contributions are embedded in the X-ray microanalysis sections.

Dale E. Newbury
Nicholas W.M. Ritchie
John Henry J. Scott
Gaithersburg, MD, USA

Joseph R. Michael
Albuquerque, NM, USA

David C. Joy
Knoxville, TN, USA

The original version of this book was revised. Index has been updated.

Scanning Electron Microscopy and Associated Techniques: Overview

Imaging Microscopic Features

The scanning electron microscope (SEM) is an instrument that creates magnified images which reveal microscopic-scale information on the size, shape, composition, crystallography, and other physical and chemical properties of a specimen. The principle of the SEM was originally demonstrated by Knoll (1935; Knoll and Theile 1939) with the first true SEM being developed by von Ardenne (1938). The modern commercial SEM emerged from extensive development in the 1950s and 1960s by Prof. Sir Charles Oatley and his many students at the University of Cambridge (Oatley 1972). The basic operating principle of the SEM involves the creation of a finely focused beam of energetic electrons by means of emission from an electron source. The energy of the electrons in this beam, E_0, is typically selected in the range from $E_0 = 0.1$ to 30 keV). After emission from the source and acceleration to high energy, the electron beam is modified by apertures, magnetic and/or electrostatic lenses, and electromagnetic coils which act to successively reduce the beam diameter and to scan the focused beam in a raster (x-y) pattern to place it sequentially at a series of closely spaced but discrete locations on the specimen. At each one of these discrete locations in the scan pattern, the interaction of the electron beam with the specimen produces two outgoing electron products: (1) backscattered electrons (BSEs), which are beam electrons that emerge from the specimen with a large fraction of their incident energy intact after experiencing scattering and deflection by the electric fields of the atoms in the sample; and (2) secondary electrons (SEs), which are electrons that escape the specimen surface after beam electrons have ejected them from atoms in the sample. Even though the beam electrons are typically at high energy, these secondary electrons experience low kinetic energy transfer and subsequently escape the specimen surface with very low kinetic energies, in the range 0–50 eV, with the majority below 5 eV in energy. At each beam location, these outgoing electron signals are measured using one or more electron detectors, usually an Everhart–Thornley "secondary electron" detector (which is actually sensitive to both SEs and BSEs) and a "dedicated backscattered electron detector" that is insensitive to SEs. For each of these detectors, the signal measured at each individual raster scan location on the sample is digitized and recorded into computer memory, and is subsequently used to determine the gray level at the corresponding X-Y location of a computer display screen, forming a single picture element (or pixel). In a conventional-vacuum SEM, the electron-optical column and the specimen chamber must operate under high vacuum conditions ($<10^{-4}$ Pa) to minimize the unwanted scattering that beam electrons as well as the BSEs and SEs would suffer by encountering atoms and molecules of atmospheric gasses. Insulating specimens that would develop surface electrical charge because of impact of the beam electrons must be given a conductive coating that is properly grounded to provide an electrical discharge path. In the variable pressure SEM (VPSEM), specimen chamber pressures can range from 1 Pa to 2000 Pa (derived from atmospheric gas or a supplied gas such as water vapor), which provides automatic discharging of uncoated insulating specimens through the ionized gas atoms and free electrons generated by beam, BSE, and SE interactions. At the high end of this VPSEM pressure range with modest specimen cooling (2–5 °C), water can be maintained in a gas–liquid equilibrium, enabling direct examination of wet specimens.

SEM electron-optical parameters can be optimized for different operational modes:

1. A small beam diameter can be selected for high spatial resolution imaging, with extremely fine scale detail revealed by possible imaging strategies employing high beam energy, for example, ◘ Fig. 1a ($E_0 = 15$ keV) and low beam energy, ◘ Fig. 1b ($E_0 = 0.8$ keV), ◘ Fig. 1c ($E_0 = 0.5$ keV), and ◘ Fig. 1d ($E_0 = 0.3$ keV). However, a negative consequence of choosing a small beam size is that the beam current is reduced as the inverse square of the beam diameter. Low beam current means that visibility is compromised for features that produce weak contrast.

2. A high beam current improves visibility of low contrast objects (e.g., ◘ Fig. 2). For any combination of beam current, pixel dwell time, and detector efficiency there is always a threshold contrast below which features of the specimen will not be visible. This threshold contrast depends on the relative size and shape of the feature of interest. The visibility of large objects and extended linear objects persists when small objects have dropped below the visibility

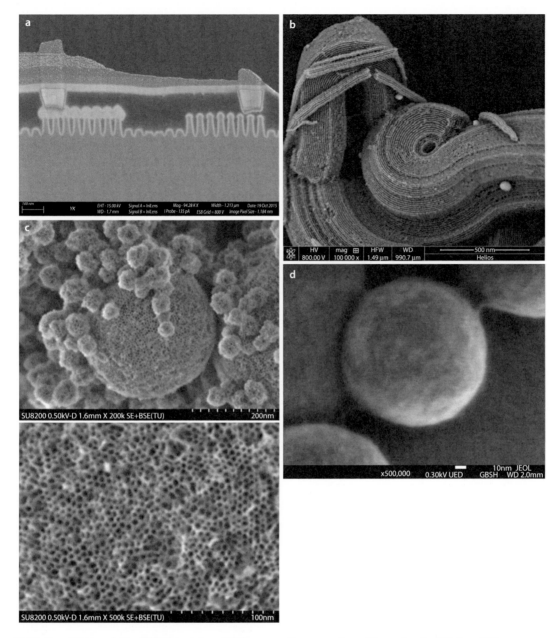

Fig. 1 **a** High resolution SEM image taken at high beam energy ($E_0 = 15$ keV) of a finFET transistor (16-nm technology) using an in-lens secondary electron detector. This cross section was prepared by inverted Ga FIB milling from backside (Zeiss Auriga Cross beam; image courtesy of John Notte, Carl Zeiss); Bar = 100 nm. **b** High resolution SEM image taken at low beam energy ($E_0 = 0.8$ keV) of zeolite (uncoated) using a through-the-lens SE detector (image courtesy of Trevan Landin, FEI); Bar = 500 nm. **c** Mesoporous silica nanosphere showing 5-nm-diameter pores imaged with a landing energy of 0.5 keV (specimen courtesy of T. Yokoi, Tokyo Institute of Technology; images courtesy of A. Muto, Hitachi High Technologies); Upper image Bar = 200 nm, Lower image Bar = 100 nm. **d** Si nanoparticle imaged with a landing energy of 0.3 keV; Bar = 10 nm (image courtesy V. Robertson, JEOL)

threshold. This threshold can only be lowered by increasing beam current, pixel dwell time, and/or detector efficiency. Selecting higher beam current means a larger beam size, causing resolution to deteriorate. Thus, there is a dynamic contest between resolution and visibility leading to inevitable limitations on feature size and feature visibility that can be achieved.

3. The beam divergence angle can be minimized to increase the depth-of-field (e.g., ◘ Fig. 3). With optimized selection of aperture size and specimen-to-objective lens distance (working distance), it is generally possible to achieve small beam convergence angles and therefore effective focus along the beam axis that is at least equal to the horizontal width of the image.

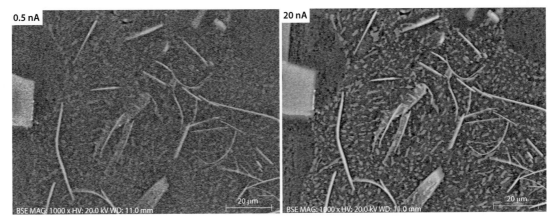

■ **Fig. 2** Effect of increasing beam current (at constant pixel dwell time) to improve visibility of low contrast features. Al-Si eutectic alloy; $E_0 = 20$ keV; semiconductor BSE detector (sum mode): (*left*) 0.5 nA; (*right*) 20 nA; Bar = 20 μm

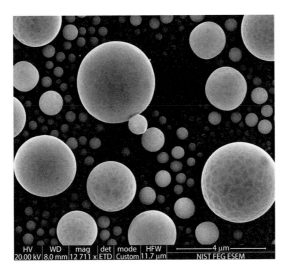

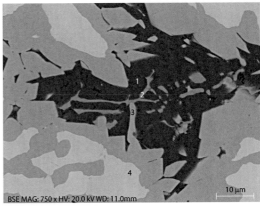

■ **Fig. 3** Large the depth-of-focus; Sn spheres; $E_0 = 20$ keV; Everhart–Thornley (positive bias) detector; Bar = 4 μm (Scott Wight, NIST)

■ **Fig. 4** Atomic number contrast with backscattered electrons. Raney nickel alloy, polished cross section; $E_0 = 20$ keV; semiconductor BSE (sum mode) detector. Note that four different phases corresponding to different compositions can be discerned; Bar = 10 μm

A negative consequence of using a small aperture to reduce the convergence/divergence angle is a reduction in beam current.

Vendor software supports collection, dynamic processing, and interpretation of SEM images, including extensive spatial measurements. Open source software such as ImageJ-Fiji, which is highlighted in this textbook, further extends these digital image processing capabilities and provides the user access to a large microscopy community effort that supports advanced image processing.

General specimen property information that can be obtained from SEM images:

1. Compositional microstructure (e.g., ■ Fig. 4). Compositional variations of 1 unit difference in average atomic number (Z) can be observed generally with BSE detection, with even greater sensitivity ($\Delta Z = 0.1$) for low ($Z = 6$) and intermediate ($Z = 30$) atomic numbers. The lateral spatial resolution is generally limited to approximately 10–100 nm depending on the specimen composition and the beam energy selected.

2. Topography (shape) (e.g., ■ Fig. 5). Topographic structure can be imaged with variations in local surface inclination as small as a few degrees. The edges of structures can be localized with a spatial resolution ranging from the incident beam diameter (which can be 1 nm or less, depending on the electron source) up to 10 nm or greater, depending on the material and the geometric nature of the edge (vertical, rounded, tapered, re-entrant, etc.).

3. Visualizing the third dimension (e.g., ■ Fig. 6). Optimizing for a large depth-of-field permits visualizing the three-dimensional structure of a specimen. However, in conventional *X-Y* image presentation, the resulting image is a projection of the three dimensional information onto a two dimensional plane, suffering

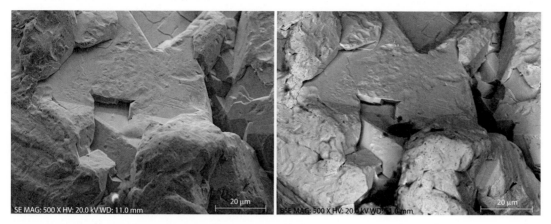

SE MAG: 500 X HV: 20.0 kV WD: 11.0 mm   20 µm   BSE MAG: 500 X HV: 20.0 kV WD: 11.0 mm   20 µm

◘ Fig. 5 Topographic contrast as viewed with different detectors: Everhart–Thornley (positive bias) and semiconductor BSE (sum mode); silver crystals; $E_0 = 20$ keV; Bar = 20 µm

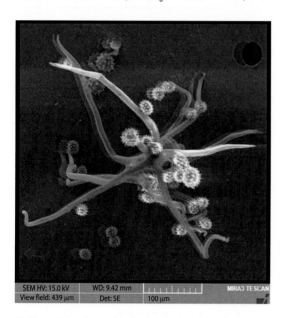

SEM HV: 15.0 kV   WD: 9.42 mm   MIRA3 TESCAN
View field: 439 µm   Det: SE   100 µm

◘ Fig. 6 Visualizing the third dimension. Anaglyph stereo pair (red filter over left eye) of pollen grains on plant fibers; $E_0 = 15$ keV; coated with Au-Pd; Bar = 100 µm

spatial distortion due to foreshortening. The true three-dimensional nature of the specimen can be recovered by applying the techniques of stereomicroscopy, which invokes the natural human visual process for stereo imaging by combining two independent views of the same area made with small angular differences.

4. Other properties which can be accessed by SEM imaging: (1) crystal structure, including grain boundaries, crystal defects, and crystal deformation effects (e.g., ◘ Fig. 8); (2) magnetic microstructure, including magnetic domains and interfaces; (3) applied electrical fields in engineered microstructures; (4) electron-stimulated optical emission (cathodoluminescence), which is sensitive to low energy electronic structure.

Measuring the Elemental Composition

The beam interaction with the specimen produces two types of X-ray photon emissions which compose the X-ray spectrum: (1) characteristic X-rays, whose specific energies provide a fingerprint that is specific to each element, with the exception of H and He, which do not emit X-rays; and (2) continuum X-rays, which occur at all photon energies from the measurement threshold to E_0 and form a background beneath the characteristic X-rays. This X-ray spectrum can be used to identify and quantify the specific elements (excepting H and He, which do not produce X-rays) present within the beam-excited interaction volume, which has dimensions ranging from approximately 100 nm to 10 µm depending on composition and beam energy, over a wide range of concentrations (C, expressed in mass fraction):

"Major constituent": $0.1 < C \leq 1$
"Minor constituent": $0.01 \leq C \leq 0.1$
"Trace constituent": $C < 0.01$

The X-ray spectrum is measured with the semiconductor energy dispersive X-ray spectrometer (EDS), which can detect photons from a threshold of approximately 40 eV to E_0 (which can be as high as 30 keV). Vendor software supports collection and analysis of spectra, and these tools can be augmented significantly with the open source software National Institute of Standards and Technology DTSA II for quantitative spectral processing and simulation, discussed in this textbook.

Analytical software supports qualitative X-ray microanalysis which involves assigning the characteristic peaks recognized in the spectrum to specific elements. Qualitative analysis presents significant challenges because of mutual peak interferences that can occur between certain combinations of elements, for example, Ti and Ba; S, Mo, and Pb; and many others, especially when the peaks of major constituents interfere with the peaks of minor or trace constituents. Operator knowledge of the physical rules governing X-ray generation and detection is needed to perform a careful review of software-generated peak identifications, and this careful review must always be performed to achieve a robust measurement result.

After a successful qualitative analysis has been performed, quantitative analysis can proceed. The characteristic intensity for each peak is automatically determined by peak fitting procedures, such as the multiple linear least squares method. The intensity measured for each element is proportional to the concentration of that element, but that intensity is also modified by all other elements present in the interaction volume through their influence on the electron scattering and retardation ("atomic number" matrix effect, Z), X-ray absorption within the specimen ("absorption" matrix effect, A), and X-ray generation induced by absorption of X-rays ("secondary fluorescence" matrix effects, F, induced by characteristic X-rays and c, induced by continuum X-rays). The complex physics of these "ZAFc" matrix corrections has been rendered into algorithms by a combined theoretical and empirical approach. The basis of quantitative electron-excited X-ray microanalysis is the "k-ratio protocol": measurement under identical conditions (beam energy, known electron dose, and spectrometer performance) of the characteristic intensities for all elements recognized in the unknown spectrum against a suite of standards containing those same elements, producing a set of k-ratios, where

$$k = I_{Unknown} / I_{Standard} \qquad (1)$$

for each element in the unknown. Standards are materials of known composition that are tested to be homogeneous at the microscopic scale, and preferably homogeneous at the nanoscale. Standards can be as simple as pure elements—e.g., C, Al, Si, Ti, Cr, Fe, Ni, Cu, Ag, Au, etc.—but for those elements that are not stable in a vacuum (e.g., gaseous elements such as O) or which degrade during electron bombardment (e.g., S, P, and Ga), stable stoichiometric compounds can be used instead, e.g., MgO for O; FeS_2 for S; and GaP for Ga and P. The most accurate analysis is performed with standards measured on the same instrument as the unknown(s), ideally in the same measurement campaign, although archived standard spectra can be effectively used if a quality measurement program is implemented to ensure the constancy of measurement conditions, including spectrometer performance parameters. With such a standards-based measurement protocol and ZAFc matrix corrections, the accuracy of the analysis can be expressed as a relative deviation from expected value (RDEV):

$$RDEV(\%) = [(Measured - True) / True] \times 100\% \qquad (2)$$

Based on extensive testing of homogeneous binary and multiple component compositions, the distribution of RDEV values for major constituents is such that a range of ±5 % relative captures 95 % of all analyses. The use of stable, high integrated count spectra (>1 million total counts from threshold to E_0) now possible with the silicon drift detector EDS (SDD-EDS), enables this level of accuracy to be achieved for major and minor constituents even when severe peak interference occurs and there is also a large concentration ratio, for example, a major constituent interfering with a minor constituent. Trace constituents that do not suffer severe interference can be measured to limits of detection as low as $C = 0.0002$ (200 parts per million) with spectra containing >10 million counts. For interference situations, much higher count spectra (>100 million counts) are required.

An alternative "standardless analysis" protocol uses libraries of standard spectra ("remote standards") measured on a different SEM platform with a similar EDS spectrometer, ideally over a wide range of beam energy and detector parameters (resolution). These library spectra are then adjusted to the local measurement conditions through comparison of one or more key spectra (e.g., locally measured spectra of particular elements such as Si and Ni). Interpolation/extrapolation is used to supply estimated spectral intensities for elements not present in or at a beam energy not represented in the library elemental suite. Testing of the standardless analysis method has shown that an RDEV range of ±25 % relative is needed to capture 95 % of all analyses.

High throughput (>100 kHz) EDS enables collection of X-ray intensity maps with gray scale representation of different concentration levels (e.g., ◘ Fig. 7a). Compositional mapping by spectrum imaging (SI) collects a full EDS spectrum at each pixel of an x-y array, and after applying the quantitative analysis procedure at each pixel, images are created for each element where the gray (or color) level is assigned based on the measured concentration (e.g., ◘ Fig. 7b).

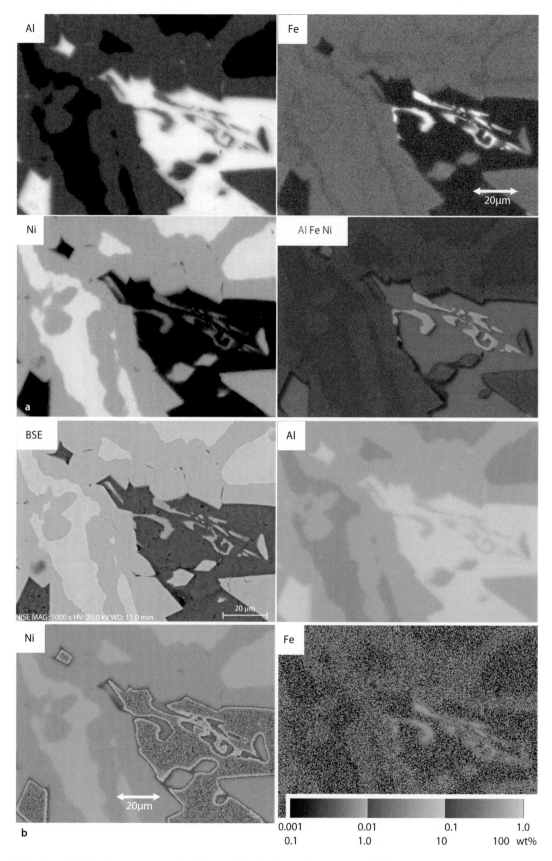

◘ **Fig. 7**   **a** EDS X-ray intensity maps for Al, Fe, and Ni and color overlay; Raney nickel alloy; $E_0 = 20$ keV. **b** SEM/BSE (sum) image and compositional maps corresponding to **a**

Measuring the Crystal Structure

An electron beam incident on a crystal can undergo electron channeling in a shallow near-surface layer which increases the initial beam penetration for certain orientations of the beam relative to the crystal planes. The additional penetration results in a slight reduction in the electron backscattering coefficient, which creates weak crystallographic contrast (a few percent) in SEM images by which differences in local crystallographic orientation can be directly observed: grain boundaries, deformations bands, and so on (e.g., ◻ Fig. 8).

The backscattered electrons exiting the specimen are subject to crystallographic diffraction effects, producing small modulations in the intensities scattered to different angles that are superimposed on the overall angular distribution that an amorphous target would produce. The resulting "electron backscatter diffraction (EBSD)" pattern provides extensive information on the local orientation, as shown in ◻ Fig. 8b for a crystal of hematite. EBSD pattern angular separations provide measurements of the crystal plane spacing, while the overall EBSD pattern reveals symmetry elements. This crystallographic information combined with elemental analysis information obtained simultaneously from the same specimen region can be used to identify the crystal structure of an unknown.

Dual-Beam Platforms: Combined Electron and Ion Beams

A "dual-beam" instrument combines a fully functional SEM with a focused ion beam (FIB), typically gallium or argon. This combination provides a flexible platform for *in situ* specimen modification through precision ion beam milling and/or ion beam mediated material deposition with sequential or simultaneous electron beam technique characterization of the newly revealed specimen surfaces. Precision material removal enables detailed study of the third dimension of a specimen with nanoscale resolution along the depth axis. An example of ion beam milling of a directionally solidified Al-Cu is shown in ◻ Fig. 9, as imaged with the SEM column on the dual-beam instrument. Additionally, ion-beam induced secondary electron emission provides scanning ion microscopy (SIM) imaging to complement SEM imaging. For imaging certain specimen properties, such as crystallographic structure, SIM produces stronger contrast than SEM. There is also an important class of stand-alone SIM instruments, such as the helium ion microscope (HIM), that are optimized for high resolution/high depth-of-field imaging performance (e.g., the same area as viewed by HIM is also shown in ◻ Fig. 9).

BSE MAG: 400 x HV: 20.0 kV WD: ↑1.0 mm   40 μm

◻ **Fig. 8** **a** Electron channeling contrast revealing grain boundaries in Ti-alloy (nominal composition: Ti-15Mo-3Nb-3Al-0.2Si); $E_0 = 20$ keV. **b** Electron backscatter diffraction (EBSD) pattern from hematite at $E_0 = 40$ keV

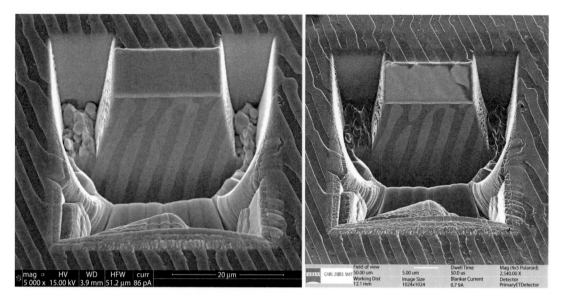

▣ Fig. 9 Directionally-solidified Al-Cu eutectic alloy after ion beam milling in a dual-beam instrument, as imaged by the SEM column (*left image*); same region imaged in the HIM (*right image*)

Modeling Electron and Ion Interactions

An important component of modern Scanning Electron Microscopy and X-ray Microanalysis is modeling the interaction of beam electrons and ions with the atoms of the specimen and its environment. Such modeling supports image interpretation, X-ray microanalysis of challenging specimens, electron crystallography methods, and many other issues. Software tools for this purpose, including Monte Carlo electron trajectory simulation, are discussed within the text. These tools are complemented by the extensive database of Electron-Solid Interactions (e.g., electron scattering and ionization cross sections, secondary electron and backscattered electron coefficients, etc.), developed by Prof. David Joy, can be found in chapter 3 on SpringerLink: http://link.springer.com/chapter/10.1007/978-1-4939-6676-9_3.

References

Knoll M (1935) Static potential and secondary emission of bodies under electron radiation. Z Tech Physik 16:467

Knoll M, Theile R (1939) Scanning electron microscope for determining the topography of surfaces and thin layers. Z Physik 113:260

Oatley C (1972) The scanning electron microscope: part 1, the instrument. Cambridge University Press, Cambridge

von Ardenne M (1938) The scanning electron microscope. Theoretical fundamentals. Z Physik 109:553

Contents

Electron Beam—Specimen Interactions: Interaction Volume

© Springer Science+Business Media LLC 2018
J. Goldstein et al., *Scanning Electron Microscopy and X-Ray Microanalysis*,
https://doi.org/10.1007/978-1-4939-6676-9_1

1.1 What Happens When the Beam Electrons Encounter Specimen Atoms?

By selecting the operating parameters of the SEM electron gun, lenses, and apertures, the microscopist controls the characteristics of the focused beam that reaches the specimen surface: energy (typically selected in the range 0.1–30 keV), diameter (0.5 nm to 1 µm or larger), beam current (1 pA to 1 µA), and convergence angle (semi-cone angle 0.001–0.05 rad). In a conventional high vacuum SEM (typically with the column and specimen chamber pressures reduced below 10^{-3} Pa), the residual atom density is so low that the beam electrons are statistically unlikely to encounter any atoms of the residual gas along the flight path from the electron source to the specimen, a distance of approximately 25 cm.

■■ The initial dimensional scale

With a cold or thermal field emission gun on a high-performance SEM, the incident beam can be focused to 1 nm in diameter, which means that for a target such as gold (atom diameter ~ 288 pm), there are approximately 12 gold atoms in the first atomic layer of the solid within the area of the beam footprint at the surface.

At the specimen surface the atom density changes abruptly to the very high density of the solid. The beam electrons interact with the specimen atoms through a variety of physical processes collectively referred to as "scattering events." The overall effects of these scattering events are to transfer energy to the specimen atoms from the beam electrons, thus setting a limit on their travel within the solid, and to alter the direction of travel of the beam electrons away from the well-defined incident beam trajectory. These beam electron–specimen interactions produce the backscattered electrons (BSE), secondary electrons (SE), and X-rays that convey information about the specimen, such as coarse- and fine-scale topographic features, composition, crystal structure, and local electrical and magnetic fields. At the level needed to interpret SEM images and to perform electron-excited X-ray microanalysis, the complex variety of scattering processes will be broadly classified into "inelastic" and "elastic" scattering.

1.2 Inelastic Scattering (Energy Loss) Limits Beam Electron Travel in the Specimen

"Inelastic" scattering refers to a variety of physical processes that act to progressively reduce the energy of the beam electron by transferring that energy to the specimen atoms through interactions with tightly bound inner-shell atomic electrons and loosely bound valence electrons. These energy loss processes include ejection of weakly bound outer-shell atomic electrons (binding energy of a few eV) to form secondary electrons; ejection of tightly bound inner shell atomic electrons (binding energy of hundreds to thousands of eV) which subsequently results in emission of characteristic X-rays; deceleration of the beam electron in the electrical field of the atoms producing an X-ray continuum over all energies from a few eV up to the beam's landing energy (E_0) (*bremsstrahlung* or "braking radiation"); generation of waves in the free electron gas that permeates conducting metallic solids (plasmons); and heating of the specimen (phonon production). While energy is lost in these inelastic scattering events, the beam electrons only deviate slightly from their current path. The energy loss due to inelastic scattering sets an eventual limit on how far the beam electron can travel in the specimen before it loses all of its energy and is absorbed by the specimen.

To understand the specific limitations on the distance traveled in the specimen imposed by inelastic scattering, a mathematical description is needed of the rate of energy loss (incremental dE, measured in eV) with distance (incremental ds, measured in nm) traveled in the specimen. Although the various inelastic scattering energy loss processes are discrete and independent, Bethe (1930) was able to summarize their collective effects into a "continuous energy loss approximation":

$$dE / ds\,(eV / nm) = -7.85\,(Z\rho / AE)\ln(1.166\,E / J) \quad \textbf{(1.1a)}$$

where E is the beam energy (keV), Z is the atomic number, ρ is the density (g/cm³), A is the atomic weight (g/mol), and J is the "mean ionization potential" (keV) given by

$$J\,(keV) = (9.76\,Z + 58.5\,Z^{-0.19}) \times 10^{-3} \quad \textbf{(1.1b)}$$

The Bethe expression is plotted for several elements (C, Al, Cu, Ag, Au) over the range of "conventional" SEM operating energies, 5–30 keV in ◘ Fig. 1.1. This figure reveals that the rate of energy loss dE/ds increases as the electron energy decreases and increases with the atomic number of the target. An electron with a beam energy of 20 keV loses energy at approximately 10 eV/nm in Au, so that if this rate was constant, the total path traveled in the specimen would be approximately 20,000 eV/(10 eV/nm) = 2000 nm = 2 µm. A better estimate of this electron "Bethe range" can be made by explicitly considering the energy dependence of dE/ds through integration of the Bethe expression, Eq. 1.1a, from the incident energy down to a lower cut-off energy (typically ~ 2 keV due to limitations on the range of applicability of the Bethe expression; see further discussion below). Based on this calculation, the Bethe range for the selection of elements is shown in ◘ Fig. 1.2. At a particular incident beam energy, the Bethe range decreases as the atomic number of the target increases, while for a particular target, the Bethe range increases as the incident beam energy increases.

Fig. 1.1 Bethe continuous energy loss model calculations for dE/ds in C, Al, Cu, Ag, and Au as a function of electron energy

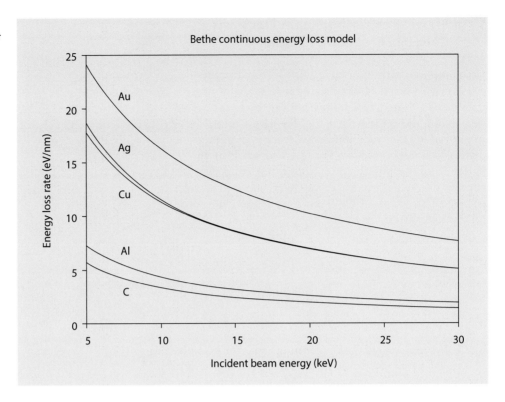

Fig. 1.2 Bethe range calculation from the continuous energy loss model by integrating over the range of energy from E_0 down to a cut-off energy of 2 keV

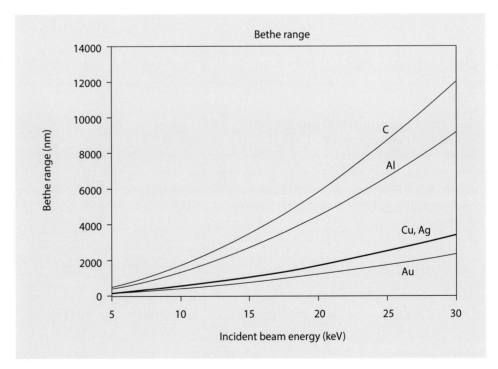

■■ **Note the change of scale**

The Bethe range for Au with an incident beam energy of 20 keV is approximately 1200 nm, a linear change in scale of a factor of 1200 over an incident beam diameter of 1 nm. If the beam–specimen interactions were restricted to a cylindrical column with the circular beam entrance footprint as its cross section and the Bethe range as its altitude, the volume of a cylinder 1 nm in diameter and 1200 nm deep would be approximately 940 nm^3, and the number of gold atoms it contained would be approximately 7.5×10^4, which can be compared to the incident beam footprint surface atom count of approximately 12.

1

1.3   Elastic Scattering: Beam Electrons Change Direction of Flight

Simultaneously with inelastic scattering, "elastic scattering" events occur when the beam electron is deflected by the electrical field of an atom (the positive nuclear charge as partially shielded by the negative charge of the atom's orbital electrons), causing the beam electron to deviate from its previous path onto a new trajectory, as illustrated schematically in ☐ Fig. 1.3a. The probability of elastic scattering depends strongly on the nuclear charge (atomic number Z) and the energy of the electron, E (keV) and is expressed mathematically as a cross section, Q:

$$Q_{\text{elastic}(>\phi_0)} = 1.62 \times 10^{-20} \left(Z^2 / E^2 \right) \cot^2 \left(\phi_0 / 2 \right)$$
$$\left[\text{events} > \phi_0 / \left[\text{electron} \left(\text{atom} / \text{cm}^2 \right) \right] \right] \quad \textbf{(1.2)}$$

where ϕ_0 is a threshold elastic scattering angle, for example, 2°. Despite the angular deviation, the beam electron energy is effectively unchanged in energy. While the average elastic scattering event causes an angular change of only a few degrees, deviations up to 180° are possible in a single elastic scattering event. Elastic scattering causes beam electrons to deviate out of the narrow angular range of incident trajectories defined by the convergence of the incident beam as controlled by the electron optics.

1.3.1   How Frequently Does Elastic Scattering Occur?

The elastic scattering cross section, Eq. 1.2, can be used to estimate how far the beam electron must travel on average to experience an elastic scattering event, a distance called the "mean free path," λ:

$$\lambda_{\text{elastic}} \left(\text{cm} \right) = A / \left[N_0 \rho Q_{\text{elastic}(>\phi_0)} \right] \quad \textbf{(1.3a)}$$

$$\lambda_{\text{elastic}} \left(\text{nm} \right) = 10^7 A / \left[N_0 \rho Q_{\text{elastic}(>\phi_0)} \right] \quad \textbf{(1.3b)}$$

where A is the atomic weight (g/mol), N_0 is Avogadro's number (atoms/mol), and ρ is the density (g/cm³). ☐ Figure 1.4 shows a plot of λ_{elastic} for various elements as a function of electron energy, where it can be seen that the mean free path

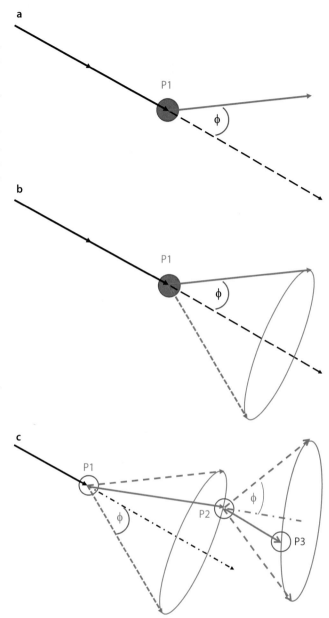

☐ **Fig. 1.3**   **a** Schematic illustration of elastic scattering. An energetic electron is deflected by the electrical field of an atom at location P1 through an angle ϕ_{elastic}. **b** Schematic illustration of the elastic scattering cone. The energetic electron scatters elastically at point P1 and can land at any location on the circumference of the base of the cone with equal probability. **c** Schematic illustration of a second scattering step, carrying the energetic electron from point P2 to point P3

is of the order of nm. Elastic scattering is thus likely to occur hundreds to thousands of times along a Bethe range of several hundred to several thousand nanometers.

Fig. 1.4 Elastic mean free path
as a function of electron kinetic
energy for various elements

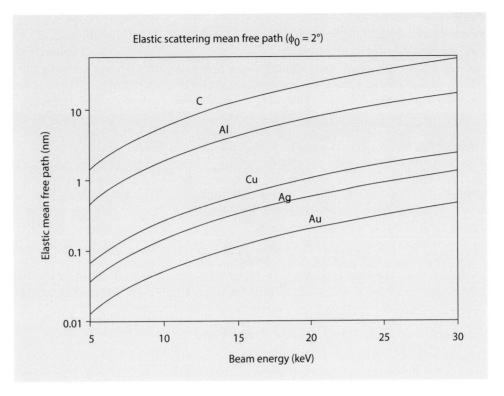

Fig. 1.4 Elastic mean free path as a function of electron kinetic energy for various elements

1.4 Simulating the Effects of Elastic Scattering: Monte Carlo Calculations

Inelastic scattering sets a limit on the total distance traveled by the beam electron. The Bethe range is an estimate of this distance and can be found by integrating the Bethe continuous energy loss expression from the incident beam energy E_0 down to a low energy limit, for example, 2 keV. Estimating the effects of elastic scattering on the beam electrons is much more complicated. Any individual elastic scattering event can result in a scattering angle within a broad range from a threshold of a fraction of a degree up to 180°, with small scattering angles much more likely than very large values and an average value typically in the range 5–10°. Moreover, the electron scattered by the atom through an angle ϕ in **Fig. 1.3a** at point P1 can actually follow any path along the surface of the three-dimensional scattering cone shown in **Fig. 1.3b** and can land anywhere in the circumference of the base of the scattering cone (i.e., the azimuthal angle in the base of the cone ranges from 0 to 360° with equal probability), resulting in a three-dimensional path. The length of the trajectory along the surface of the scattering cone depends on the frequency of elastic events with distance traveled and can be estimated from Eq. 1.3a for the elastic scattering mean free path, $\lambda_{elastic}$. The next elastic scattering event P2 causes the electron to deviate in a new direction, as

shown in **Fig. 1.3c**, creating an increasingly complex path. Because of the random component of scattering at each of many steps, this complex behavior cannot be adequately described by an algebraic expression like the Bethe continuous energy loss equation. Instead, a stepwise simulation of the electron's behavior must be constructed that incorporates inelastic and elastic scattering. Several simplifications are introduced to create a practical "Monte Carlo electron trajectory simulation":

1. All of the angular deviation of the beam electron is ascribed to elastic scattering. A mathematical model for elastic scattering is applied that utilizes a random number (hence the name "Monte Carlo" from the supposed randomness of gambling) to select a properly weighted value of the elastic scattering angle out of the possible range (from a threshold value of approximately 1° to a maximum of 180°). A second random number is used to select the azimuthal angle in the base of the scattering cone in **Fig. 1.1c**.

2. The distance between elastic scattering events, s, which lies on the surface of the scattering cone in **Fig. 1.3b**, is calculated from the elastic mean free path, Eq. 1.3b.

3. Inelastic scattering is calculated with the Bethe continuous energy loss expression, Eq. 1.1b. The specific energy loss, ΔE, along the path, s, in the surface of the scattering cone, **Fig. 1.3b**, is calculated with the Bethe continuous energy loss expression: $\Delta E = (dE/ds)*s$

1

Given a specific set of these parameters, the Monte Carlo electron trajectory simulation utilizes geometrical expressions to calculate the successive series of locations P1, P2, P3, etc., successively determining the coordinate locations (x, y, z) that the energetic electron follows within the solid. At each location P, the newly depreciated energy of the electron is known, and after the next elastic scattering angle is calculated, the new velocity vector components v_x, v_y, v_z are determined to transport the electron to the next location. A trajectory ends when either the electron energy falls below a threshold of interest (e.g., 1 keV), or else the path takes it outside the geometric bounds of the specimen, which is determined by comparing the current location (x, y, z) with the specimen boundaries. The capability of simulating electron beam interactions in specimens with complex geometrical shapes is one of the major strengths of the Monte Carlo electron trajectory simulation method.

Monte Carlo electron trajectory simulation can provide visual depictions as well as numerical results of the beam–specimen interaction, creating a powerful instructional tool for studying this complex phenomenon. Several powerful Monte Carlo simulations appropriate for SEM and X-ray microanalysis applications are available as free resources:

CASINO [▶ http://www.gel.usherbrooke.ca/casino/What.html]
Joy Monte Carlo [▶ http://web.utk.edu/~srcutk/htm/
simulati.htm]
NIST DTSA-II [▶ http://www.cstl.nist.gov/div837/837.02/
epq/dtsa2/index.html]

While the static images of Monte Carlo simulations presented below are useful instructional aids, readers are encouraged to perform their own simulations to become familiar with this powerful tool, which in more elaborate implementations is an important aid in understanding critical aspects of SEM imaging.

1.4.1 What Do Individual Monte Carlo Trajectories Look Like?

Perform a Monte Carlo simulation (CASINO simulation) for copper with a beam energy of 20 keV and a tilt of 0° (beam perpendicular to the surface) for a small number of trajectories, for example, 25. ◼ Figure 1.5a, b show two simulations of 25 trajectories each. The trajectories are actually determined in three dimensions (x-y-z, where x-y defines the surface plane and z is perpendicular to the surface) but for plotting are rendered in two dimensions (x-z), with the third

dimension y projected onto the x-z plane. (An example of the true three-dimensional trajectories, simulated with the Joy Monte Carlo, is shown in ◼ Fig. 1.6, in which a small number of trajectories (to minimize overlap) have been rendered as an anaglyph stereo representation with the convention left eye = red filter. Inspection of this simulation shows the y motion of the electrons in and out of the x-z plane.) The stochastic nature of the interaction imposed by the nature of elastic scattering is readily apparent in the great variation among the individual trajectories seen in ◼ Fig. 1.5a, b. It quickly becomes clear that individual beam electrons follow a huge range of paths and simulating a small number of trajectories does not provide an adequate view of the electron beam specimen interaction.

1.4.2 Monte Carlo Simulation To Visualize the Electron Interaction Volume

To capture a reasonable picture representation of the *electron interaction volume*, which is the region of the specimen in which the beam electrons travel and deposit energy, it is necessary to calculate many more trajectories. ◼ Figure 1.5c shows the simulation for copper, $E_0 = 20$ keV at 0° tilt extended to 500 trajectories, which reveals the full extent of the electron interaction volume. Beyond a few hundred trajectories, superimposing the three-dimensional trajectories to create a two-dimensional representation reaches diminishing returns due to overlap of the plotted lines. While simulating 500 trajectories provides a reasonable qualitative view of the electron interaction volume, Monte Carlo calculations of numerical properties of the interaction volume and related processes, such as electron backscattering (discussed in the backscattered electron module), are subject to statistically predictable variations because of the use of random numbers to select the elastic scattering parameters. Variance in repeated simulations of the same starting conditions is related to the number of trajectories and can be described with the properties of the Gaussian (normal) distribution. Thus the precision, p, of the calculation of a parameter of the interaction is related to the total number of simulated trajectories, n, and the fraction, f, of those trajectories that produce the effect of interest (e.g., backscattering):

$$p = \left(f\, n \right)^{1/2} / \left(f\, n \right) = \left(f\, n \right)^{-1/2} \qquad (1.4)$$

Fig. 1.5 **a** Copper, $E_0 = 20$ keV; 0 tilt; 25 trajectories (CASINO Monte Carlo simultion). **b** Copper, $E_0 = 20$ keV; 0 tilt; another 25 trajectories. **c** Copper, $E_0 = 20$ keV; 0 tilt; 200 trajectories

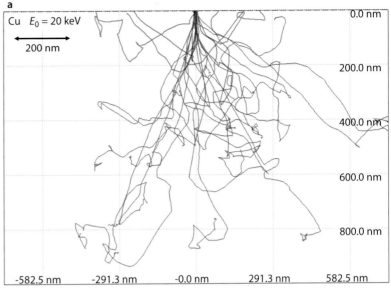

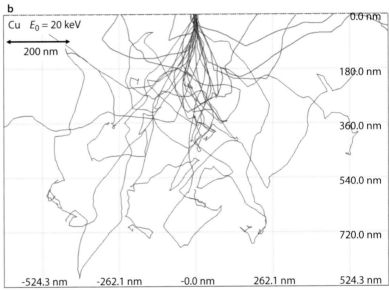

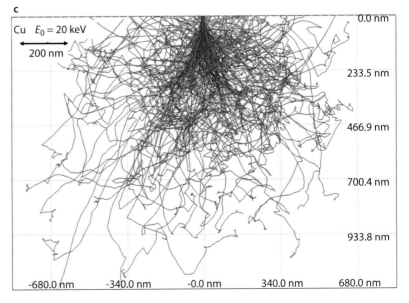

1

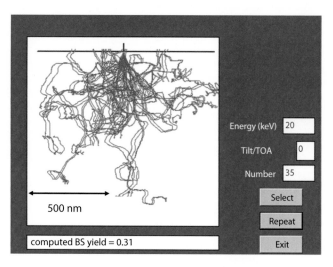

Energy (keV)   20

Tilt/TOA   0

Number   35

Select

Repeat

500 nm

computed BS yield = 0.31   Exit

◼ **Fig. 1.6**   Three-dimensional representation of a Monte Carlo simulation (Cu, 20 keV, 0° tilt) using the anaglyph stereo method (left eye = red filter) (Joy Monte Carlo)

1.4.3   Using the Monte Carlo Electron Trajectory Simulation to Study the Interaction Volume

What Are the Main Features of the Beam Electron Interaction Volume?

In ◼ Fig. 1.5c, the beam electron interaction volume is seen to be a very complex structure with dimensions extending over hundreds to thousands of nanometers from the beam impact point, depending on target material and the beam energy. At 0° tilt, the interaction volume is rotationally symmetric around the beam. While the electron trajectories provide a strong visual representation of the interaction volume, more informative numerical information is needed. The Monte Carlo simulation can provide detailed information on many aspects of the electron beam–specimen interaction. The color-encoding of the energy deposited along each trajectory, as implemented in the Joy Monte Carlo shown in ◼ Fig. 1.11, creates a view that reveals the general three-dimensional complexity of energy deposition within the interaction volume. The CASINO Monte Carlo provides an even more detailed view of energy deposition, as shown in ◼ Fig. 1.7. The energy deposition per unit volume is greatest just under the beam impact location and rapidly falls off as the periphery of the interaction volume is approached. This calculation reveals that a small cylindrical volume under the beam impact point, shown in more detail in ◼ Fig. 1.7b, receives half of the total energy deposited by the beam in the specimen (that is, the volume within the 50% contour), with the

balance of the energy deposited in a strongly non-linear fashion in the much larger portion of the interaction volume.

How Does the Interaction Volume Change with Composition?

◼ Figure 1.8 shows the interaction volume in various targets, C, Si, Cu, Ag, and Au, at fixed beam energy, $E_0 = 20$ keV, and 0° tilt. As the atomic number of the target increases, the linear dimensions of the interaction volume decrease. The form also changes from pear-shaped with a dense conical region below the beam impact for low atomic number targets to a more hemispherical shape for high atomic number targets.

◼◼ **Note the dramatic change of scale**

Approximately 12 gold atoms were encountered within the footprint of a 1-nm diameter at the surface. Without considering the effects of elastic scattering, the Bethe range for Au at an incident beam energy of 20 keV limited the penetration of the beam to approximately 1200 nm and a cylindrical volume of approximately 940 nm^3, containing approximately 5.6×10^4 Au atoms. The effect of elastic scattering is to create a three-dimensional hemispherical interaction volume with a radius of approximately 600 nm and a volume of 4.5×10^8 nm^3, containing 2.7×10^{10} Au atoms, an increase of nine orders-of-magnitude over the number of atoms encountered in the initial beam footprint on the surface.

How Does the Interaction Volume Change with Incident Beam Energy?

◼ Figure 1.9 shows the interaction volume for copper at 0° tilt over a range of incident beam energy from 5 to 30 keV. The shape of the interaction volume is relatively independent of beam energy, but the size increases rapidly as the incident beam energy increases.

How Does the Interaction Volume Change with Specimen Tilt?

◼ Figure 1.10 shows the interaction volume for copper at an incident beam energy of 20 keV and a series of tilt angles. As the tilt angle increases so that the beam approaches the surface at a progressively more shallow angle, the shape of the interaction volume changes significantly. At 0° tilt, the interaction volume is rotationally symmetric around the beam, but as the tilt angle increases the interaction volume becomes asymmetric, with the dense portion of the distribution shifting progressively away from the beam impact point. The maximum penetration of the beam is reduced as the tilt angle increases.

◻ **Fig. 1.7** **a** Isocontours of energy loss showing fraction remaining; Cu, 20 keV, 0° tilt; 50,000 trajectories (CASINO Monte Carlo simulation). **b** Expanded view of high density region of 1.7**a**

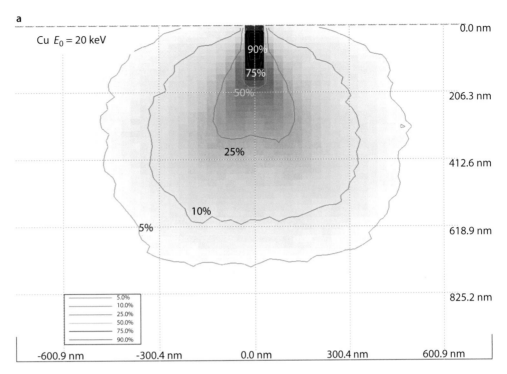

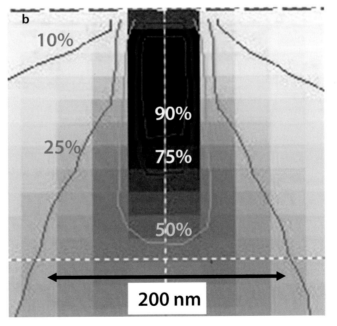

1

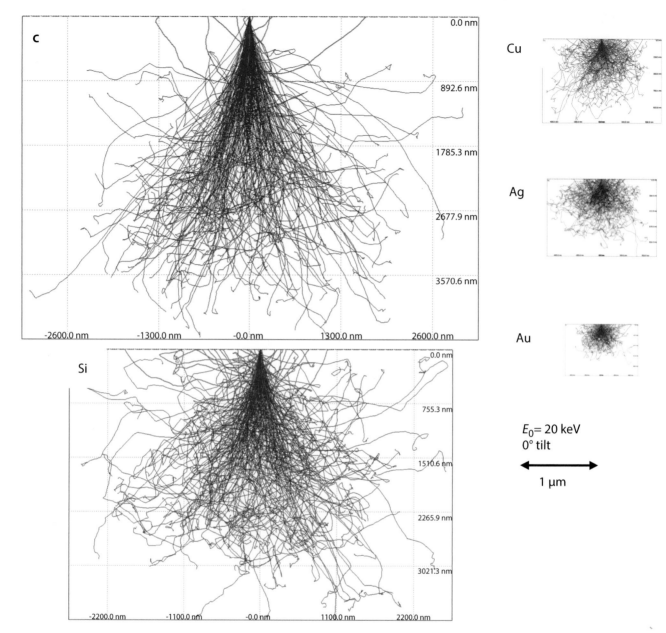

■ **Fig. 1.8**   Monte Carlo simulations for an incident beam energy of 20 keV and 0° tilt for C, Si, Cu, Ag, and Au, all shown at the same scale (CASINO Monte Carlo simulation)

■ **Fig. 1.9** Monte Carlo simulations for Cu, 0° tilt, incident beam energies 5, 10, 20, and 30 keV (CASINO Monte Carlo simulation)

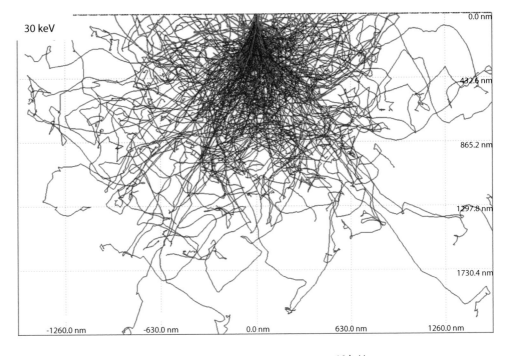

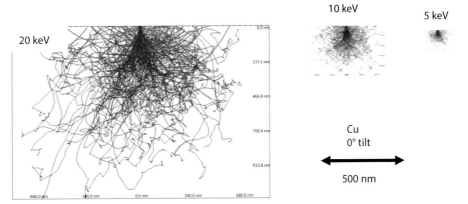

1

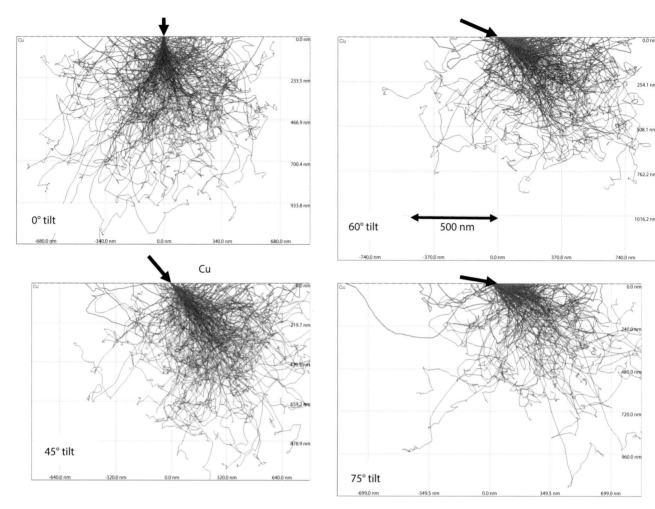

■ **Fig. 1.10**   Monte Carlo simulations for Cu, 20 keV, with various tilt angles (CASINO Monte Carlo simulation)

1.5   A Range Equation To Estimate the Size of the Interaction Volume

While the Monte Carlo simulation is a powerful tool to depict the complexity of the electron beam specimen interactions, it is often useful to have a simple estimate of the size. The Bethe range gives the maximum distance the beam electron can travel in the specimen, but this distance is measured along the complex trajectory that develops because of elastic scattering. Kanaya and Okayama (1972) developed a range equation that considered both inelastic and elastic scattering to give an estimate of the interaction volume as the radius of a hemisphere centered on the beam impact point that contained at least 95% of the trajectories:

$$R_{K-O}(nm) = 27.6 \left(A / Z^{0.89} \rho \right) E_0^{1.67} \qquad (1.5)$$

■ **Table 1.1**   Kanaya–Okayama range

	5 keV (nm)	10 keV	20 keV	30 keV (μm)
C	450 nm	1.4 μm	4.5 μm	8.9 μm
Al	413 nm	1.3 μm	4.2 μm	8.2 μm
Fe	159 nm	505 nm	1.6 μm	3.2 μm
Ag	135 nm	431 nm	1.4 μm	2.7 μm
Au	85 nm	270 nm	860 nm	1.7 μm

where A is the atomic weight (g/mol), Z is the atomic number, ρ is the density (g/cm^3), and E_0 is the incident beam energy (keV). Calculations of the Kanaya–Okayama range are presented in ■ Table 1.1. The Kanaya–Okayama range

E_0 = 20 keV; 0° tilt

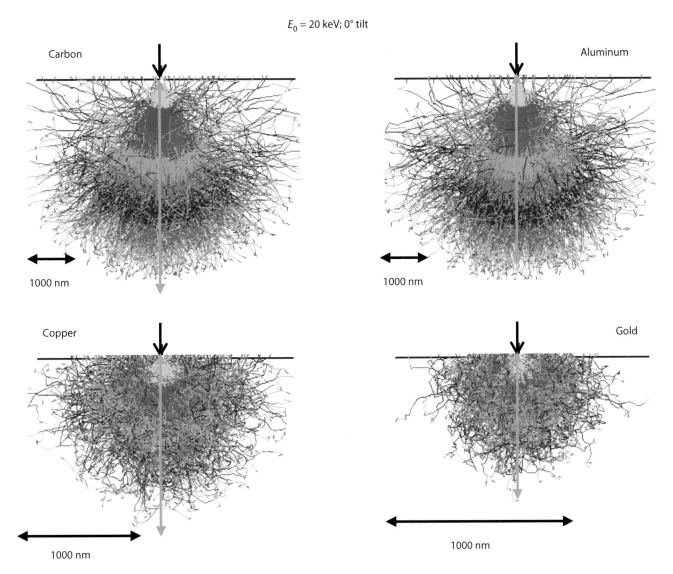

☐ **Fig. 1.11** Kanaya–Okayama range (*gold arrow*) superimposed on the interaction volume for C, Al, Cu, and Au at $E0$ = 20 keV and 0° tilt (Joy Monte Carlo simulation)

is shown superimposed on the Monte Carlo simulation of the interaction volume in ☐ Fig. 1.11 and is plotted graphically in ☐ Fig. 1.12. It is, of course, simplistic to use a single numerical value of the range to describe such a complex phenomenon as the electron interaction volume with its varying contours of energy deposition, and thus the range equation should only be considered as a "gray" number useful for estimation purposes. Nevertheless, the Kanaya–Okayama range is useful as a means to provide scaling to describe the spatial distributions of the signals produced within the interaction volume: secondary electrons, backscattered electrons, and X-rays.

1

■ **Fig. 1.12** Kanaya–Okayama range plotted for C, Al, Cu, Ag and Au as a function of E_0

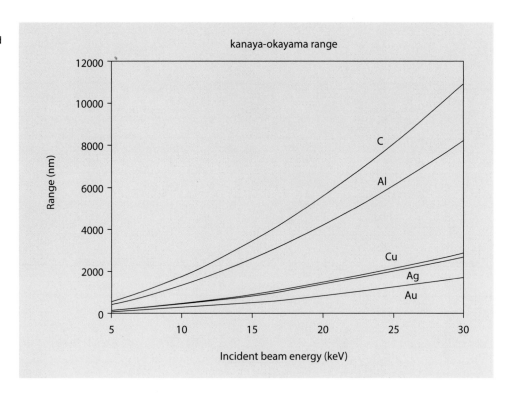

References

Bethe H (1930) Theory of the transmission of corpuscular radiation through matter. Ann Phys Leipzig 5:325

CASINO ▶ http://www.gel.usherbrooke.ca/casino/What.html

Joy Monte Carlo ▶ http://web.utk.edu/~srcutk/htm/simulati.htm

Kanaya K, Okayama S (1972) Penetration and energy-loss theory of electrons in solid targets. J Phys D Appl Phys 5:43

NISTDTSA-II ▶ http://www.cstl.nist.gov/div837/837.02/epq/dtsa2/index.html

Backscattered Electrons

© Springer Science+Business Media LLC 2018
J. Goldstein et al., *Scanning Electron Microscopy and X-Ray Microanalysis*,
https://doi.org/10.1007/978-1-4939-6676-9_2

2

2.1 Origin

Close inspection of the trajectories in the Monte Carlo simulation of a flat, bulk target of copper at 0° tilt shown in ◘ Fig. 2.1 reveals that a significant fraction of the incident beam electrons undergo sufficient scattering events to completely reverse their initial direction of travel into the specimen, causing these electrons to return to the entrance surface and exit the specimen. These beam electrons that escape from the specimen are referred to as "backscattered electrons" (BSE) and constitute an important SEM imaging signal rich in information on specimen characteristics. The BSE signal can convey information on the specimen composition, topography, mass thickness, and crystallography. This module describes the properties of backscattered electrons and how those properties are modified by specimen characteristics to produce useful information in SEM images.

2.1.1 The Numerical Measure of Backscattered Electrons

Backscattered electrons are quantified with the "backscattered electron coefficient," η, defined as

$$\eta = N_{BSE} / N_B \tag{2.1}$$

where N_B is the number of beam electrons that enter the specimen and N_{BSE} is the number of those electrons that subsequently emerge as backscattered electrons.

2.2 Critical Properties of Backscattered Electrons

2.2.1 BSE Response to Specimen Composition (η vs. Atomic Number, Z)

Use the CASINO Monte Carlo simulation software, which reports η in the output, to examine the dependence of electron backscattering on the atomic number of the specimen.

Simulate at least 10,000 trajectories at an incident energy of $E_0 = 20$ keV and a surface tilt of 0° (i.e., the beam is perpendicular to the surface). Note that statistical variations will be observed in the calculation of η due to the different selections of the random numbers used in each simulation. Repetitions of this calculation will give a distribution of results, with a precision $p = (\eta N)^{1/2}/\eta N$, so that for $N = 10,000$ trajectories and $\eta \sim 0.15$ (Si), p is expected to be 2.5 %. ◘ Figure 2.2 shows the simulation of 500 trajectories in carbon, silicon, copper, and gold with an incident energy of $E_0 = 20$ keV and a surface tilt of 0°, showing qualitatively the increase in the number of backscattered electrons with atomic number.

Detailed experimental measurements of the backscattered electron coefficient as a function of the atomic number, Z, in highly polished, flat pure element targets confirm a generally monotonic increase in η with increasing Z, as shown in ◘ Fig. 2.3a, where the classic measurements made by Heinrich (1966) at a beam energy of 20 keV are plotted. The slope of η vs. Z is highest for low atomic number targets up to approximately $Z = 14$ (Si). As Z continues to increase into the range of

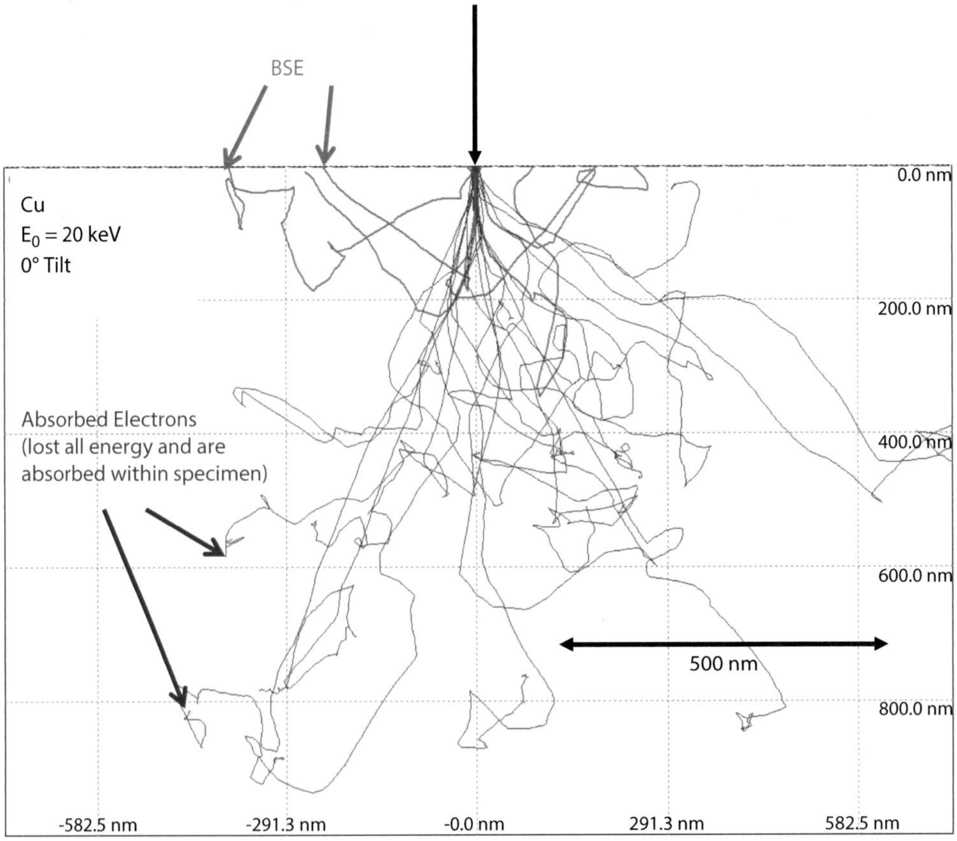

◘ **Fig. 2.1**   Monte Carlo simulation of a flat, bulk target of copper at 0° tilt. Red trajectories lead to backscattering events

BSE

Cu
$E_0 = 20$ keV
0° Tilt

Absorbed Electrons
(lost all energy and are
absorbed within specimen)

500 nm

0.0 nm

200.0 nm

400.0 nm

600.0 nm

800.0 nm

-582.5 nm   -291.3 nm   -0.0 nm   291.3 nm   582.5 nm

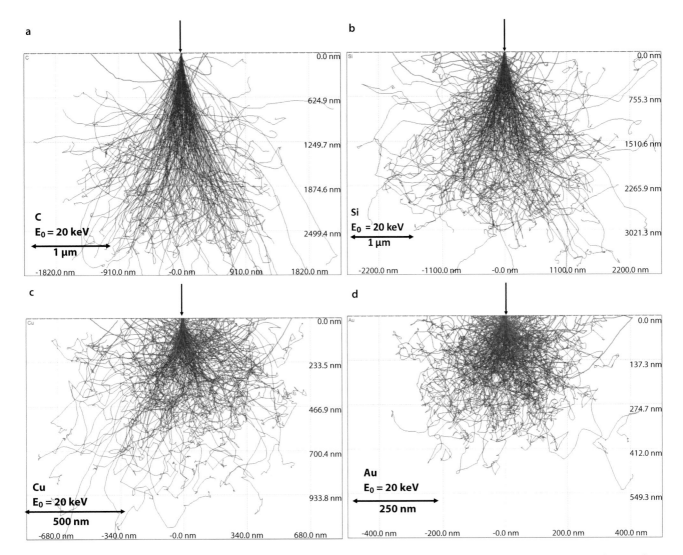

a

C
$E_0 = 20$ keV
1 μm

0.0 nm
624.9 nm
1249.7 nm
1874.6 nm
2499.4 nm

-1820.0 nm   -910.0 nm   -0.0 nm   910.0 nm   1820.0 nm

b

Si
$E_0 = 20$ keV
1 μm

0.0 nm
755.3 nm
1510.6 nm
2265.9 nm
3021.3 nm

-2200.0 nm   -1100.0 nm   -0.0 nm   1100.0 nm   2200.0 nm

c

Cu
$E_0 = 20$ keV
500 nm

0.0 nm
233.5 nm
466.9 nm
700.4 nm
933.8 nm

-680.0 nm   -340.0 nm   -0.0 nm   340.0 nm   680.0 nm

d

Au
$E_0 = 20$ keV
250 nm

0.0 nm
137.3 nm
274.7 nm
412.0 nm
549.3 nm

-400.0 nm   -200.0 nm   -0.0 nm   200.0 nm   400.0 nm

■ **Fig. 2.2**   **a** Monte Carlo simulation of 500 trajectories in carbon with an incident energy of $E_0 = 20$ keV and a surface tilt of 0° (CASINO Monte Carlo simulation). **b** Monte Carlo simulation of 500 trajectories in silicon with an incident energy of $E_0 = 20$ keV and a surface tilt of 0°. **c** Monte Carlo simulation of 500 trajectories in copper with an incident energy of $E_0 = 20$ keV and a surface tilt of 0°. **d** Monte Carlo simulation of 500 trajectories in gold with an incident energy of $E_0 = 20$ keV and a surface tilt of 0°. Red trajectories = backscattering

the transition elements, e.g., $Z = 26$ (Fe), the slope progressively decreases until at very high Z, e.g., the region around $Z = 79$ (Au), the slope becomes so shallow that there is very little change in η between adjacent elements. Plotted in addition to the experimental measurements in ■ Fig. 2.3a is a mathematical fit to the 20 keV data developed by Reuter (1972):

$$\eta = -0.0254 + 0.016\,Z - 1.86 \times 10^{-4}\,Z^2 + 8.3 \times 10^{-7}\,Z^3 \qquad (2.2)$$

This fit provides a convenient estimate of η for those elements for which direct measurements do not exist.

Experimental measurements (Heinrich 1966) have shown that the backscattered electron coefficient of a mixture of atoms that is homogeneous on the atomic scale, such as a stoichiometric compound, a glass, or certain metallic alloys, can be accurately predicted from the mass concentrations of the elemental constituents and the values of η for those pure elements:

$$\eta_{\text{mixture}} = \Sigma \eta_i C_i \qquad (2.3)$$

where C is the mass (weight) fraction and i is an index that denotes all of the elements involved.

When measurements of η vs. Z are made at different beam energies, combining the experimental measurements of Heinrich and of Bishop in ■ Fig. 2.3b, little dependence on the beam energy is found from 5 to 49 keV, with all of the measurements clustering relatively closely to the curve for the 20 keV data shown in ■ Fig. 2.3a. This result is perhaps surprising in view of the strong dependence of the dimensions of the interaction volume on the incident beam energy. The weak dependence of η upon E_0 despite the strong dependence of the beam penetration upon E_0 can be understood as a near balance between the increased energy available at higher E_0, the lower rate of loss, dE/ds, with higher E_0, and the increased penetration. Thus, although a beam electron may penetrate more deeply at high E_0, it started with more

2

◻ **Fig. 2.3   a** Electron backscatter coefficient as a function of atomic number for pure elements (Data of Heinrich 1966; fit of Reuter 1972). **b** Electron backscatter coefficient as a function of atomic number for pure elements for incident beam energies of 5 keV (data of Bishop 1966); 10 keV to 49 keV (Data of Heinrich 1966); Reuter's fit to Heinrich's 20 keV data, (1972))

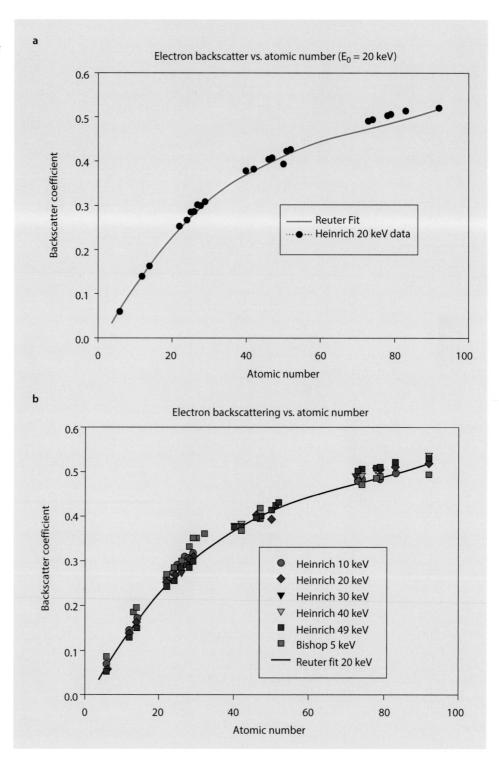

energy and lost that energy at a lower initial rate than an electron at a lower incidence energy. Thus, a higher incidence energy electron, despite penetrating deeper in the specimen, retains more energy and can continue to scatter and progress through the target to escape.

SEM Image Contrast with BSE: "Atomic Number Contrast"

Whenever a signal that can be measured in the SEM, such as backscattered electrons, follows a predictable response to a specimen property of interest, such as composition, the physical basis for a "contrast mechanism" is established. Contrast, C_{tr}, is defined as

$$C_{tr} = (S_2 - S_1) / S_2 \text{ with } S_2 > S_1 \qquad (2.4)$$

where S is the signal measured at any two locations of interest in the image field. As shown in ◘ Fig. 2.4, examples include the contrast between an object P_1 and the general background P_2 or between two objects that share an interface, P_3 and P_4. By this definition, contrast can range numerically from 0 to 1.

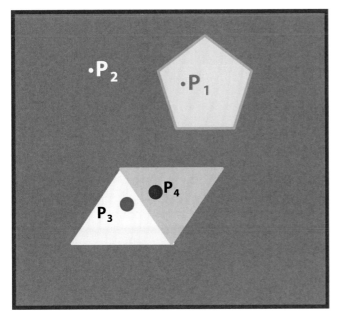

◘ **Fig. 2.4** Illustration of some possible contrast situations of interest, e.g., an object *P1* and the general background *P2* or between two objects that share an interface, *P3* and *P4*

◘ **Fig. 2.5** Backscattered electron atomic number contrast for a polished flat surface of Raney nickel (nickel-aluminum) alloy. Numbered locations identify phases with distinctly different compositions

The monotonic behavior of η vs. Z establishes the physical basis for "atomic number contrast" (also known as "Z-contrast" and "compositional contrast"). When an SEM BSE image is acquired from a flat specimen (i.e., no topography is present, at least on a scale no greater than about 5 % of the Kanaya–Okayama range for the particular material composition and incident beam energy), then local differences in composition can be observed as differences in the BSE intensity, which can be used to construct a meaningful gray-scale SEM image. The compositionally-different objects must have dimensions that are at least as large as the Kanaya-Okayama range for each distinct material so that a BSE signal characteristic of the particular composition can be measured over at least the center portion of the object. The BSE signal at beam locations on the edge of the object may be affected by penetration into the neighboring material(s).

From the definition of contrast, C_{tr}, atomic number contrast can be predicted between two materials with backscatter coefficients η_1 and η_2 when the measured signal S is proportional to η:

$$C_{tr} = (\eta_2 - \eta_1) / \eta_2 \text{ with } \eta_2 > \eta_1 \qquad (2.5)$$

An example of atomic number contrast from a polished cross section of an aluminum-nickel alloy (Raney nickel) is shown in ◘ Fig. 2.5. At least four distinct gray levels are observed, which correspond to three different Al/Ni phases with different Al-Ni compositions (labeled "1," "3," and "4" in ◘ Fig. 2.5) and a fourth phase that consists of Al-Fe-Ni (labelled "2"), with the phase containing the highest nickel concentration appearing brightest in the BSE image.

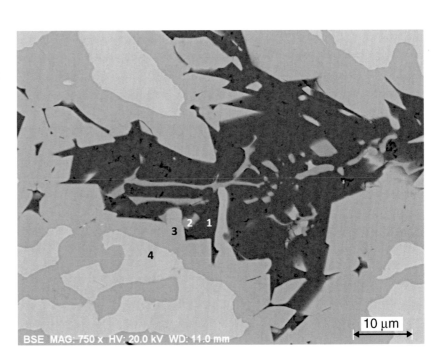

BSE MAG: 750 x HV: 20.0 kV WD: 11.0 mm

2

2.2.2 BSE Response to Specimen Inclination (η vs. Surface Tilt, θ)

Model the effect of the angle of inclination of the specimen surface to the incident beam with the Monte Carlo simulation. Select a particular element and incident beam energy, e.g., copper and $E_0 = 20$ keV, and vary the angle of incidence. Calculate at least 10,000 trajectories to obtain adequate simulation precision.

◘ Figure 2.6 shows simulations for aluminum with an incident beam energy of 15 keV at various inclinations calculated with 200 trajectories, which qualitatively reveals the increase in backscattering in a forward direction (i.e., continuing in the general direction of the incident beam) with increasing tilt angle. A more extensive series of simulations for aluminum at $E_0 = 15$ keV with 25,000 trajectories covering a greater range of specimen tilts is presented in ◘ Table 2.1, where the backscatter coefficient shows a strong dependence on the surface inclination.

◘ **Table 2.1** Backscatter vs. tilt angle for aluminum at $E_0 = 15$ keV (25,000 trajectories calculated with the CASINO Monte Carlo simulation)

Tilt (degrees)	η
0	0.129
15	0.138
30	0.169
45	0.242
60	0.367
75	0.531
80	0.612
85	0.706
88	0.796
89	0.826

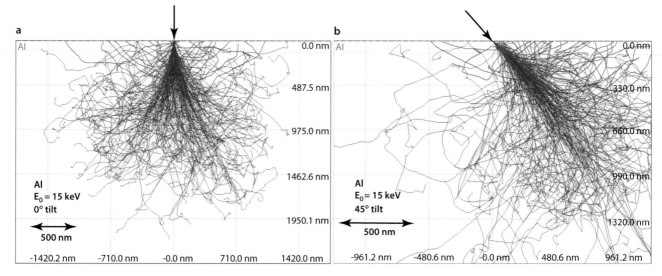

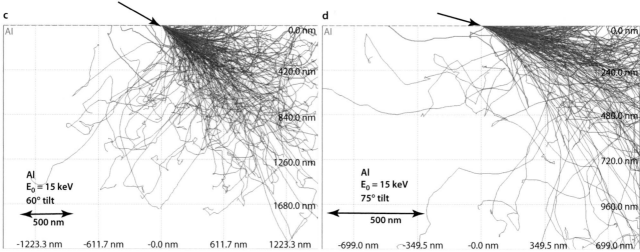

◘ **Fig. 2.6**   **a** Monte Carlo simulation for aluminum at $E_0 = 15$ keV for a tilt angle of 0°. **b** Monte Carlo simulation for aluminum at $E_0 = 15$ keV for a tilt angle of 45°. **c** Monte Carlo simulation for aluminum at $E_0 = 15$ keV for a tilt angle of 60°. **d** Monte Carlo simulation for aluminum at $E_0 = 15$ keV for a tilt angle of 75°

Figure 2.7 shows the results of similar Monte Carlo simulations for various elements as a function of surface inclination. As the surface tilt increases, η increases for all elements, converging toward unity at high tilt and grazing incidence for the incident beam.

SEM Image Contrast: "BSE Topographic Contrast—Number Effects"

This regular behavior of η vs. θ provides the basis for a contrast mechanism by which differences in the relative numbers of backscattered electrons depend on differences in the local surface inclination, which reveals the surface topography. Figure 2.8a shows an example of a pure material (polycrystalline silver) with grain faces inclined at various angles. The higher the inclination of the local surface to the incident beam, the higher will be the BSE signal, so that highly inclined surfaces appear bright, while dark surfaces are those nearly perpendicular to the beam. This image was prepared with a backscattered electron detector (discussed in the Electron Optics—Detectors module), which has a very large solid angle, so that back-scattered electrons are collected with high efficiency regardless of the direction that they travel after leaving the specimen.

Fig. 2.7 Monte Carlo calculations of electron backscattering from various tilted pure element bulk targets

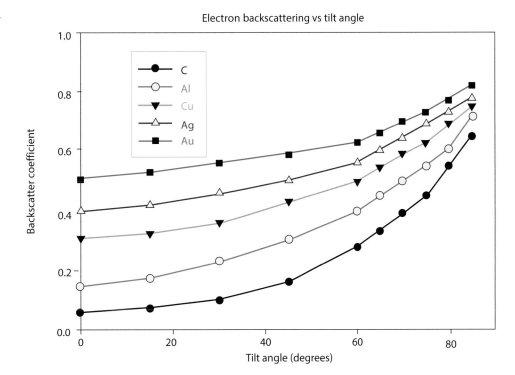

Fig. 2.8 **a** SEM backscattered electron image of a topographically irregular surface of pure silver prepared with a large collection angle BSE detector. **b** SEM backscattered electron image of the same area, prepared with a small collection angle BSE detector placed at the top of the image looking down

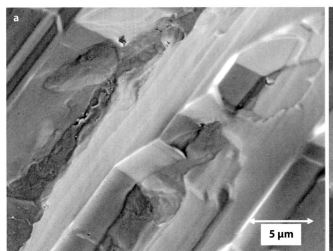

2

2.2.3 Angular Distribution of Backscattering

Beam Incident Perpendicular to the Specimen Surface (0° Tilt)

For a flat, bulk target, backscattered electrons emerge through the surface along a wide range of possible angular paths measured relative to the surface normal. When the incident beam is perpendicular to the specimen surface (0° tilt), experimental measurements and Monte Carlo simulations show that the angular distribution of the trajectories is such that the fraction along any given angle of emission is proportional to the cosine of that angle of emission, φ, between the electron trajectory and the surface normal, as shown in ◘ Fig. 2.9a:

$$\eta(\varphi) \sim \cos(\varphi) \qquad (2.6)$$

Thus, the largest number of BSEs follow a path parallel to the surface normal ($\varphi = 0°$, cosine = 1), while virtually no BSEs exit along a trajectory nearly parallel to the surface ($\varphi = 90°$, cosine = 0). The angular distribution seen in ◘ Fig. 2.8a is also rotationally symmetric around the beam: the same cosine shape would be found in any section through the distribution in any plane perpendicular to surface containing the beam vector and surface normal.

Beam Incident at an Acute Angle to the Specimen Surface (Specimen Tilt > 0°)

When a flat, bulk target is tilted so that the beam is incident at an acute angle to the surface, the angular distribution of backscattered electrons changes from the rotationally symmetric cosine function of ◘ Fig. 2.9a to the asymmetric distribution

◘ **Fig. 2.9**   **a** Cosine angular distribution observed for the directionality of backscattering from a bulk target at normal incidence (0° tilt; beam perpendicular to surface). **b** Angular distribution observed for the directionality of backscattering from a bulk target inclined (60° tilt; beam 30° above surface) (Data of Seidel quoted by Niedrig 1978)

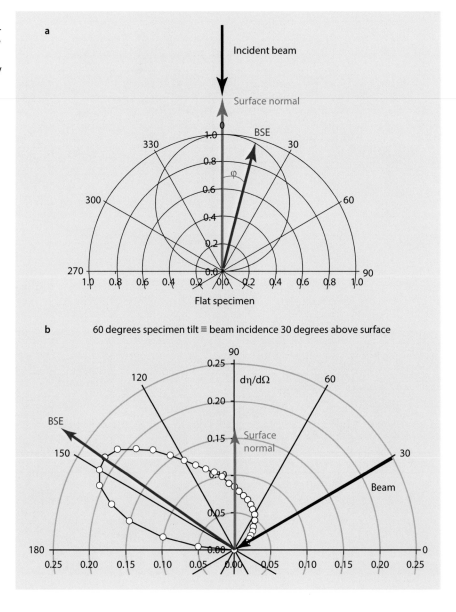

seen in ◨ Fig. 2.9b, with this distribution measured for a tilt of 60° (angle of incidence = 30°). The angular distribution is peaked in the forward direction away from the incident beam direction, with the maximum BSE emission occurring at an angle above the surface close to the value of the angle of incidence above the surface of the beam. This angular asymmetry develops slowly for tilt angles up to approximately 30°, but the asymmetry becomes increasingly pronounced with further increases in the specimen tilt. Moreover, the rotational symmetry of the 0° tilt case is also progressively lost with increasing tilt, with the asymmetric distribution seen in ◨ Fig. 2.9b becoming much narrower in the direction out of the plotting plane. See ▸ Chapter 29 for effects of crystal structure on backscattering angular distribution.

SEM Image Contrast: "BSE Topographic Contrast—Trajectory Effects"

The overall effects of specimen tilt are to increase the number of backscattered electrons and to create directionality in the backscattered electron emission, and both effects become increasingly stronger as the tilt increases. The "trajectory effects" create a very strong component of topographic contrast when viewed with a backscattered electron detector that has limited size and is placed preferentially on one side of the specimen. ◨ Figure 2.8b shows the same area as ◨ Fig. 2.8a imaged with a small solid angle detector, located at the top center of the image. Very strong contrast is created between faces tilted toward the detector, i.e., facing upward, and those tilted away, i.e., facing downward. These effects will be discussed in detail in the Image Interpretation module.

2.2.4 Spatial Distribution of Backscattering

Model a small number of trajectories (~25) so that the individual trajectories can be distinguished; e.g., for a copper target with an incident beam energy of 20 keV and 0° tilt, as seen in ◨ *Fig. 2.10 (Note: because of the random number sampling, repeated simulations will differ from each other and will be different from the printed example.) By following a number of trajectories from the point of incidence to the point of escape through the surface as backscattered electrons, it can be seen that the trajectories of beam electrons that eventually emerge as BSEs typically traverse the specimen both laterally and in depth.*

Depth Distribution of Backscattering

By performing detailed Monte Carlo simulations for many thousands of trajectories and recording for each trajectory the maximum depth of penetration into the specimen before the electron eventually escaped as a BSE, we can determine the contribution to the overall backscatter coefficient as a function of the depth of penetration, as shown for a series of elements in ◨ Fig. 2.11a. To compare the different elements, the horizontal axis of the plot is the depth normalized by the Kanaya–Okayama range for each element. From the depth distribution data in ◨ Fig. 2.11a, the cumulative backscattering coefficient as a function of depth can be calculated, and as shown in ◨ Fig. 2.11b, this distribution follows an S-shaped curve. To capture 90 % of the total backscattering, which corresponds to the region where the slope of the plot is rapidly decreasing, the backscattered electrons are found to travel a

◨ **Fig. 2.10** Monte Carlo simulation of a few trajectories in copper with an incident beam energy of 20 keV and 0° tilt to show effect of penetration depth of backscattered electrons

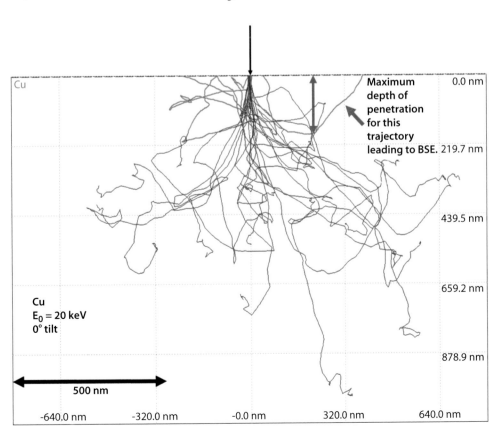

2

■ **Fig. 2.11** **a** Distribution of depth penetration of backscattered electrons in various elements. **b** Cumulative backscattering coefficient as a function of the depth of penetration in various elements, showing determination of 90 % total backscattering depth

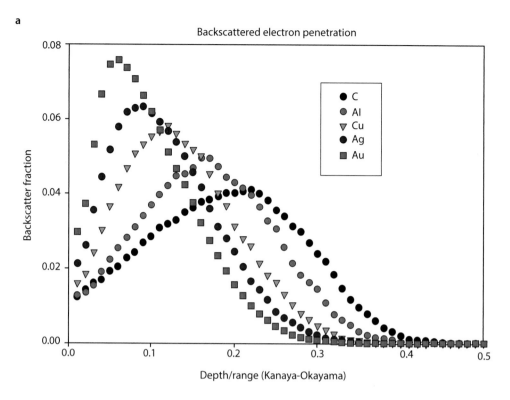

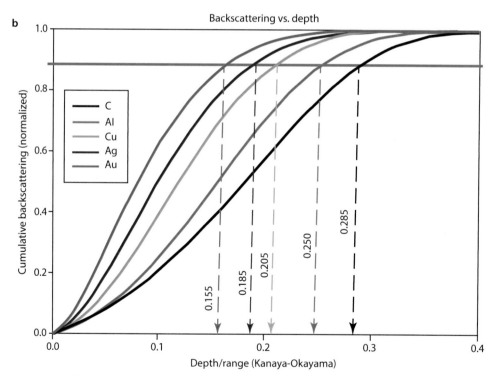

significant fraction of the Kanaya–Okayama range into the target. Strong elastic scattering materials with high atomic number such as gold sample a smaller fraction of the range than the weak elastic scattering materials such as carbon. ■ Table 2.2 lists the fractional range to capture 90 % of backscattering at normal beam incidence (0° tilt) and for a similar Monte Carlo study performed for a target at 45° tilt. For a tilted target, all materials show a slightly smaller fraction of the Kanaya–Okayama range to reach 90 % backscattering compared to the normal incidence case.

When the beam energy is increased for a specific material, the strong dependence of the total range on the incident

	0° tilt	45° tilt
C	0.285	0.23
Al	0.250	0.21
Cu	0.205	0.19
Ag	0.185	0.17
Au	0.155	0.15

beam energy leads to a strong dependence of the sampling depth of backscattered electrons, as shown in the depth distributions of backscattered electrons for copper over a wide energy range in ☐ Fig. 2.12. The substantial sampling depth of backscattered electrons combined with the strong dependence of the electron range on beam energy provides a useful tool for the microscopist. By comparing a series of images of a given area as a function of beam energy, subsurface details can be recognized. An example is shown in ☐ Fig. 2.13 for an engineered semiconductor electronic device with three-dimensional layered features, where a systematic increase in the beam energy reveals progressively deeper structures.

Radial Distribution of Backscattered Electrons

The Monte Carlo simulation can record the x-y location at which a backscattered electron exits through the surface plane, and this information can be used to calculate the radial distribution of backscattering relative to the beam impact location. The cumulative radial distribution is shown in ☐ Fig. 2.14 for a series of elements, as normalized by the Kanaya–Okayama range for each element, and an S-shaped curve is observed. ☐ Table 2.3 gives the fraction of the range necessary to capture 90 % of the total backscattering. The radial distribution is steepest for high atomic number elements, which scatter strongly compared to weakly scattering low atomic number elements. However, even for strongly scattering elements, the backscattered electrons emerge over a significant fraction of the range. This characteristic impacts the spatial resolution that can be obtained with backscattered electron images. An example is shown in ☐ Fig. 2.15 for an interface between an aluminum-rich phase and a copper-rich phase ($CuAl_2$) in directionally solidified aluminum-copper eutectic alloy. The interfaces are perpendicular to the surface and are atomically sharp. The backscattered electron signal response as the beam is scanned across the interface is more than an order-of-magnitude broader (~300 nm) due to the lateral spreading of backscattering than would be predicted from the incident beam diameter alone (10 nm).

☐ **Fig. 2.12** Backscattered electron depth distributions at various energies in copper at 0° tilt

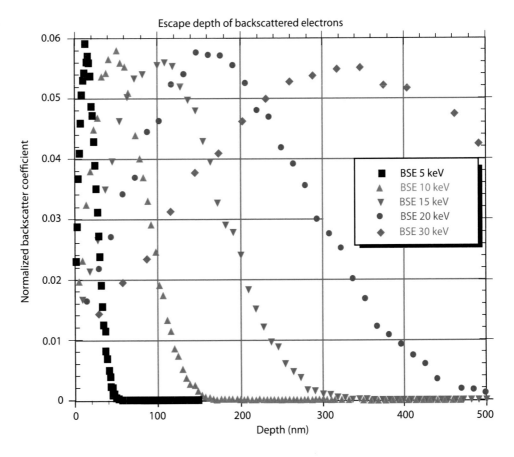

Escape depth of backscattered electrons

	BSE 5 keV
	BSE 10 keV
	BSE 15 keV
	BSE 20 keV
	BSE 30 keV

2

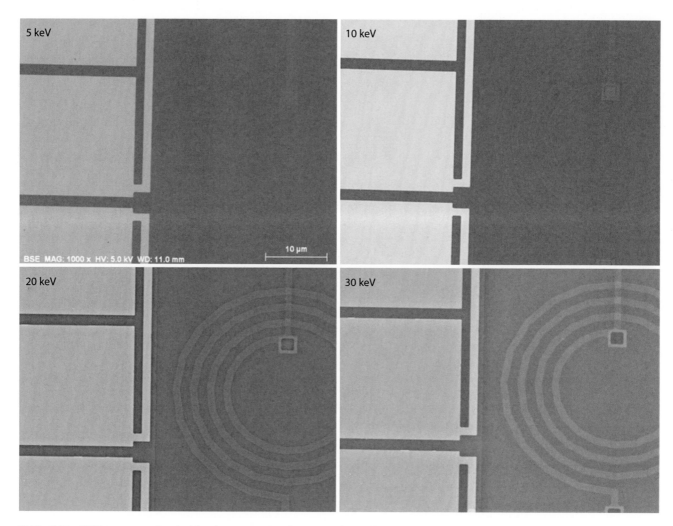

□ **Fig. 2.13** BSE images at various incident beam energies of a semiconductor device consisting of silicon and various metallization layers at different depths

□ **Fig. 2.14** Cumulative radial distribution of backscattered electrons in various bulk pure elements at 0° tilt showing determination of 90 % total backscattering radius

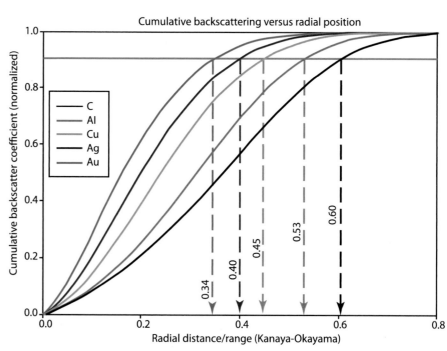

■ **Table 2.3** Fraction of the BSE radial distribution (r/R_{K-O}) to capture 90 % of backscattering	
C	0.60
Al	0.53
Cu	0.45
Ag	0.40
Au	0.34

2.2.5 Energy Distribution of Backscattered Electrons

As a beam electron travels in the specimen, inelastic scattering progressively diminishes the energy. When the trajectory of a beam electron intersects a specimen surface so that backscattering occurs, the backscattered electron will have lost a portion of the initial beam energy, E_0, with the amount lost depending on the length of the path within the specimen. The Monte Carlo simulation can record the exit energy of each backscattered electron, and from this data the energy distribution of BSE can be calculated, as shown in ■ Fig. 2.16a. The energy distribution is seen to extend from the incident beam energy down to zero energy. The energy distribution is sharply peaked at high fractional energy for a strong elastic scattering material such as gold, but the energy distribution is much broader and flatter for a weak elastic scattering material such as carbon. The backscattered electron energy spectra of ■ Fig. 2.16a can be used to calculate the cumulative backscattering distribution as a function of the fractional energy retained, E/E_0, as shown in ■ Fig. 2.16b. It is worth noting that even for weakly scattering carbon, more than half of the backscattered electrons retain at least half of the incident beam energy. The retained energy is a critical property that impacts the design of detectors for backscattered electrons.

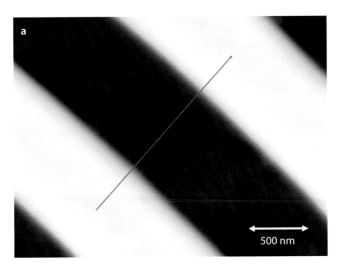

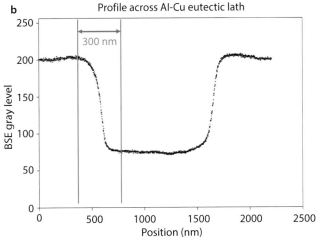

Profile across Al-Cu eutectic lath

2.3 Summary

Backscattered electrons form an important imaging signal for the SEM. A general understanding of the major properties of BSE provides the basis for interpreting images:
1. η vs. Z (atomic number)
2. η vs. θ (specimen tilt)
3. $\eta(\theta)$ vs. φ (emission angle relative to surface normal)
4. η vs. sampling depth
5. η vs. radial distance from beam
6. $\eta(E)$ vs. Z, energy distribution of BSE (■ Fig. 2.16)

■ **Fig. 2.15** **a** Backscattered electron image of a directionally solidified aluminum-copper eutectic alloy showing two phases: $CuAl_2$ (bright) and an Al-rich solid solution with copper. **b** Trace along the vector indicated in ■ Fig. 2.15a showing BSE signal profile

2

☐ Fig. 2.16 a Monte Carlo simulation of the energy of backscattered electrons for various pure elements at $E_0 = 20$ keV and 0° tilt. **b** Cumulative backscattered electron energy distribution for various pure elements at $E_0 = 20$ keV and 0° tilt

a

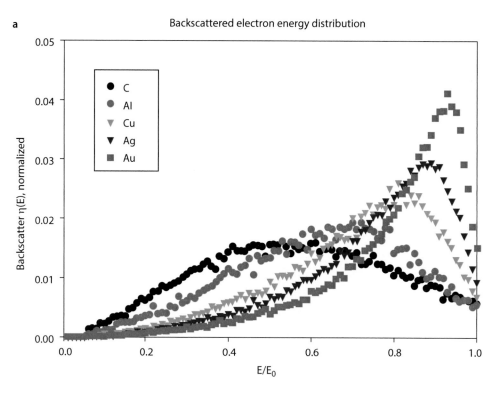

Backscattered electron energy distribution

b

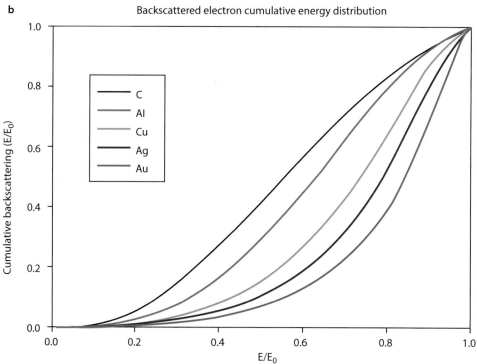

Backscattered electron cumulative energy distribution

References

Bishop H (1966) Some electron backscattering measurements for solid targets. In: Castaing R, Deschamps P, Philibert J (eds) Proceeding 4th international conferences on x-ray optics and microanalysis. Hermann, Paris, p 153

Heinrich KFJ (1966) Electron probe microanalysis by specimen current measurement. In: Castaing R, Deschamps P, Philibert J (eds)

Proceeding 4th international conferences on x-ray optics and microanalysis. Hermann, Paris, p 159

Niedrig H (1978) Physical background of electron backscattering. Scanning 1:17

Reuter W (1972) Electron backscattering as a function of atomic number. In: Shinoda G, Kohra K, Ichinokawa T (eds) Proceeding 6th International Cong x-ray optics and microanalysis. University of Tokyo Press, Tokyo, p 121

Secondary Electrons

Electronic supplementary material The online version of this chapter
(https://doi.org/10.1007/978-1-4939-6676-9_3) contains supplementary material,
which is available to authorized users.

3

3.1   Origin

Secondary electrons (SE) are created when inelastic scattering of the beam electrons ejects weakly bound valence electrons (in the case of ionically or covalently bonded materials) or conduction band electrons (in the case of metals), which have binding energies of ~ 1–15 eV to the parent atom(s). Secondary electrons are quantified by the parameter δ, which is the ratio of secondary electrons *emitted* from the specimen, N_{SE}, to the number of incident beam (primary) electrons, N_B:

$$\delta = N_{SE} / N_B \tag{3.1}$$

3.2   Energy Distribution

The most important characteristic of SE is their extremely low kinetic energy. Because of the large mismatch in relative velocities between the primary beam electron (incident energy 1–30 keV) and the weakly bound atomic electrons (1–15 eV ionization energy), the transfer of kinetic energy from the primary electron to the SE is relatively small, and as a result, the SE are ejected with low kinetic energy. After ejection, the SE must propagate through the specimen while undergoing inelastic scattering, which further decreases their kinetic energy. SE are generated along the complete trajectory of the beam electron within the specimen, but only a very small fraction of SE reach the surface with sufficient kinetic energy to exceed the surface energy barrier and escape. The energy spectrum of the secondary electrons that escape is peaked at only a few eV, as shown in ◘ Fig. 3.1a for a measurement of a copper target and an incident beam energy of $E_0 = 1$ keV. Above this peak, the intensity falls rapidly at higher kinetic energy (Koshikawa and Shimizu 1973). ◘ Figure 3.1b shows the cumulative intensity as a function of energy: 67 % of the secondary electrons from copper are emitted with less than 4 eV, and 90 % have less than 8.4 eV. Secondary electron production is considered to cease for kinetic energies above 50 eV, an arbitrary but reasonable value considering how sharply the energy distribution of ◘ Fig. 3.1a is skewed toward low energy. Inspection of the literature of secondary electrons confirms that the distribution for copper is generally representative of a large range of metals and other materials (e.g., Kanaya and Ono 1984).

3.3   Escape Depth of Secondary Electrons

The kinetic energy of SE is so low that it has a strong influence on the depth from which SE can escape from the specimen. While some inelastic scattering processes are absent because of the low kinetic energy of SE, nevertheless SE suffer rapid energy loss with distance traveled, limiting the range of an SE

to a few nanometers rather than the hundreds to thousands of nanometers for the energetic beam electrons and BSE. Thus, although SE are generated along the entire trajectory of a beam electron scattering in the target, only those SE generated close to a surface have a significant chance to escape. The probability of escape depends on the initial kinetic energy, the depth of generation, and the nature of the host material. Since there is a spectrum of initial kinetic energies, each energy represents a different escape probability and depth sensitivity. This complex behavior is difficult to measure directly, and instead researchers have made use of the Monte Carlo simulation to characterize the escape depth. ◘ Figure 3.2a shows the relative intensity of secondary electrons (over the energy range 0–50 eV) that escape from a copper target as a function of the depth of generation in the solid (Koshikawa and Shimizu 1974). ◘ Figure 3.2b shows this same data in the form of the cumulative secondary electron intensity as a function of initial generation depth. For copper, virtually no secondary electron escapes if it is created below approximately 8 nm from the surface, and 67 % of the secondary emission originates from a depth of less than 2.2 nm and 90 % from less than 4.4 nm. Kanaya and Ono (1984) modeled the mean secondary electron escape depth, d_{esc}, in terms of various material parameters:

$$d_{esc}\,(\text{nm}) = 0.267\, A\, I / \left(\rho Z^{0.66} \right) \tag{3.2}$$

where A is the atomic weight (g/mol), ρ is the density (g/cm^3), Z is the atomic number, and I is the first ionization potential (eV). When this model is applied to the solid elements of the Periodic Table, the complex behavior seen in ◘ Fig. 3.3 results. The mean escape depth varies from a low value of ~ 0.25 nm for Ce to a high value of 9 nm for Li. For copper, d_{esc} is calculated to be 1.8 nm, which can be compared to the 50 % escape value of 1.3 nm from the Monte Carlo simulation study in ◘ Fig. 3.2b. Systematic behavior of the atomic properties in Eq. 3.1 leads to systematic trends in the mean escape depth, with the low density alkali metals showing the largest values for the escape depth, while minima occur for the highest density elements in each period.

3.4   Secondary Electron Yield Versus Atomic Number

◘ Figure 3.4 shows a plot of the secondary electron coefficient as a function of atomic number for an incident beam energy of $E_0 = 5$ keV with data taken from *A Database of Electron-Solid Interactions* of Joy (2012). The measurements of δ are chaotic and inconsistent. For example, the values of δ for gold reported by various workers range from approximately 0.4 to 1.2. Oddly, all of these measured values may be "correct" in the sense that a valid, reproducible measurement was made on the particular specimen used. This behavior is really an indication of how difficult it is to make a

◻ **Fig. 3.1** **a** Secondary electron energy spectrum for copper with an incident beam energy of $E_0 = 1$ keV (Koshikawa and Shimizu 1973). **b** (Data from ◻ Fig. 3.1a replotted as the cumulative energy distribution)

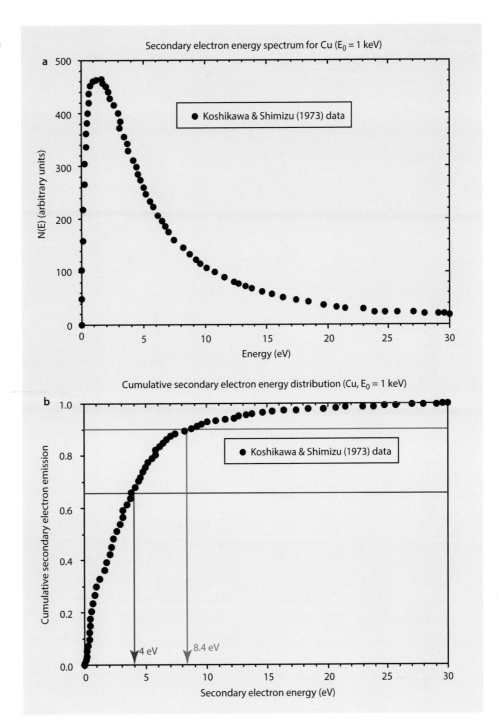

representative measurement of a property that results from very low energy electrons generated within and escaping from a very shallow layer below the surface. Thus, a surface modified by accumulations of oxide and contamination (e.g., adsorbed water, chemisorbed water, hydrocarbons, etc.) is likely to produce a value of δ that is different from the "ideal" pure element or pure compound value. If the specimen is pre-cleaned by ion bombardment in an ultra-high vacuum electron beam instrument (chamber pressure maintained below 10^{-8} Pa) which preserves the clean surface, and if the surface composition is confirmed to be that of the pure element or compound by a surface-specific

3

◼ **Fig. 3.2   a** Escape of secondary electrons from copper as a function of generation depth from Monte Carlo simulation (Koshikawa and Shimizu 1974). **b** (Data from ◼ Fig. 3.2a replotted to show the cumulative escape of secondary electrons as a function of depth of generation)

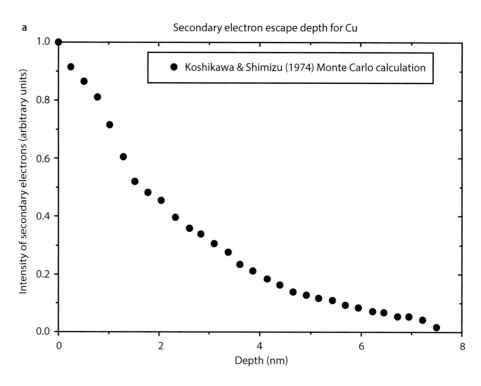

a   Secondary electron escape depth for Cu

● Koshikawa & Shimizu (1974) Monte Carlo calculation

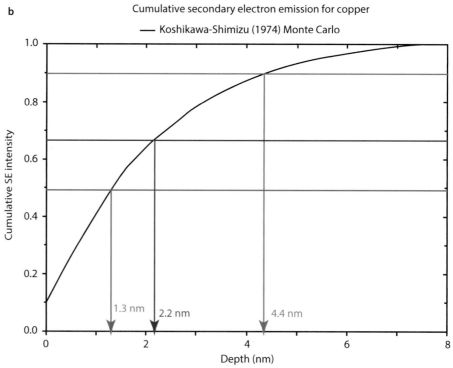

b   Cumulative secondary electron emission for copper

— Koshikawa-Shimizu (1974) Monte Carlo

measurement method such as Auger electron spectroscopy or X-ray photoelectron spectroscopy, then the measured secondary electron coefficient is likely to be representative of the pure substance. However, the surfaces of most specimens examined in the conventional-vacuum SEM (chamber pressure $\sim 10^{-4}$ Pa) or a variable pressure SEM (chamber pressure from 10^{-4} Pa to values as high as 2500 Pa) are not likely to be that of pure substances, but are almost inevitably covered with a complex mixture of oxides, hydrocarbons, and chemisorbed water molecules that quickly redeposit

■ **Fig. 3.3** Mean secondary electron escape depth for various materials as modeled by Kanaya and Ono (1984)

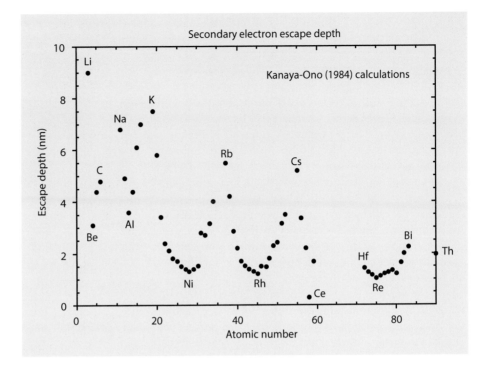

■ **Fig. 3.4** Secondary electron coefficient as a function of atomic number for $E_0 = 5$ keV (Data from the secondary electron database of Joy (2012))

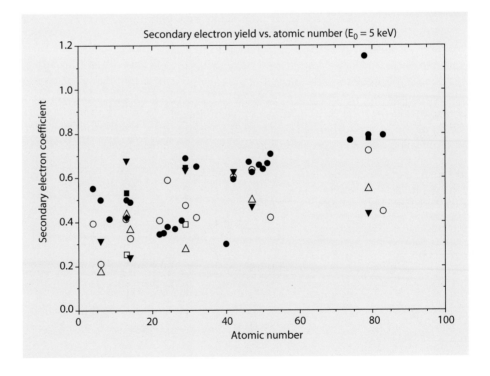

at such elevated pressures even when ion beam cleaning is utilized to expose the "true" surface. The effective secondary electron coefficient of a "real" material under typical SEM or VP-SEM vacuum conditions is unlikely to produce a consistent, predictable response as a function of the composition of the nominal substance under examination.

Thus, while compositionally dependent secondary electron signals may be occasionally observed, they are generally not predictable and reproducible, which is the critical basis for establishing a useful contrast mechanism such as that found for backscattered electrons as a function of atomic number.

3

3.5   Secondary Electron Yield Versus Specimen Tilt

When the secondary electron coefficient is measured as a function of the specimen tilt angle, θ (i.e., the specimen inclination to the beam, where a tilt of 0° means that the beam is perpendicular to the surface), a monotonic increase with tilt is observed, as shown for copper at two different incident beam energies in ◻ Fig. 3.5, which is taken from the measurements of Koshikawa and Shimizu (1973). This increase in δ with θ can be understood from the geometric argument presented schematically in ◻ Fig. 3.6. As the primary beam enters the specimen, the rate of secondary electron production is effectively constant along the path that lies within the shallow secondary electron escape depth because the beam electrons have not yet undergone sufficient scattering to modify their energies or trajectories. The length of the primary beam path within the depth of escape, d_{esc}, increases as the secant of the tilt angle. Assuming that the number of secondary electrons that eventually escape will be proportional to the number produced in this near surface region, the secondary electron coefficient is similarly expected to rise with the secant of the tilt angle. As shown in ◻ Fig. 3.5, the measured dependence of δ upon θ does not rise as fast as the secant relation that the simple geometric argument predicts. This deviation from the secant function model in ◻ Fig. 3.6 is due to the large contribution of secondary electrons produced by the exiting backscattered electrons which follow different trajectories through the escape layer, as discussed below.

The monotonic dependence of the secondary electron coefficient on the local surface inclination is an important factor in producing topographic contrast that reveals the shape of an object.

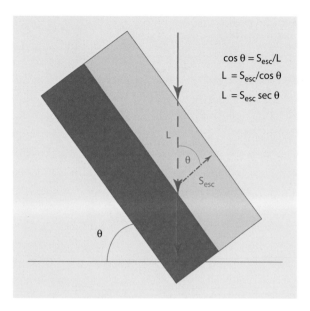

◻ **Fig. 3.6**   Simple geometric argument predicting that the secondary electron coefficient should follow a secant function of the tilt angle

3.6   Angular Distribution of Secondary Electrons

When a secondary electron is generated within the escape depth below the surface, as shown in ◻ Fig. 3.7a, the shortest path to the surface, s, lies along the direction parallel to the local surface normal. For any other trajectory at an angle φ relative to this surface normal, the path length increases in length as $s/\cos\varphi$. The probability of secondary electron escape decreases as the escape path length increases, so that the angular distribution of emitted

◻ **Fig. 3.5**   Behavior of the secondary electron coefficient as a function of surface tilt (Data of Koshikawa and Shimizu (1973)) showing a monotonic increase with tilt angle but at a much slower rate than would be predicted by a secant function

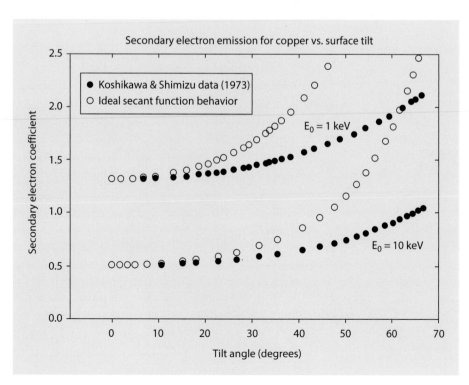

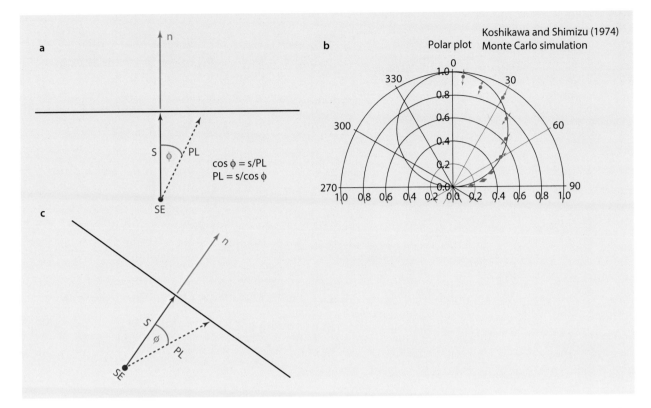

Fig. 3.7 **a** Dependence of the secondary electron escape path length on the angle relative to the surface normal. The probability of escape decreases as this path length increases. **b** Angular distribution of secondary electrons as a function of the angle relative to the surface normal as simulated by Monte Carlo calculations (Koshikawa and Shimizu 1974) compared to a cosine function. **c** The escape path length situation of **◻** Fig. 3.7a for the case of a tilted specimen. A cosine dependence relative to the surface normal is again predicted

secondary electrons is expected to follow a cosine relation with the emergence angle relative to the local surface normal. Behavior close to a cosine relation is seen in the Monte Carlo simulation of Koshikawa and Shimizu (1974) in **◻** Fig. 3.7b.

Even when the surface is highly tilted relative to the beam, the escape path length situation for a secondary electron generated below the surface is identical to the case for normal beam incidence, as shown in **◻** Fig. 3.7c. Thus, the secondary electron trajectories follow a cosine distribution relative to the local surface normal regardless of the specimen tilt.

3.7 Secondary Electron Yield Versus Beam Energy

The secondary electron coefficient increases as the incident beam energy decreases, as shown for copper in **◻** Fig. 3.8a for the conventional beam energy range ($5 \text{ keV} \leq E_0 \leq 30 \text{ keV}$) and in **◻** Fig. 3.8b for the low beam energy range ($E_0 < 5 \text{ keV}$). This behavior arises from two principal factors: (1) as the beam electron energy decreases, the rate of energy loss, dE/ds, increases so that more energy is deposited per unit of beam electron path length leading to more secondary electron generation per unit of path length; and (2) the range of the beam

electrons is reduced so more of that energy is deposited and more secondary electrons are generated in the near surface region from which secondary electrons can escape. This is a general behavior found across the Periodic Table, as seen in the plots for C, Al, Cu, Ag, and Au in **◻** Fig. 3.8c.

3.8 Spatial Characteristics of Secondary Electrons

As the beam electrons enter the sample surface, they begin to generate secondary electrons in a cylindrical volume whose cross section is defined by the footprint of the beam on the entrance surface and whose height is the escape depth of the SE, as shown schematically in **◻** Fig. 3.9 These entrance surface SE, designated the SE_1 class, preserve the lateral spatial resolution information defined by the dimensions of the focused beam and are similarly sensitive to the properties of the near surface region due to the shallow scale of their origin. As the beam electrons move deeper into the solid, they continue to generate SE, but these SE rapidly lose their small initial kinetic energy and are completely reabsorbed within an extremely short range. However, for those beam electrons that subsequently undergo enough scattering to return to the entrance surface to emerge as backscattered electrons (or reach any

3

◘ **Fig. 3.8** **a** Behavior of the secondary electron coefficient as a function of incident beam energy for the conventional beam energy range, $E_0 = 5$–30 keV (Data of Moncrieff and Barker (1976)). **b** Behavior of the secondary electron coefficient as a function of incident beam energy for the low beam energy range, $E_0 < 5$ keV (data) (Data of Bongeler et al. (1993)). **c** Dependence of the secondary electron coefficient on incident beam energy for C, Al, Cu, Ag, and Au (Reimer and Tolkamp 1980)

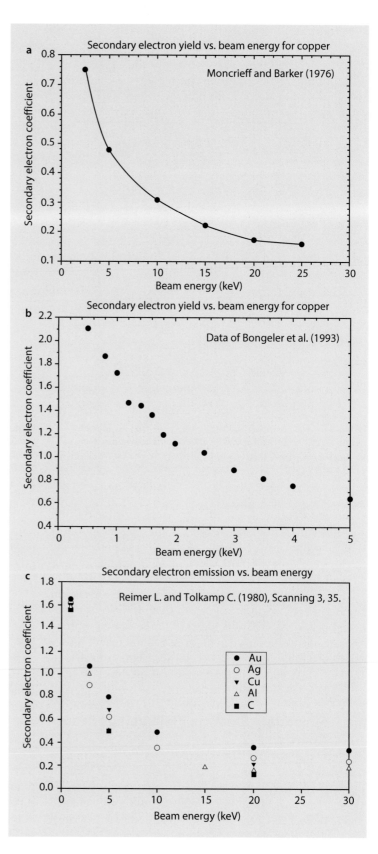

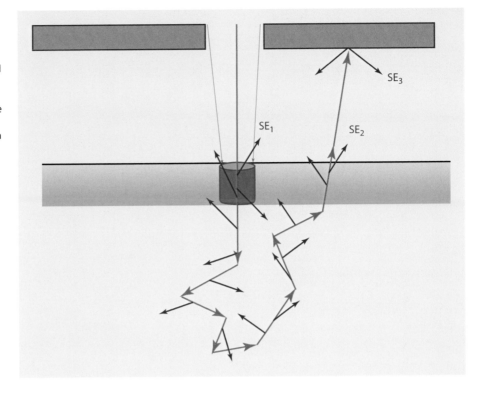

■ **Fig. 3.9** Schematic diagram showing the origins of the SE$_1$, SE$_2$, and SE$_3$ classes of secondary electrons. The SE$_1$ class carries the lateral and near-surface spatial information defined by the incident beam, while the SE$_2$ and SE$_3$ classes actually carry backscattered electron information. The blue rectangle represents the escape depth for SE, and the cylinder represents the volume from which the SE$_1$ escape

other surface for specimens with more complex topography than a simple flat bulk target), the SE that they continue to generate as they approach the surface region will escape and add to the total secondary electron production, as shown in ■ Fig. 3.9. This class of SE is designated SE$_2$ and they are indistinguishable from the SE$_1$ class based on their energy and angular distributions. However, because of their origin from the backscattered electrons, the SE$_2$ class actually carries the degraded lateral spatial distribution of the BSE: because the relative number of SE$_2$ rises and falls with backscattering, the SE$_2$ signal actually carries the same information as BSE. That is, the relative number of the SE$_2$ scales with whatever specimen property affects electron backscattering. Finally, the BSE that leave the specimen are energetic, and after traveling millimeters to centimeters in the specimen chamber, these BSE are likely to hit other metal surfaces (objective lens polepiece, chamber walls, stage components, etc.), generating a third set of secondary electrons designated SE$_3$. The SE$_3$ class again represents BSE information, including the degraded spatial resolution, not true SE$_1$ information and resolution. The SE$_1$ and SE$_2$ classes represent an inherent property of a material, while the SE$_3$ class depends on the details of the SEM specimen chamber. Peters (1984) measured the three secondary electron classes for thin and thick gold targets to estimate the relative populations of each class:

Incident beam footprint, high resolution, SE$_1$ (9 %)
BSE generated at specimen, low resolution, SE$_2$ (28 %)
BSE generated remotely on lens, chamber walls, SE$_3$ (61 %)

A small SE contribution designated the SE$_4$ class arises from pre-specimen instrumental sources such the final aperture (2 %) that depends in detail on the instrument construction (apertures, magnetic fields, etc.). These measurements show

that for gold the sum of the SE$_2$ and SE$_3$ classes which actually carry BSE is nearly ten times larger than the high resolution, high surface sensitivity SE$_1$ component. These three classes of secondary electrons influence SEM images of compositional structures and topographic structures in complex ways. The appearance of the SE image of a structure depends on the details of the secondary electron emission and the properties of the secondary electron detector used to capture the signal, as discussed in detail in the image formation module.

References

Bongeler R, Golla U, Kussens M, Reimer L, Schendler B, Senkel R, Spranck M (1993) Electron-specimen interactions in low voltage scanning electron microscopy. Scanning 15:1
Joy D (2012) Can be found in chapter 3 on SpringerLink: http://link.springer.com/chapter/10.1007/978-1-4939-6676-9_3
Kanaya K, Ono S (1984) Interaction of electron beam with the target in scanning electron microscope. In: Kyser DF, Niedrig H, Newbury DE, Shimizu R (eds) Electron interactions with solids. SEM, Inc, Chicago, pp 69–98
Koshikawa T, Shimizu R (1973) Secondary electron and backscattering measurements for polycrystalline copper with a retarding-field analyser. J Phys D Appl Phys 6:1369
Koshikawa T, Shimizu R (1974) A Monte Carlo calculation of low-energy secondary electron emission from metals. J Phys D Appl Phys 7:1303
Peters K-R (1984) Generation, collection and properties of an SE-I enriched signal suitable for high resolution SEM on bulk specimens. In: Kyser DF, Niedrig H, Newbury DE, Shimizu R (eds) Electron beam interactions with solids. SEM, Inc, AMF O'Hare, p 363
Moncrieff DA, Barker PR (1976) Secondary electron emission in the scanning electron microscope. Scanning 1:195
Reimer L, Tolkamp C (1980) Measuring the backscattering coefficient and secondary electron yield inside a scanning electron microscope. Scanning 3:35

X-Rays

© Springer Science+Business Media LLC 2018
J. Goldstein et al., *Scanning Electron Microscopy and X-Ray Microanalysis*,
https://doi.org/10.1007/978-1-4939-6676-9_4

4

4.1   Overview

Energetic beam electrons stimulate the atoms of the specimen to emit "characteristic" X-ray photons with sharply defined energies that are specific to each atom species. The critical condition for generating characteristic X-rays is that the energy of the beam electron must exceed the electron binding energy, the critical ionization energy E_c, for the particular atom species and the K-, L-, M-, and/or N- atomic shell(s). For efficient excitation, the incident beam energy should be at least twice the critical excitation energy, $E_0 > 2 E_c$. Characteristic X-rays can be used to identify and quantify the elements present within the interaction volume. Simultaneously, beam electrons generate *bremsstrahlung*, or braking radiation, which creates a continuous X-ray spectrum, the "X-ray continuum," whose energies fill the range from the practical measurement threshold of 50 eV to the incident beam energy, E_0. This continuous X-ray spectrum forms a spectral background beneath the characteristic X-rays which impacts accurate measurement of the characteristic X-rays and determines a finite concentration limit of

detection. X-rays are generated throughout a large fraction of the electron interaction volume. The spatial resolution, lateral and in-depth, of electron-excited X-ray microanalysis can be roughly estimated with a modified Kanaya–Okayama range equation or much more completely described with Monte Carlo electron trajectory simulation. Because of their generation over a range of depth, X-rays must propagate through the specimen to reach the surface and are subject to photoelectric absorption which reduces the intensity at all photon energies, but particularly at low energies.

4.2   Characteristic X-Rays

4.2.1   Origin

The process of generating characteristic X-rays is illustrated for a carbon atom in ◘ Fig. 4.1. In the initial ground state, the carbon atom has two electrons in the K-shell bound to the nucleus of the atom with an "ionization energy" E_c (also known as the "critical excitation energy," the "critical

◘ **Fig. 4.1**   Schematic diagram of the process of X-ray generation: inner shell ionization by inelastic scattering of an energetic beam electron that leaves the atom in an elevated energy state which it can lower by either of two routes involving the transition of an L-shell electron to fill the K-shell vacancy: (1) the Auger process, in which the energy difference $E_K - E_L$ is transferred to another L-shell electron, which is ejected with a characteristic energy: $E_K - E_L - E_L$; (2) photon emission, in which the energy difference $E_K - E_L$ is expressed as an X-ray photon of characteristic energy

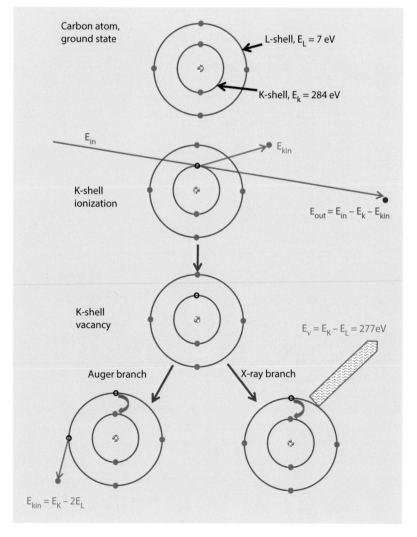

absorption energy," and the "K-edge energy") of 284 eV and four electrons in the L-shell, two each in the L_1 and the L_2 subshells bound to the atom, with an ionization energy of 7 eV. An incident energetic beam electron having initial kinetic energy $E_{in} > E_c$ can scatter inelastically with a K-shell atomic electron and cause its ejection from the atom, providing the beam electron transfers to the atomic electron kinetic energy at least equal to the ionization energy, which is the minimum energy necessary to promote the atomic electron out of the K-shell beyond the effective influence of the positive nuclear charge. The total kinetic energy transferred to the K-shell atomic electron can range up to half the energy of the incident electron. The outgoing beam electron thus suffers energy loss corresponding to the carbon K-shell ionization energy $E_K = 284$ eV plus whatever additional kinetic energy is imparted:

$$E_{out} = E_{in} - E_K - E_{kin} \qquad (4.1)$$

The ionized carbon atom is left with a vacancy in the K-shell which places it in a raised energy state that can be lowered through the transition of an electron from the L-shell to fill the K-vacancy. The difference in energy between these shells must be expressed through one of two possible routes:

1. The left branch in ◼ Fig. 4.1 involves the transfer of this K–L inter-shell transition energy difference to another L-shell electron, which is then ejected from the atom with a specific kinetic energy:

$$E_{kin} = E_K - E_L - E_L = 270 \, \text{eV} \qquad (4.2a)$$

This process leaves the atom with two L-shell vacancies for subsequent vacancy-filling transitions. This ejected electron is known as an "Auger electron," and measure-ment of its characteristic kinetic energy can identify the atom species of its origin, forming the physical basis for "Auger electron spectroscopy."

2. The right branch in ◼ Fig. 4.1 involves the creation of an X-ray photon to carry off the inter-shell transition energy:

$$E_v = E_K - E_L = 277 \, \text{eV} \qquad (4.2b)$$

Because the energies of the atomic shells of an element are sharply defined, the shell difference is also a sharply defined quantity, so that the resulting X-ray photon has an energy that is characteristic of the particular atom species and the shells involved and is thus designated as a "characteristic X-ray." Characteristic X-rays are emitted uniformly in all directions over the full unit sphere with 4π steradians solid angle. Extensive tables of characteristic X-ray energies for elements with $Z \geq 4$ (beryllium) are provided in the database embedded within the DTSA-II software. The characteristic X-ray photon energy has a very narrow range of just a few electronvolts depending on atomic number, as shown in ◼ Fig. 4.2 for the K–L_3 transition.

4.2.2 Fluorescence Yield

The Auger and X-ray branches in ◼ Fig. 4.1 are not equally probable. For a carbon atom, characteristic X-ray emission only occurs for approximately 0.26 % of the K-shell ionizations. The fraction of the ionizations that produce photons is known as the "fluorescence yield," ω. Most carbon K-shell ionizations thus result in Auger electron emission. The fluorescence yield is strongly dependent on the atomic number of the atom, increasing rapidly with Z, as shown in ◼ Fig. 4.3a for K-shell ionizations. L-shell and M-shell fluorescence

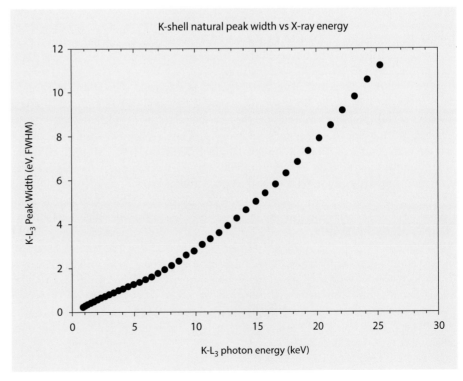

◼ **Fig. 4.2** Natural width of K-shell X-ray peaks up to 25 keV photon energy (Krause and Oliver 1979)

K-shell natural peak width vs X-ray energy

4

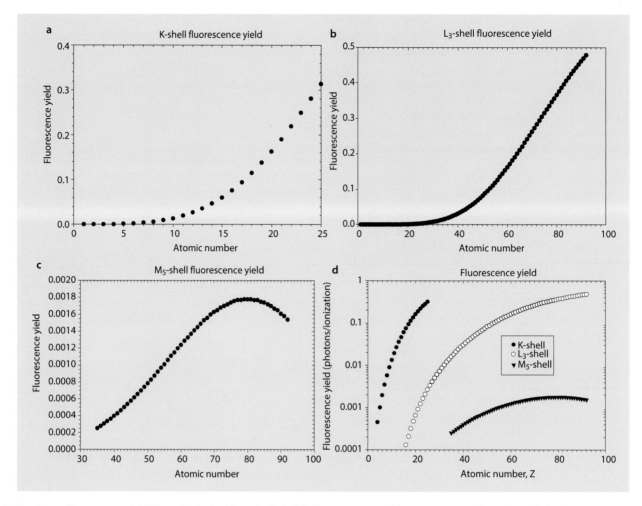

Fig. 4.3 **a** Fluorescence yield (X-rays/ionization) from the K-shell. **b** Fluorescence yield (X-rays/ionization) from the L$_3$-shell. **c** Fluorescence yield (X-rays/ionization) from the M$_5$-shell. **d** Comparison of fluorescence yields from the K-, L$_3$- and M$_5$- shells (Crawford et al. 2011)

yields are shown in ▪ Fig. 4.3b, c; and K-, L-, and M-shell yields are compared in ▪ Fig. 4.3d (Crawford et al. 2011). From ▪ Fig. 4.3d, it can be observed that, when an element can be measured with two different shells, $\omega_K > \omega_L > \omega_M$.

The shell transitions for carbon are illustrated in the shell energy diagram shown in ▪ Fig. 4.4a. Because of the small number of carbon atomic electrons, the shell energy values are limited, and only one characteristic X-ray energy is possible for carbon with a value of 277 eV. (The apparent possible transition from the L$_1$-shell to the K-shell is forbidden by the quantum mechanical rules that govern these inter-shell transitions.)

4.2.3 X-Ray Families

As the atomic number increases, the number of atomic electrons increases and the shell structure becomes more complex. For sodium, the outermost electron occupies the M-shell, so that a K-shell vacancy can be filled by a transition from the L-shell or the M-shell, producing two different characteristic X-rays, designated

$$"K - L_{2,3}" ("K\alpha") E_X = E_K - E_L = 1041\,\text{eV} \quad \textbf{(4.3a)}$$

$$"K - M" ("K\beta") E_X = E_K - E_M = 1071\,\text{eV} \quad \textbf{(4.3b)}$$

For atoms with higher atomic number than sodium, additional possibilities exist for inter-shell transitions, as shown in ▪ Fig. 4.4b, leading to splitting of the $K-L_{2,3}$ into $K-L_3$ and $K-L_2$ ($K\alpha$ into $K\alpha_1$ and $K\alpha_2$), and similarly for $K\beta$ into $K\beta_1$ and $K\beta_2$, which can be observed with energy dispersive spectrometry for X-rays with energies above 20 keV.

As these additional inter-shell transitions become possible, increasingly complex "families" of characteristic X-rays are created, as shown in the energy diagrams of ▪ Fig. 4.4c for L-shell X-rays, and 4.4d for M-shell X-rays. Only transitions that lead to X-rays that are measurable on a practical basis with energy dispersive X-ray spectrometry are shown. (There are, for example, at least 25 L-shell transitions that are possible for a heavy element such as gold, but most are of such low abundance or are so close in energy to a more abundant transition as to be undetectable by energy dispersive X-ray spectrometry.)

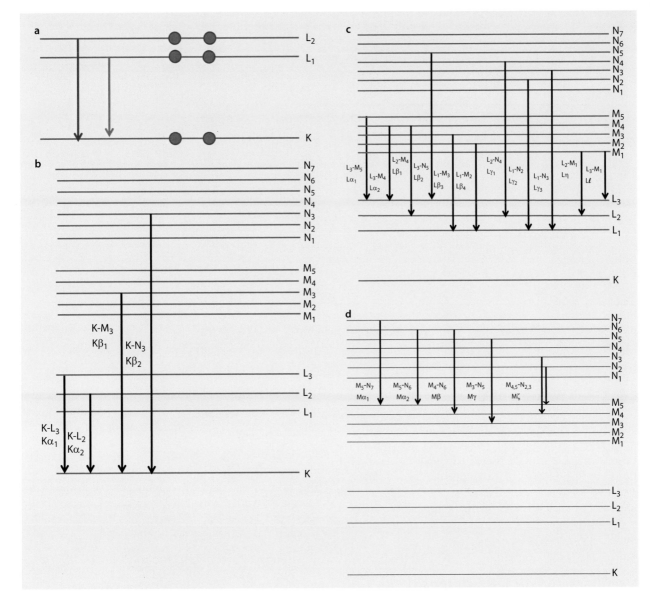

Fig. 4.4 **a** Atomic shell energy level diagram for carbon illustrating the permitted shell transition K–L$_2$ (shown in *green*) and the forbidden transition K–L$_1$ (shown in *red*). **b** Atomic shell energy level diagram illustrating possible K-shell vacancy-filling transitions. **c** Atomic shell energy level diagram illustrating possible L-shell vacancy-filling transitions. **d** Atomic shell energy level diagram illustrating some possible M-shell vacancy-filling transitions

4.2.4 X-Ray Nomenclature

Two systems are in use for designating X-rays. The traditional but now archaic Siegbahn system lists the shell where the original ionization occurs followed by a Greek letter or other symbol that suggests the order of the family members by their relative intensity, $\alpha > \beta > \gamma > \eta > \zeta$. For closely related members, numbers are also attached, for example, Lβ_1 through Lβ_{15}. Additionally, Latin letters are used for the complex minor L-shell family members: l, s, t, u, and v. While still the predominant labeling system used in commercial X-ray microanalysis software systems, the Siegbahn system has been officially replaced by the International Union of Pure and Applied Chemistry (IUPAC) labeling protocol in which the first term denotes the shell or subshell where the original ionization occurs while the second term indicates the subshell from which the electron transition occurs to fill the vacancy; for example, Kα_1 is replaced by K-L$_3$ for a K-shell ionization filled from the L$_3$ subshell. ◘ Table 4.1 gives the correspondence between the Siegbahn and IUPAC labeling schemes for the characteristic X-rays likely to be detected by energy dispersive X-ray spectrometry. Note that for the M-shell, there are minor family members detectable by EDS for which there are no Siegbahn designations.

Table 4.1 Correspondence between the Siegbahn and IUPAC nomenclature protocols (restricted to characteristic X-rays observed with energy dispersive X-ray spectrometry and photon energies from 100 eV to 25 keV)

Siegbahn	IUPAC	Siegbahn	IUPAC	Siegbahn	IUPAC
$K\alpha_1$	$K\text{-}L_3$	$L\alpha_1$	$L_3\text{-}M_5$	$M\alpha_1$	$M_5\text{-}N_7$
$K\alpha_2$	$K\text{-}L_2$	$L\alpha_2$	$L_3\text{-}M_4$	$M\alpha_2$	$M_5\text{-}N_6$
$K\beta_1$	$K\text{-}M_3$	$L\beta_1$	$L_2\text{-}M_4$	$M\beta$	$M_4\text{-}N_6$
$K\beta_2$	$K\text{-}N_{2,3}$	$L\beta_2$	$L_3\text{-}N_5$	$M\gamma$	$M_3\text{-}N_5$
		$L\beta_3$	$L_1\text{-}M_3$	$M\zeta$	$M_{4,5}\text{-}N_{2,3}$
		$L\beta_4$	$L_1\text{-}M_2$		$M_3\text{-}N_1$
		$L\gamma_1$	$L_2\text{-}N_4$		$M_2\text{-}N_1$
		$L\gamma_2$	$L_1\text{-}N_2$		$M_3\text{-}N_{4,5}$
		$L\gamma_3$	$L_1\text{-}N_3$		$M_3\text{-}O_1$
		$L\gamma_4$	$L_1\text{-}O_4$		$M_3\text{-}O_{4,5}$
		$L\eta$	$L_2\text{-}M_1$		$M_2\text{-}N_4$
		Ll	$L_3\text{-}M_1$		

4.2.5 X-Ray Weights of Lines

Within these families, the relative abundances of the characteristic X-rays are not equal. For example, for sodium the ratio of the $K\text{-}L_{2,3}$ to $K\text{-}M$ is approximately 150:1, and this ratio is a strong function of the atomic number, as shown in ◘ Fig. 4.5a for the K-shell (Heinrich et al. 1979). For the L-shell and M-shell, the X-ray families have more members, and the relative abundances are complex functions of atomic number, as shown in ◘ Fig. 4.5b, c.

4.2.6 Characteristic X-Ray Intensity

Isolated Atoms

When isolated atoms are considered, the probability of an energetic electron with energy E (keV) ionizing an atom by ejecting an atomic electron bound with ionization energy E_c (keV) can be expressed as a cross section, Q_I:

$$Q_I\left(\text{ionizations}/\left[e^-\left(\text{atom}/\text{cm}^2\right)\right]\right)$$
$$= 6.51\times10^{-20}\left[\left(n_s\,b_s\right)/E\,E_c\right]\log_e\left(c_s E/E_c\right) \quad (4.4)$$

where n_s is the number of electrons in the shell or subshell (e.g., $n_K = 2$), and b_s and c_s are constants for a given shell (e.g., $b_K = 0.35$ and $c_K = 1$) (Powell 1976). The behavior of the ionization cross section for the silicon K-shell as a function of the energy of the energetic beam electron is shown in ◘ Fig. 4.6. Starting with a zero value at 1.838 keV, the K-shell ionization energy for silicon, the cross section rapidly increases to a peak value, and then slowly decreases with further increases in the beam energy.

The relationship of the energy of the exciting electron to the ionization energy of the atomic electron is an important parameter and is designated the "overvoltage," U:

$$U = E/E_c \quad (4.5a)$$

The overvoltage that corresponds to the incident beam energy, E_0, which is the maximum value because the beam electrons subsequently lose energy due to inelastic scattering as they progress through the specimen, is designated as U_0:

$$U_0 = E_0/E_c \quad (4.5b)$$

For ionization to occur followed by X-ray emission, $U > 1$. With this definition for U, Eq. (4.4) can be rewritten as

$$Q_I\left(\text{ionizations}/\left[e^-\left(\text{atom}/\text{cm}^2\right)\right]\right)$$
$$= 6.51\times10^{-20}\left[\left(n_s\,b_s\right)/U\,E_c^2\right]\log_e\left(c_s U\right) \quad (4.6)$$

The critical excitation energy is a strong function of the atomic number of the element and of the particular shell, as shown in ◘ Fig. 4.7. Thus, for a specimen that consists of several different elements, the initial overvoltage U_0 will be different for each element, which will affect the relative generation intensities of the different elements.

X-Ray Production in Thin Foils

Thin foils may be defined as having a thickness such that most electrons pass through the foil without suffering elastic scattering out of the ideal beam cylinder (defined by the circular beam footprint on the entrance and exit surfaces and the foil thickness) and without suffering significant

◘ **Fig. 4.5** **a** Relative abundance of the K-L$_{2,3}$ to K-M (Kα to Kβ) (Heinrich et al. 1979). **b** Relative abundance of the L-shell X-rays, L$_3$-M$_{4,5}$ (Lα$_{1,2}$) = 1 (Crawford et al. 2011). **c** Relative abundance of the M-shell X-rays, M$_5$-N$_{6,7}$ (Mα) = 1 (Crawford et al. 2011)

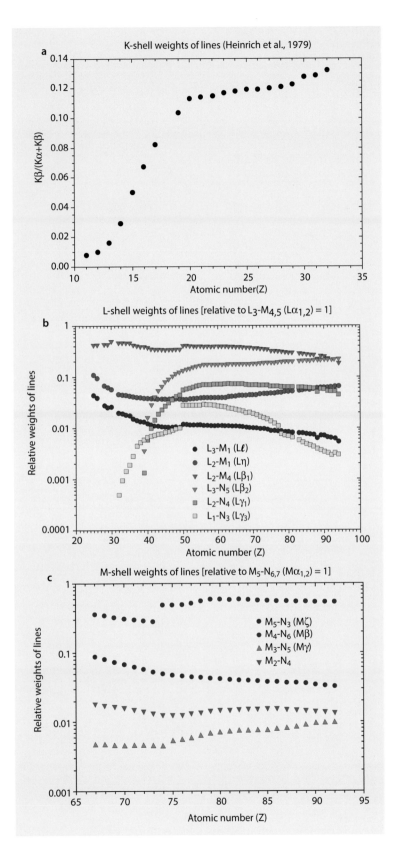

4

☐ Fig. 4.6 Ionization cross section for the silicon K-shell calculated with Eq. 4.4

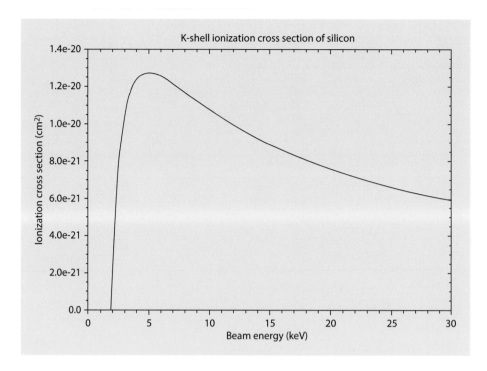

☐ Fig. 4.7 Critical ionization energy for the K-, L-, and M-shells

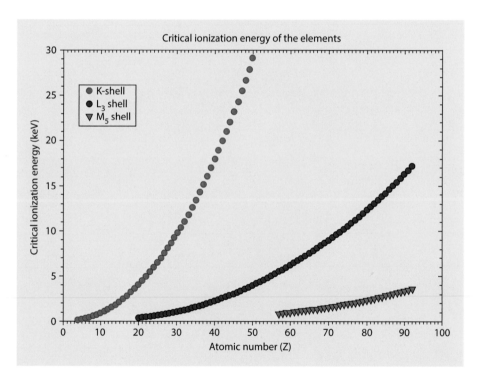

energy loss. The X-ray production in a thin foil of thickness t can be estimated from the cross section by calculating the effective density of atom targets within the foil:

$$n_X \left[\text{photons} / e^- \right] = Q_I \left[\text{ionizations} / e^- \left(\text{atom} / \text{cm}^2 \right) \right]$$
$$\times \omega \left[\text{X-rays} / \text{ionization} \right] \times N_0 \left[\text{atoms} / \text{mole} \right]$$
$$\times \left(1 / A \right) \left[\text{moles} / \text{g} \right] \times \rho \left[\text{g} / \text{cm}^3 \right]$$
$$\times t \left[\text{cm} \right] = Q_I \times \omega \times N_0 \times \rho \times t / A \quad \text{(4.7)}$$

where A is the atomic weight and N_0 is Avogadro's number.

When several elements are mixed at the atomic level in a thin specimen, the relative production of X-rays from different elements depends on the cross section and fluorescence yield, as given in Eq. 4.7, and also on the partitioning of the X-ray production among the various possible members of the X-ray families, as plotted in ☐ Fig. 4.5a–c. The relative production for the most intense transition in each X-ray family is plotted in ☐ Fig. 4.8 for $E_0 = 30$ keV. ☐ Figure 4.8 reveals

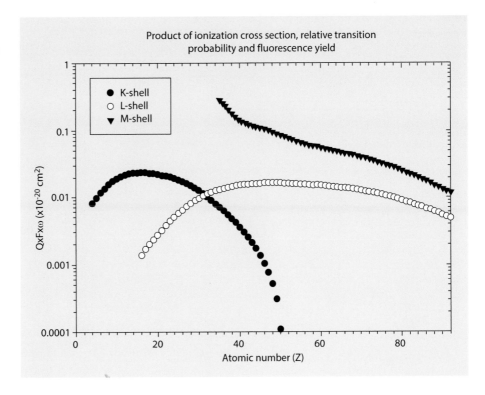

Fig. 4.8 Product of the ionization cross section, the fluorescence yield, and the relative weights of lines for the most intense member of the K-, L-, and M-shells for $E_0 = 30$ keV

strong differences in the relative abundance of the X-rays produced by different elements. This plot also reveals that over certain atomic number ranges, two different atomic shells can be excited for each element, for example, K and L for $Z = 16$ to $Z = 50$, and L and M for $Z = 36$ to $Z = 92$. For lower values of E_0, these atomic number ranges will be diminished.

X-Ray Intensity Emitted from Thick, Solid Specimens

A thick specimen is one with sufficient thickness so that it contains the full electron interaction volume, which generally requires a thickness of at least a few micrometers for most choices of composition and incident beam energy. Within the interaction volume, the complete range of elastic and inelastic scattering events occur. X-ray generation for each atom species takes place across the full energy range of the ionization cross section from the initial value corresponding to the energy of the incident beam as it enters the specimen down to the ionization energy of each atom species. Based upon experimental measurements, the X-ray intensity emitted from thick targets is found to follow an expression of the form

$$I \approx i_p \left[\left(E_0 - E_c \right) / E_0 \right]^n \approx i_p \left[U - 1 \right]^n \quad \text{(4.8)}$$

where i_p is the beam current, and n is a constant depending on the particular element and shell (Lifshin et al. 1980). The value of n is typically in the range 1.5–2.0. Equation 4.8 is plotted for an exponent of $n = 1.7$ in ◨ Fig. 4.9. The intensity

rises rapidly from a zero value at $U = 1$. For a reasonably efficient degree of X-ray excitation, it is desirable to select E_0 so that $U_0 > 2$ for the highest value of E_c among the elements of interest.

4.3 X-Ray Continuum (*bremsstrahlung*)

Simultaneously with the inner shell ionization events that lead to characteristic X-ray emission, a second physical process operates to generate X-rays, the "braking radiation," or *bremsstrahlung*, process. As illustrated in ◨ Fig. 4.10, because of the repulsion that the beam electron experiences in the negative charge cloud of the atomic electrons, it undergoes deceleration and loses kinetic energy, which is released as a photon of electromagnetic radiation. The energy lost due to deceleration can take on any value from a slight deceleration involving the loss of a few electron volts up to the loss of the total kinetic energy carried by the beam electron in a single event. Thus, the *bremsstrahlung* X-rays span all energies from a practical threshold of 100 eV up to the incident beam energy, E_0, which corresponds to an incident beam electron suffering total energy loss by deceleration in the Coulombic field of a surface atom as the beam electron enters the target and before it has lost any energy in any other inelastic scattering events. The braking radiation process thus forms a continuous energy spectrum, also referred to as the "X-ray continuum," from 100 eV to E_0, which is the so-called Duane–Hunt limit. The X-ray continuum forms a background beneath any characteristic X-rays produced by the atoms. The *bremsstrahlung* process is anisotropic, being somewhat peaked in the

Fig. 4.9 Characteristic X-ray intensity emitted from a thick specimen; exponent $n = 1.7$

4

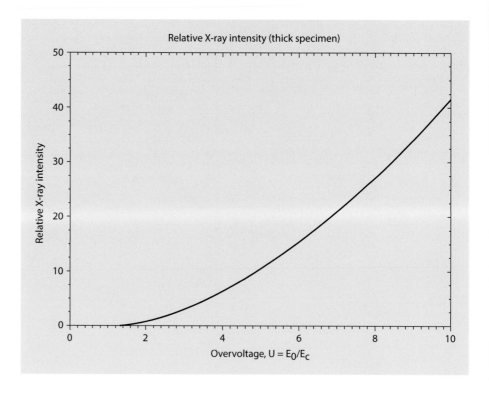

Fig. 4.10 Schematic illustration of the braking radiation (*bremsstrahlung*) process giving rise to the X-ray continuum

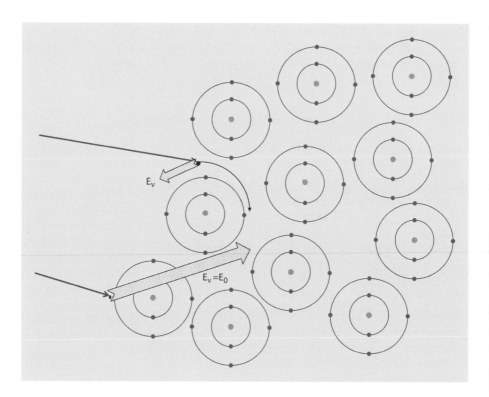

direction of the electron travel. In thin specimens where the beam electron trajectories are nearly aligned, this anisotropy can result in a different continuum intensity in the forward direction along the beam relative to the backward direction. However, in thick specimens, the near-randomization of the beam electron trajectory segments by elastic scattering effectively smooths out this anisotropy, so that the X-ray continuum is effectively rendered isotropic.

4.3.1 X-Ray Continuum Intensity

The intensity of the X-ray continuum, I_{cm}, at an energy E_ν is described by Kramers (1923) as

$$I_{cm} \approx i_p Z \left[\left(E_0 - E_\nu \right) / E_\nu \right] \tag{4.9}$$

where i_p is the incident beam current and Z is the atomic number. For a particular value of the incident energy, E_0, the intensity of the continuum decreases rapidly relative to lower photon energies as E_ν approaches E_0, the Duane–Hunt limit.

An important parameter in electron-excited X-ray microanalysis is the ratio of the characteristic X-ray intensity to the X-ray continuum intensity at the same energy, $E_{ch} = E_\nu$, often referred to as the "peak-to-background, P/B." The P/B can be estimated from Eqs. (4.8) and (4.9) with the approximation that $E_\nu \approx E_c$ so that Eq. (4.9) can be rewritten as—

$$I_{cm} \approx i_p Z \left[\left(E_0 - E_c \right) / E_c \right] \approx i_p Z \left(U - 1 \right) \tag{4.10}$$

Taking the ratio of Eqs. (4.8) and (4.10) gives

$$P / B \approx \left(1 / Z \right) \left(U - 1 \right)^{n-1} \tag{4.11}$$

The P/B is plotted in ◘ Fig. 4.11 with the assumption that $n = 1.7$, where it is seen that at low overvoltages, which are often used in electron-excited X-ray microanalysis, the characteristic intensity is low relative to higher values of U, and the intensity rises rapidly with U, while the P/B increases rapidly at low overvoltage but then more slowly as the overvoltage increases.

4.3.2 The Electron-Excited X-Ray Spectrum, As-Generated

The electron-excited X-ray spectrum generated within the target thus consists of characteristic and continuum X-rays and is shown for pure carbon with $E_0 = 20$ keV in ◘ Fig. 4.12, as calculated with the spectrum simulator in NIST Desktop Spectrum Analyzer (Fiori et al. 1992), using the Pouchou and Pichoir expression for the K-shell ionization cross section and the Kramers expression for the continuum intensity (Pouchou and Pichoir 1991; Kramers 1923). Because of the energy dependence of the continuum given by Eq. 4.10, the generated X-ray continuum has its highest intensity at the lowest photon energy and decreases at higher photon energies, reaching zero intensity at E_0. By comparison, the energy span of the characteristic C–K peak is its natural width of only 1.6 eV, which is related to the lifetime of the excited K-shell vacancy. The energy width for K-shell emission up to 25 keV photon energy is shown in ◘ Fig. 4.2 (Krause and Oliver 1979). For photon energies below 25 keV, the characteristic X-ray peaks from the K-, L-, and M- shells have natural widths less than 10 eV. In the calculated spectrum of ◘ Fig. 4.12, the C–K peak is therefore plotted as a narrow line. (X-ray peaks are often referred to as "lines" in the literature, a result of their appearance in high-energy resolution measurements of X-ray spectra by diffraction-based X-ray spectrometers.) The X-ray spectra as-generated in the target for carbon, copper, and gold are compared in ◘ Fig. 4.13, where it can be seen that at all photon energies the intensity of the X-ray continuum increases with Z, as given by Eq. 4.9. The increased complexity of the characteristic X-rays at higher Z is also readily apparent.

◘ **Fig. 4.11** X-ray intensity emitted from a thick specimen and P/B, both as a function of overvoltage with exponent $n = 1.7$

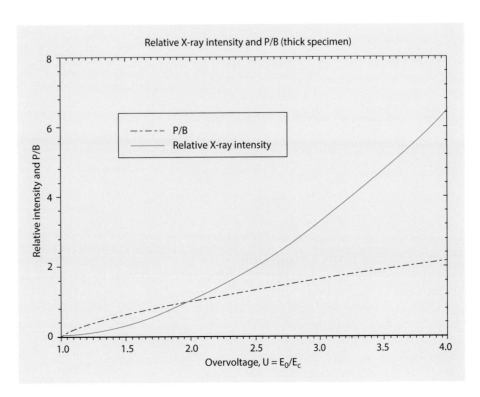

Fig. 4.12 Spectrum of pure carbon as-generated within the target calculated for $E_0 = 20$ keV with the spectrum simulator in Desktop Spectrum Analyzer

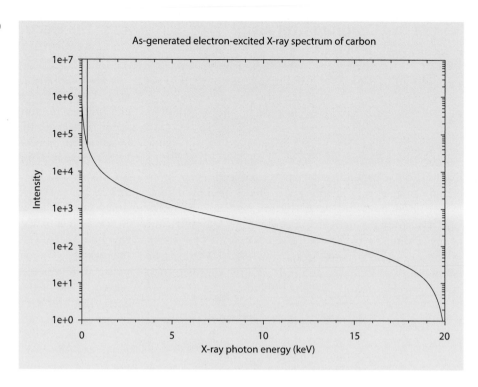

Fig. 4.13 Spectra of pure carbon, copper, and gold as-generated within the target calculated for $E_0 = 20$ keV with the spectrum simulator in Desktop Spectrum Analyzer (Fiori et al. 1992)

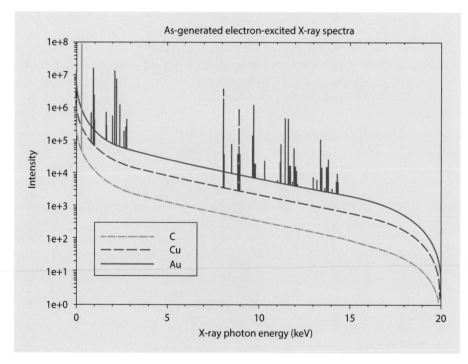

4.3.3 Range of X-ray Production

As the beam electrons scatter inelastically within the target and lose energy, inner shell ionization events can be produced from U_0 down to $U = 1$, so that depending on E_0 and the value(s) of E_c represented by the various elements in the target, X-rays will be generated over a substantial portion of the interaction volume. The "X-ray range," a crude estimate of the limiting range of X-ray generation, can be obtained from a simple modification of the Kanaya–Okayama range equation (IV-5) to compensate for the portion of the electron range beyond which the energy of beam electrons has deceased below a specific value of E_c:

$$R_{K-O}(nm) = 27.6\left(A/Z^{0.89}\rho\right)\left[E_0^{1.67} - E_c^{1.67}\right] \tag{4.12}$$

■ Table 4.2 lists calculations of the range of generation for copper K-shell X-rays ($E_c = 8.98$ keV) produced in various host elements, for example, a situation in which copper is present at a low level so it has a negligible effect on the overall Kanaya–Okayama range. As the incident beam energy decreases to $E_0 = 10$ keV, the range of production of copper K-shell X-rays decreases to a few hundred nanometers because of the very low overvoltage, $U_0 = 1.11$. The X-ray range in various matrices for the generation of various characteristic X-rays spanning a wide range of E_c is shown in ■ Fig. 4.14a–d.

4.3.4 Monte Carlo Simulation of X-Ray Generation

The X-ray range given by Eq. 4.12 provides a single value that captures the limit of the X-ray production but gives no information on the complex distribution of X-ray production within the interaction volume. Monte Carlo electron simulation can provide that level of detail (e.g., Drouin et al., 2007; Joy, 2006; Ritchie, 2015), as shown in ■ Fig. 4.15a, where the electron trajectories and the associated emitted photons of Cu K-L_3 are shown superimposed. For example, the limit of production of Cu K-L_3 that occurs when energy loss causes the beam electron energy to fall below the Cu K-shell excitation energy (8.98 keV) can be seen in the electron trajectories (green) that extend beyond the region of X-ray production (red). The effects of the host element on the

■ **Table 4.2**   Range of Cu K-shell ($E_c = 8.98$ keV) X-ray generation in various matrices

Matrix	25 keV	20 keV	15 keV	10 keV
C	6.3 µm	3.9 µm	1.9 µm	270 nm
Si	5.7 µm	3.5 µm	1.7 µm	250 nm
Fe	1.9 µm	1.2 µm	570 nm	83 nm
Au	1.0 µm	630 nm	310 nm	44 nm

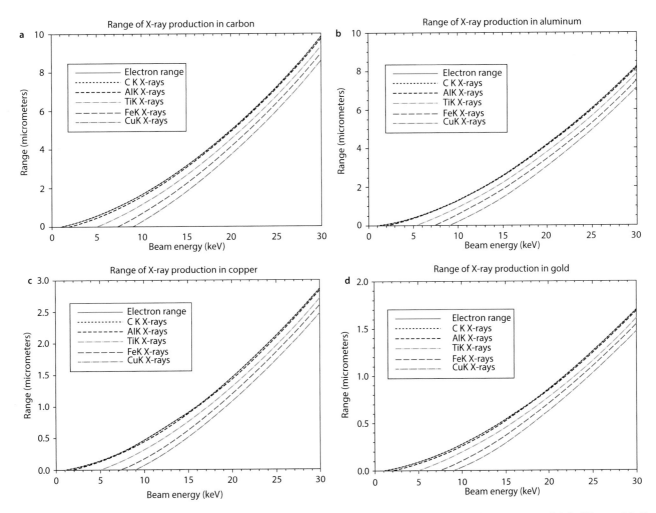

■ **Fig. 4.14**   **a** X-ray range as a function of E_0 for generation of K-shell X-rays of C, Al, Ti, Fe, and Cu in a C matrix. **b** X-ray range as a function of E_0 for generation of K-shell X-rays of C, Al, Ti, Fe, and Cu in an Al matrix. **c** X-ray range as a function of E_0 for generation of K-shell X-rays of C, Al, Ti, Fe, and Cu in a Cu matrix. **d** X-ray range as a function of E_0 for generation of K-shell X-rays of C, Al, Ti, Fe, and Cu in an Au matrix

4

■ **Fig. 4.15** **a** Monte Carlo simulation (DTSA-II) of electron trajectories and associated Cu K-shell X-ray generation in pure copper; $E_0 = 20$ keV. **b** Monte Carlo simulation (DTSA-II) of the distribution of Cu K-shell and L-shell X-rays in a Cu matrix with $E_0 = 10$ keV showing the X-rays that escape. **c** Monte Carlo simulation (DTSA-II) of the distribution of Cu K-shell and L-shell X-rays that escape in Au-1 % Cu with $E_0 = 10$ keV. **d** Monte Carlo simulation (DTSA-II) of the distribution of Cu K-shell and L-shell X-rays that escape in C-1 % Cu with $E_0 = 10$ keV (Ritchie 2015)

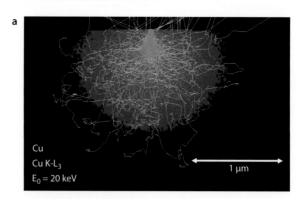

a

Cu
Cu K-L$_3$
$E_0 = 20$ keV

1 μm

b   Monte Carlo (DTSA-II) simulation of Cu X-ray production in Cu; $E_0 = 10$ keV

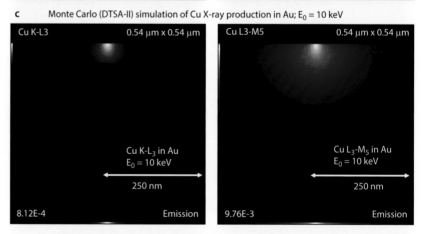

Cu K-L3   0.92 μm x 0.92 μm

Cu K-L$_3$ in Cu
$E_0 = 10$ keV
250 nm

0.089   Emission

Cu L3-M5   092 μm x 0.92 μm

Cu L$_3$-M$_5$ in Cu
$E_0 = 10$ keV
250 nm

1.000   Emission

c   Monte Carlo (DTSA-II) simulation of Cu X-ray production in Au; $E_0 = 10$ keV

Cu K-L3   0.54 μm x 0.54 μm

Cu K-L$_3$ in Au
$E_0 = 10$ keV
250 nm

8.12E-4   Emission

Cu L3-M5   0.54 μm x 0.54 μm

Cu L$_3$-M$_5$ in Au
$E_0 = 10$ keV
250 nm

9.76E-3   Emission

d   Monte Carlo (DTSA-II) simulation of Cu X-ray production in C; $E_0 = 10$ keV

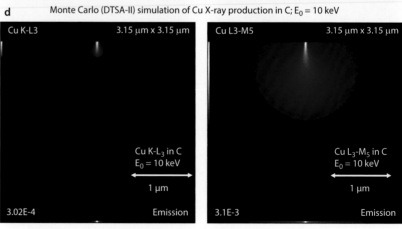

Cu K-L3   3.15 μm x 3.15 μm

Cu K-L$_3$ in C
$E_0 = 10$ keV
1 μm

3.02E-4   Emission

Cu L3-M5   3.15 μm x 3.15 μm

Cu L$_3$-M$_5$ in C
$E_0 = 10$ keV
1 μm

3.1E-3   Emission

emission volumes for Cu K-shell and L-shell X-ray generation in three different matrices—C, Cu, and Au—is shown in ◻ Fig. 4.15a–c using DTSA-II (Ritchie 2015). The individual maps of X-ray production show the intense zone of X-ray generation starting at and continuing below the beam impact point and the extended region of gradually diminishing X-ray generation. In all three matrices, there is a large difference in the generation volume for the Cu K-shell and Cu L-shell X-rays as a result of the large difference in overvoltage at $E_0 = 10$ keV: CuK $U_0 = 1.11$ and CuL $U_0 = 10.8$.

4.3.5 X-ray Depth Distribution Function, $\phi(\rho z)$

The distribution of characteristic X-ray production as a function of depth, designated "$\phi(\rho z)$" in the literature of quantitative electron-excited X-ray microanalysis, is a critical parameter that forms the basis for calculating the compositionally dependent correction ("A" factor) for the loss of X-rays due to photoelectric absorption. As shown in ◻ Fig. 4.16 for Si with $E_0 = 20$ keV, Monte Carlo electron trajectory simulation provides a

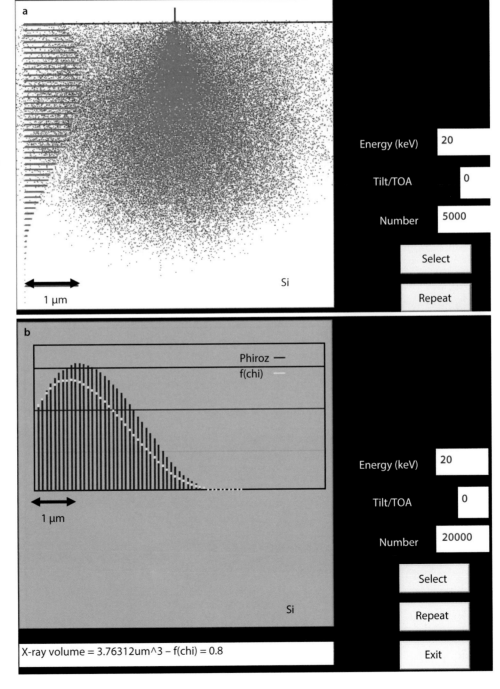

◻ **Fig. 4.16** **a** Monte Carlo calculation of the interaction volume and X-ray production in Si with $E_0 = 20$ keV. The histogram construction of the X-ray depth distribution $\phi(\rho z)$ is illustrated. (Joy Monte Carlo simulation). **b** $\phi(\rho z)$ distribution of generated Si K-L$_3$ X-rays in Si with $E_0 = 20$ keV, and the effect of absorption from each layer, giving the fraction, $f(\chi)_{depth}$, that escapes from each layer. The cumulative escape from all layers is $f(\chi) = 0.80$. (Joy Monte Carlo simulation) (Joy 2006)

method to model $\phi(\rho z)$ by dividing the target into layers of constant thickness parallel to the surface, counting the X-rays produced in each layer, and then plotting the intensity as a histogram. The intensity in each layer is normalized by the intensity produced in a single unsupported layer which is sufficiently thin so that no significant elastic scattering occurs: the electron trajectories pass through such a thin layer without deviation. The $\phi(\rho z)$ distribution has several important characteristics. For a thick specimen, the intensity produced in the first layer exceeds that of the unsupported reference layer because in addition to the X-ray intensity produced by the passage of all of the beam electrons through the first layer, elastic scattering from deeper in the specimen creates backscattered electrons which pass back through the surface layer to escape the target, producing additional X-ray generation. The intensity produced in the first layer, designated ϕ_0, thus always exceeds unity because of this extra X-ray production due to backscattering. Below the surface layer, $\phi(\rho z)$ increases as elastic scattering increases the path length of the electrons that pass obliquely through each layer, compared to the relatively unscattered passage of the incident electrons through the outermost layers before elastic scattering causes significant deviation in the trajectories. The reverse passage of backscattered electrons also adds to the generation of X-rays in the shallow layers. Eventually

a peak value in $\phi(\rho z)$ is reached, beyond which the X-ray intensity decreases due to cumulative energy loss, which reduces the overvoltage, and the relative number of backscattering events decreases. The $\phi(\rho z)$ distribution then steadily decreases to a zero intensity when the electrons have sustained sufficient energy loss to reach overvoltage $U = 1$. The limiting X-ray production range is given by Eq. 4.12.

4.4 X-Ray Absorption

The Monte Carlo simulations shown in ◘ Fig. 4.15b–d are in fact plots of the X-rays *emitted* from the sample. To escape the sample, the X-rays must pass through the sample atoms where they can undergo the process of photoelectric absorption. An X-ray whose energy exceeds the binding energy (critical excitation energy) for an atomic shell can transfer its energy to the bound electron, ejecting that electron from the atom with a kinetic energy equal to the X-ray energy minus the binding energy, as shown in ◘ Fig. 4.17, which initiates the same processes of X-ray and Auger electron emission as shown in ◘ Fig. 4.1 for inner shell ionization by energetic electrons. The major difference in the two processes is that the X-ray is annihilated in photoelectric absorption and its entire energy

◘ **Fig. 4.17** Schematic diagram of the process of X-ray generation: inner shell ionization by photoabsorption of an energetic X-ray that leaves the atom in an elevated energy state which it can lower by either of two routes involving the transition of an L-shell electron to fill the K-shell vacancy: *1* the Auger process, in which the energy difference $E_K - E_L$ is transferred to another L-shell electron, which is ejected with a characteristic energy: $E_K - E_L - E_L$; (*2*) photon emission, in which the energy difference $E_K - E_L$ is expressed as an X-ray photon of characteristic energy

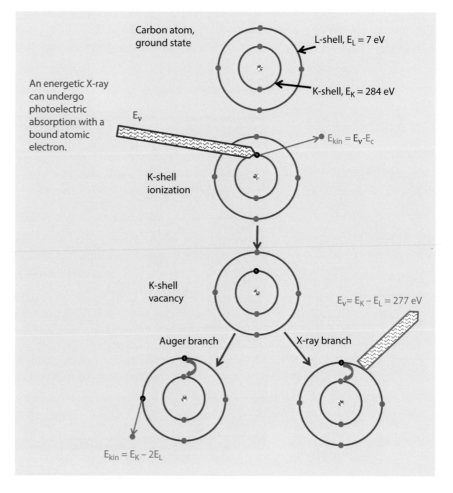

transferred to the ejected electron. Photoelectric absorption is quantified by the "mass absorption coefficient," μ/ρ, which determines the fraction of X-rays that pass through a thickness s of a material acting as the absorber:

$$I / I_0 = \exp\left[-\left(\mu/\rho\right)\rho s\right] \qquad (4.13)$$

where I_0 is the initial X-ray intensity and I is the intensity after passing through a thickness, s, of a material with density ρ.

The dimensions of the mass absorption coefficient are cm^2/g. For a given material, mass absorption coefficients generally decrease with increasing photon energy, as shown for carbon in ◻ Fig. 4.18a. The exception is near the critical excitation energy for the atomic shells of a material. The region near the C K-shell excitation energy of 284 eV is shown expanded in ◻ Fig. 4.18b, where an abrupt increase in μ/ρ occurs. An X-ray photon whose energy is just slightly greater than the critical excitation energy for an atomic shell can very

◻ **Fig. 4.18** **a** Mass absorption coefficient for C as a function of photon energy. **b** Mass absorption coefficient for C as a function of photon energy near the C critical excitation energy

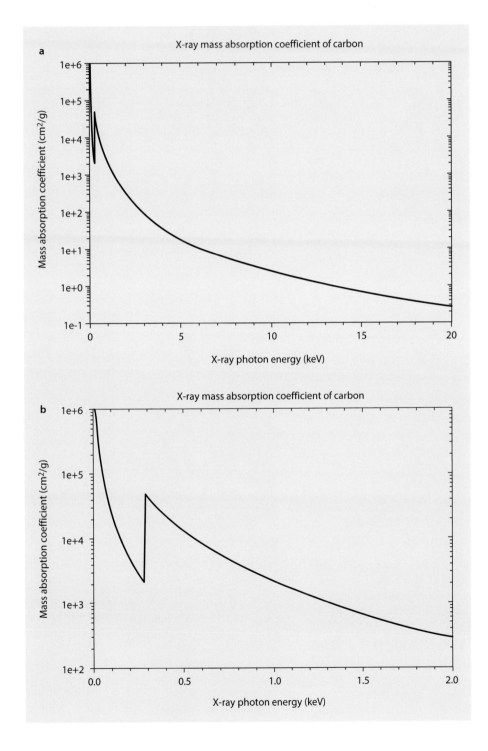

4

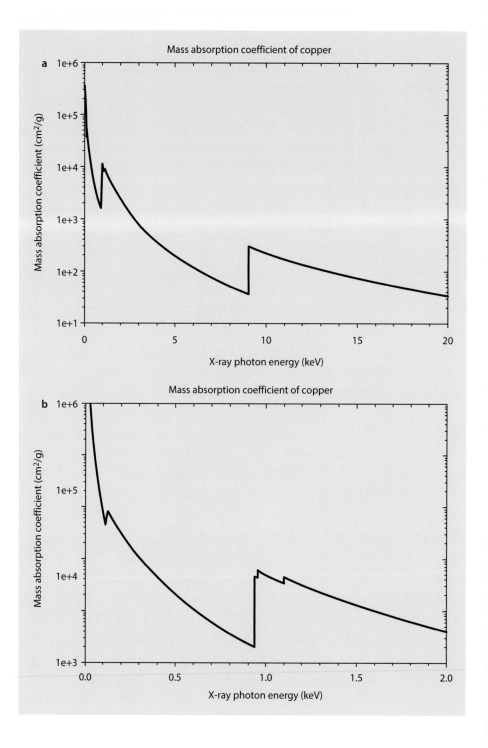

Fig. 4.19 **a** Mass absorption coefficient for Cu as a function of photon energy. **b** Mass absorption coefficient for Cu as a function of photon energy near the Cu L-shell critical excitation energies

efficiently couple its energy to the bound electron, resulting in a high value of μ/ρ. With further increases in photon energy, the efficiency of the coupling of the photon energy to the bound electron decreases so that μ/ρ also decreases. For more complex atoms with more atomic shells, the mass absorption coefficient behavior with photon energy becomes more complicated, as shown for Cu in ◘ Fig. 4.19a, b, which shows the region of the three Cu L-edges. For Au, ◘ Fig. 4.20a–c shows the regions of the three Au L-edges and the five Au M-edges.

When a material consists of an atomic-scale mixture of two or more elements, the mass absorption for the mixture is calculated as

$$\left(\mu/\rho\right)^{i}_{mix} = \Sigma_{j}\left(\mu/\rho\right)^{i}_{j} C_{j} \qquad (4.14)$$

where $(\mu/\rho)^{i}_{j}$ is the mass absorption coefficient for the X-rays of element i by element j, and C_{j} is the mass concentration of element j in the mixture.

4

Fig. 4.20 **a** Mass absorption coefficient for Au as a function of photon energy. **b** Mass absorption coefficient for Au as a function of photon energy near the Au L-shell critical excitation energies. **c** Mass absorption coefficient for Au as a function of photon energy near the Au M-shell critical excitation energies

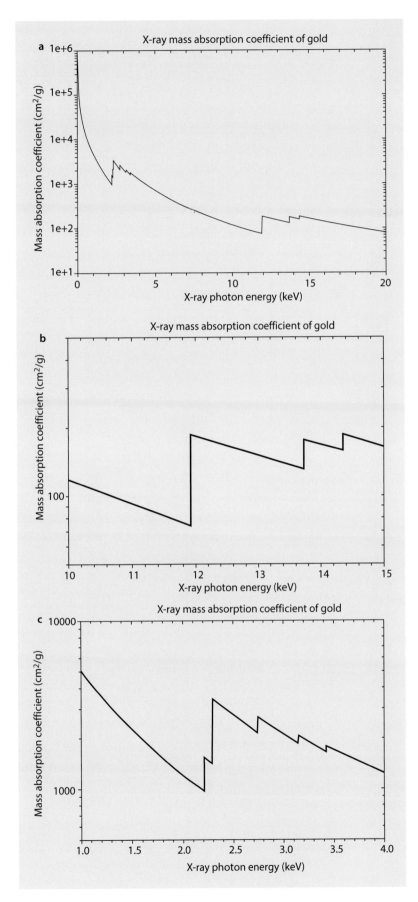

4

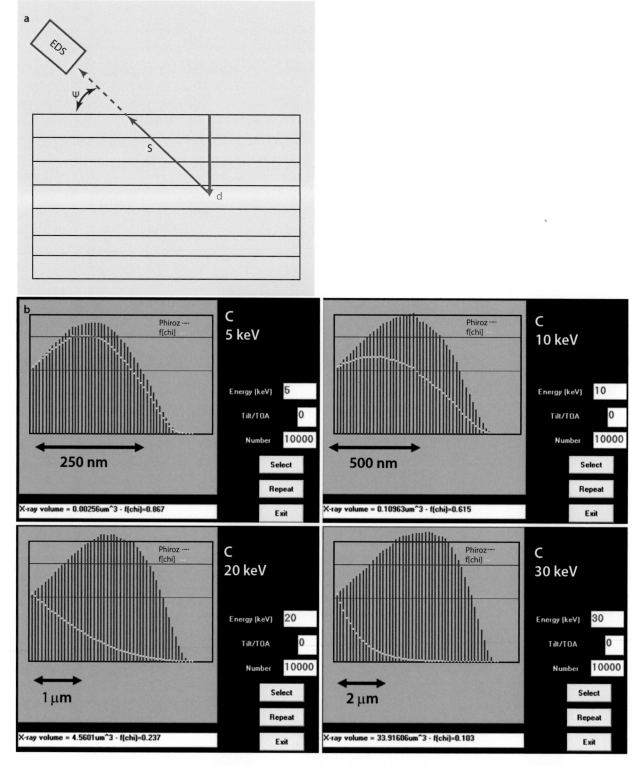

■ **Fig. 4.21**   **a** Determination of the absorption path length from a layer of the φ(ρz) distribution located at depth, d, in the direction of the X-ray detector. **b** Monte Carlo determination of the φ(ρz) distribution and absorption for carbon at various beam energies. (Joy Monte Carlo simulation)

Photoelectric absorption reduces the X-ray intensity that is emitted from the target at all photon energies. The absorption loss from each layer of the φ(ρz) distribution is calculated using Eq. 4.15 with the absorption path, s, determined as shown in ■ Fig. 4.21a from the depth, d, of the histogram slice, and the cosecant of the X-ray detector "take-off angle," ψ, which is the elevation angle of the detector above the target surface:

Table 4.3 Self-absorption of carbon K-shell X-rays as a function of beam energy

E_0	$f(\chi)$
2 keV	0.974
5	0.867
10	0.615
20	0.237
30	0.103

$$s = d\csc\psi \qquad (4.15)$$

Normalizing by the intensity generated in each layer, the ϕ(ρz) histogram gives the probability, with a value between 0 and 1, that a photon generated in that layer and emitted into the solid angle of the EDS detector will escape and reach the detector, as shown in ◘ Fig. 4.16b for each histogram bin of the silicon ϕ(ρz) distribution. The escape probability of X-rays integrated over the complete ϕ(ρz) histogram gives the parameter designated "$f(\chi)$," which is the overall escape probability, between 0 and 1, for an X-ray generated anywhere in the ϕ(ρz) distribution.

◘ Figure 4.21b shows a sequence of calculations of the C K ϕ(ρz) distribution and subsequent absorption as a function of incident beam energy. As the incident beam energy increases, the depth of electron penetration increases so that carbon characteristic X-rays are produced deeper in the target. For pure carbon with $E_0 = 5$ keV, the cumulative value of $f(\chi) = 0.867$; that is, 86.7 % of all carbon X-rays that are generated escape, while 13.3 % are absorbed. As the C X-rays are produced deeper with increasing beam energy, the total X-ray absorption increases so that the value of $f(\chi)$ for C K decreases sharply with increasing beam energy, as shown in ◘ Fig. 4.21b and in ◘ Table 4.3.

Thus, with $E_0 = 2$ keV, 97.4 % of the carbon X-rays escape the specimen, while at $E_0 = 30$ keV, nearly 90 % of the carbon X-rays generated in pure carbon are absorbed before they can exit the specimen.

When the parameter $f(\chi)$ is plotted at every photon energy from the threshold of 100 eV up to the Duane–Hunt limit of the incident beam energy E_0, X-ray absorption is seen to sharply modify the X-ray spectrum that is emitted from the target, as illustrated for carbon (◘ Fig. 4.22), copper (◘ Fig. 4.23), and gold (◘ Fig. 4.24). The high relative intensity of the X-ray continuum at low photon energies compared to higher photon energies in the generated spectrum is greatly diminished in the emitted spectrum because of the higher absorption suffered by low energy photons. Discontinuities

in $f(\chi)$ are seen at the critical ionization energy of the K-shell in carbon, the K- and L-shells in copper, and the M- and L-shells in gold, corresponding to the sharp increase in μ/ρ just above the critical ionization energy. Because the X-ray continuum is generated at all photon energies, the continuum is affected by every ionization edge represented by the atomic species present, resulting in abrupt steps in the background. An abrupt decrease in X-ray continuum intensity is observed just above the absorption edge energy due to the increase in the mass absorption coefficient. The characteristic peaks in these spectra are also diminished by absorption, but because a characteristic X-ray is always lower in energy than the ionization edge energy from which it originated, the mass absorption coefficient for characteristic X-rays is lower than that for photons with energies just above the shell ionization energies. Thus an element is relatively transparent to its own characteristic X-rays because of the decrease in X-ray absorption below the ionization edge energy.

4.5 X-Ray Fluorescence

As a consequence of photoelectric absorption shown in ◘ Fig. 4.17, the atom will subsequently undergo de-excitation following the same paths as is the case for electron ionization in ◘ Fig. 4.1. Thus, the primary X-ray spectrum of characteristic and continuum X-rays generated by the beam electron inelastic scattering events gives rise to a secondary X-ray spectrum of characteristic X-rays generated as a result of target atoms absorbing those characteristic and continuum X-rays and emitting lower energy characteristic X-rays. Because continuum X-rays are produced up to E_0, the Duane–Hunt limit, all atomic shells present with $E_c < E_0$ will be involved in generating secondary X-rays, which is referred to as "secondary X-ray fluorescence" by the X-ray microanalysis community. Generally, at any characteristic photon energy the contribution of secondary fluorescence is only a few percent or less of the intensity produced by the direct electron ionization events. However, there is a substantial difference in the spatial distribution of the primary and secondary X-rays. The primary X-rays must be produced within the interaction volume of the beam electrons, which generally has limiting dimensions of a few micrometers at most. The secondary X-rays can be produced over a much larger volume because the range of X-rays in a material is typically an order-of-magnitude (or more) greater than the range of an electron beam with E_0 from 5 to 30 keV. This effect is shown in ◘ Fig. 4.25 for an alloy of Ni-10 % Fe for the secondary fluorescence of Fe K-shell X-rays ($E_K = 7.07$ keV) by the electron-excited Ni K-$L_{2,3}$ X-rays (7.47 keV). The hemispherical volume that contains 99 % of the secondary Fe K-$L_{2,3}$ X-rays has a radius of 30 μm.

Fig. 4.22 **a** Absorption parameter $f(\chi)$ as a function of photon energy for carbon and an incident beam energy of $E_0 = 20$ keV. Note the abrupt decrease for photons just above the ionization energy of carbon at 0.284 keV. **b** Expansion of the region from 0 to 5 keV. Note the abrupt decrease for photons just above the ionization energy of carbon at 0.284 keV. **c** Comparison of the generated (*black*) and emitted (*red*) X-ray spectra for carbon with an incident beam energy of $E_0 = 20$ keV

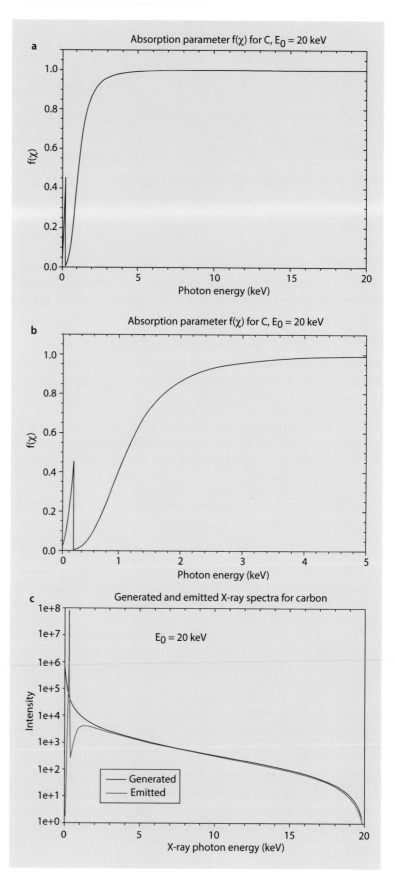

■ **Fig. 4.23** **a** Absorption parameter $f(\chi)$ as a function of photon energy for copper and an incident beam energy of $E_0 = 20$ keV. Note the abrupt decrease just above the ionization energies of the three L-shells near 0.930 keV and the K-shell ionization energy at 8.98 keV. **b** Comparison of the generated (*red*) and emitted (*blue*) X-ray spectra for copper with an incident beam energy of $E_0 = 20$ keV. **c** Comparison of the generated and emitted X-ray spectra for copper with an incident beam energy of $E_0 = 20$ keV; expanded to show the region of the Cu L-shell and Cu K-shell ionization energies

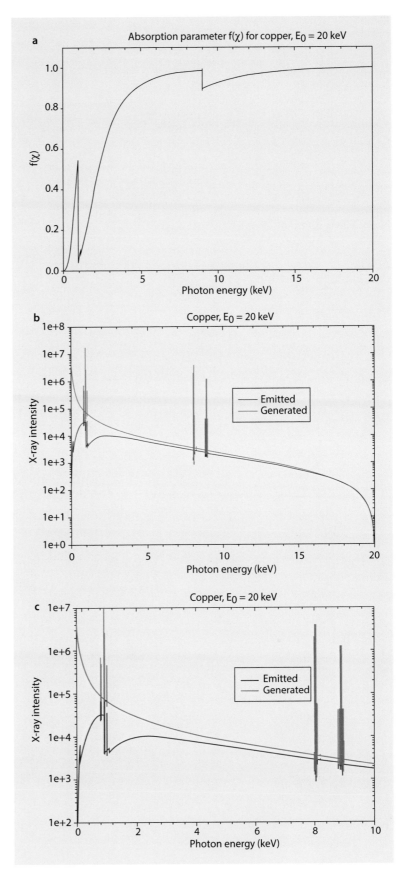

4

Fig. 4.24 **a** Absorption parameter $f(\chi)$ as a function of photon energy for gold and an incident beam energy of $E_0 = 20$ keV. Note the abrupt decrease in $f(\chi)$ for photons just above the ionization energies of the gold M-shell and gold L-shell. **b** Comparison of the generated (*red*) and emitted (*blue*) X-ray spectra for gold with an incident beam energy of $E_0 = 20$ keV. **c** Comparison of the generated (*red*) and emitted (*blue*) X-ray spectra for gold with an incident beam energy of $E_0 = 20$ keV; expansion of the region around the gold M-shell ionization edges

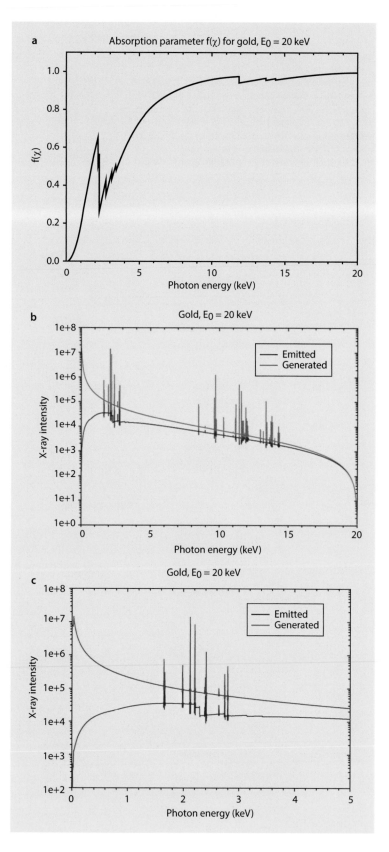

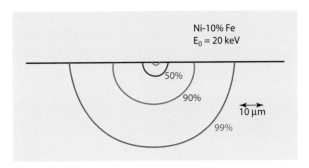

Fig. 4.25 Range of secondary fluorescence of Fe K-L$_3$ ($E_K = 7.07$ keV) by Ni K-L$_3$ X-rays (7.47 keV). *Red arc* = extent of direct electron excitation of Ni K-L$_3$ and Fe K-L$_3$. *Blue arc* = range for 50 % of total secondary fluorescence of Fe K-L$_3$ by Ni K-L$_3$; *green arc* = 90 %; *magenta arc* = 99 %

References

Crawford J, Cohen D, Doherty G, Atanacio A (2011) Calculated K, L and M-shell X-ray line intensities for light ion impact on selected targets from Z = 6 to 100. ANSTO/E-774, Australian Nuclear Science and Technology Organisation (September, 2011)

Drouin D, Couture A, Joly D, Tastet X, Aimez V, Gauvin R (2007) CASINO V2.42 – a fast and easy-to-use modeling tool for scanning electron microscopy and microanalysis users. Scanning 29:92–101

Fiori CE, Swyt CR, Myklebust RL (1992) Desktop Spectrum Analyzer (DTSA), a comprehensive software engine for electron-excited X-ray spectrometry. National Institute of Standards and Technology (NIST) – National Institutes of Health (NIH). United States Patent 5,299,138 (March 29, 1994) (Standard Reference Data Program, NIST, Gaithersburg, MD 20899)

Heinrich KFJ, Fiori CE, Myklebust RL (1979) Relative transition probabilities for the X-ray lines from the K level. J Appl Phys 50:5589

Joy D (2006) Monte Carlo simulation. Available at ► http://www.lehigh.edu/~maw3/link/mssoft/mcsim.html

Kramers H (1923) On the theory of X-ray absorption and the continuous X-ray spectrum. Phil Mag 46:836

Krause M, Oliver J (1979) Natural widths of atomic K and L levels, Kα X-ray lines and several KLL Auger lines. J Phys Chem Ref Data Monogr 8:329

Lifshin E, Ciccarelli MF, Bolon RB (1980) In: Beaman DR, Ogilvie RE, Wittry DB (eds) Proceedings 8th international conference on X-ray optics and microanalysis, vol 141. Pendell, Midland

Pouchou JL, Pichoir F (1991) Quantitative analysis of homogeneous or stratified microvolumes applying the model 'PAP'. In: Heinrich KFJ, Newbury DE (eds) Electron probe quantitation, vol 31. Plenum Press, New York

Powell C (1976) Use of Monte Carlo Calculations. In: Heinrich KFJ, Newbury DE, Yakowitz H (eds) Electron probe microanalysis and scanning electron microscopy. National Bureau of standard Special Publication 460. U.S. Government Printing Office, Washington, DC, 97

Ritchie NWM (2015) NIST DTSA-II software, including tutorials, is available free at: ► www.cstl.nist.gov/div837/837.02/epq/dtsa2/index.html

Scanning Electron Microscope (SEM) Instrumentation

© Springer Science+Business Media LLC 2018
J. Goldstein et al., *Scanning Electron Microscopy and X-Ray Microanalysis*,
https://doi.org/10.1007/978-1-4939-6676-9_5

5

5.1   Electron Beam Parameters

This chapter addresses essential topics: the quantitative attributes of an electron beam, well-known widely-used SEM modes, and electron detectors.

■   **Why Learn About Electron Optical Parameters?**

As we mentioned in the introduction to the book, the main goal of the text is to help users understand how to operate the SEM and its accessories, and how to be effective at using these powerful tools for materials characterization and analysis. It is a fair question, then, to ask why an operator of the microscope needs to understand electron optics and the optical parameters of the beam. Clearly an SEM design engineer needs to be conversant in these subjects, but why learn these concepts as an end user? The simplest answer is that while all SEMs have knobs, switches, and controls, in the end it is the electron optical beam parameters that the operator is controlling, and a basic understanding of what is being changed by those knobs is essential to becoming a skilled user. Whether the knobs and dials are "old-school" analog hardware devices or purely virtual objects that exist only in a software user interface, the operator cannot use the SEM to best advantage without a clear picture of how those knobs are changing the beam.

5.2   Electron Optical Parameters

◘ Figure 5.1 shows the basic features of an electron beam in a scanning electron microscope after it emerges from the final aperture of the objective lens and before it impacts the sample surface. While changes to the beam inside the electron gun and inside the electron column are also important to the SEM operator, a thorough understanding of the attributes of the beam in the chamber is absolutely essential to mastery of the instrument.

5.2.1 Beam Energy

One of the fundamental beam parameters is the energy of the electrons in the beam, measured in electronvolts (eV) and often represented by the symbol E, or E_0. This parameter represents the *initial* energy of the electrons as they enter the SEM chamber or the sample. Beam energy has a direct effect on many important aspects of SEM operation, such as the size of the excitation volume in the sample and the intensity of the X-rays emitted, so it is necessary to choose this parameter carefully and set it to an appropriate value before acquiring data. Frequently the beam energy is several thousand electronvolts or higher, so the kilo-electronvolt is the most common unit of beam energy, abbreviated keV. One keV is equal to 1000 eV, and many SEMs are capable of generating electron beams up to 30 keV (equal to 30,000 eV), or in a few cases even higher.

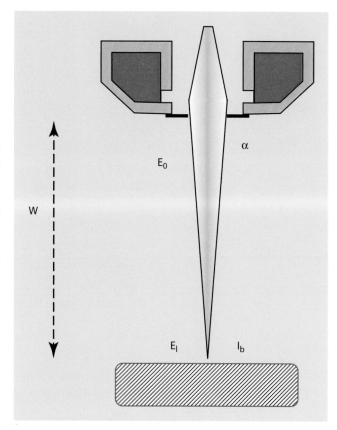

◘ **Fig. 5.1**   Basic elements of the electron beam in an SEM

If you have any experience with electronics or electrical engineering, the electronvolt as a unit of energy may be confusing at first since it sounds more like a measure of voltage, unlike the more common units of energy such as the Joule, the calorie, or the erg. The terms *electron volt* and the related SI unit *electronvolt* are related to the method used by the SEM to impart energy to the electrons that emerge from the electron source. Typically, the electrons are accelerated from low energy to high energy using an electrostatic potential difference generated by a high-voltage power supply. Negatively charged electrons are repelled from surfaces with negative electrical potential and attracted to surfaces with positive potential, and the potential difference is measured in volts. One electronvolt is simply the energy acquired by an electron when it is accelerated through a potential difference of one volt; similarly, an electron that drops through a voltage difference of 20 kilovolts (20 kV) emerges with an energy of 20 keV.

This underlying connection between the accelerating voltage used by the microscope and the resulting beam energy can help make sense of the different terminology often used in the SEM community regarding "beam energy." On many microscopes, you set the accelerating voltage with a knob or by using a graphical user interface on a computer. On these microscopes, you would select 30 kV for the accelerating voltage if you wanted to work at high beam energy, or you might select 1 kV if you wished to work at low voltage. Other microscope interfaces allow you to select the beam energy directly instead of the accelerating voltage, so the corresponding settings would be 30 keV for high beam energy work and 1 keV for low-voltage operation. In informal conversation it is common to hear 30 kV and 30 keV used interchangeably to mean the same beam setting, and usually no confusion arises from this practice. However, in written documents such as reports of analyses or academic publications, the common error of describing the beam energy using units of kilovolts or of recording the accelerating voltage in units of kilo-electronvolts should be avoided.

Landing Energy

Aside from this possible confusion between beam energy and accelerating potential, there are other subtleties in the proper characterization of the beam energy in the SEM. Depending on the technology used by the microscope manufacturer, the electrons in the microscope may change energy more than once during their path from the electron source to the surface of the sample. Some microscopes seek to improve imaging performance by modifying the electrons' energy during the mid-portion of the optical column. On more recent microscope models it is increasingly common to see beam deceleration options, which decrease the beam energy just before the electrons emerge from the objective lens.

Also common on modern instruments is the option to apply a voltage bias to the sample itself, thus allowing the SEM operator to increase the energy of the electrons as they approach the sample (in the case of a positive sample bias), or decrease the energy of the electrons (in the case of a negative sample bias). For example, if the electron beam emerges from the objective lens into the SEM sample chamber with a beam energy of 5 keV, but the sample has a negative voltage bias of 1 kV applied, the electrons will be decelerated to an energy of 4 keV when they impact the specimen.

The term used to describe the electron beam energy at the point of impact on the sample surface is *landing energy*, usually denoted by the symbol E_l. The physics of beam–specimen interaction depends only on the landing energy of the electrons, not on their energy at points further up the optical path. Critically important phenomena such as the size of the excitation volume in the specimen, the number of characteristic X-ray peaks available for use in compositional measurements, or the high energy limit of continuum X-rays emitted (the Duane–Hunt limit) are all functions of the landing energy, not the initial beam energy. Because of this, it is very important for the SEM operator to understand when landing energy differs from the beam energy at the objective lens final aperture, and how to control the value of the landing energy. The details of such subtleties vary from one vendor to another, and even from one microscope model to the next, but they are invariably described in the user documentation for every instrument. Seek help from your microscope's customer support team or an application engineer if you are not absolutely clear on how to control the landing energy on your microscope. In many situations, particularly when working with older microscopes, this distinction is not important and the terms *beam energy* and *landing energy* may be used

interchangeably without a problem, but when the distinction matters it can be crucial to accurate analysis and proper communication or reporting.

5.2.2 Beam Diameter

Another important electron beam characteristic under the control of the SEM operator is the diameter of the electron beam, which in most cases refers to the diameter of the beam as it impacts the sample surface. Beam diameter has units of length and is frequently measured in nanometers, Ångstroms, or micrometers, depending on the size of the beam. For most SEM applications the beam diameter will fall within the broad range of 1 nm to 1 μm. It is commonly represented by the symbol d, or a subscripted variant such as d_{probe} or d_p.

Before developing an understanding of the importance of beam–specimen interactions, many SEM operators naively assume that the resolution of their SEM images is dictated solely by the beam diameter. While this may be true in some situations, more often the relationship between the beam diameter and the resolution is a complex one. Perhaps this explains why the exact definition of beam diameter is not always provided, even in relatively careful writing or formal contexts. The simplest model of an electron beam is one where the beam has a circular cross section at all times, and that the electrons are distributed with uniform intensity everywhere inside the beam diameter and are completely absent outside the beam diameter. In this trivial case, the beam has hard boundaries and is the same size no matter which azimuth you use to measure it. In reality the electron beam in an SEM is much more complicated. Even if you assume that the cross section is circular, electron beams exhibit a gradient of electron density from the core of the beam out to the edges, and in many cases have a tail of faint intensity that extends quite far from the central flux. It is still possible in these case to define the meaning of beam diameter in a precise way, in terms of the full width at half-maximum of the intensity for example, or the full width at tenth-maximum if the tails are pronounced. More careful statistical models of the beam will specify the radial intensity distribution function—a Gaussian or Lorentzian distribution, for example—and will allow for non-circularity. In most situations where such precision is not warranted, however, it will suffice to assume that the beam diameter is a single number that characterizes the width in nanometers of that portion of the beam that gives rise to the most important fraction of the contrast or sample excitation as measured at the surface of the specimen.

5.2.3 Beam Current

Of all the electron beam parameters that matter to the SEM operator, beam current is perhaps highest on the list. Fortunately it is a relatively simple parameter to understand since it is entirely analogous to electrical currents of the kind found in wires, electronics, or electrical engineering. Beam current at the sample surface is a measure of the number of electrons per second that impact the specimen. It is usually measured in fractions of an ampere, such as microamperes (μA), nanoamperes (nA), or picoamperes (pA). A typical SEM beam current is about 1 nA, which corresponds to 6.25 × 10⁹ electrons per second, or approximately one electron striking the sample every 160 ps . The usual symbol used to represent beam current is I, or i, or a subscripted variant such as I_{probe}, I_{beam}, I_p, or I_b.

5.2.4 Beam Current Density

Similar to the beam current, the concept of current density is relatively easy to understand and corresponds directly with the same concept in electrical engineering or electrical design. Current density in an electron beam is defined as the beam current per unit area, and it is usually represented by the symbol J, or J_{beam}. In standard units this quantity is expressed in A/m², but there are also derived units better suited to the SEM such as nA/nm² or similar. The most important thing to understand about current density is that it is an *areal* measure, not an absolute measure; this means the current depends directly on and varies linearly with the area of the region through which the stated current density passes.

To make this concrete, let's consider an example calculation of the current density in an electron beam. ◻ Figure 5.2 shows a circular beam with a diameter of 5 nm; and the total current inside the circular beam spot is 1 nA. The area of the beam is

$$A_{circle} = \pi r^2 = \pi \left(\frac{d}{2}\right)^2 = \pi \left(\frac{5\,nm}{2}\right)^2 = 19.6\,nm^2 \quad (5.1)$$

and therefore the current density in the round beam is

$$J_{circle} = \frac{I_{circle}}{A_{circle}} = \frac{1\,nA}{19.6\,nm^2} = 0.0509\,\frac{nA}{nm^2} = 50.9\,pA\,/\,nm^2 \quad (5.2)$$

◻ Figure 5.2b shows the situation if you decrease the diameter of the beam by half, from 5 nm to 2.5 nm, yet keep the same total current in the beam. Now the area has gotten smaller, yet the current is unchanged, so the current density has increased. The new current density is

$$J_{circle} = \frac{I_{circle}}{A_{circle}} = \frac{1\,nA}{4.91\,nm^2} = 0.204\,\frac{nA}{nm^2} = 204\,pA\,/\,nm^2 \quad (5.3)$$

Since focusing the electron beam in an SEM changes the diameter of the beam but does not change the beam current, the current density must change. As can be seen in ◻ Fig. 5.2, shrinking the beam width by a factor of two results in a fourfold increase in beam current density.

5.2.5 Beam Convergence Angle, α

One of the fundamental characteristics of the electron beam found in all SEM instruments is that the shape of the beam as seen from the side is not a parallel-sided cylinder like a pencil, but rather a cone. The beam is wide where it exits the final aperture of the objective lens, and narrows steadily until (if the sample is in focus) it converges to a very small spot when it enters the specimen. A schematic of this cone is shown in ◻ Fig. 5.3. The point where the beam lands on the sample is denoted S at the bottom of the cone, and the beam-defining aperture is shown in perspective as a circle at the top of the cone, with line segment AB equal to the diameter of that aperture, d_{apt}. The vertical dashed line represents the optical axis of the SEM column, which ideally passes through the center of the final aperture, is perpendicular to the plane of that aperture, and extends down through the chamber into the sample and beyond. The sides of the cone are defined by the "edge" of the electron beam. As mentioned above in the definition of the beam diameter, this notion of a hard-edged

a Current density = 50.9 pA/nm²

b density = 204 pA/nm²

1 nA

1 nA

5 nm

2.5 nm

◻ **Fig 5.2** **a** Current density in a circular electron beam; **b** current density if the beam diameter is reduced by a factor of two with the same current

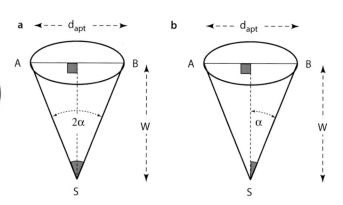

◻ **Fig. 5.3** **a** Definition of beam cone opening angle 2α; **b** definition of beam convergence (half) angle α

beam may not be physically realistic, but it is simple to understand and works well for our purposes here of understanding basic beam parameters.

In geometry, the opening angle of a cone is defined as the vertex angle ASB at the point of the cone, as shown in ◻ Fig. 5.3a. When working with the electron optics of an SEM, by convention we use the term *convergence angle* to describe how quickly the electron beam narrows to its focus as it travels down the optic axis. This convergence angle is shown in ◻ Fig. 5.3b as α, which is half of the cone opening angle. In some cases, the beam convergence angle is referred to as the *convergence half-angle* to emphasize that only half of the opening angle is intended.

Generally the numerical value of the beam convergence angle in the SEM is quite small, and the electron beam cones are much sharper and narrower than the cones used for schematic purposes in ◻ Fig. 5.3. In fact, if you ground down and reshaped the sides of a sewing needle so that it was a true cone instead of a cylinder sharpened only at the tip, you would then have a cone whose size and shape is reasonably close to the dimensions found in the SEM.

Estimating the value of the convergence angle of an electron beam is not difficult using the triangles drawn in ◻ Fig. 5.3. The length of the vertical dashed line along the optical axis is called the *working distance*, usually denoted by the symbol W. It is merely the distance from the bottom of the objective lens pole piece (taken here to be approximately the same plane as the final aperture) to the point at which the beam converges, which is typically also the surface of the sample if the sample is in focus. In practical SEM configurations this distance can be as small as a fraction of a millimeter or as large as tens of millimeters or a few centimeters, but in most situations W will be between 1 mm and 5 mm or so. The diameter of the wide end of the cone, line segment AB, is the aperture diameter, d_{apt}. This can also vary widely depending on the SEM model and the choices made by the operator, but it is certainly no larger than a fraction of a millimeter and can be much smaller, on the order of micrometers. For purposes of concreteness, let's assume W is 5 mm (i.e., 5000 μm) and the aperture is 50 μm in diameter (25 μm in radius, denoted r_{apt}).

From ◻ Fig. 5.3b we can see that triangle ASB is composed of two back-to-back right triangles. The rightmost of these has its vertex angle labeled α. The leg of that triangle adjacent to α is the working distance W, the opposite leg is the aperture radius r_{apt}, and the hypotenuse of the right triangle is the slant length of the beam cone, SB. From basic trigonometry we know that the tangent of the angle is equal to the length of its opposite leg divided by its adjacent leg, or

$$\tan \alpha = \frac{r_{apt}}{W}, \tag{5.4}$$

$$\alpha = \tan^{-1}\left(\frac{r_{apt}}{W}\right) = \tan^{-1}\left(\frac{25\,\mu m}{5000\,\mu m}\right)$$
$$= \tan^{-1}(0.005) = 0.005\,\text{radians} = 5\,\text{mrad}.$$

It is no coincidence that the arc tangent of 0.005 is almost exactly equal to 0.005 radians, since a well-known approximation in trigonometry is that

$$\tan^{-1}\theta = \theta. \tag{5.5}$$

Since in every practical case encountered in SEM imaging the angle will be sufficiently small to justify this approximation, we can write our estimate of the convergence angle in a much simpler form that does not require any trigonometric functions,

$$\alpha = \frac{r_{apt}}{W}, \text{ or } \alpha = \frac{d_{apt}}{2W} \tag{5.6}$$

As mentioned earlier, this angle is quite small, approximately equal to 0.25° or about 17 arc minutes.

5.2.6 Beam Solid Angle

In the previous section we defined the beam convergence angle in terms of 2D geometry and characterized it by a planar angle measured in the dimensionless units of radians. However, the electron beam forms a 3D cone, not a 2D triangle, so in reality it subtends a *solid angle*. This is a concept used in 3D geometry to describe the angular spread of a converging (or diverging) flux. The usual symbol for solid angle is Ω, and its units of measure are called *steradians*, abbreviated *sr*. Usually when solid angles are discussed in the context of the SEM they are used to describe the acceptance angle of an X-ray spectrometer, or sometimes a backscattered electron detector, but they are also important in fully describing the electron optical parameters of the primary beam in the SEM as well as the properties of electron guns.

◻ Figure 5.4 shows the conical electron beam in 3D, emerging from the circular beam-defining aperture at the top

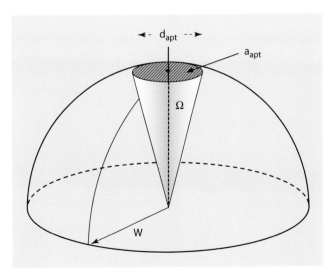

◻ **Fig. 5.4** Definition of beam solid angle, Ω. The vertical dashed line represents the optical axis of the SEM, and the distance from the aperture plane to the beam impact point is the working distance, W. This is also the radius of the imaginary hemisphere used to visualize the solid angle

5

of the figure and converging on the sample surface. The diameter of that aperture is d_{apt}, and the area of the aperture is a_{apt}. As discussed earlier, the distance the beam travels vertically from the final aperture to the point where it is focused to a spot is the working distance, W. Now imagine a complete sphere centered on the beam impact point, and with spherical radius equal to W. The upper hemisphere of this imaginary dome is depicted in ◘ Fig. 5.4 as well, and since its radius is W, the beam-defining aperture will lie on the surface of this sphere.

The key to understanding the meaning of solid angles and their numerical measure using units of steradians is to consider such a complete sphere and the fractional surface area of that sphere that is occupied by the object of interest. Every complete sphere, *regardless of diameter*, subtends exactly 4π steradians of solid angle. It follows that every hemisphere represents a solid angle of 2π steradians, no matter how small or how large the hemisphere might measure in meters. On the other hand, the surface area of a sphere A_{sphere} most certainly depends on the radius r, and can be calculated using the ancient formula

$$A_{sphere} = 4\pi r^2. \tag{5.7}$$

For the imaginary sphere and electron beam aperture shown in ◘ Fig. 5.4, we can assume realistic numbers for this calculation by adopting the values used in the beam convergence angle discussion above: $W = 5000$ μm, $d_{apt} = 50$ μm, and $r_{apt} = 25$ μm. With these values we can calculate the surface area of the complete sphere as

$$A_{sphere} = 4\pi W^2 = 4\pi \left(5000\,\mu m\right)^2 = 3.14\times10^8 mm^2. \tag{5.8}$$

and we can calculate the area of the beam-defining aperture as

$$a_{apt} = \pi r_{apt}^2 = \pi \left(25\,\mu m\right)^2 = 1.96\times10^3\,\mu m^2. \tag{5.9}$$

It is obvious from the diagram that our aperture subtends only a small fraction of the sphere upon which it rests, and it is a simple matter to calculate the value of that fraction,

$$\frac{a_{apt}}{A_{sphere}} = \frac{1.96\times10^3\,\mu m^2}{3.14\times10^8\,\mu m^2} = 6.25\times10^{-6}. \tag{5.10}$$

The important step is to realize that if the aperture occupies 6 parts in a million of the whole sphere's surface area, then it must also subtend 6 parts in a million of the 4π steradian solid angle of that whole sphere, so we can calculate the numerical solid angle of the beam by multiplying by this areal fraction

$$\Omega_{beam} = \frac{a_{apt}}{A_{sphere}} \cdot \Omega_{sphere} = \left(6.25\times10^{-6}\right)\cdot\left(4\pi\ steradian\right)$$
$$= 7.85\times10^{-5}\ sr. \tag{5.11}$$

Unless you work with solid angle calculations on a regular basis, this value probably has little physical meaning to you, and you have no sense of how big or how small 78 microsteradians are in real life. To provide some perspective, consider that both the Moon and the Sun subtend about this same solid angle when viewed from the surface of the Earth using the naked eye. The exact angular diameters (and therefore also the solid angles) of both the Sun and the Moon vary slightly during their orbits, depending on how far away they are at any given moment, but this variation is small and oscillates around average values:

$$2\alpha_{Sun} = 9.35\,mrad,\quad \alpha_{Sun} = 4.68\,mrad$$

$$2\alpha_{Moon} = 9.22\,mrad,\quad \alpha_{Moon} = 4.61\,mrad$$

$$\Omega_{Sun} = 68.7\,\mu sr$$

$$\Omega_{Moon} = 66.7\,\mu sr.$$

Of course the Sun is much, much larger than the Moon in diameter, but it is also much farther away, so the two celestial bodies appear to be about the same angular size from the perspective of the Earthbound viewer. This similarity in angular size is a coincidence, and it is the reason that during a solar eclipse that the Moon almost perfectly occludes the Sun for a short time. This analogy is instructive for the SEM operator because it helps explain how a small final beam aperture combined with a short working distance can produce the same convergence angle (and therefore depth of field) as a configuration that uses a large aperture and a long working distance. Likewise, an energy-dispersive X-ray spectrometer (EDS detector) with a small area of 10 mm² can subtend the same solid angle (and therefore collect the same number of X-rays) as a much larger 100-mm² detector sitting at a more distant detector-to-sample position.

5.2.7 Electron Optical Brightness, β

In practice the beam solid angle described in the previous section is an obscure and little-used parameter, and it is not that important for most SEM operators to understand fully. However, the concept of beam solid angle and the units of steradians affect the SEM operator much more directly through the concept of electron optical *brightness*, β. The main reason that field emission gun SEMs (FEG SEMs) enjoy drastically improved performance over SEMs that use thermionic tungsten electron sources is because of the much larger electron optical brightness of the FEG electron source. Further, the brightness of the beam when it lands on the sample is the central mathematical variable in one of the key equations of SEM operation, the brightness equation. This equation relates the beam brightness to the beam diameter, the beam current, and the convergence angle, and it is an invaluable tool that lets the SEM operator predict and manage the tradeoff between probe size and beam current.

Because of this central role in practical use of the SEM, it is worth struggling with the mathematics until you understand these concepts and can apply them in your work.

Because the term *brightness* is used in everyday language, most people have an intuitive sense that if one source of energy (say, the Sun) is brighter than another source (say your flashlight or torch) then the brighter source is emitting "more light." In other words, the flux is higher on the receiving end (i.e., at the sensor). Electron optical brightness is similar, but it is more precisely defined, considers current density instead of just total current, and factors in the change in angular divergence of the beam as it is focused or defocused by the electron lenses in the SEM column. Using the terms and concept defined in the sections above, brightness can be succinctly defined as current density per unit solid angle, and it is measured in units of A m^{-2} sr^{-1} (i.e., amperes per square meter per steradian). Based on a quick analysis of the units, it becomes obvious that if two electron beams have exactly the same current and same beam diameter at their tightest focus (and therefore the same current density), but they have different convergence angles, the beam with the smaller convergence angle will have the higher brightness. This is a result of the sr^{-1} term in the units, meaning the solid angle is in the denominator of the definition of brightness, and therefore larger solid angles result in smaller brightnesses (all other things being equal). In the case of visible light, this is why a 1-W laser is far "brighter" than a 200-W light bulb. This simultaneous dependence on current density and angular spread is also the reason for one of the most important properties of brightness as defined above: it is not changed as the electron beam is acted upon by the lenses in the SEM. In other words, to a very good approximation, the brightness of the electron beam is constant as it travels down the SEM from the electron source to the surface of the sample; and if you can estimate its value at one location along the beam you know it everywhere. One variable that does affect the brightness, however, is beam energy. In the SEM the brightness of all electron sources increases linearly with beam energy, and this change must be taken into account if you compare the brightness of beams at different energies.

Brightness Equation

One of the most valuable equations for understanding the behavior of electron beams in the SEM is the brightness equation, which relates the three parameters that define the beam:

$$\beta = \frac{4i}{\pi^2 d^2 \alpha^2} \tag{5.12}$$

If you know the numerical value of the brightness of the beam, measured in A m^{-2} sr^{-1}, then the brightness equation can provide a rough estimate of other parameters such as beam diameter, current, and convergence angle. This can be useful for explaining (quantitatively) the observed performance increase of a FEG SEM over a thermionic instrument,

for example. However, even without knowing the numerical value of the brightness β, the functional form of the equation can provide very useful information about changes in the beam.

Because the brightness, even if its value is unknown, is a constant and does not change as you change lens settings from one imaging condition to the next, the left-hand side of the equation is constant and has a fixed value. This implies the right-hand side of the equation is also fixed, so that any changes in one variable must be offset by equivalent changes in the other variables to maintain the constant value. The multiplier "4" in the numerator is a constant, as is π in the denominator. That means that the ratio of i to the product $d^2\alpha^2$ is also constant. Note that the brightness equation constrains the selection of the beam parameters such that all three parameters cannot be independently chosen. For example, this means that if the current i is increased by a factor of 9 but the convergence angle is unchanged, the beam diameter will increase by a factor of 3 to maintain the equality. Alternatively, if the convergence angle is increased by a factor of 2 (say, by decreasing the working distance by moving the sample closer to the objective lens) then the current can be increased by a factor of 4 without changing the beam size. Even more complex changes in the beam parameters can be understood and predicted in this way, so careful study of this equation and its implications will pay many dividends during your study of the SEM.

5.2.8 Focus

One of the first skills taught to new SEM operators is how to focus the image of the sample. From a practical perspective, all that is required is to observe the image produced by the SEM, and adjust the focus setting on the microscope until the image appears sharp (not blurry) and contains as much fine detail as possible. From the perspective of electron optics, it is not quite as straightforward to understand what happens during the focusing operation, especially if you remember that the SEM image is not formed using the action of a lens as would be the case in a light microscope, but rather by rastering a conical beam across the surface of the sample. ◘ Figure 5.5 shows the three basic focus conditions: overfocus, correct focus, and underfocus.

In the SEM the focus of the microscope is changed by altering the electrical current in the objective lens, which is almost always a round, electromagnetic lens. The larger the electrical current supplied to the objective lens, the more strongly it is excited and the stronger its magnetic field. This high magnetic field produces a large deflection in the electrons passing through the lens, causing the beam to be focused more strongly, so that the beam converges to crossover quickly after leaving the lens and entering the SEM chamber. In other words, a strongly excited objective lens has a shorter focal length than a weakly excited lens. On the left in ◘ Fig. 5.5, a strongly excited objective lens (short focal

Fig. 5.5 Schematic of the conical electron beam as it strikes the surface of the sample, showing overfocus (*left*), correct focus (*center*), and underfocus (*right*). From this view it is clear that if the beam converges to crossover above the surface of the sample (*left*) or below the surface (*right*), the beam diameter is wider at the sample than the diameter of an in-focus beam (*center*)

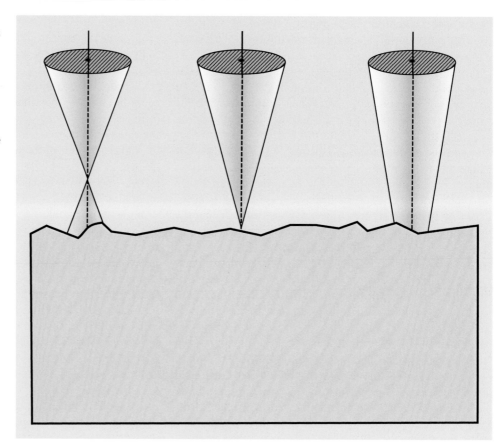

length) causes the beam to converge to focus before the electrons reach the surface of the sample. Since the electrons then begin to diverge from this crossover point, the beam has broadened beyond its narrowest waist and is wider than optimal when it strikes the surface of the sample, thus producing an out-of-focus image. Conversely, the right-hand portion of **Fig. 5.5** shows the beam behavior in an underfocused condition. Here the magnetic field is too weak, and the beam is not fully brought to crossover before it strikes the surface of the sample, and again the beam diameter is broader than optimal, resulting in an out-of-focus image.

Using these schematics as a guide, it is easier to understand what is happening electron optically when the SEM user focuses the image. Changes in the focus control result in changes in the electrical current in the objective lens, which results in raising or lowering the crossover of the electron beam relative to the surface of the sample. The distance between the objective lens exit aperture and this beam crossover point is displayed on the microscope as the working distance, W. On most microscopes you can see the working distance change numerically on the screen as you make gross changes in the focus setting, reflecting this vertical motion of the beam crossover in the SEM chamber. It is important to note that the term *working distance* is also used by some microscopists when referring to the distance between the objective lens pole piece and the surface of the sample. The value of W displayed on the microscope will accurately reflect this lens-to-sample distance if the sample is in focus.

Astigmatism

The pointy cones drawn in **Fig. 5.5** are a useful fiction for representing the large-scale behavior of a focused electron beam, but if we consider the beam shape carefully near the beam crossover point this conical model of the beam breaks down. **Figure 5.6** is a more realistic picture of the beam shape as it converges to its narrowest point and then begins to diverge again below that plane. For a variety of reasons, mostly the effects of lens aberrations and other imperfections, even at its narrowest point the beam retains a finite beam diameter. In other words, it can never be focused to a perfect geometrical point. The left side of **Fig. 5.6** shows the beam narrowing gently but never reaching a sharp point, reflecting this reality. Ideally, cross sections through the beam at different heights will all be circles, as shown in the right of **Fig. 5.6**. If the beam is underfocused or overfocused, as shown in **Fig. 5.5**, the consequence is a blurry image caused by the larger-diameter beam (larger blue circles in **Fig. 5.6** above and below the narrowest point).

In real SEMs the magnetic fields created in the electron optics are never perfectly symmetric. Although the manufacturers strive for ideal circular symmetry in round lenses, invariably there are defects in the lens yoke, the electrical windings, the machining of the pole pieces, or other problems that lead to asymmetries in the lens field and ultimately to distortions in the electron beam. Dirt or contamination buildup on the apertures in the microscope can also be an important source of distorted beam shapes. Since the dirt on the aperture is electrically non-conductive, it can

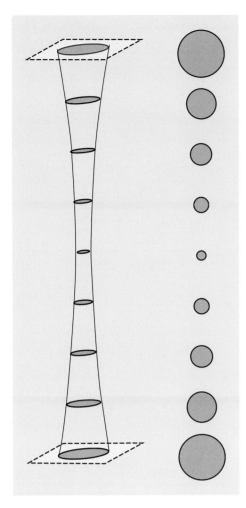

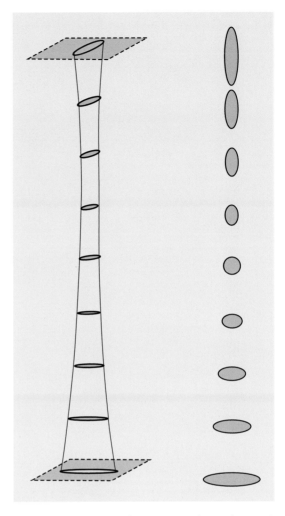

Fig. 5.6 Perspective view of the electron beam as it converges to focus and subsequently diverges (*left*), and a series of cross-sectional areas from the same beam as it travels along the optic axis (*right*). Note that although this beam does not exhibit any astigmatism it still does not focus to a point at its narrowest waist

Fig. 5.7 Perspective view of an astigmatic electron beam as it converges to focus and subsequently diverges (*left*), and a series of cross-sectional areas from the same beam as it travels along the optic axis (*right*). Because this beam exhibits significant astigmatism, the cross sections are not circular and their major axis changes direction after passing through focus

accumulate an electric charge when the beam electrons strike it, and the resulting electrostatic fields can warp and bend the beam into odd and complex shapes that no longer have a circular cross section.

By far the most important of these distortions is called *two-fold astigmatism*, which in practice is often referred to just as *astigmatism*. In this specific distortion the magnetic field that focuses the electrons is stronger in one direction than in the orthogonal direction, resulting in a beam with an elliptical cross section instead of a circular one. In beams exhibiting astigmatism the electrons come to closest focus in the x-direction at a different height than the y-direction, consistent with the formation of elliptical cross sections. These effects are shown schematically in **Fig. 5.7**. Similar to **Fig. 5.6**, the focused beam is shown in perspective on the left side of the diagram, while a series of cross sections of the beam are shown on the right of the figure. In the case shown here, as the beam moves down the optical axis of the SEM, the

cross section changes from an elongated ellipse with long axis in the y-direction, to a circle (albeit with a larger diameter than the equivalent circle in **Fig. 5.6**), and then to another elongated ellipse, but this time with its long axis oriented in the x-direction. This progression from a near-line-focus to a broader circular focus and then to a near-line-focus in an orthogonal direction is the hallmark of a beam exhibiting astigmatism.

This behavior is also easily visible in the images produced by rastering the beam on a sample. When the beam cross section is highly elongated at the surface of the sample, the image resolution is degraded badly in one direction, producing a blurring effect with pronounced linear asymmetry. It appears as if the image detail is sheared or stretched in one direction but not the other. If the focus knob is adjusted when the beam is astigmatic, a point can be reached when this image shearing or linear asymmetry is eliminated or at least greatly reduced. This is the best focus obtainable without

5

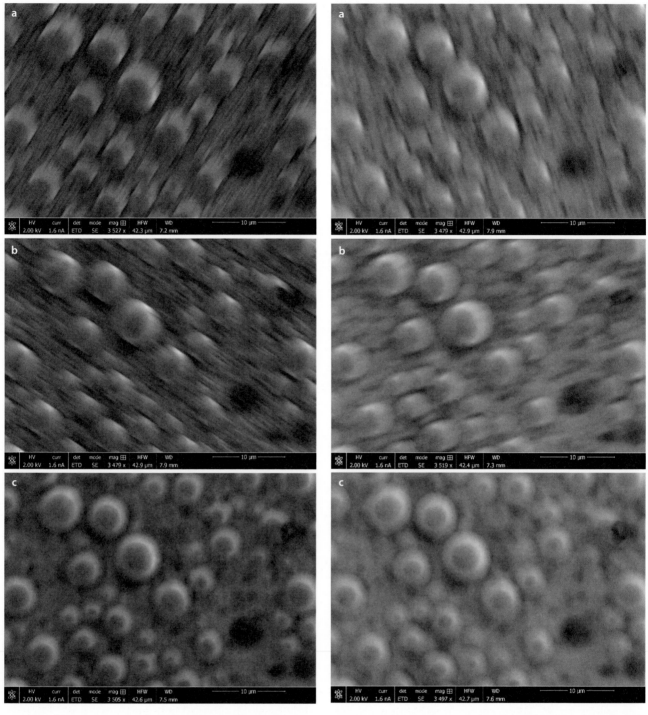

☐ **Fig. 5.8** Three SEM micrographs showing strong astigmatism in the X direction, when the sample is **a** overfocused, **b** underfocused, and **c** near focus. Note that the shearing or "tearing" appearance of fine detail in **a** appears to be in a direction perpendicular to the effect in **b**

☐ **Fig. 5.9** Three additional SEM micrographs from the same field of view shown in Fig. 5.8 above. Here the beam shows strong astigmatism in the Y-direction, when the sample is **a** overfocused, **b** underfocused, and **c** near focus

correcting the astigmatism, although it generally produces an image that is far inferior to the in-focus images obtainable with a properly stigmated beam. If the focus is further adjusted, past this point of symmetry, the image will again exhibit large amounts of shearing and stretching of the fine details, but in a different direction. This sequence of effects can be seen in ☐ Fig. 5.8. In ☐ Fig. 5.8a the sample is shown

when the objective lens is overfocused, with the beam crossover occurring above the sample surface, corresponding to the left diagram in ☐ Fig. 5.7. In panel ☐ Fig. 5.8b, the same sample with the same astigmatic beam is shown in underfocus, the right side diagram in ☐ Fig. 5.6. ☐ Figure 5.8c shows the best achievable focus; here, the shearing and stretching is minimized (or at least balanced), suggesting the cross section

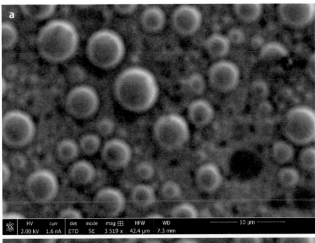

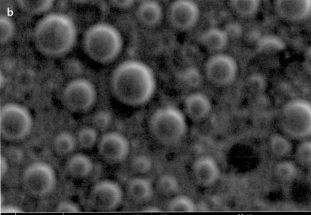

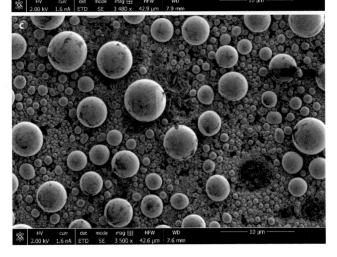

□ **Fig. 5.10** Three additional SEM micrographs from the same field of view seen in □ Figs. 5.8 and 5.9 above. Here the sample is imaged with a fully corrected beam, so neither the overfocused image **a** nor the underfocused image **b** shows significant anisotropic fine detail. Further, the in-focus image in panel **c** is much sharper then the best-focus images obtained in panel **c** of □ Figs. 5.8 and 5.9

of the electron beam is approximately circular. □ Figure 5.9 shows the sample field of view seen in □ Fig. 5.8, but imaged using a beam with pronounced astigmatism in the Y direction. In general the SEM beam will be astigmatic in both X and Y, and the operator must correct this beam distortion along both axes using the X and Y stigmators. When this is

performed correctly, a series of image like those in □ Fig. 5.10 can be obtained. In □ Fig. 5.10, both overfocused and underfocused images show loss of fine detail, but no directional distortion is present. The other significant improvement in □ Fig. 5.10c over □ Figs. 5.8c and 5.9c is that the best-focus image is much sharper when the image is properly stigmated. While this last benefit is the real reason to master the art of image stigmation, the characteristic appearance of images like those in □ Figs. 5.8a, b and 5.9a, b are very handy when adjusting the stigmation controls on the microscope.

Learning how to properly adjust the stigmation coils on an SEM can be one of the most challenging and frustrating skills to develop when first learning to use the instrument. However, as can be seen in the previous figures, being able to quickly and accurately minimize astigmatism in your electron micrographs is an essential milestone along the journey to becoming an expert scanning electron microscopy and X-ray microanalyst.

5.3 SEM Imaging Modes

SEMs are very flexible instruments, and the SEM operator has control over a large number of electron beam, detector, and stage parameters. Consequently, the number of different imaging conditions that may be employed to analyze any given sample is nearly infinite, and it is the job of the analyst to choose these conditions wisely to obtain useful information to meet the needs of the analysis. Fortunately, in many situations these choices can be narrowed to using one of the four basic modes of SEM operation: (1) High Depth-of-Field Mode, (2) Resolution Mode, (3) High-Current Mode, and (4) Low-Voltage Mode.

Below you will find practical information on how to control the fundamental electron optical parameters described earlier in the text and specific guidance for operating the SEM in the four basic modes just mentioned. Experienced SEM operators will have mastered these four modes and will be comfortable moving between them as needed. Choosing any one of these modes is a compromise, since each of them sacrifices microscope performance in some areas to achieve other imaging goals. Appreciating the strengths and weaknesses of each mode is essential to understanding when each mode is warranted. Of course some analyses will demand imaging conditions that do not fall neatly into one of these four basic modes, and the expert SEM operator will use the full flexibility of the instrument when required.

5.3.1 High Depth-of-Field Mode

Anyone familiar with compound light microscopes (LMs) understands that they have a very limited Depth-of-Field (DoF), meaning there is a limited range of vertical heights on the sample surface that will all appear to be in focus simultaneously. Parts of the sample that fall outside this range appear blurry. One of the advantages of the SEM over the light microscope is that it is capable of a much deeper depth-of-field than the LM.

5

◘ **Fig. 5.11** Schematic showing why an SEM has finite depth-of-field D_f and how it is defined

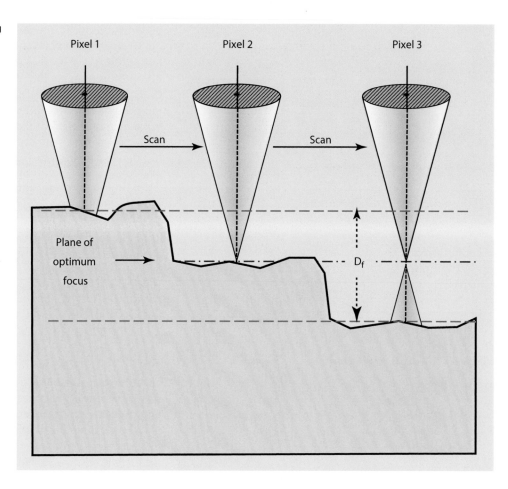

The shallow depth-of-field in the LM arises from the properties of its glass lenses, but SEMs don't use lenses to form direct images; instead they rely on lenses to focus the beam and then scan this beam from pixel to pixel to image the sample.[1] Nonetheless, they suffer from limited DoF because of the effect shown schematically in ◘ Fig. 5.11. Here the electron beam is shown striking the sample in three different locations, producing three different pixels in the image. For all three pixels the vertical position of the electron beam crossover is the same; this height is called the plane of optimum focus and is represented in ◘ Fig. 5.11 as a horizontal green dot-dashed line. For the case of pixel 2, this plane coincides with the surface of the sample. For pixel 1, the electron beam has not yet reached crossover when it

strikes the surface of the sample, at a height denoted by the upper red dotted line. This is equivalent to underfocusing the beam, with the same effect: the diameter of the probe at the sample surface is larger than optimal. If this increase in probe size is large enough, it will degrade the sharpness of the image. The height at which this blurring becomes measureable, denoted by the upper red dotted line, is the upper limit of the DoF for this beam. Similarly, for pixel 3 the sample surface is lower (i.e., further from the objective lens) than the middle case. This is analogous to overfocus because the beam comes to crossover and begins to diverge again before it lands on the sample. As before, this can degrade the sharpness of the image, and the height at which this degradation is noticeable is the lower limit of the height range that defines the DoF. The distance between these two dotted red lines is labeled D_f in ◘ Fig. 5.11, denoting the depth-of-field.

Because the definition of DoF requires the resulting blur to be noticeable (or at least measureable), it depends on many factors and can be somewhat subjective. For example, since in most cases sub-pixel blurring is not a concern to the SEM analyst, the effective DoF will often improve as the magnification decreases and the pixel size increases. However, as the magnification decreases, much more of the sample is visible in the field of view, increasing the chances that pronounced topography will lead to blurring. In these cases the DoF is

---

1   Glass lenses and transmission electron microscope lenses also have a related property known as *depth-of-focus*, a term that is often confused with depth-of-field. Depth-of-field refers to the range of heights in simultaneous focus *on the sample* (i.e., the observed field). In contrast, depth-of-focus refers to the range of positions near the imaging plane of the lens where the image is in focus. This determines, for example, how far away from the ideal imaging plane of the lens you can place a piece of film, or a CCD detector, and still capture an in-focus image. Because SEMs capture images via scanning action, the term depth-of-focus is not relevant.

◻ Fig. 5.12 **a** Diagram of the electron beam emerging from the final aperture in the objective lens and striking the sample under typical imaging conditions; a relatively large aperture diameter and short working distance create a large convergence angle and therefore a shallow depth-of-field. **b** *High Depth-of-Field Mode.* Here a small aperture diameter and long working distance *W* combine to create a small convergence angle and therefore a large depth-of-field

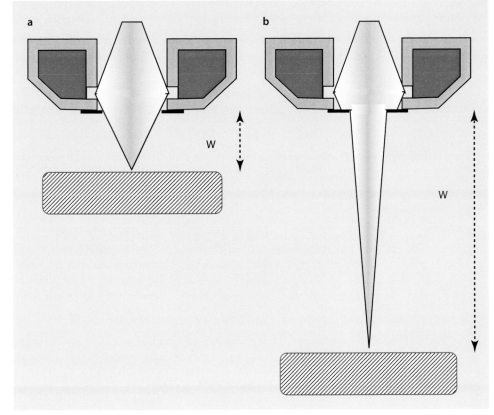

still increased at low magnification, but parts of the sample are still blurred in the image because of the large range of height visible in the expanded field of view. Nonetheless, operating the SEM in High Depth-of-Field Mode at medium to low magnifications is perhaps the most often used imaging condition for routine SEM work.

The basic idea behind High Depth-of-Field Mode is to create a set of imaging conditions where the convergence angle of the beam is small, producing a narrow pencil-like electron beam that does not change diameter rapidly with height above the sample. ◻ Figure 5.12 shows what this looks like schematically. ◻ Figure 5.12a represents typical imaging conditions, with short working distance W and normal aperture diameter. ◻ Figure 5.12b shows the imaging conditions used in High Depth-of-Field Mode, where the working distance has been increased significantly and a smaller diameter aperture is inserted. These two changes decrease the convergence angle of the electron beam and therefore increase the DoF. The effects of the aperture and working distance are independent of each other, meaning either one can improve the depth-of-field by itself.

For best results in Depth-of-Field Mode, determine the lowest stage position available (largest working distance), and drive the sample to that location. Changing the working distance is straightforward on most SEMs. Those microscopes with a manual stage will often have a physical knob on the chamber door for changing the height of the sample. Motorized stages are sometimes controlled by a hand panel,

joystick, or stand-alone stage controller, especially on older microscopes. Recent models typically use a graphical user interface, requiring the operator to enter a destination height (or "Z position") in millimeters and then executing the move. Some also allow the stage height to be changed continuously using the mouse.

Depth-of-Field Mode is also optimized by selecting a relatively small final beam aperture. The mechanisms used to change the diameter of the final aperture, and to center it on the optical axis of the microscope, vary widely from one SEM model to the next. In fact, some SEMs are designed to use a fixed or semi-fixed final aperture and do not provide an easy method of altering the aperture size. Many microscopes have manual aperture controls mounted on the outside of the SEM column (◻ Fig. 5.13). Other microscopes use a graphical user interface (GUI) to allow the operator to select one of several available apertures for insertion. Following this selection, motors driven by an X/Y- motion controller physically move the selected aperture into place and recall from memory the X- and Y- positions needed to center it. In either case the apertures themselves are arrayed linearly as a series of circular holes in a long, thin aperture strip.

A few microscopes permit you to configure Depth-of-Field Mode directly by selecting this option in the instrument control software. ◻ Figure 5.14 shows an example screenshot from one manufacturer's user interface where the operator can select a dedicated "DEPTH" setting, automatically optimizing the instrument for a small convergence angle.

5

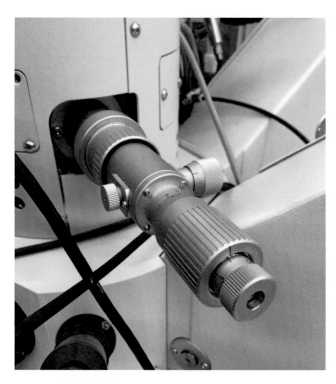

◘ Fig. 5.13 Manual aperture control mounted on the outside of an electron optical column. This mechanism allows the operator to select one of several different discrete apertures and adjust the X- and Y-positions of the aperture to center it on the optical axis of the microscope

Info Panel		
Continual	Single	Acquire
Scan Mode:	DEPTH	
HV:	10.00 kV	
Magnification:	491 x	
View field:	309.1 μm	
Speed:	1 (0.10 μs/pxl)	
WD:	17.133 mm	
Depth of Focus:	249.68 μm	
Stigmator:	2.0 % / –0.8 %	
Shift:	0.0 μm / –0.0 μm	
Rotation:	0.00 deg	
Beam Intensity:	13.04	
Absorb. Curr:	< 1 pA	
Spot Size:	8.5 nm	

◘ Fig. 5.14 Graphical user interface from one manufacturer's instrument control software that allows the user to select Depth-of-Field Mode directly. On this microscope, once the Scan Mode is set to "DEPTH," the electron column is configured automatically to create a small convergence angle and a large depth-of-field

5.3.2 High-Current Mode

Like the High Depth-of-Field Mode described above, the High-Current Mode of operating the SEM is used frequently. In many common imaging situations it delivers excellent feature visibility, useful and informative materials contrast, and adequate resolution and depth-of-field . It is particularly useful when the native contrast of the sample is low, such as when neighboring materials phases exhibit approximately equal average atomic number. It is also invaluable when performing X-ray microanalysis since the higher beam current translates directly into higher X-ray count rates. This can help by shortening the acquisition time needed to acquire individual X-ray spectra with an adequate number of counts for quantitative analysis, but it is even more important when acquiring X-ray maps or spectrum image datasets with full spectra at every spatial pixel. In all the cases mentioned above, feature visibility and count rate (both enabled by high current) are more important than spatial resolution or depth-of-field .

The basic idea behind High-Current Mode is to increase the current in the probe to boost both the signal reaching the detectors and the signal-to-noise ratio. Regardless of the electron detector in use (e.g., Everhart–Thornley detector, dedicated backscatter detector, through-the-lens detector, etc.), the signal reaching the detector scales with the signal generated at the sample, and this in turn scale directly with the current in the electron probe.

Unfortunately, the controls used to vary the electron beam current vary widely from one SEM model to the next, and different SEM manufacturers use discordant or conflicting terminology to describe these controls. As dictated by the brightness equation, the probe diameter must increase with an increase in probe current, so some manufacturers call the control "Spot Size." On some microscopes Spot Size 1 is a small spot (corresponding to a low beam current) and Spot Size 10 is a large spot (high current); a different vendor, however, may have adopted the convention where Spot Size 1 is a large spot and Spot Size 10 is a small spot. Other companies use the term *Spot Size,* but specify it in nanometers in an attempt to represent the nominal diameter of the probe. An approach growing in popularity with more modern instruments is to allow the operator to set the nominal probe current itself instead of Spot Size. As discussed above, this can be done either in discrete steps or continuously. In either case, the current steps can be labeled with arbitrary numbers (e.g., 7), they can reflect the nominal probe current (e.g., 100 pA), or sometimes they are specified as a percentage of the maximum current (e.g., 30 %). This dizzying variety of methods for labeling the desired probe current on SEMs can be confusing when switching from one instrument to another.

◘ Figure 5.15 shows two different varieties of physical knob configuration that might be encountered on the control panel of SEMs and electron probe microanalyzers (EPMAs). These analog controls are very intuitive to use because turning the knob changes the beam current in an immediate and continuous manner, allowing fine control of this parameter. Large

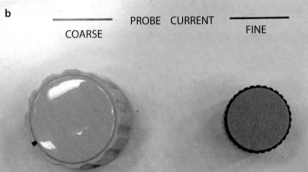

■ **Fig. 5.15** Examples of physical knobs on SEMs and electron probe microanalyzers (EPMAs) used by the operator to adjust the beam current. In both cases the operator has access to a coarse and a fine adjustment, either using one knob and a coarse/fine selector button **a**, or dedicated knobs for coarse and fine control **b**

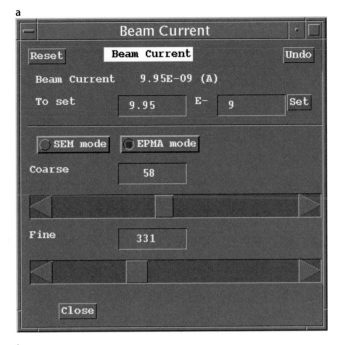

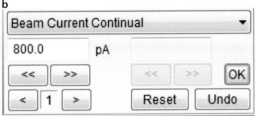

■ **Fig. 5.16** Examples of graphical user interface controls present in different manufacturers that allow the operator to control the beam current continuously. In **a**, the operator has a choice of setting the nominal current in pA digitally, or by using coarse and fine slider controls expressed in arbitrary units. In **b**, the operator also has a nominal beam current control expressed in pA as well as buttons to increase to decrease the value

changes in beam current can be made quickly by using the coarse setting of the knob shown in ■ Fig. 5.15a, or the coarse knob shown in ■ Fig. 5.15b. In both cases fine control is also possible for smaller adjustments. ■ Figure 5.16 shows two examples of computer-based beam current controls of the type found in graphical user interfaces. In both cases the operator can change the beam current using the mouse and keyboard. In ■ Fig. 5.16a this can be accomplished either by entering an exact digital value for the beam current and clicking the "Set" button, by dragging one of the two the slider controls to the left or right, or by clicking the arrow buttons to increase to decrease the current setting. Note that on this microscope, the slider positions are expressed digitally using arbitrary units (58 units for the coarse slider and 331 units for the fine slider). While these numbers are not true current values, these arbitrary settings can be useful for returning the microscope to a specific current. ■ Figure 5.16b shows a similar GUI window from the

user interface written by a different manufacturer. In this case, the operator also has access to a numerical beam current setting, nominally calibrated in true current measured in pA, as well as buttons that when clicked will increase or decrease the beam current incrementally. Finally, ■ Fig. 5.17 shows screenshots from a graphical user interface based on Spot Size instead of beam current. The operator is asked to select a specific Spot Size using a quick access pull-down menu (■ Fig. 5.17a) or a more flexible combination of a pull-down menu and up/down buttons (■ Fig. 5.17b). While these figures provide a sampling of the large variety of terms and interface layouts that the operator might encounter in the field, there are many more variations in practice than can be shown here.

Regardless of how any given SEM allows the operator to change the probe current, the most important tasks for the operator are to know how to increase and decrease current, and how to measure the current accurately once set. Even on those instruments that let the operator select a numerical probe current (e.g., 1 nA) via the user interface, it is unwise

5

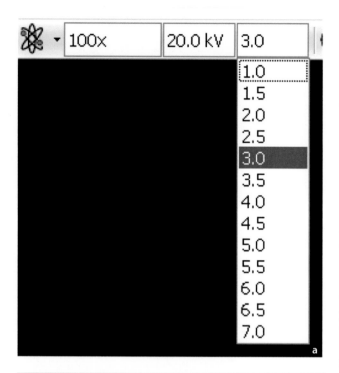

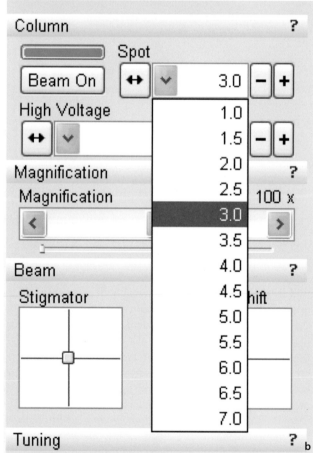

Fig. 5.17 Examples of graphical user interface controls that allow the operator to control the beam current in discrete steps expressed as changes in Spot Size. In **a**, the can choose any of several Spot Size values from a pull-down menu. In **b**, from the same microscope, the operator can access Spot Size via a pull-down menu or buttons that increase or decrease the value

to assume this setting will reliably produce the displayed value. Well-equipped SEMs have a built-in picoammeter that can be automatically inserted into the beam path to measure probe current. Getting a reading in these cases is as simple as triggering the insertion of the meter's cup and reading the value off the screen. Alternatively, a stage-mounted Faraday cup (either purchased commercially or homemade) attached through an electrical feedthrough to a benchtop picoammeter can be used instead.

Since the basic idea of High-Current Mode is to deliver sufficient probe current to the sample to generate superior signal-to-noise ratio, optimum results are obtained at medium to low magnifications, and often a larger final aperture is useful. Frame the field-of-view desired, focus the beam, and increase probe current until high-quality images can be obtained within a relatively short frame time, say, a few seconds to a minute. Check that any low-contrast features needed for analysis are sufficiently visible, and increase probe current further if they are not. For many situations, this high-current imaging approach will yield excellent images quickly and with little wasted time. If you are performing X-ray microanalysis, the approach to High-Current Mode is very similar to that for imaging, but the choice of current is dictated not by image quality but by X-ray count rate or, more suitably, the dead time percentage of the X-ray spectrometer's pulse processor.

5.3.3 Resolution Mode

Resolution Mode is probably the most demanding of the four basic SEM operational modes, chiefly because the microscope is driven at or near its limits of performance. It challenges the operator mentally, since choosing optimum imaging parameters requires deeper knowledge of electron optics and the physics of electron beams, although suitable images can be obtained with a basic understanding of the principles. It also demands more skill in operating the microscope, since small misalignments (e.g., residual stigmatism, imperfectly centered aperture) are more apparent. In fact, good alignment of the entire column is necessary to get the best resolution from the scope, while small misalignments are often tolerated in High-Current or High Depth-of-Field Mode. Resolution Mode also expects more from the microscope's environment. Mechanical vibrations, electronic noise, and AC magnetic fields near the microscope are some of the many sources of image degradation that, while generally unnoticeable, become obvious when operating in Resolution Mode. Poor sample preparation, such as overly thick evaporated metal coatings or insufficient metallographic polishing, for example, is also more evident at high magnifications. Most of these challenges are greatly reduced at lower magnifications, but the larger pixel sizes that result from low magnification obviate the need for Resolution Mode. In short, the same imaging conditions that enable Resolution Mode also highlight any shortcomings in the operator's technique, the laboratory environment, and the sample preparation.

Although every one of the basic SEM operational modes requires some compromise, in Resolution Mode the pursuit of high spatial resolution often involves compromise across the board. Small probe diameters require very low beam currents, thereby reducing the signal generated and lengthening the frame times needed. Depth-of-field is also reduced, although this is often not noticeable at high magnification, and detector choice is usually limited to the one or two channels optimized for this purpose (e.g., through-the-lens detectors).

The basic idea in Resolution Mode is to (1) minimize the probe diameter by raising the beam energy and reducing the beam current, (2) emphasize the collection of the resolution-preserving SE_1 secondary electrons generated at the beam footprint, and (3) minimize the myriad sources of image degradation by using the shortest working distance possible. Raising the beam energy helps produce smaller probe sizes because it increases the brightness of the gun. For thin samples, such as small particles sitting on an ultrathin film substrate, this produces the highest resolution. Likewise for very high-Z samples, even high landing energies have short electron ranges and therefore small excitation volumes. However, for thick samples with low atomic number, better resolution may be obtained at lower landing energies if the size of the excitation volume is the limiting factor. For any given beam energy, smaller currents always yield smaller probe sizes, as demanded by the brightness equation, so operating at tens of picoamps is not uncommon in this mode. Choice of signal carrier and detector can be crucial for obtaining high spatial resolution. Since backscattered electrons emerge from a disc comparable in size to the electron range, it is very hard to realize high resolution by using backscattered electrons (BSE) directly or BSE-generated secondaries such as SE_2 secondary electrons (generated at the sample surface by emerging BSE) or SE_3 secondary electrons (generated at great distance from the sample by BSE that strike microscope components). The highest resolution is obtained from SE_1 secondary electrons, because these emerge from the very narrow electron probe footprint on the sample surface, comparable in diameter to the probe itself. Microscopes equipped with immersion objective lenses or snorkel lenses and through-the-lens detectors (TTLs) are best at efficient collection of SE_1 electrons. Finally, bringing the sample very close to the objective lens, even less than 1 mm if practical, can improve resolution significantly. SE_1 collection is maximized by this proximity, and a short working distance (WD) can minimize the length over which beam perturbations such as AC fields can act.

The practical steps needed to configure the SEM for operation in Resolution Mode follow from the basic requirements outlined above. Get the sample as close to the objective lens as possible by carefully shortening the working distance. Computer-controlled SEMs will frequently have a software interlock designed to reduce the chances that the sample will physically impact the pole piece. Learn how this feature functions and use it effectively but carefully; high resolution is useful, but a scratched or dented pole piece can be a very expensive mistake! Also, beware that many microscopes possess more than one objective lens mode. Invariably the lens mode needed for best resolution will be the one that creates the highest magnetic field at the sample. Coupled with the proximity of a short working distance, these high magnetic field modes may lift your sample off the stage unless it consists of a non-magnetic material. Select the TTL detector if available, or other detector that preferentially utilizes SE_1 secondary electrons for imaging. Increase the accelerating voltage on the SEM to its highest setting, usually 30 kV or higher, and reduce the beam current to as low a value as practical while still maintaining visibility of the sample as noise increases. Moving to a slower frame time, longer dwell time, or enabling frame averaging will help mitigate the effects of reduced signal at low probe currents. Finally, select the optimal objective lens aperture diameter for best resolution. This can be tricky because of competing effects. Small apertures can limit the resolution due to diffraction effects, so the larger the aperture the less likely that these effects will be a problem. However, large apertures quickly amplify the effects of objective lens aberrations, especially spherical aberration, so the smallest aperture size available is ideal for reduction of aberrations. Clearly these requirements conflict with one another, and every objective lens has an intermediate aperture diameter that delivers the best resolution for any given beam energy and working distance. Some SEMs inform the operator of this optimal aperture size, while others are less helpful and leave it up to the operator to determine the best choice. In these cases, contact the SEM manufacturer's application engineer for advice or test a variety of aperture diameters on high quality imaging test specimens to understand how to manage this tradeoff.

5.3.4 Low-Voltage Mode

Of the four basic SEM modes, Low-Voltage Mode is probably the most esoteric and challenging, regarding both instrumentation and specimen issues. Reducing the landing energy of the beam is useful in many situations, and varying the beam energy should be considered when operating in High-Current Mode or Depth-of-Field Mode as needed. However, operating with landing energies below 5 keV, and especially below 1 keV, is qualitatively different than using higher energies. The performance of the SEM's entire electron optical chain, from the electron gun to the objective lens, is much worse at 1 keV than at high beam energy. While modern thermionic SEMs are often quite good performers in Low-Voltage Mode, not many years ago a field emission electron source (FEG) was considered a de facto requirement for low voltage work, and most older thermionic SEMs produce such poor images at 1 keV that they are almost useless.

For all electron sources the gun's brightness will be much lower at 1 keV than at 30 keV, which limits the current density in the probe because of the brightness equation. This in turn means the operator must work at much larger probe sizes to obtain sufficient current for imaging. Here field emission sources have a big advantage over tungsten or LaB_6 thermionic filaments because they are much brighter intrinsically, and so remain bright enough at low voltage for decent imaging. Another important concern that arises at these low beam energies is the chromatic aberration of the objective lens. This aberration causes beam electrons at different energies to be focused in different planes, reducing the current density. Although this aberration is a flaw in the lens itself and not the electrons in the beam, lower beam energies make the problem more apparent, in part because they have a larger fractional energy spread. In fact, the effects of this aberration would not be noticeable at all in a monochromatic electron beam, where all the electrons have exactly the same energy. Similarly, electron sources with naturally narrow energy spreads, such as cold field emission sources, suffer from these problems much less than sources with large energy spreads like thermionic guns. Whatever their cause, these reductions in image quality, both lower resolution and lower current density, explain why Low-Voltage Mode is commonly employed at low magnifications. Operators with expensive, high-performing field emission microscopes designed for low voltage operation will be able to work at low voltage and high magnification—even more so if the microscope is equipped with a beam monochromator, an accessory designed to artificially narrow the energy spread of the electron beam, thus reducing the effect of chromatic aberration even at very high magnifications.

Another unwanted consequence of using very low beam energies is that the resulting electron trajectories are less "stiff," meaning the electrons are more easily deflected from their intended paths by stray electric or magnetic fields near the beam. At 1 keV landing energy and below, the electrons are moving relatively slowly and are more susceptible to electrical charging in the sample, AC electric or magnetic fields in the microscope room, and electrical noise on the microscope's scan coils. These are some of the many challenges of imaging in Low-Voltage Mode.

The main advantages of Low-Voltage Mode are the much-reduced excitation volume and the resulting change in contrast for most sample materials. The range of primary beam electrons in most materials drops very rapidly as the landing energy is reduced, so the region in the sample emitting signal-carrying electrons can be very small, improving resolution in cases where it is limited by this range. The resulting surface sensitivity of the signal also tends to flatten the image contrast and it de-emphasizes materials contrast in favor of topography. Because the view of the sample in Low-Voltage Mode is often dramatically different than the equivalent image at normal beam energy, this mode often reveals important features in the sample that might be missed using routine imaging conditions.

It is possible to perform X-ray microanalysis at low voltage, but it presents special challenges and should not be attempted unless it is unavoidable. The very short electron range means the X-rays produced in the sample are generated close to the surface and very near the beam impact point. This is a good thing, because both lateral and depth resolution are improved, and absorption losses are reduced for outgoing X-rays. However, the low landing energies severely limit the number of X-ray lines that are efficiently excited, and many elements are either inaccessible, or the analyst is forced to use M- or N-shell lines with poorly measured cross sections or absorption coefficients. Complicating

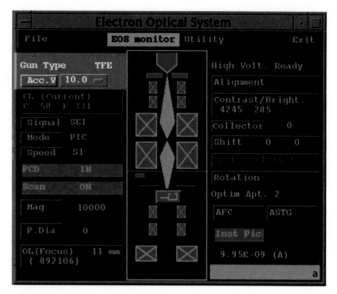

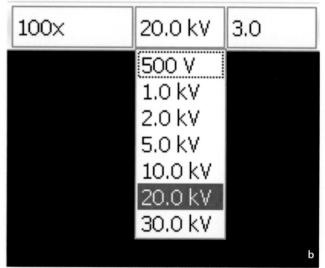

Fig. 5.18 Graphical user interface controls that allow the operator to control the beam energy. The instrument control software shown in **a** utilizes a pull-down menu on the *upper left* of the window to allow the operator to select the accelerating voltage in kilovolts (and thus the beam energy in kilo-electronvolts). The control is currently set to 10 kV. The interface in the screenshot in **b** shows a drop-down menu, allowing the SEM operator to select one of several discrete accelerating voltages between 500 V and 30 keV. In most cases, including the two shown above, the microscope allows the user to select values between the discrete settings shown in the screenshots, via a different mechanism (not shown)

matters further, the reduced brightness at low voltage means probe currents are low and X-ray count rates can be anemic.

The basic idea behind low voltage mode is simple: reduce the landing energy of the beam. The practical advice for configuring this mode is equally straightforward, since changing the beam energy on most microscopes is controlled by a dedicated knob or can be achieved by selecting the desired energy on a graphical user interface. ◻ Figure 5.18 shows two examples of GUI controls from different instruments. In screenshots the controls are expressed in accelerating voltage measured in kV; this is equivalent to controlling the beam energy in keV.

In some cases the SEM may allow the operator to apply a sample bias or use another form of beam deceleration, thus permitting the electron landing energy to differ from the beam energy. In these situations the manufacturer's instrument manual should be consulted for exact configuration guidance, and it is important to remember that it is the landing energy (not the energy of the beam as it leaves the objective lens) that governs both the electron range and the X-ray generation physics.

5.4 Electron Detectors

The SEM is equipped with one or more detectors that are sensitive to BSE, SE, or a combination of BSE and SE that emerge from the specimen as a result of the interaction of the primary electron beam. By measuring the response of BSE and SE as a function of beam location, various properties of the specimen, including composition, thickness, topography, crystallographic orientation, and magnetic and electrical fields, can be revealed in SEM images.

5.4.1 Important Properties of BSE and SE for Detector Design and Operation

Abundance
The total yield per incident beam electron of BSE or SE is sensitive to specimen properties such as the average atomic number (BSE), the chemical state (SE), local specimen inclination (BSE and SE), crystallographic orientation (BSE), and local magnetic field (BSE and SE). However, the total electron signal is not what is measured by most electron detectors in common use for SEM imaging. The actual response of a particular detector is further complicated by its limited angular range of acceptance as well as its sensitivity to the energy of the electrons being detected. The only detector which is exclusively sensitive to the number of BSE and/or SE (and not emitted trajectory or energy distributions) is the specimen itself when the specimen current is used as an imaging signal.

Angular Distribution
Knowledge of the trajectories of BSE and SE after leaving the specimen is important for placing a detector to intercept the useful signals. For a beam incident perpendicularly to a surface (i.e., the beam is parallel to the normal to the surface), BSE and SE are emitted with the same angular distribution which approximately follows the cosine function: the relative abundance along any direction is proportional to the cosine of the angle between the surface normal and that direction. Thus, the most abundant emission is along the direction parallel to the surface normal (i.e., back along the beam, where the angle $= 0°$ and $\cos 0° = 1.0$), while relatively few BSE or SE are emitted close to the surface (e.g., along a direction 1° above the surface is 89° from the surface normal, $\cos 89° = 0.017$, so that only 1.7 % is emitted compared to the

intensity emitted back along the beam). When a surface is highly inclined to the beam, the angular distribution of the SE still follows the cosine distribution, but the BSE follow a distribution that becomes progressively more asymmetric with tilt and is peaked in the forward (down slope) direction. For local surface inclinations above approximately 45°, the most likely direction of BSE emission is at an angle above the surface that is similar to the beam incidence angle above the surface. The directionality of BSE emission becomes more strongly peaked as the inclination further increases.

Kinetic Energy Response
BSE and SE have sharply differing kinetic energies. BSE retain a significant fraction of the incident energy of the beam electrons from which they originate, with typically more than 50 % of the BSE escaping while retaining more than 0.5 E_0. The BSE coefficient, the relative abundance of energetic BSE, and the peak BSE energy all increase with the atomic number of the target. Thus, for an incident beam energy of $E_0 = 20$ keV, a large fraction of the BSE will escape with a kinetic energy of 10 keV or more. By comparison, SE are much lower in kinetic energy, being emitted with less than 50 eV (by arbitrary definition). In fact, most SE exit the specimen with less than 10 eV, and the peak of the SE kinetic energy distribution is in the range 2 eV to 5 eV. Methods of detecting electrons include (1) charge generation during inelastic scattering of an energetic electron within semiconductor devices and (2) scintillation, the emission of light when an energetic electron strikes a suitably sensitive material, which includes inorganic compounds (e.g., CaF_2 with a minor dopant of the rare earth element Eu), certain glasses incorporating rare earth elements, and organic compounds (e.g., certain plastics). Both charge generation in semiconductors and scintillation require that electrons have elevated kinetic energy, typically above several kilo-electronvolts, to initiate the detection process, and the strength of the detection effect generally increases with increasing kinetic energy. Thus, most BSE produced by a beam in the conventional energy range of 10–30 keV can be directly detected with semiconductor and scintillation detectors, while these same detectors are not sensitive to SE because of their much lower kinetic energy. To detect SE, post-specimen acceleration must be applied to boost the kinetic energy of SE into the detectable range.

5.4.2 Detector Characteristics

Angular Measures for Electron Detectors
▪▪ Key Fact

Knowledge of the location of electron detectors is critical for proper interpretation of SEM images, especially of topographic features. Apparent illumination in the SEM image appears to come from the detector, while the observer's view appears to be along the incident electron beam, as discussed in detail in the Image Formation module.

5

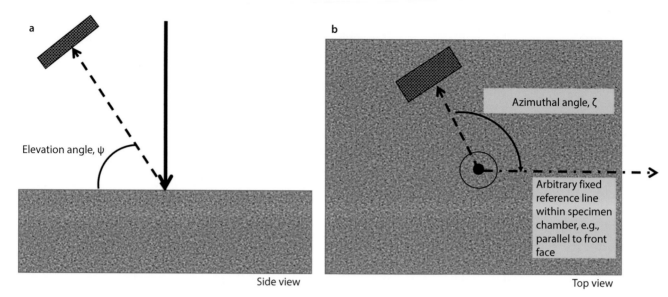

a

b

Elevation angle, ψ

Azimuthal angle, ζ

Arbitrary fixed reference line within specimen chamber, e.g., parallel to front face

Side view

Top view

☐ **Fig. 5.19**   **a** Detector take-off angle, ψ. **b** Detector azimuthal angle around beam, ζ

Elevation (Take-Off) Angle, ψ, and Azimuthal Angle, ζ

The effective position of a detector is specified by two angles. The *elevation* ("*take-off*") *angle*, designated ψ, is the angle above a horizontal plane perpendicular to the beam axis and the vector that joins the center of the detector to the beam impact position on the specimen, as shown in ☐ Fig. 5.19a. (Alternatively, the take-off angle can be measured as the complement of the angle between the beam axis and a line perpendicular to the detector face extended to the beam optic axis.) The "*azimuthal angle*," ζ, of the detector is the rotational angle around the beam to the detector line, measured relative to some arbitrary but fixed reference, such as the front face of the specimen chamber, as shown in ☐ Fig. 5.19b. When an SEM image is created, it is critical for the user to understand the relative position of the detector in the scanned image, as given by the azimuthal angle, since the illumination of the image will apparently come from the detector. Note that the "scan rotation" function, which permits the user to arbitrarily choose the angular orientation for the presentation of the image on the display, also varies the apparent angular location of the detector. It is therefore critical for the user to establish what setting of scan rotation corresponds to the correct known value of the detector azimuthal angle.

■ ■ Good Practice

Make a drawing (top view and side view) of the SEM chamber showing the physical locations of all detectors (electron, X-ray, and cathodoluminescence) and mark the values of the elevation angle, ψ, and azimuthal angle, ζ.

Solid Angle, Ω

As shown in ☐ Fig. 5.20, the effective size of the detector with an active area A placed at a distance r from the beam impact point on the specimen is given by the solid angle, Ω (Greek omega, upper case), which is defined as

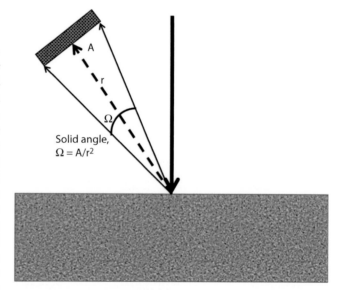

A

r

Ω

Solid angle, $\Omega = A/r^2$

☐ **Fig. 5.20**   Detector solid angle, Ω

$$\Omega = A / r^2 \ \left(\text{steradians, sr} \right) \qquad (5.13)$$

Note the strong dependence of Ω upon the distance of the detector from the beam impact on the specimen.

As an estimate of the overall geometric efficiency, ε, the solid angle of the detector can be compared to the solid angle of the hemisphere (2π sr) into which all electrons leaving a thick target are emitted:

$$\varepsilon = \Omega / 2\pi \qquad (5.14)$$

ε provides only an estimate of efficiency because the simple definition in Eq. (5.14) does not consider the non-uniform distribution in the emission of electrons from the specimen, for example, the cosine distribution of BSE at normal incidence.

Energy Response

The response of a detector may be sensitive to the kinetic energy of the striking electron. Generally an electron detector exhibits an energy threshold below which it has no response, usually a consequence of an insensitive surface layer such as a metallic coating, needed to dissipate charging, through which the incident electron must penetrate. Above this threshold, the detector response typically increases with increasing electron energy, making the detector output signal more sensitive to the high energy fraction of the electrons.

Bandwidth

The act of creating an SEM image involves scanning the beam in a time-serial fashion to dwell at a series of discrete beam locations (pixels) on the specimen, with the detector measuring the signal of interest at each location. The signal stream can thus be thought of as changing with a maximum spatial frequency defined by the speed within which successive pixels are sampled. "Bandwidth" is a general term used to describe the range from the lowest to the maximum spatial frequency that can be measured and transmitted through the amplification system. To achieve sufficiently fast scanned imaging to create the illusion of a continuous image ("flicker free") to a human observer, the imaging system must be capable of producing approximately 30 distinct image frames per second.

Ideally, the measurements of successive pixel locations are independent, with the detector returning to its quiescent state before measuring the next pixel. In reality, detectors typically require a finite decay time to dissipate the electron charge accumulated before measuring the next pixel. Thus, as the scanning speed increases so that the time separation of the pixel samples decreases, a limit will eventually be reached where the detector retains a sufficiently high fraction of the signal from the previous pixel so as to interfere with the useful measurement of the signal at the next pixel, producing a visible degradation of the image. When this situation occurs, the detector acts as a bandwidth-limiting device. For the discussion of detector performance characteristics below, detector bandwidth will be broadly classified as "high" (e.g., capable of achieving flicker-free imaging) or "low" (slow scan speeds required).

5.4.3 Common Types of Electron Detectors

Backscattered Electrons

Passive Detectors

Because a large fraction of the BSE emitted from the specimen under conventional operating conditions ($E_0 > 5$ keV) retain 50 % or more of the incident energy, they can be detected with a passive detector that does not apply any post-specimen acceleration to the BSE. Passive detectors include scintillation-based detectors and semiconductor charge-deposition based detectors.

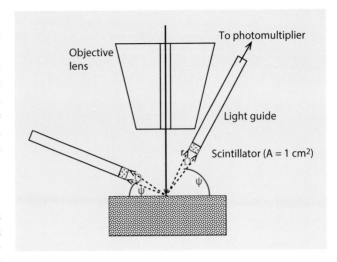

☐ **Fig. 5.21** Passive scintillator detectors for BSE. High take-off angle configuration and low take-off angle configuration

Scintillation Detectors

Energetic electrons that strike certain optically active materials cause the emission of light. Optical materials are selected that produce a high signal response that decays very rapidly, thus enabling high bandwidth operation. The emitted light is collected and passed by total internal reflection through a light guide to a photomultiplier, where the light is converted into an electrical signal with very high gain and very rapid time decay, thus preserving the high bandwidth of the original detector signal. Depending on the design, scintillator detectors can vary widely in solid angle. ☐ Figure 5.21 shows a small solid-angle design consisting of a small area scintillator (e.g., $A = 1$ cm^2) on the tip of a light guide placed at a distance of 4 cm from the beam impact, giving a solid angle of $\Omega = 0.0625$ sr and a geometric efficiency of $\varepsilon = 0.01$ or 1 %. Both a high take-off angle and a low take-off angle arrangement are illustrated.

▪▪ Adjustable Controls

Passive BSE detectors on rigid light guides have no user-adjustable operating parameters. (In operation, the "brightness" and "contrast" parameters match the amplified signal from the detector photomultiplier to the acceptable input range of the digitizer.) A passive BSE detector that employs a flexible light guide enables the microscopist to change the take-off angle, azimuthal angle, and the solid angle.

Very large solid angle scintillator-BSE detectors are possible. An example of a large solid angle design is shown in ☐ Fig. 5.22 that almost entirely surrounds the specimen with an aperture to permit the access of the beam. For a planar sample set normal to the beam, this detector spans a large range of take-off angles. The scintillator also serves as the light guide, so that a BSE that strikes anywhere on the detector surface can be detected. Due to its large area and close proximity to the specimen, the solid angle approaches 2π sr in size with a geometric efficiency greater than 90 % (Wells 1957; Robinson 1975).

5

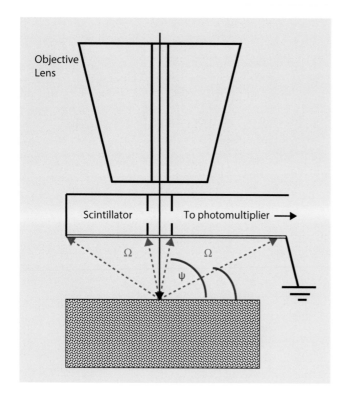

Fig. 5.22 Large solid-angle passive BSE detector

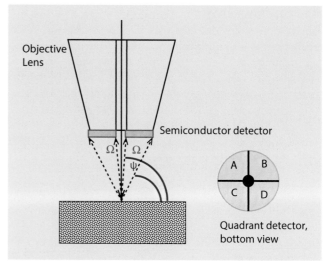

Fig. 5.23 Semiconductor annular detector, quadrant design with four separately selectable sections

■■ **Adjustable Controls**

The Wells–Robinson scintillation BSE detector is often mounted on an externally controlled, motorized retractable arm. In typical use the detector would be fully inserted to maximize the solid angle. A partial insertion that does not interrupt the beam access to the specimen can be used to intentionally provide an asymmetric detector placement to give an apparent illumination from one side.

Semiconductor BSE Detectors

Certain semiconductor devices can detect energetic electrons that penetrate into the active region of the device where they undergo inelastic scattering. One product of this energy deposition in the semiconductor is the promotion of loosely bound valence shell electrons (each leaving behind a vacancy or positively-charged "hole") into the empty conduction band where they can freely move through the semiconductor in response to an applied potential bias. By applying a suitable electrical field, these free electrons can be collected at a surface electrode and measured. For silicon, this process requires 3.6 eV of energy loss per free electron generated, so that a 15-keV BSE will generate about 4000 free electrons. Thus a BSE current of 1 nA entering the detector will create a collected current of about 4 μA as input for the next amplification stage. The collection electrodes are located on the entrance and back surfaces of the planar wafer detector, which is shown in a typical mounting as an annular detector in ◘ Fig. 5.23. The semiconductor BSE detector has the advantage of being thin, so that it can be readily mounted

under the objective lens where it will not interfere with other detectors. The size and proximity to the specimen provide a large solid angle and a high take-off angle. As shown in ◘ Fig. 5.23, the semiconductor detector can also be assembled from segments, each of which can be used as a separate detector that provides a selectable apparent illumination of the SEM image, or the signals from any combination of the detectors can be added. Semiconductor detectors can also be placed at various locations around the specimen, similar to the arrangement shown for scintillator detectors in ◘ Fig. 5.21.

The semiconductor BSE detector has an energy threshold typically in the range 1 keV to 3 keV because of energy loss suffered by the BSE during penetration through the entrance surface electrode. Above this threshold, the response of the detector increases linearly with increasing electron energy, thus providing a greater gain from the high energy fraction of BSE.

■■ **Adjustable Controls**

The semiconductor BSE detector does not have any user-adjustable parameters, with the exception of the choice of the individual components of a composite multi-detector. In some systems, the individual quadrants or halves can be selected in various combinations, or the sum of all detectors can be used. Some SEMs add an additional semiconductor detector that is placed asymmetrically away from the electron beam to enhance the effect of apparent oblique illumination.

5.4.4 Secondary Electron Detectors

Everhart–Thornley Detector

The most commonly used SEM detector is the Everhart–Thornley (E–T) detector, almost universally referred to as the "secondary electron detector." Everhart and Thornley

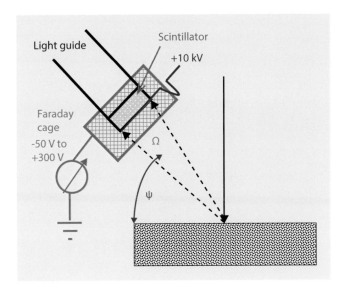

Fig. 5.24 Schematic of Everhart–Thornley detector showing the scintillator with a thin metallic surface electrode (*blue*) with an applied bias of positive 10 kV surrounded by an electrically isolated Faraday cage (*red*) which has a separate bias supply variable from negative 50 V to positive 300 V

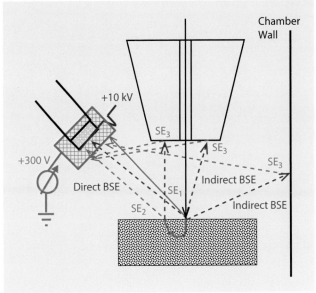

Fig. 5.25 Schematic of electron collection with a +300 V Faraday cage potential. Signals collected: direct BSE that enter solid angle of the scintillator; SE_1 produced within beam entrance footprint; SE_2 produced where BSE emerge from specimen; SE_3 produced where BSE strike the pole piece and chamber walls. SE_2 and SE_3 collection actually represents the remote BSE that could otherwise be lost

(1957) solved the problem of detecting very low energy secondary electrons by using a scintillator with a thin metal coating to which a large positive potential, 10 kV or higher, is applied. This post-specimen acceleration of the secondary electrons raises their kinetic energy to a sufficient level to cause scintillation in an appropriate material (typically plastic or glass doped with an optically active compound) after penetrating through the thin metallization layer that is applied to discharge the insulating scintillator. To protect the primary electron beam from any degradation due to encountering this large positive potential asymmetrically placed in the specimen chamber, the scintillator is surrounded by an electrically isolated "Faraday cage" to which is applied a modest positive potential of a few hundred volts (in some SEMs, the option exists to select the bias over a range typically from −50 to +300 V), as shown in **Fig. 5.24**. The primary beam is negligibly affected by exposure to this much lower potential, but the secondary electrons can still be collected with great efficiency to the vicinity of the Faraday cage, where they are then accelerated by the much higher positive potential applied to the scintillator.

While the E–T detector does indeed detect the secondary electrons emitted by the sample, the nature of the total collected signal is actually quite complicated because of the different sources of secondary electrons, as illustrated in **Fig. 5.25**. The SE_1 component generated within the landing footprint of the primary beam on the specimen cannot be distinguished from the SE_2 component produced by the exiting BSE since they are produced spatially within nanometers to micrometers and they have the same energy and angular distributions. Since the SE_2 production

depends on the BSE, rising and falling with the local effects on backscattering, the SE_2 signal actually carries BSE information. Moreover, the BSE are sufficiently energetic that while they are not significantly deflected and collected by the low Faraday cage potential, the BSE continue along their emission trajectory until they encounter the objective lens pole piece, stage components, or sample chamber walls, where they generate still more secondary electrons, designated SE_3. Although SE_3 are generated centimeters away from the beam impact, they are collected with high efficiency by the Faraday cage potential, again constituting a signal carrying BSE information since their number depends on the number of BSE ("indirect BSE"). Finally, those BSE emitted by the specimen into the solid angle defined by the E-T scintillator disk are detected ("direct BSE"). This complex mixture of signals plays an important role in creating the apparent illumination of the "secondary electron image."

■■ Adjustable Controls

On some SEMs the Faraday cage bias of the Everhart–Thornley detector can be adjusted, typically over a range from a negative potential of −100 V or less to a positive potential of a few hundred volts. When the Faraday cage potential is set to zero or a few volts negative, secondary electron collection is almost entirely suppressed, so that only the direct BSE are collected, giving a scintillator BSE detector that is of relatively small solid angle and asymmetrically placed on one side of the specimen. When the Faraday cage potential is set to the maximum positive value available, the

5

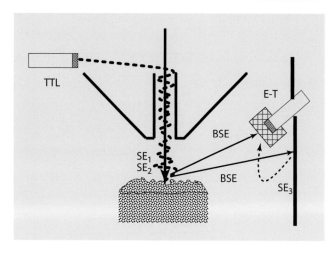

● **Fig. 5.26** "Through-the-lens" (TTL) secondary electron detector

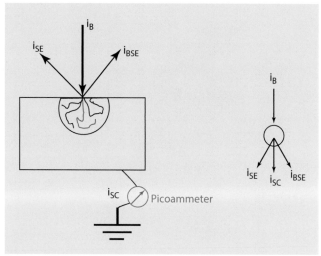

● **Fig. 5.27** Currents flowing in and out of the specimen and the electrical junction equivalent

complete suite of SE_1, SE_2, SE_3, and the direct BSE is collected, creating a complex mix of BSE and true SE image contrast effects.

Through-the-Lens (TTL) Electron Detectors

TTL SE Detector

In SEMs where the magnetic field of the objective lens projects into the specimen chamber, a "through-the-lens" (TTL) secondary electron detector can be implemented, as illustrated schematically in ● Fig. 5.26. SE_1 from the incident beam footprint and SE_2 emitted within the BSE surface distribution are captured by the magnetic field and spiral up through the lens. After emerging from the top of the lens, the secondary electrons are then attracted to an Everhart–Thornley type biased scintillator detector. The advantage of the TTL SE detector is the near complete exclusion of direct BSE and the abundant SE_3 class generated by BSE striking the chamber walls and pole piece. Since these remote SE_3 are generated on surfaces far from the optic axis of the SEM, they are not efficiently captured by the lens field. Because the SE_3 class actually represents low resolution BSE information, removing SE_3 from the overall SE signal actually improves the sensitivity of the image to the true SE_1 component, which is still diluted by the BSE-related SE_2 component. A further refinement of the through-the-lens detector is the introduction of energy filtering which allows the microscopist to select a band of SE kinetic energy.

TTL BSE Detector

For a flat specimen oriented normal to the beam, the cosine distribution of BSE creates a significant flux of BSE that pass up through the bore of the objective lens. A TTL BSE detector is created by providing either a direct scintillation BSE detector or a separate surface above the lens for BSE-to-SE conversion and subsequent detection with another E-T type detector.

5.4.5 Specimen Current: The Specimen as Its Own Detector

■ **The Specimen Can Serve as a Perfect Detector for the Total Number of BSE and SE Emitted**

Consider the interaction of the beam electrons to produce BSE and SE. For a 20-keV beam incident on copper, about 30 out of 100 beam electrons are backscattered ($\eta = 0.3$). The remaining 70 beam electrons lose all their energy in the solid, are reduced to thermal energies, and are captured. Additionally, about 10 units of charge are ejected from copper as secondary electrons ($\delta = 0.1$). This leaves a total of 60 excess electrons in the target. What is the fate of these electrons? To understand this, an alternative view is to consider the electron currents, defined as charge per unit time, which flow in and out of the specimen. Viewed in this fashion, the specimen can be treated as an electrical junction, as illustrated schematically in ● Fig. 5.27, and is subject to the fundamental rules which govern junctions in circuits. By Thevinin's junction theorem, the currents flowing in and out of the junction must exactly balance, or else there will be net accumulation or loss of electrical charge, and the specimen will charge on a macroscopic scale. If the specimen is a conductor or semiconductor and if there is a path to ground from the specimen, then electrical neutrality will be maintained by the flow of a current, designated the "specimen current" (also referred to as the "target current" or "absorbed current"), either to or from ground, depending on the exact conditions of beam energy and specimen composition. What is the magnitude of the specimen current?

Considering the specimen as a junction, the current flowing into the junction is the beam current, i_B, and the currents flowing out of the junction are the backscattered electron current, i_{BS}, and the secondary electron current, i_{SE}. For charge balance to occur, the specimen current, i_{SC}, is given by

$$i_{SC} = i_B - i_{BS} - i_{SC} \qquad (5.15)$$

For the copper target, the BSE current will be $i_{BSE} = \eta \times i_B = 0.3 \, i_B$ and the SE current will be $i_{SE} = \delta \times i_B = 0.1 \, i_B$. Substituting these values in Eq. (5.15) gives the result that the specimen current will be $i_{SC} = 0.6 \, i_B$, double the largest of the conventional emitted imaging currents, the BSE signal. If a path to ground is not provided so that the specimen current can flow, the specimen will rapidly charge.

Note that in formulating Eq. (5.15) no consideration is given to the large difference in energy carried by the BSE and SE. Since current is the passage of charge per unit time, the ejection of a 1 eV SE from the specimen carries the same weight as a 10 keV BSE in affecting the specimen current signal. Moreover, the specimen current is not sensitive to the direction of emission of BSE and SE, or to their subsequent fate in the SEM specimen chamber, as long as they do not return to the specimen as a result of re-scattering. Thus, specimen current constitutes a signal that is sensitive only to number effects, that is, the total numbers of BSE and SE leaving the specimen.

The specimen serves as its own collector for the specimen current. As such, the specimen current signal is readily available just by insulating the specimen from electrical ground and then measuring the specimen current flowing to ground through a wire to ground. Knowledge of the actual specimen current is extremely useful for establishing consistent operating conditions, and is critical for dose-based X-ray microanalysis. The original beam current itself be measured by creating a "Faraday cup" in the specimen or specimen stage by drilling a blind hole and directing the incident beam into the hole: since no BSE or SE can escape the Faraday cup, the measured specimen current then must equal the beam current. But by measuring the specimen current as a function of the beam position during the scan, an image can be formed that is sensitive to the total emission of BSE and SE regardless of the direction of emission and their subsequent fate interacting with external detectors, the final lens pole piece, and the walls of the specimen chamber. Does the specimen current signal actually convey useful information? As described below under contrast formation, the specimen current signal contains exactly the same information as that carried by the BSE and SE currents. Since external electron detectors measure a convolution of backscattered and/or secondary current with other characteristics such as energy and/or directionality, the specimen current signal can give a unique view of the specimen (Newbury 1976).

To make use of the specimen current signal, the current must be routed through an amplifier on its way to ground. The difficulty is that we must be able to work with a current similar in magnitude to the beam current, without any high gain physical amplification process such as electron-hole pair production in a solid state detector or the electron cascade in an electron multiplier. To achieve acceptable bandwidth at the high gains necessary, most current amplifiers take the form of a low input impedance operational amplifier (Fiori et al. 1974). Such amplifiers can operate with currents as low as 10 pA and still provide adequate bandwidth to view acceptable images at slow visual scan rates (one 500-line frame/s).

5.4.6 A Useful, Practical Measure of a Detector: Detective Quantum Efficiency

The geometric efficiency is just one factor in the overall performance of a detector, and while this quantity is relatively straightforward to define in the case of a passive BSE detector, as shown in ▫ Fig. 5.20, it is much more difficult to describe for an E–T detector because of the mix of BSE and direct SE_1 and SE_2 signal components and the complex conversion and collection of the remote SE_3 component produced where BSE strike the objective lens, BSE detector, and chamber walls. A second important factor in detector performance is the efficiency with which each collected electron is converted into useful signal. Thirdly, noise may be introduced at various stages in the amplification process to the digitization which creates the final intensity recorded in the computer memory for the pixel at which the beam dwells.

All of these factors are taken into account by the detective quantum efficiency (DQE). The DQE is a robust measure of detector performance that can be used in the calculation of limitations imposed on imaging through the threshold current/contrast equation (Joy et al. 1996).

The DQE is defined as (Jones 1959)

$$DQE = (S/N)^2_{experimental} \, / \, (S/N)^2_{theoretical} \qquad (5.16)$$

where S is the signal and N is the noise. Determining the DQE for a detector requires measurement of the experimental S/N ratio as produced under defined conditions of specimen composition, beam current and pixel dwell time that enable an estimate of the corresponding theoretical S/N ratio. This measurement can be performed by imaging a featureless specimen that ideally produces a fixed signal response which translates into a single gray level in the digitally recorded image, giving a direct measure of the signal, S. The corresponding noise, N, is determined from the measured width of the distribution of gray levels around the average value.

Measuring the DQE: BSE Semiconductor Detector

Joy et al. (1996) describe a procedure by which the experimental S/N ratio can be estimated from a digital image of a specimen that produces a unique gray level, so that the broadening observed in the image histogram of the ideal gray level is a quantitative measure of the various noise sources that are inevitable in the total measurement process that produces the image. Thus, the first requirement is a specimen with a highly polished featureless surface that will produce unique values of η and δ and which does not contribute any other sources of contrast (e.g., topography, compositional differences, electron channeling, or most problematically, changing δ and η values due to the accumulation of contamination). A polished silicon wafer provides a suitable sample, and with careful pre-cleaning, including plasma cleaning in the SEM airlock if available, the contamination problem can be minimized satisfactorily during the sequence of measurements required. As an alternative to silicon, a metallographically polished (but not etched) pure metallic element (metallic) surface, such as nickel, molybdenum, gold, etc., will be suitable. Because calculation of the theoretical S/N ratio is required for the DQE calculation with Eq. 5.16, the beam current must be accurately measured. The SEM must thus be equipped with a picoammeter to measure the beam current, and if an in-column Faraday cup is not available, then a specimen stage Faraday cup (e.g., a blind hole covered with a small [e.g., <100-μm-diameter] aperture) is required to completely capture the beam without loss of BSE or SE so that a measurement of the specimen current equals the beam current.

Because the detector will have a "dark current," i.e., a response with no beam current, it is necessary to make a series of measurements with changing beam current. It is also important to defeat any automatic image gain scaling that some SEMs provide as a "convenience" feature for the user that acts to automatically compensate for changes in the incident beam current by adjusting the imaging amplifier gain to maintain a steady mid-range gray level.

Measurement sequence

1. Choose a beam current which will serve as the high end of the beam current range, and using the image histogram function, adjust the imaging amplifier controls (often designated "contrast" and "brightness") to place the average gray level of the specimen near the top of the range, being careful that the upper tail of the gray level distribution of the image of the specimen does not saturate ("clip") at the maximum gray level (255 for an 8-bit image, 65,535 for a 16-bit image).

2. Keeping the same values for the image amplifier parameters (autoscaling of the imaging amplifier must be defeated before beginning the measurement process), choose a beam current that places the average gray level of the specimen near low end of the gray level range, checking to see that the gray level distribution of the image is completely within the histogram range—that is, there is no clipping of the distribution at the bottom (black) of the range.

3. With the minimum and maximum of the current range established, record a sequence of images with different beam currents between the low and high values and use the image histogram tool to determine the average gray level for each beam current.

4. A graphical plot of data measured with a semiconductor BSE detector for a polished Mo target produces the result illustrated in ◘ Fig. 5.28, where the y-axis intercept value is a measure of the dark current gray level intensity, GL_{DC} (corresponding to zero beam current) of this particular BSE detector.

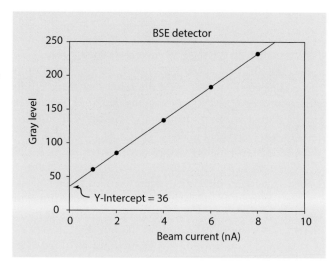

◘ **Fig. 5.28** Plot of measured gray level versus incident beam current for a BSE detector. $E_0 = 20$ keV; Mo target

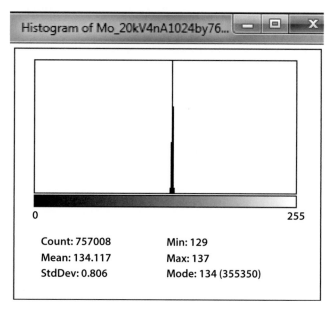

◘ **Fig. 5.29** Output of image histogram from IMAGE-J for the 4 nA image from Fig. 5.28

5. Choose an image recorded within this range and determine the mean gray level, G_{mean}, and the variance, S_{var} (the square of the standard deviation) using the image histogram function, as shown in ◘ Fig. 5.29.

Calculation sequence:

$$\left(S/N\right)_{experimental} = \left(GL_{mean} - GL_{DC}\right)/S_{var} \qquad (5.17)$$

where is S_{var} the variance (the square of the standard deviation) of the gray level distribution. For the values in ◘ Fig. 5.29 for the 4 nA data point obtained with ImageJ-Fiji,

$$\left(S/N\right)_{experimental} = \left(134.3 - 36\right)/0.575^2 = 297.3 \qquad (5.18)$$

The corresponding theoretical S/N ratio is estimated from the number n of BSE produced, which depends on the incident beam current I_B, the BSE coefficient η, and the dwell time per pixel τ:

$$n = 6.24\, I_B\, \eta\, \tau \tag{5.19}$$

where the coefficient 6.24 is appropriate for beam current expressed in pA and the dwell time expressed in μs.

Because the image pixels are independent and uncorrelated, if a mean number n of BSE is produced at each pixel the expected variance is $n^{1/2}$:

$$\left(S/N\right)_{\text{theory}} = n/n^{1/2} = n^{1/2} = \left(6.24\, I_B\, \eta\, \tau\right)^{1/2} \tag{5.20}$$

For $I_B = 4000$ pA, $\eta_{\text{Mo}} = 0.38$, and $\tau = 64$ μs

$$\left(S/N\right)_{\text{theory}} = \left(6.24\, I_B\, \eta\, \tau\right)^{1/2} = 779.1 \tag{5.21}$$

The DQE for this particular detector is thus

$$\begin{aligned} \text{DQE} &= \left(S/N\right)^2_{\text{experimental}} / \left(S/N\right)^2_{\text{theoretictal}} \\ &= 297.3^2 / 779.1^2 = 0.146 \end{aligned} \tag{5.22}$$

A similar study for an Everhart–Thornley SE-BSE detector on an electron probe X-ray microanalyzer is shown in ▣ Fig. 5.30, for which the DQE is calculated as 0.0016. ▣ Table 5.1 lists values of the DQE for various detectors, demonstrating that a large range in values is encountered, even among detectors of a specific class, for example, the E–T detector.

▣ **Table 5.1** DQE of electron detectors from different manufacturers (Joy et al. 1996)

SE detector	DQE
Everhart–Thornley	0.56
Everhart–Thornley	0.17
Everhart–Thornley	0.12
Everhart–Thornley	0.017
Everhart–Thornley	0.0008
High performance SEM:	
Everhart–Thornley (lower)	0.18
Everhart–Thornley (TTL)	0.76
Microchannel plate	0.029
BSE detector	
Scintillator BSE	0.043
Scintillator BSE	0.005
E–T BSE mode (negative bias)	0.001
E–T BSE mode (negative bias)	0.004
Microchannel plate BSE	0.058
Microchannel plate BSE	0.026

References

Everhart T, Thornley R (1960) Wide-band detector for micro-microampere low-energy electron currents. J Sci Instrum 37:246

Fiori C, Yakowitz H, Newbury D (1974) Some techniques of signal processing in scanning electron microscopy. In: Johari O (ed) SEM/1974. IIT Research Institute, Chicago, p 167

Jones R (1959) Phenomenological description of the response and detecting ability of radiation detectors. Adv Electr Electron Opt 11:88

Joy DC, Joy CS, Bunn RD (1996) Measuring the performance of scanning electron microscope detectors. Scanning 18:533

Newbury DE (1976) "The utility of specimen current imaging in the scanning electron microscope" SEM/1976/I. IIT Research Inst, Chicago, p 111

Robinson V (1975) "Backscattered electron imaging" SEM/1975, I. IIT Research Inst, Chicago, p 51

Wells OC (1957) The construction of a scanning electron microscope and its application to the study of fibres. Ph. D. Diss., Cambridge University, Cambridge

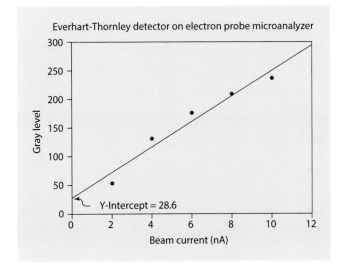

▣ **Fig. 5.30** Average gray level versus beam current for an Everhart–Thornley detector on an electron probe microanalyzer. Specimen: Si; $E_0 = 10$ keV

Image Formation

© Springer Science+Business Media LLC 2018
J. Goldstein et al., *Scanning Electron Microscopy and X-Ray Microanalysis*,
https://doi.org/10.1007/978-1-4939-6676-9_6

6.1 Image Construction by Scanning Action

After leaving the electron source, the beam follows the central (optic) axis of the lens system and is sequentially defined by apertures and focused by the magnetic and/or electrostatic fields of the lens system. Within the final (objective) lens, a system of scan coils acts to displace the beam off the optic axis so that it can be addressed to a location on the specimen, as illustrated schematically for single deflection scanning in ◘ Fig. 6.1.

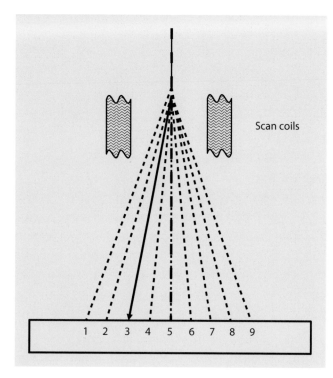

◘ **Fig. 6.1**   Scanning action to produce a sequence of discrete beam locations on the specimen

At any particular time, there is only one ray path (solid line) through the scanning system and the beam reaches only one location on the specimen, for example, position 3 in ◘ Fig. 6.1. The SEM image is a geometric construction created under computer control by addressing the focused beam to a sequence of discrete x-y locations on the specimen and measuring the effect of the interaction of the beam with the specimen at each location. For a single gray-scale SEM image, this interaction could be the output from a single electron detector, such as the Everhart–Thornley detector. It is also possible to measure the output from more than one detector simultaneously while the beam is addressed to a single x-y location. When this is done, multiple gray-scale SEM images are built up at the same time during the scan. It is essential to realize that even when these multiple signals are being collected simultaneously and multiple images are produced, only a single scan is needed; the parallel nature of the acquisition arises from parallel detection, not parallel scanning. Note that no "true image" actually exists within the SEM in the same sense as the image created in a light optical microscope, where actual ray paths extend from each point on the specimen through the lens system to a corresponding point on the image recording medium, whether that is the eye of a human viewer or the positionally sensitive detector of a digital camera. In the SEM, at each location sampled by the incident electron beam, each signal is measured with an appropriate detector and the analog measurement is converted to an equivalent digital value (using an analog-to-digital converter, ADC). The beam x-y location and the intensity(ies) I_j of the signal(s) of interest generate a digital stream of data packets (x, y, I_j), where the index j represents the various signals available: backscattered electrons (BSE), secondary electrons (SE), absorbed current, X-rays, cathodoluminescence, etc.

A simple description of this "scanning action" to create an image is shown schematically in ◘ Fig. 6.2, where an area with equal edge dimensions l being scanned on the specimen is effectively divided into an x-y grid of square picture ele-

◘ **Fig. 6.2**   Scanning action in two dimensions to produce an x-y raster, and the corresponding storage and display of image information by scan location

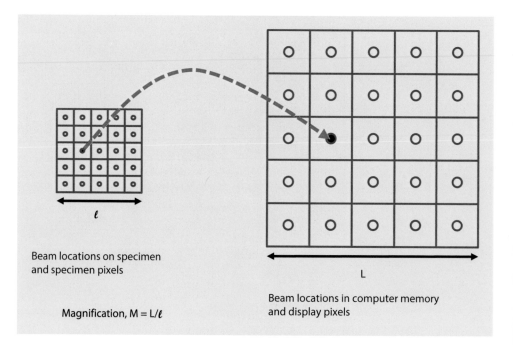

Beam locations on specimen and specimen pixels

Magnification, M = L/ℓ

Beam locations in computer memory and display pixels

6.3 · Making Dimensional Measurements With the SEM: How Big Is That Feature?

95

6

ments or "pixels" of number **n** along an edge. The specimen pixel edge dimension is given by

$$\text{Specimen pixel dimension} = l \, / \, n \qquad (6.1)$$

With equal values of the scan l along the x- and y-dimensions, the pixels will be square. Strictly, the pixel is the geometric center of the area defined by the edges given by Eq. (6.1), and the center-to-center spacing or pitch is given by Eq. (6.1). In creating an SEM image, the center of the beam is placed in the center of a specific pixel, dwells for a specific time, t_p, and the signal information I_{sig} from various sources "j"—e.g., SE, BSE, X-ray, etc.—collected during that time at that (x, y) location is stored at a corresponding location in a data matrix with a minimum of three dimensions (x, y, I_j). The final image viewed by the microscopist is created by reading the stored data matrix into a corresponding pattern of (x, y) display pixels with a total edge dimension L, and adjusting the display brightness ("gray level", varying from black to white) according to the relative strength of the measured signal(s).

6.2 Magnification

"Magnification" in such a scanning system is given by the ratio of the edge dimensions of the specimen area and the display area:

$$M = L \, / \, l \qquad (6.2)$$

Since the final display size is typically fixed, increasing the magnification in this scanning system means that the edge dimension of the area scanned on the specimen is reduced.

6.2.1 Magnification, Image Dimensions, and Scale Bars

One of the most important pieces of information that the microscopist seeks is the size of objects of interest. The first step in determining the size of an object is knowledge of the parameters in Eq. (6.2): the linear edge length of the area scanned on the specimen and on the display. The nominal SEM magnification appropriate to the display as viewed by the microscopist is typically embedded in the alphanumeric record that appears with the image as presented on most SEMs, as in the example of ◘ Fig. 6.3, and as recorded with the metadata associated with the digital record of the image. "Magnification" only has a useful meaning for the display on which the original image was viewed, since this is the display for which L in Eq. (6.2) is strictly valid. If the image is transferred to another display with a different value of L, for example, projected on a large screen, then the specific magnification value embedded in the metadata bar becomes meaningless. Much more meaningful are the x- and y-image dimensions, which are the lengths of the orthogonal boundaries of the scanned square area on the specimen,

◘ **Fig. 6.3** SEM-SE image of silver crystals showing a typical information bar specifying the electron detector, the nominal magnification, the accelerating voltage, and a scale bar

l, in ◘ Fig. 6.2, or for rectangular images, the dimensions in orthogonal directions, l by k (dimensions: millimeters, micrometers, or nanometers, as appropriate). While the image dimension(s) is a much more robust term that automatically scales with the presentation of the image, this term is also vulnerable to inadvertent mistakes, such as might happen if the image is "cropped," either digitally or manually in hard copy and the appropriate reduction in size is not recorded by modifying l (and k, if rectangular) appropriately. The most robust measure in terms of image integrity is the dimensional scale bar, which shows the length that corresponds to a specific millimeter, micrometer, or nanometer measure. Because this feature is embedded directly in the image (as well as in the metadata associated with the image), it cannot be lost unless the image is severely (and obviously) cropped. Such a scale bar automatically enlarges or contracts as the image size is modified for subsequent publication or projection.

6.3 Making Dimensional Measurements With the SEM: How Big Is That Feature?

6.3.1 Calibrating the Image

The validity of the dimensional marker displayed on the SEM image should not be automatically assumed (Postek et al. 2014). As part of a laboratory quality-assurance program, the dimensional marker and/or the x- and y-dimensions of the scanned field should be calibrated and the calibration periodically confirmed. This can be accomplished with a "scale calibration artifact," a specimen that contains features with various defined spacings whose dimensions are traceable to the fundamental primary length standard through a national measurement institution. An example of such a scale calibration artifact suitable for SEM is Reference Material RM 8820 (Postek et al. 2014; National Institute of Standards and

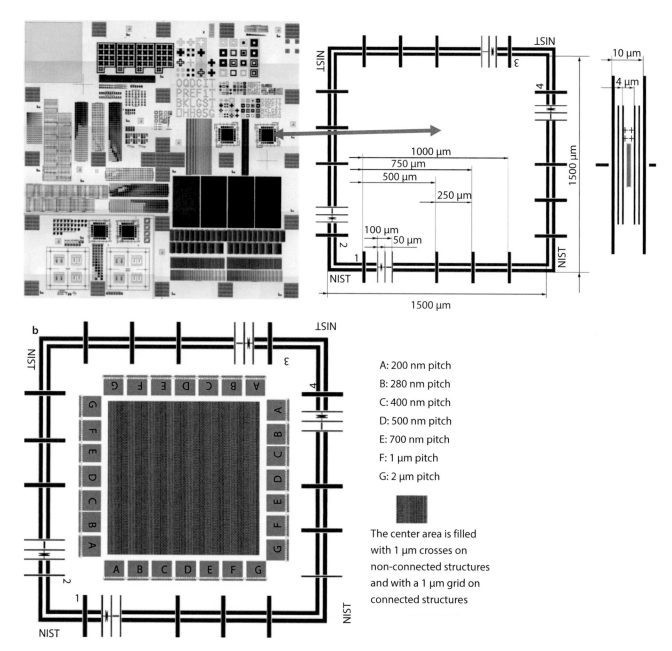

NIST RM 8820 SEM scale calibration artifact:
(lithographically patterned silicon chip, 20 mm by 20 mm)

A: 200 nm pitch

B: 280 nm pitch

C: 400 nm pitch

D: 500 nm pitch

E: 700 nm pitch

F: 1 μm pitch

G: 2 μm pitch

The center area is filled with 1 μm crosses on non-connected structures and with a 1 μm grid on connected structures

Fig. 6.4 **a** Scale calibration artifact Reference Material 8820 (National Institute of Standards and Technology, U.S.) (From Postek et al. 2014). **b** Detail within the feature noted in Fig. 6.4a (From Postek et al. 2014)

Technology [USA]), shown in Fig. 6.4. This scale calibration artifact consists of an elaborate collection of linear features produced by lithography on a silicon substrate. It is important to calibrate the SEM over the full range of magnifications to be used for subsequent work. RM 8820 contains large-scale structures suitable for low and intermediate magnifications, for example, a span of 1500 μm (1.5 mm) as indicated by the red arrows in Fig. 6.4a, that permit calibration of scan fields ranging up to 1 × 1 cm (e.g., a nominal

magnification of 10× on a 10 x 10-cm display). Scanned fields as small as 1 × 1 μm (e.g., a nominal magnification of 100,000×) can be calibrated with the series of structures with pitches of various repeat distances shown in Fig. 6.4b. The structures present in RM 8820 enable simultaneous calibration along the x- and y-axes of the image so that image distortion can be minimized. Accurate calibration in orthogonal directions is critical for establishing "square pixels" in order to avoid introducing serious distortions into the scanned image.

97   6

6.3 · Making Dimensional Measurements With the SEM: How Big Is That Feature?

Fig. 6.5 **a** Careful calibration of the x- and y-scans produces square pixels, and a faithful reproduction of shapes lying in the scan plane perpendicular to the optic axis. **b** Distortion in the display of an object caused by non-square pixels in the image scan

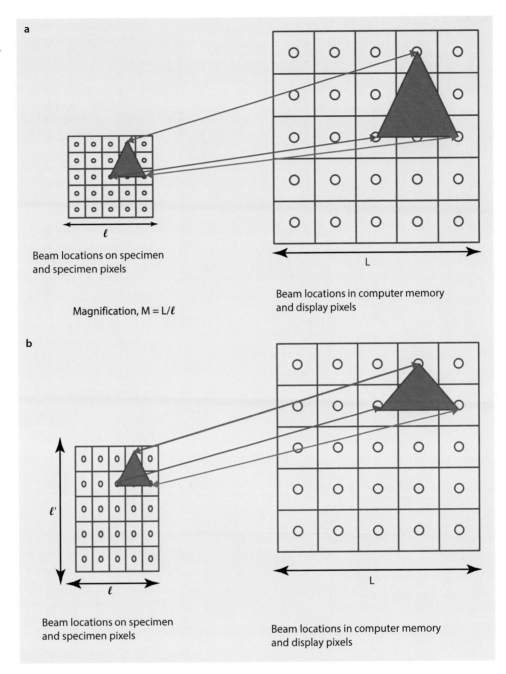

Beam locations on specimen and specimen pixels

Magnification, M = L/ℓ

Beam locations in computer memory and display pixels

Beam locations on specimen and specimen pixels

Beam locations in computer memory and display pixels

With square pixels, the shape of an object is faithfully transferred, as shown in **Fig. 6.5a**, while non-square pixels in the specimen scan result in distortion in the displayed image, **Fig. 6.5b**.

Note that for all measurements the calibration artifact must be placed normal to the optic axis of the SEM to eliminate image foreshortening effects (see further discussion below).

Using a Calibrated Structure in ImageJ-Fiji

The image-processing software engine ImageJ-Fiji includes a "Set Scale" function that enables a user to transfer the image calibration to subsequent measurements made with various functions. As shown in **Fig. 6.6a**, starting with an image of a primary or secondary calibration artifact (i.e., where "secondary" refers to a commercial vendor artifact that is traceable to a primary national measurement calibration artifact) that contains a set of defined distances, the user can specify a vector that spans a particular pitch to establish the calibration at that magnification setting. This calibration procedure should then be repeated to cover the range of magnification settings to be used for subsequent measurements of unknowns. Note that the calibration that has been performed is only strictly valid for the SEM working distance at which the calibration artifact has been imaged. When a

6

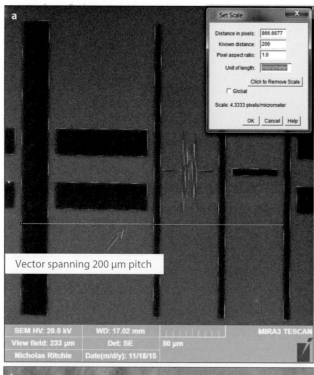

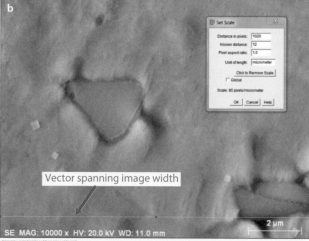

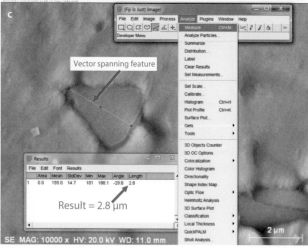

◻ Fig. 6.6 **a** ImageJ (Fiji) "Set Scale" calibration function applied to an image of NIST RM 8820. **b** ImageJ (Fiji) "Set Scale" calibration function applied to an image of an unknown (alloy IN100). **c** After "Set Scale" image calibration, subsequent use of ImageJ (Fiji) "Measure" function to determine the size of a feature of interest

different working distance (i.e., objective lens strength) is used subsequently to image the unknown specimen, the SEM software is likely to make automatic adjustments for different lens strength and scan dimensions that alter the effective magnification. For the most robust measurement environment, the user should use the calibration artifact to determine the validity of the SEM software specified scale at other working distances to develop a comprehensive calibration.

Alternatively, if the SEM magnification calibration has already been performed with an appropriate calibration artifact, then subsequent images of unknowns will be recorded with accurate dimensional information in the form of a scale bar and/or specified scan field dimensions. This information can be used with the "Set Scale" function in ImageJ-Fiji as shown for a specified field width in ◻ Fig. 6.6b where a vector (yellow) has been chosen that spans the full image width. The "Set Scale" tool will record this length and the user then specifies the "Known Distance" and the "Unit of Length." To minimize the effect of the uncertainty associated with selecting the endpoints when defining the scale for this image, the larger of the two dimensions reported in the vendor software was chosen, in this case the full horizontal field width of 12 μm rather than the much shorter embedded length scale of 2 μm.

Making Routine Linear Measurements With ImageJ-Fiji (Flat Sample Placed Normal to Optic Axis)

For the case of a flat sample placed normal to the optic axis of the SEM, linear measurements of structures can be made following a simple, straightforward procedure after the image calibration procedure has been performed. Typical image-processing software tools directly available in the SEM operational software or in external software packages such as ImageJ-Fiji enable the microscopist to make simple linear measurements of objects. With the calibration established, the "Line" tool is used to define the particular linear measurement to be made, as shown in ◻ Fig. 6.6c, and then the "Measure" tool is selected, producing the "Results" table that is shown. Repeated measurements will be accumulated in this table.

6.4   Image Defects

6.4.1   Projection Distortion (Foreshortening)

The calibration of the SEM image must be performed with the planar surface of the calibration artifact placed perpendicular to the optic axis (i.e., x- and y-axes at right angles relative to the z-axis), and only measurements that are made on planar objects that are similarly oriented will be valid. When the specimen is tilted around an axis, for example, the x-axis, the resulting SEM image is subject to projection distortion causing foreshortening along the y-axis. Foreshortening occurs because the effective magnification is reduced along the y-axis relative to the x-axis (tilt axis), as illustrated in ◻ Fig. 6.7. For nominal magnifications exceeding 100×, the

Fig. 6.7 **a** Schematic illustration of projection distortion of tilted surfaces. **b** Illustration of foreshortening of familiar objects, paper clips (*upper*) Large area image at 0 tilt; (*lower*) large area image at 70° tilt around a horizontal tilt axis. Note that parallel to the tilt axis, the paper clips have the same size, but perpendicular to the tilt axis severe foreshortening has occurred. The magnification also decreases significantly down the tilted surface, so the third paper clip appears smaller than the first (Images courtesy J. Mershon, TESCAN)

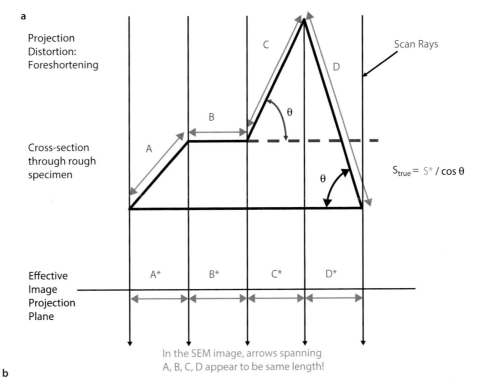

a

Projection
Distortion:
Foreshortening

Scan Rays

Cross-section
through rough
specimen

$$S_{true} = S^* / \cos \theta$$

Effective
Image
Projection
Plane

A*   B*   C*   D*

In the SEM image, arrows spanning
A, B, C, D appear to be same length!

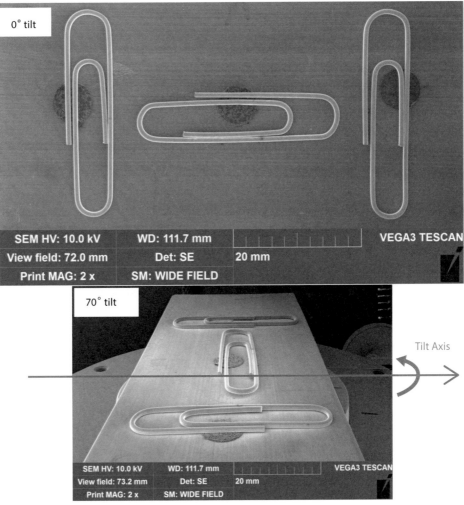

b

0° tilt

SEM HV: 10.0 kV	WD: 111.7 mm		VEGA3 TESCAN
View field: 72.0 mm	Det: SE	20 mm	
Print MAG: 2 x	SM: WIDE FIELD		

70° tilt

Tilt Axis

SEM HV: 10.0 kV	WD: 111.7 mm		VEGA3 TESCAN
View field: 73.2 mm	Det: SE	20 mm	
Print MAG: 2 x	SM: WIDE FIELD		

6

successive scan rays of the SEM image have such a small angular spread relative to the optic axis that they create a nearly parallel projection to create the geometric mapping of the specimen three-dimensional space to the two-dimensional image space. As shown in ◘ Fig. 6.7, a linear feature of length L_{true} lying in a plane tilted at an angle, θ, (where θ is defined relative to a plane perpendicular to the optic axis) is foreshortened in the SEM image according to the relation

$$L_{image} = L_{true} * \cos\theta \qquad (6.3)$$

For the situation shown in ◘ Fig. 6.7a, all four linear objects would have the same apparent size in the SEM image, but only one, object B, would be shown with the correct length since it lies in a plane perpendicular to the optic axis, while the true lengths of the other linear objects would be significantly underestimated. For the most severe case, object D, which lies on the most highly tilted surface with $\theta = 75°$, the object is a factor of 3.9 longer than it appears in the image. The effect of foreshortening is dramatically illustrated in ◘ Fig. 6.7b, where familiar objects, paper clips, are seen in a wide area SEM image at 0° tilt and 70° tilt. At high tilt, the length of the first paper clip parallel to the tilt axis remains the same, while the second paper clip that is perpendicular to the tilt axis is highly foreshortened (Note that the third paper clip, which also lies parallel to the tilt axis, appears shorter than the first paper clip because the effective magnification decreases down the tilted surface). As shown schematically in ◘ Fig. 6.8, foreshortening causes a square to appear as a rectangle. The effect of foreshortening is shown for an SEM image of a planar copper grid in ◘ Fig. 6.9, where the square openings of the grid are correctly imaged at $\theta = 0°$ in ◘ Fig. 6.9a. When the specimen

plane is tilted to $\theta = 45°$, the grid appears to have rectangular openings, as shown in ◘ Fig. 6.9b, with the shortened side of the true squares running parallel to the y-axis, while the correctly sized side runs parallel to the x-axis, which is the axis of tilt. Some SEMs are equipped with a "tilt correction" feature in which the y-scan perpendicular to the tilt axis is decreased to compensate for the extended length (relative to the x-scan along the tilt axis) of the scan excursion on the tilted specimen, as shown schematically in ◘ Fig. 6.9c. Tilt correction creates the same magnification (i.e., the same pixel dimension) along orthogonal x- and y-axes, which restores the proper shape of the squares, as seen in ◘ Fig. 6.9c. However, this scan transformation is only correct for objects that lie in the plane of the specimen. ◘ Figure 6.9c also contains a spherical particle, which appears to be circular at $\theta = 0°$ and at $\theta = 45°$ without tilt correction, since the normal scan projects the intersection of the plane of the scan sphere as a circle. However, when tilt correction is applied at $\theta = 45°$, the sphere now appears to be a distorted ovoid. Thus, applying tilt correction to the image of an object with three-dimensional features of arbitrary orientation will result in image distortions that will increase in severity with the degree of local tilt.

6.4.2 Image Defocusing (Blurring)

The act of focusing an SEM image involves adjusting the strength of the objective lens to bring the narrowest part of the focused beam cross section to be coincident with the surface. If the specimen has a flat, planar surface placed normal to the beam, then the situation illustrated in ◘ Fig. 6.10a will exist at sufficiently low magnification. ◘ Figure 6.10a

◘ **Fig. 6.8**   Effect of foreshortening of objects in a titled plane to distort square grid openings into *rectangles*

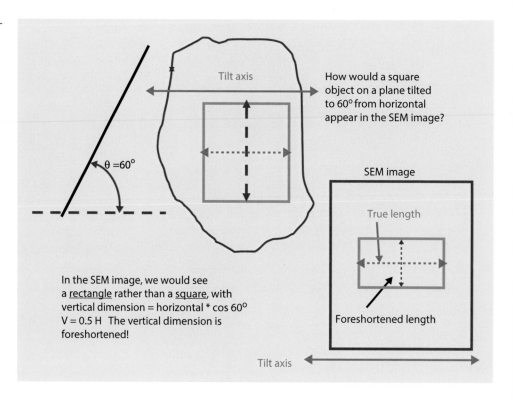

Tilt axis

How would a square object on a plane tilted to 60° from horizontal appear in the SEM image?

$\theta = 60°$

SEM image

True length

In the SEM image, we would see a rectangle rather than a square, with vertical dimension = horizontal * cos 60° V = 0.5 H   The vertical dimension is foreshortened!

Foreshortened length

Tilt axis

□ Fig. 6.9 **a** SEM/E–T (positive) image of a copper grid with a poly-styrene latex sphere; tilt = 0° (grid normal to electron beam). **b** Grid tilted to 45°; note the effect of foreshortening distorts the *square grid*

openings into *rectangles*. **c** Grid tilted to 45°; "tilt correction" applied, but note that while the square grid openings are restored to the proper shape, the *sphere* is highly distorted

shows the locations of the beam at the pixel centers in the middle of the squares and the effective sampling footprint. The sampling footprint consists of the contribution of the incident beam diameter (in this case finely focused to a diameter <10 nm) and the surface emergence area of the BSE and SE, which is controlled by the interaction volume. □ Figure 6.10a considers a situation of a low beam energy (e.g., 5 keV) and a high atomic number (e.g., Au). For these conditions, the beam sampling footprint only occupies a small fraction of each pixel area so that there is no possibility of overlap, i.e., sampling of adjacent pixels. Now consider what happens as the magnification is increased, i.e., the length l in Eq. (6.1) decreases while the pixel number, n, remains the same: the distance between pixel centers decreases, but the beam sampling footprint remains the same size for this particular material and beam landing energy. The situation shown in □ Fig. 6.10b for the original beam sampling footprint relative to the pixel spacing and

6

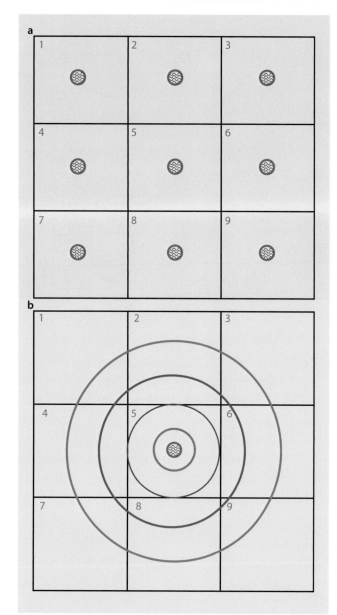

○ **Fig. 6.10** **a** The beam sampling footprint relative to the pixel spacing for a low magnification image with a low energy finely focused beam and a high atomic number target. **b** As the magnification is increased with fixed beam energy and target material, the beam sampling footprint (diameter and BSE-SE convolved) eventually fills the pixel and progressively leaks into adjacent pixels

four successive increases in magnification reveals that as the pixel spacing becomes smaller, the beam sampling footprint eventually leaks into the surrounding pixels, so that the beam no longer samples exclusively the region of a single pixel. Eventually, when enough pixels overlap, the observer will perceive this leakage as image defocusing or blurring. The reality and limitations of this situation become obvious when the microscopist seeks to perform high spatial resolution microscopy, a topic which will be covered in more depth in module 10 on high resolution SEM.

The effects of blurring are also encountered in the trivial case when the objective lens is strengthened or weakened, which moves the minimum beam convergence along the vertical axis (either up or down), as shown schematically in ○ Fig. 6.11a, increasing the size of the beam that encounters the specimen surface. The beam diameter that encounters the specimen surface will be larger in either case because of the finite convergence angle, α. As the beam samples progressively more adjacent pixels just due to the increase in beam size, and not dependent on the BSE-SE sampling footprint, the observer will eventually perceive the defocusing, and hopefully correct the situation!

Defocusing is also encountered when the specimen has features that extend along the optic axis. For example, defocusing may be encountered when planar specimens are tilted or rough topographic specimens are examined, even at low magnifications, i.e., large scanned areas, as illustrated schematically in ○ Fig. 6.11b, c. In these situations, the diameter of the converged beam that encounters the specimen depends on the distance of the feature from the bottom of the objective lens and the convergence angle of the beam, α. Because the beam is focused to a minimum diameter at a specific distance from the objective lens, the working distance W, any feature of the specimen that the scanned beam encounters at any other distance along the optic axis will inevitably involve a larger beam diameter, which can easily exceed the sampling footprint of the BSE and SE. ○ Figure 6.11d shows an image of Mt. St. Helens volcanic ash particles where the top of the large particle is in good focus, but the focus along the sides of the particles deteriorates into obvious blurring, as also occurs for the small particles dispersed around the large particle on the conductive tape support. This defocus situation can only be improved by reducing the convergence angle, α, as described in Depth-of-Field Mode operation.

6.5 Making Measurements on Surfaces With Arbitrary Topography: Stereomicroscopy

By operating in Depth-of-Field Mode, which optimizes the choice of the beam convergence angle, α, a useful range of focus along the optic axis can be established that is sufficient to render effective images of complex three-dimensional objects. ○ Figure 6.12 shows an example of a specimen (metal fracture surface) with complex surface topography. The red arrows mark members of a class of flat objects. If the microscopist's task is to measure the size of these objects, the simple linear measurement that is possible in a single SEM image is subject to large errors because the local tilt of each feature is different and unknown, which corresponds to the situation illustrated in ○ Fig. 6.7. Although lost in a single two-dimensional image, the third dimension of an irregular surface can be recovered by the technique of stereomicroscopy.

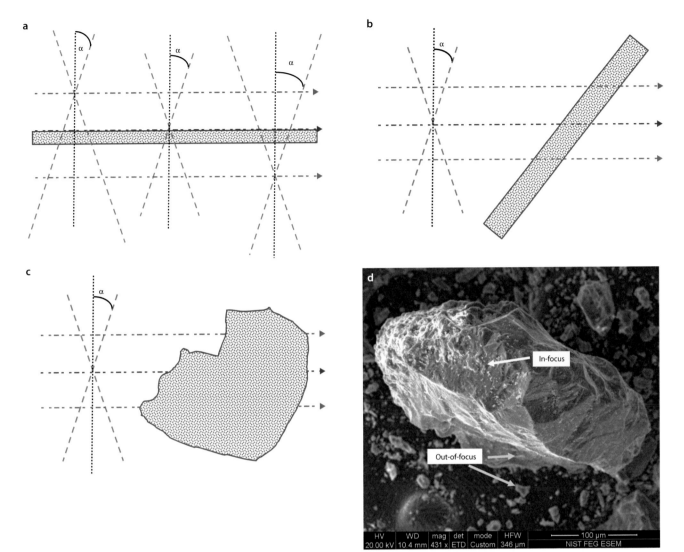

Fig. 6.11 **a** Trivial example of optimal lens strength (focused at *blue plane*) and defocusing caused by selecting the objective lens strength too high (focused at *green plane*) and too low (focused at *magenta plane*) relative to the specimen surface. **b** Effect of a tilted planar surface. The beam is scanned with fixed objective lens strength, so that different beam diameters encounter the specimen at different distances along the optic axis. **c** Effects similar to **Fig. 6.11b** but for a three-dimensional specimen of arbitrary shape. **d** An imaging situation corresponding to **Fig. 6.11c**: coated fragments of Mt. St. Helens ash mounted on conducting tape and imaged under high vacuum at $E_0 = 20$ keV with an E–T (positive) detector

6.5.1 Qualitative Stereomicroscopy

The human visual process creates the perception of depth and the three-dimensional character of objects by combining the separate two-dimensional views provided by the left eye and the right eye to create a fused image, as shown in **Fig. 6.13**. The angular difference between the eyes creates two distinct views containing parallax information, which is the horizontal shift (relative to a vertical axis) of a feature common to the two separate views. The parallax is the critical information that the brain uses to create the sensation of depth in the fused image: the larger the parallax, the closer the object is to the viewer. To create a similar sensation of depth, SEM stereomicroscopy operates by mimicking the human visual process and creating two angularly separated views of the specimen with parallax information. In SEM imaging, the electron beam is the "eye of the observer" (see the "Image Interpretation" module), so the two required images for stereo imaging must be obtained by either changing the orientation of the beam relative to the specimen (beam rocking method), or by changing the orientation of the specimen relative to the fixed electron beam (specimen tilting method). An appropriate image separation method such as the anaglyph technique (e.g., using red and cyan filters to view color-coded images) then presents the each member of the image pair to the left or right eye so that the viewer's natural imaging process will create a fused image that reveals the third dimension. (Note that there is a

6

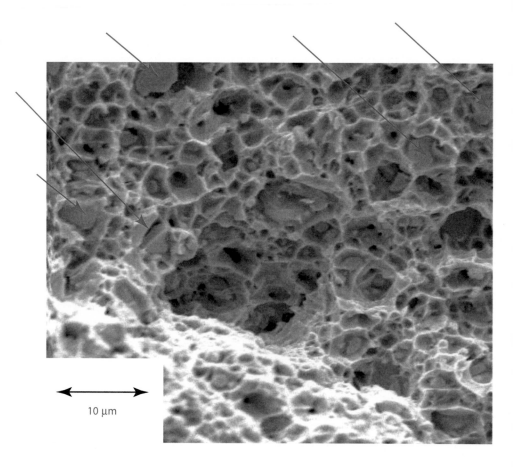

Fig. 6.12 SEM/E–T (positive) image of a metal fracture surface. The red arrows designate members of a class of flat objects embedded in this surface

10 μm

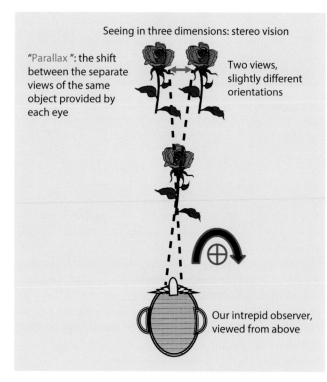

Seeing in three dimensions: stereo vision

"Parallax": the shift between the separate views of the same object provided by each eye

Two views, slightly different orientations

Our intrepid observer, viewed from above

Fig. 6.13 Schematic illustration of an observer's creation of a stereo view of an object. Note that the parallax (shift between the two views) is across a vertical axis

fraction of the population for whom this process is not effective at creating the sense of viewing a three-dimensional object.)

Fixed beam, Specimen Position Altered

Parallax can be created by changing the specimen tilt relative to the optic axis (beam) by recording two images with a difference in tilt angle ranging from 2° to 10°. The specific value depends on the degree of topography of the specimen, and the optimum choice may require a trial-and-error approach. Weak topography will generally require a larger tilt difference to create a suitable three-dimensional effect. However, if the tilt angle difference between the images is made too large, it may not be possible for a viewer to successfully fuse the images and visualize the topography, especially for large-scale topography.

A suitable procedure to achieve SEM stereomicroscopy with a fixed beam by tilting the specimen has the following steps:

1. Determine where the tilt axis lies in the SEM image. The eventual images must be presented to the viewer with horizontal parallax (i.e., all the shift between the two images must be across a vertical axis), so the tilt axis must be oriented vertically. The images can be recorded and rotated appropriately within image processing software such as ImageJ-Fiji, or the scan rotation function of the SEM can be used to orient the tilt axis

☐ Fig. 6.14 Illustration of the main page of the "Anaglyph Maker STEREOEYE" software (► http://www.stereoeye.jp/index_e.html) showing the windows where the left (low tilt) and right (high tilt) SEM/E–T (positive) images are selected and the resulting anaglyph (convention: red filter for the left eye). Note that brightness and contrast and fine position adjustments are available to the user. Specimen: ceramic fibers, coated with Au-Pd; $E_0 = 5$ keV

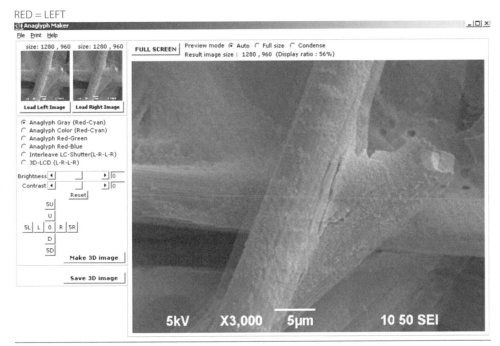

along the vertical. In either case, note the location of the Everhart–Thornley detector in the image, which will provide the general sense of illumination. Ideally, the position of the E–T detector should be at the top of the image. However, after image rotation to orient the tilt axis vertically, the effective position of the Everhart-Thornley detector is likely to be different from this ideal 12-o'clock position (top center of image).

2. Record an image of the area of interest at the low tilt angle, for example, stage tilt = 0°.

3. Using this image as a reference, increase the tilt angle to the desired value, e.g., stage tilt = 5°, while maintaining the location of the field of view. Depending on the mechanical sophistication of the specimen stage, changing the tilt may cause the field of view to shift laterally, requiring continual relocation of the desired field of view during the tilting process to avoid losing the area of interest, especially at high magnification on specimens with complex topography.

4. The vertical position of the specimen may also shift during tilting. To avoid introducing rotation in the second (high tilt image) by changing the objective lens strength to re-focus, the vertical stage motion (z-axis) should be used to refocus the image. After careful adjustment of the x-y-z position of the stage using the low tilt image to locate the area of interest, record this high tilt image.

5. Within the image-processing software, assign the low tilt image to the RED image channel and the high tilt image to the CYAN (GREEN-BLUE channels combined, or the individual GREEN or BLUE image channels, depending on the type of anaglyph viewing filters available). Apply the image fusion function to create the stereo image, and view this image display with appropriate red (left eye)

and blue (right eye) glasses. Note: The image-processing software may allow fine scale adjustments (shifts and/or rotations) to improve the registration of the images. This procedure is illustrated in ☐ Fig. 6.14 for the "Anaglyph Maker STEREOEYE" software (► http://www.stereoeye.jp/index_e.html). Examples of "stereo pairs" created in this manner are shown in ☐ Fig. 6.15 (a particle of ash from the Mt. St. Helens eruption) and ☐ Fig. 6.16 (gypsum crystals from cement).

▪▪ Note

While usually successful, this SEM stereomicroscopy "recipe" may not produce the desired stereo effect on your particular instrument. Because of differing conventions for labeling tilt motions or due to unexpected image rotation applied in the software, the sense of the topography may be reversed (e.g., a topographic feature that is an "inner" falsely becomes an "outer" and vice versa). It is good practice when first implementing stereomicroscopy with an SEM to start with a simple specimen with known topography such as a coin with raised lettering or a scratch on a flat surface. Apply the procedure above and inspect the results to determine if the proper sense of the topography has been achieved in the resulting stereo pair. If not, be sure the parallax shift is horizontal, that is, across a vertical axis (if necessary, use software functions to rotate the images to vertically orient the tilt axis). If the tilt axis is vertical but the stereo pair still shows the wrong sense of the topography, try reversing the images so the "high tilt" image is now viewed by the left eye and the "low tilt" image viewed by the right eye. Once the proper procedure has been discovered to give the correct sense of the topography on a known test structure, follow this convention for future stereomicroscopy work. (Note: A small but significant fraction of observers find it difficult to fuse the images to form a stereo image.)

6

□ **Fig. 6.15** Anaglyph stereo presentation of SEM/E–T(positive) images (E_0 = 20 keV) of a grain of Mt. St. Helens volcanic ash prepared by the stage tilting stereo method

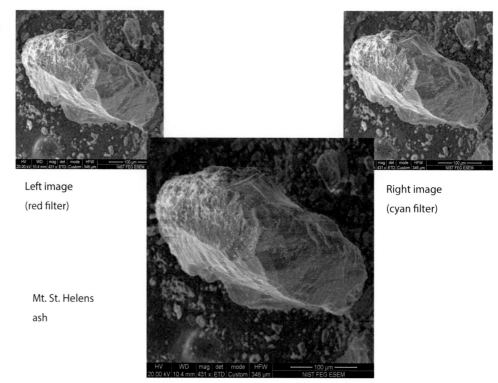

Left image
(red filter)

Right image
(cyan filter)

Mt. St. Helens

ash

□ **Fig. 6.16** Anaglyph stereo presentation of SEM/E–T(positive) images (E_0 = 20 keV) of a grain of gypsum crystals prepared by the stage tilting stereo method

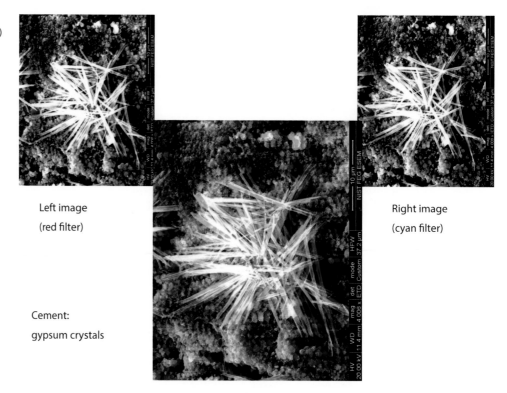

Left image
(red filter)

Right image
(cyan filter)

Cement:

gypsum crystals

Fixed Specimen, Beam Incidence Angle Changed

The beam incidence angle relative to the specimen can be changed by a small value by means of a deflection in the final stage of the scan to create the two distinct views needed to achieve the stereo effect. An example of a stereo pair created in this manner is shown in □ Fig. 6.17 for a

fractured fragment of galena. By applying the two beam tilts to alternate image scans at high rate, "live" 3D SEM imaging can be achieved that is nearly "flicker free." By eliminating the need for mechanical stage motion as well as avoiding problems which frequently occur due to shifting of the area of interest during mechanical tilting, the speed of the beam tilting method makes it very powerful for studying complex

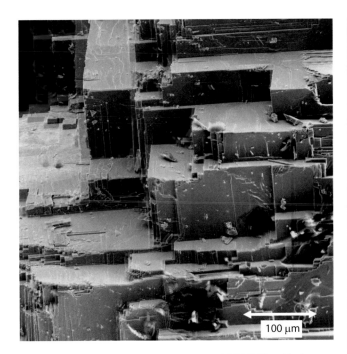

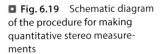

◘ Fig. 6.17 Anaglyph stereo presentation of SEM/E–T(positive) images ($E_0 = 15$ keV) of a fractured galena crystal prepared by the beam tilting stereo method

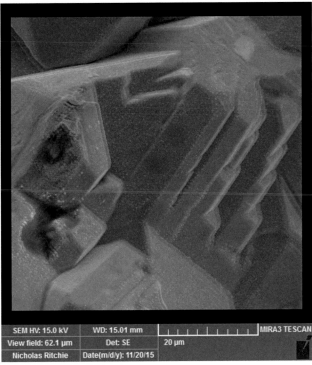

SEM HV: 15.0 kV	WD: 15.01 mm		MIRA3 TESCAN
View field: 62.1 µm	Det: SE	20 µm	
Nicholas Ritchie	Date(m/d/y): 11/20/15		

◘ Fig. 6.18 Anaglyph stereo presentation of SEM/E–T(positive) images ($E_0 = 15$ keV) of a silver crystal prepared by the beam tilting stereo method

◘ Fig. 6.19 Schematic diagram of the procedure for making quantitative stereo measurements

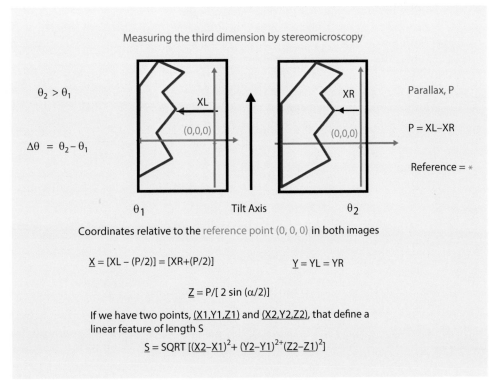

Measuring the third dimension by stereomicroscopy

$\theta_2 > \theta_1$

$\Delta\theta = \theta_2 - \theta_1$

XL

(0,0,0)

θ_1

Tilt Axis

XR

(0,0,0)

θ_2

Parallax, P

P = XL–XR

Reference = *

Coordinates relative to the reference point (0, 0, 0) in both images

$\underline{X} = [XL - (P/2)] = [XR+(P/2)]$ $\underline{Y} = YL = YR$

$\underline{Z} = P/[\,2 \sin(\alpha/2)]$

If we have two points, (X1,Y1,Z1) and (X2,Y2,Z2), that define a linear feature of length S

$\underline{S} = SQRT\,[(\underline{X2}-\underline{X1})^2 + (\underline{Y2}-\underline{Y1})^{2+}(\underline{Z2}-\underline{Z1})^2]$

topography. This is especially true at high magnification when the act of mechanical stage tilting is more likely to cause significant lateral shifting of the specimen, rendering the mechanical stage tilt stereo method tedious. An example of a stereo pair for a silver crystal produced with the beam tilt method at higher magnification is shown in ◘ Fig. 6.18.

6.5.2 Quantitative Stereomicroscopy

Quantitation of the topography in SEM micrographs can be carried out by calculating the Z-coordinate of the feature from measurements of the x- and y-coordinates in the members of a stereo pair, as illustrated schematically in ◘ Fig. 6.19

6

(Boyde 1973, 1974a,b; Wells 1974). This procedure can be accomplished even if the operator is not personally able to perceive the qualitative stereo effect using the anaglyph or other methods to present the two images.

1. The first step is to record a stereo pair with tilt angles θ_1 and θ_2 and with the tilt axis placed in a vertical orientation in the images. The difference in tilt angle between the members of the stereo pair is a critical parameter:

$$\Delta\theta = \theta_2 - \theta_1 \tag{6.4}$$

2. A set of orthogonal axes is centered on a recognizable feature, as shown in the schematic example in ◻ Fig. 6.19. This point will then be arbitrarily assigned the X-, Y-, Z-coordinates $(0, 0, 0)$ and all subsequent height measurements will be with respect to this point. The axes are selected so that the y-axis is parallel to the tilt axis and the x-axis is perpendicular to the tilt axis.

3. For the feature of interest, the (X, Y)-coordinates are measured in the Left (X_L, Y_L) and Right (X_R, Y_R) members of the stereo pair using the calibrated distance marker. The parallax, P, of a feature is given by

$$P = (X_L - X_R) \tag{6.5}$$

With this convention, points lying above the tilt axis will have positive parallax values P. Note that as an internal consistency check, $Y_L = Y_R$ if the y-axis has been properly aligned with the tilt axis.

4. For SEM magnifications above a nominal value of 100×, the scan angle will be sufficiently small that it can be assumed that the scan is effectively moving parallel to the optic axis, which enables the use of simple formulas for quantification. With reference to the fixed point $(0, 0, 0)$, the three-dimensional coordinates X_3, Y_3, Z_3 of the chosen feature are given by

$$Z_3 = P / \left[2\sin(\Delta\theta / 2) \right] \tag{6.6}$$

$$X_3 = (P/2) + X_L = X_R - (P/2) \tag{6.7}$$

(Note that Eq. (6.7) provides a self-consistency check for the X_3 coordinate.)

$$Y_3 = Y_L = Y_R \tag{6.8}$$

Note that if the measured coordinates y_L and y_R are not the same then this implies that the tilt axis is not accurately parallel to Y and the axes must then be rotated to correct this error.

By measuring any two points with coordinates, (X_M, Y_M, Z_M) and (X_N, Y_M, Z_M), the length L of the straight line connecting the points is given by

$$L = \mathrm{SQRT}\left[(X_M - X_N)^2 + (Y_M - Y_N)^2 + (Z_M - Z_N)^2 \right] \tag{6.9}$$

Measuring a Simple Vertical Displacement

The stereo pair in ◻ Fig. 6.20a illustrates a typical three-dimensional measurement problem: for this screw thread, how far above or below is the feature circled in green relative to the feature circled in yellow? The left image (low tilt, $\theta = 0°$) and right image (high tilt, $\theta = 5°$) are prepared according to the convention described above and oriented so that the tilt axis is vertical. It is good practice to inspect the stereo pair with the anaglyph method shown in ◻ Fig. 6.14 to ensure that the stereo pair is properly arranged, and to qualitatively assess the nature of the topography, i.e., determine how features are arranged relative to each other, as shown for this image of the screw thread in ◻ Fig. 6.20a. In ◻ Fig. 6.20b, a set of x- (horizontal) and y- (vertical) axes are established in each image centered on the feature in the yellow circle, which is assigned the origin of coordinates $(0, 0, 0)$. Using this coordinate system, measurements are made of the feature of interest (within the green circle) in the left $(X_L = 144 \ \mu m, Y_L = -118 \ \mu m)$ and right $(X_R = 198 \ \mu m, Y_R = -118 \ \mu m)$ images. The parallax P is then

$$P = X_L - X_R = 144\,\mu m - 198\,\mu m = -54\,\mu m \tag{6.10}$$

Note that the sign of the parallax is negative, which means that the green circle feature is below the yellow circle feature, a result that is confirmed by the qualitative inspection of the stereo pair in ◻ Fig. 6.20a. Inserting these values into Eq. (6.6), the Z-coordinate of the end of the green circle feature relative to the yellow circle feature is calculated to be:

$$\begin{aligned} Z_3 &= P / \left[2\sin(\Delta\theta / 2) \right] \\ &= -54\,\mu m / \left[2\sin(5°/2) \right] \\ &= -619\,\mu m \end{aligned} \tag{6.11}$$

Thus, the feature in the green circle is 619 μm below the feature in the yellow circle at the origin of coordinates. The uncertainty budget for this measurement consists of the following components:

1. Scale calibration error: with the careful use of a primary or secondary dimensional artifact, this uncertainty contribution can be reduced to 1 % relative or less.

2. Measurement of the feature individual coordinates: The magnitude of this uncertainty contribution depends on how well the position of a feature can be recognized and on the separation of the features of interest. By selecting a magnification such that the features whose vertical separation is to be measured span at least half of the image field, the uncertainty in the individual coordinates should be approximately 1 % relative, and in the difference of X-coordinates $(X_L - X_R)$ about 2 % relative. For closely spaced features, the magnitude of this uncertainty contribution will increase.

3. Uncertainty in the individual tilt settings: The magnitude of this uncertainty is dependent on the degree of backlash in the mechanical stage motions. Backlash

Fig. 6.20 **a** Stereo pair of a machined screw thread—SEM/E–T(positive) images; $E_0 = 20$ keV. **b** Stereo pair with superimposed axes for measurement of coordinates needed for quantitative stereomicroscopy calculations

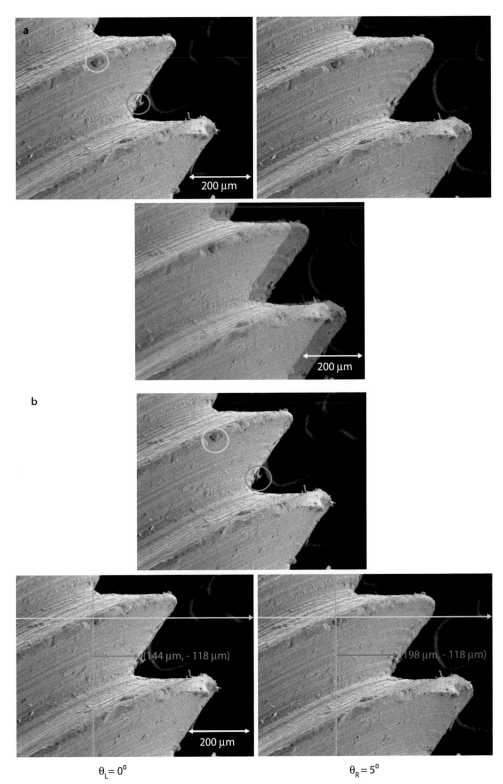

effects can be minimized by selecting the initial (low) tilt value to correspond to a well-defined detent position if the mechanical stage is so designed, such as a physical stop at 0° tilt. With a properly maintained mechanical stage, the uncertainty in the tilt angle difference $\Delta\theta$ is

estimated to be approximately 2 % for $\Delta\theta = 5^0$, with the relative uncertainty increasing for smaller values of $\Delta\theta$.

4. Considering all of these sources of uncertainty, the measurement should be assigned an overall uncertainty of ±5 % relative.

References

Boyde A (1973) Quantitative photogrammetric analysis and qualitative stereoscopic analysis of SEM images. J Microsc 98:452

Boyde A (1974a) A stereo-plotting device for SEM micrographs and a real time 3-D system for the SEM. In: Johari O (ed) SEM/1974. IIT Research Institute, Chicago, p 93

Boyde A (1974b) Photogrammetry of stereo pair SEM images using separate measurements from the two images. In: Johari O (ed) SEM/1974. IIT Research Institute, Chicago, p 101

Postek MT, Vladar AE, Ming B, Bunday B (2014) Documentation for Reference Material (RM) 8820: a versatile, multipurpose dimensional metrology calibration standard for scanned particle beam, scanned probe, and optical microscopy. NIST Special Publication 1170 ► https://www-s.nist.gov/srmors/view_detail.cfm?srm=8820

Wells O (1974) Scanning electron microscopy. McGraw-Hill, New York

SEM Image Interpretation

© Springer Science+Business Media LLC 2018
J. Goldstein et al., *Scanning Electron Microscopy and X-Ray Microanalysis*,
https://doi.org/10.1007/978-1-4939-6676-9_7

7.1   Information in SEM Images

Information in SEM images about specimen properties is conveyed when contrast in the backscattered and/or secondary electron signals is created by differences in the interaction of the beam electrons between a specimen feature and its surroundings. The resulting differences in the backscattered and secondary electron signals (S) convey information about specimen properties through a variety of contrast mechanisms. Contrast (C_{tr}) is defined as

$$C_{tr} = \left(S_{max} - S_{min} \right) / S_{max} \qquad (7.1)$$

where is S_{max} is the larger of the signals. By this definition, $0 \le C_{tr} \le 1$.

Contrast can be conveyed in the signal by one or more of three different mechanisms:

1. Number effects. Number effects refer to contrast which arises as a result of different numbers of electrons leaving the specimen at different beam locations in response to changes in the specimen characteristics at those locations.
2. Trajectory effects. Trajectory effects refer to contrast resulting from differences in the paths the electrons travel after leaving the specimen.
3. Energy effects. Energy effects occur when the contrast is carried by a certain portion of the backscattered electron or secondary electron energy distribution. For example, the high-energy backscattered electrons are generally the most useful for imaging the specimen using contrast mechanisms such as atomic number contrast or crystallographic contrast. Low-energy

secondary electrons are likely to escape from a shallow surface region of a specimen and convey surface information.

7.2   Interpretation of SEM Images of Compositional Microstructure

7.2.1   Atomic Number Contrast With Backscattered Electrons

The monotonic dependence of electron backscattering upon atomic number (η vs. Z, shown in ◘ Fig. 2.3) constitutes a number effect with predictable behavior that enables SEM imaging to reveal the compositional microstructure of a specimen through the contrast mechanism variously known as "atomic number contrast," "compositional contrast," "material contrast," or "Z-contrast." Ideally, to observe unobscured atomic number contrast, the specimen should be flat so that topography does not independently modify electron backscattering. An example of atomic number contrast observed in a polished cross section of Raney nickel alloy using signal collected with a semiconductor backscattered electron (BSE) detector is shown in ◘ Fig. 7.1, where four regions with progressively higher gray levels can be identified. The systematic behavior of η versus Z allows the observer to confidently conclude that the average atomic number of these four regions increases as the average gray level increases. SEM/EDS microanalysis of these regions presented in ◘ Table 7.1 gives the compositional results and calculated average atomic number, Z_{av}, of each phase. The Z_{av} values correspond to the trend of the gray levels of the phases observed in ◘ Fig. 7.1.

◘ **Fig. 7.1**   Raney nickel; $E_0 = 20$ keV; semiconductor BSE detector (SUM mode)

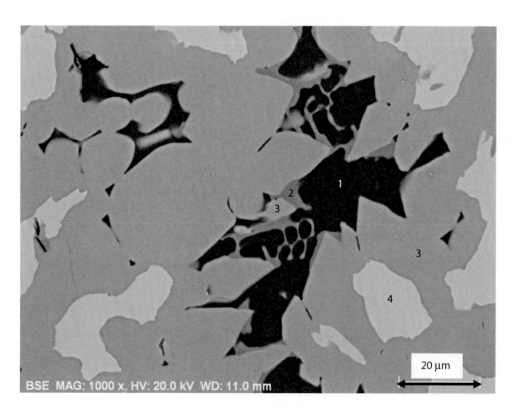

◘ Table 7.1 Raney nickel alloy (measured composition, calculated average atomic number, backscatter coefficient, and atomic number contrast across the boundary between adjacent phases)

Phase	Al (mass frac)	Fe (mass frac)	Ni (mass frac)	Z_{av}	Calculated, η	Contrast
1	0.9874	0.0003	0.0123	13.2	0.155	
2	0.6824	0.0409	0.2768	17.7	0.204	1–2 0.24
3	0.5817	0.0026	0.4155	19.3	0.22	2–3 0.073
4	0.4192	0.0007	0.5801	21.7	0.243	3–4 0.095

7.2.2 Calculating Atomic Number Contrast

An SEM is typically equipped with a "dedicated backscattered electron detector" (e.g., semiconductor or passive scintillator) that produces a signal, S, proportional to the number of BSEs that strike it and thus to the backscattered electron coefficient, η, of the specimen. Note that other factors, such as the energy distribution of the BSEs, can also influence the detector response.

If the detector responded only to the number of BSEs, the contrast C_{tr}, can be estimated as

$$C_{tr} = \left(S_{max} - S_{min}\right)/S_{max} = \left(\eta_{max} - \eta_{min}\right)/\eta_{max} \quad (7.2)$$

Values of the backscatter coefficient for $E_0 \geq 10$ keV can be conveniently estimated using the fit to η versus Z (Eq. 2.2). Note that for mixtures that are uniform at the atomic level (e.g., alloy solid solutions, compounds, glasses, etc.), the backscattered electron coefficient can be calculated from the mass fraction average of the atomic number inserted into Eq. 2.2 (as illustrated for the Al-Fe-Ni phases listed in ◘ Table 7.1), or alternatively, from the mass fraction average of the pure element backscatter coefficients.

The greater the difference in atomic number between two materials, the greater is the atomic number contrast. Consider two elements with a significant difference in atomic number, for example, Al ($Z = 13$, $\eta = 0.152$) and Cu ($Z = 29$, $\eta = 0.302$). From Eq. (7.1), the atomic number contrast between Al and Cu is estimated to be

$$C_{tr} = \left(\eta_{max} - \eta_{min}\right)/\eta_{max}$$
$$= \left(0.302 - 0.152\right)/0.302 = 0.497 \quad (7.3)$$

When the contrast is calculated between elements separated by one unit of atomic number, much lower values are found, which has an important consequence on establishing visibility, as discussed below. Note that the slope of η versus Z decreases as Z increases, so that the contrast (which is the slope of η vs. Z) between adjacent elements ($\Delta Z = 1$) also decreases. For example, the contrast between Al ($Z = 13$, $\eta = 0.152$) and Si ($Z = 14$, $\eta = 0.164$) where the slope of η versus Z is relatively high is

$$C_{tr} = \left(\eta_{max} - \eta_{min}\right)/\eta_{max}$$
$$= \left(0.164 - 0.152\right)/0.164 = 0.073 \quad (7.4)$$

A similar calculation for Cu ($Z = 29$, $\eta = 0.302$) and Zn ($Z = 30$, $\eta = 0.310$) where the slope of η versus Z is lower gives

$$C_{tr} = \left(0.310 - 0.302\right)/0.310 = 0.026 \quad (7.5)$$

For high atomic number elements, the slope of η versus Z approaches zero, so that a calculation for Pt ($Z = 78$, $\eta = 0.484$) and Au ($Z = 79$, $\eta = 0.487$) gives a very low contrast:

$$C_{tr} = \left(0.487 - 0.484\right)/0.487 = 0.0062 \quad (7.6)$$

◘ Figure 7.2 summarizes this behavior in a plot of the BSE atomic number contrast for a unit change in Z as a function of Z.

7.2.3 BSE Atomic Number Contrast With the Everhart–Thornley Detector

The appearance of atomic number contrast for a polished cross section of Al-Cu aligned eutectic, which consists of an Al-2% Cu solid solution and the intermetallic $CuAl_2$, is shown as viewed with a semiconductor BSE detector in ◘ Fig. 7.3a and an Everhart–Thornley detector (positively biased) in ◘ Fig. 7.3b. The E–T detector is usually thought of as a secondary electron detector, and while it captures the SE_1 signal, it also captures BSEs that are directly emitted into the solid angle defined by the scintillator. Additionally, BSEs are also represented in the E–T detector signal by the large contribution of SE_2 and SE_3, which are actually BSE-modulated signals. Thus, although the SE_1 signal of the E–T detector does not show predictable variation with composition, the BSE components of the E–T signal reveal the atomic number contrast seen in ◘ Fig. 7.3b. It must be noted, however, that because of the sensitivity of the E–T detector to edge effects and topography, these fine-scale features are much more visible in ◘ Fig. 7.3b than in ◘ Fig. 7.3a.

For both the dedicated semiconductor BSE detector and the E–T detector, the higher atomic number regions appear brighter than the lower atomic number regions, as independently confirmed by energy dispersive X-ray spectrometry of both materials. However, the semiconductor BSE detector

7

Fig. 7.2 Atomic number contrast for pure elements with $\Delta Z = 1$

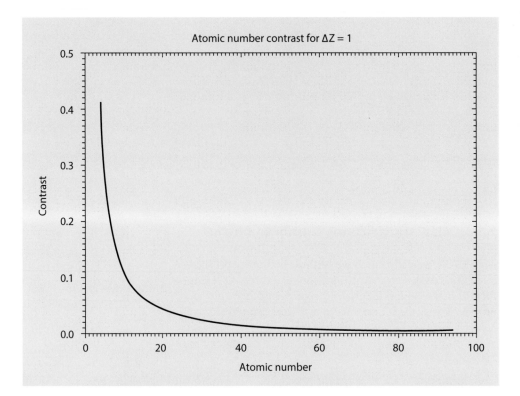

Fig. 7.3 Aligned Al-Cu eutectic; $E_0 = 20$ keV: **a** semiconductor BSE detector (SUM mode); **b** Everhart–Thornley detector (positive bias)

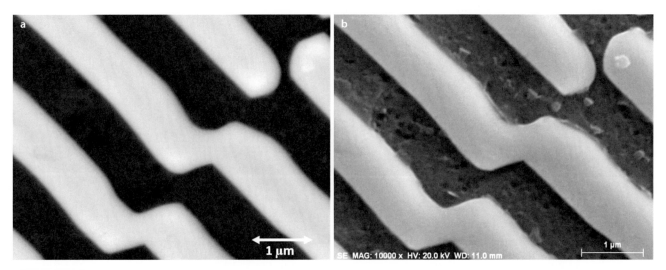

actually enhances the atomic number contrast over that estimated from the composition (Al-0.02Cu, $\eta = 0.155$; CuAl$_2$, $\eta = 0.232$, which gives $C_{tr} = 0.33$). The semiconductor detector shows increased response from higher energy backscattered electrons, which are produced in greater relative abundance from Cu compared to Al, thus enhancing the difference in the measured signals. The response of the Everhart–Thornley detector (positive bias) to BSEs is more complex. The BSEs that directly strike the scintillator produce a greater response with increasing energy. However, this component is small compared to the BSEs that strike the objective lens pole piece and chamber walls, where they are converted to SE$_3$ and subsequently collected. For these remote BSEs, the lower energy fraction actually create SEs more efficiently.

7.3 Interpretation of SEM Images of Specimen Topography

Imaging the topographic features of specimens is one of the most important applications of the SEM, enabling the microscopist to gain information on size and shape of features. Topographic contrast has several components arising from both backscattered electrons and secondary electrons:

1. The backscattered electron coefficient shows a strong dependence on the surface inclination, η versus θ. This effect contributes a number component to the observed contrast.
2. Backscattering from a surface perpendicular to the beam (i.e., 0° tilt) is directional and follows a cosine

distribution $\eta(\varphi) \approx \cos \varphi$ (where φ is an angle measured from the surface normal) that is rotationally symmetric around the beam. This effect contributes a trajectory component of contrast.

3. Backscattering from a surface tilted to an angle θ becomes more highly directional and asymmetrical as θ increases, tending to peak in the forward scattering direction. This effect contributes a trajectory component of contrast.

4. The secondary electron coefficient δ is strongly dependent on the surface inclination, $\delta(\theta) \approx \sec \theta$, increasing rapidly as the beam approaches grazing incidence. This effect contributes a number component of contrast.

Imaging of topography should be regarded as qualitative in nature because the details of the image such as shading depend not only on the specimen characteristics but also upon the response of the particular electron detector as well as its location and solid angle of acceptance. Nevertheless, the interpretation of all SEM images of topography is based on two principles regardless of the detector being used:

1. Observer's Point-of-View: The microscopist views the specimen features as if looking along the electron beam.
2. Apparent Illumination of the Scene:
   a. The apparent major source of lighting of the scene comes from the position of the electron detector.
   b. Depending on the detector used, there may appear to be minor illumination sources coming from other directions.

7.3.1 Imaging Specimen Topography With the Everhart–Thornley Detector

SEM images of specimen topography collected with the Everhart–Thornley (positive bias) detector (Everhart and Thornley 1960) are surprisingly easy to interpret, considering how drastically the imaging technique differs from ordinary human visual experience: A finely focused electron beam steps sequentially through a series of locations on the specimen and a mixture of the backscattered electron and secondary electron signals, subject to the four number and trajectory effects noted above that result from complex beam–specimen interactions, is used to create the gray-scale image on the display. Nevertheless, a completely untrained observer (even a young child) can be reasonably expected to intuitively understand the general shape of a three-dimensional object from the details of the pattern of highlights and shading in the SEM/E–T (positive bias) image. In fact, the appearance of a three-dimensional object viewed in an SEM/E–T (positive bias) image is strikingly similar to the view that would be obtained if that object were viewed with a conventional light source and the human eye, producing the so-called "light-optical analogy." This situation is quite remarkable, and the relative ease with which SEM/E–T (positive bias) images can be utilized is a major source of the utility and popularity of the SEM. It is important to understand the origin of this SEM/E–T (positive bias) light-optical analogy and what pathological effects can occur to diminish or destroy the effect, possibly leading to incorrect image interpretation of topography.

The E–T detector is mounted on the wall of the SEM specimen chamber asymmetrically off the beam axis, as illustrated schematically in ◻ Fig. 7.4. The interaction of the beam

◻ **Fig. 7.4** Schematic illustration of the various sources of signals generated from topography: BSEs, SE_1 and SE_2, and remote SE_3 and collection by the Everhart–Thornley detector

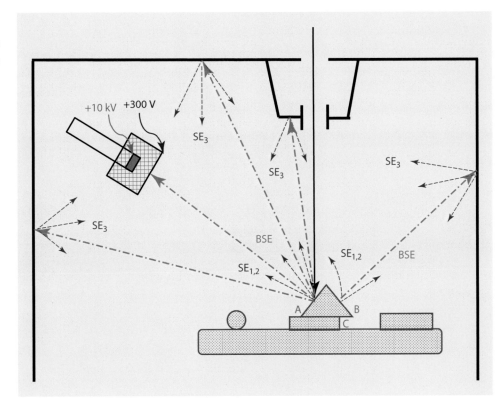

7

with the specimen results in backscattering of beam electrons and secondary electron emission (type SE_1 produced by the beam electrons entering the specimen and type SE_2 produced by the exiting BSEs). Energetic BSEs carrying at least a few kilo-electronvolts of kinetic energy that directly strike the E–T scintillator are always detected, even if the scintillator is passive with no positive accelerating potential applied. In typical operation the E–T detector is operated with a large positive accelerating potential (+10 kV or higher) on the scintillator and a small positive bias (e.g., +300 V) on the Faraday cage which surrounds the scintillator. The small positive bias on the cage attracts SEs with high efficiency to the detector. Once they pass inside the Faraday cage, the SEs are accelerated to detectable kinetic energy by the high positive potential applied to the face of the scintillator. In addition to the SE_1 and SE_2 signals produced at the specimen, the E–T (positive bias) detector also collects some of the remotely produced SE_3 which are generated where the BSEs strike the objective lens and the walls of the specimen chamber. Thus, in ◘ Fig. 7.4 a feature such as face "A," which is tilted toward the E–T detector, scatters some BSEs directly to the scintillator, which add to the SE_1, SE_2, and SE_3 signals that are also collected, making "A" appear especially bright compared to face "B." Because "B" is tilted away from the E–T (positive bias) detector, it does not make a direct BSE contribution, but some SE_1 and SE_2 signals will be collected from "B" by the Faraday cage potential, which causes SEs to follow curving trajectories, while remote SE_3 signals from face "B" will also be collected. Only features the electron beam does not directly strike, such as the re-entrant feature "C," will fail to generate any collectable signal and thus appear black.

7.3.2 The Light-Optical Analogy to the SEM/E–T (Positive Bias) Image

The complex mix of direct BSEs, SE_1 and SE_2, and remote SE_3 illustrated in ◘ Fig. 7.4 effectively illuminates the specimen in a way similar to the "real world" landscape scene illustrated schematically in ◘ Fig. 7.5 (Oatley 1972). A viewer in an airplane looks down on a hilly landscape that is directionally illuminated by the Sun at a shallow (oblique) angle, highlighting sloping hillsides such as "A," while a general pattern of diffuse light originates from scattering of sunlight by clouds and the atmosphere that illuminates all features, including those not in the direct path of the sunlight, such as hillside "B," while the cave "C" receives no illumination. To establish this light-optical analogy, we must match components with similar characteristics:

1. The human observer's eye, which has a very sharply defined line-of-sight, is matched in characteristic by the electron beam, which presents a very narrow cone angle of rays: thus, the observer of an SEM image is effectively looking along the beam, and what the beam can strike is what can be observed in an image.

2. The illumination of an outdoor scene by the Sun consists of a direct component (direct rays that strongly light those surfaces that they strike) and an indirect component (diffuse scattering of the Sun's rays from clouds and the atmosphere, weakly illuminating the scene from all angles). For the E–T detector (positive bias), there is a direct signal component that acts like the Sun (BSEs emitted by the specimen into the solid angle defined by the scintillator, as well as SE_1 and SE_2 directly collected

◘ **Fig. 7.5** Human visual experience equivalent to the observer position and lighting situation of the Everhart–Thornley (positive bias) detector

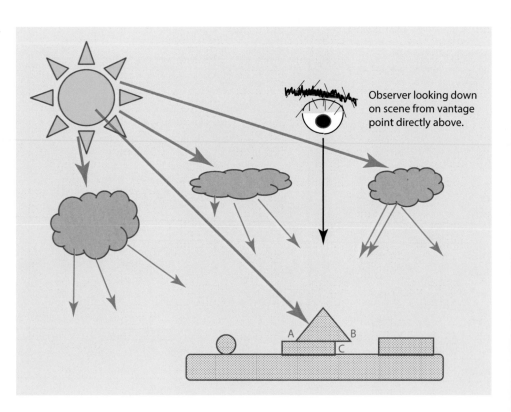

from the specimen) and an indirect component that acts like diffuse illumination (SE$_3$ collected from all surfaces struck by BSEs).

Though counterintuitive, in the SEM the detector is the apparent source of illumination while the observer looks along the electron beam.

Establishing a Robust Light-Optical Analogy

The human visual process has developed in a world of top lighting ( Fig. 7.6): sunlight comes from above in the outdoors; our indoor environment is illuminated from light sources on the ceiling or lamp fixtures placed above our comfortable reading chair. We instinctively expect that brightly illuminated features must be facing upward to receive light from the source above, while poorly illuminated features are facing away from the light source. Thus, to establish the strongest possible light-optical analogy for the SEM/E–T (positive bias) image, we need to create a situation of apparent top lighting. Because the strong source of apparent illumination in an SEM image appears to come from the detector (direct BSEs, SE$_1$, and SE$_2$ for an E–T [positive bias] detector), by ensuring that the effective location of the E–T detector is at the top of the SEM image field as it is presented to the viewer, any feature facing the E–T detector will appear bright, thus establishing that the apparent lighting of the scene presented to the viewer will be from above. All features that can be reached by the electron beam will produce some signal, even those facing away from the E–T (positive bias) detector or that are screened by local topography, through the

collection of the SE$_3$ component. Thus, if we imagine the specimen scene to be illuminated by a primary light source, then that light source occupies the position of the E–T detector and the viewer of that scene is looking along the electron beam. The SE$_3$ component of the signal provides a general diffuse secondary source of illumination that appears to come from all directions.

Getting It Wrong: Breaking the Light-Optical Analogy of the Everhart–Thornley (Positive Bias) Detector

If the microscopist is not careful, it is possible to break the light-optical analogy of the Everhart-Thornley (positive bias) detector. This situation can arise through improper collection of the image by misuse of the feature called "Scan Rotation" (or in subsequent off-line image modification with image processing software). "Scan Rotation" is a commonly available feature of nearly all SEM systems that allows the microscopist to arbitrarily orient an image on the display screen. While this may seem to be a useful feature that enables the presentation of the features of a specimen in a more aesthetically pleasing manner (e.g., aligning a fiber along the long axis of a rectangular image), scan rotation changes the apparent position of the E–T detector (indeed, of all detectors) in the image with potentially serious consequences that can compromise the light-optical analogy of the E–T (positive bias) detector. The observer is naturally accustomed to having top illumination when interpreting images of topography, that is, the apparent source of illumination coming from the top of the field-of-view and shining

Fig. 7.6 We have evolved in a world of top lighting. Features facing the Sun are brightly illuminated, while features facing away are shaded but receive some illumination from atmospheric scattering. Bright = facing upwards

7

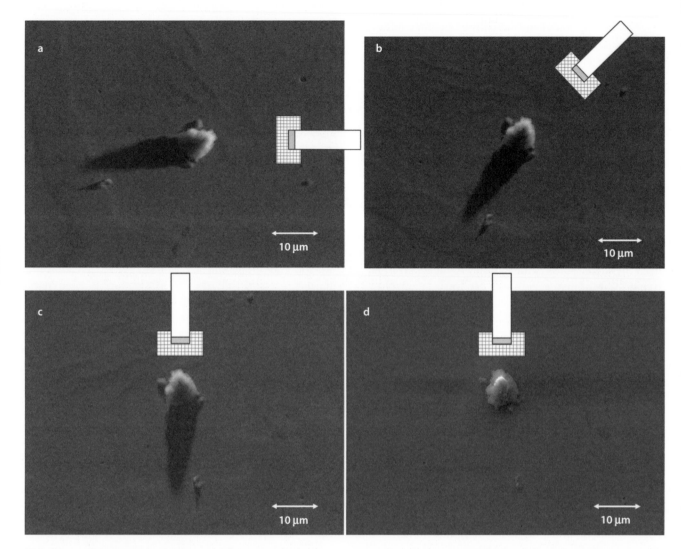

■ **Fig. 7.7** **a** SEM image of a particle on a surface as prepared with the E–T (negative bias) detector in the 90° clockwise position shown; $E_0 = 20$ keV. Note strong shadowing pointing away from E–T. **b** SEM image of a particle on a surface as prepared with the E–T (negative bias) detector in the 45° clockwise position shown; $E_0 = 20$ keV. Note strong shadowing pointing away from E–T. **c** SEM image of a particle on a surface as prepared with the E–T (negative bias) detector in the 0° clockwise (12 o'clock) position shown; $E_0 = 20$ keV. Note strong shadowing pointing away from E–T. **d** SEM image of a particle on a surface as prepared with the E–T (positive bias) detector in the 0° clockwise (12 o'clock) position shown; $E_0 = 20$ keV. Note lack of shadowing but bright surface facing the E–T (positive bias) detector

down on the features of the specimen. When the top lighting condition is violated and the observer is unaware of the alteration of the scene illumination, then the sense of the topography can appear inverted. Arbitrary scan rotation can effectively place the E–T detector, or any other asymmetrically placed (i.e., off-axis) detector, at the bottom or sides of the image, and if the observer is unaware of this situation of unfamiliar illumination, misinterpretation of the specimen topography is likely to result. This is especially true in the case of specimens for which there are limited visual clues. For example, the SEM image of an insect contains many familiar features—e.g., head, eyes, legs, etc.—that make it almost impossible to invert the topography regardless of the

apparent lighting. By comparison, the image of an undulating surface of an unknown object may provide no clues that cause the proper sense of the topography to "click in" for the observer. Having top illumination is critical in such cases. When a microscopist works in a multi-user facility, the possibility must always be considered that a previous user may have arbitrarily adjusted the scan rotation. As part of a personal quality-assurance plan, the careful microscopist should confirm that the location of the E–T detector is at the top center of the image. ■ Figure 7.7 demonstrates a procedure that enables unambiguous location of the E–T detector. Some (but not all) implementations of the E–T detector enable the user to "deconstruct" the E–T detector image by selectively excluding the SE component of the total signal—either by changing the Faraday cage voltage to negative values to reject the very low energy SEs (e.g., −50 V cage bias) or by eliminating the high potential on the scintillator so that SEs cannot be accelerated to sufficient kinetic energy to excite scintillation. Even without the high potential applied to the scintillator, the E–T detector remains sensitive to the high energy BSEs generated by a high energy primary beam, for example., $E_0 \geq 20$ keV, which creates a large fraction of BSEs with energy >10 keV. As a passive scintillator or with the negative Faraday cage potential applied, the E–T (negative bias) detector only collects the small fraction of high energy BSEs scattered into the solid angle defined by the E–T scintillator. When the direct BSE mode of the E–T (negative bias) detector is selected, debris on a flat surface is found to create distinct shadows that point away from the apparent source of illumination, the E–T detector. By using the scan rotation, the effective position of the E–T detector can then be moved to the top of the image, as shown in the sequence of ◘ Fig. 7.7a–c, thus achieving the desired top-lighting situation. When the conventional E–T (positive bias) is used to image this same field of view (◘ Fig. 7.7d), the strong shadow of the particle disappears because of the efficient collection of SEs, particularly the SE_3 component, and now has a bright edge along the top which reinforces the impression that it rises above the general surface.

Note that physically rotating the specimen stage to change the angular relation of the specimen relative to the E–T (or any other) detector does not change the location of the apparent source of illumination in the displayed image. Rotating the specimen stage changes which specimen features are directed toward the detector, but the scan orientation on the displayed image determines the relative position of the detector in the image presented to the viewer and the apparent direction of the illumination.

Deconstructing the SEM/E–T Image of Topography

It is often useful to examine the separate SE and BSE components of the E–T detector image. An example of a blocky fragment of pyrite (FeS_2) imaged with a positively-biased E–T detector is shown in ◘ Fig. 7.8a. In this image, the effective position of the E–T detector relative to the presentation of the image is at the top center. ◘ Figure 7.8b shows the same field of view with the Faraday cage biased negatively to exclude SEs so that only direct BSEs contribute to the SEM image. The image contrast is now extremely harsh, since topographic features facing toward the detector are illuminated, while those facing away are completely lost. Comparing ◘ Fig. 7.8a, b, the features that appear bright in the BSE-only image are also brighter in the full BSE + SE image obtained with the positively biased E–T detector, demonstrating the presence of the direct-BSE component. The much softer contrast of nearly all surfaces seen in the BSE + SE image of

◘ Fig. 7.8a demonstrates the efficiency of the E–T detector for collection of signal from virtually all surfaces of the specimen that the primary beam strikes.

7.3.3 Imaging Specimen Topography With a Semiconductor BSE Detector

A segmented (A and B semicircular segments) semiconductor BSE detector placed directly above the specimen is illustrated schematically in ◘ Fig. 7.9. This BSE detector is mounted below the final lens and is placed symmetrically around the beam, so that in the summation mode it acts as an annular detector. A simple topographic specimen is illustrated, oriented so that the left face directs BSEs toward the A-segment, while the right face directs BSEs toward the B-segment. This A and B detector pair is typically arranged so that one of the segments, "A," is oriented so that it appears to illuminate from the top of the image, while the "B" segment appears to illuminate from the bottom of the image. The segmented detector enables selection of several modes of operation: SUM mode (A + B), DIFFERENCE mode (A − B), and individual detectors A or B) (Kimoto and Hashimoto 1966).

SUM Mode (A + B)

The two-segment semiconductor BSE detector operating in the summation (A + B) mode was used to image the same pyrite specimen previously imaged with the E–T (positive bias) and E–T (negative bias), as shown in ◘ Fig. 7.8c. The placement of the large solid angle BSE is so close to the primary electron beam that it creates the effect of apparent wide-angle illumination that is highly directional along the line-of-sight of the observer, which would be the light-optical equivalent of being inside a flashlight looking along the beam. With such directional illumination along the observer's line-of-sight, the brightest topographic features are those oriented perpendicular to the line-of-sight, while tilted surfaces appear darker, resulting in a substantially different impression of the topography of the pyrite specimen compared to the E–T (positive bias) image in ◘ Fig. 7.8a. The large solid angle of the detector acts to suppress topographic contrast, since local differences in the directionality of BSE emission caused by differently inclined surfaces are effectively eliminated when the diverging BSEs are intercepted by another part of the large BSE detector.

Another effect that is observed in the A + B image is the class of very bright inclusions which were subsequently determined to be galena (PbS) by X-ray microanalysis. The large difference in average atomic number between FeS_2 ($Z_{av} = 20.7$) and PbS ($Z_{av} = 73.2$) results in strong atomic number (compositional) between the PbS inclusions and the FeS_2 matrix. Although there is a significant BSE signal component in the E–T (positive bias) image in ◘ Fig. 7.8a, the

☐ Fig. 7.8 **a** SEM/E–T (positive bias) image of a fractured fragment of pyrite; $E_0 = 20$ keV. **b** SEM/E–T (negative bias) image of a fractured fragment of pyrite; $E_0 = 20$ keV. **c** SEM/BSE (A + B) SUM-mode image of a fractured fragment of pyrite; $E_0 = 20$ keV. **d** SEM/BSE (A segment) image (detector at top of image field) of a fractured fragment of pyrite (FeS$_2$); $E_0 = 20$ keV. **e** SEM/BSE (B segment) image (detector at bottom of image field) of a fractured fragment of pyrite (FeS$_2$); $E_0 = 20$ keV. **f** SEM/BSE (A–B) image (detector DIFFERENCE image) of a fractured fragment of pyrite (FeS$_2$); $E_0 = 20$ keV. **g** SEM/BSE (B–A) image (detector DIFFERENCE image) of a fractured fragment of pyrite (FeS$_2$); $E_0 = 20$ keV

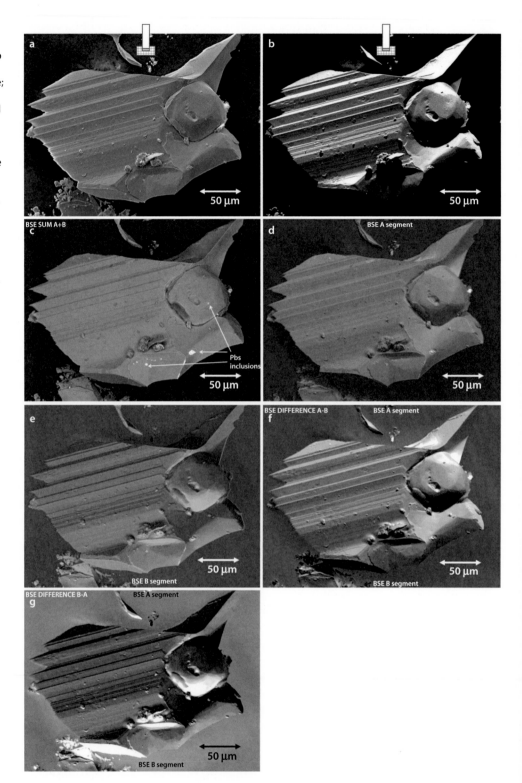

topographic contrast is so strong that it overwhelms the compositional contrast.

Examining Images Prepared With the Individual Detector Segments

Some semiconductor BSE detector systems enable the microscopist to view BSE images prepared with the signal derived from the individual components of a segmented detector. As illustrated in ☐ Fig. 7.9 for a two-segment BSE detector, the individual segments effectively provide an off-axis, asymmetric illumination of the specimen. Comparing the A-segment and B-segment images of the pyrite crystal in ☐ Fig. 7.8d, e, the features facing each detector can be discerned and a sense of the topography can be obtained by comparing the two images. But note the strong effect of the apparent inversion of the sense of the topography in the

Fig. 7.9 Schematic illustration of a segmented annular semiconductor BSE detector

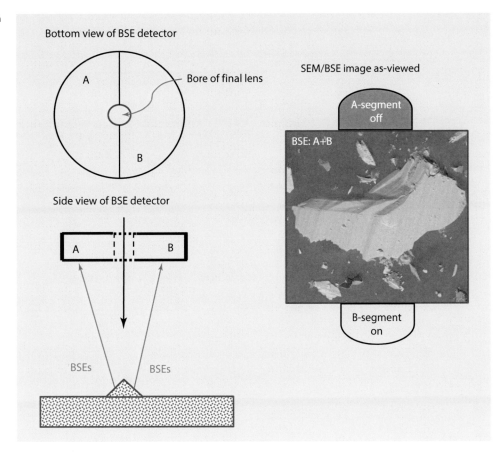

Bottom view of BSE detector

A

Bore of final lens

B

SEM/BSE image as-viewed

A-segment off

BSE: A+B

B-segment on

Side view of BSE detector

A   B

BSEs   BSEs

B-segment image, where the illumination comes from the bottom of the field, compared to the A-segment image, where the illumination comes from the top of the field of view.

DIFFERENCE Mode (A−B)

The signals from the individual BSE detector segments "A" and "B" can be subtracted from each other, producing the image seen in **Fig. 7.8f. Because the detector segments "A" and "B" effectively illuminate the specimen from two different directions, as seen in **Fig. 7.8d, e, taking the difference A–B between the detector signals tends to enhance these directional differences, producing the strong contrast seen in **Fig. 7.8f.

Note that when subtracting the signals the order of the segments in the subtraction has a profound effect on appearance of the final image. **Figure 7.8g shows the image created with the order of subtraction reversed to give B–A. Because the observer is so strongly biased toward interpreting an image as if it must have top lighting, bright features are automatically interpreted as facing upward. This automatic

assumption of top lighting has the effect for most viewers of **Fig. 7.8g to strongly invert the apparent sense of the topography, so that protuberances in the A–B image become concavities in the B–A image. If BSE detector difference images are to be at all useful and not misleading, it is critical to determine the proper order of subtraction. A suitable test procedure is to image a specimen with known topography, such as the raised lettering on a coin or a particle standing on top of a flat surface.

References

Everhart TE, Thornley RFM (1960) Wide-band detector for micro-microampere low-energy electron currents. J Sci Instr 37:246

Kimoto S, Hashimoto H (1966) Stereoscopic observation in scanning microscopy using multiple detectors. In: McKinley T, Heinrich K, Wittry D (eds) The electron microprobe. Wiley, New York, p 480

Oatley CW (1972) The scanning electron microscope, Part I, the instrument. Cambridge University Press, Cambridge

The Visibility of Features in SEM Images

© Springer Science+Business Media LLC 2018
J. Goldstein et al., *Scanning Electron Microscopy and X-Ray Microanalysis*,
https://doi.org/10.1007/978-1-4939-6676-9_8

The detection in SEM images of specimen features such as compositional differences, topography (shape, inclination, edges, etc.), and physical differences (crystal orientation, magnetic fields, electrical fields, etc.), depends on satisfying two criteria: (1) establishing the minimum conditions necessary to ensure that the contrast created by the beam–specimen interaction responding to differences in specimen features is statistically significant in the imaging signal (backscattered electrons [BSE], secondary electrons [SE], or a combination) compared to the inevitable random signal fluctuations (noise); and (2) applying appropriate signal processing and digital image processing to render the contrast information that exists in the signal visible to the observer viewing the final image display.

8.1 Signal Quality: Threshold Contrast and Threshold Current

An SEM image is constructed by addressing the beam to a specific location on the specimen for a fixed dwell time, τ, during which a number of beam electrons are injected through the focused beam footprint into the specimen. The resulting beam–specimen interactions cause the emission of BSE and SE, a fraction of which will be detected and measured with appropriate electron detectors. This measured BSE and/or SE signal is then assigned to that pixel as it is digitally stored and subsequently displayed as a gray-level image. Both the incident electron beam current and the measured BSE and/or SE signals, S_i, involve discrete numbers of electrons: n_B, n_{BSE}, and n_{SE}. The emission of the incident beam current from the electron gun and the subsequent BSE/SE generation due to elastic and inelastic scattering in the specimen are stochastic processes; that is, the mechanisms are subject to random variations over time. Thus, repeated sampling of any imaging signal, S, made at the same specimen location with

the same nominal beam current and dwell time will produce a range of values distributed about a mean count n, with the standard deviation of this distribution described by $\bar{n}^{1/2}$. This natural variation in repeated samplings of the signal S is termed "noise," N. The measure of the signal quality is termed the "signal-to-noise ratio," S/N, given by

$$\frac{S}{N} = \bar{n} / \bar{n}^{1/2} = \bar{n}^{1/2} \tag{8.1}$$

Equation (8.1) shows that as the mean number of collected signal counts increases, the signal quality S/N improves as the random fluctuations become a progressively smaller fraction of the total signal.

◻ Figure 8.1 shows schematically the result of repeated scans over a series of pixels that cross a feature of interest. The signal value S changes in response to the change in the specimen property (composition, topography, etc.), but the repeated scans do not produce exactly the same response due to the inevitable noise in the signal generation processes. When an observer views a scanned image, this noise is superimposed on the legitimate changes in signal (contrast) of features in the image, reducing the visibility. Rose (1948) made an extensive study of the ability of observers viewing scanned television images to detect the contrast between objects of different size and the background in the presence of various levels of noise. Rose found that for the average observer to distinguish small objects with dimensions about 5 % of the image width against the background, the change in signal due to the contrast, ΔS, had to exceed the noise, N, by a factor of 5:

$$\Delta S > 5\,N \tag{8.2}$$

Synthesized digital images in ◻ Figs. 8.2 and 8.3 demonstrate how the visibility is affected by noise and the relative size of objects. ◻ Figure 8.2a shows a synthesized object from the

◻ **Fig. 8.1** Schematic representation of signal response across a specimen feature with the underlying long integration time average (*smooth line*)

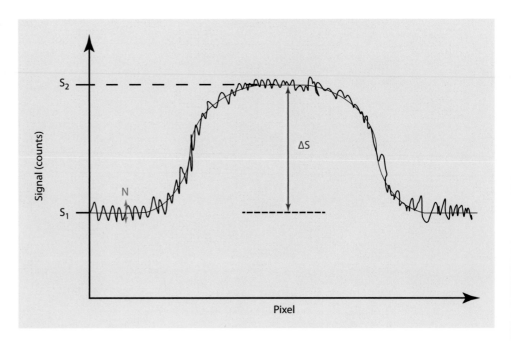

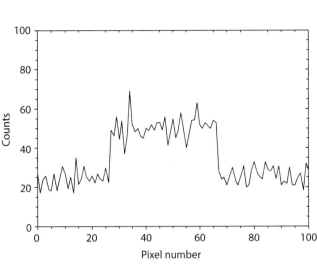

◘ Fig. 8.2 **a** Synthesized image from the template shown in **◘** Fig. 8.3a. **b** Trace of the signal across the circular object

template shown in **◘** Fig. 8.3a with a specified signal and superimposed random noise, and **◘** Fig. 8.2b shows a plot of the signal through one of the test objects. **◘** Figure 8.3 shows synthesized images for various levels of the S and ΔS relative to N. In **◘** Fig. 8.3b $\Delta S = 5 = N$; **◘** Fig. 8.3c $\Delta S = 10 = 2\,N$; and **◘** Fig. 8.3d $\Delta S = 25 = 5\,N$, which just matches the Rose criterion. While the large-scale features are visible in all three images, the fine-scale objects are completely lost in image **◘** Fig. 8.3b, and only fully visible when the Rose criterion is satisfied in image **◘** Fig. 8.3d.

The Rose visibility criterion can be used as the basis to develop the quantitative relation between the threshold contrast, that is, the minimum level of contrast potentially visible in the signal, and the beam current. The noise can be considered in terms of the number of signal events, $N = n^{1/2}$:

$$\Delta S > 5\bar{n}^{1/2} \tag{8.3}$$

Equation (8.3) can be expressed in terms of contrast (defined as $C = \Delta S / S$) by dividing through by the signal:

$$\frac{\Delta S}{S} = C > \frac{5\bar{n}^{1/2}}{S} = \frac{5\bar{n}^{1/2}}{\bar{n}} \tag{8.4}$$

$$C > \frac{5}{\bar{n}^{1/2}} \tag{8.5}$$

$$\bar{n} > \left(\frac{5}{C}\right)^2 \tag{8.6}$$

Equation (8.6) indicates that in order to observe a specific level of contrast, C, a mean number of signal carriers, given

by $(5/C)^2$, must be collected per picture element. Considering electrons as signal carriers, the number of electrons which must be collected per picture element in the dwell time, τ, can be converted into a signal current, i_s

$$i_s = \frac{\bar{n}e}{\tau} \tag{8.7}$$

where e is the electron charge $(1.6 \times 10^{-19}\ \mathrm{C})$. Substituting Eq. (8.6) into Eq. (8.7) gives the following result:

$$i > \frac{25e}{C^2\tau} \tag{8.8}$$

The signal current, i_s, differs from the beam current, i_B, by the fractional signal generation per incident beam electron (η for BSE and δ for SE or a combination for a detector which is simultaneously sensitive to both classes of electrons) and the efficiency with which the signal is converted to useful information for the image. This factor is given by the detective quantum efficiency (DQE) (Joy et al. 1996) and depends on the solid angle of collection and the response of the detector (see the full DQE description in the Electron Detectors module):

$$i_s = i_B(\eta,\delta)\,\mathrm{DQE} \tag{8.9}$$

Combining Eqs. (8.8) and (8.9) yields

$$i_B > \frac{25\left(1.6\times10^{-19}\ \mathrm{C}\right)}{(\eta,\delta)\,\mathrm{DQE}\,C^2\tau} \tag{8.10}$$

8

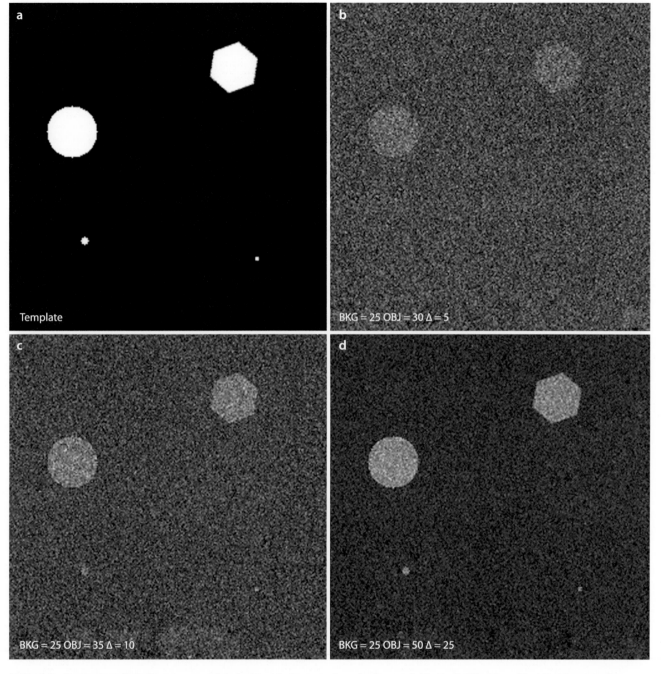

◻ **Fig. 8.3** Synthesized digital images: **a** template; **b** object $S=5$ counts above background, $\Delta S=5=N$, $S/B=1.2$; **c** object $S=10$ counts above background, $\Delta S=10=2N$, $S/B=1.4$; **d** object $S=25$ counts above background, $\Delta S=25=5N$, $S/B=2$

The picture element dwell time, τ, can be replaced by the time to scan a full frame, t_F, from the relation

$$\tau = \frac{t_F}{N_{PE}} \tag{8.11}$$

where N_{PE} is the number of pixels in the entire image. Substituting Eq. (8.11) into Eq. (8.10),

$$i_B > \frac{(4\times10^{-18})N_{PE}}{(\eta,\delta)\,\mathrm{DQE}\,C^2 t_F}\,(\mathrm{coulomb/s=amperes}) \tag{8.12}$$

For an image with 1024×1024 picture elements, Eq. (8.12) can be stated as

$$i_B > \frac{(4\times10^{-12}\,A)}{(\eta,\delta)\,\mathrm{DQE}\,C^2 t_F} \tag{8.13}$$

Equation (8.12) is referred to as the "Threshold Equation" (Oatley et al. 1965; Oatley 1972) because it defines the minimum beam current, the "threshold current," necessary to observe a specified level of contrast, C, with a signal production efficiency specified by η and/or δ and the detector

Fig. 8.4 Plot of the threshold contrast vs. frame time for an image with 1024 by 1024 pixels and an overall signal conversion efficiency of 0.25. Contours of constant current from 1 μA to 1 pA are shown

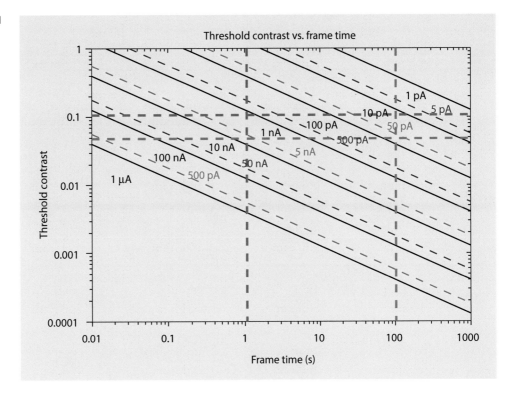

Fig. 8.5 Plot of the threshold current vs. frame time for an image with 1024 by 1024 pixels and an overall signal conversion efficiency of 0.25. Contours of constant contrast from 1 to 0.001 are shown

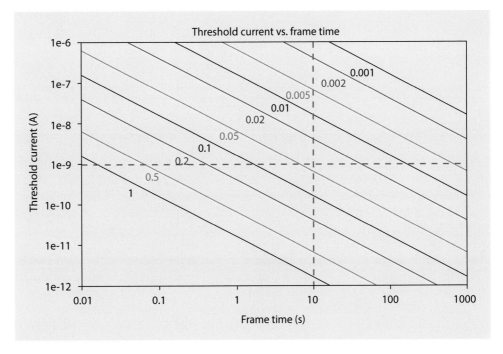

performance described by the DQE (Joy et al. 1996). Alternatively, if we measure the current which is available in the beam that reaches the specimen (e.g., with a Faraday cup and specimen current picoammeter), then we can calculate the minimum contrast, the so-called "threshold contrast," which can be observed in an image prepared under these conditions. Objects in the field of view that do not produce this threshold contrast cannot be distinguished from the noise of random background fluctuations. Equations (8.12 and 8.13) lead to the following *critical limitation* on SEM imaging performance:

For any particular selection of operating parameters—beam current, signal (backscattered electrons, secondary electrons, or a combination), detector performance (DQE), image pixel density and dwell time—there is always a level of contrast below which objects *cannot be detected*. Objects producing contrast below this threshold contrast cannot be recovered by applying any post-collection image-processing algorithms.

The graphical plots shown in ☐ Fig. 8.4 (threshold contrast vs. frame time for various values of the beam current) and ☐ Fig. 8.5 (threshold current vs. frame time for various values

8

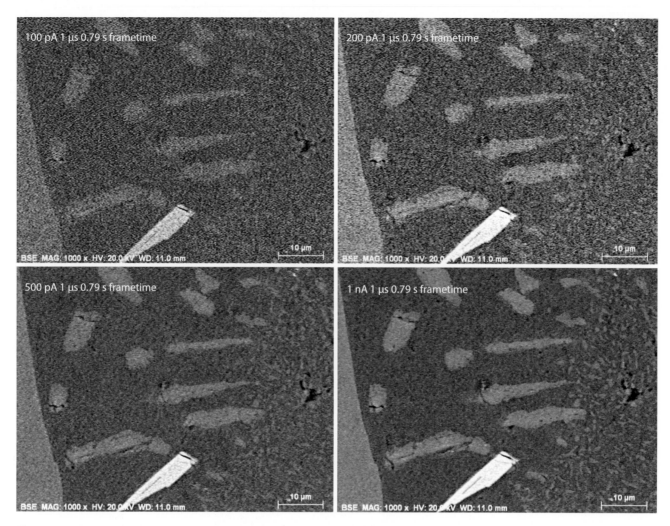

100 pA 1 µs 0.79 s frametime

BSE MAG: 1000 x HV: 20.0 kV WD: 11.0 mm   10 µm

200 pA 1 µs 0.79 s frametime

BSE MAG: 1000 x HV: 20.0 kV WD: 11.0 mm   10 µm

500 pA 1 µs 0.79 s frametime

BSE MAG: 1000 x HV: 20.0 kV WD: 11.0 mm   10 µm

1 nA 1 µs 0.79 s frametime

BSE MAG: 1000 x HV: 20.0 kV WD: 11.0 mm   10 µm

Fig. 8.6   Al-Si eutectic alloy. BSE images (1024 by 784 pixels; 1-µs pixel dwell) at various beam currents

of the contrast) provide a useful way to understand the relationships of the parameters of the Threshold Equation. These plots have been derived from Eq. (8.13) with the assumptions that the image has 1024 by 1024 pixels and the overall signal generation/collection efficiency (the product of η and/or δ and the DQE) is 0.25; that is, one signal-carrying electron (backscattered and/or secondary) is registered in the final image for every four beam electrons that strike the specimen. This collection efficiency is a reasonable assumption for a target, such as gold, which has high backscattering and secondary electron coefficients, when the electrons are detected with an efficient positively-biased E–T detector. ☐ Figure 8.4 reveals that imaging a contrast level of $C = 0.10$ (10 %) with a frame time of 1 s (a pixel dwell time of ~1 µs for a 1024 × 1024-pixel scan) requires a beam current in excess of 1 nA, whereas if 100 s is used for the frame time (pixel dwell time of ~100 µs), the required beam current falls to about 10 pA. If the specimen only produces a contrast level of 0.05 (5 %), a beam current above 5 nA must be used. Conversely, if a particular value of the beam current is selected, ☐ Fig. 8.5 demonstrates that there will always be a level of contrast below which objects will be effectively invisible. For example, if a beam current of 1 nA is used for a 10-s frame time, all objects producing con-

trast less than approximately 0.05 (5 %) against the background will be lost. Once the current required to image a specific contrast level is known from the Threshold Equation, the minimum beam size that contains this current can be estimated with the Brightness Equation. A severe penalty in minimum probe size is incurred when the contrast is low because of the requirement for high beam current needed to exceed the threshold current. Moreover, this ideal beam size will be increased due to the aberrations that degrade electron optical performance.

The Rose criterion is actually a conservative estimate of visibility threshold conditions since it is appropriate for small discrete features with linear dimensions down to a few percent of the image width or small details on larger structures. For objects that constitute a large fraction of the image or which have an extended linear nature, such as an edge or a fiber, the ability of an observer's visual process to effectively combine information over many contiguous pixels actually relaxes the visibility criterion, as illustrated in the synthesized images in ☐ Fig. 8.3 (Bright et al. 1998). The effect of the size of a feature on visibility of real features can be seen in ☐ Figs. 8.6 and 8.7, which show BSE images (semiconductor detector) of a commercial aluminum–silicon eutectic casting

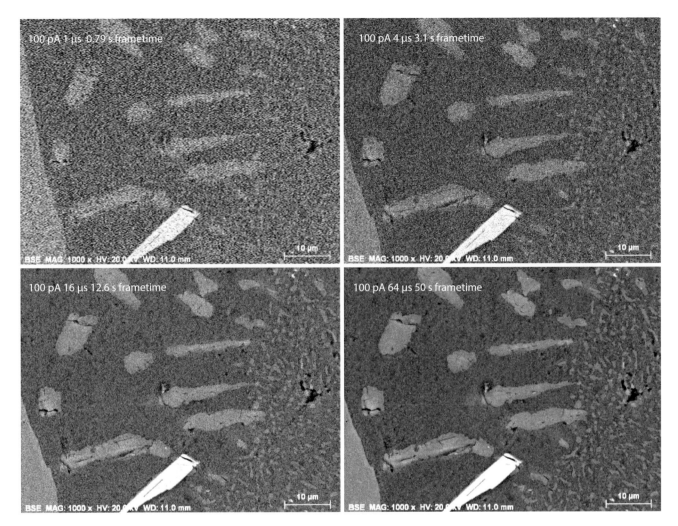

□ **Fig. 8.7** Al-Si eutectic alloy. BSE images (1024 by 784 pixels; 100-pA beam current) pixel dwell at various frame times

alloy under various conditions. The two principal phases of this material are nearly pure Al and Si, which produce a contrast based on the respective BSE coefficients of $C = \Delta\eta/\eta_{max} = (0.14-0.13)/0.14 \approx 0.07$ or 7 % contrast. As the beam current is decreased with fixed frame time (□ Fig. 8.6) or the frame time is decreased with fixed beam current (□ Fig. 8.7), the visibility of the fine-scale features at the right-hand side of the image diminishes and these small features are eventually lost, whereas the large-scale features on the left-hand side of the image remain visible over the range of experimental parameters despite having the same compositional difference and thus producing the same contrast.

While the Threshold Equation provides "gray numbers" for the threshold parameters due to the variability of the human observer and the relative size of objects, the impact of the Threshold Equation must be considered in developing imaging strategy. Unfortunately, poor imaging strategy can render the SEM completely ineffective in detecting the features of interest. A careful imaging strategy will first estimate the likely level of contrast from the objects of interest (or assume the worst possible case that the features being sought produce very low contrast, e.g., <0.01) and then select instrument parameters capable of detecting that contrast. An

example is shown in □ Fig. 8.8, which shows a sequence of images of a polished carbon planchet upon which a droplet containing a dilute salt was deposited by inkjet printing. The images were prepared at constant beam current but with increasing pixel dwell time, which represents a section through the Threshold Equation plot shown in □ Fig. 8.9. Even the largest-scale features are lost in the image prepared with the shortest dwell time. Careful study of these images reveals that new information is being added throughout the image sequence, and likely there would be additional information recovered with further increases in the pixel time or by increasing the beam current.

Finally, it must be recognized that there is a substantial "observer effect" for objects producing contrast near the threshold of visibility: different observers may have substantially different success in detecting features in images (Bright et al. 1998). Thus the threshold current or threshold contrast calculated with Eq. (8.12) should be considered a "fuzzy number" rather than an absolute threshold, since visibility depends on several factors, including the size and shape of the features of interest as well as the visual acuity of the particular observer and his/her experience in evaluating images. The limitations imposed by the threshold equation and the

8

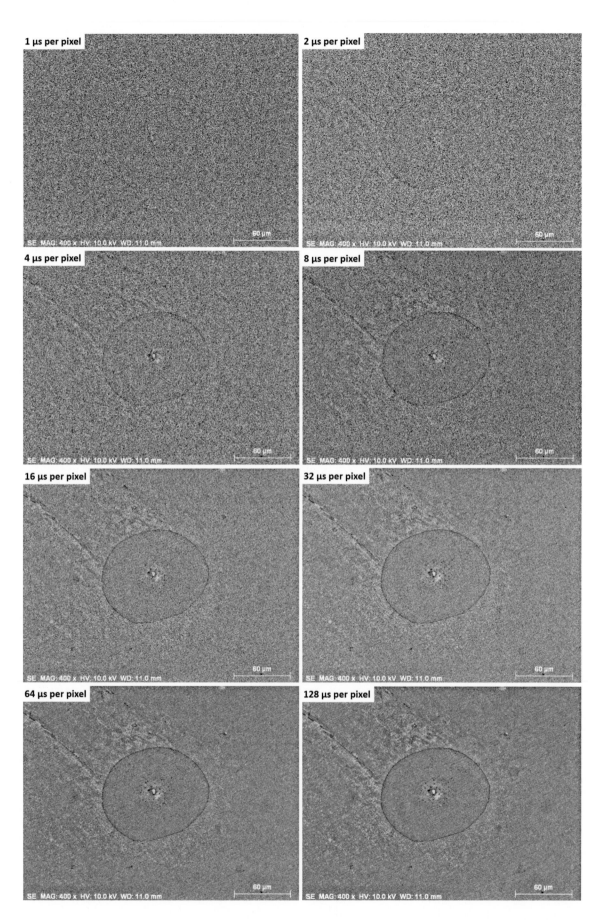

□ **Fig. 8.8** Threshold imaging visibility; image sequence with increasing pixel dwell time at constant beam current. Inkjet deposited droplet on carbon planchet; $E_0 = 10$ keV; Everhart–Thornley (positive bias) detector. Post-collection processing with ImageJ (FIJI) CLAHE function

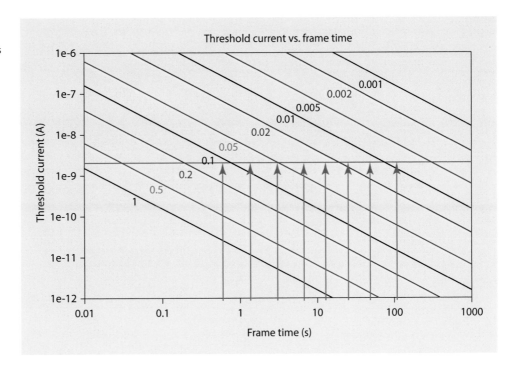

Fig. 8.9 Threshold current plot showing time sequence at constant beam current. Contours of constant contrast from 1 to 0.001 are shown

observer effect mean that a negative result in an SEM study, that is, the failure to find an expected feature in an image, may occur because of the choice of imaging conditions and the observer's limitations, not because the feature does not exist in the specimen under study. Thus, best practices in SEM imaging of low contrast features must include a comprehensive strategy to systematically vary the imaging parameters, including beam current and dwell time, to be sure that the visibility threshold is adequately exceeded before an object can be declared to be absent with a high degree of confidence.

References

Bright DS, Newbury DE, Steel EB (1998) Visibility of objects in computer simulations of noisy micrographs. J Micros 189:25–42

Joy DC, Joy CS, Bunn RD (1996) Experimental measurements of the efficiency, and the static and dynamic response of electron detectors in the scanning electron microscope. Scanning 18:181

Oatley CW (1972) The scanning electron microscope: Part 1 the instrument. Cambridge University Press, Cambridge

Oatley CW, Nixon WC, Pease RFW (1965) Scanning electron microscopy. In: Advances in electronics and electron physics. Academic Press, New York, p 181

Rose A (1948) Television pickup tubes and the problem of vision. In: Marton L (ed) Advances in electronics and electron physics, vol 1. Academic Press, New York, p 131

Image Defects

© Springer Science+Business Media LLC 2018
J. Goldstein et al., *Scanning Electron Microscopy and X-Ray Microanalysis*,
https://doi.org/10.1007/978-1-4939-6676-9_9

SEM images are subject to defects that can arise from a variety of mechanisms, including charging, radiation damage, contamination, and moiré fringe effects, among others. Image defects are very dependent on the specific nature of the specimen, and often they are anecdotal, experienced but not reported in the SEM literature. The examples described below are not a complete catalog but are presented to alert the microscopist to the possibility of such image defects so as to avoid interpreting artifact as fact.

9.1   Charging

Charging is one of the major image defects commonly encountered in SEM imaging, especially when using the Everhart–Thornley (positive bias) "secondary electron" detector, which is especially sensitive to even slight charging.

9.1.1   What Is Specimen Charging?

The specimen can be thought of as an electrical junction into which the beam current, i_B, flows. The phenomena of backscattering of the beam electrons and secondary electron emission represent currents flowing out of the junction, i_{BSE} ($= \eta\, i_B$) and i_{SE} ($= \delta\, i_B$). For a copper target and an incident beam energy of 20 keV, η is approximately 0.3 and δ is approximately 0.1, which together account for 0.4 or 40 % of the charges injected into the specimen by the beam current. The remaining beam current must flow from the specimen to ground to avoid the accumulation of charge in the junction (Kirchoff's current law). The balance of the currents for a non-charging junction is then given by

$$\sum i_{in} = \sum i_{out}$$
$$i_B = i_{BSE} + i_{SE} + i_{SC} \qquad (9.1)$$

where i_{SC} is the specimen (or absorbed) current. For the example of copper, $i_{SC} = 0.6\, i_B$.

The specimen stage is typically constructed so that the specimen is electrically isolated from electrical ground to permit various measurements. A wire connection to the stage establishes the conduction path for the specimen current to travel to the electrical ground. This design enables a current meter to be installed in this path to ground, allowing direct measurement of the specimen current and enabling measurement of the true beam current with a Faraday cup (which captures all electrons that enter it) in place of the specimen. Moreover, this specimen current signal can be used to form an image of the specimen (see the "Electron Detectors" module) However, if the electrical path from the specimen surface to ground is interrupted, the conditions for charge balance in Eq. (9.1) cannot be established, even if the specimen is a metallic conductor. The electrons injected into the specimen by the beam will then accumulate, and the specimen will develop a high negative electrical charge

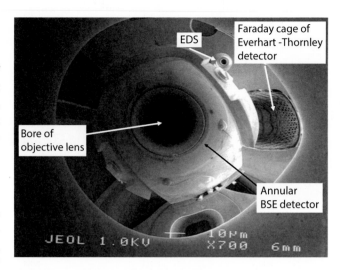

Fig. 9.1   SEM image (Everhart–Thornley detector, positive bias) obtained by disconnecting grounding wire from the specimen stage and reflecting the scan from a flat, conducting substrate; $E_0 = 1$ keV

relative to ground. The electrical field from this negative charge will decelerate the incoming beam electrons, and in extreme cases the specimen will actually act like an electron mirror. The scanning beam will be reflected before reaching the surface, so that it actually scans the inside of the specimen chamber, creating an image that reveals the objective lens, detectors, and other features of the specimen chamber, as shown in ￼ Fig. 9.1.

If the electrical path to ground is established, then the excess charges will be dissipated in the form of the specimen current provided the specimen has sufficient conductivity. Because all materials (except superconductors) have the property of electrical resistivity, ρ, the specimen has a resistance R ($R = \rho\, L/A$, where L is the length of the specimen and A is the cross section), and the passage of the specimen current, i_{SC}, through this resistance will cause a potential drop across the specimen, $V = i_{SC}\, R$. For a metal, ρ is typically of the order of $10^{-6}\,\Omega$-cm, so that a specimen 1-cm thick with a cross-sectional area of 1 cm^2 will have a resistance of $10^{-6}\,\Omega$, and a beam current of 1 nA (10^{-9} A) will cause a potential of about 10^{-15} V to develop across the specimen. For a high purity (undoped) semiconductor such as silicon or germanium, ρ is approximately 10^4 to $10^6\,\Omega$-cm, and the 1-nA beam will cause a potential of 1 mV (10^{-3} V) or less to develop across the 1-cm cube specimen, which is still negligible. The flow of the specimen current to ground becomes a critical problem when dealing with non-conducting (insulating) specimens. Insulating specimens include a very wide variety of materials such as plastics, polymers, elastomers, minerals, rocks, glasses, ceramics, and others, which may be encountered as bulk solids, porous solids, foams, particles, or fibers. Virtually all biological specimens become non-conducting when water is removed by drying, substitution with low vapor pressure polymers, or frozen in place. For an insulator such as an oxide, ρ is very high, 10^6 to $10^{16}\,\Omega$-cm, which prevents the smooth motion of the electrons injected by the beam through the specimen to ground; electrons accumulate in the immediate vicinity of the beam impact, raising the local potential and creating a range of phenomena described as "charging."

9.1.2 Recognizing Charging Phenomena in SEM Images

Charging phenomena cover a wide range of observed behaviors in SEM images of imperfectly conducting specimens. Secondary electrons (SEs) are emitted with very low energy, by definition $E_{SE} < 50$ eV, with most carrying less than 5 eV. Such low energy, slow-moving SEs can be strongly deflected by local electrical fields caused by charging. The Everhart–Thornley (positive bias) detector collects SEs by means of a positive potential of a few hundred volts (e.g., +300 V) applied to the Faraday cage at a distance of several centimeters (e.g., 3 cm) from the specimen, creating an electrical field at the specimen of approximately 10^4 V/m. SEs emitted from a conducting specimen are strongly attracted to follow the field lines from the grounded specimen surface to the positively biased Faraday cage grid, and thus into the high voltage field applied to the face of the scintillator of the Everhart–Thornley (E–T) detector. If the specimen charges locally to develop even a few volts' potential, the local electrical field from the charged region relative to nearby uncharged areas of the specimen a few micrometers away or to the grounded stub a few millimeters away likely to be much stronger (10^5 to 10^7 V/m) than the field imposed by the E–T detector. Depending on the positive or negative character, this specimen field may have either a repulsive or an attractive effect. Thus, depending on the details of the local electrical field, the collection of SEs by the E–T detector may be enhanced or diminished. Negatively charging areas will appear bright, while in positively charging areas the SEs are attracted back to the specimen surface or to the stub so that such a region appears dark. Thus, the typical

appearance of an isolated insulating particle undergoing charging on a conducting surface is a bright, often saturated signal (possibly accompanied by amplifier overloading effects due to signal saturation) surrounded by a dark halo that extends over surrounding conducting portions of the specimen where the local field induced by the charging causes the SEs to be recollected. This type of voltage contrast must be regarded as an artifact, because it interferes with and overwhelms the regular behavior of secondary electron (SE) emission with local surface inclination that we depend upon for sensible image interpretation of specimen topography with the E–T detector. ◻ Figure 9.2 shows examples of charging effects observed when imaging insulating particles on a conducting metallic substrate with the E–T (positive bias) detector. There are regions on the particles that are extremely bright due to high negative charging that increases the detector collection efficiency surrounded by a dark "halo" where a positive mirror charge develops, lowering the collection efficiency. Often these charging effects, while extreme in the E–T (positive bias) image due to the disruption of SE trajectories, will be negligible in a backscattered electron (BSE) image simultaneously collected from the same field of view, because the much higher energy BSEs are not significantly deflected by the low surface potential. An example is shown in ◻ Fig. 9.3, where the SE image, ◻ Fig. 9.3a, shows extreme bright-dark regions due to charging while the corresponding BSE image, ◻ Fig. 9.3b, shows details of the structure of the particle. In more extreme cases of charging, the true topographic contrast image of the specimen may be completely overwhelmed by the charging effects, which in the most extreme cases will actually deflect the beam causing image discontinuities. An example is shown in ◻ Fig. 9.4, which compares images (Everhart–Thornley detector, positive bias) of an uncoated calcite crystal at $E_0 = 1.5$ keV, where the true shape of the object can be seen, and at $E_0 = 5$ keV, where charging completely overwhelms the topographic contrast.

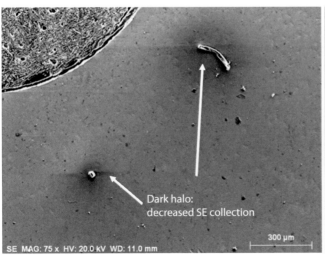
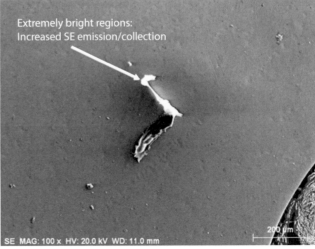

◻ **Fig. 9.2** Examples of charging artifacts observed in images of dust particles on a metallic substrate. $E_0 = 20$ keV; Everhart–Thornley (positive bias) detector

9

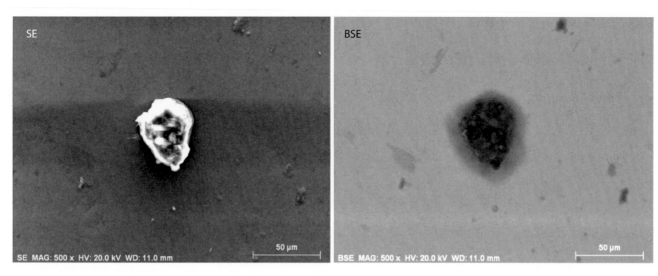

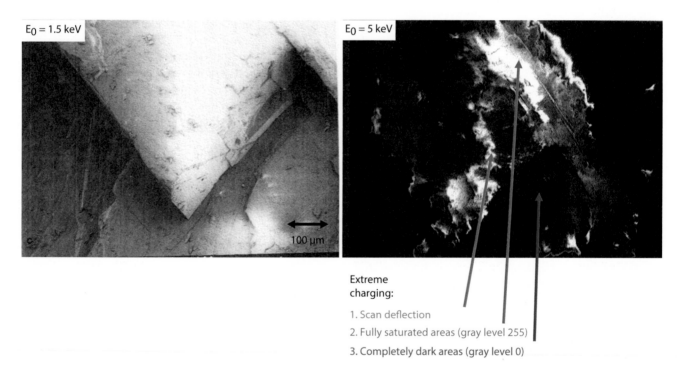

Extreme
charging:

1. Scan deflection
2. Fully saturated areas (gray level 255)
3. Completely dark areas (gray level 0)

■ **Fig. 9.4** Comparison of images of an uncoated calcite crystal viewed at (*left*) $E_0 = 1.5$ keV, showing topographic contrast; (*right*) $E_0 = 5$ keV, showing extreme charging effects; Everhart–Thornley (positive bias) detector

Charging phenomena are incompletely understood and are often found to be dynamic with time, a result of the time-dependent motion of the beam due to scanning action and due to the electrical breakdown properties of materials as well as differences in surface and bulk resistivity. An insulating specimen acts as a local capacitor, so that placing the beam at a pixel causes a charge to build up with an RC time constant as a function of the dwell time, followed by a decay of that charge when the beam moves away. Depending on the exact material properties, especially the surface resistivity which is often much lower than the bulk resistivity, and the beam conditions (beam energy, current, and scan rate), the injected charge may only partially dissipate before the beam returns in the scan cycle, leading to strong effects in SEM images. Moreover, local specimen properties may cause charging effects to vary with position in the same image. A time-dependent charging situation at a pixel is shown schematically in ■ Fig. 9.5, where the surface potential at a particular pixel accumulates with the dwell time and then decays until the beam returns. In more extreme behavior, the accumulated charge may cause local electrical breakdown and abruptly discharge to ground. The time dependence of

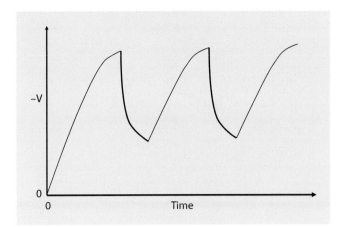

Fig. 9.5 Schematic illustration of the potential developed at a pixel as a function of time showing repeated beam dwells

charging can result in very different imaging results as the pixel dwell time is changed from rapid scanning for surveying a specimen to slow scanning for recording images with better signal-to-noise. An example of this phenomenon is shown in ◻ Fig. 9.6, where an image of an uncoated mineral fragment taken with $E_0 = 1$ keV appears to be free of charging artifacts with a pixel dwell time of 1.6 µs, but longer dwell times lead to the in-growth of a bright region due to charging. Charging artifacts can often be minimized by avoiding slow scanning through the use of rapid scanning and summing repeated scans to improve the signal-to-noise of the final image.

Charging of some specimens can create contrast that can easily be misinterpreted as specimen features. An example is shown in ◻ Fig. 9.7, where most of the polystyrene latex spheres (PSLs) imaged at $E_0 = 1$ keV with the Everhart–Thornley (positive bias) detector show true topographic details, but five of the PSLs have bright dots at the center, which might easily be mistaken for high atomic number inclusions or fine scale topographic features rising above the spherical surfaces. Raising the beam energy to 1.5 keV and higher reveals progressively more extensive and obvious evidence of charging artifacts. The nature of this charging artifact is revealed in ◻ Fig. 9.8, which compares an image of the PSLs at higher magnification and $E_0 = 5$ keV with a low

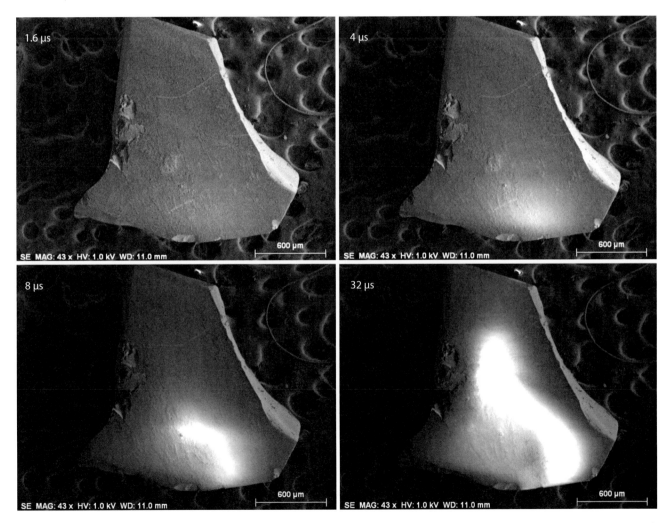

Fig. 9.6 Sequence of images of an uncoated quartz fragment imaged at $E_0 = 1$ keV with increasing pixel dwell times, showing development of charging; Everhart–Thornley (positive bias) detector

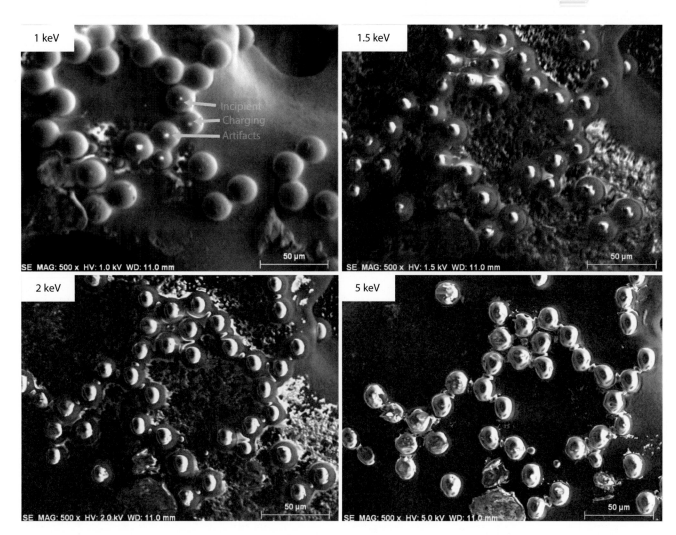

Fig. 9.7 Polystyrene latex spheres imaged over a range of beam energy, showing development of charging artifacts; Everhart–Thornley (positive bias) detector

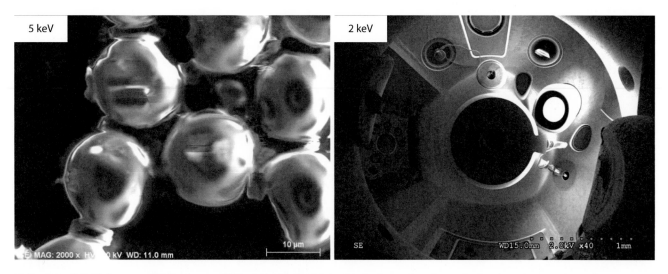

Fig. 9.8 (*left*) Higher magnification image of PSLs at $E_0 = 5$ keV; (*right*) reflection image from large plastic sphere that was charged at $E_0 = 10$ keV and then imaged at $E_0 = 2$ keV; Everhart–Thornley (positive bias) detector

magnification image of a large plastic sphere (5 mm in diameter) that was first subjected to bombardment at $E_0 = 10$ keV, followed by imaging at $E_0 = 2$ keV where the deposited charge acts to reflect the beam and produce a "fish-eye" lens view of the SEM chamber. Close examination of the higher magnification PSL images shows that each of these microscopic spheres is acting like a tiny "fish-eye lens" and producing a highly distorted view of the SEM chamber.

9.1.3 Techniques to Control Charging Artifacts (High Vacuum Instruments)

Observing Uncoated Specimens

To understand the basic charging behavior of an uncoated insulator imaged with different selections of the incident beam energy, consider ◘ Fig. 9.9, which shows the behavior of the processes of backscattering and secondary electron emission as a function of beam energy. For beam energies above 5 keV, generally $\eta + \delta < 1$, so that more electrons are injected into the specimen by the beam than leave as BSEs and SEs, leading to an accumulation of negative charge in an insulator. For most materials, especially insulators, as the beam energy is lowered, the total emission of BSEs and SEs increases, eventually reaching an upper cross-over energy, E_2 (which typically lies in the range 2–5 keV depending on the material) where $\eta + \delta = 1$, and the charge injected by the beam is just balanced by the charge leaving as BSEs and SEs. If a beam energy is selected just above E_2 where $\eta + \delta < 1$, the local build-up of negative charge acts to repel the subsequent incoming beam electrons while the beam remains at that pixel, lowering the effective kinetic energy with which the beam strikes the surface eventually reaching the E_2 energy and a dynamically stable charge balance. For beam energies below the E_2 value and above the lower cross-over energy E_1 (approximately

0.5–2 keV, depending on the material), the emission of SE can actually reach very large values for insulators with δ_{max} ranging from 2 to 20 depending on the material. Thus, in this beam energy region $\eta + \delta > 1$, resulting in positive charging which increases the kinetic energy of the incoming beam electrons until the E_2 energy is reached and charge balance occurs. This dynamic charge stability enables uncoated insulators to be imaged, as shown in the example of the uncoated mineral particle shown in ◘ Fig. 9.10, where a charge-free image is obtained at $E_0 = 1$ keV, but charging effects are observed at $E_0 \geq 2$ keV. Achieving effective "dynamic charge balance microscopy" is sensitive to material and specimen shape (local tilt as it affects BSEs and particularly SE emission), and success depends on optimizing several instrument parameters: beam energy, beam current, and scan speed. Note that the uncoated mineral specimen used in the beam energy sequence in ◘ Fig. 9.10 is the same used for the pixel dwell time sequence at $E_0 = 1$ keV in ◘ Fig. 9.6 where charging is observed when longer dwell times are used, demonstrating the complex response of charging to multiple variables.

Coating an Insulating Specimen for Charge Dissipation

Conductive coatings can be deposited by thermal evaporation with electron beam heating (metals, alloys) or with resistive heating (carbon), by high energy ion beam sputtering (metals, alloys), or by low energy plasma ion sputtering (alloys). The coating must cover all of the specimen, including complex topographic shapes, to provide a continuous conducting path across the surface to dissipate the charge injected into the specimen by the electron beam. It is important to coat all surfaces that are directly exposed to the electron beam or which might receive charge from BSEs, possibly after re-scattering of those BSEs. Note that applying a conductive coating alone may not be sufficient to achieve efficient charge dissipation. Many specimens may be so thick that the sides may not actually receive an adequate amount of the coating material, as illustrated in ◘ Fig. 9.11, even with rotation during the coating process. It is necessary to complete the path from the coating to the electrical ground with a conducting material that exhibits a low vapor pressure material that is compatible with the microscope's vacuum requirement, such as a metal wire, conducting tape, or metal foil.

It is desirable to make the coating as thin as possible, and for many samples an effective conducting film can be 2–10 nm in thickness. A beam with $E_0 > 5$ keV will penetrate through this coating and 10–100 times (or more) deeper depending on material and the incident beam energy, thus depositing most of the charge in the insulator itself. However, the presence of a ground plane and conducting path nanometers to micrometers away from the implanted charge creates a very high local field gradient, $>10^6$ V/m, apparently leading to continuous breakdown and discharging. The strongest evidence that a continuous discharge situation is established

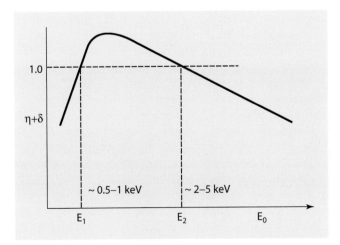

◘ **Fig. 9.9** Schematic illustration of the total emission of backscattered electrons and secondary electrons as a function of incident beam energy; note upper and lower cross-over energies where $\eta + \delta = 1$

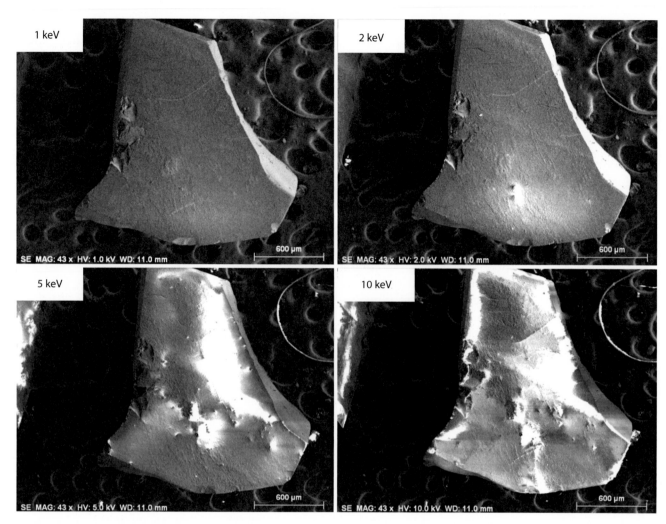

◘ Fig. 9.10 Beam energy sequence showing development of charging as the energy is increased. Specimen: uncoated quartz fragment; 1.6 µs per pixel dwell time; Everhart–Thornley (positive bias) detector

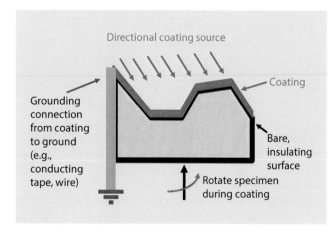

◘ Fig. 9.11 Schematic diagram showing the need to provide a grounding path from a surface coating due to uncoated or poorly coated sides of a non-conducting specimen

that avoids the build-up of charge is the behavior of the Duane–Hunt energy limit of the X-ray continuum. As the beam electrons are decelerated in the Coulombic field of the atoms, the energy lost is emitted as photons of electromagnetic radiation, termed *bremsstrahlung*, or braking radiation, and forming a continuous spectrum of photon energies up to the incident beam energy, which is the Duane–Hunt energy limit. Examination of the upper limit with a calibrated EDS detector provides proof of the highest energy with which beam electrons enter the specimen. When charging occurs, the potential that is developed serves to decelerate subsequent beam electrons and reduce the effective E_0 with which they arrive, lowering the Duane–Hunt energy limit. ◘ Figure 9.12 illustrates such an experiment. The true beam energy should first be confirmed by measuring the Duane–Hunt limit with a conducting high atomic number metal such as tantalum or gold, which produces a high continuum intensity since I_{cm}

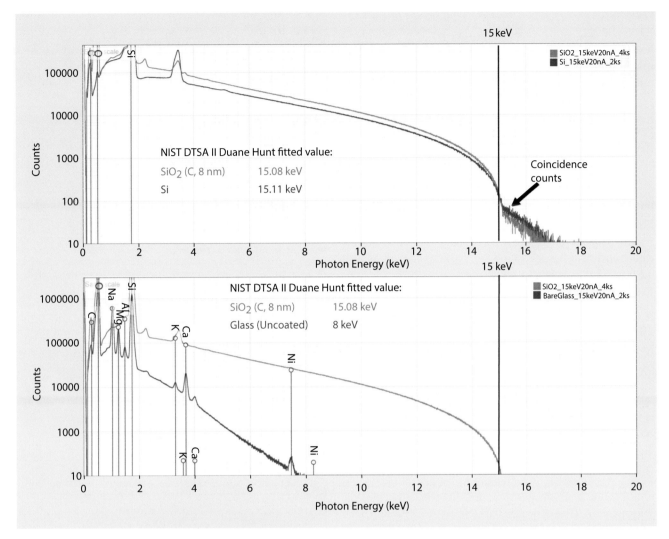

Fig. 9.12 Effects of charging on the Duane–Hunt energy limit of the X-ray continuum: (*upper*) comparison of silicon and coated (C, 8 nm) SiO$_2$ showing almost identical values; (*lower*) comparison of coated (C, 8 nm) SiO$_2$ and uncoated glass showing significant depression of the Duane–Hunt limit due to charging

scales with the atomic number. Note that because of pulse coincidence events, there will always be a small number of photons measured above the true Duane–Hunt limit. The true limit can be estimated with good accuracy by fitting the upper energy range of the continuum intensity, preferably over a region that is several kilo-electronvolts in width and that does not contain any characteristic X-ray peaks, and then finding where the curve intersects zero intensity to define the Duane–Hunt limit. NIST DTSA II performs such a fit and the result is recorded in the metadata reported for each spectrum processed. Once the beam energy is established on a conducting specimen, then the experiment consists of measuring a coated and uncoated insulator. In **Fig. 9.12** (upper plot) spectra are shown for Si (measured Duane–Hunt limit = 15.11 keV) and coated (C, 8 nm) SiO$_2$ (measured Duane–Hunt limit = 15.08 keV), which indicates

there is no significant charging in the coated SiO$_2$. When an uncoated glass slide is bombarded at $E_0 = 15$ keV, the charging induced by the electron beam causes charging and thus severely depresses the Duane–Hunt limit to 8 keV, as seen in **Fig. 9.12** (lower plot), as well as a sharp difference in the shape of the X-ray continuum at higher photon energy.

Choosing the Coating for Imaging Morphology

The ideal coating should be continuous and featureless so that it does not interfere with imaging the true fine-scale features of the specimen. Since the SE$_1$ signal is such an important source of high resolution information, a material that has a high SE coefficient should be chosen. Because the SE$_1$ signal originates within a thin surface layer that is a few nanometers in thickness, having this layer consist of a high atomic number material such as gold that has a high SE coefficient will increase the relative abundance of the high resolution SE_1 signal, especially if the specimen consists of much lower atomic number materials, such as biological material. By using the thinnest possible coating, there is only a vanishingly small contribution to electron backscattering which would tend to degrade high resolution performance.

Although gold has a high SE coefficient, pure gold tends to form discontinuous islands whose structure can interfere with visualizing the desired specimen fine scale topographic structure. This island formation can be avoided by using alloys such as gold-palladium, or other pure metals, for example, chromium, platinum, or iridium, which can be deposited by plasma ion sputtering or ion beam sputtering. The elevated pressure in the plasma coater tends to randomize the paths followed by the sputtered atoms, reducing the directionality of the deposition and coating many re-entrant surfaces. For specimens which are thermally fragile, low deposition rates combined with specimen cooling can reduce the damage.

9.2   Radiation Damage

Certain materials are susceptible to radiation damage ("beam damage") under energetic electron bombardment. "Soft" materials such as organic compounds are especially vulnerable to radiation damage, but damage can also be observed in "hard" materials such as minerals and ceramics, especially if water is present in the crystal structure, as is the case for hydrated minerals. Radiation damage can occur at all length scales, from macroscopic to nanoscale. Radiation damage may manifest itself as material decomposition in which mass is actually lost as a volatile gas, or the material may change density, either collapsing or swelling. On an atomic scale, atoms may be dislodged creating vacancies or interstitial atoms in the host lattice.

An example of radiation damage on a coarse scale is illustrated in ◻ Fig. 9.13, which shows a conductive double-sided sticky polymer tab of the type that is often used as a substrate for dispersing particles. This material was found to be extremely sensitive to electron bombardment. As the magnification was successively reduced in a series of 20-s scans, radiation damage in the form of collapse of the structure at the previous higher magnification scan was readily apparent after a single 20-s scan (20 keV and 10 nA). Note that when this tab is used as a direct support for particles, the susceptibility of the tab material to distortion due radiation damage can lead to unacceptable image drift. Instability in the position of the target particle occurs due to changes in the support tape immediately adjacent to the particle of interest where electrons strike, directly at the edges of the image raster and as a result of backscattering off the particle. Other support materials are less susceptible to radiation damage. ◻ Figure 9.14 shows a detail on a different conductive sticky tape material. After a much higher

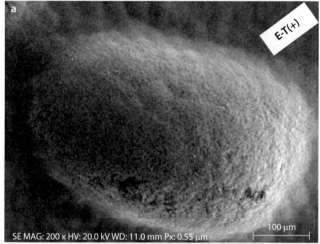

◻ **Fig. 9.13**   SEM Everhart–Thornley (positive bias) image of double-sticky conducting tab

◻ **Fig. 9.14**   Conducting tape: **a** Initial image. **b** Image after a dose of 15 min exposure at higher magnification (20 keV and 10 nA); Everhart–Thornley (positive bias)

dose (15 min of bombardment at 20 keV and 10 nA), a much less significant collapse crater is seen to have formed. It is prudent to examine the behavior of the support materials under electron bombardment prior to use in a particle preparation.

If radiation damage occurs and interferes with successful imaging of the structures of interest, the microscopist has several possible strategies:

1. Follow a minimum dose microscopy strategy.
   a. Radiation damage scales with dose. Use the lowest possible beam current and frame time consistent with establishing the visibility of the features of interest. It may be necessary to determine these parameters for establishing visibility for the particular specimen by operating initially on a portion of the specimen that can be sacrificed.
   b. Once optimum beam current and frame time have been established, the SEM can be focused and stigmated on an area adjacent to the features of interest, and the stage then translated to bring the area of interest into position. After the image is recorded using the shortest possible frame time consistent with establishing visibility, the beam should be blanked (ideally into a Faraday cup) to stop further electron bombardment while the stored image is examined before proceeding.
2. Change the beam energy
   Intuitively, it would seem logical to lower the beam energy to reduce radiation damage, and depending on the particular material and the exact mechanism of radiation damage, a lower beam energy may be useful. However, the energy deposited per unit volume actually increases significantly as the beam energy is lowered! From the Kanaya–Okayama range, the beam linear beam penetration scales approximately as $E_0^{1.67}$ so that the volume excited by the beam scales as $(R_{K-O})^3$ or E_0^5. The energy deposited per unit volume scales as E_0/E_0^5 or $1/E_0^4$. Thus, the volume density of energy deposition increases by a factor of 10^4 as the beam energy decreases from $E_0 = 10$ keV to $E_0 = 1$ keV. Raising the beam energy may actually be a better choice to minimize radiation damage.
3. Lower the specimen temperature
   Radiation damage mechanisms may be thermally sensitive. If a cold stage capable of achieving liquid nitrogen temperature or lower is available, radiation damage may be suppressed, especially if low temperature operation is combined with a minimum dose microscopy strategy.

9.3 Contamination

"Contamination" broadly refers to a class of phenomena observed in SEM images in which a foreign material is deposited on the specimen as a result of the electron beam bombardment. Contamination is a manifestation of radiation damage in which the material that undergoes radiation damage is unintentionally present, usually as a result of the original environment of the specimen or as a result of inadequate cleaning during preparation. Contamination typically arises from hydrocarbons that have been previously deposited on the specimen surface, usually inadvertently. Such compounds are very vulnerable to radiation damage. Hydrocarbons may "crack" under electron irradiation into gaseous components, leaving behind a deposit of elemental carbon. While the beam can interact with hydrocarbons present in the area being scanned, electron beam induced migration of hydrocarbons across the surface to actually increase the local contamination has been observed (Hren 1986). Sources of contamination can occur in the SEM itself. However, for a modern SEM that has been well maintained and for which scrupulous attention has been paid to degreasing and subsequently cleanly handling all specimens and stage components, contamination from the instrument itself should be negligible. Ideally, an instrument should be equipped with a vacuum airlock to minimize the exposure of the specimen chamber to laboratory air and possible contamination during sample exchange. A plasma cleaner that operates in the specimen airlock during the pump down cycle can greatly reduce specimen-related contamination by decomposing the hydrocarbons, provided the specimen itself is not damaged by the active oxygen plasma that is produced.

A typical observation of contamination is illustrated in ◘ Fig. 9.15a, where the SEM was first used to image an area at certain magnification and the magnification was subsequently reduced to scan a larger area. A "scan rectangle" is observed in the lower magnification image that corresponds to the area previously scanned at higher magnification. Within this scan rectangle, the SE coefficient has changed because of the deposition of a foreign material during electron bombardment, most likely a carbon-rich material which has a lower SE coefficient. Note that the contamination is most pronounced at the edge of the scanned field, where the beam is briefly held stationary before starting the next scanned line so that the greatest electron dose is applied along this edge.

"Etching," the opposite of contamination, can also occur (Hren 1986). An example is shown in ◘ Fig. 9.15b, where a bright scan rectangle is observed in an image of an aluminum stub after reducing the magnification following scanning for several minutes at higher magnification. In this case, the radiation damage has actually removed an overlayer of contamination on the specimen, revealing the underlying aluminum with its native oxide surface layer (~4 nm thick), which has an increased SE coefficient compared to the carbon-rich contamination layer.

Contamination is usually dose-dependent, so that the high dose necessary for high resolution microscopy, for example, a small scanned area (i.e., high magnification) with a high current density beam from a field emission gun, is likely to encounter contamination effects. This situation is illustrated

Fig. 9.15 **a** Contamination area observed after a higher magnification scan; Everhart–Thornley (positive bias). The extent of the contamination is visible upon lowering the magnification of the scan, thus increasing the scanned area. **b** Etching of a surface contamination layer observed during imaging of an aluminum stub; Everhart–Thornley (positive bias); 10 keV and 10 nA

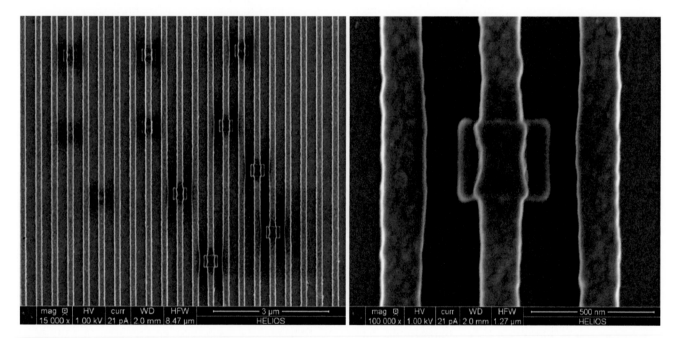

Fig. 9.16 Contamination observed during dimensional measurements performed under high resolution conditions on a patterned silicon substrate (Postek and Vladar 2014). Note broadening of the structure (*right*) due to contamination

in ◘ Fig. 9.16, which shows contamination in scanned areas on a patterned silicon sample used for dimensional metrology (Postek and Vladar 2014). The contamination in this case was so severe that it actually altered the apparent width of the measured features. To perform successful measurements, the authors developed an aggressive cleaning procedure that minimized contamination effects for this class of specimens. Their strategy may prove useful for other materials as well (Postek and Vladar 2014).

9.4   Moiré Effects: Imaging What Isn't Actually There

An SEM image appears to be continuous, but it is constructed as a regular repeating two-dimensional pattern of pixels. Thus, the viewer is effectively looking at the specimen through a two-dimensional periodic grid, and if the specimen itself has a structure that has a regularly repeating pattern, then a moiré pattern of fringes can form between

Fig. 9.17 Moiré fringe effects observed for the periodic structures in NIST RM 8820 (magnification calibration artifact). Note the different moiré patterns in the different calibration regions; Everhart–Thornley (positive bias) detector

the two patterns. The form of the moiré interference fringes depends on the spacing and orientation of the specimen periodic pattern and the scan pattern. Moiré patterns are maximized when the spatial frequencies of the two patterns are similar or an integer multiple of each other (i.e., they are commensurate). The formation of moiré patterns is illustrated in ◘ Fig. 9.17, which shows various etched patterns in the NIST RM 8820 magnification calibration artifact. The structures have different spacings in each of the fields viewed at the lowest magnification so that different moiré patterns are observed in each field. As the magnification is increased the scan field decreases in size so that the SEM pattern changes its periodicity (spatial frequency), causing

the moiré pattern to change. Finally, at sufficiently high magnification, the specimen periodic structure becomes sufficiently different from the scan pattern that the moiré fringes are lost.

Moiré effects can be very subtle. The periodic bright flares at fine edges, as seen in ◘ Fig. 9.18, are moiré patterns created when the fine scale structure approaches the periodicity of the scan grid.

To avoid interpreting moiré effects as real structures, the relative position and/or rotation of the specimen and the scan grid should be changed. A real structure will be preserved by such an action, while the moiré pattern will change.

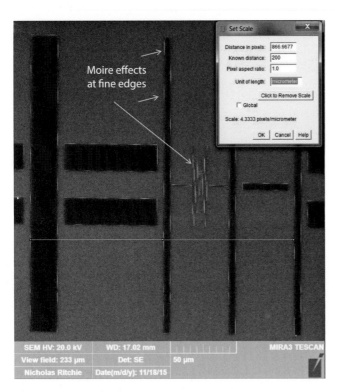

Fig. 9.18 Moiré effects seen as periodic bright flares at the edge of fine structures in NIST RM 8820 (magnification calibration artifact); Everhart–Thornley (positive bias) detector

References

Hren J (1986) Barriers to AEM: contamination and etching. In: Joy D, Romig A, Goldstein J (eds) Principles of analytical electron microscopy. Plenum, New York, p 353

Postek M, Vladar A (2014) Does your SEM really tell the truth? How would you know? Part 2. Scanning 36:347

9

High Resolution Imaging

© Springer Science+Business Media LLC 2018
J. Goldstein et al., *Scanning Electron Microscopy and X-Ray Microanalysis*,
https://doi.org/10.1007/978-1-4939-6676-9_10

10

10.1 What Is "High Resolution SEM Imaging"?

"I know high resolution when I see it, but sometimes it doesn't seem to be achievable!"

"High resolution SEM imaging" refers to the capability of discerning fine-scale spatial features of a specimen. Such features may be free-standing objects or structures embedded in a matrix. The definition of "fine-scale" depends on the application, which may involve sub-nanometer features in the most extreme cases. The most important factor determining the limit of spatial resolution is the footprint of the incident beam as it enters the specimen. Depending on the level of performance of the electron optics, the limiting beam diameter can be as small as 1 nm or even finer. However, the ultimate resolution performance is likely to be substantially poorer than the beam footprint and will be determined by one or more of several additional factors: (1) delocalization of the imaging signal, which consists of secondary electrons and/or backscattered electrons, due to the physics of the beam electron – specimen interactions; (2) constraints imposed on the beam size needed to satisfy the Threshold Equation to establish the visibility for the contrast produced by the features of interest; (3) mechanical stability of the SEM; (4) mechanical stability of the specimen mounting; (5) the vacuum environment and specimen cleanliness necessary to avoid contamination of the specimen; (6) degradation of the specimen due to radiation damage; and (7) stray electromagnetic fields in the SEM environment. Recognizing these factors and minimizing or eliminating their impact is critical to achieving optimum high resolution imaging performance. Because achieving satisfactory high resolution SEM often involves operating at the performance limit of the instrument as well as the technique, the experience may vary from one specimen type to another, with different limiting factors manifesting themselves in different situations. Most importantly, because of the limitations on feature visibility imposed by the Threshold Current/Contrast Equation, for a given choice of operating conditions, there will always be a level of feature contrast below which specimen features will not be visible. Thus, there is always a possible "now you see it, now you don't" experience lurking when we seek to operate at the limit of the SEM performance envelope.

10.2 Instrumentation Considerations

High resolution SEM requires that the instrument produce a finely focused, astigmatic beam, in the extreme 1 nm or less in diameter, that carries as much current as possible to maximize contrast visibility. This challenge has been solved by different vendors using a variety of electron optical designs. The electron sources most appropriate to high resolution work are (1) cold field emission, which produces the highest brightness among possible sources (e.g., ~10^9 A/(cm^{2}sr^{-1}) at $E_0 = 20$ keV) but which suffers from emission current

instability with a time constant of seconds to minutes and (2) Schottky thermally assisted field emission, which produces high brightness (e.g., ~10^8 A/(cm^{2}sr^{-1}) at $E_0 = 20$ keV) and high stability both over the short term (seconds to minutes) and long term (hours to days).

10.3 Pixel Size, Beam Footprint, and Delocalized Signals

The fundamental step in recording an SEM image is to create a picture element (pixel) by placing the focused beam at a fixed location on the specimen and collecting the signal(s) generated by the beam–specimen interaction over a specific dwell time. The pixel is the smallest unit of information that is recorded in the SEM image. The linear distance between adjacent pixels (the pixel pitch) is the length of edge of the area scanned on the specimen divided by the number of pixels along that edge. As the magnification is increased at fixed pixel number, the area scanned on the specimen decreases and the pixel pitch decreases. Each pixel represents a unique sampling of specimen features and properties, provided that the signal(s) collected is isolated within the area represented by that pixel. "Resolution" means the capability of distinguishing changes in specimen properties between contiguous pixels that represent a fine-scale feature against the adjacent background pixels or against pixels that represent other possibly similar nearby features. Resolution degrades when the signal(s) collected delocalizes out of the area represented by a pixel into the area represented by adjacent pixels so that the signal no longer exclusively samples the pixel of interest. Signal delocalization has two consequences, the loss of spatial specificity and the diminution of the feature contrast, which affects visibility. Thus, when the lateral leakage becomes sufficiently large, the observer will perceive blurring, and less obviously the feature contrast will diminish, possibly falling below the threshold of visibility.

How closely spaced are adjacent pixels of an image? ▣ Table 10.1 lists the distance between pixels as a function of the nominal magnification (relative to a 10 x 10-cm display) for a 1000 x 1000 pixel scan. For low magnifications, for example, less than a nominal value of 100×, the large scan fields result in pixel-to-pixel distances that are large enough (pixel pitch >1 μm) to contain nearly all of the possible information-carrying backscattered electrons (BSE) and secondary electrons (SE$_1$, SE$_2$, and SE$_3$) that result from the beam electron–specimen interactions, despite the lateral delocalization that occurs within the interaction volume for the BSE (SE$_3$) and SE$_2$ signals.

▣ Table 10.1 reveals that the footprint of a 1-nm focused beam will fit inside a single pixel up to a nominal magnification of 100,000×. However, as discussed in the "Electron Beam–Specimen Interactions" module, the BSE and the SE$_2$ and SE$_3$ signals, which are created by the BSE and carry the same spatial information, are subject to substantial lateral delocalization because of the scattering of the beam electrons giving rise to the beam interaction volume, which is beam

◻ Table 10.1 Relationship between nominal magnification and pixel dimension

Nominal magnification (10 × 10-cm display)	Edge of scanned area (μm)	Pixel pitch (1000 × 1000-pixel scan)
40×	2500	2.5 μm
100×	1000	1 μm
200×	500	500 nm
400×	250	250 nm
1000×	100	100 nm
2000×	50	50 nm
4000×	25	25 nm
10,000×	10	10 nm
20,000×	5	5 nm
40,000×	2.5	2.5 nm
100,000×	1	1 nm
200,000×	0.5	500 pm
400,000×	0.25	250 pm
1,000,000×	0.1	100 pm

◻ Table 10.2 Diameter of the area at the surface from which 90 % of BSE (SE$_3$) and SE$_2$ emerge

E_0	C	Cu	Au
30 keV	11.8 μm	2.6 μm	1.2 μm
20 keV	6.0 μm	1.4 μm	590 nm
10 keV	1.9 μm	410 nm	180 nm
5 keV	590 nm	130 nm	58 nm
2 keV	128 nm	28 nm	12 nm
1 keV	41 nm	8.8 nm	3.9 nm
0.5 keV	12.7 nm	2.8 nm	1.2 nm
0.25 keV	4.0 nm	0.9 nm	0.39 nm
0.1 keV	0.86 nm	0.19 nm	0.08 nm

energy and composition dependent. ◻ Table 10.2 gives the diameter of the footprint of the area that contains 90 % of the BSE, SE$_2$, and SE$_3$ emission, which is compositionally dependent, as calculated from the cumulative radial spreading plotted in ◻ Fig. 2.14. The radial spreading is surprisingly large when compared to the distance between pixels in ◻ Table 10.1. For a beam energy of 10 keV, the BSE (SE$_3$) and SE$_2$ signals will delocalize out of a single pixel at very low magnifications, approximately 40× for C, 200× for Cu, and 1000× for Au. Even allowing for the fact that the average observer viewing an SEM

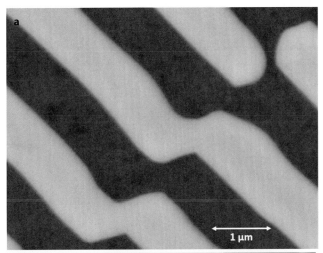

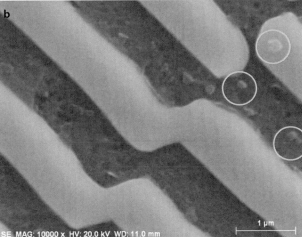

◻ Fig. 10.1 Aluminum-copper eutectic alloy, directionally solidified. The phases are CuAl$_2$ and an Al(Cu) solid solution. Beam energy = 20 keV. **a** Two-segment semiconductor BSE detector, sum mode (A + B). **b** Everhart–Thornley detector(positive bias)

image prepared with a high pixel density scan will only perceive blurring when several pixels effectively overlap, these are surprisingly modest magnification values. Considering that high resolution SEM performance is routinely expected and is apparently delivered, this begs the question: Is such poor resolution actually encountered in practice and why does it not prevent useful high resolution applications of the SEM? ◻ Figure 10.1a shows an example of degraded resolution observed in BSE imaging at $E_0 = 20$ keV of what should be nearly atomically sharp interfaces in directionally solidified Al-Cu eutectic. This material contains repeated interfaces (which were carefully aligned to be parallel to the incident beam) between the two phases of the eutectic, CuAl$_2$ intermetallic, and Al(Cu) solid solution. A similar image is shown in ◻ Fig. 2.14 with a plot of the BSE signal (recorded with a large solid angle semiconductor detector) across the interface. The BSE signal changes over approximately 300 nm rather than being limited by the beam size, which is approximately 5 nm for this image. The same area is imaged with the Everhart–Thornley detector(positive bias) in ◻ Fig. 10.1b and shows finer-scale details, that is, "better resolution." The positively

10

biased Everhart–Thornley (E–T) detector collects a complex mixture of BSE and SE signals, including a large BSE component (Oatley 1972). The BSE component consists of a relatively small contribution from the BSEs that directly strike the scintillator (because of its small solid angle) but this direct BSE component is augmented by a much larger contribution of indirectly collected BSEs from the relatively abundant SE_2 (produced as all BSEs exit the specimen surface) and SE_3 (created when the BSEs strike the objective lens pole piece and specimen chamber walls). For an intermediate atomic number target such as copper, the SE_2 class created as the BSEs emerge constitutes about 45 % of the total SE signal collected by the E–T(positive bias) detector (Peters 1984, 1985). The SE_3 class from BSE-to-SE conversion at the objective lens pole piece and specimen chamber walls constitutes about 40 % of the total SE intensity. The SE_2 and SE_3, constituting 85 % of the total SE signal, respond to BSE number effects and create most of the atomic number contrast seen in the E–T(positive bias) image. However, the SE_2 and SE_3 are subject to the same lateral delocalization suffered by the BSEs and result in a similar loss of edge resolution. Fortunately for achieving useful high resolution SEM, the E–T (positive bias) detector also collects the SE_1 component (about 15 % of the total SE signal for copper) which is emitted from the footprint of the incident beam. The SE_1 signal component thus retains high resolution spatial information on the scale of the beam, and that information is superimposed on the lower resolution spatial information carried by the BSE, SE_2, and SE_3 signals. Careful inspection of ◘ Fig. 10.1b reveals several examples of discrete fine particles which appear in much sharper focus than the boundaries of the Al-Cu eutectic phases. These particles are distinguished by

bright edges and uniform interiors and are due in part to the dominance of the SE_1 component that occurs at the edges of structures but which are lost in the pure BSE image of ◘ Fig. 10.1a.

10.4 Secondary Electron Contrast at High Spatial Resolution

The secondary electron coefficient responds to changes in the local inclination (topography) of the specimen approximately following a secant function:

$$\delta(\theta) = \delta_0 \sec\theta \qquad (10.1)$$

where δ_0 is the secondary electron coefficient at normal beam incidence, i.e., $\theta = 0°$. The contrast between two surfaces at different tilts can be estimated by taking the derivative of Eq. 10.1:

$$d\delta(\theta) = \delta_0 \sec\theta \tan\theta \, d\theta \qquad (10.2)$$

The contrast for a small change in tilt angle $d\theta$ is then

$$C \sim d\delta(\theta)/\delta(\theta) = \delta_0 \sec\theta \tan\theta \, d\theta / \delta_0 \sec\theta$$
$$= \tan\theta \, d\theta \qquad (10.3)$$

As the local tilt angle increases, the contrast between two adjacent planar surfaces with a small difference in tilt angle, $d\theta$, increases as the average tilt angle, θ, increases, as shown in ◘ Fig. 10.2 for surfaces with a difference in tilt of $d\theta = 1°, 5°$

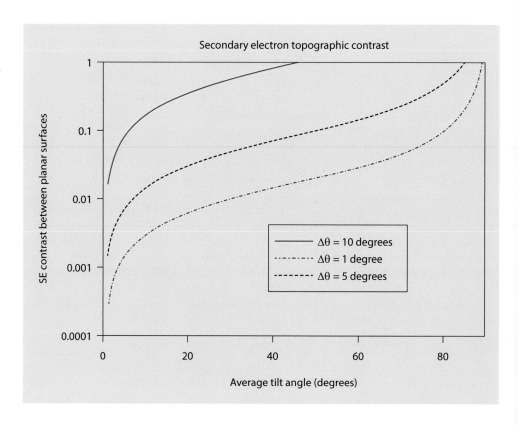

◘ **Fig. 10.2** Plot of secondary electron topographic contrast between two flat surfaces with a difference in tilt angle of 1°, 5°, and 10°

and 10°. Superimposed on this broad scale secondary electron topographic contrast are strong sources of contrast associated with situations where the range of SEs dominates leading to enhanced SE escape:

1. When the beam strikes nearly tangentially, that is, grazing incidence when θ approaches 90° and sec θ reaches very high values, as the beam travels near the surface and a high SE signal is produced, an effect that is seen in the calculated contrast at high tilt angles in ◘ Figs. 10.2 and 10.3 show an example of a group of particles imaged with an E–T(positive bias) detector. High SE signals occur where the beam strikes the edges of the particles at grazing incidence, compared to the interior of the particles where the incidence angle is more nearly normal.

2. At feature edges, especially edges that are thin compared to the primary electron range. These mechanisms result in a very noticeable "bright edge effect."

10.4.1 SE Range Effects Produce Bright Edges (Isolated Edges)

Because of their extremely low kinetic energy of a few kiloelectronvolts, SEs have a short range of travel in a solid and thus can only escape from a shallow depth. The mean escape depth (SE range) is approximately 10 nm for a conductor. When the beam is located in bulk material well away from edges, as shown schematically in ◘ Fig. 10.4, the surface area from which SEs can escape is effectively constant as the beam is scanned, and the SE emission (SE_1, SE_2, and SE_2) is thus constant and equal to the bulk SE coefficient appropriate to the target material at the local inclination angle. However, when the beam approaches an edge of a feature, such as the vertical wall shown in ◘ Fig. 10.4, the escape of SEs is enhanced by the proximity of additional surface area that lies within the SE escape range. As the incident beam travels nearly parallel to the vertical face, the proximity of the surface along an extended portion of the beam path further enhances the escape of SEs, resulting in a very large

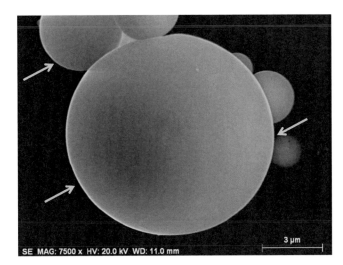

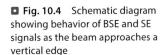

SE MAG: 7500 x HV: 20.0 kV WD: 11.0 mm

◘ **Fig. 10.3**   SEM image of SRM 470 (Glass K -411) micro-particles prepared with an Everhart–Thornley detector(positive bias) and $E_0 = 20$ keV. Note bright edges where the beam strikes tangentially

◘ **Fig. 10.4** Schematic diagram showing behavior of BSE and SE signals as the beam approaches a vertical edge

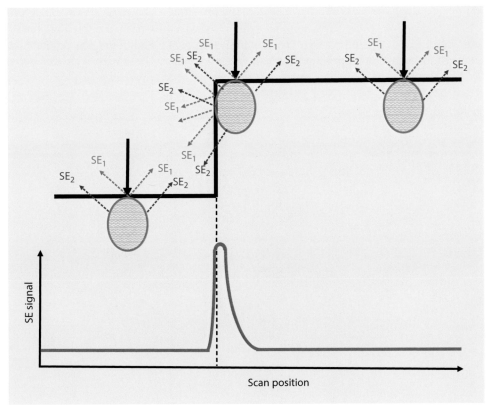

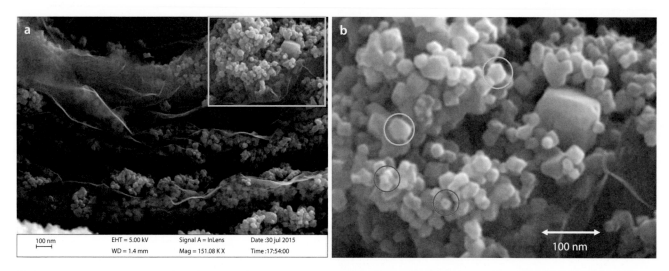

Fig. 10.5   a SEM image at $E_0 = 5$ keV of TiO_2 particles using a through-the-lens detector for SE_1 and SE_2 (Bar = 100 nm). **b** Note bright edge effects and convergence of bright edges for the smallest particles (Example courtesy John Notte, Zeiss)

excess of SEs compared to the bulk interior. In addition, there will be enhanced escape of BSEs near the edge, and these BSEs will likely strike other nearby specimen and instrument surfaces, producing even more SEs. All of the signals collected when the beam is placed at a given scan location are assigned to that location in the image no matter where on the specimen or SEM chamber those signals are generated. The apparent SE emission coefficient when the beam is placed near an edge is thus greatly increased over the bulk interior value, often by a factor of two to ten depending on the exact geometric circumstances. The edges of an object will appear very bright relative to the interior of the object, as shown in ■ Fig. 10.5 (e.g., objects in yellow circles in ■ Fig. 10.5b) for particles of TiO_2. Since the edges are often the most important factor in defining a feature, a contrast mechanism that produces such an enhanced edge

signal compared to bulk is a significant advantage. This is especially true considering the limitations that are imposed on high resolution performance by the demands of the Threshold Current/Contrast Equation, as discussed below.

10.4.2   Even More Localized Signal: Edges Which Are Thin Relative to the Beam Range

The enhanced SE escape near an edge shown in ■ Fig. 10.4 is further increased when the beam approaches a feature edge that is thin enough for penetration of the beam electrons. As shown schematically in ■ Fig. 10.6, not only are additional SEs generated as the beam electrons emerge as "BSEs" through the bottom and sides of the thin edge

Fig. 10.6   Schematic diagram of the enhanced BSE and SE production at an edge thin enough for beam penetration. BSEs may strike multiple surfaces, creating several generations of SEs

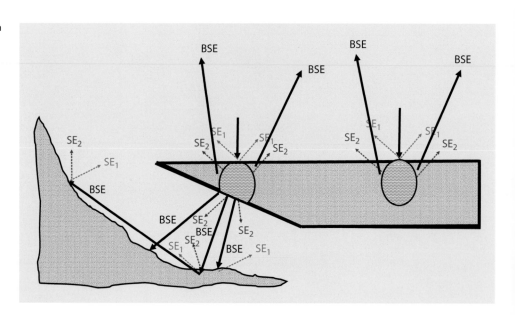

structure, but these energetic BSEs will continue to travel, backscattering off other nearby specimen surfaces and the SEM lens and chamber walls, producing additional generations of SEs at each surface they strike. These additional SEs will be collected with significant efficiency by the E–T (positive bias) detector and assigned to each pixel as the beam approaches the edge, further increasing the signal at a thin edge relative to the interior and thus increasing the contrast of edges.

10.4.3 Too Much of a Good Thing: The Bright Edge Effect Can Hinder Distinguishing Shape

As the dimensions of a free-standing object such as a particle or the diameter of a fiber approach the secondary electron escape length, the bright edge effects from two or more edges will converge, as shown schematically in ◘ Fig. 10.7 and in the image of TiO2 particles (e.g., objects in magenta circles)

◘ **Fig. 10.7** Convergence of bright edges as feature dimensions approach the SE escape distance. **a** Object edges separated by several multiples of the SE escape distance so that edge effects are distinct; **b** object edges sufficiently close for edge effects to begin to merge

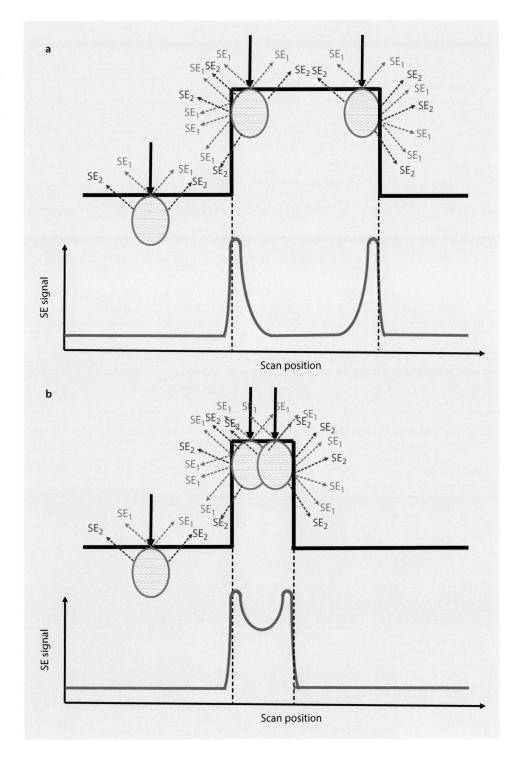

shown in ■ Fig. 10.5b. While the object will appear in high contrast as a very bright feature against the background, making it relatively easy to detect, as an object decreases in size it becomes difficult and eventually impossible to discern the true shape of an equiaxed object and to accurately measure its dimensions.

10.4.4 Too Much of a Good Thing: The Bright Edge Effect Hinders Locating the True Position of an Edge for Critical Dimension Metrology

While the enhanced SE escape at an edge is a great advantage in visualizing the presence of an edge, the extreme signal excursion and its rapid change with beam position make it difficult to locate the absolute position of the edge within the SE range, which can span 10 nm or even more for insulating materials. For advanced metrology applications such as semiconductor manufacturing critical dimension measurements where nanometer to sub-nanometer accuracy is required, detailed Monte Carlo modeling, as shown in ■ Fig. 10.8, of the beam electron, backscattered electron, and secondary electron trajectories as influenced by the specific geometry of the edge, is needed to deconvolve the measured signal profile as a function of scan position so as to recover the best estimate of the true edge location and object shape (NIST JMONSEL: Villarrubia et al. 2015). An example of an SEM signal profile across a structure and the shape recovered after deconvolution through modeling is shown in ■ Fig. 10.9. An application of this approach is shown in ■ Fig. 10.10, where a three-dimensional photoresist line was first imaged in a top-down SEM view (■ Fig. 10.10a). Monte Carlo modeling applied to the signal profiles obtained from the top-down

view enabled a best fit estimate of the shape and dimensions of the line. The structure was subsequently cross-sectioned by ion beam milling to produce the SEM view shown in ■ Fig. 10.10b. The best estimate of the structure obtained from the top-down imaging and modeling (red trace) is shown superimposed on the direct image of the cross-section edges (blue trace), showing excellent correspondence with this approach.

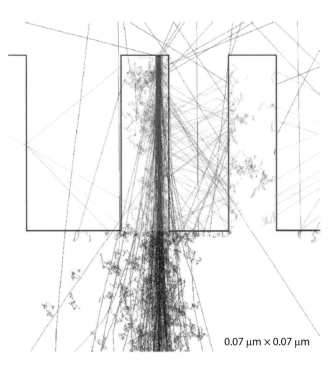

0.07 μm × 0.07 μm

■ **Fig. 10.8** Monte Carlo electron trajectory simulation of complex interactions at line-width structures as calculated with the J-MONSEL code (Villarrubia et al. 2015)

■ **Fig. 10.9** Application of J-MONSEL Monte Carlo simulation to measured SEM profile data and the estimated shape that best fits the data (Villarrubia et al. 2015)

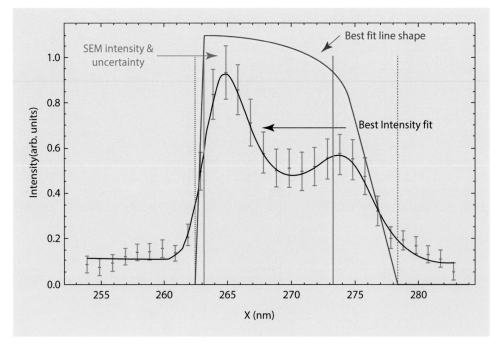

Fig. 10.10 **a** Top-down SEM image of line-width test structures; $E_0 = 15$ keV. **b** Side view of structures revealed by focused ion beam milling showing estimated shape from modeling of the top-down image (*red trace*) compared with the edges directly found in the cross sectional image (*blue*) (Villarrubia et al. 2015)

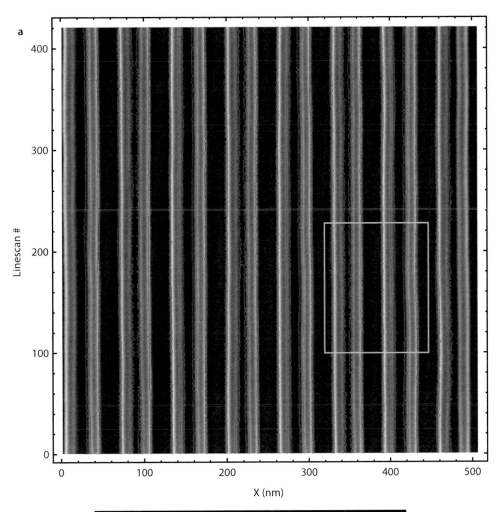

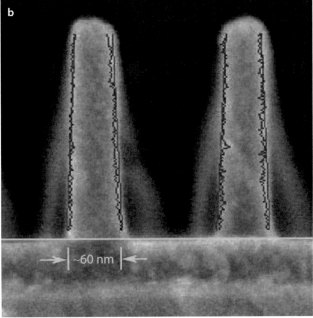

10.5   Achieving High Resolution with Secondary Electrons

Type 1 secondary electrons (SE_1), which are generated within the footprint of the incident beam and from the SE escape depth of a few nanometers, constitute an inherently high spatial resolution signal. SE_1 are capable of responding to specimen properties with lateral dimensions equal to the beam size as it is made progressively finer. Unfortunately, with the conventional Everhart–Thornley (positive bias) detector, the SE_1 are difficult to distinguish from the SE_2 and SE_3 signals which are created by the emerging BSEs, which effectively carry BSE information, and which are thus subject to the same long range spatial delocalization as BSEs. Strategies to improve high resolution imaging with SEs seek to modify the spatial characteristics and/or relative abundance of the SE_2 and SE_3 compared to the SE_1.

10.5.1   Beam Energy Strategies

◘ Figure 10.11a shows schematically the narrow spatial distribution of the SE_1 emitted from a finely focused beam superimposed on the broader spatial distribution of the of the SE_2 and SE_3 that are created from the BSE distribution that would arise from a beam of intermediate energy, for example, 10 keV, on a material of intermediate atomic number, for example, Cu. While the beam can be focused to progressively smaller sizes within the limitations of the electron-optical system and the SE_1 will follow the beam footprint as it is reduced, the BSE-SE_2-SE_3 distributions remain at a fixed size defined by the extent of the interaction volume, which depends primarily on the composition and the beam energy and is insensitive to small beam size. For the situation shown in ◘ Fig. 10.11a, the SE_1 distribution can interact over a short spatial range with a feature that has dimensions similar to the focused beam footprint, but the extended BSE-SE_2-SE_3 distribution interacts with this feature over a longer range. The BSE-SE_2-SE_3 create a long, gradually decreasing signal tail, so that a sharp feature appears blurred. There are two different strategies for improving the resolution by choosing the beam energy at the extreme limits of the SEM range.

Low Beam Energy Strategy

As the beam energy is lowered, the electron range decreases rapidly, varying approximately as $E_0^{1.67}$. Since the BSE-SE_2-SE_3 distributions scale with the range, when the beam

◘ **Fig. 10.11**   **a** Schematic diagram of the SE_1 and SE_2 spatial distributions for an intermediate beam energy, e.g., $E_0 = 5$–10 keV. **b** Schematic diagram of the SE_1 and SE_2 spatial distributions for low beam energy, e.g., $E_0 = 1$ keV. **c** Schematic diagram of the SE_1 and SE_2 spatial distributions for high beam energy, e.g., $E_0 = 30$ keV

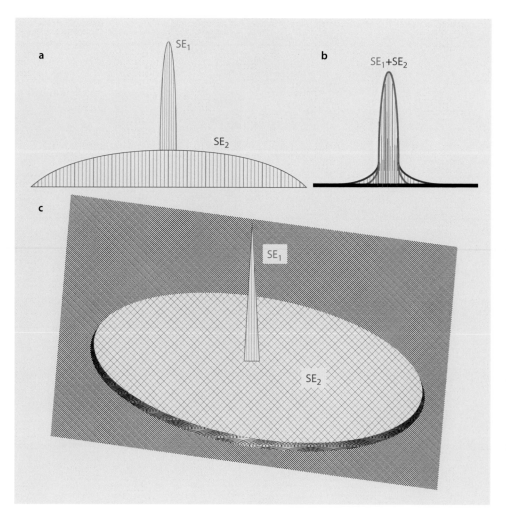

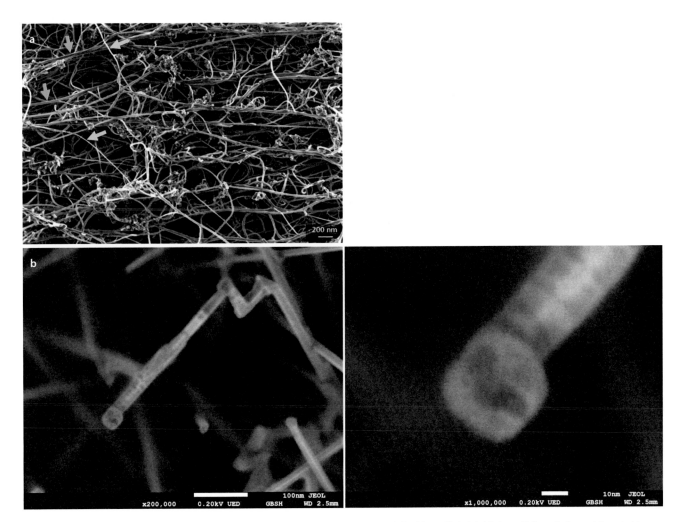

Fig. 10.12 **a** High resolution achieved at low beam energy, $E_0 = 1$ keV: image of carbon nanofibers. Note broad fibers (*cyan arrows*) with bright edges and darker interiors and thin fibers (*yellow arrows*) for which the bright edge effects converge (Bar = 200 nm) (Example courtesy John Notte, Zeiss). **b** High spatial resolution achieved at low landing energy: SnO$_2$ whisker imaged with a landing energy of 0.2 keV (left, Bar = 100 nm) (right, Bar = 10 nm) (Images courtesy V. Robertson, JEOL)

energy is reduced so that $E_0 \leq 2$ keV, the situation illustrated in ◻ Fig. 10.11b is reached for a finely focused beam (Joy 1984; Pawley 1984). The BSE-SE$_2$-SE$_3$ distributions collapse onto the SE$_1$ distribution, and all the signals now represent high spatial resolution information. An example of carbon nanofibers imaged at $E_0 = 1$ keV to achieve high resolution is shown in ◻ Fig. 10.12a. In ◻ Fig. 10.12a, the edges of the wider fibers appear bright (e.g., blue arrows) relative to the interior, as shown schematically in ◻ Fig. 10.7a. ◻ Figure 10.12 also illustrates the convergence of the bright edges of the narrow fibers (e.g., yellow arrows), as illustrated in ◻ Fig. 10.12b, to produce a very bright object against the background.

By applying a negative potential to the specimen, the landing energy can be reduced even further while preserving high spatial resolution, as shown in ◻ Fig. 10.12b for tin oxide whiskers imaged with a TTL SE detector at a landing energy of $E_0 = 0.2$ keV.

There are limitations of low beam energy operation that must be acknowledged (Pawley 1984). An inevitable consequence of low beam energy operation is the linear reduction in source brightness, which reduces the current that is contained in the focused probe which in turn affects feature visibility. Low energy beams are also more susceptible to interference from outside sources of electromagnetic radiation.

High Beam Energy Strategy

As the beam energy is increased, the electron range increases rapidly as $E_0^{1.67}$, broadening the spatial distribution of the BSE-SE$_2$-SE$_3$ signals while the SE$_1$ distribution remains fixed to the beam footprint. For example, when the beam energy is increased from 10 to 30 keV, the range increases by a factor of 6.3. With sufficient broadening, the spatial distributions of the BSE-SE$_2$-SE$_3$ signals do not significantly respond during beam scanning to small-scale features to which the SE$_1$ are sensitive. The BSE-SE$_2$-SE$_3$ signals then represent

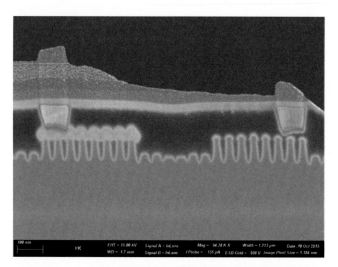

Fig. 10.13 High resolution achieved at high beam energy, $E_0 = 15$ keV: finFET transistor (16-nm technology) using the in lens SE detector in the Zeiss Auriga Cross beam. This cross section was prepared by inverted Ga FIB milling from backside (Bar = 100 nm) (Image courtesy of John Notte, Carl Zeiss)

a background noise component that, while it reduces the overall signal-to-noise, does not significantly alter the signal profiles across features. An advantage of operating at high beam energy is that the source brightness is increased, thus enabling more current to be obtained in a given focused probe size, which can help to compensate for the reduced signal-to-noise caused by the remote BSE-SE_2-SE_3 signals. An example of high beam energy imaging to achieve high resolution is shown in **Fig. 10.13**.

10.5.2 Improving the SE_1 Signal

Since the SE_1 Signal Is So Critical To Achieving High Resolution, What Can Be Done To Improve It?

Excluding the SE_3 Component

For a bulk specimen, the high resolution SE_1 component only forms 5–20 % of the total SE signal collected by the E–T (positive bias) detector, while the lower resolution SE_2 and SE_3 components of roughly similar strength form the majority of the SE signal. While the SE_1 and SE_2 components are generated within 1 to 10 μm, the SE_3 are produced millimeters to centimeters away from the specimen when the BSEs strike instrument components. This substantial physical separation is exploited by the class of "through-the-lens" (TTL) detectors, which utilize the strong magnetic field of the objective lens to capture the SE_1 and SE_2 which travel up the bore of the lens and are accelerated onto a scintillator-photomultiplier detector. Virtually all of the SE_3 are excluded by their points of origin being outside of the lens magnetic field. For an SE_1 component of 10 % and SE_2 and SE_3 components of 45 %, the ratio of high resolution/low resolution signals thus changes from 0.1 for the E–T (positive bias) detector to 0.22 for the TTL detector.

Making More SE_1: Apply a Thin High-δ Metal Coating

Because SEs are generated within a thin surface layer, the SE coefficient δ of the first few atomic layers will dominate the SE emission of the specimen. For specimens that consist of elements such as carbon with a low value of δ, the SE_1 signal can be increased by applying a thin coating (one to a few nanometers) of a high SE emitter such as gold-palladium (rather than pure gold, which deposits as islands that can be mistaken for specimen structure), or platinum-family metals. While such a coating can also serve to dissipate charging from an insulating specimen, even for conducting carbonaceous materials the heavy-metal coating increases the surface SE_1 emission of the specimen while not significantly increasing the scattering of beam electrons due to its minimal thickness so that BSE, SE_2, and SE_3 signals are not affected. As shown schematically in **Fig. 10.14a**, the SE signal across an uncoated particle shows an increase at the edge due to the grazing beam incidence, but after a thin high-δ metal coating

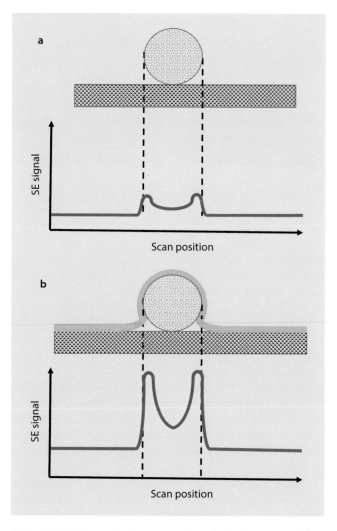

Fig. 10.14 Schematic illustration of the effect of heavy metal, high δ coating to increase contrast from low-Z targets: **a** SE signal trace from an uncoated particle; **b** signal trace after coating with thin Au-Pd

is applied (▪ Fig. 10.14b), the SE signal at the edges of features will be substantially enhanced.

Making Fewer BSEs, SE$_2$, and SE$_3$ by Eliminating Bulk Scattering From the Substrate

For the important class of specimens such as nanoscale particles which have such small mass thickness that the beam electrons penetrate through the particle into the underlying bulk substrate, the large BSE, SE$_2$ and SE$_3$ components that dominates the E–T(positive bias) signal respond to substrate properties and don't actually represent specimen information at all. Thus, the high resolution imaging situation can be significantly improved by eliminating the bulk substrate. The particles are deposited on an ultrathin (~10-nm) carbon film supported on a metal (Cu, Ni, etc.) grid. This grid is placed over a deep blind hole drilled in a block of carbon that will serve as a Faraday cup for the beam electrons that pass though the particles, as shown schematically in ▪ Fig. 10.15a. An example of this preparation is shown for particles of SRM470

▪ **Fig. 10.15   a** Schematic illustration of specimen mounting strategy to minimize background by eliminating the bulk substrate. **b** Scanning transmission electron microscopy (STEM) two component detector for high energy electrons: on-axis bright-field detector and surrounding annular dark-field detector

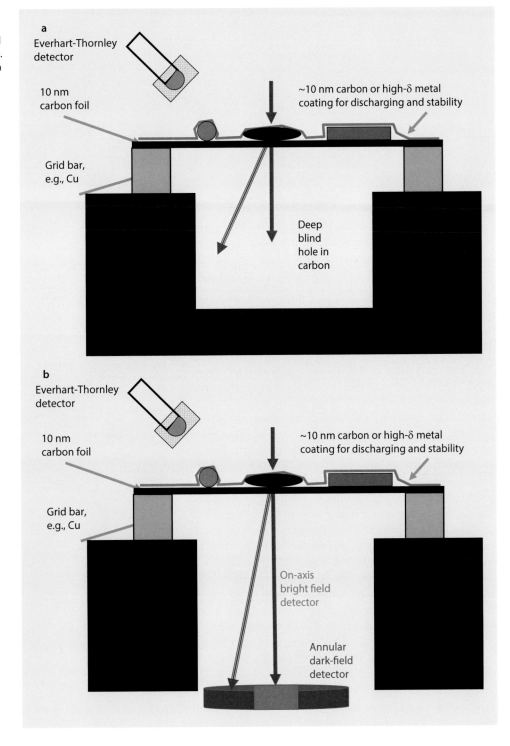

K-309 particle shards on thin carbon

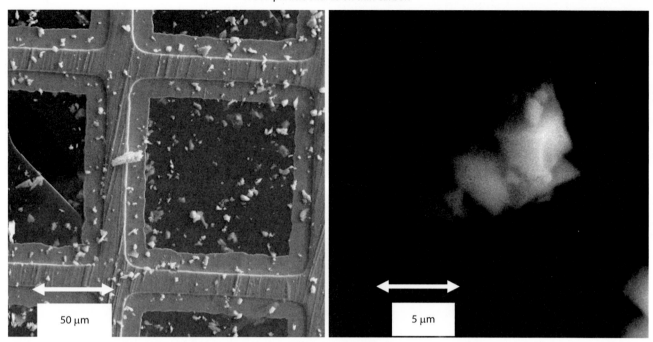

Conventional Everhart-Thornley (+bias) detector above specimen

◨ **Fig. 10.16**   SEM imaging glass shards deposited on a thin (~ 10-nm carbon) at $E_0 = 20$ keV and placed over a deep blind hole in a carbon block

(K411 glass) in ◨ Fig. 10.16. By selecting operation at the highest beam energy available, for example, 20–30 keV, backscattering will be minimized along with the SE_2 and SE_3 signals.

Scanning Transmission Electron Microscopy in the Scanning Electron Microscope (STEM-in-SEM)

The "thin film" support method for nanoscale particles and other thin specimens (either inherently thin or prepared as thin sections by ion beam milling) can be further exploited by collecting the beam electrons that transmit through the specimen to create a scanning transmission electron microscope (STEM) image, as illustrated in ◨ Fig. 10.15b. To create the STEM image, an appropriate detector, such as a passive scintillator-photomultiplier, is placed below the specimen grid on the optical axis. The size of this detector is such that it accepts only electrons traveling close to the optical axis that pass through the specimen unscattered. Those electrons that experience even a small angle elastic scattering event that causes an angular deviation of a few degrees will miss the detector. Thus, the regions of the specimen with minimal scattering will appear bright, while those with sufficient mass to cause elastic scattering will appear dark, creating a "bright-field" image. A more elaborate STEM detector array can include an annular ring detector co-mounted with the central on-axis bright-field detector to capture the elastically scattered transmitted electrons from the specimen, as illustrated in ◨ Fig. 10.15b. This off-axis annular detector produces a "dark-field" image since the thin regions such as the support film that do not produce significant scattering events

will appear dark. Portions of the specimen that do scatter sufficiently will appear bright. Since elastic scattering depends strongly on local atomic number, compositional effects can be observed in the dark field STEM image. An example of a high resolution STEM-in-SEM image created with an annular off-axis detector is shown in ◨ Fig. 10.17.

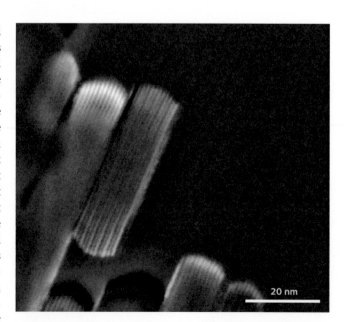

◨ **Fig. 10.17**   Dark-field annular detector STEM image of $BaFe_{12}O_{19}$ nanoparticles; $E_0 = 22$ keV using oriented dark-field detector in the Zeiss Gemini SEM. The 1.1-nm (002) lattice spacing is clearly evident (Image courtesy of John Notte, Carl Zeiss. Image processed with ImageJ-Fiji CLAHE function)

■ **Fig. 10.18** Schematic cross section of a STEM-in-SEM detector that makes use of the Everhart–Thornley(positive bias) detector to form a bright-field STEM image

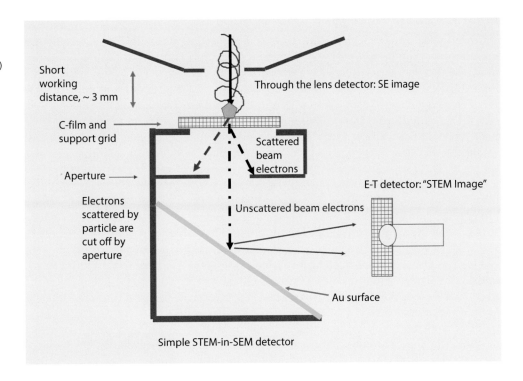

Aerosol particles collected on lacey carbon
25 keV cold-FEG-SEM

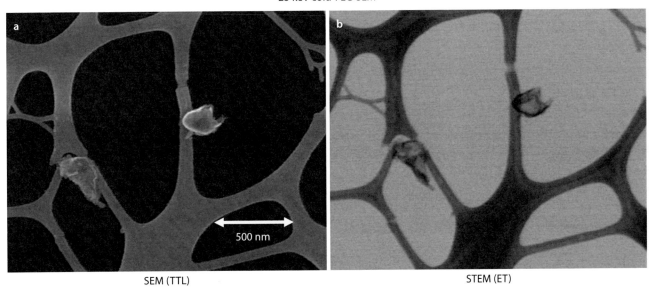

SEM (TTL)   STEM (ET)

■ **Fig. 10.19** Aerosol contamination particles deposited on lacey-carbon film and simultaneously imaged with a TTL detector for SE_1 and the STEM-in-SEM detector shown in ■ Fig. 10.18 (Example courtesy John Small, NIST)

A simple STEM-in-SEM bright-field detector can be created as shown in ■ Fig. 10.18. The grid carrying the thin specimen is placed over an aperture that serves to stop electrons that have suffered an elastic scattering event in the specimen. The unscattered beam electrons pass through this aperture and strike a gold-covered surface below, where they generate strong SE emission, which is then attracted to the E–T(positive bias) detector, creating a bright-field image. If the SEM is also equipped with a TTL detector, the nearly pure SE_1 image that arises from a thin specimen can be collected with the TTL detector simultaneously with the bright-field STEM image collected with the E–T(positive bias) detector, as shown for particles supported on a lacey-carbon film in ■ Fig. 10.19.

10.5.3 Eliminate the Use of SEs Altogether: "Low Loss BSEs"

BSEs are usually considered a low resolution signal because of the substantial delocalization that results from multiple elastic scattering of the beam electrons at conventional beam

10

energy, for example, $E_0 \geq 10$ keV. However, high resolution SEM can be achieved by eliminating the use of SEs as the imaging signal and instead relying on the BSEs, specifically those that have lost very little of the initial beam energy. Because of the energy loss due to inelastic scattering that occurs for high energy beam electrons at a nearly constant rate, dE/ds, with distance traveled in the specimen, low loss BSEs represent beam electrons that have emerged from the specimen after traveling very short paths through the specimen. These low loss electrons are thus sensitive to specimen scattering properties very close to the entrance beam footprint and from a very shallow surface region, thus constituting a high resolution signal. Wells (1974a, b) first demonstrated the utility of this approach by using an energy filter to select the "low loss" backscattered electrons (LL BSE) that had lost less than a specified fraction, for example, 5%, of the initial beam energy. At normal beam incidence, the LL BSE fraction of the total BSE population is very low, and their trajectories are spread over a wide angular range, the 2π azimuth around the beam, making their efficient collection difficult. The population of LL BSE can be increased, and their angular spread greatly decreased, by tilting the specimen to a high angle, for example, 70° or higher. As shown schematically in ◻ Fig. 10.20, at this tilt angle a single elastic

scattering event greater than 20°, which also has a suitable azimuthal angular component along the trajectory, can carry the beam electron out of the specimen as a low loss BSE after traveling along a short path within the specimen. The energy filter with an applied potential V + ΔV then serves to decelerate and exclude BSEs that have lost more than a specified ΔE of the incident energy. Since the electrons that pass through the filter have been retarded to a low kinetic energy, the detector following the filter must include an acceleration field, such as that of the Everhart–Thornley detector, to raise the kinetic energy to a detectable level for detection.

An example comparing TTL SE and LL BSE (10% energy window) images of etched photoresist at low beam energy ($E_0 = 2$ keV) is shown in ◻ Fig. 10.21 (Postek et al. 2001). Note the enhanced surface detail visible on the top of the resist pattern in the LL BSE image compared to the SE image. The extreme directionality of the LL BSE detector leads to loss of signal on surfaces not tilted toward the detector, resulting in

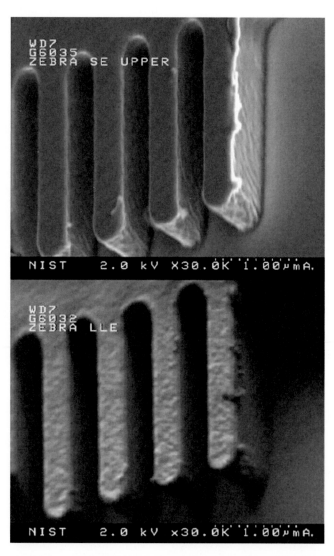

◻ **Fig. 10.20** Schematic illustration of low loss BSE imaging from a highly tilted specimen using an energy filter

◻ **Fig. 10.21** SE (*upper*) and low loss BSE (*lower*) images of photoresist at $E_0 = 2$ keV. Note the enhanced detail visible on the surface of the LL-BSE image compared to the SE image (Postek et al. 2001)

poor signal collection on the sides of the steps, which are illuminated in the TTL SE detector image.

While the example in ◘ Fig. 10.21 illustrates the utility of LL BSE imaging at low beam energy, LL BSE imaging also enables operation of the SEM at high beam energy (Wells 1971), thus maximizing the electron gun brightness to enable a small beam with maximum current. Low loss images provide both high lateral spatial resolution and a shallow sampling depth.

10.6 Factors That Hinder Achieving High Resolution

10.6.1 Achieving Visibility: The Threshold Contrast

High resolution SEM involves working with a finely focused beam which even when optimized to minimize the effects of aberrations inevitably carries a small current, often as low as a few picoamperes, because of the restrictions imposed by the Brightness Equation. The inevitable consequence of operating with low beam current is the problem of establishing the visibility of the features of interest because of the restrictions imposed by the Threshold Equation. For a given selection of operating parameters, including beam current, detector solid angle, signal conversion efficiency, and pixel dwell time, there is always a threshold of detectable contrast. Features producing contrast below this threshold contrast will not be visible at the pixel density selected for the scan, even with post-processing of the image with various advanced image manipulation algorithms. It is important to understand that a major consequence of the Threshold Equation is that the absence of a feature in an SEM image is not a guarantee of the absence of that feature on the specimen: the feature may not be producing sufficient contrast to exceed the threshold contrast for the particular operating conditions chosen. Because of the action of the "bright edge effect" in high resolution SE images to produce very high contrast, approaching unity, between the edges of a feature and its interior, the ready visibility of the edges of features, while obviously useful and important, can give a false sense of security with regard to the absence of topographic details within the bulk of a feature. In fact, those weaker topographic features may be producing contrast that is below the threshold of visibility. To perform "due diligence" and explore the possibility of features lurking below the threshold of visibility, the threshold contrast must be lowered:

$$i_{\mathrm{B}} > \frac{\left(4 \times 10^{-18}\right) N_{PE}}{(\eta,\delta)\, DQEC^2 t_F} \left(\mathrm{coulomb}/s = \mathrm{amperes}\right) \quad \textbf{(10.4)}$$

where N_{PE} is the number of pixels in the image scan, η and δ are the backscatter or secondary electron coefficients as

appropriate to the signal selected, DQE is the detective quantum efficiency, which includes the solid angle of collection for the electrons of interest and the conversion into detected signal, C is the contrast that the feature produces, and t_{F} is the frame time. Equation 10.4 reveals the constraints the microscopist faces: if the beam current is determined by the requirement to maintain a certain beam size and the detector has been optimized for the signal(s) that the features of interest are likely to produce, then the only factor remaining to manipulate to lower the threshold contrast is to extend the dwell time per pixel ($t_{\mathrm{F}}/N_{\mathrm{PE}}$). While using longer pixel dwell times is certainly an important strategy that should be exploited, other factors may limit its utility, including specimen drift, contamination, and damage due to increased dose. Thus, performing high resolution SEM almost always a dynamic tension when establishing the visibility of low contrast features between the electron dose needed to exceed the threshold of visibility and the consequences of that electron dose to the specimen.

10.6.2 Pathological Specimen Behavior

The electron dose needed for high resolution SEM even with an optimized instrument can exceed the radiation damage threshold for certain materials, especially "soft" materials such as biological materials and other weakly bonded organic and inorganic substances. Damage may be readily apparent in repeated scans, especially when the magnification is lowered after recording an image. If such specimen damage is severe, a "minimum-dose" strategy may be necessary, including such procedures as focusing and optimizing the image on a nearby area, blanking the beam, translating the specimen to an unexposed area, and then exposing the specimen for a single imaging frame.

Another possibility is to explore the sensitivity of the specimen to damage over a wide range of beam energy. It may seem likely that operating at low beam energy should minimize specimen damage, but this may not be the case. Because the electron range scales as $E_0^{1.67}$ and the volume as $(E_0^{1.67})^3$ while the energy deposited scales as E_0, the energy deposited per unit volume scales roughly as

$$\mathrm{Energy}/\mathrm{unit\ volume} = E_0 / \left(E_0^{1.67}\right)^3 = 1/E_0^4 \quad \textbf{(10.5)}$$

Thus, increasing the beam energy from 1 to 10 keV lowers the energy deposited per unit volume by a factor of approximately 10,000. This simplistic argument obviously ignores the substantial variation in the energy density within the interaction volume as well as the possibility that some damage mechanisms have an energy threshold for activation that may be avoided by lowering the beam energy. Nevertheless, Eq. 10.5 suggests that examining the material susceptibility to damage over a wide range of beam energy may be a useful strategy.

10.6.3 Pathological Specimen and Instrumentation Behavior

Contamination

A modern SEM that is well maintained should not be the source of any contamination that is observed. The first requirement of avoiding contamination is a specimen preparation protocol that minimizes the incorporation of or retention of contaminating compounds when processing the specimen. This caution includes the specimen as well as the mounting materials such as sticky conductive tape. A specimen airlock that minimizes the volume brought to atmosphere for specimen exchange as well as providing the important capability of pre-pumping the specimen to remove volatile compounds prior to insertion in the specimen chamber is an important capability for high resolution SEM. The specimen airlock can also be equipped with a "plasma cleaner" that generates a low energy oxygen ion stream for destruction and removal of organic compounds that produce contamination. If contamination is still observed after a careful preparation and insertion protocol has been followed, it is much more likely that the source of contamination remains the specimen itself and not the SEM vacuum system.

Instabilities

Unstable imaging conditions can arise from several sources. (1) Drift and vibration: The specimen preparation, the method of attachment to the substrate, the attachment of the specimen mount to the stage, and the stage itself must all have high stability to avoid drift, which is most noticeable at high magnification, and isolation from sources of vibration. Note that some mounting materials such as sticky tape may be subject to beam damage and distortion when struck by the beam electrons, leading to significant drift. One of the most stable mechanical stage designs is to be mounted within the bore of the objective lens, although such designs severely limit the size of the specimen and the extent of lateral motion that can be achieved. (2) Electromagnetic radiation interference: A periodic distortion is sometimes observed that is a result of interference from various sources of electromagnetic radiation, including emissions from 60-Hz AC sources, including emissions from fluorescent lighting fixtures. Rather than being random, this type of interference can synchronize with the scan and can be recorded. An example of this type of image defect is shown in ◨ Fig. 10.22. Eliminating this type of interference and the resulting image defects can be extremely challenging.

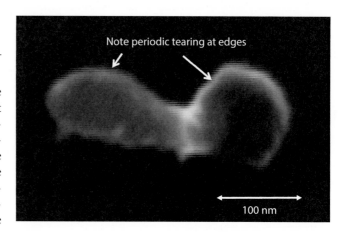

◨ **Fig. 10.22** SEM image of nanoparticles showing tearing at the particle edges caused by some source of electromagnetic interference whose frequency is constant and apparently locked to the 60 Hz AC power

References

Joy DC (1984) Beam interactions, contrast and resolution in the SEM. J Microsc 136:241

Oatley CW (1972) The scanning electron microscope, part 1, the instrument. Cambridge University Press, Cambridge

Pawley JB (1984) Low voltage scanning electron microscopy. J Microsc 136:45

Peters K-R (1984) Generation, collection, and properties of an SE-1 enriched signal suitable for high resolution SEM on bulk specimens. In: Kyser DF, Niedrig H, Newbury DE, Shimizu R (eds) Electron beam interactions with solids for microscopy, microanalysis, and microlithography. SEM, Inc, AMF O'Hare, p 363

Peters K-R (1985) "Working at higher magnifications in scanning electron microscopy with secondary and backscattered electrons on metal coated biological specimens" SEM/1985. SEM, Inc, AMF O'Hare, p 1519

Postek MT, Vladar AE, Wells OC, Lowney JL (2001) Application of the low-loss scanning electron microscope image to integrated circuit technology. Part 1 – applications to accurate dimension measurements. Scanning 23:298

Villarrubia J, Vladar A, Ming B, Kline R, Sunday D, Chawla J, List S (2015) Scanning electron microscope measurement of width and shape of 10 nm patterned lines using a JMONSEL-modeled library. Ultramicroscopy 154:15–28

Wells OC (1971) Low-loss image for surface scanning electron microscopy. Appl Phys Lett 19:232

Wells OC (1974a) Scanning electron microscopy. McGraw-Hill, New York

Wells OC (1974b) Resolution of the topographic image in the SEM. SEM/1974. IIT Research Inst, Chicago, p 1

Low Beam Energy SEM

Electronic supplementary material The online version of this chapter
(https://doi.org/10.1007/978-1-4939-6676-9_11) contains supplementary material,
which is available to authorized users.

The incident beam energy is one of the most useful parameters over which the microscopist has control because it determines the lateral and depth sampling of the specimen properties by the critical imaging signals. The Kanaya–Okayama electron range varies strongly with the incident beam energy:

$$R_{\text{K-O}}\,(\text{nm}) = \left(27.6\,A\,/\,Z^{0.89}\,\rho\right)E_0^{1.67} \tag{11.1}$$

where A is the atomic weight (g/mol), Z is the atomic number, ρ is the density (g/cm³), and E_0 (keV) is the incident beam energy, which is shown graphically in ◻ Fig. 11.1a–c.

11.1   What Constitutes "Low" Beam Energy SEM Imaging?

The rapid but continuous decrease of the range with E_0 shown in ◻ Fig. 11.1a raises the question, Where does "low" beam energy SEM imaging begin? That is, what value of E_0 constitutes the upper bound of "low" beam energy microscopy? As

will be discussed below, useful SEM imaging can now be accomplished down to remarkably low arrival energies at the specimen surface, less than 100 eV. The upper bound for E_0 is arbitrary, but a reasonable limit is the value discussed in the "Low Beam Energy X-Ray Microanalysis" module, where it is found that $E_0 = 5$ keV is the lowest beam energy for which a useful characteristic X-ray peak can be excited for all elements of the periodic table, excepting H and He, which do not produce characteristic X-rays. Thus, the plot of the range for $E_0 \leq 5$ keV shown in ◻ Fig. 11.1b will be taken to define the range for low beam energy SEM.

11.2   Secondary Electron and Backscattered Electron Signal Characteristics in the Low Beam Energy Range

The characteristics of the secondary electron (SE) and backscattered electron (BSE) signals observed in conventional SEM imaging performed at high beam energy ($E_0 \geq 10$ keV) can be summarized as follows: (1) For most elements, $\eta > \delta$.

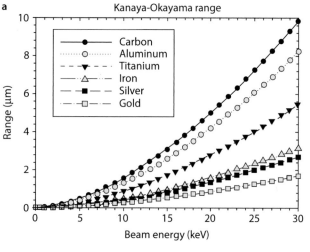

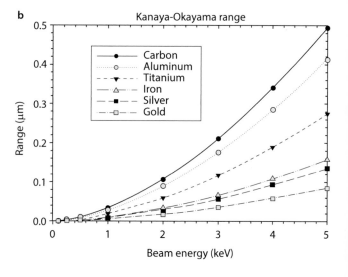

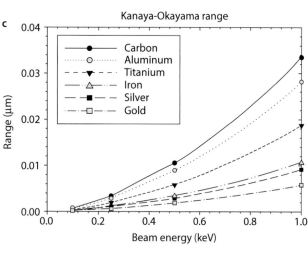

◻ **Fig. 11.1**   Plot of the Kanaya–Okayama range for various elements: **a** 0–30 keV; **b** 0–5 keV; **c** 0–1 keV

Fig. 11.2 Secondary electron coefficient, δ, as a function of beam energy for C, Al, Cu, and Au, taken from the data of Bongeler et al. (1993)

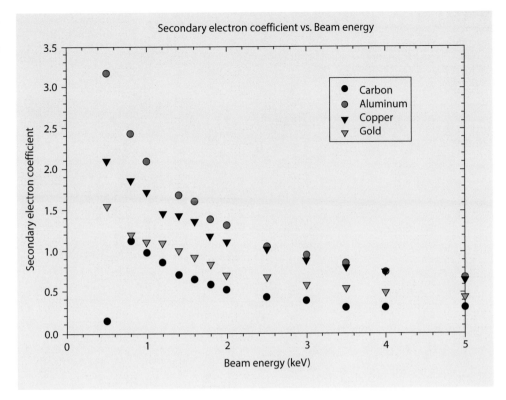

(2) Although the SE_1 are sensitive to surface characteristics within the escape depth of ~10 nm (metals), this surface sensitivity is diluted by the more numerous SE_2 and SE_3, which compose about 75–85 % of the total SE signal. SE_2 and SE_3 carry BSE information since they are created by the exiting BSEs at the specimen surface and on the chamber walls. Because the BSEs escape from approximately 15 % (high Z) to 30 % (low Z) of $R_{K–O}$, BSE depth sensitivity in turn determines the effective sampling of sub-surface information carried by the SE_2 and the SE_3, which is one to two orders of magnitude greater than the ~10 nm of the SE_1.

As E_0 is reduced into the low beam energy range below 5 keV, the rapid reduction in the electron range given by equation 11.1, as shown in ◘ Fig. 11.1 b, strongly influences the secondary electron coefficient: (1) The fraction of the incident energy lost by the beam electrons near the surface increases, which in turn increases the production of SEs, so that δ increases as the beam energy is reduced, as shown in ◘ Fig. 11.2 for several elements for measurements conducted in one laboratory. Because of this significant increase in SE production in the low beam energy range, generally δ > η, as shown for Au in ◘ Fig. 11.3. In low beam energy SEM, backscattering still occurs, but due to their much greater abundance SEs generally dominate the signal collected by the Everhart–Thornley (E-T)(positive bias) detector. (2) As the beam energy decreases, the collapse of the lateral and depth ranges increases the fraction of the SE_2 and SE_3 that carry surface information equivalent to the SE_1. This trend makes the SE image increasingly sensitive to

the surface characteristics of the material as the beam energy is reduced. However, the surface of a material is often unexpectedly complex. Upon exposure to the atmosphere, most "pure" elements form a thin surface oxide layer, for example, approximately 4 nm of Al_2O_3 forms on Al. Moreover, this surface layer may incorporate water chemically to form hydroxide and/or carbon dioxide to form carbonate, or there may be physical adsorption of these and other compounds from the environment which may not evaporate under vacuum. Additionally, there may be unexpected contamination from hydrocarbons deposited on the specimen surface which generally arise from the environment to which the specimen was exposed prior to the SEM. In some cases such contamination may be deposited from the SEM vacuum system if sufficient care has not been previously taken to eliminate sources of volatile contamination by rigorous specimen cleaning and by prepumping in an airlock prior to transferring into the specimen chamber. Complex surface composition is the likely reason for the wide range of δ values reported by various researchers measuring a nominally common target, as illustrated in ◘ Fig. 11.4 for aluminum, where reported values of δ span a factor of 4 or more. This is a common result across the periodic table, as seen in the SE database compiled by Joy (2012). The strong surface sensitivity of the SE and BSE signals at low beam energy to the condition of the specimen surface means that SEM image interpretation of "real" as-received specimens will be challenging. *In situ* cleaning by ion beam milling in a "dual beam" platform may

■ **Fig. 11.3** Secondary electron coefficient, δ, and backscatter electron coefficient, η, as a function of beam energy for Au, taken from the data of Bongeler et al. (1993)

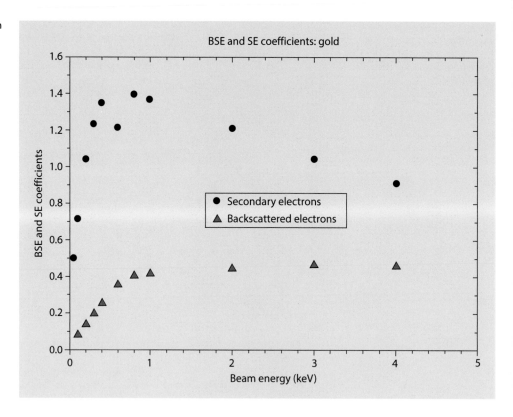

■ **Fig. 11.4** Secondary electron coefficient, δ, as a function of beam energy for Al (Taken from the data of various authors)

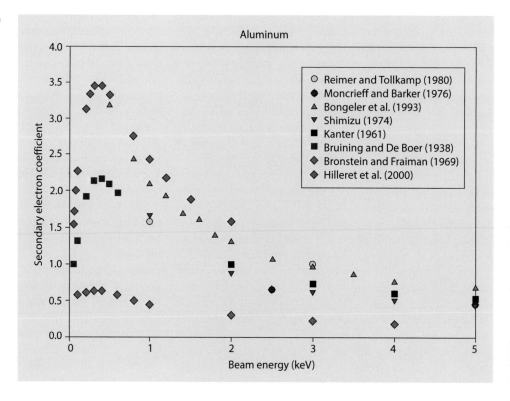

11

remove such artifacts. However, even with ion beam cleaning, it must be recognized that at the vacuum levels of the conventional "high vacuum" SEM, for example, 10^{-4} Pa (10^{-6} torr), the partial pressure of oxygen is sufficiently high that a monolayer of oxide will form on a reactive surface such as Al in a matter of seconds. Thus, while ion beam milling may successfully remove contamination, oxide formation at least at the monolayer level may be unavoidable

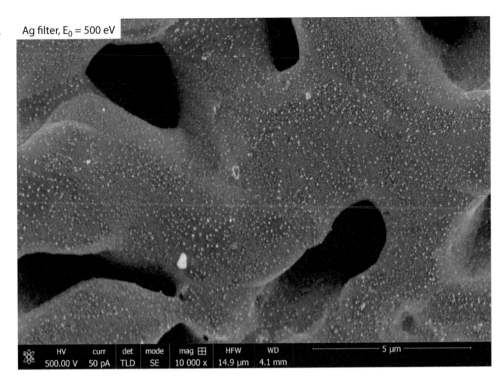

■ **Fig. 11.5** SEM image of a silver filter obtained at $E_0 = 0.5$ keV with a through-the-lens secondary electron detector; Bar = 5 μm (Image courtesy of Keana Scott, NIST)

Ag filter, $E_0 = 500$ eV

	HV	curr	det	mode	mag ⊞	HFW	WD	———— 5 μm ————
	500.00 V	50 pA	TLD	SE	10 000 x	14.9 μm	4.1 mm	

unless an ultrahigh vacuum instrument is used, where the chamber pressure is $<10^{-8}$ Pa (10^{-10} torr).

11.3 Selecting the Beam Energy to Control the Spatial Sampling of Imaging Signals

11.3.1 Low Beam Energy for High Lateral Resolution SEM

The electron range controls the lateral spatial distribution of the backscattered electrons: 90 % of BSEs escape radially from approximately 30 % R_{K-O} (high Z) to 60 % R_{K-O} (low Z). The lateral spatial distribution of the SE_2, which is created as the BSE escape through the surface, and the SE_3, which is the BSE-to-SE conversion signal that results when BSE strike the objective lens pole piece, the stage components, and the chamber walls, effectively sample the same spatial range as the BSE. As the incident beam energy is lowered, the BSE (SE_3) and SE_2 signal lateral distributions collapse onto the SE_1 distribution, which is restricted to the beam footprint, so that at sufficiently low beam energy all of these signals carry high spatial resolution information similar to the SE_1. With a modern high performance SEM equipped with a high brightness source, for example, a cold field emission gun or a Schottky thermally assisted field emission gun, capable of delivering a useful beam current into a nanometer or sub-nanometer diameter beam, low beam energy SEM operation has become the

most frequent choice to achieve high lateral spatial resolution imaging of bulk specimens, as discussed in detail in the "High Resolution SEM" module. An example of high spatial resolution achieved at low beam energy is shown in ■ Fig. 11.5 for a silver filter material imaged at $E_0 = 0.5$ keV with a "through-the-lens" secondary electron detector. Unfortunately, there is no simple rule like η vs. Z at high beam energy for interpreting the contrast seen in this image. For example, why does the population of nanoscale particles appear extremely bright against the general mid-level gray of the bulk background of the silver structure. These features may appear bright because of local compositional differences such as thicker oxides or there may be a physical change such as increased surface area for SE emission due to nanoscale roughening.

11.3.2 Low Beam Energy for High Depth Resolution SEM

The strong exponential dependence of the beam penetration on the incident energy controls the sampling of sub-surface specimen properties by the BSEs and SEs, which can provide insight on the depth dimension. Observing a specimen as the beam energy is progressively lowered to record systematic changes can reveal lateral heterogeneities in surface composition. ■ Fig. 11.6 shows such a sequence of images from high beam energy (30 keV) to low beam energy (1 keV) prepared with an E–T(positive bias) detector where the specimen is an aluminum stub upon which was deposited approximately

11

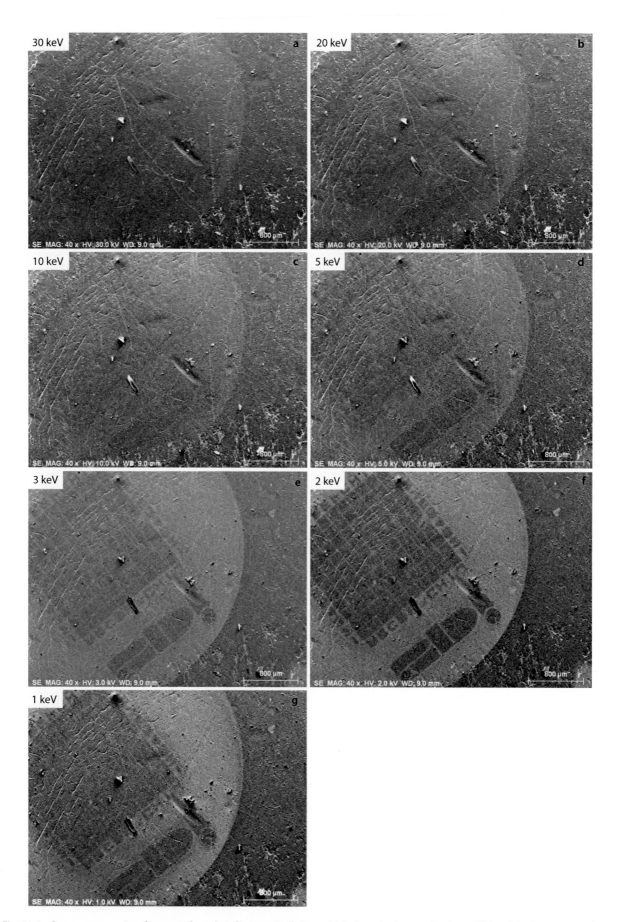

■ **Fig. 11.6**   Beam energy series of images of a carbon film, nominally 7 nm thick, deposited on an aluminum SEM stub in the as-received condition prepared with an E–T(positive bias) detector: **a** 30 keV; **b** 20 keV; **c** 10 keV; **d** 5 keV; **e** 3 keV; **f** 2 keV; **g** 1 keV; Bar = 800 μm

7 nm of carbon shadowed through a grid. The contrast between the carbon and the aluminum behaves in a complex fashion. The C-Al contrast is only weakly visible above $E_0 = 5$ keV despite a high electron dose, long image integration, and post-acquisition image processing for contrast enhancement. The C-Al contrast increases sharply as the beam energy decreases below 5 keV, reaching a maximum at $E_0 = 2$ keV and then decreasing below this energy. The increase in contrast below 5 keV is consistent with the increasing separation between the values of δ for C and Al seen in ◘ Fig. 11.2. The eventual decrease in the C-Al contrast below $E_0 = 2$ keV is not consistent with the measurements plotted in ◘ Fig. 11.2, where the difference between δ for C and Al actually increases below $E_0 = 2$ keV, which should increase the contrast. Despite the difficulty in interpreting these trends in contrast, this example demonstrates that lateral differences in the surface can be detected, provided care is taken to fully explore the image response to changing the beam energy parameter.

11.3.3 Extremely Low Beam Energy Imaging

High performance SEMs typically operate down to beam energies below 0.5 keV, with the lower limit depending on the vendor and the particular model. Ultralow beam energies below 0.1 keV can be achieved through different electron-optical techniques, including biasing the specimen to −V. Specimen biasing acts to decelerate the beam electrons emitted at energy E_0 from the column so that the landing energy, that is, the kinetic energy remaining when the beam electrons reach the specimen surface, is $E_0 - eV$, where e represents the electronic charge. Ultralow beam energy imaging is illustrated in ◘ Fig. 11.7, where the surface of a silica (SiO_2) specimen is imaged at a landing energy of 0.030 keV (30 eV). ◘ Figure 11.8 shows gold islands on carbon imaged with a landing energy of 0.01 keV (10 eV). At such low incident energy, only the outermost atomic/molecular layers are probed by the beam.

◘ **Fig. 11.7** Extremely low landing energy ($E_0 = 0.030$ keV) SEM image of silica (SiO_2) prepared with an Everhart-Thornley E–T(positive bias) detector and a beam current of 250 pA revealing fine-scale texture and surface topography; Bar = 2 μm (Image courtesy of Carl Zeiss)

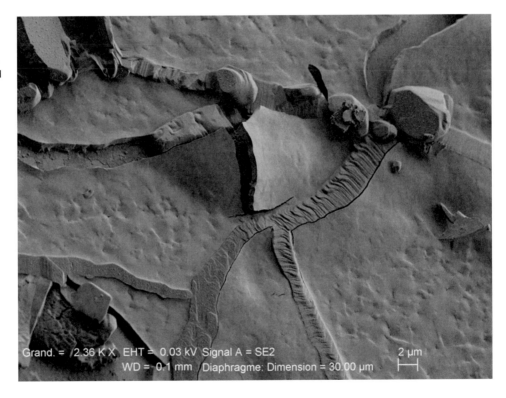

Fig. 11.8 Extremely low
landing energy ($E_0 = 0.010$ keV)
image of gold islands evaporated
on carbon; TTL SE detector;
Bar = 100 nm (Image courtesy
of V. Robertson, JEOL)

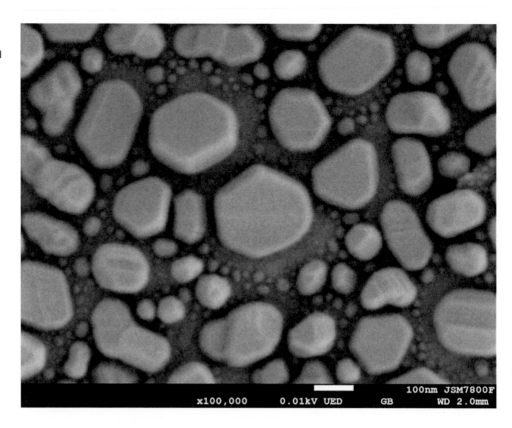

11

References

Bongeler R, Golla U, Kussens M, Reimer L, Schendler B, Senkel R, Spranck M (1993) Electron-specimen interactions in low-voltage scanning electron microscopy. Scanning 15:1

Bronstein IM, Fraiman BS (1969) "Secondary electron emission" Vtorichnaya elektronnaya emissiya. Nauka, Moskva, p 340

Bruining H, De Boer JM (1938) Secondary electron emission: Part 1 secondary electron emission of metals. Phys Ther 5:17p

Hilleret N, Bojko J, Grobner O, Henrist B, Scheuerlein C, and Taborelli M (2000) "The secondary electron yield of technical materials and its Variation with surface treatments" in Proc.7th European Particle Accelerator Conference, Vienna

Joy D (2012) Can be found in chapter 3 on SpringerLink: http://link.springer.com/chapter/10.1007/978-1-4939-6676-9_3

Kanter M (1961) Energy dissipation and secondary electron emission in solids. Phys Rev 121:1677

Moncrieff DA, Barker PR (1976) Secondary electron emission in the scanning electron microscope. Scanning 1:195

Reimer L, Tolkamp C (1980) Measuring the backscattering coefficient and secondary electron yield inside a scanning electron microscope. Scanning 3:35

Shimizu R (1974) Secondary electron yield with primary electron beam of keV. J Appl Phys 45:2107

Variable Pressure Scanning Electron Microscopy (VPSEM)

© Springer Science+Business Media LLC 2018
J. Goldstein et al., *Scanning Electron Microscopy and X-Ray Microanalysis*,
https://doi.org/10.1007/978-1-4939-6676-9_12

12

12.1   Review: The Conventional SEM High Vacuum Environment

The conventional SEM must operate with a pressure in the sample chamber below ~10^{-4} Pa (~10^{-6} torr), a condition determined by the need to satisfy four key instrumental operating conditions:

12.1.1   Stable Electron Source Operation

The pressure in the electron gun must be maintained below 10^{-4} Pa (~10^{-6} torr) for stable operation of a conventional thermal emission tungsten filament and below 10^{-7} Pa (~10^{-9} torr) for a thermally assisted field emission source. Although a separate pumping system is typically devoted to the electron source to maintain the proper vacuum, if the specimen chamber pressure in a conventional SEM is allowed to rise, gas molecules will diffuse to the gun, raising the pressure and causing unstable operation and early failure.

12.1.2   Maintaining Beam Integrity

An electron emitted from the source that encounters a gas atom along the path to the specimen will scatter elastically, changing the trajectory and causing the electron to deviate out of the focused beam. To preserve the integrity of the beam, the column and chamber pressure must be reduced to the point that the number of collisions between the beam electrons and the residual gas molecules is negligible along the entire path, which typically extends to 25 cm or more.

12.1.3   Stable Operation of the Everhart–Thornley Secondary Electron Detector

To serve as a detector for secondary electrons, the Everhart–Thornley secondary electron detector must be operated with a bias of +10,000 volts or more applied to the face of the scintillator to accelerate the SE and raise their kinetic energy sufficiently to cause light emission. If the chamber pressure exceeds approximately 100 mPa (~10^{-3} torr), electrical discharge events will begin to occur due to gas ionization between the scintillator (+10,000 V) and the Faraday cage (+250 V), which is located in close proximity, initially increasing the noise and thus degrading the signal-to-noise ratio. As the chamber pressure is further increased, electrical arcing will eventually cause total operational failure.

12.1.4   Minimizing Contamination

A major source of specimen contamination during examination arises from the cracking of hydrocarbons by the electron beam. A critical factor in determining contamination rates is the availability of hydrocarbon molecules for the beam electrons to hit. To achieve a low contamination environment, the pumping system must be capable of achieving low ultimate operating pressures. A specimen exchange airlock can pump off most volatiles, minimizing the exposure of the specimen chamber, and the airlock can be augmented with a plasma cleaning system to actively destroy volatiles. Finally, the vacuum system can be augmented with careful cold surface trapping of any remaining volatiles from the specimen or those that can backstream from the pump so as to minimize the partial pressure of hydrocarbons. Most importantly, to avoid introducing unnecessary sources of contamination, the microscopist must be very careful in handling instrument parts and specimens to avoid inadvertently depositing highly volatile hydrocarbons, such as those associated with skin oils deposited in fingerprints, into the conventional SEM. With this level of operational care when operating in a well maintained modern instrument, beam-induced contamination when observed almost always results from residual hydrocarbons on the specimen which remain from incomplete cleaning rather than from hydrocarbons from the vacuum system itself.

A significant price is paid to operate the SEM with such a "clean" high vacuum. The specimen must be prepared in a condition so as not to evolve gases in the vacuum environment. Many important materials, such as biological tissues, contain liquid water, which will rapidly evaporate at reduced pressure, distorting the microscopic details of a specimen and disturbing the stable operating conditions of the microscope. This water, and any other volatile substances, must be removed during sample preparation to examine the specimen in a "dry" state, or the water must be immobilized by freezing to low temperatures ("frozen, hydrated samples"). Such specimen preparation is both time-consuming and prone to introducing artifacts, including the redistribution of "diffusible" elements, such as the alkali ions of salts.

12.2   How Does VPSEM Differ From the Conventional SEM Vacuum Environment?

The development of the variable pressure scanning electron microscope (VPSEM) has enabled operation with elevated specimen chamber pressures in the range ~1–2500 Pa (~0.01–20 torr) while still maintaining a high level of SEM imaging performance (Danilatos 1988, 1991). The VPSEM utilizes "differential pumping" with several stages to obtain the desired elevated pressure in the specimen chamber while simultaneously maintaining a satisfactory pressure for stable operation of the electron gun and protection of the beam electrons from encountering elevated gas pressure along most of the flight path down the column. Differential pumping consists of establishing a series of regions of successively lower pressure, with each region separated by small apertures from the regions on either side and each region having its own dedicated pumping path. The probability of gas molecules moving from one region to the next is limited by the area of the aperture. In the VPSEM, these differential pumping apertures also serve as the beam-defining apertures. A typical vacuum design consists of separate pumping systems for the specimen chamber, one region for each lens, and finally the electron gun. Such a vacuum system can maintain a pressure differential of six orders of magnitude or more between the specimen chamber and the electron gun, enabling use of both conventional thermionic sources and thermally-assisted field emission sources. A wide variety of gases can be used in the elevated pressure sample chamber, including oxygen, nitrogen, argon, and water vapor. Because the imaging conditions are extremely sensitive to the sample chamber pressure, careful regulation of the pressure and of its stability for extended periods is required.

12.3 Benefits of Scanning Electron Microscopy at Elevated Pressures

There are several special benefits to performing scanning electron microscopy at elevated pressures.

12.3.1 Control of Specimen Charging

Insulating materials suffer charging in the conventional high vacuum SEM because the high resistivity of the specimen prevents the migration of the charges injected by the beam, as partially offset by charges that leave the specimen as backscattered and secondary electrons, to reach an electrical ground. Consequently, there develops a local accumulation of charge. Depending on the beam energy, the material properties, and the local inclination of the specimen to the beam, negative or positive charging can occur. Charging phenomena can be manifest in many ways in SEM images, ranging at the threshold from diminished collection of secondary electrons which reduces the signal-to-noise ratio to more extreme effects where the local charge accumulation is high enough to cause actual displacement of the position of the beam, often seen as discontinuities in the scanned image. In the most extreme cases, the charge may be sufficient for the specimen to act as a mirror and deflect the beam entirely. In conventional SEM operation, charging is typically eliminated or at least minimized by applying a thin conducting coating to an insulating specimen and connecting the coating layer to electrical ground.

In the VPSEM, incident beam electrons, BSE and SE can scatter inelastically with gas atoms near the specimen, ionizing those gas atoms to create free low kinetic energy electrons and positive ions. Areas of an insulating specimen that charge will attract the appropriate oppo-

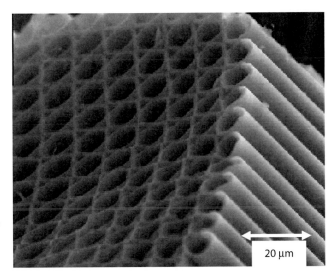

Fig. 12.1 Uncoated glass polycapillary as imaged in a VPSEM (conditions: 20 keV; 500 Pa water vapor; gaseous secondary electron detector)

sitely charged species from this charge cloud, the positively ionized gas atoms or the free electrons, leading to local dynamic charge neutralization, enabling insulating materials to be examined without a coating. Moreover, the environmental gas, the ionized gas atoms, and the free electrons can penetrate into complex geometric features such as deep holes, features which would be very difficult to coat to establish a conducting path for conventional high vacuum SEM. An example of VPSEM imaging of a very complex insulating object is shown in ◘ Fig. 12.1, which is an array of glass microcapillaries examined without any coating. No charging is observed in this secondary electron VPSEM image with $E_0 = 20$ keV (prepared with a gaseous secondary electron detector, as described below) despite the very deep recesses in the structure. Another example is shown in ◘ Fig. 12.2a, which shows a comparison of images of a complex polymer foam imaged in high vacuum SEM at a low beam energy of $E_0 = 4$ keV with an Everhart–Thornley (E–T) detector, showing the development of charging, and in VPSEM mode with $E_0 = 20$ keV and an off-axis backscattered electron (BSE) detector, showing no charging effects. A challenging insulating sample with a complex surface is shown in ◘ Fig. 12.2b, which depicts fresh popcorn imaged under VPSEM conditions with a BSE detector.

Achieving suppression of charging for such complex insulating objects as those shown in ◘ Figs. 12.1 and 12.2 involves careful control of the usual parameters of beam energy, beam current, and specimen tilt. In VPSEM operation the additional critical variables of environmental gas species and partial pressure must be carefully explored. Additionally, the special detectors for SE that have been developed for VPSEM operation can also play a role in charge suppression.

12

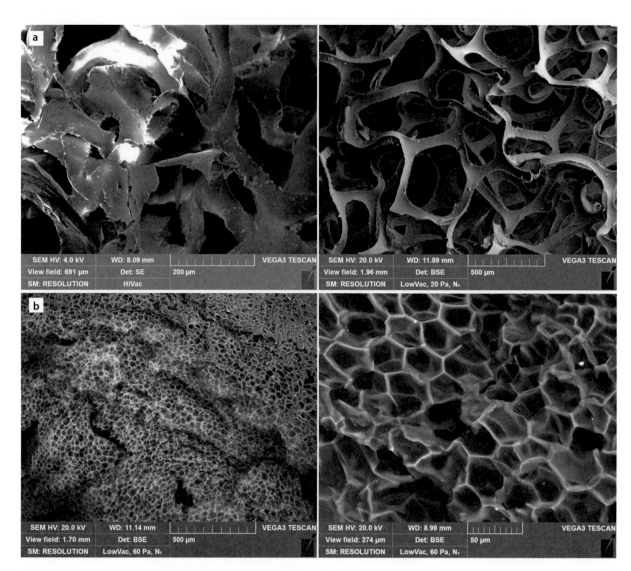

◻ Fig. 12.2   a Uncoated polymer foam imaged (left) with high vacuum SEM, $E_0 = 4$ keV, E-T(+) detector (bar = 200 μm); and (right) VPSEM, $E_0 = 20$ keV, off-axis BSE detector (bar = 500 μm) (Images courtesy J. Mershan, TESCAN). **b** (left and right) Uncoated freshly popped popcorn imaged with VPSEM, $E_0 = 20$ keV, BSE detector; 60 Pa N_2 (left, bar = 500 μm) (right, bar = 50 μm) (Images courtesy J. Mershon, TESCAN; sample source: Lehigh Microscopy School)

12.3.2   Controlling the Water Environment of a Specimen

Careful control and preservation of the water content of a specimen can be critical to recording SEM images that are free from artifacts or suffer only minimal artifacts. Additionally, when there is control of the partial pressure of water vapor in the specimen chamber to maintain liquid water in equilibrium with the gas phase, it becomes possible to observe chemical reactions that are mediated by water.

By monitoring and controlling the relative humidity, it is possible to add water by condensation or remove it by evaporation. ◻ Figure 12.3 shows the pressure–temperature phase diagram for water. The pressure–temperature conditions to maintain liquid water, ice, and water vapor in equilibrium can be achieved at the upper end of the operating pressure range of certain VPSEMs when augmented

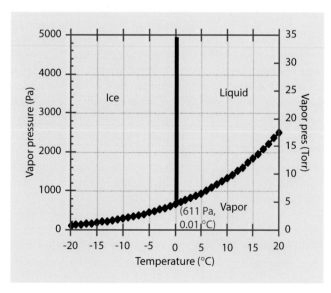

◻ Fig. 12.3   Phase diagram for water

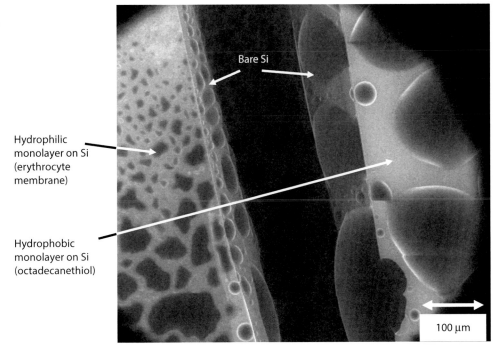

Fig. 12.4 VPSEM imaging of water condensed in situ on silicon treated with a hydrophobic layer (octadecanethiol), a hydrophilic layer (erythrocyte membrane), and bare, uncoated silicon (nearly vertical fracture surfaces) (Example courtesy Scott Wight, NIST)

Bare Si

Hydrophilic monolayer on Si (erythrocyte membrane)

Hydrophobic monolayer on Si (octadecanethiol)

100 µm

with a cooling stage capable of reaching −5 °C to 5 °C. With careful control of both the pressure of water vapor added to the specimen chamber and of the specimen temperature, the microscopist can select the relative humidity in the sample chamber so that water can be evaporated, condensed, or maintained in liquid–gas or solid-gas equilibrium. In addition to direct examination of water-containing specimens, experiments can be performed in which the presence and quantity amount of water is controlled as a variable, enabling a wide range of chemical reactions to be observed. ◘ Figure 12.4 shows an example of the condensation of water on a silicon wafer, one side of which was covered with a hydrophobic layer while the other was coated with a hydrophilic layer, directly revealing the differences in the wetting behavior on the two applied layers, as well as the bare silicon exposed by fracturing the specimen.

12.4 Gas Scattering Modification of the Focused Electron Beam

The differential pumping system achieves vacuum levels that minimize gas scattering and preserve the beam integrity as it passes from the electron source through the electron-optical column. As the beam emerges from the high vacuum of the electron column through the final aperture into the elevated pressure of the specimen chamber, the volume density of gas atoms rapidly increases, and with it the probability that elastic scattering events with the gas atoms will occur. Although the volume density of the gas atoms in the chamber is very low compared to the density of a solid material, the path length that the beam electrons must travel in the elevated pressure region of the sample chamber typically ranges from

1 mm to 10 mm or more before reaching the specimen surface. As illustrated schematically in ◘ Fig. 12.5, elastic scattering events that occur with the gas molecules along this path cause beam electrons to substantially deviate out of the focused beam to create a "skirt". Even a small angle elastic event with a 1-degree scattering angle that occurs 1 mm above the specimen surface will cause the beam electron to be displaced by 17 µm radially from the focused beam.

How large is the gas-scattering skirt? The extent of the beam skirt can be estimated from the ideal gas law (the density of particles at a pressure p is given by $n/V = p/RT$, where n is the number of moles, V is the volume, R is the gas constant, and T is the temperature) and by using the cross section for elastic scattering for a single event (Danilatos 1988):

$$R_s = (0.364 Z / E)(p / T)^{1/2} L^{3/2} \qquad (12.1)$$

where R_s = skirt radius (m)
Z = atomic number of the gas
E = beam energy (keV)
p = pressure (Pa)
T = temperature (K)
L = Gas Path Length (GPL) (m)

◘ Figure 12.6 plots the skirt radius for a beam energy of 20 keV as a function of the gas path length through oxygen at several different chamber pressures. For a pressure of 100 Pa and a gas path length of 5 mm, the skirt radius is calculated to be 30 µm. Consider the change in scale from the focused beam to the skirt that results from gas scattering. The high vacuum beam footprint that gives the lateral extent of the BSE, SE, and X-ray production can be estimated with the

◻ Fig. 12.5 Schematic diagram showing gas scattering leading to development of the skirt surrounding the unscattered beam at the specimen surface

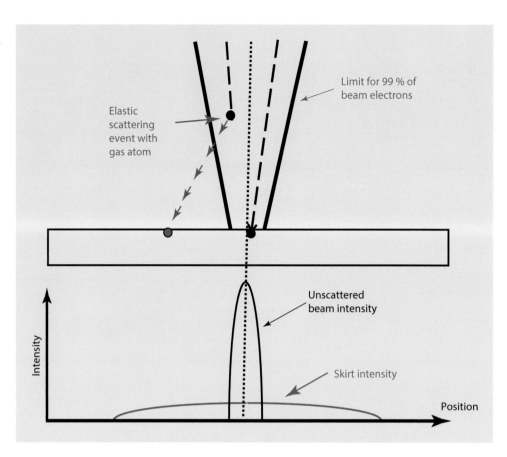

◻ Fig. 12.6 Development of beam skirt for 20-keV electrons passing through oxygen at various pressures as calculated with Eq. 12.1

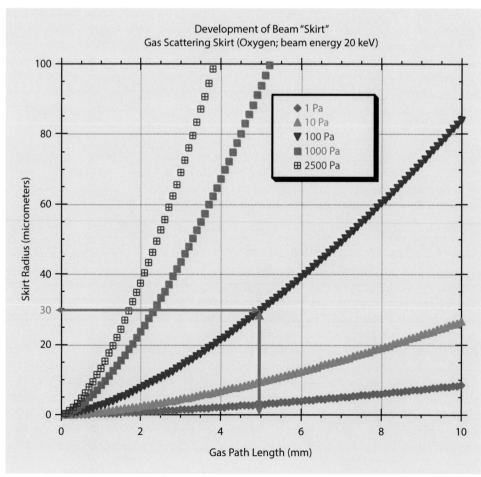

Kanaya–Okayama range equation. For a copper specimen and $E_0 = 20$ keV, the full range $R_{K-O} = 1.5$ µm, which is also a good estimate of the diameter of the interaction volume projected on the entrance surface. With a beam/interaction volume footprint radius of 0.75 µm, the gas scattering skirt of 30-µm radius is thus a factor of 40 larger in linear dimension, and the skirt is a factor of 1600 larger in area than that due the focused beam and beam·specimen interactions. Considering just a 10-nm incident beam diameter (5-nm radius), the gas scattering skirt is 6000 times larger.

While Eq. 12.1 is useful to estimate the extent of the gas scattering skirt under VPSEM conditions, it provides no information on the relative fraction of the beam that remains unscattered or on the distribution of gas-scattered electrons within the skirt. The Monte Carlo simulation embedded in NIST DTSA-II enables explicit treatment of gas scattering to provide detailed information on the unscattered beam electrons as well as the spatial distribution of electrons scattered

into the skirt. The VPSEM menu of DTSA-II allows selection of the critical variables: the gas path length, the gas pressure, and the gas species (He, N_2, O_2, H_2O, or Ar). ◻ Table 12.1 gives an example of the Monte Carlo output for the electron scattering out of the beam for a 5-mm gas path length through 100 Pa of water vapor. In addition to the radial distribution, the DTSA II Monte Carlo reports the unscattered fraction that remains in the focused beam, a value that is critical for estimating the likely success of VPSEM imaging, as described below.

◻ Figure 12.7a plots the gas scattering predicted by the Monte Carlo simulation for a gas path length of 5 mm and 100 Pa of O_2, presented as the cumulative electron intensity as a function of radial distance out to 50 µm from the beam center. For these conditions the unscattered beam retains about 0.70 of the beam intensity that enters the specimen chamber. The skirt out to a radius of 30 µm contains a cumulative intensity of 0.84 of the incident beam

◻ **Table 12.1** NIST DTSA-II Monte Carlo simulation for 20-keV electrons passing through 5 mm of water vapor at 100 Pa

Ring	Inner Radius, µm	Outer radius, µm	Ring area, µm²	Electron count	Electron fraction	Cumulative (%)
Undeflected	—	—	—	42,279	0.661	—
1	0.0	2.5	19.6	46,789	0.731	73.1
2	2.5	5.0	58.9	2431	0.038	76.9
3	5.0	7.5	98.2	1457	0.023	79.2
4	7.5	10.0	137.4	1081	0.017	80.9
5	10.0	12.5	176.7	834	0.013	82.2
6	12.5	15.0	216.0	730	0.011	83.3
7	15.0	17.5	255.3	589	0.009	84.2
8	17.5	20.0	294.5	554	0.009	85.1
9	20.0	22.5	333.8	490	0.008	85.9
10	22.5	25.0	373.1	393	0.006	86.5
11	25.0	27.5	412.3	395	0.006	87.1
12	27.5	30.0	451.6	341	0.005	87.6
13	30.0	32.5	490.9	271	0.004	88.1
14	32.5	35.0	530.1	309	0.005	88.5
15	35.0	37.5	569.4	274	0.004	89.0
16	37.5	40.0	608.7	248	0.004	89.4
17	40.0	42.5	648.0	224	0.004	89.7
18	42.5	45.0	687.2	217	0.003	90.0
19	45.0	47.5	726.5	204	0.003	90.4
20	47.5	50.0	765.8	191	0.003	90.7

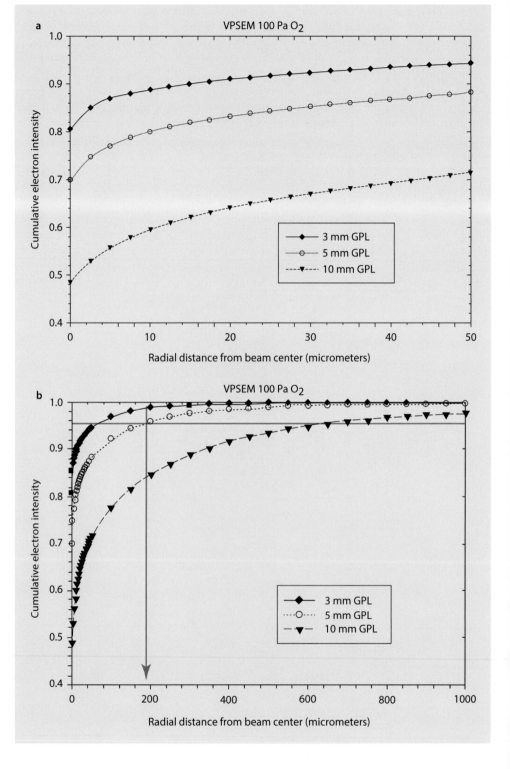

Fig. 12.7 **a** Cumulative electron intensity as a function of distance (0–50 μm) from the beam center for 20-keV electrons passing through 100 Pa of oxygen as calculated with NIST DTSA-II. **b** Cumulative electron intensity as a function of distance (0–1000 μm) from the beam center for 20-keV electrons passing through 100 Pa of oxygen as calculated with NIST DTSA-II (GPL = Gas Path Length)

current. To capture 0.95 of the total beam current requires a radial distance to approximately 190 μm, as shown in Fig. 12.7b, and the last 0.05 of the beam electrons are distributed out to 1000 μm (1 mm). The strong effect of the gas path length on the skirt radius, which follows a 3/2 exponent in the scattering Eq. 12.1, can be seen in Fig. 12.7 by comparing the plots for 3-, 5-, and 10-mm gas path lengths.

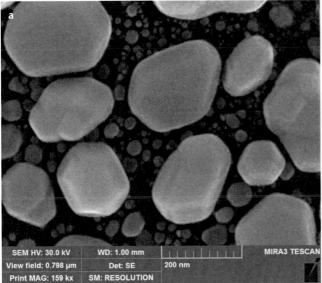

SEM HV: 30.0 kV | WD: 1.00 mm | 200 nm | MIRA3 TESCAN
View field: 0.798 µm | Det: SE |
Print MAG: 159 kx | SM: RESOLUTION |

SEM HV: 30.0 kV | WD: 6.54 mm | 200 nm | MIRA3 TESCAN
View field: 1.000 µm | Det: BSE |
Print MAG: 127 kx | LowVac, 300 Pa, N₂ |

☐ **Fig. 12.8** **a** High resolution SEM imaging of gold deposited on carbon in conventional SEM; $E_0 = 30$ keV; E–T (positive bias) detector (bar = 200 nm) (image courtesy J. Mershon, TESCAN). **b** High resolution SEM imaging of gold deposited on carbon in VPSEM; 300 Pa N₂; $E_0 = 30$ keV; BSE detector (bar = 200 nm) (Image courtesy J. Mershon, TESCAN)

12.5 VPSEM Image Resolution

Remarkably, despite the strong gas scattering and the development of the skirt around the focused beam, the image resolution that can be achieved in VPSEM operation is very similar to that for the same specimen imaged at the same incident beam energy in a conventional high vacuum SEM. A comparison of high vacuum SEM and VPSEM imaging performance for gold islands on carbon using a modern thermal field emission gun SEM is shown in ☐ Fig. 12.8, showing comparable spatial resolution, as originally demonstrated by Danilatos (1993). This extraordinary imaging performance in the VPSEM can be understood by recognizing that elastic scattering is a stochastic process. As beam electrons encounter the gas molecules and atoms in the elevated pressure region, elastic scattering events occur, but not every electron suffers elastic scattering immediately. There remains an unscattered fraction of electrons that follows the expected path defined by the objective lens field and lands in the focused beam footprint identical to the situation at high vacuum but with reduced intensity due to the gas scattering events that rob the beam of some of the electrons. As the gas scattering path, which is a product of working distance and the gas pressure, increases, the unscattered fraction of the beam decreases and eventually reaches zero intensity. The fraction of unscattered electrons that remain in the beam can be calculated by the Monte Carlo simulation in DTSA-II, and an example is plotted in ☐ Fig. 12.9. For 20-keV electrons passing through 10 mm of water vapor at 200 Pa, approximately 20 %

of the original beam current reaches the specimen surface unscattered and contained within the focused beam. The electrons that remain in the beam behave exactly as they would in a high vacuum SEM, creating the same interaction volume and generating secondary and backscattered electrons with exactly the same spatial distributions. The electrons that land in the scattering skirt also generate secondary and backscattered electrons in response to the local specimen characteristics they encounter, for example, surface inclination, roughness, composition, an so on, which may be different from the region sampled by the focused beam. Because these skirt electron interactions effectively arise from a broad, diffuse area rather than a focused beam, they cannot respond to fine-scale spatial details of the specimen as the beam is scanned. The skirt electrons interact over such a broad area that effectively they only contribute increased noise to the measurement, degrading the signal-to-noise ratio of the useful high resolution signal generated by the unscattered electrons that remain in the focused beam. This degraded signal-to-noise does degrade the visibility threshold, which can be compensated by increasing the beam current and/or by increasing the pixel dwell time. The degradation of feature visibility due to gas scattering has the most impact at the short pixel dwell times (high scan rates) that are typically selected for rapidly surveying a specimen to search for features of interest. The prudent VPSEM microscopist will always use long pixel dwell times to reduce the contrast visibility threshold to ensure that a low-contrast feature of interest can be observed.

Fig. 12.9 Fraction of a 20-keV beam that remains unscattered after passage through 10 mm of water vapor at various pressures

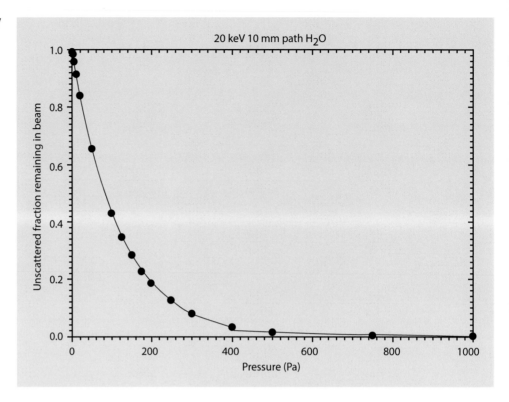

12.6 Detectors for Elevated Pressure Microscopy

12.6.1 Backscattered Electrons—Passive Scintillator Detector

As noted above, the E–T detector, or any other detector which employs a high accelerating voltage post-specimen, such as the channel plate multiplier, cannot be used at elevated VPSEM pressures due to ionization of the gas atoms leading to large-scale electrical breakdown. The passive backscattered electron detectors, including the semiconductor and scintillator detectors, are suitable for elevated pressure, since the backscattered electrons suffer negligible energy loss while transiting the environmental gas and thus retain sufficient energy to activate the scintillator without post-specimen acceleration. In fact, an added advantage of elevated pressure VPSEM operation is that the gas discharging allows the bare scintillator to be used without the metallic coating required for conventional high vacuum operation. An example of a VPSEM image of polymer foam prepared with a large symmetric BSE detector placed symmetrically above the specimen is shown in ◘ Fig. 12.10 (left).

   ◘ Figure 12.11a shows an example of a BSE image of polished Raney nickel alloy obtained with a passive scintillator detector in water vapor at a pressure of 500 Pa (3.8 torr) with a beam energy of 20 keV. This BSE image shows compositional contrast similar to that observed under high vacuum conventional SEM imaging.

12.6.2 Secondary Electrons–Gas Amplification Detector

To utilize the low energy secondary electrons in the VPSEM, a special elevated pressure SE detector that utilizes ionization of the environmental gas (gaseous secondary electron detector, GSED) has been developed (Danilatos 1990). As shown schematically in ◘ Fig. 12.12, an electrode (which may also serve as the final pressure limiting aperture) in close proximity to the electrically grounded specimen is maintained at a modest accelerating voltage of a few hundred volts positive. The exact value of this applied voltage is selected so as not to exceed the breakdown voltage for the gas species and pressure being utilized. The SE emitted from the specimen are accelerated toward this electrode and undergo collisions with the gas molecules, ionizing them and creating positive ions and more free electrons. The mean free path for this process is a few tens of micrometers, depending on the gas pressure and the accelerating voltage, so that multiple generations of ionizing collisions can occur between SE emission from the specimen and collection at the positive electrode. Moreover, the electrons ejected from the gas atoms are also accelerated toward the wire and ionize other gas atoms, resulting in a cascade of increasing charge carriers, progressively amplifying the current collected at the electrode by a factor up to several hundred compared to the SE current originally emitted from the specimen. While BSE can also contribute to the total signal collected at the electrode by collisions with gas molecules,

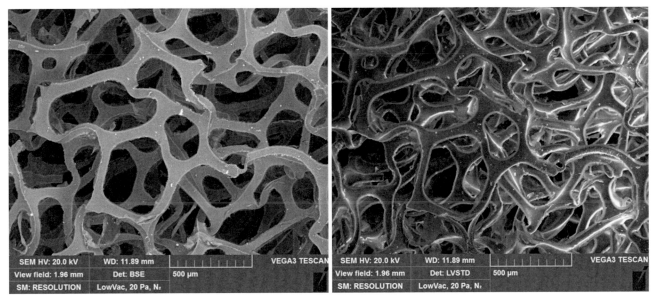

◘ Fig. 12.10   Uncoated polymer foam imaged under VPSEM conditions at $E_0 = 20$ keV: (*left*) large solid angle symmetric BSE detector placed above the specimen; (bar = 500 μm) (*right*) same area with induced field SE detector (bar = 500 μm) (Images courtesy J. Mershan, TESCAN)

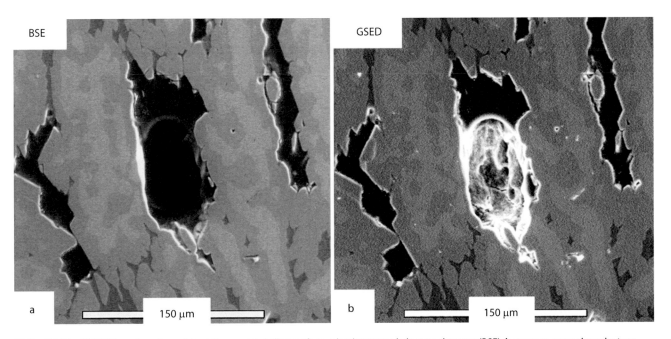

◘ Fig. 12.11   VPSEM imaging of a polished Raney nickel alloy surface. **a** backscattered electron detector (BSE). **b** gaseous secondary electron detector (GSED). Note the details visible in the shrinkage cavity in the GSED image

the mean free path for gas collisions increases rapidly with increasing electron energy thus decreasing the frequency of gas ionizations by the BSE. The contribution of the high energy BSE to the current amplification cascade is much less than that of the SE. To make simultaneous use of both the BSE and SE signals, a detector array such as that shown in ◘ Fig. 12.13 can be utilized, combining an annular scintillator BSE detector with the GSED. An example of the same area of the polished Raney nickel alloy simultaneously

obtained with the GSED is shown in ◘ Fig. 12.11b, operating under VPSEM conditions with water vapor at a pressure of 600 Pa.

Other variants of the GSED have been developed that make use of other physical phenomena that occur in the complex charged particle environment around the beam impact on the specimen, including the magnetic field induced by the motion of the charged particles and the cathodoluminescence of certain environmental gases induced by the SE

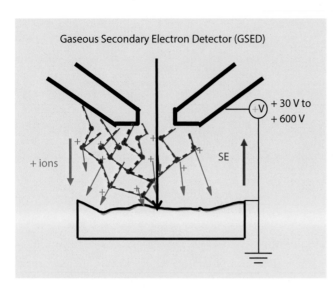

● **Fig. 12.12**   Schematic diagram showing principle of operation of the gaseous secondary electron detector (GSED)

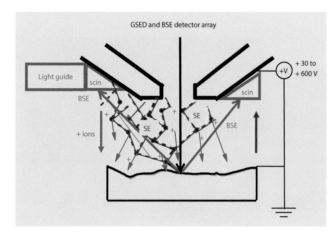

● **Fig. 12.13**   Co-mounted BSE and GSED detectors, showing positions relative to the electron beam

and BSE. An example of an induced-field SE detector image is shown in ● Fig. 12.10 (right).

12.7   Contrast in VPSEM

In general, the contrast mechanisms for the BSE and SE signals that are familiar from conventional high vacuum SEM operate in a similar fashion in VPSEM. For example, in the BSE detector image shown in ● Fig. 12.11a, most of the region of the Raney nickel alloy being viewed consists of a flat polished surface. Close examination of this image reveals atomic

number (compositional) contrast from the flat surface that is consistent with what would be observed for this specimen with the BSE signal in a conventional high vacuum SEM operating at the same beam energy. This same atomic number contrast can be observed in the simultaneously recorded GSED SE image in ● Fig. 12.11b. Atomic number contrast appears in the GSED SE image because of the atomic number dependence of the SE_2 class of secondary electrons that are generated by the exiting BSEs and are thus subject to the same contrast mechanisms as the BSEs. This is again familiar contrast behavior equivalent to high vacuum SEM imaging experience with the E–T detector. An important difference in the VPSEM case is the loss of the large contribution to atomic number contrast made by the SE_3 class in a high vacuum SEM. The SE_3 contribution is not a significant factor in the VPSEM since the SE_3 are generated on the chamber walls and objective lens outside of the accelerating field of the GSED and thus do not contribute to the SE signal.

Most BSE and SE images can be interpreted from the experience of high vacuum SEM, but as in all SEM image interpretation, the microscopist must always consider the apparent illumination situation provided by the detector in use. The GSED class of detectors is effectively located very close to the incident beam and thus provide apparent illumination along the line-of-sight. Moreover, the degree of amplification increases with distance of the surface from the GSED detector. These characteristics of the GSED lead to an important difference between ● Fig. 12.11a, b. The deep cavity is much brighter in the GSED image compared to the BSE image. The cavity walls and floor are fully illuminated by the electron beam and fine scale features can be captured at the bottom, as shown in the progressive image sequence in ● Fig. 12.14. The differing contrast in these images is a result of the relative positions and signal responses of the BSE and GSED detectors. Both detectors are annular, but the GSED detector is effectively looking along the beam and produces apparent lighting along the viewer's direction of sight. The annular BSE detector intercepts BSEs traveling at a minimum angle of approximately 20 degrees to the beam so that the effective lighting appears to come from outside the viewer's direction of sight. The cavity appears dark in the BSE image because although the primary beam strikes the walls and floor, there is no line-of-sight from cavity surfaces to the BSE detector. The BSEs are strongly reabsorbed by the walls and/or scattered out of the line-of-sight collection of the BSE detector. Because the environmental gas penetrates into the holes, as long as the primary beam can strike a surface and cause it to emit secondary electrons, the positive collection potential on the final pressure limiting aperture will attract electrons from the ionization cascade and generate a measurable SE signal, as shown schematically in ● Fig. 12.15 (Newbury 1996). Because of the added ionization path represented by the depth

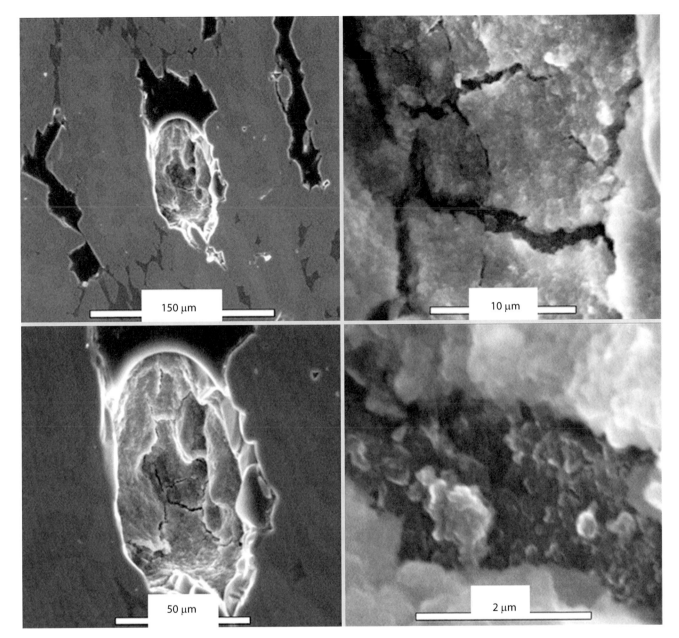

Fig. 12.14 Sequence of VPSEM images showing details revealed at the bottom of a cavity in polished Raney nickel alloy

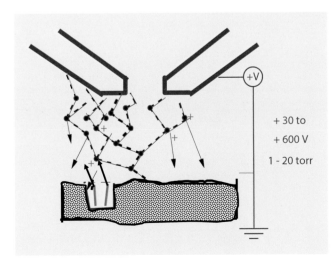

Fig. 12.15 Effect of extended gas path length created by a cavity

of the cavity, the SEs generated deep in the cavity create additional generations of cascade multiplication, increasing the signal compared to the flat surface of the specimen, making the cavities appear bright relative to the flat surface.

References

Danilatos GD (1988) Foundations of environmental scanning electron microscopy. Adv Electronics Electron Opt 71:109

Danilatos GD (1990) Mechanism of detection and imaging in the ESEM. J Microsc 160:9

Danilatos GD (1991) Review and outline of environmental SEM at present. J Microsc 162:391

Danilatos DD (1993) Introduction to the ESEM instrument. Microsc Res Tech 25:354

Newbury D (1996) Imaging deep holes in structures with gaseous secondary electron detection in the environmental scanning electron microscope. Scanning 18:474

ImageJ and Fiji

© Springer Science+Business Media LLC 2018
J. Goldstein et al., *Scanning Electron Microscopy and X-Ray Microanalysis*,
https://doi.org/10.1007/978-1-4939-6676-9_13

Software is an essential tool for the scanning electron microscopist and X-ray microanalyst (SEMXM). In the past, software was an important optional means of augmenting the electron microscope and X-ray spectrometer, permitting powerful additional analysis and enabling new characterization methods that were not possible with bare instrumentation. Today, however, it is simply not possible to function as an SEMXM practitioner without using at least a minimal amount of software. A graphical user interface (GUI) is an integral part of how the operator controls the hardware on most modern microscopes, and in some cases it is the only interface. Even many seemingly analog controls such as focus knobs, magnification knobs, or stigmators are actually digital interfaces mounted on hand-panel controllers that connect to the microscope control computer via a USB interface.

In addition to its role in data acquisition, software is now indispensable in the processing, exploration, and visualization of SEMXM data and analysis results. Fortunately, most manufacturers provide high-quality commercial software packages to support the hardware they sell and to aid the analyst in the most common materials characterization tasks. Usually this software has been carefully engineered, often at great cost, and smart analysts will take advantage of this software whenever it meets their needs. However, closed-source commercial software suffers from several limitations. Because the source code is not available for inspection, the procedures and algorithms used by the software cannot be checked for accuracy or completeness, and must be accepted as a "black box." Further, it is often very difficult to modify closed source software, either to add missing features needed by the analyst or to customize the workflow to meet specific job requirements. In this regard, open source software is more flexible and more extensible. The cost of commercial software packages can also be a downside, especially in an academic or teaching environment or in any situation where many duplicate copies of the software are required. Clearly a no-cost, open source solution is preferable to a high-cost commercial application if you need to install 50 copies for instructional purposes.

One of the most popular free and open source software packages for SEM image analysis is ImageJ, a Java program that has grown over the decades from a small application started at the National Institutes of Health (NIH) into a large international collaboration with hundreds of contributors and many, many thousands of users (▶ http://imagej.net).

13.1 The ImageJ Universe

ImageJ has grown into a large and multifaceted suite of related tools, and how all these parts fit together (and which are useful for SEM and X-ray microanalysis) may not be immediately obvious. The project began in the late 1970s when Wayne Rasband, working at NIH, authored a simple image processing program in the Pascal programming language that he called *Image*. This original application ran only on the PDP-11, but in 1987 when the Apple Macintosh II was becoming popular, Rasband undertook the development of a Mac version of the tool called *NIH Image*. Largely to enable cross-platform compatibility and to allow non-Macintosh users to run the program, it was again rewritten, this time using the Java programming language. The result was the first version of *ImageJ* in 1997 (Schneider et al. 2012, 2015).

The availability of ImageJ on the Microsoft PC and Unix platforms as well as Macintosh undoubtedly added to its popularity, but just as important was the decision to create an open software architecture that encouraged contributions from a large community of interested software developers. As a result, ImageJ benefitted from a prodigious number of code submissions in the form of macros and plugins as well as edits to the core application itself. Partly to manage this organic growth of the package, partly to reorganize the code base, and in part to introduce improvements that could not be added incrementally, NIH funded the *ImageJ2* project in 2009 to overhaul this widely useful and very popular program, and to create a more robust and more capable foundation for future enhancements (▶ http://imagej.net/ImageJ2).

Both ImageJ and ImageJ2 have benefitted from independent software development projects that interoperate with these programs. The *Bio-Formats* file I/O library as well as other related projects led by the Laboratory for Optical and Computational Instrumentation (LOCI) at the University of Wisconsin (▶ https://loci.wisc.edu) are important resources in the ImageJ universe and have added valuable functionality. The Bio-Formats project responded to the community's need for software that would read and write the large number of vendor-supplied image file formats, mostly for light microscopy (LM). Today the Bio-Formats library goes well beyond LM vendor formats and encompasses 140 different file types, including many useful for SEMXM, such FEI and JEOL images, multi-image TIFFs (useful for EDS multi-element maps), movie formats like AVI for SEM time-lapse imaging, etc. A follow-on LOCI project called SCIFIO aims to extend the I/O library's scope to include N-dimensional files (Hiner et al. 2016). Both projects are closely associated with the Open Microscopy Environment (OME) project and the OME consortium (▶ http://www.openmicroscopy.org). Similarly, the *ImgLib2* project aims to provide a neutral, Java-based computational library for processing N-dimensional scientific datasets of the kind targeted by SCIFIO (Pietzsch et al. 2012).

Given the complexity of this rapidly evolving ecosystem of interrelated and interoperable tools that support ImageJ, it is not surprising that some users find it difficult to understand how all the pieces fit together and how to exploit all the power available in this software suite. Fortunately, there is a simple way to access much of this power: by installing *Fiji*.

13.2 Fiji

Fiji, which is a recursive acronym that stands for "Fiji Is Just ImageJ," is a coherent distribution of ImageJ2 that is easy to install and comes pre-bundled with a large collection of useful plugins and enhancements to the bare ImageJ2 application (Schindelin et al. 2012). It is often thought of as "ImageJ with Batteries Included." The Fiji website provides several convenient installation packages for both the 32-bit and 64-bit versions of Fiji for common operating systems such as Microsoft Windows (currently Windows XP, Vista, 7, 8, and 10) and Linux (on amd64 and x86 architectures). Pre-built and tested versions for Mac OS X 10.8 (Mountain Lion) and later are also available. By default, these bundles include a version of the Java Runtime Environment (JRE) configured for Fiji's use that can coexist with other instances of Java on the host computer, but "bare" distributions of Fiji are available that will attempt to utilize your computer's existing JRE if that is preferred. Of course, as an Open Source software project, all of the source code for Fiji can be downloaded.

Installation of Fiji is straightforward because it has been configured as a *portable application*, meaning it is designed to run from its own directory as a standalone application. Installation is as simple as downloading the distribution and unpacking it; Fiji does not use an installer, does not copy shared libraries into destination directories scattered around the file system, and it does not store configuration information in system databases (e.g., the Windows registry). Because of this design, once installed it can be moved or copied simply by moving the directory tree. This portability also means it runs quite well from a USB flash drive or removable hard drive.

After launching Fiji you will be presented with the Fiji main window (◘ Fig. 13.1a), which contains the Menu bar, the Tools bar, and a Status bar for messages and other application feedback to the user. Selecting "Update…" or "Update Fiji" on the Help menu will trigger the updater, one of the most useful features of Fiji. Because Fiji is configured by default to start the updater immediately after program launch, for many new users this is the first piece of Fiji

◘ **Fig. 13.1 a** Fiji main window. **b** Fiji updater

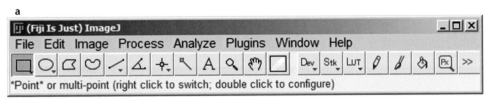

13

functionality they encounter. Upon activation the updater will scan your local Fiji installation and calculate checksums for everything to see if any components are out-of-date, or if new features have been added since it was last run. It will then confer with the global Fiji code repositories to look for updated Java Archive files (.jar files) and offer to download and install them for the user. □ Figure 13.1b shows an example of this, where the updater has located numerous changes in the ImageJ, Fiji, and Bio-Formats repositories. By selecting the "Apply changes" button the software will fetch the latest code and apply all the patches to the user's local Fiji installation. □ Figure 13.2 shows a window listing a selection of available Fiji update sites illustrating the rich community resources.

13.3 Plugins

One of the most powerful features of Fiji is the enormous collection of plugins, macros, and other extensions that have been developed by third-party contributors in the scientific

community. Fiji comes with some of the most useful plugins pre-installed, and these are accessible from the Plugins menu item. Hundreds of powerful features are accessible this way, exposed to the user in a series of cascading menus and submenus. Such a large set of choices can be overwhelming at first, but many of the plugins are meant for light microscopy, so the SEM analyst may find it simpler to ignore some of them. However, the Non-local means denoising plugin, the Optic flow plugin, and the myriad of morphological operations under the Plugins|Process menu are all useful for SEM microscopists, as are the dozens of features in the Registration, Segmentation, Stacks, Stitching, Transform, and Utilities submenus.

Sometimes the appearance of a plugin as a single entry in the Fiji menu structure belies the full power of that plugin. Indeed, some of the most impressive plugins available for Fiji might be considered entire image processing packages in their own right. An example of this is the Trainable WEKA Classifier plugin that appears as a single entry on the Segmentation submenu of the Plugins menu. WEKA is an acronym that stands for "Waikato Environment for Knowledge Analysis," a tool developed by the Machine Learning Group at the University of Waikato in New Zealand (Hall et al. 2009). WEKA is a full-featured and very popular open source software suite written in Java for machine learning (ML) researchers. It provides an open, cross-platform workbench for common ML tasks such as data mining, feature selection, clustering, classification, and regression, going well beyond just image analysis. The Fiji plugin is a gateway into this large array of tools and provides a convenient interface for processing SEM images using a modern machine learning framework (▶ http://imagej.net/Trainable_Weka_Segmentation).

Some of the most widely used and powerful plugins in Fiji have been back-ported into ImageJ itself, and are available directly from the main application's menu structure. An example of this is the Process menu option known as "Contrast Limited Adaptive Histogram Equalization," or CLAHE (Zuiderveld 1994). First developed in 1994, this algorithm has been implemented in a wide variety of image processing tools. It is designed to amplify local contrast by performing histogram equalization on small subsets (tiles) within the source image, but to limit the allowed amplification to reduce the tendency to magnify the noise in relatively homogeneous patches. ◻ Figure 13.3a shows a scanning electron micrograph of microfabricated features on silicon, acquired at 20 keV using an Everhart–Thornley detector. Because of slight misalignment of the raster with the linear features, Moiré contrast is evident in the image as bright edges on some features, and there are pure white and pure black horizontal lines that have been added to simulate contrast artifacts. These extreme limits of intensity preclude the usual brightness/contrast adjustments, but the CLAHE algorithm recovers invisible details without loss of information, ◻ Fig. 13.3b.

While it is possible that the ideal software tool for your project is available in Fiji itself (e.g., CLAHE) or in one of the many plugins loaded into Fiji by default (e.g., Trainable Weka Segmentation), it is much more likely that the tool you are looking for is not in the distribution you downloaded from the Fiji website. Only a small fraction of the plugins available to the user have been installed in the menu tree. A much larger collection awaits the user who is willing to explore the many optional Fiji update sites. The Updater window shown in ◻ Fig. 13.1b has a "Manage update sites" button at the lower left. If you press this button you are presented with a list of optional plugin repositories, as shown in ◻ Fig. 13.4a. When checked, these additional update sites will be accessed and used by the Updater to find new functionality to add into the base distribution. Some of the sites shown in ◻ Fig. 13.4a only add one or two items to the Plugins menu, while others import

a much larger amount of supplemental code and capability. For example, the "Cookbook" site listed in ◻ Fig. 13.4a adds a new top-level menu item to the Fiji main window, as shown in ◻ Fig. 13.4b. This new menu contains example code to help new users follow along with a community-written tutorial introduction to ImageJ, available on the ImageJ website (▶ http://imagej.net/Cookbook).

Occasionally a set of useful plugins will be written by a researcher or contributor who is unable or unwilling to make them available as an update site. The ImageJ website offers free hosting of update sites for any author of plugins, and organizations can run their own Fiji update sites if they wish. If these are not already a selectable option on the Manage update sites list (◻ Fig. 13.4a), the "Add update site" button allows the user to manually follow a third-party update site. As a last resort, plugins may also be manually installed into the Fiji plugins directory, but they will not be automatically updated so this is discouraged.

Thus, there are really four tiers of plugins across the ImageJ universe: (1) core ImageJ plugins that are bundled into the base ImageJ package (more than 1000 plugins in 2016); (2) core Fiji plugins, included by default in the "Batteries Included" Fiji distributions (more than 1000 additional plugins in 2016); (3) plugins available from additional update sites; and (4) plugins that must be located, downloaded, and installed manually. While this last category of plugins is the most likely to be buggy and poorly supported, any plugin written by a co-worker or officemate will often fall into this category, so the code may be highly specific to your task or your organization—don't overlook these!

13.4 Where to Learn More

Learning ImageJ or Fiji can be a daunting task for the beginner, and no attempt was made here to provide even a basic introduction to opening, exploring, manipulating, and saving SEM micrographs or X-ray data. However, there are many excellent resources for learning Fiji on the web, and the community offers several support channels for those who need additional help. Fiji itself has a built-in Help menu with links to the ImageJ and Fiji websites, newsgroups, online documentation, example code, developer tools, guidance documents, etc. The ImageJ Help page maintains links to the ImageJ Forum, Chat Room, and IRC channel as well as pointers to the ImageJ tag on Stack Overflow and Reddit, popular online locations for ImageJ and Fiji questions and answers. Finally, there is a synoptic search engine for many of the above resources at ▶ http://search.imagej.net.

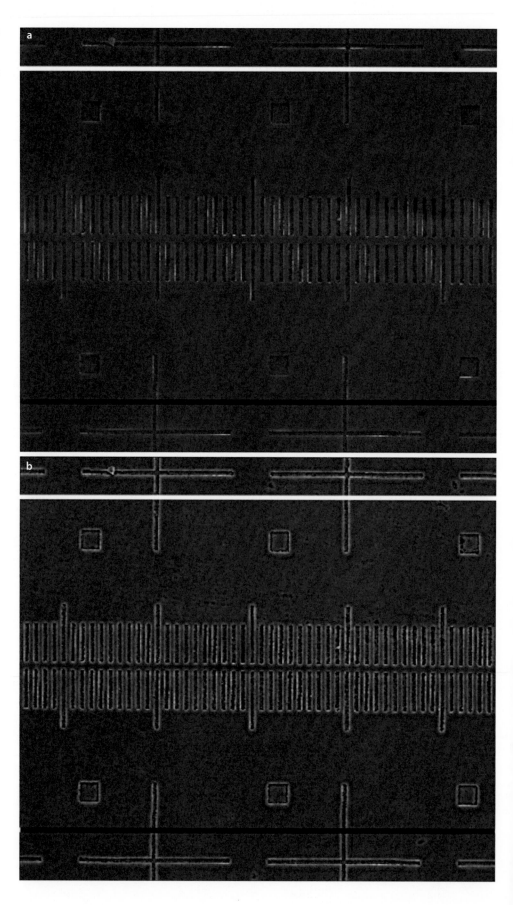

■ **Fig. 13.3** Application of Fiji's CLAHE processing to a low-contrast SEM images **a** and the resulting enhanced image **b**. The images are 256 μm wide

13

▪ Fig. 13.4 **a** Fiji's Manage
Update Sites window, showing
some of the many optional plugin
repositories available for use. **b** A
new top-level menu item called
"Cookbook" imported from the
Cookbook update site

a

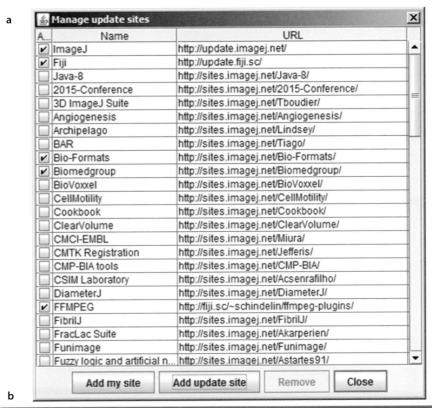

b

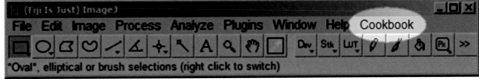

References

Hall M, Frank E, Holmes G, Pfahringer B, Reutemann P, Witten IH (2009)
The WEKA data mining software: an update. ACM SIGKDD Explor
Newsl 11(1):10–18

Hiner M, Rueden C, Eliceiri K (2016) SCIFIO [Software]. ▶ http://scif.io

Pietzsch T, Preibisch S, Tomancak P et al (2012) ImgLib2—generic image
processing in Java. Bioinformatics 28(22):3009–3011, ▶ http://
imagej.net/ImgLib2

Schindelin J, Arganda-Carreras I, Frise E et al (2012) "Fiji: an open-source
platform for biological-image analysis". Nat Methods 9(7):676–682,
▶ https://fiji.sc (Note that Fiji is not spelled FIJI)

Schindelin J, Rueden CT, Hiner MC et al (2015) The imageJ ecosystem: an
open platform for biomedical image analysis. Mol Reprod Dev
82(7–8):518–529

Schneider CA, Rasband WS, Eliceiri KW (2012) NIH Image to ImageJ: 25
years of image analysis. Nat Methods 9(7):671–675

Zuiderveld K (1994) "Contrast limited adaptive histogram equalization."
graphic gems IV. Academic Press Professional, San Diego, pp 474–485

SEM Imaging Checklist

© Springer Science+Business Media LLC 2018
J. Goldstein et al., *Scanning Electron Microscopy and X-Ray Microanalysis*,
https://doi.org/10.1007/978-1-4939-6676-9_14

14.1 Specimen Considerations (High Vacuum SEM; Specimen Chamber Pressure < 10^{-3} Pa)

14.1.1 Conducting or Semiconducting Specimens

A conducting or semiconducting specimen must maintain good contact with electrical ground to dissipate the injected beam current. Without such an electrical path, even a highly conducting specimen such as a metal will show charging artifacts, in the extreme case acting as an electron mirror and reflecting the beam off the specimen. A typical strategy is to use an adhesive such as double-sided conducting tape to both grip the specimen to a support, for example, a stub or a planchet, as well as to make the necessary electrical path connection. Note that some adhesives may only be suitable for low magnification (scanned field dimensions greater than $100 \times 100\ \mu m$, nominally less than 1,000× magnification) and intermediate magnification (scanned field dimensions between 100 μm x 100 μm, nominally less than 1,000× magnification and 10 μm × 10 μm, nominally less than 10,000× magnification) due to dimensional changes which may occur as the adhesive outgases in the SEM leading to image instability such as drift. Good practice is to adequately outgas the mounted specimen in the SEM airlock or a separate vacuum system to minimize contamination in the SEM as well as to minimize further dimensional shrinkage. Note that some adhesive media are also subject to dimensional change due to electron radiation damage during imaging, which can also lead to image drift.

14.1.2 Insulating Specimens

For SEM imaging above the low beam energy range ($E_0 \leq 5$ keV), insulating specimens must be coated with a suitable conducting layer to dissipate the charge injected by the beam and avoid charging artifacts. Note that after this layer is applied, a connection to electrical ground must be established for the coating to be effective. For tall specimens, the side of the specimen may not receive adequate coating to create a conducting path. A small strip of adhesive tape may be used for this purpose, running from the coating to the conducting stub. Note that for complex shapes, surfaces that do not receive the coating due to geometric shading may still accumulate charge even if not directly exposed to the beam due to re-scattering of backscattered electrons (BSEs).

To optimize imaging, the conductive coating should have a high secondary electron coefficient (e.g., Au-Pd, Cr, platinum-family metals). While thermally evaporated carbon is an effective, tough coating suitable for elemental X-ray microanalysis, the low secondary electron coefficient of carbon makes it a poor choice for imaging, especially for high resolution work involving high magnification where establishing visibility is critical.

The coating should be the thinnest possible that is effective at discharging the specimen, typically a few nanometers or less for ion-sputtered coatings. For high resolution imaging, the coating material should be chosen to have the least possible structure, for example, Au-Pd, which produces a continuous fine-grained layer, rather than pure Au, which tends to produce discontinuous islands.

Uncoated insulating specimens can be successfully imaged with minimum charging artifacts by carefully choosing the beam energy, typically in the range 0.1 keV–5 keV with the exact value dependent on the material, specimen topography, tilt, beam current, and scan speed to achieve a charge-neutral condition in which the charge injected by the beam is matched by the charge ejected as backscattered electrons and secondary electrons.

14.2 Electron Signals Available

14.2.1 Beam Electron Range

Beam electrons penetrate into the specimen spreading laterally through elastic scattering and losing energy through inelastic scattering creating the interaction volume (IV). The Kanaya–Okayama range equation gives the total penetration distance (for a beam incident perpendicular to the specimen surface):

$$R_{K-O}\,(\text{nm}) \;=\; (27.6\ \text{A})/(Z^{0.89}\rho)\ E_0^{1.67} \qquad (14.1)$$

where A is the atomic weight (g/mol), Z is the atomic number, ρ is the density (g/cm³), and E_0 is the incident beam energy (keV).

14.2.2 Backscattered Electrons

BSEs are beam electrons that escape the specimen after one or many elastic scattering events. The BSE coefficient increases with increasing atomic number of the target (compositional contrast) and with increasing tilt of a surface (topographic contrast). BSEs have a wide spectrum of kinetic energy, but over half retain a significant fraction, 50 % or more, of the incident beam energy. BSE sample specimen depths as great as 0.15 (high Z) to 0.3 (low Z) of R_{K-O} and spread laterally by 0.2 (high Z) to 0.5 (low Z) of R_{K-O}. From a flat surface normal to the incident beam, BSEs follow a cosine angular distribution (angle measured relative to the surface normal), while for tilted flat surfaces, the angular distribution becomes more strongly peaked in the forward direction with increasing surface tilt.

14.2.3 Secondary Electrons

Secondary electrons (SEs) are specimen electrons that are ejected through beam electron – atom interactions. SE have a distribution of kinetic energy which peaks at a few electron-volts. SEs sample only a few nanometers into the specimen due to this low kinetic energy. SE emission increases strongly with surface tilt (topographic contrast). SE emission increases as the beam energy decreases. Three classes of SEs are recognized: (1) SE_1 are produced as the beam electrons enter the specimen surface within footprint of the beam, potentially carrying high resolution information, and are sensitive to the first few nm below the surface. (2) SE_2 are produced as beam electrons exit as BSEs and are actually sensitive to BSE characteristics (lateral and depth sampling). (3) SE_3 are produced as the BSEs strike the objective lens and specimen chamber walls, and are also sensitive to BSE characteristics (lateral and depth sampling). SEs are sensitive to electrical and magnetic fields, and even a few volts of surface potential ("charging") can alter SE trajectories and eventual collection.

14.3 Selecting the Electron Detector

14.3.1 Everhart–Thornley Detector ("Secondary Electron" Detector)

Virtually all SEMs are equipped with an Everhart–Thornley detector, often referred to as the "secondary electron (SE)" detector. While SEs constitute a large fraction of the E–T signal, the E–T detector is also sensitive to BSEs directly and indirectly through the collection of SE_2 and SE_3. The E–T detector is the usual choice for imaging problems involving fine spatial details. The effective collection angle for SEs is nearly 2π sr. Some E–T detectors allow user selection of the potential applied to the SE-collecting Faraday cage so that the SE signal can be minimized or eliminated leaving a BSE signal. This BSE signal is collected over a very small solid angle, ~ 0.01 sr.

14.3.2 Backscattered Electron Detectors

Most SEMs are also equipped with a "dedicated" backscattered electron detector which has no sensitivity to SEs. Passive scintillator BSE detectors and semiconductor BSE detectors are typically placed on the bottom of the objective lens above the specimen, giving a large solid angle of collection approaching 2π sr. Both types have an energy threshold below which there is no response, the value of which depends on the particular detector in use and is typically in the range 1 keV to 5 keV. Above this threshold, the detector response increases nearly linearly with BSE energy, creating a modest energy selectivity.

14.3.3 "Through-the-Lens" Detectors

Some high performance SEMs include "through-the-lens" (TTL) detectors which use the strong magnetic field of the objective lens to capture SEs. The collection is restricted to the SE_1 and SE_2 signals, with the SE_3 component excluded. Since SE_3 actually carry lower resolution BSE information, excluding SE_3 benefits high resolution imaging. TTL BSE detectors capture the portion of the BSEs emitted into the bore of the lens. Some TTL SE and TTL BSE detectors can energy filter the signal-carrying electrons according to their energy.

14.4 Selecting the Beam Energy for SEM Imaging

The optimum beam energy depends on the nature of the imaging problem to be solved. The location of the feature (s) of interest on the surface or within the specimen; the contrast generating mechanism (s), and the degree of spatial resolution to be achieved are examples of factors to be considered.

14.4.1 Compositional Contrast With Backscattered Electrons

Choose $E_0 \geq 10$ keV: Above 5 keV, electron backscattering follows a nearly monotonic increase with atomic number, resulting in easily interpretable compositional contrast (aka "atomic number contrast"; "Z-contrast"). Because of the energy threshold of the passive scintillator BSE detector and semiconductor BSE detector (~1 keV to 5 keV), by selecting $E_0 \geq 10$ keV the BSE detector will operate reliably with the energy spectrum of BSEs produced by the specimen. For maximum compositional contrast, a flat polished specimen should be placed at 0^0 tilt (i.e., perpendicular to the beam).

14.4.2 Topographic Contrast With Backscattered Electrons

Choose $E_0 \geq 10$ keV: BSE detectors can respond strongly to variations in specimen topography, so the same beam energy conditions apply as for compositional contrast (▶ Sect. 14.4.1) to assure efficient BSE detector response. Local variations in the specimen surface tilt cause BSEs to travel in different directions. BSE topographic contrast is maximized by a small BSE detector placed on one side of the beam (e.g., Everhart–Thornley detector with zero or negative Faraday cage bias) and minimized by large BSE detectors placed symmetrically around the beam (e.g., large passive scintillator or semiconductor detector).

14.4.3 Topographic Contrast With Secondary Electrons

Choose any E_0 within the operating range: Topographic contrast is usually viewed in "secondary electron" images prepared with the E–T detector, positively biased for SE collection. The E–T detector is designed to efficiently collect and detect SEs, which are produced at all incident beam energies and are maximized at low beam energy.

14.4.4 High Resolution SEM Imaging

Two beam energy strategies optimize imaging fine-scale details by maximizing the contribution of the SE that are produced within the footprint of the focused beam:

Strategy 1

Choose the highest available beam energy, $E_0 \geq 25$ keV. The SE_1 component of the total SE signal retains the high resolution information at the scale of the beam entrance footprint. Due to lateral spreading of the interaction volume, the BSE and their associated SE_2 and SE_3 signals actually degrade spatial resolution at intermediate beam energy (e.g., 5 keV to 20 keV). As the beam energy increases, the electron range increases as $E_0^{1.67}$, causing the lateral spreading of BSEs to increase. When these signal components are spread out as much as possible by using the maximum beam energy, their contribution diminishes toward random noise, while the high resolution SE_1 contribution remains. Degraded signal-to-noise means that longer pixel dwell will be necessary to establish visibility of weak contrast. An additional advantage is the improvement in gun brightness, which increases linearly with E_0, so that more beam current can be obtained in the focused beam of a given size.

Strategy 2

Choose low beam energy, $E_0 \leq 2$ keV: as the beam energy is reduced, the electron range decreases as $E_0^{1.67}$, which collapses the BSE and associated SE_2 and SE_3 signals to dimensions approaching that of the footprint of the focused beam which defines the SE_1 distribution. These abundant BSE, SE_2 and SE_3 signals thus contribute to the high resolution signal rather than degrading it. Although there is a significant penalty in gun brightness imposed by low beam energy operation, the increased abundance of the high resolution signals partially compensates for the loss in gun brightness.

14.5 Selecting the Beam Current

14.5.1 High Resolution Imaging

Imaging fine spatial details requires a small beam diameter, which requires choosing a strong first condenser lens that inevitably restricts the beam current to a low value. Beam current (I_B), beam diameter (d), and beam divergence (α) are related through the Brightness (β) Equation:

$$\beta = 4 I_B / \left(\pi^2 d^2 \alpha^2 \right) \tag{14.2}$$

Using a small beam for high resolution inevitably restricts the beam current available. An important consequence of operating with low beam current is poor visibility of low contrast features.

14.5.2 Low Contrast Features Require High Beam Current and/or Long Frame Time to Establish Visibility

Contrast (C_{tr}), $C_{tr} = (S_2 - S_1)/S_2$, where $S_2 > S_1$, arises when the properties of a feature (e.g., composition, mass thickness, and/or surface tilt) cause a difference in the BSE (η) and/or SE (δ) thus altering the measured signal, $S_{feature} = S_2$, compared to the background signal, $S_{background} = S_1$, from adjacent parts of the

specimen. The visibility of this contrast depends on satisfying the Threshold Current Equation:

$$I_{th} > 4\,\text{pA} / \left(\delta\ \text{DQE}\ C_{tr}^2\ t_F \right) \tag{14.3a}$$

or in terms of the contrast threshold as

$$C_{th} > \text{SQRT} \left[4\,\text{pA} / \left(I_B\ \delta\ \text{DQE}\ t_F \right) \right] \tag{14.3b}$$

where δ is the secondary electron coefficient (η if imaging with backscattered electrons), DQE is the detective quantum efficiency (effectively the fraction of the collected electrons—detector solid angle and detection—that contribute to the measured signal), and t_F is the frame time (s) for a 1024 by 1024-pixel image. Lower values of C_{th} can be obtained with higher beam current and/or longer frame times. *For any selection of beam current and frame time, there is always a threshold contrast below which features will not be visible.*

14.6 Image Presentation

14.6.1 "Live" Display Adjustments

After the visibility threshold has been established for a contrast level C_{th} through appropriate selection of beam current and frame time, the imaging signal must be manipulated to properly present this contrast on the final image display. An image histogram function allows monitoring of the distribution of the displayed signal. Ideally, the signal amplification parameters (e.g., "contrast" and "gain" or other designations) are adjusted so signal variations span nearly the entire gray-scale range of the digitizer (8-bit, 0– 255) without reaching pure white (level 255) to avoid saturation or pure black (level 0) to avoid "bottoming"; both conditions cause loss of information.

14.6.2 Post-Collection Processing

Provided that the signal has been properly digitized (no saturation or bottoming), various digital image processing algorithms can be applied to the stored image to improve the displayed image, including contrast and brightness adjustment, non-linear expansion of a portion of the gray scale range, edge enhancements, and many others. ImageJ-Fiji provides a free open source platform of these software tools.

14.7 Image Interpretation

14.7.1 Observer's Point of View

The SEM image is interpreted as if the observer is looking along the incident electron beam. Your eye is the beam!

14.7.2 Direction of Illumination

The apparent source of illumination is from the position of the detector. The detector is the apparent flashlight!

14.7.3 Contrast Encoding

SEM image contrast is carried by number effects (different numbers of electrons leave the specimen because of local properties), trajectory effects (differences in the directions electrons travel after leaving the specimen), and energy effects (some contrast mechanisms are more sensitive to higher energy BSEs).

14.7.4 Imaging Topography With the Everhart–Thornley Detector

We are strongly conditioned to expect "top lighting"; that is, the illumination of a scene comes from above (e.g., sun in the sky, lighting fixtures on the ceiling). The E–T detector (positively biased to collect SE) collects a complex mix of SEs and BSEs, which produces an image of topographic surfaces that is easily interpretable if the effective position of the E–T detector is at the top of the scanned image, achieving top lighting. This condition can be achieved by adjusting the "scan rotation" control to place the E–T detector at the top (i.e., 12 o' clock position) of the scanned image (use a simple object like a particle—ideally a sphere—on a flat surface to establish the proper value of scan rotation). Brightly illuminated features then are those that face upwards. With top lighting, most viewers will properly interpret the sense of topography. Stereomicroscopy techniques can be employed to reinforce the proper interpretation of topography.

14.7.5 Annular BSE Detector (Semiconductor Sum Mode A + B and Passive Scintillator)

Because the BSE detector surrounds the electron beam symmetrically, the illumination appears to be along the viewer's line-of-sight, much like looking along a flashlight beam. Surfaces perpendicular to the beam appear bright, tilted surfaces darker. These detectors favor number contrast mechanisms such as BSE compositional contrast (atomic number contrast).

14.7.6 Semiconductor BSE Detector Difference Mode, A−B

The difference mode suppresses number effects but enhances trajectory effects such as topography.

14.7.7 Everhart–Thornley Detector, Negatively Biased to Reject SE

E–T(negative bias) collects only BSE within a small solid angle, giving the effect of strong oblique illumination (similar to a scene illuminated with a shallow sun angle and viewed from above, e.g., observer in an airplane at dawn or sunset).

14.8 Variable Pressure Scanning Electron Microscopy (VPSEM)

- Conventional SEM specimen chamber pressure < 10^{-3} Pa.
- VPSEM chamber pressure: 1 to 2000 pA (upper limit depends on specific VPSEM).

14.8.1 VPSEM Advantages

- Electron beam–BSE-SE interactions with gas atoms create ions and free electrons that discharge insulating specimens, minimizing charging artifacts.
- Water can be maintained in equilibrium (e.g., 750 Pa and 3 °C), enabling observation of biological specimens with minimum preparation as well as water-based reactions.

14.8.2 VPSEM Disadvantages

- The beam loses electrons due to gas scattering, reducing the effective useful signal generated by the electrons remaining unscattered in the focused beam while increasing noise due to scattered electron interactions. Nevertheless, nearly uncompromised high spatial resolution can be achieved. But for high resolution, compensate for loss of current in the beam by using longer frame times. For lower magnifications, compensate by using higher beam current.
- High voltage detectors such as the Everhart–Thornley secondary electron detector cannot operate due to high chamber pressure.
- VPSEM electron detectors: Gas cascade amplification detector (GSED) for SE detects SE_1 and SE_2 but avoids SE_3; passive semiconductor or scintillator detectors for BSE.

SEM Case Studies

© Springer Science+Business Media LLC 2018
J. Goldstein et al., *Scanning Electron Microscopy and X-Ray Microanalysis*,
https://doi.org/10.1007/978-1-4939-6676-9_15

15.1   Case Study: How High Is That Feature Relative to Another?

When studying the topographic features of a specimen, the microscopist has several useful software tools available. Qualitative stereomicroscopy provides a composite view from two images of the same area, prepared with different tilts relative to the optic axis, that gives a visual sensation of the specimen topography, as shown for a fractured galena crystal using the anaglyph method in ◘ Fig. 15.1 (software: Anaglyph Maker). The "3D Viewer" plugin in ImageJ-Fiji can take the same members of the stereo pair and render the three-dimensional surface, as shown in ◘ Fig. 15.2, which can then be rotated to "view" the surface from different orientations (◘ Fig. 15.3).

If the question is, How high is that feature relative to another?, as shown in ◘ Fig. 15.4 for the step height indicated by the yellow arrow, then the methodology of quantitative stereomicroscopy can be applied. First, a set of X-Y-coordinates is established by locating features common to both members of the stereo pair; for example, in ◘ Fig. 15.5a the red crosshair is placed on a feature in the lower surface which will define the origin of coordinates (0, 0). A feature is similarly identified in the upper surface, for example, the particle marked by the blue crosshair in ◘ Fig. 15.5a and the red arrow in ◘ Fig. 15.5b. The principle of the parallax measurement of the upper feature relative to the lower feature using these coordinate axes is illustrated in ◘ Fig. 15.5b, c. What is needed is the difference in the X-coordinates of the lower and upper reference features to determine the length of the X-vector from the measurement axes, which is the parallax for this feature. These measurements are conveniently made using the pixel coordinate feature in ImageJ-Fiji. By employing the expanded views presented in ◘ Fig. 15.6a, b for the left image, the individual pixels that define the reference points can be more readily seen, which improves the specificity of the feature location within the two images to ±1 pixel, minimizing this important source of measurement error. Having first calibrated the images using the "Set Scale" tool, the x-coordinate values from the pixel coordinate tool, from both the left and right images (with a tilt difference $\Delta\theta = 4^0$ with an estimated uncertainty of $\pm 1^0$), were used in the following calculations:

$$\text{Left image : X-vector length}\left(\text{red}\right)$$
$$= 214.9 \ \mu m - 137.8 \ \mu m = 77.1 \ \mu m \qquad (15.1)$$

$$\text{Right image : X-vector length}\left(\text{blue}\right)$$
$$= 187.9 \ \mu m - 121.0 \ \mu m = 66.9 \ \mu m \qquad (15.2)$$

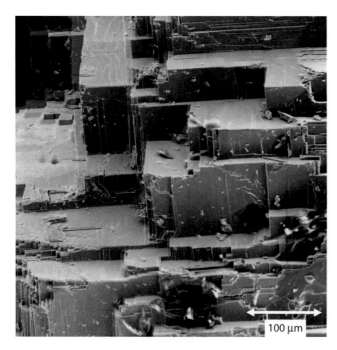

◘ **Fig. 15.1**   Anaglyph stereo pair presentation (software: Anaglyph Maker), to be viewed with the red filter over the left eye. Sample: fractured galena; Everhart–Thornley detector(positive bias); $E_0 = 20$ keV

◘ **Fig. 15.2**   "3D Viewer" plugin tool in ImageJ-Fiji operating on the same stereo pair presented as a two-image stack to create a rendering of the object surface

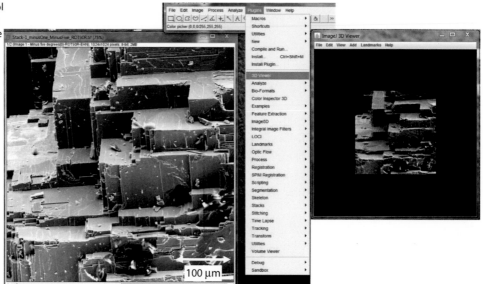

15.1 · Case Study: How High Is That Feature Relative to Another?

203

15

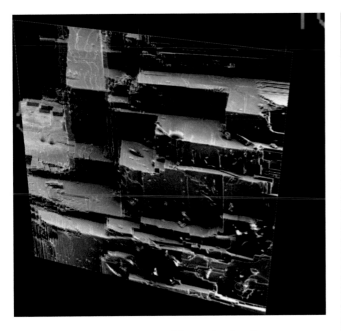

Fig. 15.3 "3D Viewer" rotation of the rendered surface

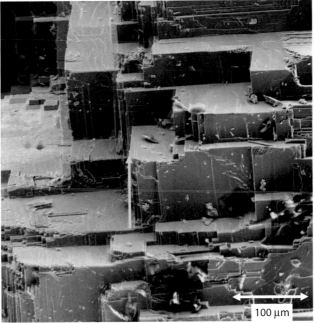

Fig. 15.4 Step height to be measured (*yellow arrow*)

Fig. 15.5 **a** Selection of lower (*red crosshair*) and upper (*blue crosshair*) features that lie on the lower and upper surfaces of the step to be measured. **b** Coordinate system established in the left-hand image relative to the lower surface feature. **c** Coordinate system established in the right-hand image relative to the lower surface feature

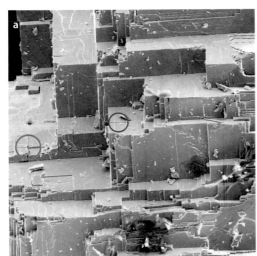

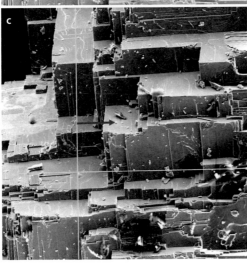

481.62x481.62 micrometer (1024x1024); 8-bit; 1MB

□ Fig. 15.6   a Use of the single pixel measurement feature in ImageJ-Fiji to select the reference pixel (center of the *red circle*) on the lower surface feature. **b** Use of the single pixel measurement feature in ImageJ-Fiji to select the reference pixel (center of the *blue circle*) on the upper surface feature

Parallax $= X_{Left} - X_{right} = 77.1$ μm $- 66.9$ μm $= 10.2$ μm (Note that the parallax has a positive sign, so the feature is *above* the reference point.)

$$Z = P / \left[2\sin(\Delta\theta)/2 \right] = 10.2 \text{ μm} / \left[2\sin\left(4^0/2\right) \right]$$
$$= 146.1 \text{ μm} \pm 6\% \tag{15.3}$$

Thus, the step represented by the yellow arrow in □ Fig. 15.4 is 146.1 μm ± 6% above the origin of the yellow arrow. The estimated uncertainty has two major components: an uncertainty of 0.1⁰ in the tilt angle difference contributes an uncertainty of ± 5% to the calculated step height. A ± 1 pixel uncertainty in selecting the same reference pixels for the lower and upper features in both images contributes ± 4% to the calculated step height.

15.2   Revealing Shallow Surface Relief

Surfaces with topographic structures that create shallow surface relief a few tens to hundreds of nanometers above the general surface provide special challenges to SEM imaging: (1) Shallow topography creates only small changes in the electron interaction volume and in the resulting emitted secondary electron (SE) and backscattered (BSE) signals as the beam is scanned across a feature, resulting in low contrast. (2) The strongest changes in the emitted signals from the weak topographic features will be found in the trajectory effects of the

BSE rather than in the numbers of the BSE or SE signals, so that an appropriate detector should be chosen that emphasizes the BSE trajectory component of topographic contrast. (3) Because the shallow relief is likely to provide very few "clues" as to the sense of the topography, it is critical to establish a condition of top lighting so that the sense of the local topography can be more easily determined. (4) Establishing the visibility of low contrast requires exceeding a high threshold current, so that careful control of beam current will be necessary. (5) The displayed image must be contrast manipulated to render the low contrast visible to the observer, which may be challenging if other sources of contrast are present.

An example of the shallow surface relief imaging problem is illustrated by a highly polished specimen with a microstructure consisting of large islands of Fe_3C (cementite) in pearlite (interpenetrating lath-like structures of Fe_3C and an iron-carbon solid solution). The strategy for obtaining a useful image of this complex specimen is based on the realization that the weak contrast from the shallow topography will be maximized with the BSE signal detected with a detector with a small solid angle of collection placed asymmetrically relative to the specimen and with a shallow detector elevation angle above the surface to produce the effect of oblique illumination. The small solid angle means that most BSE trajectories not directed into the detector will be lost, which actually increases the contrast. The asymmetric placement and shallow elevation angle ensures that the apparent illumination will come from a source that skims the surface, creating the effect of oblique illumination which creates strong shadows. The Everhart–Thornley (E–T) detector when biased negatively to reject SEs becomes a small solid angle BSE detector with these characteristics. The E–T detector is typically mounted so as to produce a shallow elevation angle relative to a specimen plane that is oriented perpendicular to the incident beam (0° tilt for a planar specimen). Before proceeding with the imaging campaign, the relative position of the E–T detector is confirmed to be at the 12-o'clock position in the image by using the "scan rotation" function and a specimen with known topography.

Figure 15.7 shows an image of the iron-carbon microstructure with the negatively biased E–T detector placed at the top of the image. A high beam current (10 nA) and a long dwell time (256 μs per pixel) were used to establish the visibility of low contrast. The displayed contrast was expanded by first ensuring that the histogram of gray levels in the raw image was centered at mid-range and did not clip at the black or white ends. The "brightness" and "contrast" functions in ImageJ-Fiji were used to spread the input BSE

intensity levels over a larger gray-scale output range. The contrast can be interpreted as follows: With the apparent illumination established as coming from the top of the image, bright edges must therefore be facing upward, and conversely, dark edges must be facing away. Thus, the topography of the Fe_3C islands can be seen to project slightly above the general surface. This situation occurs because the Fe_3C is harder than the iron-carbon solid solution, so that when this material is polished, the softer iron-carbon solid solution erodes slightly faster than the harder Fe_3C phase, which then stands in slight relief above the iron-carbon solid solution.

When this same field of view is imaged with the E–T detector positively biased, Fig. 15.8a, the same general contrast is seen, but there are significant differences in the fine-scale details. Several of these differences are highlighted in Fig. 15.8b. (1) It is much easier to discern the numerous small pits (e.g., yellow circles) in the E–T(positive bias) image because of the strong "bright edge" effects that manifest along the lip of each hole. (2) There are small objects (e.g., blue circles) which appear in the E–T(negative bias) image but which appear anomalously dark in the E–T(positive bias) image. These objects are likely to be non-conducting oxide inclusions that are charging positively, which decreases SE collection.

When this same field of view is imaged with the annular semiconductor BSE detector (sum mode, A + B) which provides apparent uniform, symmetric illumination along the beam, as shown in Fig. 15.9, the contrast from the shallow topography of the edges of the Fe_3C islands is entirely lost, whereas the compositional contrast (atomic number contrast) between the Fe_3C islands and the iron-carbon solid solution and Fe_3C lamellae is much more prominent.

Finally, when the BSE difference mode (A − B) mode is used, Fig. 15.10, the atomic number contrast is suppressed and the topographic contrast is enhanced. Note that the features highlighted in the blue circles in Fig. 15.8b are almost completely lost.

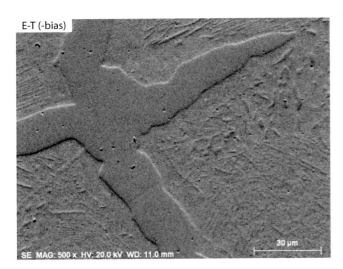

Fig. 15.7 Highly polished iron-carbon specimen imaged at $E_0 = 20$ keV and $I_p = 10$ nA with the negatively biased E–T detector. Image dimensions: 140×105 μm (Bar = 30 μm)

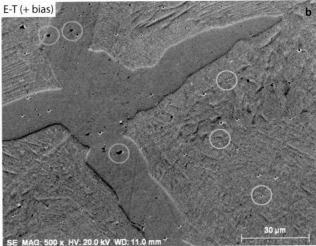

Fig. 15.8 **a** Same area imaged with a positively biased E–T detector. **b** Selected features highlighted for comparison

BSE (sum)

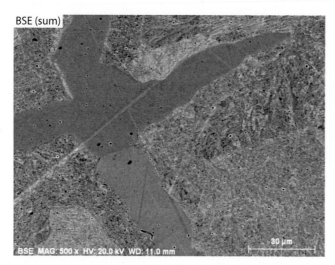

BSE MAG: 500 x HV: 20.0 kV WD: 11.0 mm   30 μm

◘ **Fig. 15.9**   Same area imaged with an annular semiconductor BSE detector (sum mode, A + B)

E_0 = 5 keV

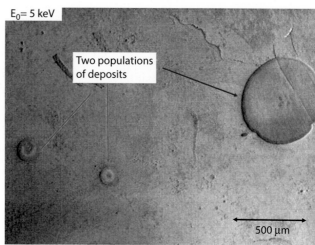

Two populations of deposits

500 μm

◘ **Fig. 15.11**   SEM-ET (positive bias) image of ink-jet deposits on a polished carbon substrate with E_0 = 5 keV; 32 μs/pixel = 25 s frame time

BSE (diff)

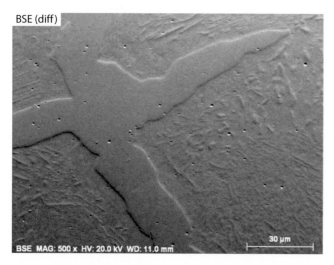

BSE MAG: 500 x HV: 20.0 kV WD: 11.0 mm   30 μm

◘ **Fig. 15.10**   Same area imaged with an annular semiconductor BSE detector (difference mode, A − B)

15.3   Case Study: Detecting Ink-Jet Printer Deposits

Ink-jet printing was used to deposit controlled quantities of reagents in individual droplets onto a polished carbon substrate in a project to create standards and test materials for instrumental microanalysis techniques such as secondary ion mass spectrometry. The spatial distribution of the dried deposits was of interest, as well as any heterogeneity within the deposits, which required elemental microanalysis.

The first critical step, detecting the ink-jet printed spots, proved to be a challenge because of the low contrast created by the thin, low mass deposit. Employing a low beam energy, $E_0 \leq 5$ keV, and secondary electron imaging using the positively biased Everhart–Thornley detector, maximized the contrast of the deposits, enabling detection of two different size classes of deposits, as seen in ◘ Fig. 15.11. When the beam energy was increased to enable the required elemental X-ray microanalysis, the visibility of the deposits diminished rapidly. Even with $E_0 = 10$ keV, which is the lowest practical beam energy to excite the K-shell X-rays of the transition metals, and a beam current of 10 nA, the deposits were not visible in high scan rate ("flicker free") imaging that is typically used when surveying large areas of a specimen to find features of interest. To reliably relocate the deposits at higher beam energy, a successful imaging strategy required both high beam current, for example, 10 nA, and long frame time, several seconds or longer, as shown in ◘ Fig. 15.12a–h. In this image series, the deposits are not visible at the shortest frame time of 0.79 s, which is similar to the visual persistence of a rapid scanned image. The class of approximately 100-μm diameter deposits is fully visible in the 6.4 s/frame image. As the frame time is successively extended to 100 s, additional features of progressively lower contrast become visible with each increase in frame time.

15

15.3 · Case Study: Detecting Ink-Jet Printer Deposits

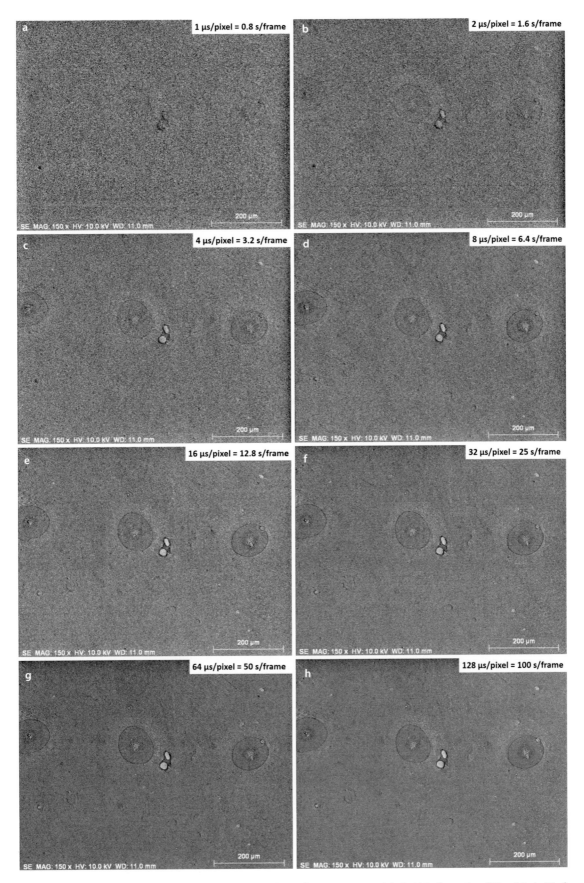

◨ **Fig. 15.12** **a** SEM-ET (positive bias) image of ink-jet deposits on a polished carbon substrate with $E_0 = 10$ keV: 1 μs/pixel = 0.8 s frame time. **b** 2 μs/pixel = 1.6 s frame time. **c** 4 μs/pixel = 3.2 s frame time. **d** 8 μs/pixel = 6.4 s

frame time. **e** 16 μs/pixel = 12.8 s frame time. **f** 32 μs/pixel = 25 s frame time. **g** 64 μs/pixel = 50 s frame time. **h** 64 μs/pixel = 100 s frame time

Energy Dispersive X-ray Spectrometry: Physical Principles and User-Selected Parameters

© Springer Science+Business Media LLC 2018
J. Goldstein et al., *Scanning Electron Microscopy and X-Ray Microanalysis*,
https://doi.org/10.1007/978-1-4939-6676-9_16

16.1 The Energy Dispersive Spectrometry (EDS) Process

As illustrated in ◘ Fig. 16.1, the physical basis of energy dispersive X-ray spectrometry (EDS) with a semiconductor detector begins with photoelectric absorption of an X-ray photon in the active volume of the semiconductor (Si). The entire energy of the photon is transferred to a bound inner shell atomic electron, which is ejected with kinetic energy equal to the photon energy minus the shell ionization energy (binding energy), 1.838 keV for the Si K-shell and 0.098 keV for the Si L-shell. The ejected photoelectron undergoes inelastic scattering within the Si crystal. One of the consequences of the energy loss is the promotion of bound outer shell valence electrons to the conduction band of the semiconductor, leaving behind positively charged "holes" in the valence band. In the conduction band, the free electrons can move in response to a potential applied between the entrance surface electrode and the back surface electrode across the thickness of the Si crystal, while the positive holes in the conduction band drift in the opposite direction, resulting in the collection of electrons at the anode on the back surface of the EDS detector. This charge generation process requires approximately 3.6 eV per electron hole pair, so that the number of charge carriers is proportional to the original photon energy, E_p:

$$n = E_p / 3.6 \, \text{eV} \tag{16.1}$$

For a Mn K-L_3 photon with an energy of 5.895 keV, approximately 1638 electron–hole pairs are created, comprising a charge of 2.6×10^{-16} coulombs. Because the detector can respond to any photon energy from a threshold of approximately 50 eV to 30 keV or more, the process has been named "energy dispersive," although in the spectrometry sense there is no actual dispersion such as occurs in a diffraction element spectrometer.

The original type of EDS was the lithium-drifted silicon [Si(Li)-EDS] detector (◘ Fig. 16.1a), with a uniform electrode on the front and rear surfaces (Fitzgerald et al. 1968). Over the last 10 years, the Si(Li)-EDS has been replaced by the silicon drift detector design (SDD-EDS), illustrated in ◘ Fig. 16.1b (Gatti and Rehak 1984; Struder et al. 1998). The SDD-EDS uses the same detection physics with a uniform front surface electrode, but the rear surface electrode is a complex pattern of nested ring electrodes with a small central anode. A pattern of potentials applied to the individual ring electrode creates an internal "collection channel," which acts to bring free electrons deposited anywhere in the detector volume to the central anode for collection. (Note that in some designs the small anode is placed asymmetrically on one side in a "teardrop" shape.)

The determination of the photon energy through the collection and measurement of this charge deposited in the detector requires an extremely sensitive and sophisticated electronic system, which operates automatically under computer control with only a limited number of parameters

under the user's control, as described below under "Best Practices." The charge measurement in the detector provides the fundamental unit of information to construct the EDS spectrum, which is created in the form of a histogram in which the horizontal axis is a series of energy bins, and the vertical axis is the number of photons whose energy fits within that bin value. As shown in ◘ Fig. 16.2, from the user's point of view this process of EDS detection can be considered simply as a "black box" which receives the X-ray photon, measures the photon energy, and increments the spectrum histogram being constructed in the computer memory by one unit at the appropriate energy bin. The typical photon energy range that can be measured by EDS starts at a threshold of 0.05 keV and extends to 30 keV or even higher, depending on the detector design.

16.1.1 The Principal EDS Artifact: Peak Broadening (EDS Resolution Function)

If the EDS detection and measurement process were perfect, all the measurements for a particular characteristic X-ray peak would be placed in a single energy bin with a very narrow width. For example, the natural energy width of Mn K-L_3 is approximately 1.5 eV. However, the number of electron–hole pairs generated from a characteristic X-ray photon that is sharply defined in energy is nevertheless subject to natural statistical fluctuations. The number of charge carriers that are created follows the Gaussian (normal) distribution, so that the variation in the number of charge carriers, n, in repeated measurements of photons of the same energy is expected to follow $1\sigma = n^{1/2}$. The 1σ value for the 1638 charge carriers for the MnK-L_3 photon is approximately 40 electron-hole pairs, which corresponds to a broadening contribution to the peak width of 0.024 (2.4 %) which can be compared to the natural width of 1.5 eV/5895 eV (from ◘ Fig. 4.2), measured as the full peak width a half-maximum height (FWHM), or 0.00025 (0.025 %), which is a broadening factor of approximately 100. The EDS peak width (FWHM) measured experimentally is a function of the photon energy, which can be estimated approximately as (Fiori and Newbury 1978)

$$\text{FWHM}(E) = \left[\left(2.5 \left(E - E_{\text{ref}} \right) + \text{FHWM}_{\text{ref}}^2 \right) \right]^{1/2} \tag{16.2}$$

where $\text{FWHM}(E)$, FHWM_{ref}, E and E_{ref} are expressed in electronvolts. The reference values for Eq. (16.2) can be conveniently taken from the values for Mn K-$L_{2,3}$ for a particular EDS system.

The EDS resolution function creates the principal artifact encountered in the measured EDS spectrum, which is the substantial broadening by a factor of 20 or more of the measured characteristic X-ray peaks, as shown in ◘ Fig. 16.3, where the peak markers (thin vertical lines) are approximately the true width of the Mn K-family characteristic X-ray peaks. Of course, all photons that are measured are

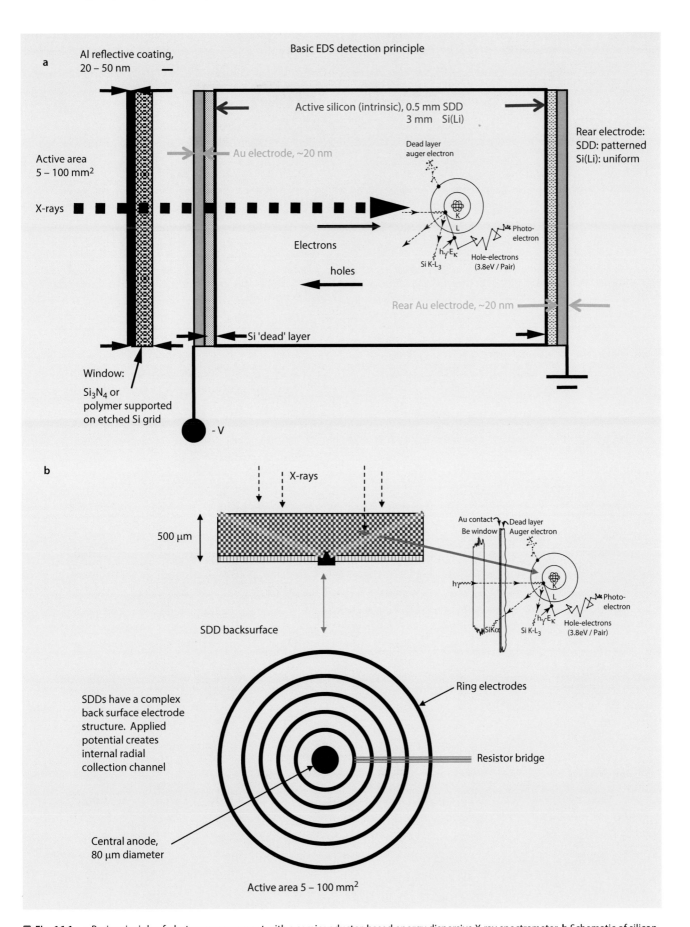

Fig. 16.1   a Basic principle of photon measurement with a semiconductor-based energy dispersive X-ray spectrometer. **b** Schematic of silicon drift detector (SDD) design, showing the complex patterned back surface electrode with a small central anode

■ **Fig. 16.2** "Black box" representation of the EDS detection and histogram binning process

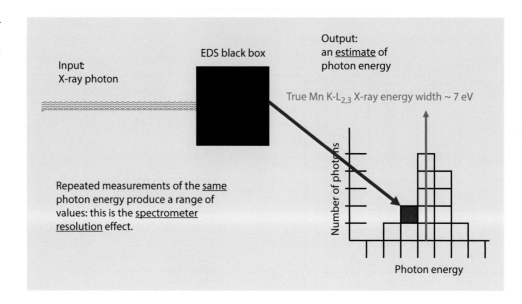

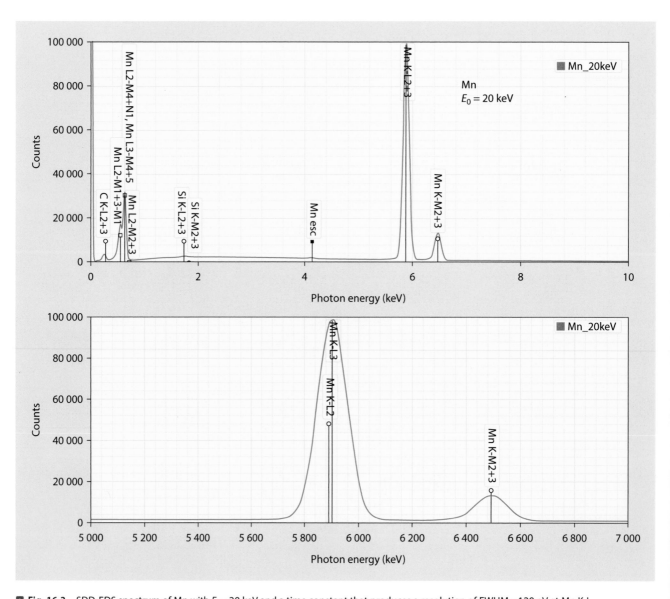

■ **Fig. 16.3** SDD-EDS spectrum of Mn with $E_0 = 20$ keV and a time constant that produces a resolution of FWHM = 129 eV at Mn K-$L_{2,3}$

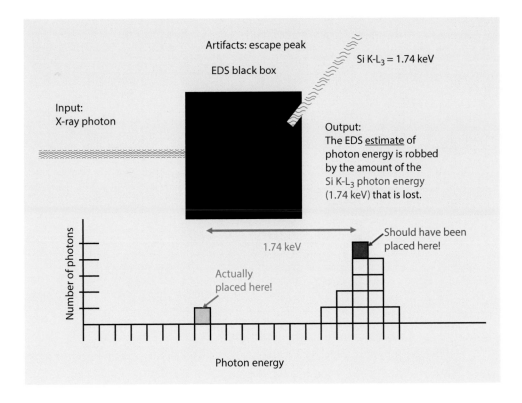

☐ Fig. 16.4 EDS "black box" representation of the Si-escape peak artifact

subject to the EDS resolution function, including the X-ray *bremsstrahlung* (continuum) background, but because the continuum is created at all energies up to E_0 and because of its slow variation with photon energy, the distortions introduced into the continuum background by the EDS resolution function are more difficult to discern.

The major impact of peak broadening is the frequent occurrence in practical analytical situations of mutually interfering peaks that arise even with pure elements, for example, Si K-L$_3$ and Si K-M$_3$ and the Fe L-family. When mixtures of elements are analyzed, Interferences are especially frequent when elements with atomic numbers above 20 are present since these elements have increasingly complex spectra of L- and M- shell X-rays that have a wide energy span. A secondary impact of peak broadening occurs when trace elements are to be measured. Peak broadening has the effect of spreading the characteristic X-rays over a wide range of the X-ray continuum background. Variance in the background sets the ultimate limit of detection.

16.1.2 Minor Artifacts: The Si-Escape Peak

After photoelectric absorption by a silicon atom in the detector, the atom is left in an ionized excited state with a vacancy in the K- or L- shell. This excited state will decay by intershell electron transitions that result in the emission of a Si Auger electron (e.g., KLL), which will undergo inelastic scattering and contribute to the free charge generation, or in about 10 % of the events, a Si K-shell X-ray. This Si K-shell X-ray will propagate in the detector and in most cases will undergo photoelectric absorption with the L-shell, ejecting

another photoelectron and further contributing to the charge generation. However, in a small number of events, as illustrated schematically in ☐ Fig. 16.4, the Si K-shell X-ray will escape from the detector, carrying with it 1.740 keV (for a Si K-L$_3$ X-ray) and robbing the original photon being captured of this amount of energy, which creates an artifact peak at an energy corresponding to:

$$\text{Escape peak energy} = \text{Parent peak energy} - 1.740\,\text{keV} \quad \textbf{(16.3)}$$

Si-escape peaks are illustrated for tin and gold in ☐ Fig. 16.5. The intensity ratio of the Si-escape peak/parent peak depends on the energy of the parent photon, with a maximum value for this ratio occurring for photon energies just above the Si K-shell ionization energy (1.838 keV) and decreasing as the photon energy increases. It is important to identify Si-escape peaks so that they are not mistaken for elements present at minor or trace levels.

16.1.3 Minor Artifacts: Coincidence Peaks

Although the EDS spectrum appears to an observer to accumulate simultaneously at all energies, the EDS system is in fact only capable of processing one photon at a time, with a duty cycle that ranges from 200 ns to several microseconds, depending on the particular EDS. If a second photon should enter the detector during this measurement period, the photon energies would be added together, producing an artifact known as a "coincidence peak" or a "sum peak," as illustrated schematically in ☐ Fig. 16.6. An "anti-coincidence function" or "fast discriminator" is incorporated in the signal processing

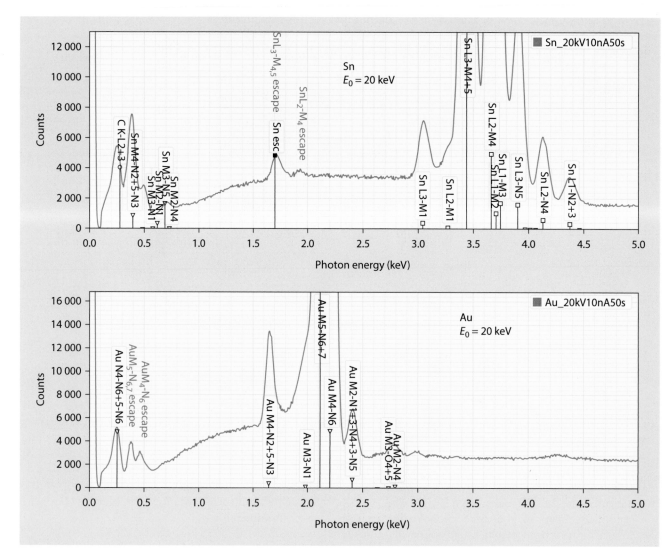

◻ **Fig. 16.5**   Si-escape peaks observed with an SDD-EDS for Sn and Au ($E_0 = 20$ keV)

◻ **Fig. 16.6**   EDS "black box" representation of photon coincidence

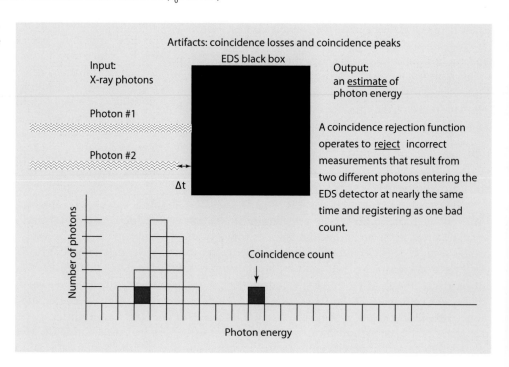

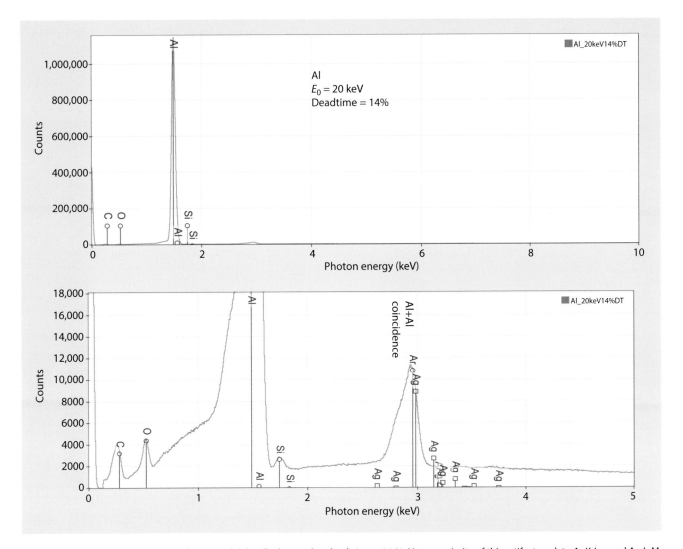

Fig. 16.7 ◘ **Fig. 16.7** Al at $E_0 = 20$ keV. Coincidence peak (Al + Al) observed at dead-time = 14 %. Note proximity of this artifact peak to Ar K-L$_{2,3}$ and Ag L-M family

chain to suppress this effect by rejecting the measurement of both photons, but as the flux of X-rays increases, an increasing frequency of events will occur in which the time separation between the two events is too short for the anti-coincidence function to recognize and reject the separate photons, resulting in an artifact sum photon. This coincidence phenomenon can occur between any two photons, for example, two characteristic X-rays, a characteristic X-ray plus a continuum X-ray, or two continuum X-rays. Coincidence produces a readily recognizable artifact peak when coincidence occurs between two photon energies that are particularly abundant, which is the case for high intensity characteristic X-ray peaks. An example is shown in ◘ Fig. 16.7, where two Al K-L$_3$ photons (1.487 keV) combine to produce a coincidence peak at 2.972 keV. Coincidence events can be formed from any two characteristic peaks, for example, O K-L + Si K-L$_2$. It is important to identify coincidence peaks so that they are not mistaken for characteristic peaks of elements present at minor or trace levels. Coincidence events involving lower energy photons will occur above the Duane–Hunt high energy limit

(which corresponds to the incident beam energy, E_0) and should not be mistaken for the true limit.

16.1.4 Minor Artifacts: Si Absorption Edge and Si Internal Fluorescence Peak

X-rays entering the EDS must pass through a window, typically a thin polymer, which is often supported on an etched silicon grid. Some X-rays will be absorbed in this grid silicon, especially those whose photon energy is just above the Si K-ionization energy (1.839 keV). In addition, there is a thin inactive Si layer ("dead-layer") just below the entrance electrode of the EDS that also acts to absorb X-rays. The X-ray mass absorption coefficient of silicon increases abruptly at the K-shell ionization energy, and this has the effect of increasing the absorption of the X-ray continuum, producing an abrupt step. However, the EDS resolution function acts to broaden all photon energies so that this sharp feature is also broadened, as seen in ◘ Fig. 16.8 (after peak fitting for Si) and made into a

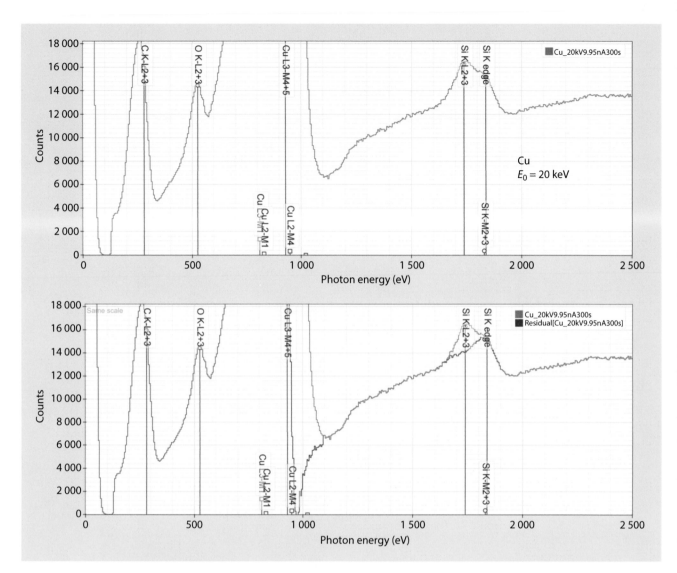

◻ Fig. 16.8   Cu at $E_0 = 20$ keV. The artifact Si peak is a combination of the Si K-absorption edge and the Si internal fluorescence peak (peak fitting in lower spectrum) created by absorption of X-rays in the Si support grid and the Si detector dead-layer, and subsequent Si X-ray emission

16

peak-like structure. The absorption of X-rays by the Si grid and Si dead-layer ionizes Si atoms and subsequently results in the emission of Si K-shell X-rays, which contribute a false Si peak to the spectrum. In the example for a copper target shown in ◻ Fig. 16.8, the apparent level of Si contributed by the internal fluorescence artifact is approximately 0.002 mass fraction.

16.2   "Best Practices" for Electron-Excited EDS Operation

While modern EDS systems are well supported by computer automation, there remain parameters whose selection is the responsibility of the user.

16.2.1   Operation of the EDS System

Before commencing any EDS microanalysis campaign, the analyst should follow an established checklist with careful attention to the measurement science of EDS operation. To establish the basis for quantitative analysis, the EDS parameters must be chosen consistently, especially if the analyst wishes to use archived spectra to serve as standards.

Choosing the EDS Time Constant (Resolution and Throughput)

The EDS amplifier time constant (a generic term which may be locally known as "shaping time," "processing time," "resolution," "count rate range," "1–6," etc.) should be checked. There are usually at least two settings, one that optimizes resolution (at the cost of X-ray throughput) and one that optimizes throughput (at the cost of resolution). Confirming the desired choice of the time constant is critical for consistent recording of spectra, especially if the analyst is using archived spectra to serve as standards for quantitative analysis. This is especially important when the EDS system is in a multi-user environment, since the previous user may have altered this parameter.

Channel Width and Number

The energy width of the histogram bins is typically chosen as 5, 10, or 20 eV. The bin energy width determines how many bins will define an X-ray peak. Since the peak width is a function of photon energy, as described by Eq. (16.2), decreasing from approximately 129 eV at Mn K-L3 (5.895 keV) to approximately 50 eV FWHM for C K-L$_2$ (0.282 keV), a selection of a 5-eV bin width is a useful choice to optimize peak fitting since this choice will provide 10 channels across C K-L$_2$. The number of bins that comprise the spectrum multiplied by the bin width gives the energy span. It is useful to capture the complete energy spectrum from a threshold of 0.1 keV to the incident beam energy, E_0. Thus, to span 0–20 keV with 5 eV bins requires 4096 channels.

Choosing the Solid Angle of the EDS

The solid angle Ω of a detector with an active area A at a distance r from the specimen is

$$\Omega = A / r^2 \qquad (16.4)$$

If the EDS is mounted on a translatable slide that can alter the detector-to-specimen distance, then the user must select a specific value for this distance for consistency with archived standard spectra if these are to be used in quantitative analysis procedures. Because of the exponent on the distance parameter r in Eq. (16.4), a small error in r propagates to a much larger error in the solid angle and a proportional deviation in the measured intensity.

Selecting a Beam Current for an Acceptable Level of System Dead-Time

X-rays are generated randomly in time with an average rate determined by the flux of electrons striking the specimen, thus scaling with the incident beam current. As discussed above, the EDS system can measure only one X-ray photon at a time, so that it is effectively unavailable if another photon arrives while the system is "busy" measuring the first photon. Depending on the separation in the time of arrival of the second photon, the anti-coincidence function will exclude the second photon, but if the measurement of the first photon is not sufficiently advanced, both photons will be excluded from the measurement and effectively lost. Due to this photon loss, the output count rate (OCR) in counts/second of the detector will always be less than the input count rate (ICR). The relation between the OCR and ICR is shown in ◻ Fig. 16.9 for a four-detector SDD-EDS. An automatic correction function measures the time increments when the detector is busy processing photons, and to compensate for possible photon loss during this "dead-time," additional time is added at the conclusion of the user-specified measurement time so that all

◻ **Fig. 16.9** Output count rate versus input count rate for a four-detector SDD-EDS

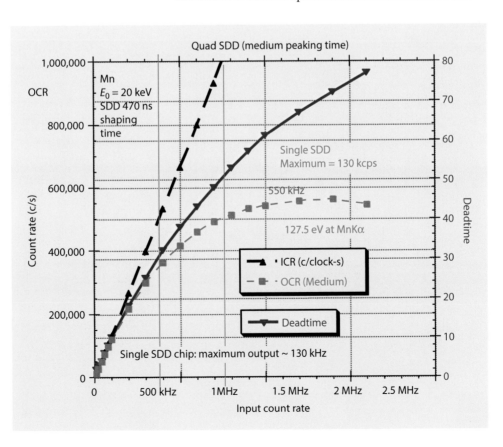

measurements are made on the basis of the same "live-time" so as to achieve constant dose for quantitative measurements. The level of activity of the EDS is reported to the user as a percentage "dead-time":

$$\text{Deadtime}(\%) = \left[(\text{ICR} - \text{OCR}) / \text{ICR} \right] \times 100\% \quad \textbf{(16.5)}$$

Dead-time increases as the beam current increases. The dead time correction circuit can correct the measurement

time over the full dead-time range to 80 % or higher. (Note that as a component of a quality measurement system, the dead-time correction function should be periodically checked by systematically changing the beam current and comparing the measured X-ray intensity with predicted.) However, as the dead-time increases and the arrival rate of X-rays at the EDS increases, coincidence events become progressively more prominent. This effect is illustrated in ◻ Fig. 16.10 for a sequence of spectra from a glass with six

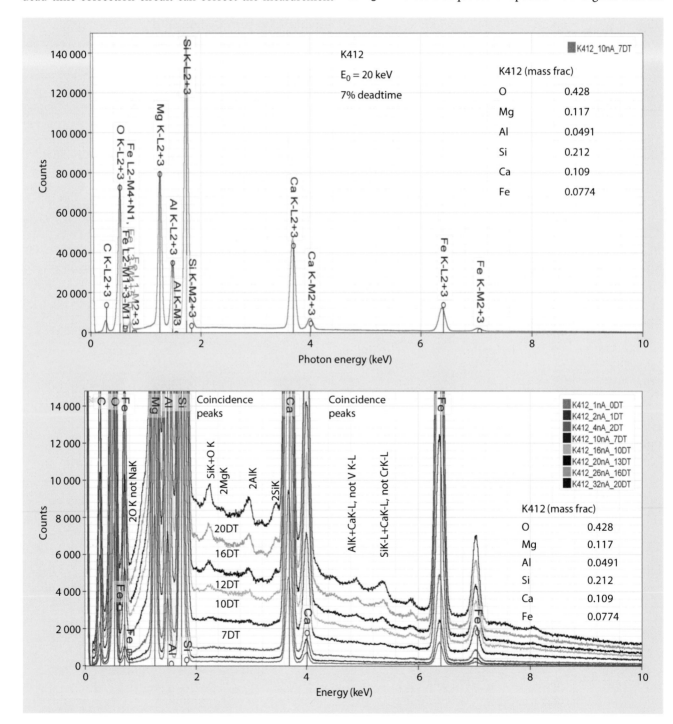

◻ **Fig. 16.10**   Development of coincidence peaks as a function of dead-time. NIST Standard Reference Material SRM 470 (Mineral Glasses) K412. (*upper*) SDD-EDS spectrum at 7 % dead-time showing the characteristic

peaks for O, Mg, Al, Si, Ca, and Fe. (*lower*) SDD-EDS spectra recorded over arrange of dead-times showing in-growth of coincidence peaks. Note elemental misidentifications that are possible major constituents (O, Mg, Al, Si, Ca, and Fe) measured at increasing dead-time. The spectra show the in-growth of a series of coincidence peaks as the dead-time increases. With the long pulses of the Si(Li) EDS technology, the pulse inspection function was effective in minimizing coincidence effects to dead-times in the range 20–30%. There is more vendor-to-vendor variability in SDD-EDS technology. Some vendors provide coincidence detection that will permit dead-times of up to 50%, while others are restricted to 10% dead-time. Since there is variability among vendors' SDD performance, it is useful to perform a measurement to determine the performance characteristic of each detector. See the sidebar for a procedure implementing such a procedure. Regardless of dead-time restrictions, an SDD-EDS is still a factor of 10 or more faster than an Si(Li)-EDS for the same resolution. In summary, as a critical step in establishing a quality measurement strategy, the beam current (for a specific EDS solid angle) should be selected to produce an acceptable rate of coincidence events in the worst-case scenario. This beam current can then be used for all measurements with reasonable expectation that the dead-time will be within acceptable limits.

■ ■ Sidebar: Protocol for Determining the Optimal Probe Current and Dead-Time

Aluminum produces one of the highest fluxes of X-rays per unit probe current: With the Al K-shell ionization energy of 1.559 keV, a modest beam energy of 15 keV provides an overvoltage of 9.6 for strong excitation. Al K-L$_2$ is of sufficient energy (1.487 keV) that it has low self-absorption, and at this energy the SDD efficiency is also relatively high. The Al K-L$_2$ energy is low enough that this peak is also quite susceptible to coincidence events. Pure aluminum thus makes an ideal sample for testing the coincidence detection performance of a detector and for determining the maximum practical probe current for a given beam energy.

1. Place a mounted, flat, polished sample of pure Al in the SEM chamber at optimal analytical working distance.
2. Mount a Faraday cup with a picoammeter in the SEM chamber.
3. Configure the detector at the desired process time.
4. Configure the SEM at the desired beam energy and an initial probe current. Measure the probe current using the Faraday cup/picoammeter.
5. Collect a spectrum from the pure Al sample with at least 10,000 counts in the Al K peak.
6. Use your vendor's software (or NIST DTSA-II) to integrate the background-corrected intensity in the Al K peak ($E = 1.486$ keV).
7. Use your vendor's software (or NIST DTSA-II) to look for and integrate the background-corrected intensity in the Al K + Al K coincidence peak ($E = 2.972$ keV).
8. Determine the ratio of the integrated intensity I(Al K + Al K)/I(Al K). We desire this ratio to be smaller than 0.01 (1%). In some trace analysis situations, it may desirable to have this ratio less than 0.001 (0.1%). Setting this limit too low will limit throughput but

setting it too high may make trace element analysis challenging.
9. If the ratio is too large, decrease the probe current and re-measure the probe current and the Al spectrum. Re-measure the ratio I(Al K + Al K)/I(Al K).
10. Repeat steps 5–10 until a suitable probe current has been determine.
11. Finally, note the suitable probe current and use it consistently at the beam energy for which it was determined.

16.3 Practical Aspects of Ensuring EDS Performance for a Quality Measurement Environment

The modern energy dispersive X-ray spectrometer is an amazing device capable of measuring the energy of tens of thousands of X-ray events per second. The spectra can be processed to extract measures of composition with a precision of a fraction of a weight-percent. However, this potential will not be realized if the detector is not performing optimally. It is important to ensure that the detector is mounted and configured optimally each time it is used. Some parameters change infrequently and need only be checked when a significant modification is made to the detector or the SEM. Other parameters and performance metrics can change from day-to-day and need to be verified more frequently. The following sections will step through a series of tests in a rationally ordered progression. The initial tests and configuration steps need only be performed occasionally, for example, when the detector is first commissioned or when a significant service event has occurred. Later steps, like ensuring proper calibration, should be performed regularly and a archival record of the results maintained.

16.3.1 Detector Geometry

In most electron-beam instruments, the EDS detector is mounted on a fixed flange to ensure a consistent sample/detector geometry with a fixed elevation angle. Almost all modern EDS detectors are mounted in a tubular snout with the crystal mounted at the end of the snout and the face of the active detector element perpendicular to the principle axis of the snout. The principal axis of the snout is oriented in the instrument such that it intersects with the electron beam axis at the "optimal working distance." This geometry is illustrated in ◻ Fig. 16.11.

Often the detector is mounted on the flange on a sliding mechanism that allows the position of the detector to translate (move in and out) along the axis of the snout. The elevation angle is nominally held fixed during the translation but the distance from the detector crystal changes and along with it the solid angle (Ω) subtended by the detector. The solid

Fig. 16.11 The elevation angle is a fixed property of the instrument/detector

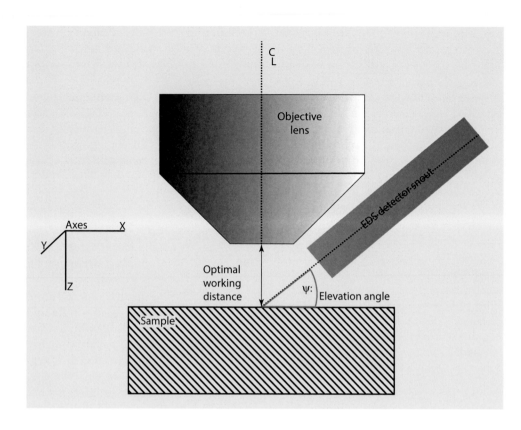

Fig. 16.12 The solid angle is a function of sample-to-detector distance and the active detector area

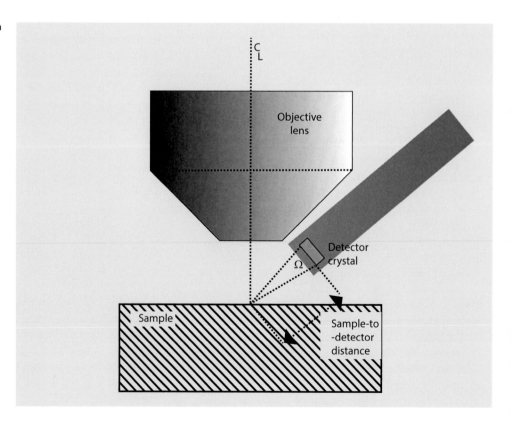

angle is illustrated in the ☐ Fig. 16.12. Moving the detector away from the sample is designed to decrease the solid angle but does not change the elevation angle or the optimal working distance.

It is important to be able to maintain a reproducible solid angle through consistent repositioning of the detector. Some slide mechanisms are motorized. Motorized mechanisms should define an "inserted" position and a "retracted" position, which you should test to ensure that the positioning is reproducible between insertions. Manual mechanisms usually provide a threaded screw with a manual crank to pull the detector in and out. The threaded screw will usually have a pair of interlocking nuts which can be positioned to define a consistent insertion position. The procedure in the sidebar below will allow you to set and maintain a constant solid angle and thus also a consistent detector collection efficiency.

■ Sidebar: Setting a Constant Detector-to-Sample Distance to Maintain Solid Angle

1. Locate the pair of lock nuts on the screw mechanism. Move the lock nuts to the inner most position on the treaded rod.
2. Insert the detector as close to the sample as possible. Ensure that the detector does not touch the interior of the microscope. The detector snout must be electrically isolated (no conductive path) from the interior of the chamber to eliminate noise caused by electrical ground loops.
3. Twist the upper lock nut to limit the motion of the detector towards the sample. Tighten the lower nut to lock the upper nut into position.

4. Test the reproducibility of the insertion point by extracting and inserting the detector and collecting a series of spectra. If the characteristic peak intensities are reproducible (to much better than a fraction of a percent) between insertions, the precision is adequate.

The take-off angle is the angle at which X-rays exit a flat sample in the direction of the detector. For a flat sample mounted perpendicular to the electron beam at the optimal working distance, the take-off angle equals the elevation angle. If the sample is tilted or the sample surface is at a slightly different working distance, then the take-off angle can be computed from the sample tilt, the working distance and the sample-to-detector distance. This is shown in ■ Fig. 16.13. Often you will hear the terms elevation angle and take-off angle used interchangeably. It is more precise however to think of the elevation angle as being a fixed property of the instrument/detector geometry and the take-off angle being dependent upon instrument-specimen configuration.

■ Check 1: Verify the Elevation Angle
It is critical that your quantitative analysis software has the correct elevation/take-off angle. Matrix correction algorithms use the take-off angle to calculate the correct absorption

■ **Fig. 16.13** The take-off angle is a function of the elevation angle, the sample tilt and the actual working distance

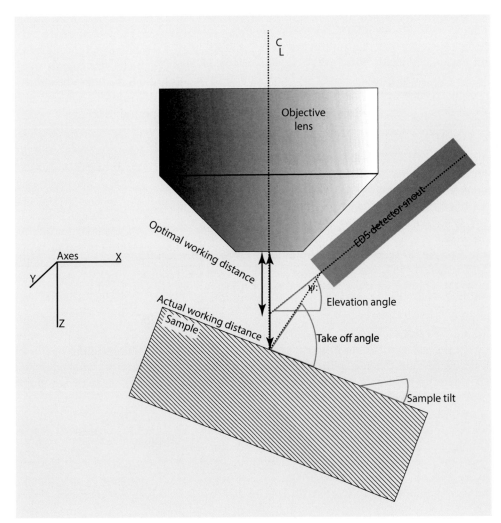

◻ **Fig. 16.14** Using a cell phone inclinometer to measure the elevation angle

correction. The elevation angle (usually sloppily called the "take-off angle") is often a configuration option that may be available to you to modify or may require a service engineer. Usually you can check the elevation angle by opening a spectrum data file in a text editor. Most spectrum files are ASCII text files and the vendors write a line containing the elevation or nominal take-off angle.

Compare the value produced by the software with the physical position of your detector relative to the beam axis. Sometimes, the correct instrument specific value of the take-off angle will be handwritten in the EDS vendor's documentation. Other times, you can extract the angle from instrument specific schematic diagrams from the SEM or EDS vendors. Regardless, it is a good idea to verify the elevation angle using a protractor or a smart phone. Many smart phones contain inclinometers which are accurate to within a degree and "bubble level apps" are available to turn the phone into a digital inclinometer. First, test the accuracy of the cell phone inclinometer using a 30-60-90 triangle or similar reference shape. Then use the cell phone to measure the angle of the snout relative to the angle of the column and calculate the elevation angle. ◻ Figure 16.14 shows that a cell phone measures the elevation angle of this detector to within half a degree (90°− 54.5° = 35.5° ~ 35° nominal value). Achieving similar accuracy with a protractor and a bubble level is difficult.

16.3.2 Process Time

The "process time" (also called the "throughput setting," the "detector time constant," the "resolution setting" or other names) determines how much time the detector electronics dedicates to processing each incoming X-ray. A longer "process time" tends to produce lower X-ray throughput but higher spectral resolution. Shorter "process times" tend to produce higher X-ray throughput but lower spectral resolution. By throughput, we mean the maximum number of X-rays that the detector can measure

per unit time. By spectral resolution we mean the width of a characteristic peak, usually taken to be the Mn K-L$_{2,3}$ (Kα) peak.

Higher throughput is desirable because it allows you to use higher probe currents to produce more X-rays and produce measured spectra with a larger number of measured X-rays. Higher resolution is desirable because it becomes easier to distinguish characteristic peaks of similar energy. Both are virtues but one is achieved at the expense of the other.

The task of selecting a process time is the task of balancing good throughput with adequate resolution. At the best resolution settings, throughput is seriously compromised and as a result the precision of quantitative analysis is also compromised. At the highest throughput settings, resolution is compromised and it becomes much more difficult to distinguish interfering peaks.

Fortunately, resolution degrades relatively slowly while throughput increases quickly. Moderate pulse process times tend to degrade the ultimate resolution by a few percent, while increasing throughput by much larger factors. Every vendor is different but typically a moderate pulse process time will produce both adequate resolution and excellent throughput.

While some modern SDD are capable of ultimate resolutions of 122–128 eV, compromising the resolution to 130–135 eV will produce an excellent compromise between resolution and throughput. Even a resolution of 140–150 eV at high throughputs can produce excellent quantitative results because the precision of spectrum fitting is more determined by the number of measured X-rays than the spectral resolution.

- **Check 2: Selecting a Process Time**
  - Ensure that your EDS detector electronics are set to a moderate process time that produces a good throughput with a resolution within 5 eV of the best resolution. Record this parameter in your electronic notebook for future reference.
  - Do not use an "adaptive process time" for standards-based quantitative analysis. Adaptive process times allow the resolution to change with throughput making spectrum fitting challenging and less accurate.

The optimization of throughput is also confounded by an another practical limitation—coincident X-rays or pulse-pileup—and will be addressed in a later section.

16.3.3 Optimal Working Distance

Nominally, the optimal working distance should be specified in instrument schematics. However, it is worth checking because it may vary slightly or there may be a mistake in design or manufacturing. From the detector geometry section, we know that the optimal working distance is the distance at which the axis along the detector snout intersects with the electron beam axis (see ▫ Fig. 16.13). This working distance should also produce the largest flux of X-rays and be the distance which is least sensitive to slight errors in vertical positioning.

- **Check 3: Measuring the Optimal Working Distance**
  1. Select a sample like a piece of copper and mount it perpendicular to the electron beam axis.
  2. Select a beam energy of 15–25 keV and a moderate probe current to produce ~ 20% dead time.
  3. Move the stage's vertical axis to place the sample close to the objective lens pole piece. Be careful not to run the sample into a detector.
  4. Focus the image and record the working distance reported by your SEM's software.
  5. Collect a 60 live-time second spectrum.
  6. Move the stage away from the objective lens pole piece in 1-mm steps.
  7. Repeat steps 4–6, taking the working distance through the nominal optimal working distance.
  8. Process the spectra. Extract the total number of counts in the energy range from about 100 eV to the beam energy and plot this number against the working distance.
  9. Determine the working distance which produces the largest number of counts. Since this working distance represents an inflection point, the slope will be minimum and thus the sensitivity with respect to working distance will also be minimized.

16.3.4 Detector Orientation

In the previous section, we assumed that the principal axis of the detector snout is oriented to intersect with the electron beam axis. In other words, the detector points towards the sample. It is usually the EDS vendor's responsibility to ensure that the mounting flange has been designed to correctly orient and position the detector. The next check will verify this.

The active face of an EDS detector is a planar area that is mounted perpendicular to the snout axis. In front of the detector element there is usually a window and an electron trap. Most windows are ultrathin layers of polymer or silicon nitride mounted on a grid for mechanical strength. Examples of two support grids are given in ▫ Fig. 16.15. While the grid may have an open area fraction of 75–80 %, the silicon or carbon grid bars are often very thick (0.38 mm) to enhance mechanical rigidity under the strain of up to one atmosphere of differential pressure. Off-axis the grid bars can occlude the direct transmission of X-rays from the sample to the detector element. Furthermore, the magnetic electron trap can also occlude X-rays from off-axis. As a result, an EDS detector is more sensitive to X-rays produced on the snout axis than slightly off the axis. The result is a position dependent efficiency which peaks on axis and decreases as the source of the X-rays is further from axis.

A wide field-of-view X-ray spectrum image can demonstrate the position sensitivity and can be used to ensure that the detector snout and detector active element are oriented correctly.

- **Check 4: Collect a Wide Field X-ray Spectrum Image**
  1. Mount a flat, polished piece of Cu in your SEM.
  2. Image the Cu at the optimal working distance and at 20–25 keV to excite both the K and L lines.
  3. Find out how wide a field-of-view your SEM can image at the optimal working distance. The example in ▫ Fig. 16.16 uses a 4-mm field-of-view.
  4. Collect a high count X-ray spectrum image from the Cu. Acquiring at a moderate-to-high probe current for an hour or more at 256 × 256-pixel image dimensions should produce sufficiently high signal-to-noise data.
  5. Process the data to extract and plot the raw intensities at each pixel in each of the Cu K-$L_{2,3}$ and Cu L-family lines.
     1. It is important to extract the raw intensities and not the normalized intensities since we are looking for variation in the raw intensity as a function of position.
     2. The open source software ImageJ-Fiji (ImageJ plus additional tools) can be used to process spectrum image data if it can be converted to a RAW format.

	AP3	AP5
Thickness (µm)	380	265
Rib width (µm)	59	45
Opening width (µm)	190	190
Open area %	76%	78%
Acceptance angle	53°	72°

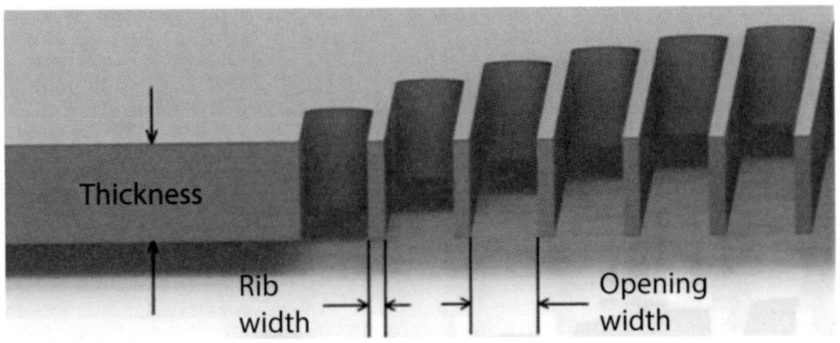

▫ **Fig. 16.15** Window support grid dimensions for two common window types (Source: MOXTEK)

6. Plot the data to demonstrate the variation of the intensity as a function of position. ◙ Figure 16.16 shows the map from a well-oriented detector plotted using a thermal color scheme in which red represents the highest intensity and blue represents zero intensity. ◙ Figure 16.17 shows traverses extracted from the ◙ Fig. 16.16 data on diagonals representing parallel to the detector axis and perpendicular to the detector axis. Verify that the most intense region in the intensity plots is in the center of the image area.

7. Note the extent of the region of uniform efficiency. Variation from ideal uniform sensitivity has consequences.
   1. Low magnification X-ray spectrum images will suffer from reduced intensity towards the edges.
   2. Point mode X-ray spectrum acquisitions collected off the optical axis will also suffer from dimi-

nished intensities leading to low analytical totals and sub-optimal quantitative results.

3. Typically, the Cu L-family peak is more sensitive due to absorption by the vacuum window support grid's Si ribs. ◙ Figure 16.15 (source: Moxtek) shows the design of two recent Moxtek support grids. The vertical sensitivity is usually minimized by orienting the grid ribs vertically.

■ **Sidebar: Processing a "RAW" Spectrum Image with ImageJ-Fiji**

1. Convert the X-ray spectrum image data into a RAW file. A RAW file is large binary representation of the data in the spectrum image. Each pixel in the spectrum image consists of a spectrum encode in an integer binary format. The pixels are organized in a continuous array row-by-row. The size of the file is typically equal to (channel depth) × (row dimension) × (column dimensions) × (2 or 4 bytes per integer value).

2. Import the RAW data file into ImageJ using the "Import → Raw" tools to create a "stack" as shown in ◙ Fig. 16.18.

3. As imported, the orientation of the stack will depend upon how the data in the RAW file is organized. Regardless of the original orientation, you will need to pivot the data a couple times using the "Image → Stack → Reslice" tool. First, to identify the range of channels that represent the Cu L-family and Cu K-$L_{2,3}$ intensities. Second, to align the spectrum data with the Z dimension so that the "Image → Stack → Z-project" tool can be used to create plots representing the intensities in the Cu L-family and Cu K-$L_{2,3}$ channels.

4. The initial view of the imported spectrum will usually show the data as shown in ◙ Fig. 16.19. In this view it is possible to identify the range of

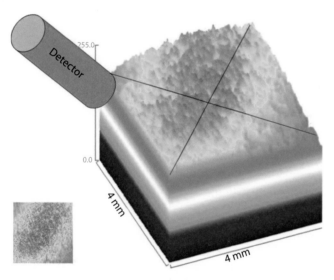
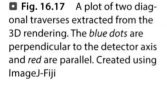

◙ **Fig. 16.16** A 3D rendering of the intensity in the Cu K line over a 4 mm by 4 mm mapped area. Created using ImageJ-Fiji

◙ **Fig. 16.17** A plot of two diagonal traverses extracted from the 3D rendering. The *blue dots* are perpendicular to the detector axis and *red* are parallel. Created using ImageJ-Fiji

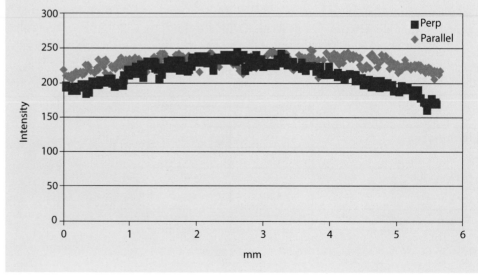

channels to sum together for the Cu L-family and Cu K-L$_{2,3}$ lines. The Cu K-L$_{2,3}$ lines are often quite dim.

5. Using the "Image → Stack → Reslice" tool (see ◘ Fig. 16.20) rotate the stack until it looks like ◘ Fig. 16.21
6. Use the "Image → Stack → Z-project" twice (◘ Fig. 16.22) to extract the range of channels identified in ◘ Fig. 16.19 as associated with the Cu L-family and Cu K-L$_{2,3}$ lines.
7. Convert the gray-scale image to a thermal scale using "Image → Lookup Tables → Thermal."
8. Convert the image to a plot using "Analyze → Surface Plot"
9. Extract traverses from the image using the straight line tool to define the traverse and then the "Analyze → Plot Profile" tool to extract and plot the data.

16.3.5 Count Rate Linearity

One of the most important circuits in an X-ray pulse processor accounts for the time during which the pulse processor is busy processing X-ray events. When the pulse processor is processing an X-ray, it is unavailable to process new incoming X-rays. This time is called "dead-time." In contrast, the time during which the processor is not busy and is available is called "live-time." The sum of "live-time" and "dead-time" is called "real-time" – the time you would measure using a wall clock.

For calculating the effective probe dose, live-time is the critical parameter. The effective probe dose consists of those electrons which could produce measurable X-rays. So the effective probe dose equals the live-time times the probe current. The effective probe dose is always less than the real probe dose, which is the product of the real time and the

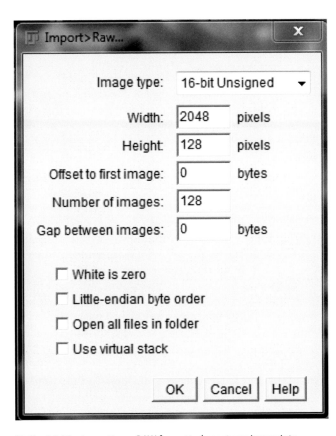

◘ **Fig. 16.18** Importing a RAW formatted spectrum image into ImageJ-Fiji

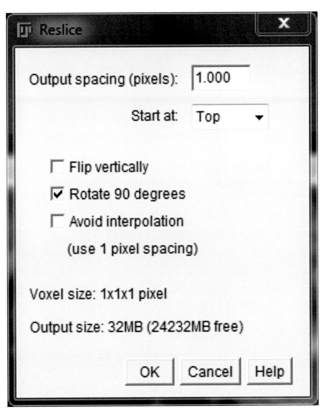

◘ **Fig. 16.20** "Reslice" tool used to rotate the stack

◘ **Fig. 16.19** The spectrum image perspective as imported into ImageJ-Fiji. The bright strip is the Cu L-family while the much fainter band is Cu K-L$_{2,3}$

◘ **Fig. 16.21** The spectrum image was collected on a 128 by 128 pixel grid representing 4 × 4 mm. After data rotation to view the x-y plane, the first energy slice, 1 out of 1024, is shown. It appears black because it contains no counts

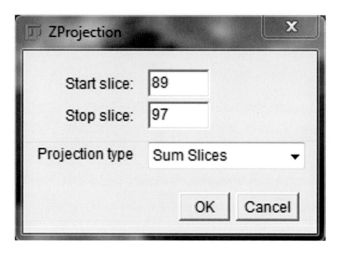

◘ **Fig. 16.22** Extracting the Cu L-family lines from the spectrum planes 89 to 97

probe current, due to the loss of useful electrons during the dead-time.

It is very important for accurate quantitative analysis that the measured X-ray intensity is linear with respect to the effective probe dose. Twice the effective probe dose should produce twice the measured intensity. Not only is this

important because it is often hard to replicate the identical probe current but it is also important because different materials measured with the same probe current will produce different dead times.

There really is no excuse for a modern X-ray detector not to be linear. However, it is worth checking because the test can expose other potential problems like a non-linear or off-set probe current meter.

- **Check 5: Count Rate Linearity with Effective Probe Current**
  Equipment:
  - Faraday cup
  - Picoammeter
  - Flat, polished copper sample
  Procedure:
  1. Mount the Faraday cup and the polished copper sample on the stage at the same nominal working distance.
  2. Image the sample at the optimal working distance and a beam energy selected in the 15–25 keV range.
  3. Start a factor of 10 or more below the optimal probe current. Measure and record the probe current using the Faraday cup and the picoammeter.
  4. Collect a 60 live-time second spectrum from the copper sample.
  5. Increase the probe current through a sequence of approximately 10–20 steps from the initial probe current to approximately 2 times the optimal probe current. Collect a 60 live-time second spectrum at each probe current and measure and record the probe current using the Faraday cup. A plot of such a measurement sequence is shown in ◘ Fig. 16.23 (upper).
  6. Integrate the total number of counts in the range of channels representing the Cu K-L$_{2,3}$ characteristic peak. (You don't need to background correct the integral.) Plot the measured intensity divided by the probe current against the measured probe current. The result should be a horizontal line, as shown in ◘ Fig. 16.23 (lower).
     1. If the line is not horizontal, the problem may be in the detector or in the probe current measurement.
     2. The probe current measurement could be non-linear meaning the plot of the measured probe current to true probe current is not a straight line.
     3. Alternatively, the probe current may have a zero offset. The zero offset can be measured by blanking the beam and recording the measured current at zero true current. Subsequent probe current measurement can be offset by this value.

16.3.6 Energy Calibration Linearity

Consistent energy calibration is critically important for reproducible quantitative analysis. Pick a nominal channel width (5-eV/channel will work fine in almost all cases) and

Fig. 16.23 The number of X-rays recorded in the spectrum should scale linearly with the dose (the product of live-time and probe current)

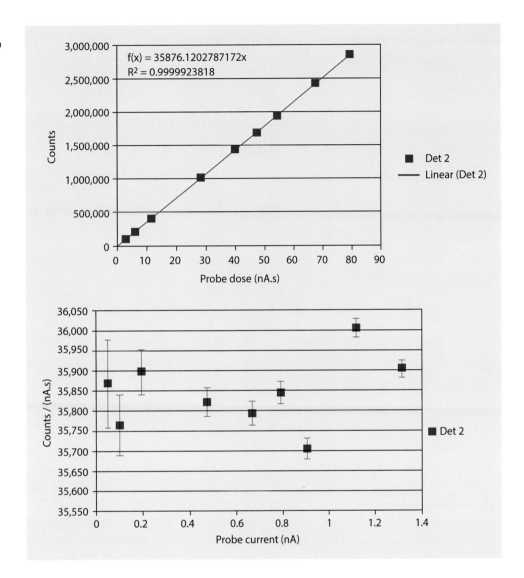

use this value for all your data acquisition. Before each day's measurements ensure that the detector is calibrated consistently by following the same protocol to check and, if necessary, recalibrate your detector.

On a modern pulse processor, calibration is usually performed using the EDS vendor's software. The software will prompt you to collect a spectrum from an established material. The software will examine the spectrum and extract the positions of various characteristic X-ray features. The software will then perform an internal adjustment to center these features in the correct channels. Most modern pulse processors perform a continuous zero offset calibration using a "zero strobe pulse" the pulse processor adds to the signal stream for diagnostic purposes. Thus the only parameter they usually adjust when performing a calibration is an electronic gain. Usually, this involves identifying a single high energy characteristic line (like the Mn K-$L_{2,3}$ or the Cu K-$L_{2,3}$) and adjusting the gain until this feature is centered on the appropriate channel. The calibration is thus a two-point calibration—either a low energy characteristic line or the zero strobe at low energy and a second characteristic line at high

energy. Two points are sufficient to unambiguously calibrate a linear function. The calibration (peak position) and resolution (peak shape) should be constant with input count rate (or dead-time), as shown in ☐ Fig. 16.24.

To a very high degree, modern EDS detectors are linear. However if you look carefully in the mid-range of energies, you many notice the KLM markers may be misaligned by a channel or two. This is evidence that your detector is not perfectly linear but this need not represent a true performance problem.

Since energy calibration is so critical but is also one of many parameters that should be measured as part of a complete EDS Quality Control (QC) program, the validation will be discussed in a later section.

16.3.7 Other Items

■ **Light Transparency and IR Cameras**

Most (but not all) modern EDS detectors have a vacuum tight window that is opaque to infrared and visible light.

Fig. 16.24 The position and widths of the characteristic peaks should not vary with probe current

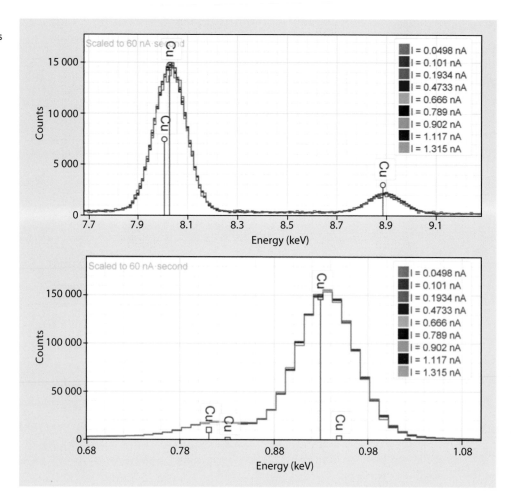

16

Usually, the window is coated with a thin layer of aluminum (or other metal) to keep light in the chamber from creating a spurious signal on the EDS detector.

Regardless, it is worthwhile to test whether your detector is sensitive to light. You may be surprised by a pinhole light leak or a window without an adequate opaque layer.

■ **Check 6: Check for SEM Light Sources**

The windows on some EDS detectors are not opaque to light and light in the chamber will produce noise counts particularly in low channels.

1. Enumerate the potential sources of light inside your SEM. Sources to consider:
   1. An IR camera
   2. Stage position sensors
   3. The tungsten filament (essentially a light bulb)
   4. Chamber windows
   5. Cathodoluminescence from samples like zinc selenide or benitoite.
2. Collect a series of spectra and examine these spectra for anomalies.
   1. When possible collect the spectra without an electron beam so there should be no source of X-rays.
   2. Collect a spectrum with the light source turned off and a spectrum with the source on. There should be no difference.

16.3.8   Setting Up a Quality Control Program

An ongoing QC program is a valuable way to demonstrate that your data and results can be trusted——yesterday, today, and tomorrow. If a client ever questions some data, it is useful to be able to go back to the day that data was collected and show that your instrument and detector were performing adequately. A well-designed QC program need not take much time. A single spectrum from a consistent sample collected under consistent conditions is sufficient to identify most common failure modes and to document the long-term performance of your detector. A well-designed QC program is likely to save time by eliminating the possibility of collecting data when the detector is miscalibrated or otherwise misbehaving.

■ **Check 7: Implement a Quick QC Program**

1. Maintain a sample consisting of a Faraday cup and a piece of Cu or Mn. Make use of this sample each day on which you intend on collecting quantitative EDS data to ensure that the detector is calibrated.
2. Image the sample at a consistent working distance (the "optimal working distance"), a consistent beam energy, and a consistent probe current.
3. Collect a spectrum from a sample for a consistent live-time.

4. Process the spectrum to extract the raw intensities in the K-L$_{2,3}$ and L-family lines, the resolution, the actual zero offset and gain, and the total number of counts. Record and plot these values on a control chart.

Using the QC Tools Within DTSA-II

While it is possible to implement a QC program by combining an EDS vendor's software with careful record-keeping, DTSA-II provides tools specifically designed to implement a basic EDS detector QC program. The tools import spectra, make some basic sanity checks, process the spectrum to extract QC metrics, archive the metrics and report the metrics.

Creating a QC Project

- A QC project is an archive of spectra collected from a specified sample under similar conditions on a specified detector/instrument. The spectra are fitted with a modeled spectrum and the resulting quantities recorded to track these values over the lifetime of the detector.
- Creating a QC Project involves specifying a detector and the conditions which are to be held consistent (■ Fig. 16.25). If you use a single detector with different process times, either select a constant process time or, better yet, create individual projects for each process time.
    - The material may be simple or complex but should be robust, durable and provide characteristic lines over a large range of energies. Copper is

ideal because it is readily available, stable and has both K-L$_{2,3}$ (~8.04 keV) and L-family (~930 eV) lines.
- The beam energy should be sufficient to adequately excite (overvoltage > 2) all the lines of interest in the sample.
- The nominal working distance should be the detector's optimal working distance. The sample should be brought into focus (using the stage Z-axis if necessary) at this working distance before collecting a spectrum.
- The nominal probe current is the probe current at which the spectra will be collected (to within a few percent). Usually, the "use probe current normalization" will be selected so that all intensities are scaled relative to the probe current measured with a in-lens cup or a Faraday cup.
- After creating the project, you will need to go through the QC tool additional times to add measured spectra.

- Adding Spectra to a QC Project
    - To add spectra to a QC project, you first need to specify which project you will be adding the spectra. Once you select the project, the "material," "beam energy," "nominal working distance," and "nominal probe current" boxes will fill with the associated information (see ■ Fig. 16.26).

■ **Fig. 16.25** The panel in DTSA-II for creating QC projects. This panel is accessed through the "Tools → QC Alien" menu item

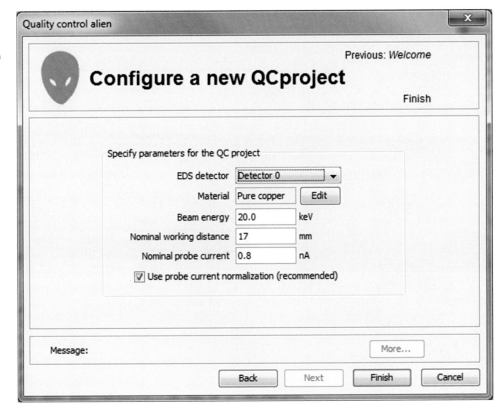

■ **Fig. 16.26** Panel for selection of detector and material parameters

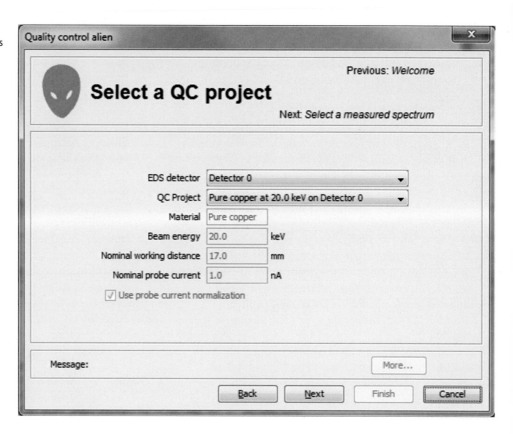

16

— Once the project has been selected, you will need to specify a spectrum to add to the project. Needless to say, the spectrum should have been collected on the correct detector under the conditions specified (see ■ Fig. 16.27).

— After selecting the spectrum, the spectrum will be processed by fitting it to a modeled spectrum shape. The resulting fit parameters will be reported and compared with the fit parameters from previous fits. The results are organized into columns associated with the current fit, the average of all fits, average of the first (up to) 10 fits and the average of the last (up to) 10 fits. Review these values to determine whether there has been short or long term drift in any of the fit parameters. The same information is shown in the DTSA-II Report table, as shown in ■ Fig. 16.28.

— You can also generate reports containing these quantities as tabular values, as shown in ■ Fig. 16.29, and plotted on control charts, as shown in ■ Fig. 16.30.

■ **Generating QC Reports**

A QC Report is a quick way to track the long-term performance of your detector. QC Reports are also generated using the QC tool accessible through the "Tools → QC Alien" menu item.

You will specify the detector and QC project along with the fit values that you wish to report. The report will be generated into a new HTML document and the result displayed in your system's default web browser. The report will look like ■ Fig. 16.30, with header information and a series of control charts. At the bottom of the report is a table containing all the data values that went into creating the control charts. All the values computed when the spectrum was added to the QC project will be available to display in the QC report.

16.3.9   Purchasing an SDD

If you were to survey EDS vendor's advertisements, you'd come to conclusion that two hardware characteristics determine the "best" EDS detector—resolution and detector area. Over the last decade, detector areas have become larger and larger and detector resolutions have improved significantly too. The performance of Si(Li) detectors scaled poorly with size because the detector capacitance scaled with size. SDD, on the other hand, perform only slightly worse (throughput and resolution) as the detector area increases. As a result, even a basic modern SDD-EDS detector is larger and performs better than the best Si(Li)-EDS detector of a decade ago.

Fig. 16.27 Panel for selection of measured spectrum for archive

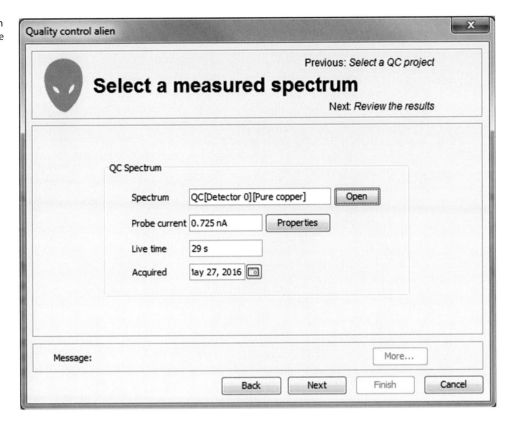

Fig. 16.28 Results of information extracted from measured reference spectrum

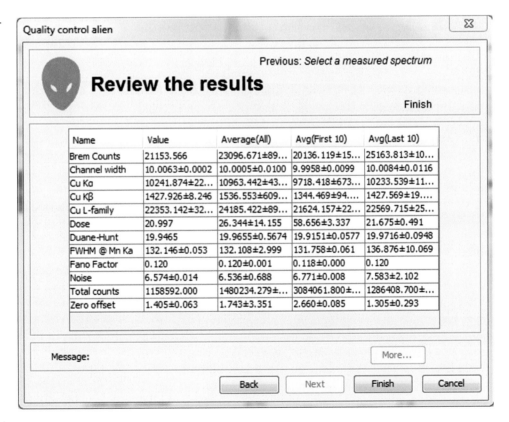

Name	Value	Average(All)	Avg(First 10)	Avg(Last 10)
Brem Counts	21153.566	23096.671±89...	20136.119±15...	25163.813±10...
Channel width	10.0063±0.0002	10.0005±0.0100	9.9958±0.0099	10.0084±0.0116
Cu Kα	10241.874±22...	10963.442±43...	9718.418±673...	10233.539±11...
Cu Kβ	1427.926±8.246	1536.553±609...	1344.469±94....	1427.569±19....
Cu L-family	22353.142±32...	24185.422±89...	21624.157±22...	22569.715±25...
Dose	20.997	26.344±14.155	58.656±3.337	21.675±0.491
Duane-Hunt	19.9465	19.9655±0.5674	19.9151±0.0577	19.9716±0.0948
FWHM @ Mn Ka	132.146±0.053	132.108±2.999	131.758±0.061	136.876±10.069
Fano Factor	0.120	0.120±0.001	0.118±0.000	0.120
Noise	6.574±0.014	6.536±0.688	6.771±0.008	7.583±2.102
Total counts	1158592.000	1480234.279±...	3084061.800±...	1286408.700±...
Zero offset	1.405±0.063	1.743±3.351	2.660±0.085	1.305±0.293

QC Measurement Recorded

Spectrum	QC [Detector 0][Pure copper]
Index	920
Timestamp	2016-05-27 10:33:00.071
Project	Pure copper at 20.0 keV on Detector 0

Name	Value	First 10	Last 10	All
Brem Counts	21153.566	20136.119±1587.620	25159.335±10725.581	23104.667±8993.435
Channel width	10.0063±0.0002	9.9958±0.0099	10.0078±0.0119	10.0005±0.0100
Cu Kα	10241.874±22.085	9718.418±673.187	10229.021±114.918	10966.411±4359.408
Cu Kβ	1427.926±8.246	1344.469±94.784	1430.549±21.392	1537.000±610.996
Cu L-family	22353.142±32.628	21624.157±2221.392	22593.581±242.852	24192.963±8918.157
Dose	20.997	58.656±3.337	21.770±0.439	26.366±14.180
Duane-Hunt	19.9465	19.9151±0.0577	19.9753±0.0945	19.9656±0.5686
FWHM @ Mn Ka	132.146±0.053	131.758±0.061	136.829±10.091	132.107±3.005
Fano Factor	0.120	0.118±0.000	0.120	0.120±0.001
Noise	6.574±0.014	6.771±0.008	7.570±2.108	6.536±0.690
Total counts	1158592.000	3084061.800±87926.172	1292007.000±230599.127	1481557.909±785814.751
Zero offset	1.405±0.063	2.660±0.085	1.288±0.292	1.744±3.358

◘ **Fig. 16.29** Report of current QC spectrum measurement parameters compared to archival values

Regardless of what the EDS vendors literature tells us, while both of these performance characteristics are important, neither is the basis of a well-considered choice of detector. Resolution and area are indirect proxies for the performance characteristics that should really drive the decision process—good throughput at an adequate resolution and a large solid angle of detection.

First, a word or two about two characteristics which are absolutely required for good quantitative analysis. Fortunately, almost all modern SDD meet these two important requirements—linearity and stability.

Linearity of Output Count Rate with Live-Time Dose

The number of X-rays measured must be proportional to the number of X-rays generated. If you generated ten times as many X-rays, you should measure ten times as many X-rays. Otherwise, the k-ratio, the basis of all quantitative analysis, would depend not only upon the composition of the material but also the probe current.

Perform the check in section Count Rate Linearity to evaluate a candidate detector's linearity performance.

Resolution and Peak Position Stability with Count Rate

The detector resolution and peak position must not change appreciably with a variation of a factor of ten or more in X-ray flux.

The same spectra used to demonstrate linearity can be used to demonstrate peak position and resolution stability. Use DTSA-II's calibration tool to fit the spectra and extract full width at half-maximum (FWHM) and channel width values for each spectrum. Plot the spectra and results as shown in ◘ Fig. 16.31.

Having ensured these two basic characteristics, the choice of next most important characteristic depends upon how your detector will be used. If signal quantity is a problem because you are limited to low probe currents, STEM mode analysis of microparticles, low beam energy analysis, or another reason why the flux of X-rays is limited, then a detector that maximizes solid angle is important. If on the other hand, you can produce a lot of X-rays, then throughput at an adequate resolution is more important. Regardless, both criteria should be part of your evaluation process.

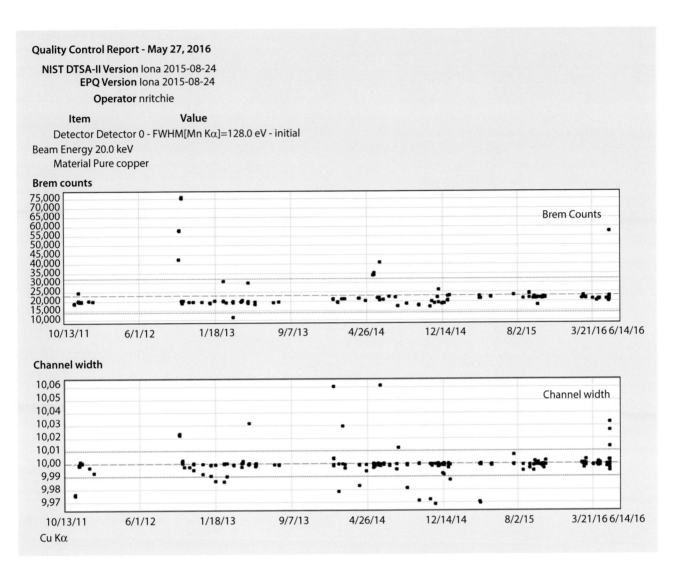

Quality Control Report - May 27, 2016

NIST DTSA-II Version Iona 2015-08-24
EPQ Version Iona 2015-08-24
Operator nritchie

Item	Value

Detector Detector 0 - FWHM[Mn Kα]=128.0 eV - initial
Beam Energy 20.0 keV
Material Pure copper

Brem counts

Channel width

Cu Kα

■ **Fig. 16.30** Quality control report charts

■ **Fig. 16.31** Neither the resolution (FWHM at Mn K-L$_{2,3}$ (Kα)) or the gain (in eV/channel) should change with probe current

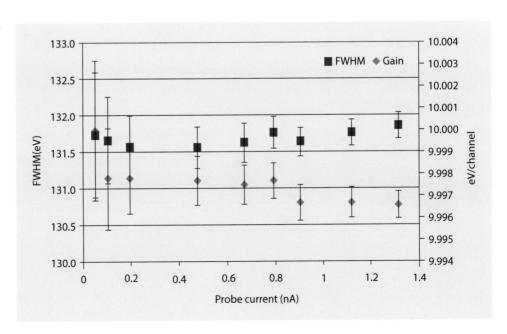

Solid Angle for Low X-ray Flux

The fraction of X-rays emitted by the sample that strike the detector is proportional to the solid angle. The solid angle is a function of both the active area of the detector and the distance from the sample to the detector. The detected fraction is linearly proportional to the active area of the detector but inversely proportional to the square of the distance from the sample to the detector so the position of the detector is critical. It is not reasonable to assume that a larger area detector will always produce a larger solid angle. Larger area detectors may require larger diameter snouts which may not be able to be positioned as close to the sample. Larger area detectors may also produce slightly poorer resolution and/or slightly lower ultimate throughput due to increases in "ballistic deficit" — the spreading of electron packets in the active detector area.

The only way to know ahead of time what solid angle you can expect is to ask they vendor to provide schematics showing how your detector will be positioned in your instrument. The critical parameters are sample-to-detector distance at the maximum insertion position, the optimal working distance, the detector area, and the elevation angle. These parameters can be used within DTSA-II to model the X-ray signal you can expect to measure from the types of samples and the probe currents you use.

Maximizing Throughput at Moderate Resolution

Modern detectors are capable of extraordinary resolutions and high throughput, though not both at the same time. The best resolutions are achieved at long pulse process times, which produce poor ultimate throughput. The highest throughputs are achieved at short pulse process times; but, while it may be possible to measure many X-rays per unit time, coincidence events (pulse pile up) limit the quantitative accuracy. The quantitative performance of the detector is three-way trade-off between throughput, resolution, and coincidence rate. Typically, this is accomplished by defining an acceptable coincidence rate as discussed in the section on process time and determining the process time that maximizes the throughput at this coincidence rate. This process time will typically be a slight compromise from the one that produces the optimal resolution but typically not by more than a few eV FWHM at Mn K-L$_{2,3}$ (Kα). A few eV of resolution degradation is usually an acceptable compromise as throughput is far more important than resolution for accurate quantitative EDS microanalysis.

- Special Case: Low Energy Sensitivity

If measuring low energy X-rays in the sub-200-eV range is particularly important to you, then you should focus your criteria on this energy region and understand that to optimize this regime will likely require compromises to throughput. Find samples similar to the ones you will commonly measure and use these samples to evaluate the performance of the candidate detectors.

References

Fiori C, Newbury D (1978) Artifacts in energy dispersive X-ray spectrometry. Scan Electron Microsc 1:401

Fitzgerald R, Keil K, Heinrich KFJ (1968) Solid-state energy-dispersive spectrometer for electron-microprobe X-ray analysis. Science 159:528

Gatti E, Rehak P (1984) Semiconductor drift chamber – an application of a novel charge transport scheme. Nucl Instr Meth A 225:608

Struder L, Fiorini C, Gatti E, Hartmann R, Holl P, Krause N, Lechner P, Longini A, Lutz G, Kemmer J, Meidinger N, Popp M, Soltau H, Van Zanthier C (1998) High-resolution high-count-rate X-ray spectroscopy with state-of-the-art silicon detectors. Mikrochim Acta Suppl 15:11

16

DTSA-II EDS Software

© Springer Science+Business Media LLC 2018
J. Goldstein et al., *Scanning Electron Microscopy and X-Ray Microanalysis*,
https://doi.org/10.1007/978-1-4939-6676-9_17

17.1 Getting Started With NIST DTSA-II

17.1.1 Motivation

Reading about a new subject is good but there is nothing like *doing* to reinforce understanding. With this in mind, the authors of this textbook have designed a number of practical exercises that reinforce the book's subject matter. Some of these exercises can be performed with software you have available to you—either instrument vendor software or a spreadsheet like MS Excel or LibreOffice/OpenOffice Calc. Other exercises require functionality which may not be present in all instrument vendor's software. Regardless, it is much easier to explain an exercise when everyone is working with the same tools.

To ensure that everyone has the tools necessary to perform the exercises, we have developed the National Institute of Standards and Technology software called DTSA-II (Ritchie, 2009, 2010, 2011a, b, c, 2012, 2017a, b). NIST DTSA-II is quantitative X-ray microanalysis software designed with education and best practices in mind. Furthermore, as an output of the US Federal Government, it is not subject to copyright restrictions and freely available to all regardless of affiliation or nationality. From a practical perspective, this means that you can install and use DTSA-II on any suitable computer. You can give it to colleagues or students. You can use it at home, in your office, or in the lab. If you are so inclined, the source code is available to allow you to review the implementation or to enhance the tool for your own special purposes.

Exercises in the textbook will be designed around the capabilities of DTSA-II. Many will take advantage of the graphical user interface to manipulate and interrogate spectra. A number will take advantage of the command line scripting interface to access low-level data or to perform advanced operations.

17.1.2 Platform

NIST DTSA-II is written in the multi-platform run-time environment Java. This means that the same program runs on Microsoft Windows (XP, Vista, 7, 8.X, 10.X), Apple OS X 10.6+, and many flavors of Linux and UNIX. The run-time environment adjusts the look-and-feel of the application to be consistent with the standards for each operating system. For Windows, Linux, and UNIX, the main menu is part of the main application window. In OS X, the main menu is at the top of the primary screen.

To make installation as easy as possible on each environment, an installer has been developed which works on Windows, OS X, and Linux/UNIX. The installer verifies that an appropriate version of Java is available and place the executable and data files in a location that is consistent with operating system guidelines. Detailed installation instructions are available on the download site.

17.1.3 Overview

DTSA-II was designed around common user interface metaphors and should feel consistent with other programs on your operating system. It has a main menu which provides a handful of high-level interactions such as file access, processing, simulation, reporting, and help. Many of these menu items lead to "wizard-style" dialogs which take you step-by-step through some more complex operation like experiment design, spectrum quantification, or spectrum simulation. The goal is to make common operations as simple as possible.

Additional tools are often available through context sensitive menus. Each region on the DTSA-II main window provides different functionality. Within these regions, you are likely to want to perform various context sensitive operations. These operations are accesses through menus that are accessed by placing the mouse over the region and issuing a "right-button click" on Windows/Linux/UNIX or an Apple-Command key + mouse click on OS X. The context-sensitive menu items may perform operations immediately, or they may request additional information through dialog boxes.

On the main screen, the bottom half of window is three tabs—the Spectrum tab, the Report tab, and the Command tab. The spectrum tab is useful for investigating and manipulating spectra. The report tab provides a record of the work completed during this invocation of the program. Since the report is in HTML (Hypertext Markup Language) and stored by date, it is possible to review old reports either in DTSA-II or in a standard web browser. Some operating systems will index HTML documents making it easy to find old designs, analyses or simulations.

The final tab contains a command line interface. The command line interface implements a Python syntax scripting environment. Python is a popular, powerful and complete scripting language. Through Python it is possible (though not necessarily easy) to perform anything that can be done through the GUI. It is also possible to do a lot more. For example, the GUI makes available some common, useful geometries for performing Monte Carlo simulations of spectrum generation. Through scripting, it is possible to simulate arbitrary sample and detector geometries. Some of the examples in the text will involve scripting. These scripts will be installed with the software so they are readily available and so you can use these as the basis for your own custom scripts.

An important foundational concept in DTSA-II is the definition of an X-ray detector. The software comes with a "default detector" which represents a typical Si(Li) detector on a typical SEM. This detector will produce adequate results for many purposes. However, it is better and more useful if you create your own detector definition or definitions to reflect the design and performance of your detector(s) and SEM. You define your own detector using the "Preferences" dialog which is access through the "File → Preferences" main menu item. To select and activate your detector, you select it in the "Default Detector" drop down lists on the middle, left side of the main DTSA-II window on the "Spectrum Tab."

17.1.4 Design

DTSA-II is, in many ways, much more like vendor software used to be. This has advantages and disadvantages. Over the years, vendors have simplified their software. They have removed many more advanced spectrum manipulation tools and they have streamlined their software to make getting an answer as straightforward as possible. If your goal is simply to collect a spectrum, press a button, and report a result, the vendor software is ideal. However, if you want to develop a more deep understanding of how spectrum analysis works, many vendors have buried the tools or removed them entirely. DTSA-II retains many of the advanced spectrum manipulation and interrogation tools.

DTSA-II is designed with Einstein's suggestion about simplicity in mind: "Everything should be made as simple as possible, but not simpler." DTSA-II was designed with the goal of making the most reliable and accurate means of quantification, standards-based quantification, as simple as possible, but not simpler. When there is a choice that might compromise reliability or accuracy for simplicity, reliability and accuracy wins out.

One such example is "auto-quant." Most microanalysis software will automatically place peak markers on spectra. Unfortunately, these markers have time and time again been demonstrated to be reliable in many but far from all cases. Users grow dependent on auto-quant and when it fails they often don't have the experience or confidence to identify the failures. The consequence is that the qualitative and then since the qualitative results are used to produce quantitative results, the quantitative results are just plain wrong.

Rather than risking being wrong, DTSA-II requires the user to perform manual peak identification. The process is more tedious and requires more understanding by the user. But no more understanding than is necessary to judge whether the vendor's auto-qual has worked correctly. If you as a user can't perform manual qualitative analysis reliably, you should not be using the vendor's auto-qual.

17.1.5 The Three -Leg Stool: Simulation, Quantification and Experiment Design

NIST DTSA-II is designed to tie together three tools which are integral to the process of performing high-quality X-ray microanalysis—simulation, quantification, and experiment design. Simulation allows you to understand the measurement process for both simple measurements and more complex materials and geometries. Quantification allows you to turn spectra into estimates of composition. Experiment design ties together simulation and quantification to allow you to develop the most accurate and reliable measurement protocols.

Simulation

Spectrum generation can be modeled either using analytical models or using Monte Carlo models. The difference is that analytical models are deterministic, they always produce the same output for the equivalent input, and they are less computationally intensive. They are limited, however, in the geometries for which we know how to perform the analytical calculation. Monte Carlo models are based on pseudo-random simulation of the physics of electron interactions and X-ray production. Individual electron trajectories are traced as they meander through the sample. Interactions like elastic scattering off the electrons and nucleus in the sample are modeled. Inelastic interactions like core-shell ionization are also modeled. Each core shell ionization is followed by either an Auger electron or an X-ray photon. The trajectories of these can also be modeled. The resulting X-rays can be collected in a modeled detector and the result presented as a dose-correct spectrum.

So, in summary, analytical models are quicker, but Montel Carlo models are more flexible. Regardless, in domains where they are both applicable, they produce similar but not identical results.

Quantification

Accurate, reliable quantification is the goal. Turning measured spectra into reliable estimates of material composition can be a challenge. Our techniques work well when we are careful to prepare our samples, collect our spectra, and process the data. However, there are many pitfalls and potential sources of error for the novice or the overconfident.

DTSA-II implements some of the most reliable algorithms for spectrum quantification. First, DTSA-II assumes that you will be comparing your unknown spectrum to spectra collected from standard materials. Standards-based quantification is the most accurate and reliable technique known. Second, DTSA-II implements robust algorithms for comparing peak intensities between standards and unknown. DTSA-II uses linear least squares fitting of background filtered spectra. This algorithm is robust, accurate, and makes very good use of the all the information present in each peak. It also provides mechanism called the *residual* to determine whether the correct elements have been identified and fit.

Fitting produces k-ratios which are the first-order estimates of composition. To extract the true composition, the k-ratios must be scaled to account for differences in absorption, atomic number, and secondary fluorescence. DTSA-II implements a handful of different matrix correction algorithms although users are encouraged to use the default algorithm ('XPP' by Pouchou and Pichoir, 1991) unless they have a compelling reason to do otherwise.

Experiment Design

One thing that has long hindered people from performing standards-based quantification is the complexity of designing an optimal standards-based measurement. The choices that go into designing a good measurement are subtle. How does one select the optimal beam energy? How does one select the best materials to use as standards? How does one determine when a reference spectrum[1] is needed in addition to the standard spectra? How long an acquisition is required to produce the desired measurement precision? What limits of detection can I expect to achieve? Do I want to optimize accuracy or simply precision? Am I interested in minimizing the total error budget or am I interested in optimizing the measurement of one (or a couple of) elements?

In fact, many of these decisions are interrelated in subtle ways. Increasing the beam energy will often improve precision (more counts) but will reduce accuracy (more absorption). The best standard for a precision measurement is likely a pure element while the best standard for an accurate measurement is likely a material similar to the unknown.

Then we must also consider subtle interactions between elements. If the emission from one element falls near in energy to the absorption edges of another element, accuracy may be reduced due to complex near edge absorption effects. All the different considerations make the mind reel and intimidate all but the most confident practitioners of the art.

DTSA-II addresses these problems through an experiment design tool. The experiment design tool calculates the uncertainty budget for an ensemble of different alternative measurement protocols. It then suggests the experiment protocol which optimizes the uncertainty budget. It outlines which spectra need to be acquired and the doses necessary to achieve the user's desired measurement precision. This is then presented in the report page as a recipe that the analyst can taking into the laboratory.

Experiment design builds upon an expert's understanding of the quantification process and makes extensive use of spectrum simulation. Through spectrum simulation, DTSA-II can understand how peak interferences and detector performance will influence the measurement process. Through spectrum simulation carefully calibrated to the performance of your detectors, the experiment optimizer can predict how much dose (probe current x time) is required. Often the result is good news. We often spend much too much time on some spectra and too little on others.

17.1.6 Introduction to Fundamental Concepts

For the most part, the functionality of DTSA-II will be introduced along with the relevant microanalytical concept. However, there are a handful of concepts which provide a skeleton around which the rest of the program is built. It is necessary to understand these concepts to use the program effectively.

DTSA-II was designed around the idea of being able simulate what you measure. With DTSA-II, it is possible to simulate the full measurement process for both simple and complex samples. You can simulate the spectrum from an unknown material and from the standard materials necessary to quantify the unknown spectrum. You can quantify the simulated spectra just like you can quantify measured spectra. This ability allows you to understand the measurement process in ways that are simply not possible otherwise. It is possible to investigate how changes in sample geometry or contamination or coatings will influence the results. It is possible to visualize the electron trajectories and X-ray production and absorption.

However to do this, it is necessary to be able to model the sample, the physics of electron transport, atomic ionization and X-ray production and transport, and the detection of X-rays. The physics of electrons and X-rays is not perfectly known, but at least it doesn't change between one instrument and another. The biggest change between instruments is the X-ray detection process. Not all detectors are created equal.

To compensate for the detection process, DTSA-II builds algorithmic models of X-ray detectors based on the properties of the detector. These models are then used to convert the simulated X-ray flux into a simulated measured spectrum. The better these models, the better DTSA-II is able to simulate and quantify spectra.

Modeled Detectors (◻ Fig. 17.1)

To make optimal use of DTSA-II, you will need to create a detector model to describe each of your X-ray detectors. Each detector model reflects the performance of a specific detector in a specific instrument at a specific resolution/throughput setting. Each physical detector should be associated with at least one detector model. A single physical detector may have more than one detector model if the detector is regularly operated at different resolution / throughput settings.

Some of the information necessary to build the detector model is readily available from product literature or from a call to the vendor. Unfortunately, some pieces of information are less easy to discover. Some require access to very specialized samples or equipment, but fortunately, accepting the default values won't overly affect utility of the simulated results. ◻ Table 17.1 identifies which values are critical and which are less critical.

Detector models are created in the "Preferences" dialog which is accessed through the "File → Preferences" main menu item. The tree view on the left side of the dialog allows you to navigate through various preference pages. By default, a root node labeled "Instruments and Detectors" is created with a branch called "Probe" and a leaf node called "Si(Li)." The branch "Probe" reflects a very basic traditional SEM/microprobe. The leaf "Si(Li)" reflects a typical lithium-drifted silicon detector with a ultra-thin window and a resolution of 132 eV at Mn Kα. You can examine the definition of this detector to determine which pieces of information are necessary to fully describe a detector.

---

[1]   Don't worry if you don't understand the difference between a reference and a standard spectrum. This will be explained later.

Fig. 17.1 The preferences dialog showing a panel containing properties of a detector

Table 17.1 This table identifies parameters that have a critical influence on simulated spectra and those that have a less critical influence. You should be able to determine the correct value of the critical parameters from vendor literature or a call to the vendor. The vendor may be able to provide the less critical values too but if they can't just accept the defaults

Critical	Less critical
Window type	Gold layer (accept the default)
Elevation angle	Aluminum layer (accept the default)
Optimal working distance	Nickel layer (accept the default)
Sample-to-detector distance (estimate)	Dead layer (accept the default)
Detector area	Zero strobe discriminator (easy to estimate)
Crystal thickness	
Number of channels	
Energy scale (nominal)	
Zero offset (nominal)	
Resolution at Mn Kα (approximate)	
Azimuthal angle	

Some pieces of information are specific to your instrument and the way the detector is mounted in the instrument. Window-type, detector area, crystal thickness, resolution, gold layer thickness, aluminum layer thickness, nickel layer thickness, and dead layer thickness are model specific properties of the detector. Elevation angle, optimal working distance, sample-to-detector distance, and azimuthal angle are determined by how the detector is mounted in your instrument. The energy scale, zero offset, and the resolution are dependent upon hardware settings that are usually configured within the vendor's acquisition software. The oldest systems may have physical hardware switches.

Window Type (◻ Fig. 17.2)

As is discussed elsewhere, most detectors are protected from contamination by an X-ray transparent window. Older windows were made of ultrathin beryllium foils or occasionally boron-nitride or diamond films. Almost all modern detectors use ultrathin polymer windows although the recently introduced silicon nitride (Si_3N_4) windows show great promise.

Each type of window has a different efficiency as a function of energy. The largest variation in efficiency is seen below 1 keV. Here the absorption edges in the elements making up the windows can lead to large jumps in efficiency over narrow energy ranges. Diamond represents an extreme example in which the absorption edge at 0.283 keV leads to a three order-of-magnitude change in efficiency.

Your vendor should be able to tell you the make and model of the window on your detector.

The Optimal Working Distance (◻ Figs. 17.3 and 17.4)

The position and orientation of your EDS detector is optimized for a certain sample position. Typically, the optimal sample position is located on the electron-beam axis at an optimal working distance. At this distance, the effective elevation angle equals the nominal elevation angle. Sometimes, the optimal working distance will be specified in the drawings the EDS vendor used to design the detector mounting hardware (◻ Fig. 17.4). Other times, it is necessary to estimate the optimal working distance finding the sample position that produces the largest X-ray flux. The optimal working distance is measured on the same scale as the focal distance since the working distance value recorded in spectrum files is typically this value.

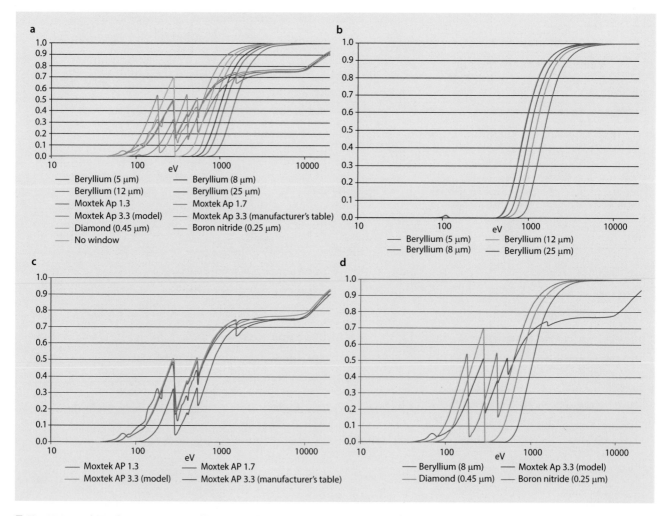

◻ **Fig. 17.2**   **a–d**: Window transmission efficiency as a function of photon energy

◘ Fig. 17.3 Definitions of geometric parameters

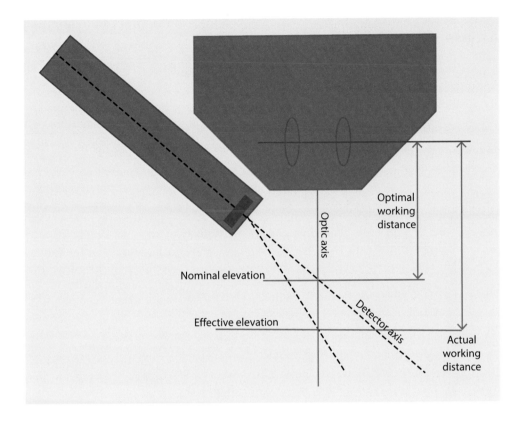

◘ Fig. 17.4 Example of vendor-supplied schematics showing the optimal working distance (17 mm) and the sample-to-detector distance (34 mm) for a microscope with multiple detectors (Source: TESCAN)

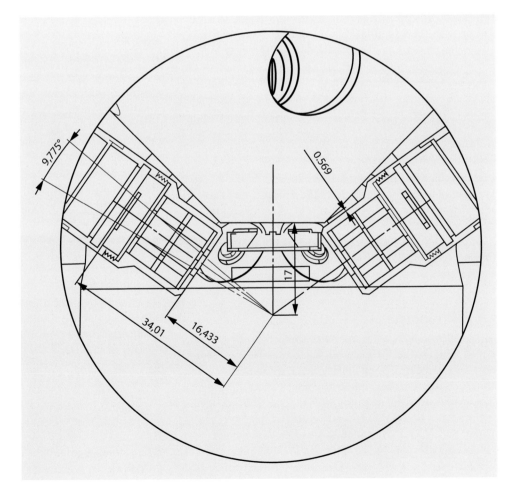

The difference between the effective elevation defined by the actual working distance and the nominal elevation as defined by the intersection of the detector axis and the optic axis. The effective elevation angle can be calculated from the actual working distance given the optimal working distance, the nominal elevation angle, and the nominal sample-to-detector distance.

Elevation Angle

The elevation angle is defined as the angle between the detector axis and the plane perpendicular to the optic axis. The elevation angle is a fixed property of the detector as it is mounted in an instrument. The elevation angle is closely related to the take-off angle. For a sample whose top surface is perpendicular to the optic axis, the take-off angle at the optimal working distance equals the elevation angle.

Elevation angles typically range between 30° and 50° with between 35° and 40° being the most common. The correct detector elevation is important for accurate quantification as the matrix correction has a strong dependence on this parameter.

Sample-to-Detector Distance

The sample-to-detector distance is the distance from front face of the detector crystal to the intersection of the optic and detector axes. The sample-to-detector distance helps to define the solid angle of acceptance for the detector. The sample-to-detector distance can often be extracted from the drawings used to design the detector mounting hardware (See ◻ Fig. 17.3).

Alternatively, you can estimate the distance and adjust the value by comparing the total integrated counts in a simulated spectrum with the total integrated counts in an equivalent measured spectrum. The simulated counts will decrease as the square in the increase of the sample-to-detector distance.

Detector Area

The detector area is the nominal surface area of the detector crystal visible (unobstructed) from the perspective of the optimal analysis point. The detector area is one of the values that detector vendors explicitly specify when describing a detector. Typical values of detector area are 5, 10, 30, 50, or 80 mm^2. The detector area does not account for area obstructed by grid bars on the window but does account for area obstructed by a collimator or other permanent pieces of hardware.

Crystal Thickness

The detection efficiency for hard (higher-energy) X-rays depends upon the thickness of the active detector crystal area. Si(Li) detectors tend to have much thicker crystals and thus measure X-rays with energies above 10 keV more efficiently. Silicon drift detectors (SDD) tend to be about an order of magnitude thinner and become increasingly transparent to X-rays above about 10 keV. DTSA-II defaults to a thickness of 5 mm for Si(Li) detectors and 0.45 mm for SDD. These values will work adequately for most purposes if a vendor specified value is unavailable.

Number of Channels, Energy Scale, and Zero Offset

A detector's energy calibration is described by three quantities—the number of bins or channels, the width of each bin (energy scale), and the offset of the zero-th bin (zero offset). The number of bins is often a power-of-two, most often 2,048 but sometimes 1,024 or 4,096. This number represents the number of individual, adjacent energy bins in the spectrum. The width of each bin is assumed to be a nice constant—typically 10 eV, 5 eV, 2.5, or occasionally 20 eV. The detector electronics are then adjusted (in older systems through physical potentiometers or in modern systems through digital calibration) to produce this width.

The zero offset allows the vendor to offset ('shift') the energy scale for the entire spectrum by a fixed energy or to compensate for a slight offset in the electronics. Some vendors don't make use of ability and the zero offset is fixed at zero. Other vendors use a negative zero offset to measure the full width of an artificial peak they intentionally insert into the data stream at 0 eV called the zero-strobe peak. The zero-strobe peak is often used to automatically correct for electronic drift. Often, DTSA-II can read these values from a vendor's spectrum file using the "Import from spectrum" tool.

You don't need to enter the exact energy scale and zero offset when you create the detector as the calibration tool can be used to refine these values.

Resolution at Mn Kα (Approximate)

The resolution is a measure of the performance of an EDS detector. Since the resolution depends upon X-ray energy in a predictable manner, the resolution is by long established standard reported as the "full width half maximum" (FWHM) of the Mn Kα peak (5.899 keV).

The full width at half-maximum is defined as the width of a peak as measured half way from the base to the peak. This is illustrated in the ◻ Fig. 17.5. The full height is measured from the level of the continuum background to the top of the peak. A line is drawn across the peak at half the full height. To account for the finite bin width, a line is drawn on each side of the peak from the center of the bin above the line to the center of the bin below the line. The intersection of this diagonal line is assumed to be the true peak edge position. The width is then measured from these intersection points and calibrated relative to the energy scale.

The graphical method for estimating the FWHM is not as accurate as numerical fitting of Gaussian line shapes. The calibration tool uses the numerical method and is the preferred method (◻ Fig. 17.6).

Azimuthal Angle

The azimuthal angle describes the angular position of the detector rotated around the optic axis. The azimuthal angle is particularly important when modeling samples that are tilted or have complex morphology.

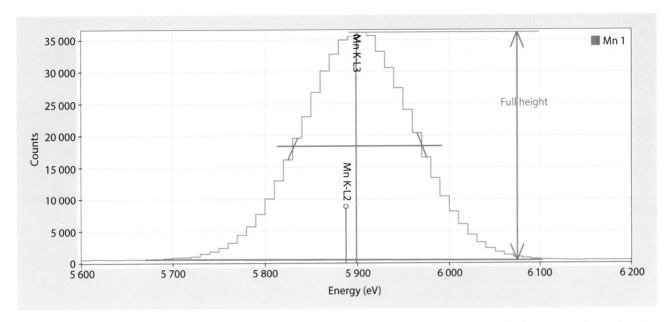

Fig. 17.5 Estimating the full width at half-maximum peak width. This peak is approximately 139 eV FWHM which you can confirm with a ruler

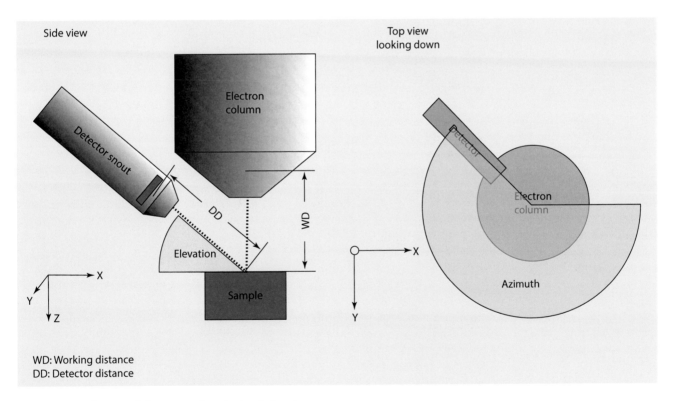

WD: Working distance
DD: Detector distance

Fig. 17.6 Definitions of elevation angle and azimuthal angle

Gold Layer, Aluminum Layer, Nickel Layer

Detector crystals typically have conductive layers on their front face to ensure conductivity. These layers can be constructed by depositing various different metals on the surface. The absorption profiles of these layers will decrease the efficiency of the detector. The layer thicknesses are particularly relevant for simulation; however, other uncertainties usually exceed the effect of the conductive layer.

Dead Layer

The dead layer is an inactive or partially active layer of silicon on the front face of the detector. The dead layer will absorb some X-rays (particularly low energy X-rays) and produce few to no electron–hole pairs. The result is a fraction of X-rays which produce no signal or a smaller signal than their energy would suggest. The result is twofold: The first effect is a diminishment of the number of low energy X-rays detected. The second effect is a low energy tail, called incomplete charge collection, on low energy X-ray peaks.

The dead layer in modern detectors is very thin and typically produces very little incomplete charge collection. Older Si(Li) detectors had thicker layers and worse incomplete charge collection.

Zero Strobe Discriminator (◘ Figs. 17.7 and 17.8)

The zero strobe discriminator is an energy below which all spectrum counts will be set to zero before many spectrum processing operations are performed.

The zero strobe is an artificial peak inserted by the detector electronics at 0 eV. The zero strobe is used to automatically determine the noise performance of the detector and to automatically adjust the offset of the detector to compensate for shifts in calibration. The zero strobe does not interfere with real X-ray events because it is located below the energies at which the detector is sensitive.

Some vendors automatically strip out the zero strobe out before presenting the spectrum. Others leave it in because it can provide useful information. When it does appear, it can negatively impact processing low energy peaks. To mitigate this problem, the zero strobe discriminator can be used to strip the zero strobe from the spectrum. The zero strobe dis-

criminator should be set to an energy just above the high energy tail of the zero strobe.

Material Editor Dialog (◘ Figs. 17.9, 17.10, 17.11, 17.12, 17.13, and 17.14)

The material editor dialog is used to enter compositional and density information throughout DTSA-II. This dialog allows you to enter compositional information either as mass fractions or atomic fractions. It also provides shortcut mechanisms for looking up definitions in a database or entering compositions using the chemical formula.

Method 1: Mass fractions (see ◘ Fig. 17.10)
Method 2: Atomic fractions (see ◘ Fig. 17.11)
Method 3: Chemical formula (see ◘ Fig. 17.12)
Method 4: Database lookup (see ◘ Fig. 17.13)
Method 5: Advanced chemical formulas (see ◘ Fig. 17.16)

So if your database contains a definition for "Albite" and you press the search button 🔍 , the table will be filled with the mass and atomic fractions and the density as recorded in the database for "Albite." The database is updated each time you select the "Ok" button. Over time, it is possible to fill the database with every material that you commonly see in your laboratory. The name "unknown" is special and is never saved to the database.

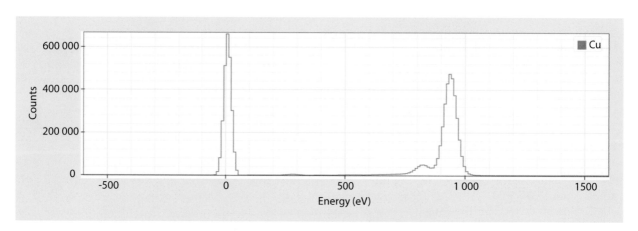

◘ **Fig. 17.7**   A raw Cu spectrum showing the zero strobe peak centered at 0 eV and the Cu L peaks centered near 940 eV

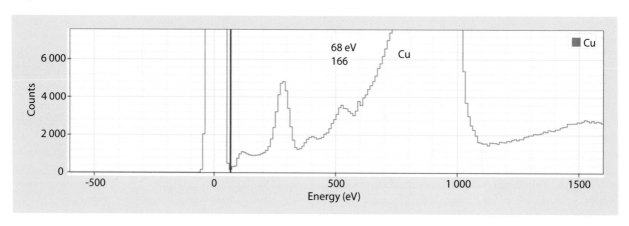

◘ **Fig. 17.8**   The blue line shows an appropriate placement of the zero strobe discriminator between the high energy edge of the zero strobe and the start of the real X-ray data

17

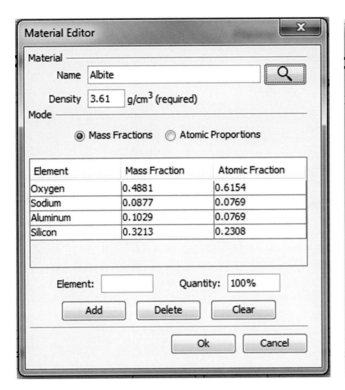

Fig. 17.9 The material editor dialog. Materials are defined by a name ("Albite"), a density ("3.61 g/cm³"), and a mapping between elements and quantities. Albite is defined as $NaAlSi_3O_8$ which is equivalent to the mass fractions and atomic fractions displayed in the table

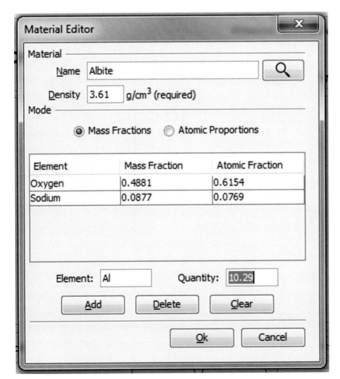

Fig. 17.10 The relative amount of each element in mass fractions may be entered manually using the "Element" and "Quantity" edit boxes and the "Add" button. Note that the mode radio button is set to "Mass Fractions" and that the quantity is entered in percent but displayed in mass fractions, so "10.29" corresponds to a mass fraction of "0.1029". The element may be specified by the common abbreviation ("Al"), the full name ("aluminum") or the atomic number ("13")

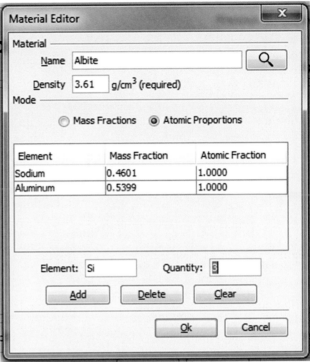

Fig. 17.11 The relative amount of each element in atomic fractions may be entered manually using the "Element" and "Quantity" edit boxes and the "Add" button. Note that the mode radio button is set to "Atomic Proportions" and that the quantity is entered as a number of atoms in a unit cell. The element may be specified by the common abbreviation ("Si"), the full name ("silicon") or the atomic number ("14")

17.2 Simulation in DTSA-II

17.2.1 Introduction

Simulation, particularly Monte Carlo simulation, is a powerful tool for understanding the measurement process. Without the ability to visualize how electrons and X-rays interact with the sample, it can be very hard to predict the significance of a measurement. Does the incident beam remain within the sample? Where are the measured X-rays coming from? Can I choose better instrumental conditions for the measurements? Without simulation, these insight can only be gained with years of experience or based on simple rules of thumb.

Too often we are asked to analyze non-ideal samples. Monte Carlo simulation is one of the few mechanisms we have to ground-truth these measurements. Consider the humble particle. When is it acceptable to consider a particle to be essentially bulk and what are the approximate errors associated with this assumption?

17.2.2 Monte Carlo Simulation

Monte Carlo models are particularly useful because they permit the simulation of arbitrarily complex sample geometries. NIST DTSA-II provides a handful of different

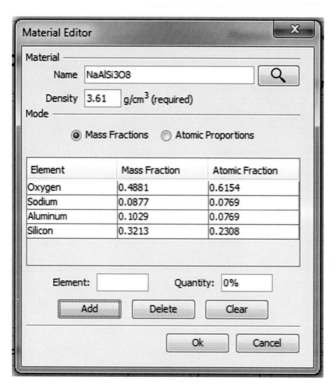

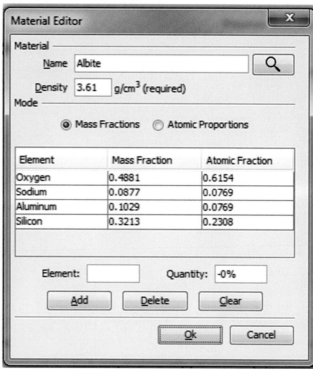

◻ **Fig. 17.12** It is possible to enter the chemical formula directly into the "Name" edit box. When the search button is pressed the chemical formula will be parsed and the appropriate mass and atomic fractions entered into the table. Capitalization of the element abbreviations is important as "CO" is very different from "Co"—one is a gas and the other a metal. More complex formulas like fluorapatite ("$Ca_5(PO_4)_3F$") can be entered using parenthesis to group terms. It is important that the formula is unambiguous. Once the formula has been parsed you may specify a new operator friendly name for the material like "Albite" of "Fluorapatite" in the "Name" edit box

◻ **Fig. 17.13** To assist the user, DTSA-II maintains a database of materials. Each time the user enters a new material or redefines an old material, the database is updated. The database is indexed by "Name"

common geometries through the graphical user interface. More complex geometries can be simulated using the scripting interface.

Monte Carlo modeling is based on simulating the trajectories of thousands of electrons and X-rays. The simulated electrons are given an initial energy and trajectory which models the initial energy and trajectory of electrons from an SEM gun. The simulated electrons scatter and lose energy as their trajectories take them through the sample. Occasionally, an electron may ionize an atom and generate a characteristic X-ray. Occasionally, an electron may decelerate and generate a *bremsstrahlung* photon. The X-rays can be tracked from the point of generation to a simulated detector. A simulated spectrum can be accumulated. If care is taken modeling the electron transport and the X-ray interactions, the simulated spectrum can mimic a measured spectrum. Other data such as emitted intensities, emission images, trajectory images and excitation volumes can be accumulated.

Multi-element materials are modeled as mass-fraction averaged mixtures of elements. Complex sample geometries can be constructed out of discrete sample shapes that include blocks, spheres, cylinders, regions bounded by planar surfaces, and sums and differences of the basic shapes. In this way, it is possible to model arbitrarily complex sample geometries.

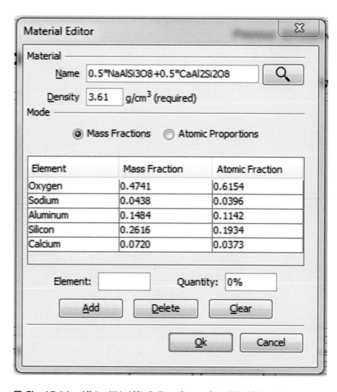

◻ **Fig. 17.14** Albite ("$NaAlSi_3O_8$") and anorthite ("$CaAl_2Si_2O_8$") represent end members of the plagioclase solid solution series. To calculate a admixture of 50% by mass albite and 50% by mass anorthite, you can enter the formula "0.5*NaAlSi3O8 + 0.5*CaAl2Si2O8". Other admixtures of stoichiometric compounds can be calculated in a similar manner. Remember to provide a user friendly name for the database

17.2.3 Using the GUI To Perform a Simulation

The simplest way to simulate one of an array of common sample geometries in through the "simulation alien." The "simulation alien" is a dialog that takes you step-by-step through the process of simulating an X-ray spectrum. The dialog requests information that defines the sample geometry, the instrument, and detector, the measurement conditions and the information to simulate. The results of the simulation include a spectrum, raw intensity data, electron trajectory images, X-ray emission images, and excitation volume information. The "simulation alien" is accessed through the "Tools" application menu (◘ Fig. 17.15).

Many different common sample geometries are available through the "simulation alien."
- Analytical model of a bulk, homogeneous material.
  - A $\varphi(\rho z)$-based analytical spectrum simulation model. This model simulates a spectrum in a fraction of a second but is only suited to bulk samples.
- Monte Carlo model of a bulk, homogeneous material.
  - The Monte Carlo equivalent of the $\varphi(\rho z)$-based analytical spectrum simulation model (see ◘ Fig. 17.16).
- Monte Carlo model of a film on a bulk, homogeneous substrate.
  - A model of a user specified thickness film on a substrate (or, optionally, unsupported) (see ◘ Fig. 17.17).
- Monte Carlo model of a sphere on a bulk, homogeneous substrate (see ◘ Fig. 17.18).

- A model of a user specified radius sphere on a substrate (or, optionally, unsupported).
- Monte Carlo model of a cube on a bulk, homogeneous substrate.
  - A model of a user specified size cube on a substrate (or, optionally, unsupported) (see ◘ Fig. 17.19).
- Monte Carlo model of an inclusion on a bulk, homogeneous substrate.
  - A model of a block inclusion of specified square cross section and specified thickness in a substrate (or, optionally, unsupported) (see ◘ Fig. 17.20).
- Monte Carlo model of a beam near an interface.
  - A model of two materials separated by a vertical interface nominally along the y-axis. The beam can be placed a distance from the interface in either material. Positive distances place the beam in the primary material and negative distances are in the secondary material (see ◘ Fig. 17.21).
- Monte Carlo model of a pyramid with a square base.
  - The user can specify the length of the base edge and the height of the pyramid (see ◘ Fig. 17.22).
- Monte Carlo model of a cylinder on its side
  - The user can specify the length and diameter of the cylinder (see ◘ Fig. 17.23).
- Monte Carlo model of a cylinder on its end
  - The user can specify the length and diameter of the cylinder (see ◘ Fig. 17.24).
- Monte Carlo model of a hemispherical cap
  - The user can specify the radius of the hemispherical cap (see ◘ Fig. 17.25).

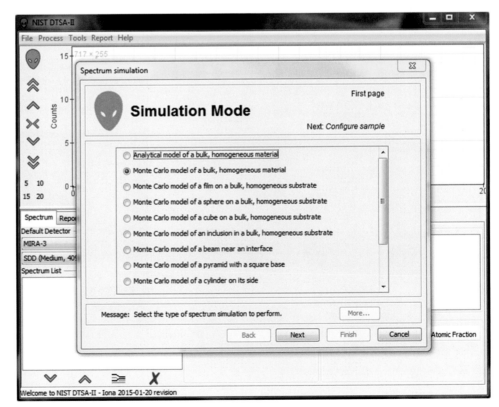

◘ Fig. 17.15 Simulation mode window in DTSA-II

- Monte Carlo model of a block
  - The user can specify the block base (square) and the height.
- Monte Carlo model of an equilateral prism
  - The user can specify the edge of the triangle and the length of the prism (◘ Figs. 17.16, 17.17, 17.18, 17.19, 17.20, 17.21, 17.22, 17.23, 17.24, 17.25, 17.26, 17.27, and 17.28).

Each simulation mode takes different parameters to configure the sample geometry. This page is for the simulation of a cube which requires two materials (substrate and cube), the dimensions of the cube and the sample rotation.

The sample rotation parameter is available for all modes for which are not rotationally invariant. The best way to understand the sample rotation parameter is to imagine rotating the sample about the optic axis at the point on the

sample at which the beam intersects (◘ Figs. 17.29, 17.30, and 17.31).

All the models require you to specify at least one material. The material editor (described elsewhere) allows you to specify the material. Since the density is a critical parameter, you must specify it (◘ Fig. 17.32).

Simulations are designed to model the spectra you could collect on your instrument with your detector. By default, the simulation "instrument configuration" page assumes that you want to simulate the "default detector" as specified on the main "Spectrum" tab. However, you can specify a different instrument, detector and calibration if you desire.

You also need to specify an incident beam energy. This is the kinetic energy with which the electrons strike the sample and is specified in kilo-electronvolts (keV).

The probe dose determines the relative intensity in the spectrum. Probe dose is specified in nano-amp seconds

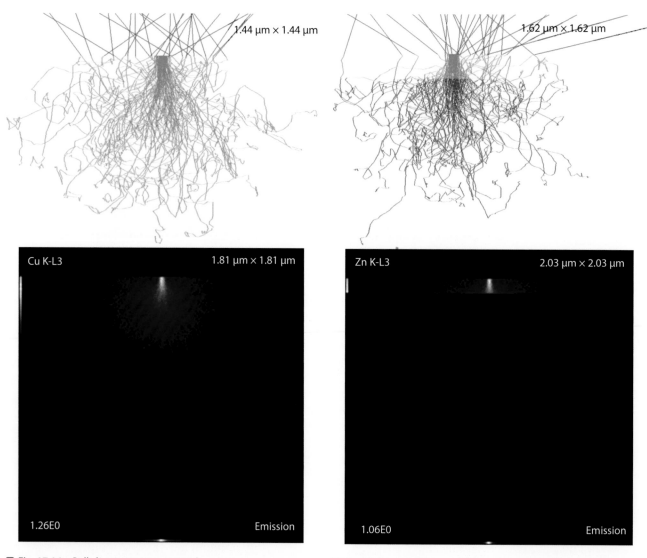

◘ **Fig. 17.16**   Bulk, homogeneous material

◘ **Fig. 17.17**   Thin film on substrate. Parameters: film thickness

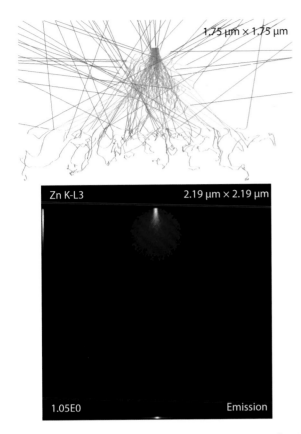

1.75 µm × 1.75 µm

Zn K-L3   2.19 µm × 2.19 µm

1.05E0   Emission

Fig. 17.18 Spherical particle on a substrate. Parameters: Sphere's radius

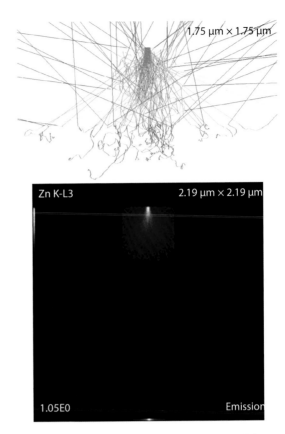

1.75 µm × 1.75 µm

Zn K-L3   2.19 µm × 2.19 µm

1.05E0   Emission

Fig. 17.19 Cubic particle on a substrate. Parameters: Cube dimension

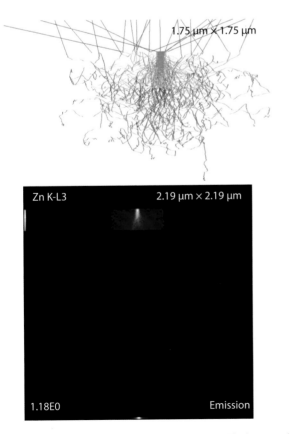

1.75 µm × 1.75 µm

Zn K-L3   2.19 µm × 2.19 µm

1.18E0   Emission

Fig. 17.20 Block inclusion in substrate. Parameters: Thickness and edge length

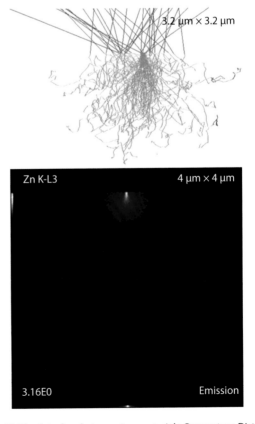

3.2 µm × 3.2 µm

Zn K-L3   4 µm × 4 µm

3.16E0   Emission

Fig. 17.21 Interface between two materials. Parameters: Distance from interface

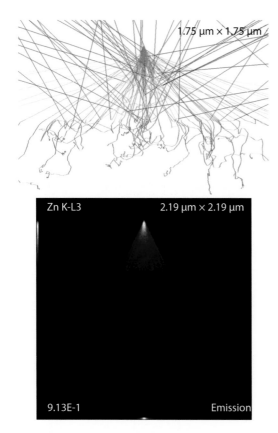

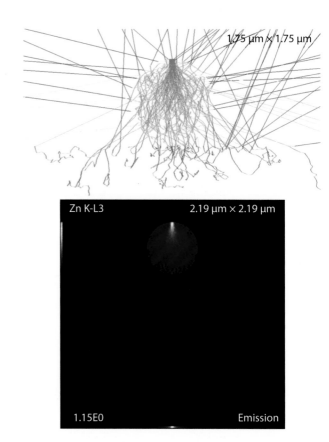

■ **Fig. 17.22**   Square pyramid on a substrate. Parameters: Height and base edge length

■ **Fig. 17.23**   Cylinder (fiber) on substrate. Parameters: Fiber diameter and length

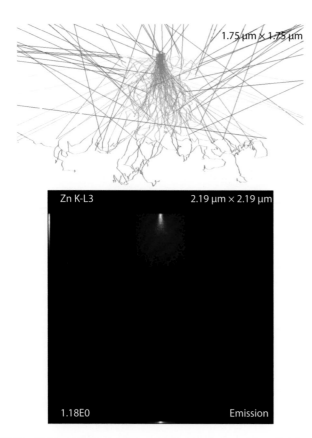

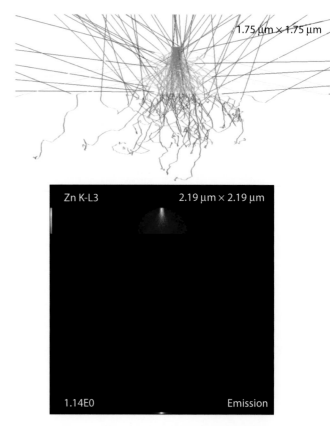

■ **Fig. 17.24**   Cylinder (can) on end on substrate. Parameters: Height and fiber diameter

■ **Fig. 17.25**   Hemispherical cap on substrate. Parameters: Cap radius

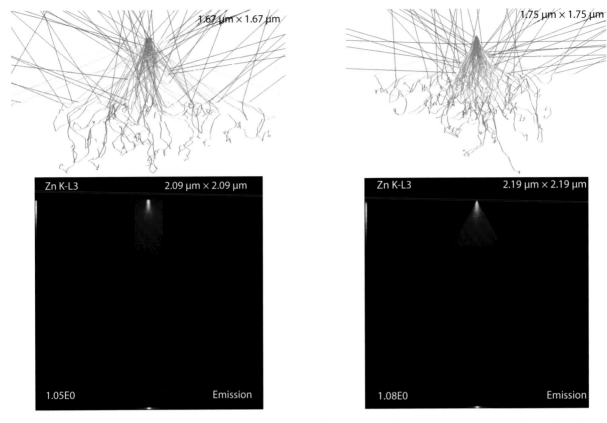

Fig. 17.26 Rectangular block on substrate. Parameters: Block height and base edge length

Fig. 17.27 Triangular prism on substrate. Parameters: Triangle edge and prism lengths

Fig. 17.28 "Configure sample" menu

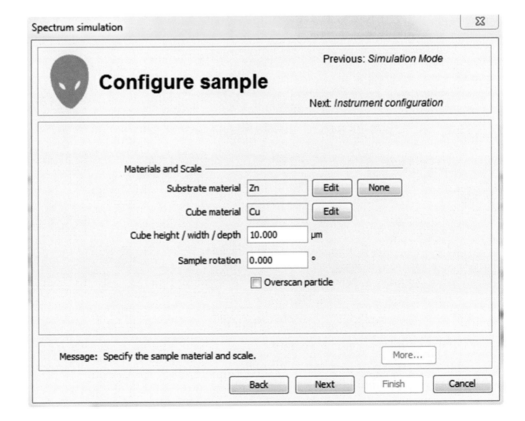

Fig. 17.29 **a** Shows how a pyramid with square base model rotates and **b** shows how a beam near an interface model rotates. Both figures take the perspective of looking down along the optic axis at the sample. The shortest distance to the interface is maintained in ☐ figure **b**

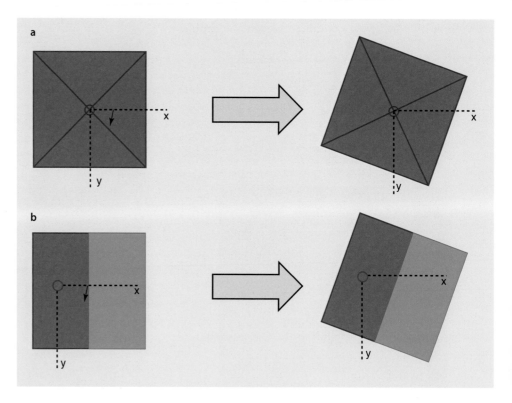

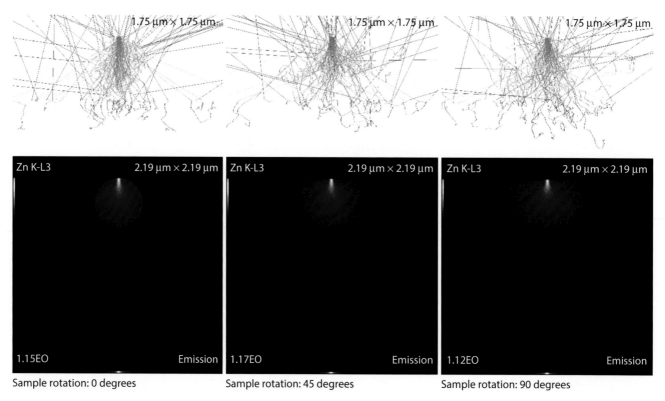

Sample rotation: 0 degrees
Sample rotation: 45 degrees
Sample rotation: 90 degrees

Fig. 17.30 A fiber rotated through 0°, 45°, and 90°

◘ Fig. 17.31 "Material editor" menu

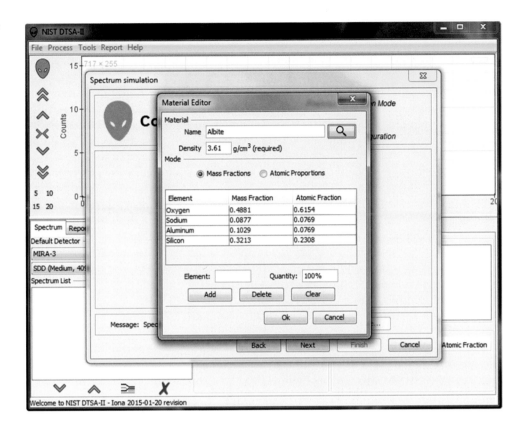

◘ Fig. 17.32 "Instrument configuration" menu

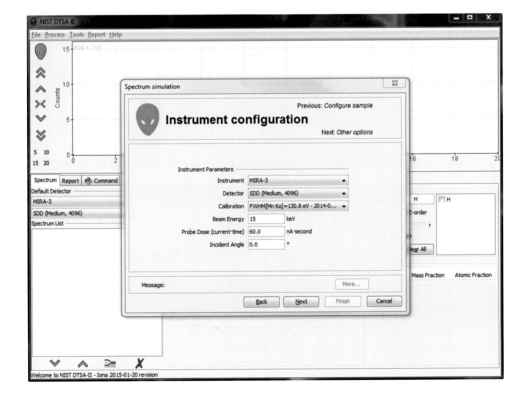

(nA · s = nC). This quantity is a product of the probe current (nA) and the spectrum acquisition live-time (seconds.) Remember the probe current is a measure of the actual number of electrons striking the sample per unit time, so the probe dose is equivalent to a number of electrons striking the sample during the measurement. Doubling the dose doubles the number of electrons striking the sample and thus also doubles the average number of X-rays generated in the sample.

The incidence angle (nominally 0°) allows you to simulate a tilted sample (◘ Fig. 17.33).

The incident angle is defined relative to the optic axis. The pivot occurs at the surface of the sample, which is placed at the detector's optimal working distance. A positive rotation is towards the X-axis (an azimuth of 0°) and a negative rotation towards the –X axis (an azimuth of 180°.) The detector displayed is at an azimuth of 180°. A negative incidence angle would tilt the sample toward the detector. Arbitrary tilts may be simulated by moving the detector around the azimuth (◘ Figs. 17.33, 17.34, 17.35, and 17.36).

The "other options" page allows you to specify whether the spectrum is simulated with or without variance due to count statistics. If you select to "apply simulated count sta-

tistics," you may also select to output multiple spectra based on the simulated spectrum but differ by pseudorandom count statistics. You may also select to run additional simulated electron trajectories. The number of simulated electron trajectories determines the simulation to simulation variance in characteristic X-ray intensities. The default number of electron trajectories typically produces about 1 % variance. The variance decreases as the square-root of the number of simulated trajectories. You may also specify which X-ray generation modes to simulate including both characteristic and *bremsstrahlung* primary emission and secondary emission due to characteristic primary emission or *bremsstrahlung* primary emission (◘ Figs. 17.37, 17.38, and 17.39).

The "configure VP" page provides an advanced option to simulate the beam scatter in a variable-pressure or environmental SEM. If the check box is selected, you may select a gas ("water," "helium," "nitrogen," "oxygen" or "argon"), a gas path length and a nominal pressure. The gas path length is the distance from the final pressure limiting aperture to the sample. 1 Torr is equivalent to 133 Pa (◘ Figs. 17.40 and 17.41).

The primary output of a simulation is a spectrum. The simulated spectrum looks and acts to the best of our ability

◘ **Fig. 17.33** Definition of angles for a tilted specimen

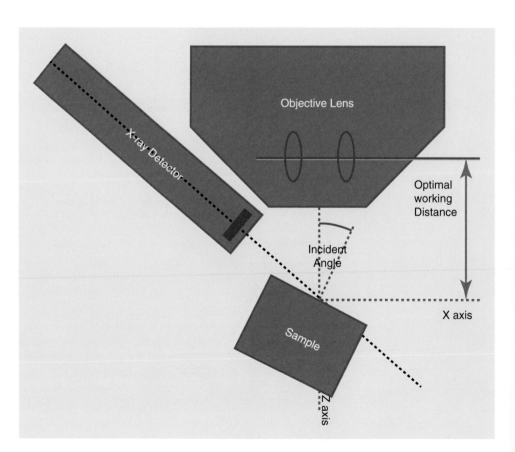

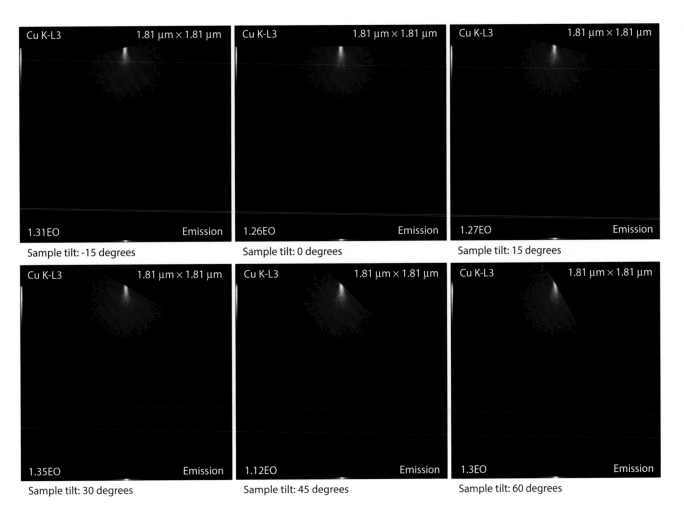

Fig. 17.34   Tilting a bulk sample

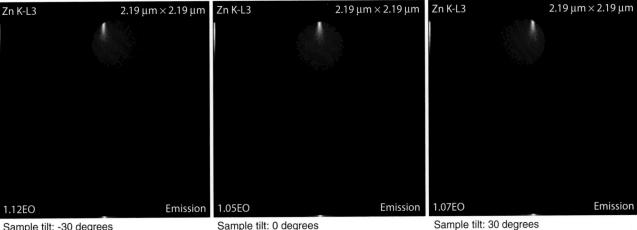

Fig. 17.35   Tilting a spherical sample

◻ Fig. 17.36 Menu for selecting number of trajectories, repetitions, and X-ray generation modes

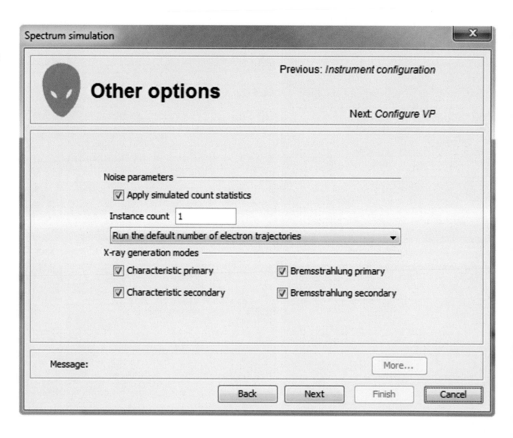

to simulate measured spectra like a spectrum collected from the specified material under the specified conditions on the specified detector. You can treat simulated spectra like measured spectra in the sense that you can simulate an "unknown" standards and reference spectra and quantify the "unknown" as though it had been measured (◻ Fig. 17.42).

The report tab contains additional details about the simulation including configuration and results information. The first table summarizes the simulation configuration parameters. This table is available for all simulation modes (◻ Fig. 17.43).

The "simulation results" table contains additional details derived from the simulation. The first two lines contain links to the simulated spectrum and a Virtual Reality Markup Language (VRML) representation of the sample and electron trajectories. The subsequent rows contain raw data detailing the generation and emission of various different kinds of characteristic X-rays. Only those modes which were selected will be available. The generated X-rays column tabulates the relative number of X-rays

generated within the sample and emitted into a milli-steradian. The emitted column tabulates the X-rays that are generated and also escape the sample in the direction of the detector. The ratio is the fraction of generated X-rays that escape the sample in the direction of the detector. The final rows compare the relative amount of characteristic fluorescence to the amount of primary characteristic and *bremsstrahlung* emission (◻ Fig. 17.44).

The emission images show where the measured X-rays were generated. Because the images only display X-rays that escape the sample, the distinction between the strongly absorbed X-rays like the O K-L3 (Kα) and the less strongly absorbed like the Si K-L3 (Kα) is evidenced by the flattened emission profile in the Si K-L3 image. The last image shows the first 100 simulated electron trajectories (down to a kinetic energy of 50 eV.) The color of the trajectory segment varies as the electron passes through the different materials present in the sample. The gray lines exiting the top of the image are backscattered electrons (trajectories in vacuum).

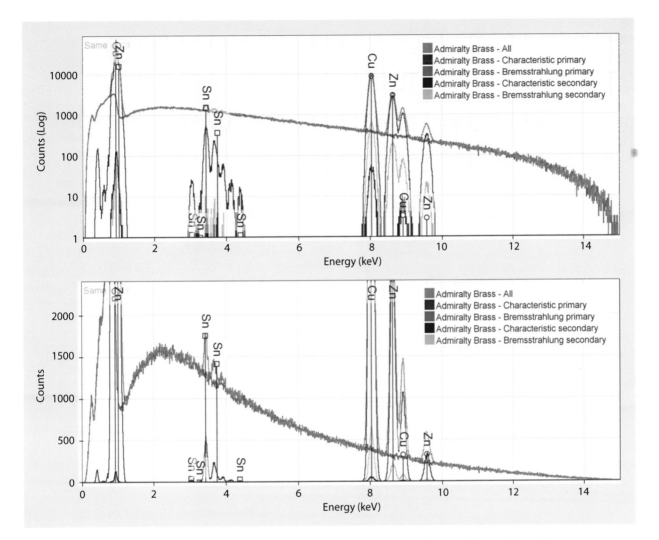

Fig. 17.37 Simulated admiralty brass (69 % Cu, 30 % Zn, and 1 % Tin) for various different selections of generation modes

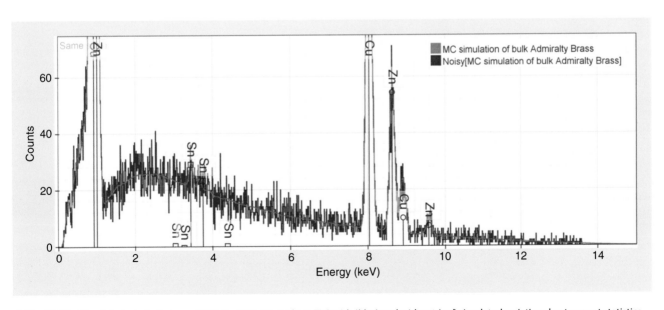

Fig. 17.38 Simulated admiralty brass (69 % Cu, 30 % Zn, and 1 % Tin) with (*blue*) and without (*red*) simulated variation due to count statistics

Fig. 17.39 Menu for selection of variable pressure SEM operating conditions

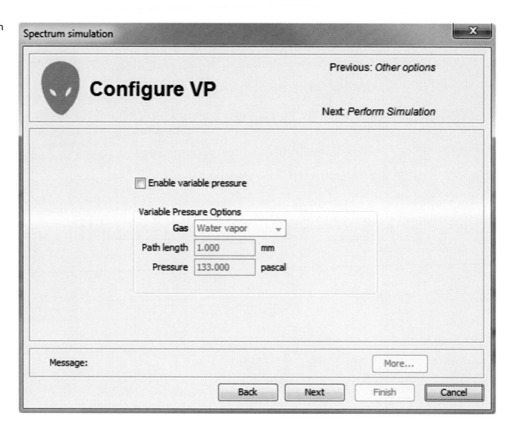

Fig. 17.40 The "perform simulation" page shows progress as the electron trajectories are simulated. When the simulation is complete the "finish" button will enable to allow you to close the dialog

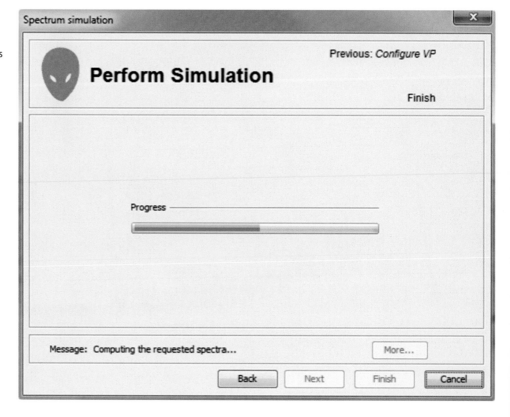

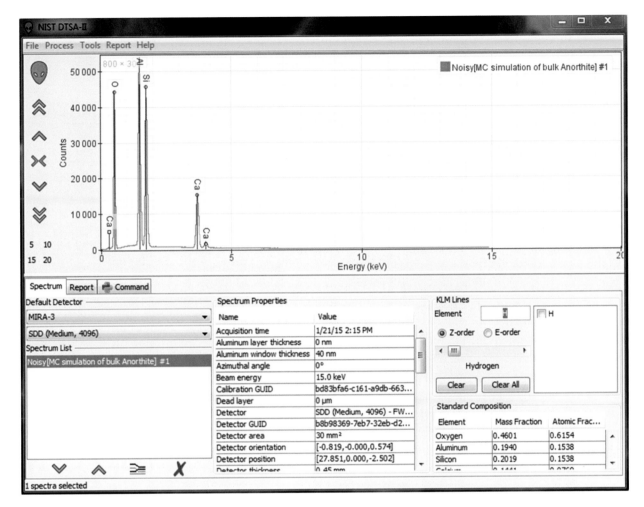

☐ **Fig. 17.41**   Simulation results: X-ray spectrum

☐ **Fig. 17.42**   Simulation results:
configuration record

Simulation Configuration				
Simulation mode	Monte Carlo model of a bulk sample			
Material	Anorthite			
	Element	Mass Fraction	Mass Fraction (normalized)	Atomic Fraction
	O	0.4601	0.4601	0.6154
	Al	0.1940	0.1940	0.1538
	Si	0.2019	0.2019	0.1538
	Ca	0.1441	0.1441	0.0769
Sample rotation	0.0°			
Sample tilt	0.0°			
Beam energy	15.0 keV			
Probe dose	60.0 nA·s			
Instrument	MIRA-3			
Detector	SDD (Medium, 4096)			
Calibration	FWHM[Mn Kα]=130.8 eV - 2014-02-18 00:00			
Overscan	False			
Vacuum conditions	High vacuum			
Replicas (with poisson noise)	1			

Fig. 17.43 Simulation results: table of X-ray intensities from various sources

Simulation Results

Result 1	Noisy[MC simulation of bulk Anorthite] #1			
Trajectory view	VRML World View File			
Characteristic	Transition	Generated 1/msR	Emitted 1/msR	Ratio (%)
	O All	91,816,279.1	20,029,188.1	21.8%
	Al Kα	28,067,811.9	20,031,447.5	71.4%
	Al Kβ	114,057.1	84,361.9	74.0%
	Si Kα	26,960,986.6	18,249,064.9	67.7%
	Si Kβ	459,584.6	325,666.6	70.9%
	Ca Kα	8,067,774.1	7,482,648.8	92.7%
	Ca Kβ	859,214.2	809,593.5	94.2%
Characteristic Fluorescence	Transition	Generated 1/msR	Emitted 1/msR	Ratio (%)
	O All	103,834.0	13,858.1	13.3%
	Al Kα	378,534.1	191,421.7	50.6%
	Si Kα	94,400.7	24,117.3	25.5%
	Si Kβ	1,069.9	298.3	27.9%
Bremsstrahulung Fluorescence	Transition	Generated 1/msR	Emitted 1/msR	Ratio (%)
	O All	52,830.4	8,065.4	15.3%
	Al Kα	108,362.6	44,802.6	41.3%
	Si Kα	137,984.3	47,597.8	34.5%
	Si Kβ	1,563.9	584.1	37.3%
	Ca Kα	190,086.6	93,453.6	49.2%
	Ca Kβ	20,244.1	11,052.8	54.6%
Comparing Characteristic to Characteristic Fluorescence	Transition	Generated (ratio)	Emitted (ratio)	
	O All	0.0011	0.0007	
	Al Kα	0.0135	0.0096	
	Al Kβ	0.0000	0.0000	
	Si Kα	0.0035	0.0013	
	Si Kβ	0.0023	0.0009	
	Ca Kα	0.0000	0.0000	
	Ca Kβ	0.0000	0.0000	
Comparing Characteristic to Bremsstrahulung Fluorescence	Transition	Generated (ratio)	Emitted (ratio)	
	O All	0.0006	0.00004	
	Al Kα	0.0039	0.00022	
	Al Kβ	0.0000	0.00000	
	Si Kα	0.0051	0.00026	
	Si Kβ	0.0034	0.00018	
	Ca Kα	0.0236	0.0125	
	Ca Kβ	0.0236	0.0137	

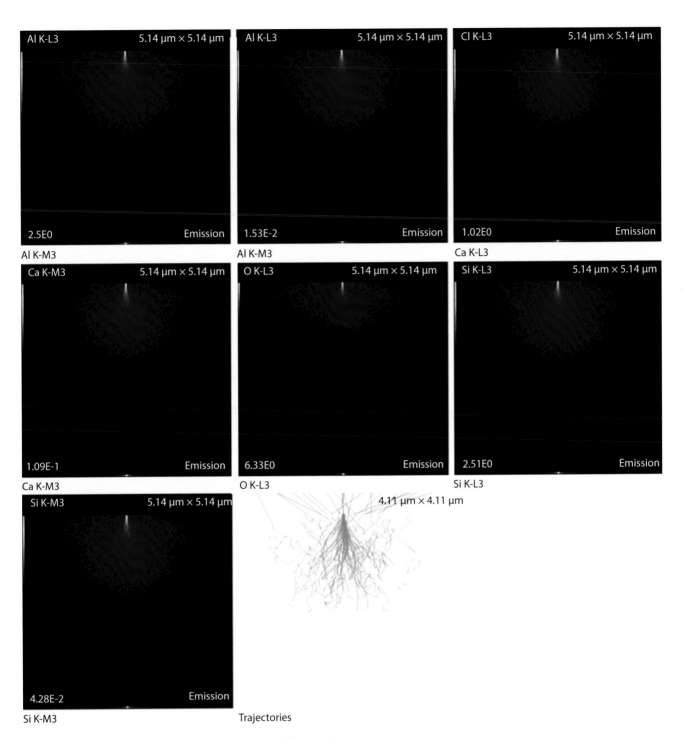

■ Fig. 17.44   Simulation results: X-ray emission images and trajectories

Fig. 17.45 Output table listing depth and volumes from which 50 % or 90 % of the measured X-rays are emitted

Fractional Emission Depths and Volumes

Ionization Edge	Ionization Energy (keV)	F(50 %) Depth (µm)	F(90 %) Depth (µm)	F(99.9 %) Depth (µm)	F(50 %) Volume (µm)³	F(90 %) Volume (µm)³	F(99.9 %) Volume (µm)³
O K	0.532	0.321	0.835	1.740	0.254	4.007	11.420
Al K	1.560	1.597	1.336	1.997	0.686	5.507	12.098
Si K	1.838	0.584	1.233	1.997	0.585	5.219	11.937
Si L1	0.148	0.064	0.193	0.347	0.017	0.618	2.220
Ca K	4.038	0.584	1.226	1.869	0.398	3.948	10.522
Ca L1	0.438	0.205	0.584	1.361	0.102	2.728	9.370
Ca L2	0.350	0.193	0.565	1.233	0.051	2.313	8.421
Ca L3	0.346	0.193	0.482	1.226	0.042	2.254	8.260

Fig. 17.46 (*upper*) Simulated trajectories for high-vacuum conditions. All the electrons strike the inclusion. (*lower*) Trajectories simulated for variable pressure mode with 1-mm gas path length through water vapor at 133 Pa. The *green* trajectories are the incident and backscattered electrons

44 µm × 44 µm

44 µm × 44 µm

17.2.4 Optional Tables

The "Fractional Emission Depths and Volumes" Table (Fig. 17.45)

When simulating a bulk sample, an additional report table shows the depth and the volumes from which 50 % or 90 % of the measured X-rays are emitted. The depth and volume are largely determined by the ionization edge energy and X-ray absorption. Low ionization edge energies emit from larger volumes but lower energy X-rays also tend to be more strongly absorbed.

The "VP Scatter Data" Table

In variable pressure mode, the gas in the chamber can scatter the incident electrons before they strike the sample. Despite the fact that these scatters tend to be small angle events, the path length is relatively long and electrons can scatter hundreds of microns to millimeters. The consequence can be demonstrated by simulating a moderate sized inclusion, shown in Fig. 17.46. In simulated high-vacuum mode, the excitation volume remains entirely within the inclusion. In simulated variable-pressure mode, the electrons can scatter out of the beam, entirely missing the inclusion and striking the surrounding matrix.

One way to understand the scatter is to consider a series of concentric rings on the surface of the sample centered at the beam axis. The "VP Scatter Data" table (Fig. 17.47), summarizes the number and number fraction of the incident electrons which intersect the various rings. In this simulation, 87 % of the initial electrons are undeflected. However, at least one electron (0.1 %) is scattered further than 700 µm and 1.2 % are scattered more than 50 µm. This qualitative information is useful because it gives us a sense of how significant beam scatter will be in variable pressure mode. It gives a sense of whether true quantitative analysis is possible and how much of an error will be introduced by the beam scatter. The consequences are evident in the spectrum from an inclusion of admiralty brass in an aluminum. The aluminum is present in significant quantities in the variable pressure mode acquisition.

Figure 17.48 shows EDS spectra calculated for a brass inclusion in an aluminum matrix under VPSEM (red) and conventional vacuum (blue) operation. The large peak for Al under VPSEM conditions reveals the extent of gas scattering outside the focused beam. Interestingly, the Al is not zero in the "high-vacuum" spectrum because of continuum generated secondary fluorescence. Increasing the size of the inclusion does not eliminate the slight Al peak but turning off the simulation of continuum fluorescence does.

Fig. 17.47 Distribution of gas-scattered electrons into a series of concentric rings

VP Scatter Data
Electron trajectory count = 1000

Ring	Inner Radius μm	Outer Radius μm	Ring area μm²	Electron Count	Electron Fraction	Cumulative
Undeflected	-	-	-	869	0.869	-
1	0.0	2.5	19.6	921	0.921	92.1%
2	2.5	5.0	58.9	15	0.015	93.6%
3	5.0	7.5	98.2	8	0.008	94.4%
4	7.5	10.0	137.4	8	0.008	95.2%
5	10.0	12.5	176.7	8	0.008	96.0%
6	12.5	15.0	216.0	7	0.007	96.7%
7	15.0	17.5	255.3	4	0.004	97.1%
8	17.5	20.0	294.5	2	0.002	97.3%
9	20.0	22.5	333.8	3	0.003	97.6%
10	22.5	25.0	373.1	3	0.003	97.9%
11	25.0	27.5	412.3	2	0.002	98.1%
12	27.5	30.0	451.6	2	0.002	98.3%
13	30.0	32.5	490.9	0	0.000	98.3%
14	32.5	35.0	530.1	0	0.000	98.3%
15	35.0	37.5	569.4	1	0.001	98.4%
16	37.5	40.0	608.7	1	0.001	98.5%
17	40.0	42.5	648.0	1	0.001	98.6%
18	42.5	45.0	687.2	0	0.000	98.6%
19	45.0	47.5	726.5	1	0.001	98.7%
20	47.5	50.0	765.8	1	0.001	98.8%
21	50.0	100.0	23561.9	8	0.008	99.6%
22	100.0	150.0	39269.9	0	0.000	99.6%
23	150.0	2000.0	54977.9	1	0.001	99.7%
24	200.0	250.0	70685.8	0	0.000	99.7%
25	250.0	300.0	86393.8	1	0.001	99.8%
26	300.0	350.0	102101.8	0	0.000	99.8%
27	350.0	400.0	117809.7	1	0.001	99.9%
28	400.0	450.0	133517.7	0	0.000	99.9%
29	450.0	500.0	149225.7	0	0.000	99.9%
30	500.0	550.0	164933.6	0	0.000	99.9%
31	550.0	600.0	180641.6	0	0.000	99.9%
32	600.0	650.0	196349.5	0	0.000	99.9%
33	650.0	700.0	212057.5	0	0.000	99.9%
34	700.0	750.0	227765.5	1	0.001	100.0%
35	750.0	800.0	243473.4	0	0.000	100.0%
36	800.0	850.0	259181.4	0	0.000	100.0%
37	850.0	900.0	274889.4	0	0.000	100.0%
38	900.0	950.0	290597.3	0	0.000	100.0%
39	950.0	1000.0	306305.3	0	0.000	100.0%

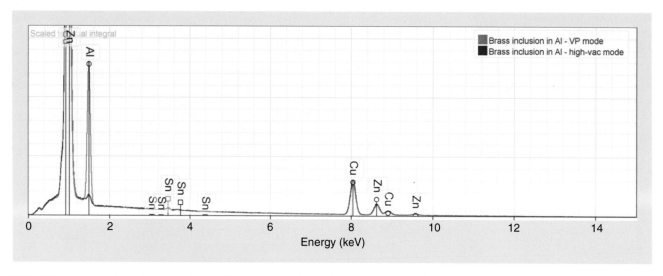

Fig. 17.48 Comparison of spectra calculated for a brass inclusion in aluminum: VPSEM (*red*) and conventional vacuum (*blue*) operation. 1-mm gas path length through water vapor at 133 Pa

References

Pouchou J-L, Pichoir P (1991) "Quantitative analysis of homogeneous or stratified microvolumes applying the model PAP". In: Heinrich K, Newbury D (eds) Electron Probe Quantitation. New York: Plenum, p163

Ritchie N (2009) Spectrum simulation in DTSA-II. Microsc Microanal 15:454

Ritchie N (2010) Using DTSA-II to simulate and interpret energy dispersive spectra from particles. Microsc Microanal 16:248

Ritchie N (2011a) "Getting Started with NIST DTSA-II." In: Lyman C (ed) Microscopy Today. Cambridge Univ. Press. 19(1):26

Ritchie N (2011b) "Manipulating Spectra with DTSA-II." In: Lyman C (ed) Microscopy Today. Cambridge Univ. Press. 19(3):34

Ritchie N (2011c) "Standards-Based Quantification in DTSA-II – Part I." In: Lyman C (ed) Microscopy Today. Cambridge Univ. Press. 19(5):30

Ritchie N (2012) "Standards-Based Quantification in DTSA-II – Part II." In: Lyman C (ed) Microscopy Today. Cambridge Univ. Press. 20(1):14

Ritchie N (2017a) "DTSA-II, a multiplatform software package for quantitative x-ray microanalysis." Available free at the National Institute of Standards and Technology website (search "DTSA-II"): ► http://www.cstl.nist.gov/div837/837.02/epq/dtsa2/index.html

Ritchie N (2017b) "Efficient simulation of secondary fluorescence Via NIST DTSA-II Monte Carlo" Microsc Microanal 23:618

17

Qualitative Elemental Analysis by Energy Dispersive X-Ray Spectrometry

© Springer Science+Business Media LLC 2018
J. Goldstein et al., *Scanning Electron Microscopy and X-Ray Microanalysis*,
https://doi.org/10.1007/978-1-4939-6676-9_18

■■ **Overview**

Qualitative elemental analysis involves the assignment of elements to the characteristic X-ray peaks recognized in the energy dispersive X-ray spectrometry (EDS) spectrum. This function is routinely performed with automatic peak identification (e.g., "AutoPeakID") software embedded in the vendor EDS system. While automatic peak identification is a valuable tool, the careful analyst will always manually identify elements by hand first and only use the automatic peak identification to confirm the manual elemental identification, even at the level of major constituents (mass concentration, $C > 0.1$), but especially for minor ($0.01 \leq C \leq 0.1$) and trace ($C < 0.1$) constituents. Using automatic peak identification before manual identification tends to lead to a cognitive flaw called confirmation bias - the tendency to interpret data in a way that confirms one's preexisting beliefs or hypotheses.

18.1 Quality Assurance Issues for Qualitative Analysis: EDS Calibration

Before attempting automatic or manual peak identification, it is critical that the EDS system be properly calibrated to ensure that accurate energy values are measured for the characteristic X-ray peaks. Follow the vendor's recommended procedure to rigorously establish the calibration. The calibration procedure typically involves measuring a known material such as copper that provides characteristic X-ray peaks at low photon energy (e.g., Cu L_3-M_5 at 0.928 keV) and at high photon energy (Cu K-L_3 at 8.040 keV). Alternatively, a composite aluminum-copper target (e.g., a copper penny partially wrapped in aluminum foil and continuously scanned so as to excite both Al and Cu) can be used to provide the Al K-L_3 (1.487 keV) as the low energy peak and Cu K-L_3 for the high energy peak. After calibration, peaks occurring within this energy range (e.g., Ti K-L_3 at 4.508 keV and Fe K-L_3 at 6.400 keV) should be measured to confirm linearity. A well-calibrated EDS should produce measured photon energies within ±2.5 eV of the ideal value. Low photon energy peaks below 1 keV photon energy should also be measured, for example, O K (e.g., from MgO) and C K. For some EDS systems, non-linearity may be encountered in the low photon energy range. ◘ Figure 18.1 shows an EDS spectrum for $CaCO_3$ in which the O K peak at 0.523 keV is found at the correct energy, but the C K peak at 0.282 keV shows a significant deviation below the correct energy due to non-linear response in this range caused by incomplete charge collection.

All calibration spectra should be stored as part of the laboratory Quality Assurance documentation, and the calibration procedure should be performed regularly, preferably weekly and especially whenever the EDS system is powered down and restarted.

18.2 Principles of Qualitative EDS Analysis

The knowledge base needed to accomplish high-confidence peak identification consists of three components: (1) the physics of characteristic X-ray generation and propagation; (2) a complete database of the energies of all critical ionization energies and corresponding characteristic peaks for all elements (except H and He, which do not produce characteristic X-rays); and (3) the artifacts inherent in EDS measurement.

18.2.1 Critical Concepts From the Physics of Characteristic X-ray Generation and Propagation

What factors determine if characteristic peaks are generated and detectable?

Exciting Characteristic X-Rays

A specific characteristic X-ray can only be produced if the incident beam energy, E_0, exceeds the critical ionization energy, E_c, for the atomic shell whose ionization leads to the emission of that characteristic X-ray. This requirement is parameterized as the overvoltage, U_0:

$$U_0 = E_0 / E_c > 1 \tag{18.1}$$

Note that for a particular element, if the beam energy is selected so that $U_0 > 1$ for the K-shell, then for higher atomic number elements with complex atomic shell structures, shells with lower values of E_c will also be ionized; for example, if Cu K-shell X-rays are created, there will also be Cu L-shell X-rays, Au L-family, and Au M-family X-rays, etc.

While $U_0 > 1$ sets the minimum beam energy criterion to generate a particular characteristic X-ray, the relative intensity of that X-ray generated from a thick target (where the thickness exceeds the electron range) depends on the overvoltage and the incident beam current, i_B:

$$I_{ch} \sim i_B (U_0 - 1)^n \tag{18.2}$$

where the exponent n is approximately 1.5. The X-ray continuum intensity, I_{cm}, that forms the spectral background at all photon energies up to E_0 (the Duane–Hunt limit), arises from the electron *bremsstrahlung* and depends on the photon energy, E_v, and the beam energy:

$$I_{cm} \sim i_B (E_0 - E_v) / E_v \tag{18.3a}$$

$$I_{cm} \sim i_B (U_0 - 1) \qquad \text{for } E_v \sim E_c \tag{18.3b}$$

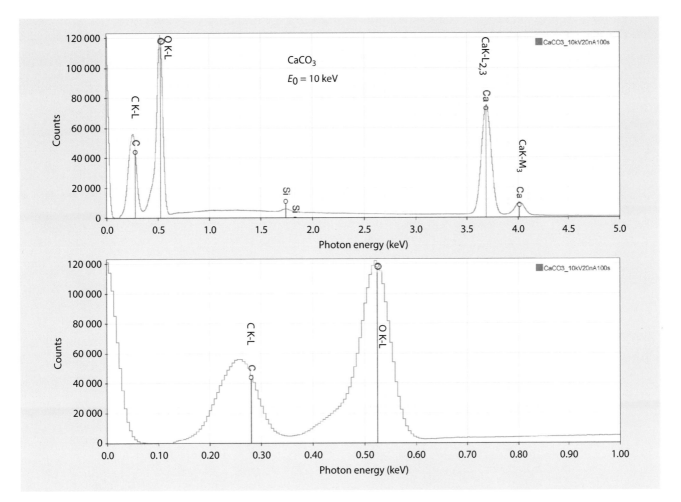

☐ **Fig. 18.1** Spectrum of calcium carbonate. Note non-linear behavior at low photon energy, e.g., the C K-shell peak is significantly shifted below the true energy value given by the marker

The characteristic peak to continuum background, P/B, which determines the visibility of peaks above the background, is found as the ratio of equations (18.2) and (18.3b):

$$P / B = I_{ch} / I_{cm} \sim (U_0 - 1)^n / (U_0 - 1)$$
$$= (U_0 - 1)^{n-1} \qquad (18.4)$$

Since the exponent $n \sim 1.5$, in the expression for P/B the value of $n - 1 \sim 0.5$, so that as U_0 is lowered, the P/B decreases dramatically, reducing the visibility of peaks, as shown in ☐ Fig. 18.2 for the K-shell peaks of silicon.

Fluorescence Yield

A second factor that affects the detectability of characteristic peaks is the fluorescence yield, the fraction of ionizations that leads to photon emission. The fluorescence yield varies sharply depending on the shells involved, with the fluorescence yields for a particular element generally trending K>L >>M. An example for barium L-shell and M-shell X-rays is shown in ☐ Fig. 18.3, where the Ba M-family X-rays are seen to have a much lower P/B than the Ba L-family X-rays, making Ba difficult to identify with high confidence if only the Ba M-family is excited, a condition that will exist for Ba if E_0 is chosen below the 5.25 keV ionization energy for the Ba L_3-shell.

X-ray Absorption

A third factor which can strongly influence the visibility and detection of peaks is absorption of characteristic X-rays as they travel through the specimen and the window and surface layers of the EDS detector. X-ray absorption along a path of length s through the specimen is a non-linear process:

$$I / I_0 = \exp\left[-(\mu / \rho)\rho s\right] \qquad (18.5)$$

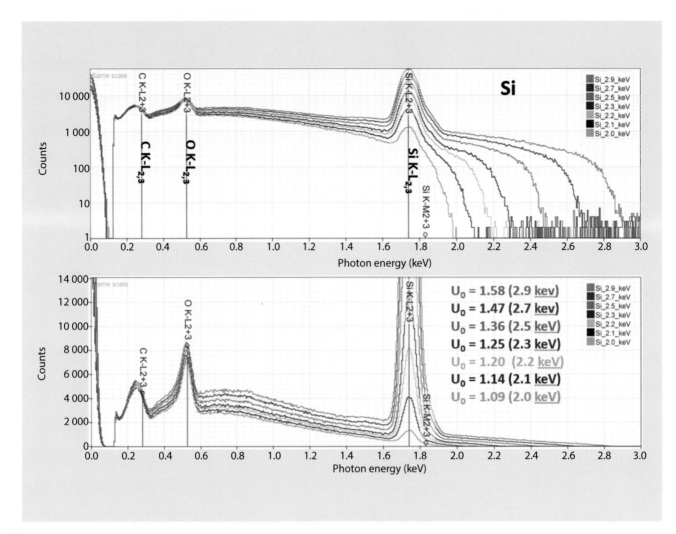

□ **Fig. 18.2**   Si at various overvoltages, showing diminishing peak visibility as the excitation decreases

where I_0 is the original intensity and I is the intensity that remains after path s through a material of density ρ having a mass absorption coefficient μ/ρ for the photon energy of interest. The mass absorption coefficient depends strongly on the photon energy and the specific elements that the photons are passing through. Generally, a photon will be strongly absorbed, i.e., there will be a large mass absorption coefficient, if the energy of the photon lies in a range of approximately 1 keV above the critical ionization energy for another element that is present in the analyzed volume. An extreme case is illustrated in □ Fig. 18.4 for SiC, where at $E_0 = 20$ keV with the spectrum scaled to the Si K-peak, the C K peak is barely visible despite C's making up half of the composition on an atomic basis. Strong absorption of the C K X-ray at 0.282 keV occurs because this energy lies just above the Si L_3 critical ionization energy at 0.110 keV, resulting in an extremely large value for the mass absorption coefficient. (There are also other factors that apply to this case, including the relative fluorescence yields, for which $\omega_C < \omega_{Si}$, and the

relative detector efficiency, $\varepsilon_{C} < \varepsilon_{Si}$, as described in the "EDS" module.) Because the absorption path length, s, depends strongly on the electron range, R_{K-O}, which scales approximately as the 1.7 power of the incident beam energy, decreasing the beam energy reduces the absorption path of the C K X-rays, making the C K-peak more prominent relative to the Si K-peak, as shown in □ Fig. 18.4 for a series of progressively lower beam energies.

The possibility of a high absorption situation for elements that must be measured with a low photon energy requires an analytical strategy such that when analyzing an unknown, the analyst should start at high beam energy, $E_0 \geq 20$ keV, and work down in beam energy. The analyst must be prepared to utilize low beam energies, $E_0 \leq 5$ keV, to evaluate the possibility of high absorption situations, such as those encountered for low atomic number elements ($Z < 10$). For these elements, the only detectable peaks have low photon energies (<1 keV) and are thus subject to high absorption when high incident beam energy is used.

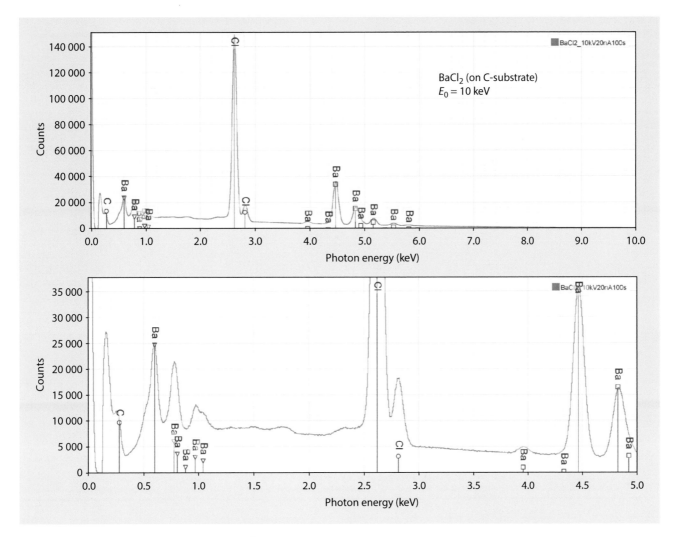

18.2.2 X-Ray Energy Database: Families of X-Rays

The X-ray energy database is typically accessed through EDS software as a display of "KLM" markers showing the position of the peak (and possibly also the corresponding critical ionization energies) and the relative peak heights of the members of each X-ray family, examples of which are seen in ■ Figs. 18.1, 18.2, 18.3, and 18.4. The underlying database must contain such all characteristic X-rays (for ionization energies up to 25 keV) for all elements (excepting H and He which do not produce characteristic X-rays). No elements should be excluded, all X-ray families with photon energies below 30 keV should be included, and no minor X-ray family members should be excluded. As an example, a section of the DTSA-II X-ray energy database that displays the information for gold is presented in ■ Table 18.1. While it is true that many of the closely spaced (in photon energy) and low abundance X-ray peaks cannot

be resolved by EDS because of the limited energy resolution, these peaks are nevertheless convolved in the measured spectrum. When a constituent is present at high concentration and is excited with adequate overvoltage, at least some of these low abundance family members will be readily detectable, for example, the L_3-M_1 (Ll) and $M_{4,5}N_{2,3}$ (Mζ) peaks, as shown for Ba in ■ Fig. 18.5 and the $AuM_{4,5}N_{2,3}$, $AuM_1N_{1,3}$ and AuM_2N_4 peaks seen in ■ Fig. 18.6. Note also that the low energy performance of the silicon drift detector (SDD)-EDS is such that the Au N-family peaks are detected.

18.2.3 Artifacts of the EDS Detection Process

The EDS detection process is subject to two principal artifacts that must be properly cataloged to avoid subsequent misidentification.

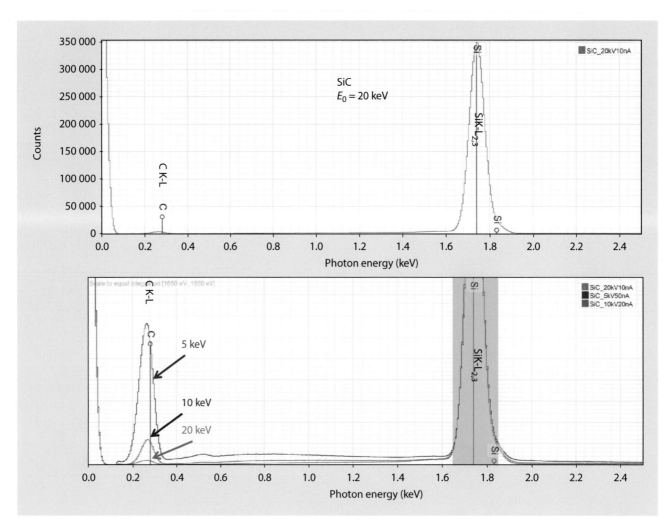

□ **Fig. 18.4**   EDS spectra of SiC: (*upper*) at $E_0 = 20$ keV, the C K-L_3 peak is barely visible; (*lower*) as the beam energy is lowered to reduce absorption, the C K-L_3 peak becomes more prominent (all spectra scaled to the Si K-L_3 region.)

□ **Table 18.1**   Comprehensive listing of all X-ray transitions for gold (DTSA-II database)

IUPAC	Siegbahn	Weight	Energy (keV)	Wavelength (Å)
Au K-N5	Au Kβ4	0.0000	80.391	0.154226
Au K-N3	Au Kβ2	0.0500	80.1795	0.154633
Au K-M5	Au Kβ5	0.0005	78.5192	0.157903
Au K-M3	Au Kβ1	0.1500	77.9819	0.158991
Au K-M2	Au Kβ3	0.1500	77.5771	0.159821
Au K-L3	Au Kα1	1.0000	68.8062	0.180193
Au K-L2	Au Kα2	0.5000	66.9913	0.185075
Au L1-O4	Au L1O4/L1O5	0.0027	14.3445	0.864333
Au L1-O3	Au Lγ4	0.0062	14.2991	0.867077
Au L1-O2	Au Lγ4p	0.0001	14.2811	0.86817
Au L1-O1	Au L1O1	0.0027	14.245	0.87037

18

◨ **Table 18.1** (continued)

IUPAC	Siegbahn	Weight	Energy (keV)	Wavelength (Å)
Au L1-N5	Au Lγ11	0.0007	14.0189	0.884407
Au L1-N4	Au L1N4	0.0001	14.0008	0.885551
Au L1-N3	Au Lγ3	0.0194	13.8074	0.897955
Au L2-O4	Au Lγ6	0.0109	13.7253	0.903326
Au L1-N2	Au Lγ2	0.0015	13.7091	0.904393
Au L2-O3	Au L2O3	0.0001	13.6799	0.906324
Au L2-O2	Au L2O2	0.0001	13.6619	0.907518
Au L2-N6	Au Lν	0.0003	13.6472	0.908495
Au L2-O1	Au Lγ8	0.0007	13.6258	0.909922
Au L1-N1	Au L1N1	0.0001	13.594	0.912051
Au L2-N5	Au L2N5	0.0001	13.3997	0.925276
Au L2-N4	Au Lγ1	0.0841	13.3816	0.926527
Au L2-N3	Au L2N3	0.0027	13.1882	0.940115
Au L2-N2	Au L2N2	0.0001	13.0899	0.947174
Au L2-N1	Au Lγ5	0.0035	12.9748	0.955577
Au L1-M5	Au Lβ9	0.0004	12.1471	1.02069
Au L1-M4	Au Lβ10	0.0054	12.0617	1.02792
Au L3-P1	Au L3P1	0.0001	11.935	1.03883
Au L3-O4	Au Lβ5	0.0438	11.9104	1.04097
Au L3-O2	Au L3O2	0.0001	11.847	1.04655
Au L3-N6	Au Lu	0.0009	11.8323	1.04785
Au L3-O1	Au Lβ7	0.0004	11.8109	1.04974
Au L1-M3	Au Lβ3	0.0690	11.6098	1.06793
Au L3-N5	Au Lβ2	0.2195	11.5848	1.07023
Au L3-N4	Au Lβ15	0.0000	11.5667	1.07191
Au L2-M5	Au L2M5	0.0001	11.5279	1.07551
Au L2-M4	Au Lβ1	0.4015	11.4425	1.08354
Au L3-N3	Au L3N3	0.0001	11.3733	1.09013
Au L3-N2	Au L3N2	0.0001	11.275	1.09964
Au L1-M2	Au Lβ4	0.0594	11.205	1.10651
Au L3-N1	Au Lβ6	0.0140	11.1599	1.11098
Au L2-M3	Au Lβ17	0.0005	10.9906	1.12809
Au L1-M1	Au L1M1	0.0001	10.9279	1.13457
Au L2-M2	Au L2M2	0.0001	10.5858	1.17123
Au L2-M1	Au Lη	0.0138	10.3087	1.20271
Au L3-M5	Au Lα1	1.0000	9.713	1.27648
Au L3-M4	Au Lα2	0.1139	9.6276	1.2878

(continued)

◘ **Table 18.1** (continued)

IUPAC	Siegbahn	Weight	Energy (keV)	Wavelength (Å)
Au L3-M3	Au Ls	0.0001	9.1757	1.35122
Au L3-M2	Au Lt	0.0012	8.7709	1.41359
Au L3-M1	Au Lℓ	0.0562	8.4938	1.4597
Au M1-N3	Au M1N3	0.0000	2.8795	4.30575
Au M2-N4	Au M2N4	0.0290	2.7958	4.43466
Au M3-O5	Au M3O5	0.0100	2.73621	4.53124
Au M3-O4	Au M3O4	0.0050	2.73469	4.53375
Au M3-O1	Au M3O1	0.0027	2.6352	4.70493
Au M3-N5	Au Mγ	0.0851	2.4091	5.14649
Au M3-N4	Au M3N4	0.0100	2.391	5.18545
Au M2-N1	Au M2N1	0.0029	2.389	5.18979
Au M4-O2	Au M4O2	0.0010	2.2194	5.58638
Au M4-N6	Au Mβ	0.5944	2.2047	5.62363
Au M5-O3	Au M5O3	0.0001	2.152	5.76135
Au M5-N7	Au Mα1	1.0000	2.1229	5.84032
Au M5-N6	Au Mα2	1.0000	2.1193	5.85024
Au M3-N1	Au M3N1	0.0290	1.9842	6.24857
Au M4-N3	Au M4N3	0.0001	1.7457	7.10226
Au M5-N3	Au Mζ1	0.0134	1.6603	7.46758
Au M4-N2	Au Mζ2	0.0451	1.6474	7.52605
Au N4-N6	Au N4N6	1.0000	0.2656	46.6808
Au N5-N6	Au N5N6/N5N7	1.0000	0.2475	50.0946

Si Escape Peak

The Si escape peak results when a Si K–L₃ X-ray ($E = 1.740$ keV), which is created following the photoionization of a silicon atom and is usually reabsorbed within the detector volume, escapes. This results in an energy loss of 1.740 keV from the parent X-ray, creating a "silicon escape peak," as shown in ◘ Fig. 18.7 (upper spectrum) for the titanium K-family X-rays. Escape peaks can only be created for parent photon energies above 1.740 keV and are formed at a fixed fraction of the parent peak, with that fraction rapidly decreasing as the parent X-ray energy increases above 1.740 keV. For parent peaks with photon energies above 6 keV, escape peaks are so small that they are difficult to detect. Escape peaks can occur from any parent peak, but as a practical matter only the major members of a family are likely to produce detectable escape peaks. Note that in the example shown for the titanium K-family in ◘ Fig. 18.7, the escape peaks for Ti K-$L_{2,3}$ and Ti K-$M_{2,3}$ are both

detected because of the high count spectrum. The EDS system should mark all possible escape peaks and not subsequently misidentify them as other elements. For example, if not properly assigned, the escape peak for Ti K-$L_{2,3}$ could be mistaken for Cl K-$L_{2,3}$. Note that some vendor software removes the escape peaks in the final processed spectrum that is displayed to the user, as shown in ◘ Fig. 18.7 (lower spectrum).

Coincidence Peaks

Although the EDS spectrum may appear to an observer to be collected simultaneously at all photon energies, in reality only one photon can be measured at a time. Because X-rays are created randomly in time, as the rate of production (input count rate) increases, the possibility of two photons entering the detector and creating an artifact coincidence event increases in probability. An inspection function continuously monitors the detector to reject such events, but at

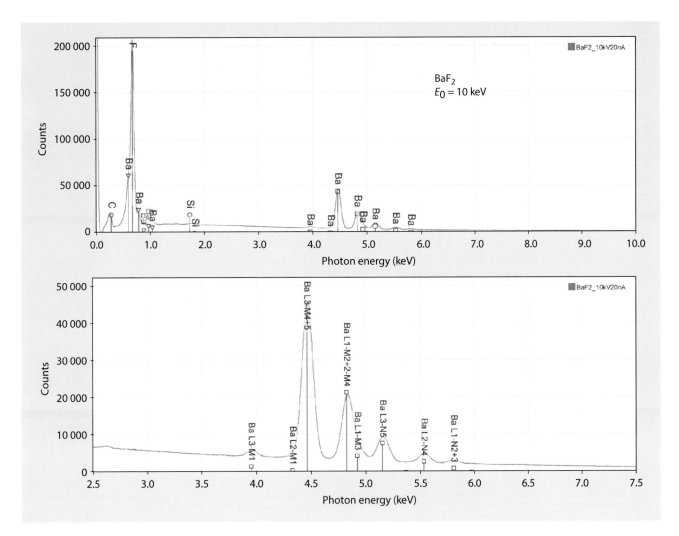

☐ **Fig. 18.5**   EDS spectrum of BaF$_2$ excited with a beam energy of 10 keV

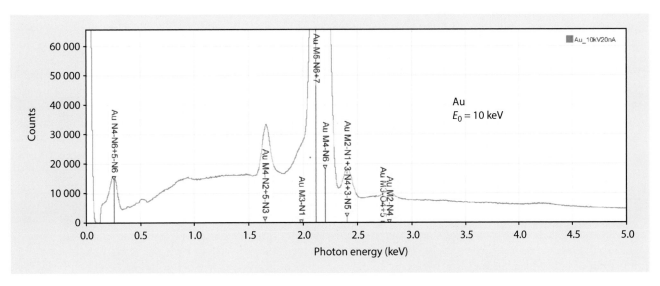

☐ **Fig. 18.6**   EDS spectrum of Au excited with a beam energy of 10 keV

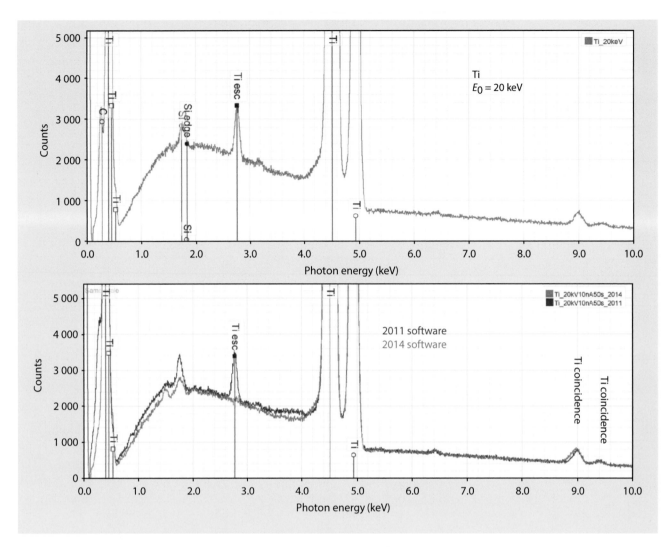

■ **Fig. 18.7**   EDS spectrum of titanium recorded at $E_0 = 20$ keV and a dead-time of 11 %

sufficiently high input count rates its finite time resolution will be overwhelmed and coincidence events will begin to populate the spectrum. This phenomenon is illustrated in ■ Fig. 18.7 for pure Ti, where coincidence peaks are seen for Ti K-$L_{2,3}$ + Ti K-$L_{2,3}$ and Ti K-$L_{2,3}$ + Ti K-$M_{2,3}$. The height of a coincidence peak relative to the parent peak depends on the arrival rate of X-rays at the detector, and thus upon the dead-time. Even at low dead-time coincidence events are likely to be observed for highly excited, low photon energy peaks such as Al and Si in pure element or high concentration targets, as shown for Si in ■ Fig. 18.8 at 12 % dead-time and in ■ Fig. 18.9 at 1 % dead-time. Operating at low dead-time can reduce the height of the coincidence peak relative to the parent peak, but coincidence can never entirely be eliminated by reducing the input count rate. For complex compositions, a wide array of coincidence peaks involving many combinations of highly excited peaks can be encountered, as shown

in ■ Fig. 18.10. As the dead-time increases, numerous coincidence peaks are observed, several of which could be misidentified as elements present as trace to minor constituents. At the highest dead-times, the coincidence peaks are seen to occupy much of the useful spectral energy range where legitimate minor and trace constituents might be measured. Coincidence can involve any two photons of any energies, but the noticeable effects in EDS spectra consist of two characteristic peak photons that originate from major constituents that are highly excited, for exmple, A + A, A + B, B + B, an so on. Since coincidence depends on the arrival rate of photons, the analyst can exert some control by operating at low dead-time (10 % or less) to minimize but not eliminate the effect. Some vendor EDS systems use statistical models of coincidence to post process the spectrum, removing the coincidence events and restoring the events to their proper parent peaks.

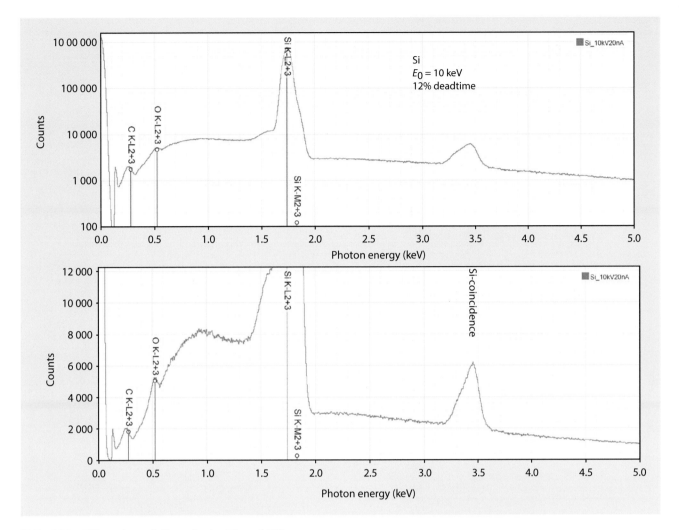

18.3 Performing Manual Qualitative Analysis

18.3.1 Why Are Skills in Manual Qualitative Analysis Important?

The automatic peak identification function supplied in all vendor software is a powerful and useful tool, but it should only be used to confirm manual identifications rather than vice versa. Studies have shown that incorrect peak assignments occur with vendor software in a few percent of analyses even for major constituents ($C > 0.1$ mass fraction, 10 weight percent) that produce prominent spectral peaks (Newbury 2005). □ Table 18.2 lists groups of elements for which incorrect peak assignments have been observed in vendor software from different sources. Extensive observations suggest that peak misidentifications occur for major

constituents in several percent of qualitative analyses of major constituents. The problem of incorrect assignments becomes even more significant for minor and trace constituents that produce peaks that inevitably occur at low peak-to-background and for which it may be difficult to recognize more than one characteristic peak (Newbury 2009). The frequency of incorrect peak assignments for minor and trace constituents can be 10 % or more, with both false positives (incorrect peak assignments) and false negatives (legitimate peaks ignored). For operation at low beam energy where the incident beam energy restricts the atomic shells which can be ionized to produce X-rays, peak identification is even more problematic at all concentration levels (Newbury 2007).

For minor and trace constituents, incorrect elemental identifications arise from incomplete identification of minor family members of X-ray families actually associated with previously identified major constituents, as well as artifact

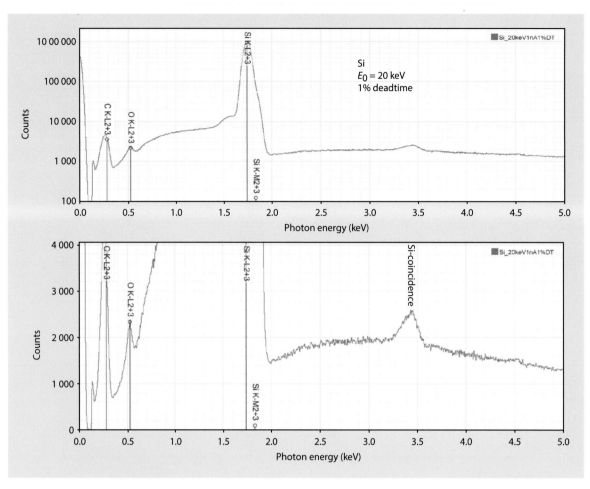

■ **Fig. 18.9**   EDS Spectrum of silicon at a dead-time of 1 %

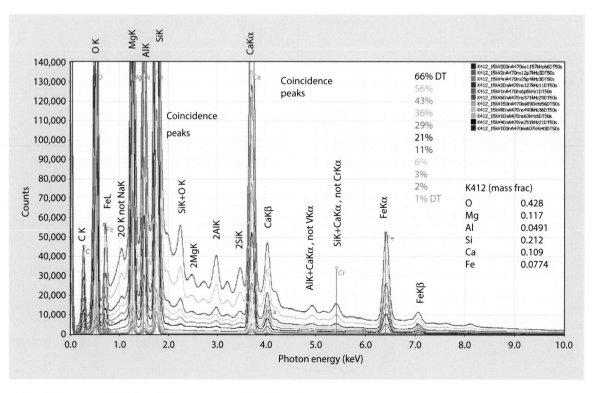

■ **Fig. 18.10**   EDS spectra of NIST glass K412 over a range of dead-times from 1 % to 66 %

◻ **Table 18.2** Characteristic X-ray peaks vulnerable to misidentification

Energy range	Elements, peaks, and photon energies
0.390–0.395 keV	N K-L_3 (0.392); Sc L_3-$M_{4,5}$ (0.395)
0.510–0.525 keV	O K-L_3 (0.523); V L_3-$M_{4,5}$ (0.511)
0.670–0.710 keV	F K-L_3 (0.677); Fe L_3-$M_{4,5}$ (0.705) (0.677); Fe L_3-$M_{4,5}$ (0.705)
0.845–0.855 keV	Ne K-L_3 (0.848); Ni L_3-$M_{4,5}$ (0.851)
1.00–1.05 keV	Na K-$L_{2,3}$ (1.041); Zn L_3-$M_{4,5}$ (1.012); Pm M_5-$N_{6,7}$ (1.032)
1.20–1.30 keV	Mg K-$L_{2,3}$ (1.253); As L_3-$M_{4,5}$ (1.282); Tb M_5-$N_{6,7}$ (1.246)
1.45–1.55 keV	Al K-$L_{2,3}$ (1.487); Br L_3-$M_{4,5}$ (1.480); Yb M_5-$N_{6,7}$ (1.521)
1.70–1.80 keV	Si K-$L_{2,3}$ (1.740); Ta M_5-$N_{6,7}$ (1.709); W M_5-$N_{6,7}$ (1.774)
2.00–2.05 keV	P K-$L_{2,3}$ (2.013); Zr L_3-$M_{4,5}$ (2.042); Pt M_5-$N_{6,7}$ (2.048)
2.10–2.20 keV	Nb L_3-$M_{4,5}$ (2.166); Au M_5-$N_{6,7}$ (2.120); Hg M_5-$N_{6,7}$ (2.191)
2.28–2.35 keV	S K-$L_{2,3}$ (2.307); Mo L_3-$M_{4,5}$ (2.293); Pb M_5-$N_{6,7}$ (2.342)
2.40–2.45 keV	Tc L_3-$M_{4,5}$ (2.424); Pb M_4-N_6 (2.443); Bi M_5-$N_{6,7}$ (2.419)
2.60–2.70 keV	Cl K-$L_{2,3}$ (2.621); Rh L_3-$M_{4,5}$ (2.696)
2.95–3.00 keV	Ar K-$L_{2,3}$ (2.956); Ag L_3-$M_{4,5}$ (2.983); Th M_5-$N_{6,7}$ (2.996)
3.10–3.20 keV	Cd L_3-$M_{4,5}$ (3.132); U M_5-$N_{6,7}$ (3.170)
3.25–3.35 keV	K K-$L_{2,3}$ (3.312); In L_3-$M_{4,5}$ (3.285); U M_4-N_6 (3.336)
4.45–4.55 keV	Ti K-$L_{2,3}$ (4.510); Ba L_3-$M_{4,5}$ (4.467)
4.90–5.00 keV	Ti K-M_3 (4.931); V K-$L_{2,3}$ (4.949)

peaks that arise from the silicon escape peak and from coincidence peaks. A particularly insidious problem occurs when automatic peak identification software delivers identifications of peaks with low peak-to-background too early in the EDS accumulation before adequate counts have been recorded. Statistical fluctuations in the continuum background create "false peaks" that may appear to correspond to minor or trace constituents. This problem can be recognized when an apparent peak identification solution for these low level peaks subsequently changes as more counts are accumulated. The danger is that the analyst may choose to stop the accumulation prematurely and be misled by the low level "peaks" that do not actually exist.

When the analyst must operate only at low beam energy ($E_0 \leq 5$ keV), the peak misidentification problem is exacerbated by the loss of the higher photon energies where X-ray family members are more widely spread and more easily identified, as well as the confidence-increasing redundancy provided by having K-L and L-M family pairs for identification of intermediate and high atomic number elements (Newbury 2009).

Even well-implemented automatic peak identification software is likely to ignore peaks with low peak-to-background that may correspond to trace constituents because the likelihood of a mistake becomes so large. Thus, if

it is important to the analyst to identify the presence of a trace element (s) with a high degree of confidence, manual peak identification will be necessary.

18.3.2 Performing Manual Qualitative Analysis: Choosing the Instrument Operating Conditions

Beam Energy

Equation 18.1 reveals that one selection of the beam energy may not be sufficient to solve a particular problem, and the analyst must be prepared to explore a range of beam energies to access desired atomic shells. The peak height relative to the spectral background increases rapidly as U_0 is increased, enabling better detection of the characteristic peak (s). Having adequate overvoltage is especially important as the concentration of an element decreases from major to minor to trace. As a general rule, it is desirable to have $U_0 > 2$ for the analyzed shells of all elements that occur in a particular analysis. For initial surveying of an unknown specimen, it is useful to select a beam energy of 20 keV or higher to provide an overvoltage of at least 2 for ionization edges up to 10 keV. Elements with intermediate atomic numbers (e.g., 22, $Ti \leq Z \leq 42$, Mo) and high atomic number (e.g., $Z \geq 56$, Ba) elements have complex atomic shell structures that produce two families of detectable characteristic X-rays (with 20 keV $\leq E_0 \leq$ 30 keV), for example, the Cu K-family and L-family; the Au L-family and M-family. A second advantage of selecting the beam energy to excite the higher energy X-ray family for an element is that it enables a high confidence identification since the peaks that form the family are more widely separated in photon energy and thus more likely to be resolved with EDS. Note that the physics of X-ray generation requires that all members of the X-ray family of a tentative elemental assignment must be present. Identifying all family members in the correct relative intensity ratios gives high confidence that the element assignment is correct as well as avoiding subsequent misidentification of these minor family members.

Choosing the EDS Resolution (Detector Time Constant)

EDS systems provide two or more choices for the detector time constant. The user has a choice of a short detector time constant that gives higher throughput (photons recorded per unit time) at the expense of poorer peak resolution or a long time constant that improves the resolution at the cost of throughput. The analyst thus has a critical choice to make: more counts per unit time or better resolution. Statham (1995) analyzed these throughput-resolution trade-offs with respect to various analytical situations and concluded that a strategy that emphasizes maximizing the number of X-ray counts rather than resolution produces the most robust results.

Choosing the Count Rate (Detector Dead-Time)

A closely related consideration is the problem of pulse coincidence creating artifact peaks, which are reduced (but not eliminated) by using lower dead-time. Note that a specific level of dead-time, for example, 10%, corresponds to a higher throughput when a shorter time constant is chosen. With the beam energy and detector time constant selected, the rate at which X-rays arrive at the EDS and subsequent output depends on two factors: (1) the detector solid angle and (2) the beam current. If the EDS detector is movable relative to the specimen, the specimen-to-detector distance should be chosen in a consistent fashion to enable subsequent return to the same operating conditions for robust standards-based quantitative analysis. A typical choice is to move the detector as close to the specimen as possible to maximize the detector solid angle, $\Omega = A/r^2$, by minimizing r, the detector-to-specimen distance, where A is the active area of the detector. Always ensure that any possible stage motions will not cause the specimen to strike the EDS. With the EDS solid angle fixed, the input count rate will then be controlled by the beam current. A useful strategy is to choose a beam current that creates an EDS dead-time of approximately 10% on a highly excited characteristic X-ray, such as Al K-$L_{2,3}$ from pure aluminum. To establish dose-corrected standards-based quantitative analysis, this same detector solid angle and beam current should be used for all measurements. It is often desirable to maximize the recorded counts per unit of real (clock) time. Higher beam current leading to higher dead-time, for example, 30–40%, can be utilized, but the spectrum is likely to have coincidence peaks like those shown in ◻ Fig. 18.10, which can greatly complicate the recognition and measurement of the peaks of minor and trace constituents. Note that some vendor software systems effectively block coincidence peaks or else remove them from the spectrum by post-processing with a stochastic model that predicts coincidence peaks based on the parent peak count rates.

Obtaining Adequate Counts

The analyst must accumulate adequate X-ray counts to distinguish a peak against the random fluctuations of the background (X-ray continuum). While it is relatively easy to record sufficient counts to recognize the principal peak for a major constituent, detection of the minor family member (s) to increase confidence in the elemental assignment may require recording a substantially greater total count. For minor or trace constituents, an even greater dose is likely to be required just to detect the principal family members, and to obtain minor family members to increase confidence in an elemental identification will require a dose greater by another factor of ten or more. A peak is considered *detectable* if it satisfies the following criterion (Currie 1968):

$$n_P > 3 n_B^{1/2} \tag{18.6}$$

where n_P is the number of peak counts and n_B is the number of background counts under the peak. Note that "detectable" does not imply optimally measureable, for example, obtaining accurate peak energy. While Eq. 18.6 defines the minimum counts to detect a peak, accurate measurement of the peak position to identify the peak may require higher counts. The effect of increasing the total spectral intensity to "develop" low relative intensity peaks from trace constituents is shown in ◻ Fig. 18.11.

▪▪ Golden Rule

If it is difficult to recognize a peak above fluctuations in the background, accumulate more counts. Patience is a virtue!

18.4   Identifying the Peaks

After a suitable spectrum has been accumulated, the analyst can proceed to perform manual qualitative analysis.

18.4.1   Employ the Available Software Tools

Manual qualitative analysis is performed using the support of available software tools such as KLM markers that show the energy positions and relative heights of X-ray family members to assign peaks recognized in the spectrum to specific elements. Before using this important software tool, the user

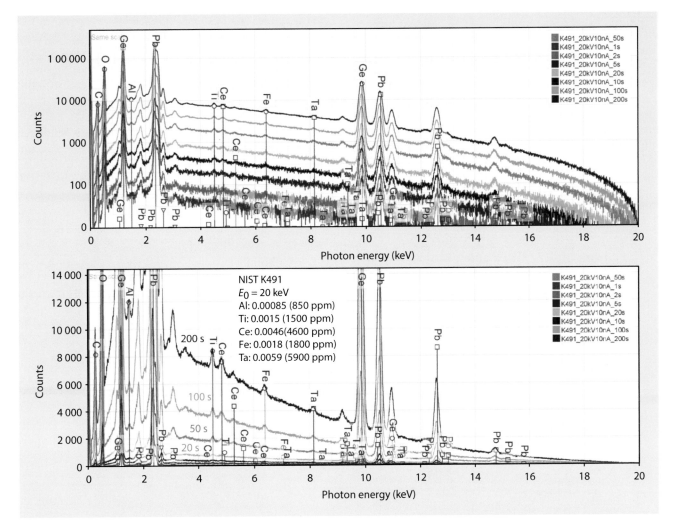

□ **Fig. 18.11** Detection of trace constituent peaks in NIST microanalysis glass K491 as the dose is increased. Integrated spectrum counts: 1 s = 0.14 million; 2 s = 0.28 million; 5 s = 0.70 million; 10 s = 1.4 million; 20 s = 2.8 million; 50 s = 7.1 million; 100 s = 14.3 million; 200 s = 28.5 million

should confirm that all elements in the periodic table are enabled and all X-ray family members are shown in the KLM markers.

As each element is tentatively identified from its major family peak, a systematic search must then be made to locate all possible peaks that *must* be associated with that element: (1) all minor family members; (2) a second X-ray family at lower energy (e.g., K and L or L and M), and for the highest atomic number elements, the N-family can also be observed, as shown in □ Fig. 18.6; and (3) any associated EDS artifact peaks (escape peaks and coincidence peaks). This careful inspection regimen and meticulous bookkeeping raises the confidence in the tentative assignment. Properly assigning the minor family members and the artifact peaks to the proper element will diminish the possibility of subsequent incorrect assignment of those peaks to other constituents that might appear to be present at minor or trace levels.

It can be helpful to think of qualitative analysis as the process of eliminating those elements which cannot possibly be present rather than the process of including those that

definitely are present. It isn't the natural perspective and it takes more thought and effort but it is much less prone to errors of omission.

Imagine that you are at a zoo. You have a list of 20 animals of various sizes and shapes and you are asked to answer the question, what animals in this list *could possibly* be in this cage. You look around and see a rhinoceros laying down near the back of the cage and no other animal. You might be tempted to say that the only animal that could possibly be in the cage is the one you see – the rhinoceros. However, the list also contains snakes, mice, fish and elephants. You can rule out elephants because they are too big to hide behind a rhino. You can rule out fish because the environment is inappropriate. You can't rule out the possibility that a mouse or snake is in the cage hidden behind the rhino. It is only by eliminating those animals that are too large (elephant) or can't survive (fish) behind the rhino that you can come to the full list of animals that could potentially be present in the cage – the rhino and any animals which could be hiding behind the rhino. If you want to be certain that you haven't missed an animal that could *possibly* be present in the cage, the process of culling animals that couldn't possibly be present is far more robust than the process of including animals that definitely are present.

Spectra are similar. Not only is it possible that the obvious elements are present but also those that could be hidden by the ones that are readily identified. Fortunately, there is a tool to help us to see through spectra and expose the hidden components – the residual spectrum. The residual spectrum is the intensity that remains in each channel after peak fitting has been performed for the specified elements. It is like being able to ask the rhino to move and then being able to see what is hidden behind – maybe a mouse or a snake or maybe nothing. An example of the utility of the residual spectrum is shown in Figure 20.8.

18.4.2   Identifying the Peaks: Major Constituents

Start with peaks located in the higher photon energy (>4 keV) region of the spectrum and work downward in energy, even if there are higher peaks in the lower photon energy region (<4 keV). The logic for this strategy is that K-shell and L-shell characteristic X-rays above 4 keV are produced in families that provide two or more peaks with distinctive relative abundances for which the energy resolution of EDS is sufficient to easily separate these peaks. Having two or more peaks to identify greatly increases the confidence with which an elemental identification can be made, enabling the analyst to achieve an unambiguous result. For each peak that is recognized, first test whether its energy corresponds closely to a particular K-L_3 (Kα) peak. The physics of X-ray generation demands that the corresponding K-M_3 (Kβ) peak must also be present in roughly a 10:1 ratio. If K-family peaks do not match the peak in question, examine L-family possibilities, noting that three or more L-peaks are likely to be detectable: L_3–$M_{4,5}$ (Lα), L_2-M_4 (Lβ), and L_2-N_4 (Lγ). Locate and mark all minor family members such as L_3-M_1 (Ll). Locate and mark the escape peaks, if any, associated with the major family members. Locate and mark, if any, the coincidence peaks associated with the major family members, which may be located at very high energy, for example, as shown for Cu K-L_3 coincidence in ◘ Fig. 18.12.

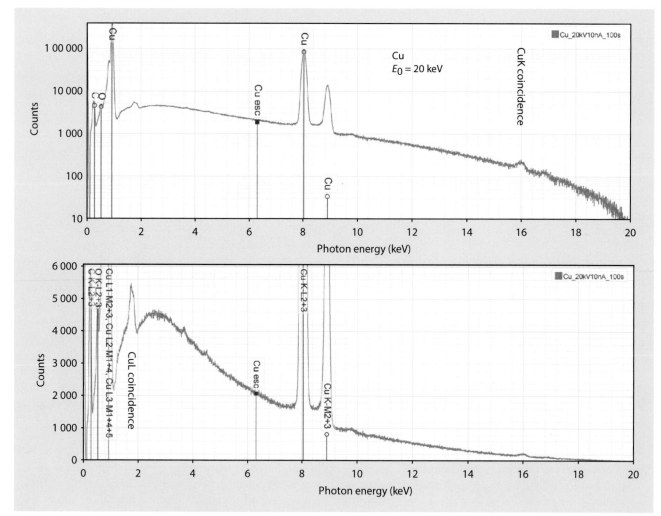

◘ **Fig. 18.12**   EDS spectrum of Cu at E_0 = 20 keV showing a coincidence peak for CuK-L_3 at 16.08 keV

18.4.3 Lower Photon Energy Region

As major spectral peaks located at lower photon energy (<4 keV) are considered, the energy separation diminishes and the relative peak heights decrease for the members of each X-ray family. EDS is no longer able to resolve these peaks, leading to a situation where only one peak is available for identification for K-family X-rays below 2 keV in energy. The K-L$_3$ peak appears symmetric since the K-M$_3$ peak has low relative intensity, as shown for Al K-L$_3$ in $\square$ Fig. 18.13a. For L- and M-family X-rays in the low photon energy range, the composite peak appears asymmetric. As shown for Br in $\square$ Fig. 18.13b, the major peaks L$_3$-M$_5$(Lα) and L$_2$-M$_4$(Lβ) occur with a ratio of approximately 2:1 and the low abundance but separated L$_3$-M$_1$ (Ll) and L2-M1 (Lη) can also aid in the identification providing the spectrum contains adequate counts. Similarly, the M$_5$-N$_{6,7}$(Mα) and M$_4$-N$_6$(Mβ) peaks occur with a ratio of 1/0.6 and the well separated minor family members W M$_{5,4}$-N$_{3,2}$ (Mζ) and W M$_3$-N$_5$ (Mγ) can be detected in a high count spectrum, as shown for W in $\square$ Fig. 18.13c.

18.4.4 Identifying the Peaks: Minor and Trace Constituents

After all major peaks and their associated minor family members and artifact peaks have been located and identified with high confidence as belonging to particular elements, the analyst can proceed to identify any remaining peaks which are now likely to be associated with minor and trace level constituents. Achieving the same degree of high confidence in the identification of lower concentration constituents is more difficult since the lower concentrations reduce all X-ray intensities so that minor family members are more difficult to detect. The situation is likely to require accumulating additional X-ray counts to improve the detectability of minor X-ray family members and increase the confidence of the assignment of elemental identification. In general, establishing the presence of a constituent at trace level is a significant challenge that requires not only collecting a high count spectrum that satisfies the limit of detection criterion but also scrupulous attention to identifying all possible minor family members and artifacts from the X-ray families of the major and minor constituents.

18.4.5 Checking Your Work

The only way to be confident that the qualitative analysis is correct to quantify the spectrum and examine the residual spectrum. When every element has been correctly identified and quantified, the analytical total should be approximately unity and there should be no obvious structure in the residual spectrum that cannot be explained through chemistry or minor chemical peak shifts. This iterative qualitative – quantitative analysis scheme to discover minor and trace elements hidden under the high intensity peaks of major constituents will covered in ▶ Chapter 19.

18.5 A Worked Example of Manual Peak Identification

Alloy IN100 is a complex mixture of transition and heavy elements that provides several challenges to manual peak identification:

1. $\square$ Figure 18.14a shows the spectrum from 0 to 20 keV excited with $E_0 = 20$ keV. Using the KLM marker tools in DTSA II, starting at high photon energy and working downward, the first high peak encountered shows a good match to Ni K-L$_3$ and the corresponding Ni K-M$_3$ is also found at the correct ratio, as well as the Ni L-family at low photon energy. The position of the Ni K-L$_3$ escape peak is marked. Inspection for possible coincidence peaks does not reveal a significant population due to the low dead-time (8 %) used to accumulate the spectrum and the large number of peaks over which the input count rate is partitioned so that even the most intense peak has a relatively low count rate and does not produce significant coincidence.

2. Working down in energy ($\square$ Fig. 18.14b), the next peak is seen to correspond to Co K-L$_3$, but the Co K-M$_3$ suffers interference from Ni K-L$_3$ and only appears as an asymmetric deviation on the high energy side. Likewise, the Co L-family is unresolved from the Ni L-family.

3. The next set of peaks match Cr, as shown in $\square$ Fig. 18.14c.

4. Continuing, $\square$ Fig. 18.14d shows a match for the peaks of Ti, but the apparent ratio of Ti K-L$_3$/Ti K-M$_3$ is approximately 5:1, whereas the true ratio is about 10:1, which suggests that another element must be present. Expansion of this region in $\square$ Fig. 18.14e reveals that V is likely to be present but with severe interference between V K-L$_3$ and Ti K-M$_3$. While the anomalous peak ratio observed for TiK-L$_3$/TiK-M$_3$ is a strong clue that another element must be present, this example shows one of the limitations of manual peak identification, namely, that peaks representing minor and trace constituents can be lost under the higher intensity peaks of higher concentration constituents as the concentration ratio becomes large. Detecting such interferences of constituents with large concentration ratios requires the careful peak-fitting procedure that is embedded in the quantitative analysis procedures described in module 19.

5. In $\square$ Fig. 18.14f, the next peak group best matches the Mo L-family. This photon energy range involves possible interferences from the S K-family, the Mo L-family, and the Pb M-family. The possibility of identifying the peak group as the Pb M-family which occurs this energy range, can be rejected because of the absence of the Pb L-family, as shown in $\square$ Fig. 18.14g. The possible presence of the S K-family ($\square$ Fig. 18.14h) is much more difficult to exclude because S cannot be effectively measured by an alternate X-ray family such as the S L-family due to the low fluorescence yield. While the shape of the peak cluster does not match S K-L$_3$ and S K-M$_3$, the presence of S can only be confidently

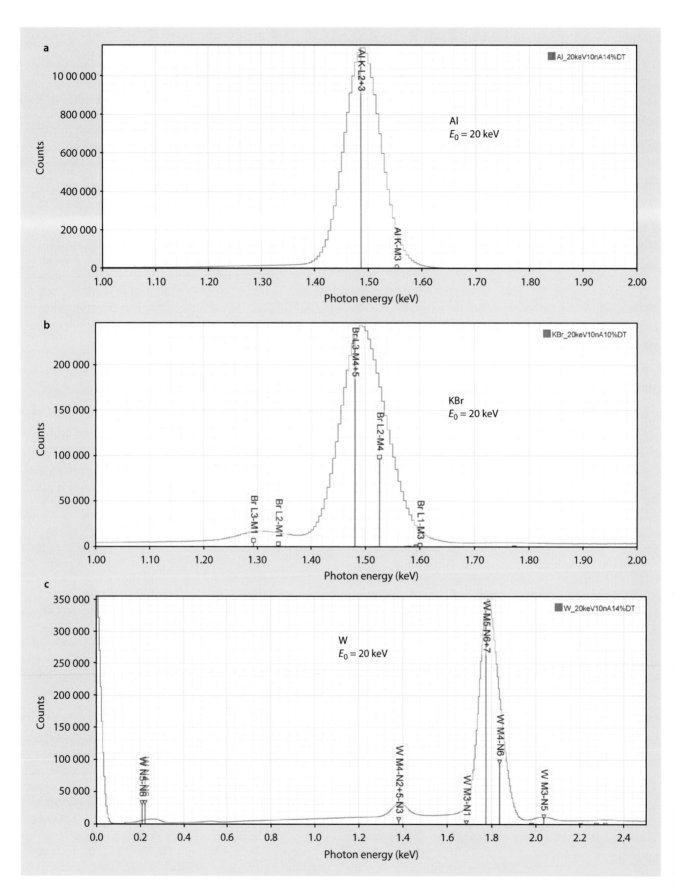

■ **Fig. 18.13** **a** EDS spectrum of Al at $E_0 = 20$ keV; note symmetry of Al K-family peaks. **b** EDS spectrum of KBr at $E_0 = 20$ keV; note asymmetry of Br L-family peaks. **c** EDS spectrum of W at $E_0 = 20$ keV; note asymmetry of W M-family peaks

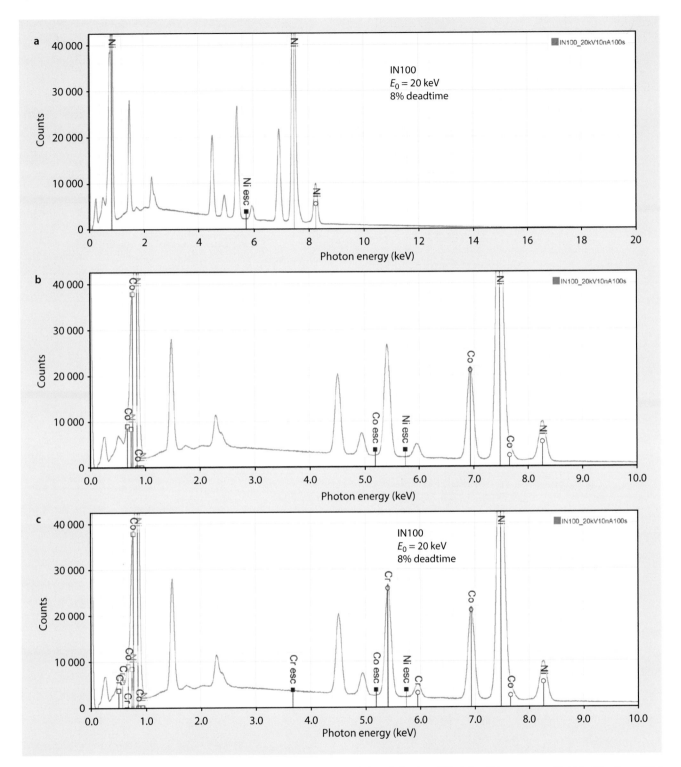

Fig. 18.14 **a** Alloy IN100 recorded with $E_0 = 20$ keV and at 8% dead-time showing identification of Ni. **b** Identification of Co. **c** Identification of Cr. **d** Identification of Ti. **e** Identification of V. **f** Identification of Mo. **g** Rejection of Pb. **h** Possible presence of S. **i** Identification of Al. **j** Rejection of Br. **k** Identification of C. **l** Identification of Si

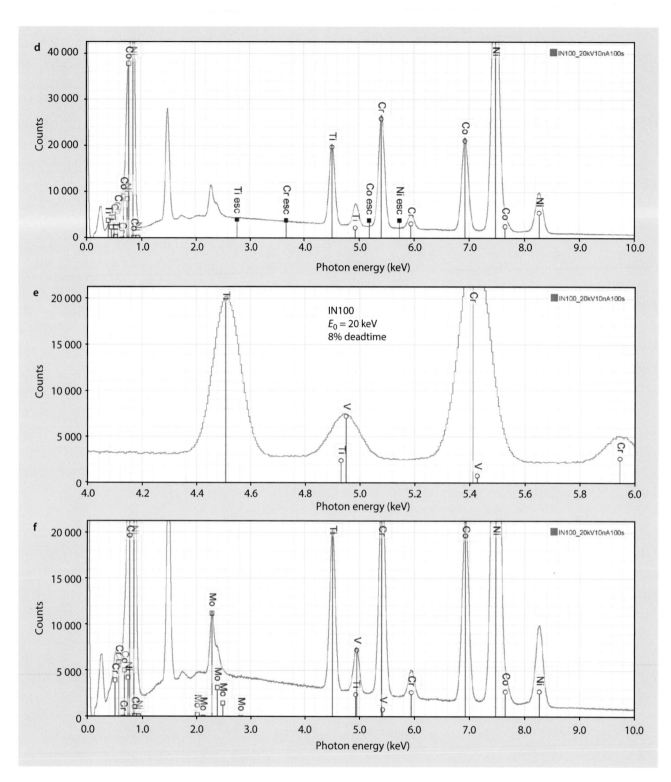

18

■ **Fig. 18.14**   (continued)

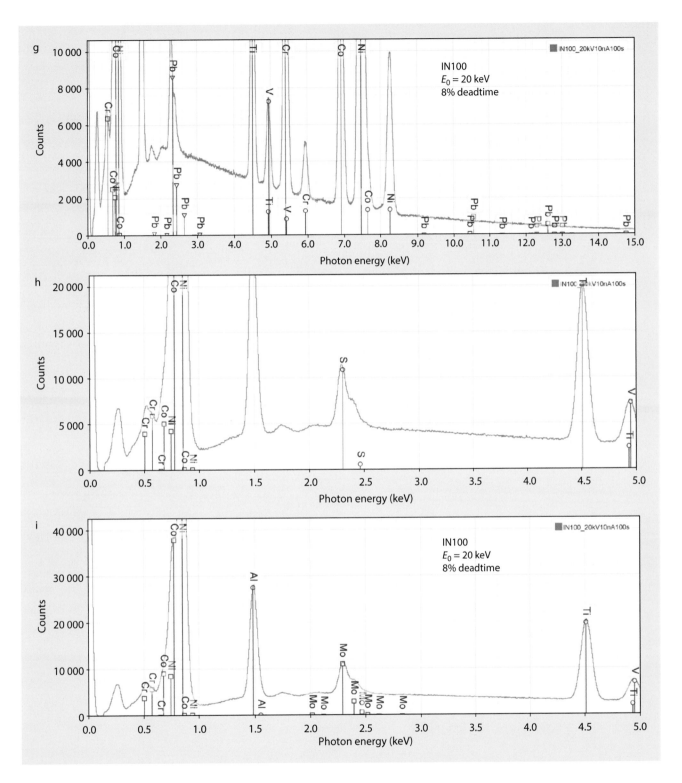

Fig. 18.14 (continued)

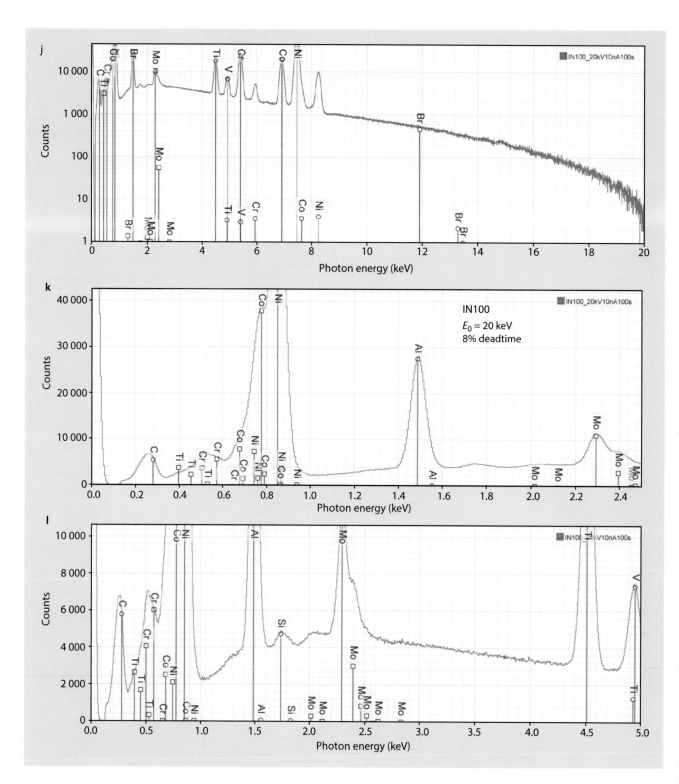

◘ **Fig. 18.14**   (continued)

confirmed by peak fitting procedures during quantitative analysis.

6. The next peak matches the Al K-family (◘ Fig. 18.14i) but in this photon energy range only one peak is available for identification. The Br L-family also fits this peak (◘ Fig. 18.14j) but Br can be dismissed because of the absence of the Br K-family.

7. The last significant peak is found to correspond to C K (◘ Fig. 18.14k) noting that due to the non-linearity of the photon energy scale for this detector below 400 eV, the peak is displaced to a lower energy from the ideal position.

8. Finally, inspection of the remaining low peak-to-background peaks reveals just one candidate, which corresponds to the Si K-family (◘ Fig. 18.14l).

References

Currie LA (1968) Limits for qualitative detection and quantitative determination. Anal Chem 40:586

Newbury D (2005) Misidentification of major constituents by automatic qualitative energy dispersive X-ray microanalysis: a problem that threatens the credibility of the analytical community. Microsc Microanal 11:545

Newbury D (2007) Mistakes encountered during automatic peak identification in low beam energy X-ray microanalysis. Scanning 29:137

Newbury D (2009) Mistakes encountered during automatic peak identification of minor and trace constituents in electron-excited energy dispersive X-ray microanalysis. Scanning 31:1

Statham P (1995) Quantifying Benefits of resolution and count rate in EDX microanalysis. In: Williams D, Goldstein J, Newbury D (eds) X-ray spectrometry in electron beam instruments. Plenum, New York, pp 101–126

Quantitative Analysis: From k-ratio to Composition

© Springer Science+Business Media LLC 2018
J. Goldstein et al., *Scanning Electron Microscopy and X-Ray Microanalysis*,
https://doi.org/10.1007/978-1-4939-6676-9_19

19.1   What Is a k-ratio?

A *k-ratio* is the ratio of a pair of characteristic X-ray line intensities, *I*, measured under similar experimental conditions for the unknown (unk) and standard (std):

$$k = I_{unk} / I_{std} \qquad (19.1)$$

The measured intensities can be associated with a single characteristic X-ray line (as is typically the case for wavelength spectrometers) or associated with a family of characteristic X-ray lines (as is typically the case for energy dispersive spectrometers.) The numerator of the k-ratio is typically the intensity measured from an unknown sample and the denominator is typically the intensity measured from a standard material—a material of known composition.

Both the numerator and the denominator of the k-ratio must be measured under similar, well-controlled instrument conditions. The electron beam energy must be the same. The probe dose, the number of electrons striking the sample during the measurement, should be the same (or the intensity scaled to equivalent dose.) The position of the sample relative to the beam and to the detector should be fixed. Both the sample and the standard(s) should be prepared to a high degree of surface polish, ideally to reduce surface relief below 50 nm, and the surface should not be chemically etched. If the unknown is non-conducting, the same thickness of conducting coating, usually carbon with a thickness below 10 nm, should be applied to both the unknown and the standard(s). Ideally, the only aspect that should differ between the measurement of the unknown and the standard are the compositions of the materials.

The k-ratio is the first estimate of material composition. From a set of k-ratios, we can estimate the unknown material composition. In many cases, to a good approximation:

$$C_{Z,unk} \sim k_Z C_{Z,std} = I_{unk} / I_{std} C_{Z,std} \qquad (19.2)$$

where $C_{Z,unk}$ and $C_{Z,std}$ are the mass fraction of element Z in the unknown and standard, respectively, and k_Z is the k-ratio measured for element Z. This relationship is called "Castaing's first approximation" after the seminal figure in X-ray microanalysis, who established the k-ratio as the basis for quantitative analysis (Castaing 1951).

By taking a ratio of intensities collected under similar conditions, the k-ratio is independent of various pieces of poorly known information.

1. The k-ratio eliminates the need to know the efficiency of the detector since both sample and unknown are measured on the same detector at the same relative position. Since the efficiency as a multiplier of the intensity is identical in the numerator and denominator of the k-ratio, the efficiency cancels quantitatively in the ratio.
2. The k-ratio mitigates the need to know the physics of the X-ray generation process if the same elements are excited under essentially the same conditions. The ionization cross section, the relaxation rates, and other poorly

known physical parameters are the same for an element in the standard and the unknown.

In the history of the development of quantitative electron-excited X-ray microanalysis, the X-ray intensities for the unknown and standards were measured sequentially with a wavelength spectrometer in terms of X-ray counts. The raw measurement contains counts that can be attributed to both the continuum background (*bremsstrahlung*) and characteristic X-rays. Since k-ratio is a function of only the characteristic X-rays, the contribution of the continuum must be estimated. Usually, this is accomplished by measuring two off-peak measurements bounding the peak and using interpolation to estimate the intensity of the continuum background at the peak position. The estimated continuum is subtracted from the measured on peak intensity to give the characteristic X-ray line intensity.

Extracting the k-ratio with an energy dispersive spectrometer can be done in a similar manner for isolated peaks. However, to deal with the peak interferences frequently encountered in EDS spectra, it is necessary to simultaneously consider all of the spectrum channels that span the mutually interfering peaks. Through a process called *linear least squares fitting*, a scale factor is computed which represents the multiplicative factor by which the integrated area under the characteristic peak from the standard must be multiplied by to equal the integrated area under the characteristic peak from the unknown. This scale factor is the k-ratio, and the fitting process separates the intensity components of the interfering peaks and the continuum background. The integrated counts measured for the unknown and for the standard for element Z enable an estimate of the precision of the measurement for that element. Linear least squares fitting is employed in NIST DTSA-II to recover characteristic X-ray intensities, even in situations with extreme peak overlaps.

A measured k-ratio of zero suggests that there is none of the associated element in the unknown. A measurement on a standard with exactly the same composition as the unknown will nominally produce a k-ratio of unity for all elements present. Typically, k-ratios will fall in a range from 0 to 10 depending on the relative concentration of element Z in the unknown and the standard. A k-ratio less than zero can occur when count statistics and the fitting estimate of the background intensity conspire to produce a slightly negative characteristic intensity. Of course, there is no such thing as negative X-ray counts, and negative k-ratios should be set to zero before the matrix correction is applied. A k-ratio larger than unity happens when the standard generates fewer X-rays than the unknown. This can happen if the standard contains less of the element and/or if the X-ray is strongly absorbed by the standard. Usually, a well-designed measurement strategy won't result in a k-ratio much larger than unity. We desire to use a standard where the concentration of element Z is high so as to minimize the contribution of the uncertainty in the amount of Z in the standard to the overall uncertainty budget of the measurement, as well as to minimize the uncertainty

19

contribution of the count statistics in the standard spectrum to the overall measurement precision. The ideal standard to maximize concentration is a pure element, but for those elements whose physical and chemical properties prohibit them from being used in the pure state, for example, gaseous elements such as Cl, F, I, low melting elements such as Ga or In, or pure elements which deteriorate under electron bombardment, for example, P, S, a binary compound can be used, for example, GaP, FeS_2, CuS, KCl, etc.

> **TIP**
>
> To get the most accurate trace and minor constituent measurements, it is best to average together many k-ratios from distinct measurements before applying the matrix correction. Don't truncate negative k-ratios before you average or you'll bias your results in the positive direction.

19.2 Uncertainties in k-ratios

All k-ratio measurements must have associated uncertainty estimates. The primary source of uncertainty in a k-ratio measurement is typically count statistics although instrumental instability also can contribute. X-ray emission is a classic example of a Poisson process or a random process described by a negative exponential distribution.

Negative exponential distributions are interesting because they are "memoryless." For a sequence of events described by a negative exponential distribution, the likelihood of an event's occurring in an interval τ is equally as likely regardless of when the previous event occurred. Just because an event hasn't occurred for a long time doesn't make an event any more likely in the subsequent time interval. In fact, the most probable time for the next event is immediately following the previous.

If X-rays are measured at an average rate R, the average number of X-rays that will be measured over a time t is $N = R \cdot t$. Since the X-ray events occur randomly dispersed in time, the actual number measured in a time t will rarely ever be exactly $N = R \cdot t$. Instead, 68.2% of the time the actual number measured will fall within the interval ($N - \Delta N$, $N + \Delta N$) where $\Delta N \sim N^{1/2}$ when N is large (usually true for X-ray counts). This interval is often called the "one sigma" interval. The one-sigma fractional uncertainty is thus $N/N^{1/2} = 1/N^{1/2}$, which for constant R decreases as t increased. This is to say that generally, it is possible to make more precise measurements by spending more time making the measurement. All else remaining constant, for example, instrument stability and specimen stability under electron bombardment, a measurement taken for a duration of $4 t$ will have twice the precision of a measurement take for t.

Poisson statistics apply to both the WDS and EDS measurement processes. For WDS, the on-peak and background measurements all have associated Poissonian statistical

uncertainties. For EDS, each channel in the spectrum has an associated Poissonian statistical uncertainty. In both cases, the statistical uncertainties must be taken into account carefully so that an estimate of the measurement precision can be associated with the k-ratio.

The best practices for calculating and reporting measurement uncertainties are described in the ISO Guide to Uncertainty in Measurement (ISO 2008). Marinenko and Leigh (2010) applied the ISO GUM to the problem of k-ratios and quantitative corrections in X-ray microanalytical measurements. In the case of WDS measurements, the application of ISO GUM is relatively straightforward and the details are in Marinenko and Leigh (2010). For EDS measurements, the process is more complicated. If uncertainties are associated with each channel in the standard and unknown spectra, the k-ratio uncertainties are obtained as part of the process of weighted linear squares fitting.

19.3 Sets of k-ratios

Typically, a single compositional measurement consists of the measurement of a number of k-ratios—typically one or more per element in the unknown. The k-ratios in the set are usually all collected under the same measurement conditions but need not be. It is possible to collect individual k-ratios at different beam energies, probe doses or even on different detectors (e.g., multiple wavelength dispersive spectrometers or multiple EDS with different isolation windows).

There may more than one k-ratio per element. Particularly when the data is collected on an energy dispersive spectrometer, more than one distinct characteristic peak per element may be present. For period 4 transition metals, the K and L line families are usually both present. In higher Z elements, both the L and M families may be present. This redundancy provides a question – Which k-ratio should be used in the composition calculation?

While it is in theory possible to use all the redundant information simultaneously to determine the composition, standard practice is to select the k-ratio which is likely to produce the most accurate measurement. The selection is nontrivial as it involves difficult to characterize aspects of the measurement and correction procedures. Historically, selecting the optimal X-ray peak has been something of an art. There are rules-of-thumb, but they involve subtle compromises and deep intuition.

This subject is discussed in more detail in Appendix 19.A. For the moment, we will assume that one k-ratio has been selected for each measured element.

$$k = \left\{ k_Z : Z \in \text{elements} \right\} \tag{19.3}$$

Our task then becomes converting this set of k-ratios into an estimate of the unknown material's composition.

$$C = \left\{ C_Z : Z \in \text{elements} \right\} \tag{19.4}$$

19.4   Converting Sets of k-ratios Into Composition

As stated earlier, the k-ratio is often a good first approximation to the composition. However, we can do better. The physics of the generation and absorption of X-rays is sufficiently well understood that we can use physical models to compensate for non-ideal characteristics of the measurement process. These corrections are called matrix corrections as they compensate for differences in the matrix (read matrix to mean "material") between the standard material and the unknown material.

Matrix correction procedures are typically divided into two classes $\varphi(\rho z)$ and ZAF-type corrections. The details will be discussed in Appendix 19.A. The distinction is primarily how the calculation is divided into independent sub-calculations. In a ZAF-type algorithm, the corrections for differences in mean atomic number (the Z term), X-ray absorption (the A term) and secondary fluorescence (the F term) are calculated separately. $\varphi(\rho z)$ matrix correction algorithms combine the Z and A terms into a single calculation. The distinction between $\varphi(\rho z)$ and ZAF is irrelevant for this discussion so the matrix correction will be described by the generic $ZAF(C_A; P)$ where this expression refers to the matrix correction associated with a material with composition C_A and measurement parameters P. The terms k_Z and C_Z refer to the k-ratio and composition of the Z-th element in the unknown.

$$k_z = \frac{ZAF\left(C_{\text{unk}};P\right)}{ZAF\left(C_{\text{std}};P\right)} \frac{C_{\text{unk}}}{C_{\text{std}}} \tag{19.5a}$$

To state the task clearly, we have measured $\{k_Z : Z \in \text{elements}\}$. We want to know which $\{C_Z : Z \in \text{elements}\}$ produces the observed set of k-ratios.

$$C_{\text{unk}} = k_z \, C_{\text{std}} \frac{ZAF\left(C_{\text{std}};P\right)}{ZAF\left(C_{\text{unk}};P\right)} \tag{19.5b}$$

However, there is a problem. Our ability to calculate k_Z depends upon knowledge of the composition of the unknown, C_{unk}. Unfortunately, we don't know the composition of the unknown. That is what we are trying to measure.

Fortunately, we can use a trick called "iteration" or successive approximation to solve this dilemma. The strategy is as follows:

1. Estimate the composition of the unknown. Castaing's First Approximation is a good place to start.
2. Calculate an improved estimate of $C_{Z,\text{unk}}$ based on the previous estimated composition.
3. Update the composition estimate based on the new calculation.
4. Test whether the resulting computed k-ratios are sufficiently similar to the measured k-ratios.
5. Repeat steps 2–5 until step 4 is satisfied.

While there is no theoretical guarantee that this algorithm will always converge or that the result is unique, in practice, this algorithm has proven to be extremely robust.

19.5   The Analytical Total

The result of the iteration procedure is a set of estimates of the mass fraction for each element in the unknown. We know these mass fractions should sum to unity—they account for all the matter in the material. However, the measurement process is not perfect and even with the best measurements there is variation around unity.

The sum of the mass fractions is called the *analytical total*. The analytical total is an important tool to validate the measurement process. If the analytic total varies significantly from unity, it suggests a problem with the measurement. Analytical totals less than one can suggest a missed element (such as an unanticipated oxidized region of the specimen), a reduced excitation volume, an unanticipated sample geometry (film or inclusion), or deviation from the measurement conditions between the unknown and standard(s). Analytic totals greater than unity likely arise because of measurement condition deviation or sample geometry issues.

19.6   Normalization

As mentioned in the previous section, the analytical total is rarely exactly unity. When it isn't, the accuracy of a measurement can often be improved by normalizing the measured mass fractions, C_i, by the analytical total of all N constituents to produce the *normalized mass fractions, $C_{i,n}$*:

$$C_{i,n} = C_i / \sum_1^N C_i \tag{19.6}$$

This procedure should be performed with care and the analytic total reported along with the normalized mass fractions. Normalization is not guaranteed to improve results and can cover up for some measurement errors like missing an element or inappropriately accounting for sample morphology. The analytical total is important information and the normalized mass fractions should never be reported without also reporting the analytical total. Any analysis which sums exactly to unity should be viewed with some skepticism.

Careful inspection of the raw analytical total is a critical step in the analytical process. If all constituents present are measured with a standards-based/matrix correction procedure, including oxygen (or another element) determined by the method of assumed stoichiometry, then the analytical total can be expected to fall in the range 0.98 to 1.02 (98 weight percent to 102 weight percent). Deviations outside this range should raise the analyst's concern. The reasons for such deviations above and below this range may include unexpected changes in the measurement conditions, such as variations in the beam current, or problems with the specimen, such as local topography such as a pit or other excursion from an ideal flat polished surface. For a deviation below the expected range, an important additional possibility is that there is at least one unmeasured constituent. For example, if a local region of oxidation is encountered while analyzing a metallic sample, the analytical total will drop to approximately 0.7 (70 weight percent) because of the significant fraction of oxygen in a metal oxide. Note that "standardless analysis" (see below) may automatically force the analytical total to unity (100 weight percent) because of the loss of knowledge of the local electron dose used in the measurement. Some vendor software uses a locally measured spectrum on a known material, e.g., Cu, to transfer the local measurement conditions to the conditions used to measure the vendor spectrum database. Another approach is to use the peak-to-background to provide an internal normalization. Even with these approaches, the analytical total may not have as narrow a range as standards-based analysis. The analyst must be aware of what normalization scheme may be applied to the results. An analytical total of exactly unity (100 weight percent) should be regarded with suspicion.

19.7 Other Ways to Estimate C_Z

k-ratios are not the only information we can use to estimate the amount of an element Z, C_Z. Sometimes it is not possible or not desirable to measure k_Z. For example, low Z elements, like H or He, don't produce X-rays or low Z elements like Li, B and Be produce X-rays which are so strongly absorbed that few escape to be measured. In other cases, we might know the composition of the matrix material and all we really care about is a trace contaminant. Alternatively, we might know that certain elements like O often combine with other elements following predictable stoichiometric relationships. In these cases, it may be better to inject other sources of information into our composition calculation algorithm.

19.7.1 Oxygen by Assumed Stoichiometry

Oxygen can be difficult to measure directly because of its relatively low energy X-rays. O X-rays are readily absorbed by other elements. Fortunately, many elements combine readily with oxygen in predictable ratios. For example, Si oxidizes to form SiO_2 and Al oxidizes to form Al_2O_3. Rather than measure O directly, it is useful to compute the quantity of other elements from their k-ratios and then compute the amount of O it would take to fully oxidize these elements. This quantity of O is added in to the next estimated composition.

NIST DTSA-II has a table of common elemental stoichiometries for calculations that invoke assumed stoichiometry. For many elements, there may be more than one stable oxidation state. For example, iron oxidizes to FeO (wüstite), Fe_3O_4 (magnetite), and Fe_2O_3 (hematite). All three forms occur in natural minerals. The choice of oxidation state can be selected by the user, often relying upon independent information such as a crystallographic determination or based upon the most common oxidation state that is encountered in nature.

The same basic concept can be applied to other elements which combine in predicable ratios.

19.7.2 Waters of Crystallization

Water of crystallization (also known as water of hydration or crystallization water) is water that occurs within crystals. Typically, water of crystallization is annotated by adding "·nH_2O" to the end of the base chemical formula. For example, $CuSO_4 \cdot 5H_2O$ is copper(II) sulfate pentahydrate. This expression indicates that five molecules of water have been added to copper sulfate. Crystals may be fully hydrated or partially hydrated depending upon whether the maximum achievable number of water molecules are associated with each base molecule. $CuSO_4$ is partially hydrated if there are fewer than five water molecules per $CuSO_4$ molecule. Some crystals hydrate in a humid environment. Hydration molecules (water) can often be driven off by strong heating, and some hydrated materials undergo loss of water molecules due to electron beam damage.

Measuring water of crystallization involves measuring O directly and comparing this measurement with the amount of water predicted by performing a stoichiometric calculation on the base molecule. Any surplus oxygen (oxygen measured but not accounted for by stoichiometry) is assumed to be in the form of water and two additional hydrogen atoms are added to each surplus oxygen atom. The resulting composition can be reported as the base molecule + "·nH_2O" where n is the relative number waters per base molecule.

19.7.3 Element by Difference

All matter consists of 100 % of some set of elements. If we were able to measure the mass fraction of N-1 of the N elements in a material with perfect accuracy then the mass fraction of the Nth element would be the difference

$$C_N = 1 - \sum_{i=1}^{N-1} C_i \tag{19.7}$$

Of course, we can't measure the N-1 elements with perfect accuracy, but we can apply the same concept to estimate the quantity of difficult to measure elements.

This approach has numerous pitfalls. First, the uncertainty in difference is the sum of the uncertainties for the mass fractions of the N-1 elements. This can be quite large particularly when N is large. Second, since we assume the total mass fraction sums to unity, there is no redundant check like the analytic total to validate the measurement.

19.8 Ways of Reporting Composition

19.8.1 Mass Fraction

The most common way to report the composition of a material is in terms of the mass fraction. To understand the mass fraction, consider a block of material containing a mixture of different atoms. Weigh the block. Now imagine separating the block into distinct piles, each pile containing all the atoms from one element in the block. Weigh each of the separated piles. The mass fraction is calculated as the ratio of the mass of pile containing element Z over the total mass of the block. Since each element in the block is represented by a pile and since none of the atoms are lost in the process of dividing the block, the sum of mass fractions equals unity. Of course, we can't really do this measurement this way for most materials but conceptually we can understand mass fraction as though we can. As a simple example, consider the mineral pyrite, FeS_2. The molecular formula combines one gram-mole of Fe (atomic weight $A = 55.85$ g/mole) with two gram moles of S ($A = 32.07$ g/mole) for a compound molecular weight of 119.99 g/mole. The mass fractions, C_w, of the constituents are thus

$$C_{w,Fe} = 55.85/119.99 = 0.4655 \quad \text{(19.8a)}$$

$$C_{w,S} = 64.14/119.99 = 0.5345 \quad \text{(19.8b)}$$

The mass fraction is the fundamental output of electron probe X-ray microanalysis measurements. All other output modes are calculated from the mass fraction.

Weight fraction, mass percent, weight percent are all commonly seen synonyms for mass fractions. Mass fraction or mass percent is the preferred nomenclature because it is independent of local gravity.

19.8.2 Atomic Fraction

If we perform the same mental experiment as was described in the mass fraction section, but instead of weighing the piles, we instead count the number of atoms in the block and each of the piles. If we then calculate the ratio of the number of atoms of element Z relative to the total number of atoms in the block, this is the atomic fraction, C_a. For the example of FeS_2, which contains a total of three atoms in the molecular formula, the atomic fractions are

$$C_{a,Fe} = 1/3 = 0.3333 \quad \text{(19.9a)}$$

$$C_{a,S} = 2/3 = 0.6667 \quad \text{(19.9b)}$$

We can calculate atomic fraction C_a from mass fraction C_w and vice versa using the atomic weights A of the elements.

$$C_{a,i} = \left(C_{w,i}/A_i\right)\Big/\sum_1^N\left(C_{w,i}/A_i\right) \quad \text{(19.10)}$$

Where N is the number of elements involved in the mixture.

Starting with the atomic fractions, the mass fractions are calculated according to the formula

$$C_{w,i} = \left(C_{a,i}*A_i\right)\Big/\sum_1^N\left(C_{a,i}*A_i\right) \quad \text{(19.11)}$$

It is appropriate to use the atomic weights as suggested by the IUPAC (► http://www.ciaaw.org/atomic_weights4.htm). These weights are based on assumed mixes of isotopes as are typically seen in terrestrial samples. Occasionally, when it is known that an element is present in a perturbed isotopic mix, it may be appropriate to use this information to calculate a more accurate atomic weight. Since the atomic fraction depends upon assumed atomic weights, the atomic fraction is less fundamental than the mass fraction.

19.8.3 Stoichiometry

Stoichiometry is closely related to atomic fraction. Many materials can be described simply in terms of the chemical formula of its most basic constituent unit. For example, silicon and oxygen combine to form a material in which the most basic repeating element consists of SiO_2. Stoichiometry can be readily translated into atomic fraction. Since our measurements are imprecise, the stoichiometry rarely works out in clean integral units. However, the measurement is often precise enough to distinguish between two or more valence states.

19.8.4 Oxide Fractions

Oxide fractions are closely related to stoichiometry. When a material such as a natural mineral is a mixture of oxides, it can make sense to report the composition as a linear sum of the oxide constituents by mass fraction.

■ Table 19.1 shows the analysis of NIST SRM470 (K412 glass) with the results reported as oxide fraction, mass fraction, and atomic fraction.

Example Calculations

Calculating the mass fraction from the oxide fraction for Al in K412 glass:

$$C_{w,Al} = \frac{2\times26.9815}{2\times26.9815+3\times15.999}0.0927 = 0.04906 \quad \text{(19.12)}$$

Calculating the atomic fraction from the mass fraction:

$$C_{a,Al} = \frac{\dfrac{0.0491}{26.9815}}{\dfrac{0.1166}{24.305}+\dfrac{0.0491}{26.9815}+\dfrac{0.2120}{28.085}+\dfrac{0.1090}{40.078}+\dfrac{0.0774}{55.845}+\dfrac{0.4275}{15.999}}$$

$$\text{(19.13)}$$

◘ Table 19.1 Three different ways to report the composition of NIST SRM 470 glass K412.

K412 Glass							
Element	Mg	Al	Si	Ca	Fe	O	Sum
Valence	2	3	1	2	2	−2	–
Atomic weight (AMU)	24.305	26.9815	28.085	40.078	55.845	15.999	
Oxide fraction	0.1933 ± 0.0020 MgO	0.0927 ± 0.0020 Al$_2$O$_3$	0.4535 ± 0.0020 SiO$_2$	0.1525 ± 0.0020 CaO	0.0996 ± 0.0020 FeO	–	0.9916
Mass fraction	0.1166	0.0491	0 2120	0.1090	0.0774	0 4276	0.9916
Atomic fraction	0.1066	0.0404	0.1678	0.0604	0.0308	0.5940	1

Note the analytic total is less than 1, indicating an imprecision in the certified value

19.9 The Accuracy of Quantitative Electron-Excited X-ray Microanalysis

19.9.1 Standards-Based k-ratio Protocol

Quantitative electron-excited X-ray microanalysis following the standards-based k-ratio protocol is a relative not an absolute analysis method. The unknown is measured relative to standards of well known composition such as pure elements and stoichiometric compounds with fixed atom ratios, for example, FeS$_2$. The accuracy of the method can only be tested by analyzing materials whose composition is known from independent (and ideally absolute) analysis methods and whose composition has been found to be homogeneous at the sub-micrometer scale. There are limited numbers of special materials that fit these strict compositional requirements to qualify as certified reference materials for electron beam X-ray microanalysis, including certain metal alloys, intermetallic compounds, and glasses. Limited numbers of these materials are available from national standards institutions, such as the National Institute of Standards and Technology (U.S.) (e.g., Marinenko et al. 1990) and the European Commission Community Bureau of Reference (e.g., Saunders et al. 2004). Certain mineral species have been characterized to serve as standards, which are of particular use to the geochemistry community, by the Smithsonian Institution. These certified reference materials and related materials such as minerals can serve directly as standards for analyses, but their other important function is to serve as challenge materials to test the quantification methods. Additional materials suitable for testing the method include stoichiometric compounds with formulae that define specific, unvarying compositions; that is, the same materials that can also be used as standards. Thus, FeS$_2$ could be used as an "unknown" for a test analysis with Fe and CuS as the standards, while CuS could be analyzed with Cu and FeS$_2$ as standards. From such analyses of certified reference materials and other test materials, the relative deviation from the expected value (RDEV)

(also referred to as "relative error") is calculated with the "expected" value taken as the stoichiometric formula value or the value obtained from an "absolute" analytical method, such as gravimetric analysis:

$$\text{RDEV} = \left[\frac{\left(\text{Analyzed value} - \text{expected value} \right)}{\text{expected value}} \right] \times 100\% \qquad \text{(19.14)}$$

Note that by this equation a positive RDEV indicates an overestimate of the concentration, while a negative RDEV indicates an underestimate. By analyzing many test materials spanning the periodic table and determining the relative deviation from the expected value (relative error), the analytical performance can be estimated. For example, early studies of quantitative electron probe microanalysis with wavelength dispersive X-ray spectrometry following the standards-based k-ratio protocol and ZAF matrix corrections produced a distribution of RDEV values (relative errors) such that 95 % of the analyses were captured in an RDEV (relative error) range of ±5 % relative, as shown in ◘ Fig. 19.1 (Yakowitz 1975).

Subsequent development and refinement of the matrix correction procedures by many researchers improved upon this level of accuracy. Pouchou and Pichoir (1991) described an advanced matrix correction model based upon extensive experimental measurements of the $\varphi(\rho z)$ description of the depth distribution of ionization. Incorporating explicit measurements for low energy photons, this approach has been especially successful for low photon energy X-rays which were subject to high absorption. A comparison of corrections of the same k-ratio dataset with their $\varphi(\rho z)$ method and with the conventional ZAF method showed significant narrowing of the RDEV distribution and elimination of significant large RDEV values, as shown in ◘ Fig. 19.2. With this improvement, approximately 95 % of analyses fall within ±2.5 % RDEV.

▶ Chapter 20 will illustrate examples of quantitative electron-excited X-ray microanalysis with silicon drift detector (SDD)-EDS performed on flat bulk specimens following the k-ratio

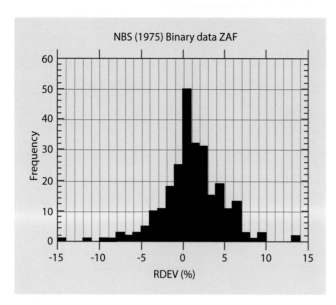

Fig. 19.1   Histogram of relative deviation from expected value (relative error) for electron probe microanalysis with wavelength dispersive spectrometry following the k-ratio protocol with standards and ZAF corrections (Yakowitz 1975)

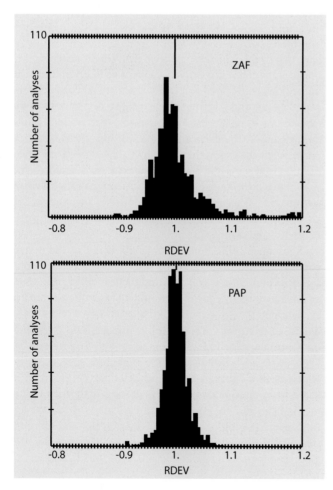

Fig. 19.2   Comparison of quantitative analysis of an EPMA-WDS k = ratio database by conventional ZAF and by the PAP $\varphi(\rho z)$ model (Pouchou and Pichoir 1991)

protocol and PAP $\varphi(\rho z)$ matrix corrections with NIST DTSA-II. The level of accuracy achieved with this SDD-EDS approach fits within the RDEV histogram achieved with EPMA-WDS for major, minor, and trace constituents, even when severe peak interference occurs. It should be noted that SDD-EDS is sufficiently stable with time that, providing a quality measurement protocol is in place to ensure that all measurements are made under identical conditions of beam energy, known dose, specimen orientation, and SDD-EDS performance, archived standards can be used without significant loss of accuracy.

19.9.2   "Standardless Analysis"

Virtually all vendor analytical software includes the option for "standardless analysis." Standardless analysis requires only the spectrum of the unknown, the list of elements identified during qualitative analysis, and the beam energy; and the software will report quantitative concentration values, including oxygen by assumed stoichiometry if desired. "Standardless analysis" is usually implemented as a "black box" tool without extensive documentation. The approach is the same as the standards-based analysis protocol: a k-ratio is the starting point, but the spectrum of the unknown only provides the numerator of the k-ratio. A "first principles physics" calculation of the standard intensity for the denominator of the k-ratio, while possible, is difficult because of the lack of accurate values of critical parameters in the equations for X-ray generation and propagation. Instead, the general approach employed throughout the EDS industry is the use of a library of remotely measured standards to provide the intensity for the denominator of the k-ratio. Pure element and binary compound standards are measured under defined conditions at several beam energies on a well characterized EDS. When standardless analysis is invoked, the appropriate elemental intensities are selected from this database of standards, and any missing elements not represented in the database are supplied by interpolation aided by the physical equations of X-ray generation and propagation. If a beam energy is requested for which reference values are not available in the database, the equations of the physics of X-ray generation are used to appropriately adjust the available intensities. Usually a reference spectrum that is locally measured on a pure element, for example, Mn or Cu, is used to compare the efficiency of the EDS on a channel-by-channel basis to the vendor EDS that was originally used to measure the standards library. Because of its simplicity of operation, standardless analysis enjoys great popularity. Probably 95 % or more of quantitative EDS analyses are performed with the standardless analysis procedure. While it is useful and is continually being improved, standardless analysis is subject to a substantially wider RDEV distribution than standards-based analysis with locally measured standards. Standardless analysis of a wide range of test materials produced the RDEV histogram shown in ▪ Fig. 19.3 (Newbury et al. 1995). This distribution is such that 95 % of all analyses fall within a range of ±25 % relative. If this level of analytical accuracy is

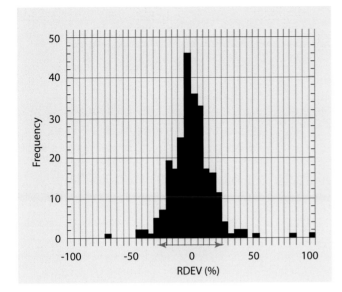

Fig. 19.3 RDEV distribution observed for a vendor standardless analysis procedure (Newbury et al. 1995)

Fig. 19.4 **a** RDEV distribution observed for a vendor standardless analysis procedures in 2016 with oxygen calculated directly for oxidized specimens. **b** RDEV distribution observed for a vendor standardless analysis procedures in 2016 with oxygen calculated by stoichiometry for oxidized specimens

sufficient, then standardless analysis is an acceptable procedure, providing this RDEV distribution, or a version appropriate to the standardless analysis software supplied by the vendor, is used to inform those who will make use of the quantitative analyses of the possible range of the results. While the standardless protocol may eventually equal the performance of standards-based analysis, recent results for current versions of standardless analysis, reported in **Fig. 19.4**, suggest that a wide RDEV distribution is still being experienced, at least from some vendors.

It should be noted that when 95 % of all analyses fall within a range of ±25 % relative, it may often not be possible to correctly determine the formula of the major constituents of a stoichiometric compound. An example of this situation is presented in **Table 19.2**, which gives the results of an SEM-EDS analysis of a $YBa_2Cu_3O_{7-X}$ single crystal by the k-ratio/standards protocol (NIST DTSA) compared to standardless analysis performed with two different vendors' software. While the proper formula is recovered with the

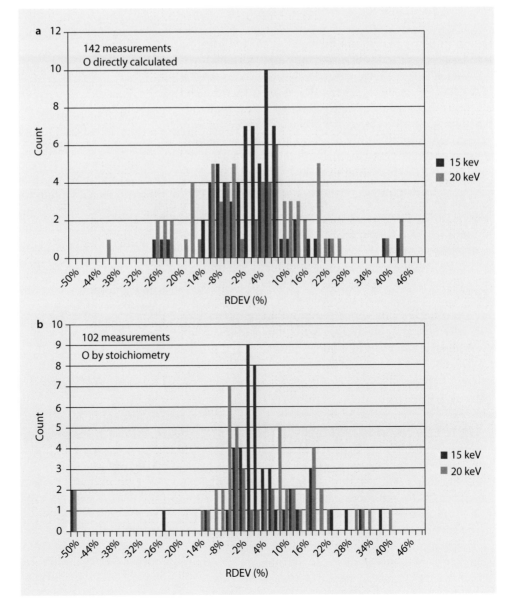

☐ **Table 19.2** SEM-EDS analysis of a $YBa_2Cu_3O_{7-x}$ single crystal (O calculated by stoichiometry)

	Y (true) 0.133 mass conc	Ba (true) 0.412	Cu (true) 0.286
k-ratio Stds ZAF	0.138 (+4%)	0.411 (−0.2%)	0.281 (−2%) Cu-K $Y_1Ba_2Cu_3O_{6.4}$

Standards: Y and Cu pure elements; Ba (NIST glass K309)

Standardless Analysis (two different vendors):

M1	0.173 (+30%)	0.400 (−3%)	0.267 (−7%) Cu-K $Y_2Ba_3Cu_4O_{10}$
M1	0.158 (+19%)	0.362 (−12%)	0.316 (+10%) Cu-L $Y_2Ba_3Cu_6O_{12}$
M2	0.165 (+24%)	0.387 (−6%)	0.287 (+0.4%) Cu-K $Y_2Ba_3Cu_5O_{11}$
M2	0.168 (+26%)	0.395 (−4%)	0.276 (−3.5%) Cu-L $Y_4Ba_6Cu_9O_{21}$

standards-based analysis, the formulae calculated from the standardless results do not match the proper formula.

Another shortcoming of standardless analysis is the loss of the information on the dose and the absolute spectrometer efficiency that is automatically embedded in the standards-based k-ratio/matrix corrections protocol. Without the dose and absolute spectrometer efficiency information, standardless analysis results must inevitably be internally normalized to unity (100%) so that the calculated concentrations have realistic meaning, thereby losing the very useful information present in the raw analytical total that is available in the standards-based k-ratio/matrix corrections protocol. It must be noted that standardless analysis results will always sum to unity, even if one or more constituents are not recognized during qualitative analysis or are inadvertently lost from the suite of elements being analyzed. If the local dose and spectrometer efficiency can be accurately scaled to the conditions used to record remote standards, then standardless analysis can determine a meaningful analytical total, but this is not commonly implemented in vendor software.

19.10 Appendix

19.10.1 The Need for Matrix Corrections To Achieve Quantitative Analysis

There has long been confusion around the definition of the expression 'ZAF' used to compensate for material differences in X-ray microanalysis measurements. There are two competing definitions. Neither is wrong and both exist in the literature and implemented in microanalysis software. However, the two definitions lead to numerical values of the matrix corrections that are related by being numerical inverses of each other.

For the sake of argument, let's call these two definitions ZAF_A and ZAF_B. In both definitions, $k = I_{unk}/I_{std}$ and C_{unk} is the mass fraction of the element in the unknown.

ZAF_A is defined by the expression:

$$k = C_{unk} ZAF_A \tag{19.15a}$$

ZAF_B is defined by the expression:

$$C_{unk} = k\,ZAF_B \tag{19.15b}$$

If we solve each equation for k/C_{unk} and equate the resulting expression, we discover that

$$ZAF_A = 1/ZAF_B \tag{19.15c}$$

Needless to say, these inconsistent definitions can cause significant confusion. Whenever interpreting matrix corrections in the literature, it is important to identify which convention the author is using.

The confusion extends to this book. Most of this book has been written using the first convention (ZAF_A) however, the previous (third) edition of this book used the second convention (ZAF_B). The following section which has been pulled from the third edition continues to use the ZAF_B convention as this was the definition favored by the writer. NIST DTSA-II and CITZAF uses the $k = C_{unk} ZAF_A$ convention. (Contribution of the late Prof. Joseph Goldstein taken from SEMXM-3, ► Chapter 9)

Upon initial examination, it would seem that quantitative analysis should be extremely simple. Just form the ratio of the characteristic X-ray intensity for a given element measured from the specimen to that measured from the standard, and that ratio should be equal to the ratio of concentrations for a given element between the specimen and the standard. As was first noted by Castaing (1951), the primary *generated* intensities are roughly proportional to the respective mass fractions of the emitting element. If other contributions to X-ray generation are very small, the *measured* intensity ratios between specimen and standard are roughly equal to the ratios of the mass or weight fractions of the emitting element. This assumption is often applied to X-ray quantitation and is called Castaing's "first approximation to quantitative analysis" and is given by

$$C_{i,unk} / C_{i,std} = I_{i,unk} / I_{i,std} = k \tag{19.16}$$

The terms $C_{i,unk}$ and $C_{i,std}$ are the composition in weight (mass) concentration of element i in the unknown and in the standard, respectively. The ratio of the measured unknown-to-standard intensities after continuum background is subtracted and peak overlaps are accounted for, $I_{i,unk}/I_{i,std}$, is the basic experimental measurement which underlies all quantitative X-ray microanalysis and is given the special designation as the "k-ratio."

Careful measurements performed on homogeneous substances of known multi-element composition compared to pure element standards reveal that there are significant sys-

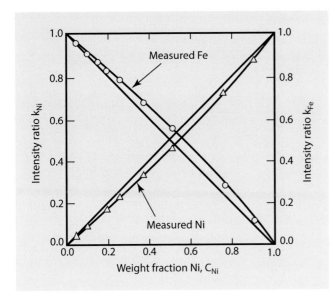

☐ **Fig. 19.5** Measured Fe K-L$_3$ and Ni K-L$_3$ k-ratios versus the weight fraction of Ni at $E_0 = 30$ keV. Curves are measured k-ratio data, while straight lines represent ideal behavior (i.e., no matrix effects)

tematic deviations between the ratio of measured intensities and the ratio of concentrations. An example of these deviations is shown in ☐ Fig. 19.5, which depicts the deviations of measured X-ray intensities in the iron-nickel binary system from the linear behavior predicted by the first approximation to quantitative analysis, Eq. (19.16). ☐ Figure 19.5 shows the measurement of $I_{i,unk}/I_{i,std} = k$ for Ni K-L$_3$ and Fe K-L$_3$ in nine well-characterized homogeneous Fe-Ni standards (Goldstein et al. 1965). The data were taken at an initial electron beam energy of 30 keV and a take-off angle $\psi = 52.5°$. The intensity ratio k_{Ni} or k_{Fe} is the $I_{i,unk}/I_{i,std}$ measurement for Ni and Fe, respectively, relative to pure element standards. The straight lines plotted between pure Fe and pure Ni indicate the relationship between composition and intensity ratio given in Eq. (19.16). For Ni K-L$_3$, the actual data fall below the linear first approximation and indicate that there is an X-ray absorption effect taking place, that is, more absorption in the sample than in the standard. For Fe K-L$_3$, the measured data fall above the first approximation and indicate that there is a fluorescence effect taking place in the sample. In this alloy the Ni K-L$_3$ radiation is heavily absorbed by the iron and the Fe K-L$_3$ radiation is increased due to X-ray fluorescence by the Ni K-L$_3$ radiation over that generated by the bombarding electrons.

These effects that cause deviations from the simple linear behavior given by Eq. (19.16) are referred to as matrix or inter-element effects. As described in the following sections, the measured intensities from specimen and standard need to be corrected for differences in electron backscatter and energy loss, X-ray absorption along the path through the solid to reach the detector, and secondary X-ray generation and emission that follows absorption, in order to arrive at the ratio of *generated* intensities and hence the value of $C_{i,unk}$. The magnitude of the matrix effects can be quite large, exceeding

factors of ten or more in certain systems. Recognition of the complexity of the problem of the analysis of solid samples has led numerous investigators to develop the theoretical treatment of the quantitative analysis scheme, first proposed by Castaing (1951).

19.10.2 The Physical Origin of Matrix Effects

What is the origin of these matrix effects? The X-ray intensity *generated* for each element in the specimen is proportional to the concentration of that element, the probability of X-ray production (ionization cross section) for that element, the path length of the electrons in the specimen, and the fraction of incident electrons which remain in the specimen and are not backscattered. It is very difficult to calculate the absolute generated intensity for the elements present in a specimen directly. Moreover, the intensity that the analyst must deal with is the *measured* intensity. The measured intensity is even more difficult to calculate, particularly because absorption and fluorescence of the generated X-rays may occur in the specimen, thus further modifying the measured X-ray intensity from that predicted on the basis of the ionization cross section alone. Instrumental factors such as differing spectrometer efficiency as a function of X-ray energy must also be considered. Many of these factors are dependent on the atomic species involved. Thus, in mixtures of elements, matrix effects arise because of differences in elastic and inelastic scattering processes and in the propagation of X-rays through the specimen to reach the detector. For conceptual as well as calculational reasons, it is convenient to divide the matrix effects into atomic number, Z_i; X-ray absorption, A_i; and X-ray fluorescence, F_i, effects.

Using these matrix effects, the most common form of the correction equation is

$$C_{i,unk}/C_{i,std} == [\mathrm{ZAF}]_i \, [\, I_{i,unk}/I_{i,std}\,] = [ZAF]_i \cdot k_i \qquad \textbf{(19.17)}$$

where $C_{i,unk}$ is the weight fraction of the element of interest in the unknown and $C_{i,std}$ is the weight fraction of i in the standard. This equation must be applied separately for *each* element present in the sample. Equation (19.17) is used to express the matrix effects and is the common basis for X-ray microanalysis in the SEM/EPMA.

It is important for the analyst to develop a good idea of the origin and the importance of each of the three major non-linear effects on X-ray measurement for quantitative analysis of a large range of specimens.

19.10.3 ZAF Factors in Microanalysis

The matrix effects Z, A, and F all contribute to the correction for X-ray analysis as given in Eq. (19.17). This section discusses each of the matrix effects individually. The combined effect of ZAF determines the total matrix correction.

Atomic Number Effect, Z (Effect of Backscattering [R] and Energy Loss [S])

One approach to the atomic number effect is to consider directly the two different factors, backscattering (R) and stopping power (S), which determine the amount of generated X-ray intensity in an unknown. Dividing the stopping power, S, for the unknown and standard by the backscattering term, R, for the unknown and standard yields the atomic number matrix factor, Z_i, for each element, i, in the unknown. A discussion of the R and S factors follows.

Backscattering, R: The process of elastic scattering in a solid sample leads to backscattering which results in the premature loss of a significant fraction of the beam electrons from the target before all of the ionizing power of those electrons has been expended generating X-rays of the various elemental constituents. From ◘ Fig. 2.3a, which depicts the backscattering coefficient as a function of atomic number, this effect is seen to be strong, particularly if the elements involved in the unknown and standard have widely differing atomic numbers. For example, consider the analysis of a minor constituent, for example, 1 weight %, of aluminum in gold, against a pure aluminum standard. In the aluminum standard, the backscattering coefficient is about 15 % at a beam energy of 20 keV, while for gold the value is about 50 %. When aluminum is measured as a standard, about 85 % of the beam electrons completely expend their energy in the target, making the maximum amount of Al K-L$_3$ X-rays. In gold, only 50 % are stopped in the target, so by this effect, aluminum dispersed in gold is actually under represented in the X-rays generated in the specimen relative to the pure aluminum standard. The energy distribution of backscattered electrons further exacerbates this effect. Not only are more electrons backscattered from high atomic number targets, but as shown in ◘ Fig. 2.16a, b, the backscattered electrons from high atomic number targets carry off a higher fraction of their incident energy, further reducing the energy available for ionization of inner shells. The integrated effects of backscattering and the backscattered electron energy distribution form the basis of the "R-factor" in the atomic number correction of the "ZAF" formulation of matrix corrections.

Stopping power, S: The rate of energy loss due to inelastic scattering also depends strongly on the atomic number. For quantitative X-ray calculations, the concept of the stopping power, S, of the target is used. S is the rate of energy loss given by the Bethe continuous energy loss approximation, Eq. (1.1), divided by the density, ρ, giving $S = -(1/\rho)(dE/ds)$. Using the Bethe formulation for the rate of energy loss (dE/ds), one observes that the stopping power is a decreasing function of atomic number. The low atomic number targets actually remove energy from the beam electron more rapidly with mass depth (ρz), the product of the density of the sample (ρ), and the depth dimension (z) than high atomic number targets.

An example of the importance of the atomic number effect is shown in ◘ Fig. 19.6. This figure shows the measurement of the intensity ratio k_{Au} and k_{Cu} for Au L-M and Cu

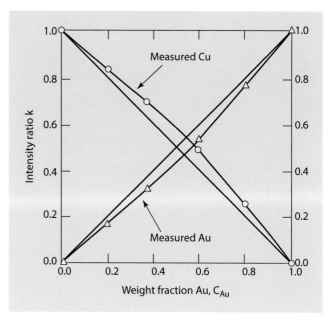

◘ **Fig. 19.6** Measured Au L$_3$-M$_5$ and Cu K-L$_3$ k-ratios versus the weight fraction of Au at $E_0 = 25$ keV. Curves are measured k-ratio data, while straight lines represent ideal behavior (i.e., no matrix effects)

K-L$_3$ for four well-characterized homogeneous Au-Cu standards (Heinrich et al. 1971). The data were taken at an initial electron beam energy of 15 keV and a take-off angle of 52.5°, and pure Au and pure Cu were used as standards. The atomic number difference between these two elements is 50. The straight lines plotted on ◘ Fig. 19.6 between pure Au and pure Cu indicate the relationship between composition and intensity ratio given in Eq. (19.17). For both Au L-M and Cu K-L$_3$, the absorption matrix effect, A_i, is less than 1 %, and the fluorescence matrix effect, F_i, is less than 2 %. For Cu K-L$_3$, the measured data fall above the first approximation and almost all the deviation is due to the atomic number effect, the difference in atomic number between the Au-Cu alloy and the Cu standard. As an example, for the 40.1 wt% Au specimen, the atomic number matrix factor, Z_{Cu}, is 1.12, an increase in the Cu K-L$_3$ intensity by 12 %. For Au L-M, the measured data fall below Castaing's first approximation and almost all the deviation is due to the atomic number effect. As an example, for the 40.1 wt % Au specimen, the atomic number effect, Z_{Au}, is 0.806, a decrease in the Au L-M intensity by 20 %. In this example, the S factor is larger and the R factor is smaller for the Cu K-L$_3$ X-rays leading to a larger S/R ratio and hence a larger Z_{Cu} effect. Just the opposite is true for the Au L-M X-rays leading to a smaller Z_{Au} effect. The effects of R and S tend to go in opposite directions and to cancel.

X-ray Generation With Depth, $\varphi(\rho z)$

A second approach to calculating the atomic number effect is to determine the X-ray generation in depth as a function of atomic number and electron beam energy. As shown in ► Chapters 1, 2, and 4, the paths of beam electrons within the specimen can be represented by Monte Carlo simulations of electron trajectories. In the Monte Carlo simulation tech-

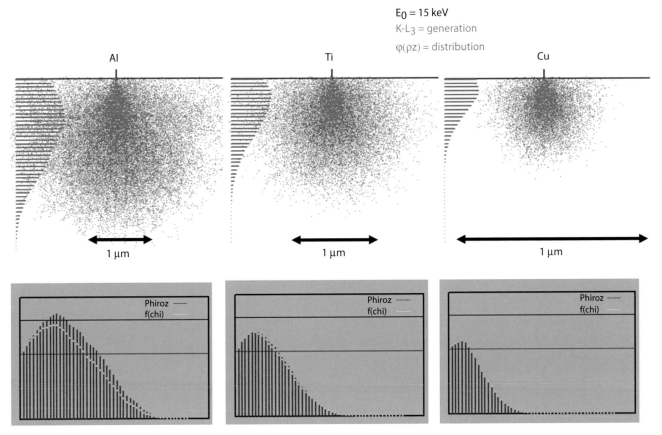

E₀ = 15 keV
K-L₃ = generation
φ(ρz) = distribution

◘ Fig. 19.7 Monte Carlo simulations (Joy Monte Carlo) of X-ray generation at E_0 = 15 keV for Al K-L₃, Ti K-L₃, and Cu K-L₃, showing (upper) the sites of X-ray generation (red dots) projected on the x-z plane, and the resulting $\varphi(\rho z)$ distribution. (lower) the $\varphi(\rho z)$ distribution is plotted with the associated f(χ) distribution showing the escape of X-rays following absorption

nique, the detailed history of an electron trajectory is calculated in a stepwise manner. At each point along the trajectory, both elastic and inelastic scattering events can occur. The production of characteristic X-rays, an inelastic scattering process, can occur along the path of an electron as long as the energy E of the electron is above the critical excitation energy, E_c, of the characteristic X-ray of interest.

◘ Figure 19.7 displays Monte Carlo simulations of the positions where K-shell X-ray interactions occur for three elements, Al, Ti, and Cu, using an initial electron energy, E_0, of 15 keV. The incoming electron beam is assumed to have a zero width and to impact normal to the sample surface. X-ray generation occurs in the lateral directions, x and y, and in depth dimension, z. The micrometer marker gives the distance in both the x and z dimensions. Each dot indicates the generation of an X-ray; the dense regions indicate that a large number of X-rays are generated. This figure shows that the X-ray generation volume decreases with increasing atomic number (Al, Z = 13; Ti, Z = 22; Cu, Z = 29) for the same initial electron energy. The decrease in X-ray generation volume is due to (1) an increase in elastic scattering with atomic number, which deviates the electron path from the initial beam direction; and (2) an increase in critical excitation energy, E_c, that gives a corresponding decrease in overvoltage U ($U = E_0/E_c$) with atomic number. This decreases the fraction of the initial electron energy available for the production of

characteristic X-rays. A decrease in overvoltage, U, decreases the energy range over which X-rays can be produced.

One can observe from ◘ Fig. 19.7 that there is a non-even distribution of X-ray generation with depth, z, for specimens with various atomic numbers and initial electron beam energies. This variation is illustrated by the histograms on the left side of the Monte Carlo simulations. These histograms plot the number of X-rays generated with depth into the specimen. In detail the X-ray generation for most specimens is somewhat higher just below the surface of the specimen and decreases to zero when the electron energy, E, falls below the critical excitation energy, E_c, of the characteristic X-ray of interest.

As illustrated from the Monte Carlo simulations, the atomic number of the specimen strongly affects the distribution of X-rays generated in specimens. These effects are even more complex when considering more interesting multi-element samples as well as the generation of L and M shell X-ray radiation.

◘ Figure 19.7 clearly shows that X-ray generation varies with depth as well as with specimen atomic number. In practice it is very difficult to measure or calculate an absolute value for the X-ray intensity generated with depth. Therefore, we follow the practice first suggested by Castaing (1951) of using a relative or a normalized generated intensity which varies with depth, called φ (ρz). The term ρz is called the mass depth and is the product of the density ρ of the sample

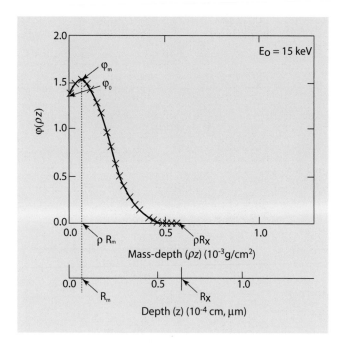

◻ **Fig. 19.8** Schematic illustration of the $\varphi(\rho z)$ depth distribution of X-ray generation, with the definitions of specific terms: φ_0, φ_m, ρR_m, R_m, ρR_X, and R_X

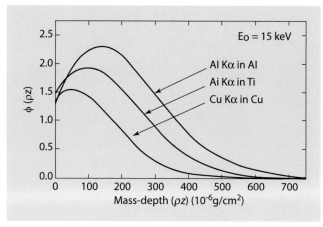

◻ **Fig. 19.9** Calculated $\varphi(\rho z)$ curves for Al K-L$_3$ in Al; Ti K-L$_3$ in Ti; and Cu K-L$_3$ in Cu at $E_0 = 15$ keV; calculated using PROZA

(g/cm^3) and the linear depth dimension, z (cm), so that the product ρz has units of g/cm^2. The mass depth, ρz, is more commonly used than the depth term, z. The use of the mass depth removes the strong variable of density when comparing specimens of different atomic number. Therefore it is important to recognize the difference between the two terms as the discussion of X-ray generation proceeds.

The general shape of the depth distribution of the generated X-rays, the $\varphi(\rho z)$ versus ρz curve, is shown in ◻ Fig. 19.8. The amount of X-ray production in any layer of the histogram is related to the amount of elastic scattering, the initial electron beam energy, and the energy of the characteristic X-ray of interest. The intensity in any layer of the $\varphi(\rho z)$ versus ρz curve is normalized to the intensity generated in an ideal thin layer, where "thin" is a thickness such that effectively no significant elastic scattering occurs and the incident electrons pass through perpendicular to the layer. As the incident beam penetrates the layers of material in depth, the length of the trajectory in each successive layer increases because (1) elastic scattering deviates the beam electrons out of the straight line path, which was initially parallel to the surface normal, thus requiring a longer path to cross the layer and (2) backscattering results in electrons, which were scattered deeper in the specimen, crossing the layer in the opposite direction following a continuous range of angles relative to the surface normal. Due to these factors, X-ray production increases with depth from the surface, $\rho z = 0$, and goes through a peak, φ_m, at a certain depth ρR_m (see ◻ Fig. 19.8). Another consequence of backscattering is that surface layer production, φ_0, is larger than 1.0 in solid samples because the backscattered electrons excite X-rays as they pass through the surface layer and leave the sample, adding to the intensity created by all of the inci-

dent beam electrons that passed through the surface layer. After the depth ρR_m, X-ray production begins to decrease with depth because the backscattering of the beam electrons reduces the number of electrons available at increasing depth ρz and the remaining electrons lose energy and therefore ionizing power as they scatter at increasing depths. Finally X-ray production goes to zero at $\rho z = \rho Rx$ where the energy of the beam electrons no longer exceeds E_c.

Now that we have discussed and described the depth distribution of the production of X-rays using the $\varphi(\rho z)$ versus ρz curves, it is important to understand how these curves differ with the type of specimen that is analyzed and the operating conditions of the instrument. The specimen and operating conditions that are most important in this regard are the average atomic number, Z, of the specimen and the initial electron beam energy, E_0 chosen for the analysis. Calculations of $\varphi(\rho z)$ versus ρz curves have been made for this Appendix using the PROZA program (Bastin and Heijligers 1990). In ◻ Fig. 19.9, the $\varphi(\rho z)$ versus ρz curves for the K-L$_3$ X-rays of pure Al, Ti, and Cu specimens at 15 keV are displayed. The shapes of the $\varphi(\rho z)$ versus ρz curves are quite different. The φ_0 values, relative to the value of φm for each curve, increase from Al to Cu due to increased backscattering which produces additional X-ray radiation. On an absolute basis, the φ_0 value for Cu is smaller than the value for Ti because the overvoltage, U_0, for the Cu K-L$_3$ X-ray at $E_0 = 15$ keV is low ($U_0 = 1.67$) and the energy of many of the backscattered electrons is not sufficient to excite Cu K-L$_3$ X-rays near the surface. The values of ρR_m and ρR_x decrease with increasing Z and a smaller X-ray excitation volume is produced. This decrease would be much more evident if we plotted $\varphi(\rho z)$ versus z, the linear depth of X-ray excitation, since the use of mass depth includes the density, which changes significantly from Al to Cu.

◻ Figure 19.10 shows calculated $\varphi(\rho z)$ versus ρz curves, using the PROZA program (Bastin and Heijligers 1990, 1991) at an initial beam energy of 15 keV for Al K-L$_3$ and Cu K-L$_3$ radiation for the pure elements Al and Cu. These curves are compared in ◻ Fig. 19.10 with calculated $\varphi(\rho z)$ versus ρz curves at 15 keV for Al K-L$_3$ and Cu K-L$_3$ in a binary sample containing Al with 3 wt % Cu. The φ_0 value of the Cu K-L$_3$

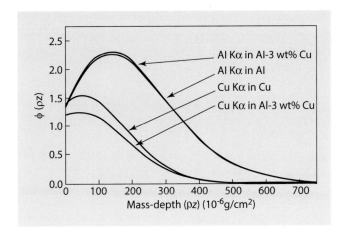

☐ **Fig. 19.10** Calculated $\varphi(\rho z)$ curves for Al K-L$_3$ and Cu K-L$_3$ in Al, Cu, and Al-3wt%Cu at $E_0 = 15$ keV; calculated using PROZA

☐ **Table 19.3** Generated X-ray intensities in Al, Cu, and Al-3wt%Cu alloy, as calculated with PROZA (Bastin and Heijligers 1990, 1991)

Sample	X-ray	$\varphi(\rho z)_{i,gen}$ Area (cm²/g)	Atomic number factor, Z_i	φ_0
Cu	Cu K-L$_3$	3.34×10^{-4}	1.0	1.39
Al	Al K-L$_3$	7.85×10^{-4}	1.0	1.33
Al-3wt%Cu	Cu K-L$_3$	2.76×10^{-4}	0.826	1.20
Al-3wt%Cu	Al K-L$_3$	7.89×10^{-4}	1.005	1.34

curve in the alloy is smaller than that of pure Cu because the average atomic number of the Al – 3 wt % Cu sample is so much lower, almost the same as pure Al. In this case, less back-scattering of the primary high energy electron beam occurs and fewer Cu K-L$_3$ X-rays are generated. On the other hand the Al K-L$_3$ $\varphi(\rho z)$ curves for the alloy and the pure element are essentially the same since the average atomic number of the specimen is so close to that of pure Al. Although the variation of $\varphi(\rho z)$ curves with atomic number and initial operating energy is complex, a knowledge of the pertinent X-ray genera-tion curves is critical to understanding what is happening in the specimen and the standard for the element of interest.

The generated characteristic X-ray intensity, I$_i$ gen, for each element, i, in the specimen can be obtained by taking the area under the $\varphi(\rho z)$ versus ρz curve, that is, by sum-ming the values of $\varphi(\rho z)$ for all the layers $\Delta(\rho z)$ in mass thickness within the specimen for the X-ray of interest. We will call this area "$\varphi(\rho z)i,_{gen}$ Area." ☐ Table 19.3 lists the cal-culated values, using the PROZA program, of the $\varphi(\rho z)i,_{gen}$ Area for the 15 kev $\varphi(\rho z)$ curves shown in ☐ Fig. 19.10 (Cu K-L$_3$ and Al K-L$_3$ in the pure elements and Cu K-L$_3$ and Al K-L$_3$ in an alloy of Al – 3 wt % Cu and the corresponding values of φ_0). A comparison of the $\varphi(\rho z)i,_{gen}$ Area values for Al K-L$_3$ in Al and in the Al – 3 wt % Cu alloy shows very similar values while a comparison of the $\varphi(\rho z)i,_{gen}$ Area val-ues for Cu K-L$_3$ in pure Cu and in the Al – 3 wt % Cu alloy shows that about 17 % fewer Cu K-L$_3$ X-rays are generated in the alloy. The latter variation is due to the different atomic numbers of the pure Cu and the Al – 3 wt% Cu alloy speci-men. The different atomic number matrices cause a change in φ_0 (see ☐ Table 19.3) and the height of the $\varphi(\rho z)$ curves.

The atomic number correction, Z_i, can be calculated by taking the ratio of $\varphi(\rho z)_{i,gen}$ Area for the standard to $\varphi(\rho z)_{i,gen}$ Area for element i in the specimen. Pure Cu and pure Al are the standards for Cu K-L$_3$ and Al K-L$_3$ respectively. The val-ues of the calculated ratios of generated X-ray intensities, pure element standard to specimen (Atomic number effect, Z_{Al}, Z_{Cu}) are also given in ☐ Table 19.3 As discussed above, it is expected that the atomic number correction for a heavy element (Cu) in a light element matrix (Al – 3 wt % Cu) is

less than 1.0 and the atomic number correction for a light element (Al) in a heavy element matrix (Al – 3 wt % Cu) is greater than 1.0. The calculated data in ☐ Table 19.3 also show this relationship.

In summary, the atomic number matrix correction, Z_i, is equal to the ratio of $Z_{i,std}$ in the standard to $Z_{i,unk}$ in the unknown. Using appropriate $\varphi(\rho z)$ curves, correction Z_i can be calculated by taking the ratio of $I_{gen,std}$ for the standard to $I_{gen,unk}$ for the unknown for each element, i, in the sample. It is important to note that the $\varphi(\rho z)$ curves for multi-element samples and elemental standards which can be used for the calculation of the atomic number effect inherently contain the R and S factors discussed previously.

X-ray Absorption Effect, A

☐ Figure 19.11 illustrates the effect of varying the initial elec-tron beam energy using Monte Carlo simulations on the positions where K-shell X-ray generation occurs for Cu at three initial electron energies, 10, 20, and 30 keV. This figure shows that the Cu characteristic X-rays are generated deeper in the specimen and the X-ray generation volume becomes larger as E_0 increases. From these plots, we can see that the sites of inner shell ionizations which give rise to characteris-tic X-rays are created over a range of depth below the surface of the specimen.

Created over a range of depth, the X-rays will have to pass through a certain amount of matter to reach the detector, and as explained in ► Chapter 4 (X-rays), the photoelectric absorption process will decrease the intensity. It is important to realize that the X-ray photons are either absorbed or else they pass through the specimen with their original energy unchanged, so that they are still characteristic of the atoms which emitted the X-rays. Absorption follows an exponential law, so as X-rays are generated deeper in the specimen, a pro-gressively greater fraction is lost to absorption.

From the Monte Carlo plots of ☐ Fig. 19.11, one recognizes that the depth distribution of ionization is a complicated func-tion. To quantitatively calculate the effect of X-ray absorption, an accurate description of the X-ray distribution in depth is

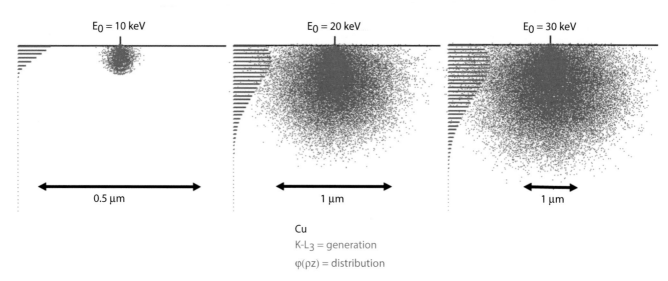

Cu
K-L$_3$ = generation
$\varphi(\rho z)$ = distribution

◘ Fig. 19.11   Monte Carlo simulations (Joy Monte Carlo) of the X-ray generation volume for Cu K-L$_3$ at E_0 = 10 keV, 20 keV and 30 keV. The sites of X-ray generation (red dots) are projected on the x-z plane, and the resulting $\varphi(\rho z)$ distribution is shown

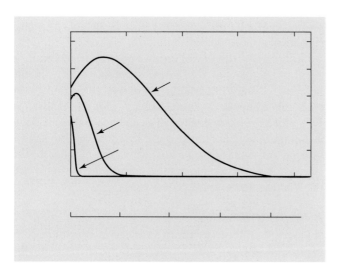

◘ Fig. 19.12   Calculated $\varphi(\rho z)$ curves for Cu K-L$_3$ in Cu at E_0 = 10 keV, 20 keV, and 30 keV; calculated using PROZA

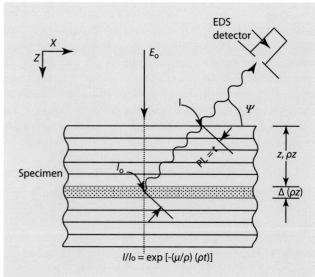

$$I/I_0 = \exp\left[-(\mu/\rho)(\rho t)\right]$$

◘ Fig. 19.13   Schematic diagram of absorption in the measurement or calculation of the $\varphi(\rho z)$ curve for emitted X-rays. PL = path length, ψ = X-ray take-off angle (detector elevation angle above surface)

needed. Fortunately, the complex three-dimensional distribution can be reduced to a one-dimensional problem for the calculation of absorption, since the path out of the specimen towards the X-ray detector only depends on depth. The $\varphi(\rho z)$ curves discussed previously give the generated X-ray distribution of X-rays in depth (See ◘ Figs. 19.8, 19.9, and 19.10). ◘ Figure 19.12 shows calculated $\varphi(\rho z)$ curves for Cu K-L$_3$ X-rays in pure Cu for initial beam energies of 10, 15, and 30 keV. The curves extend deeper (in mass depth or depth) in the sample with increasing E_0. The φ_0 values also increase with increasing initial electron beam energies since the energy of the backscattered electrons increases with higher values of E_0.

The X-rays which escape from any depth can be found by placing the appropriate path length in the X-ray absorption equation for the ratio of the measured X-ray intensity, I, to the generated X-ray intensity at some position in the sample, I_0:

$$I/I_0 = \exp\left[-(\mu/\rho)(\rho t)\right] \qquad (19.18)$$

The terms in the absorption equation are (μ/ρ), the mass absorption coefficient; ρ, the specimen density; and t, the path length (PL) that the X-ray traverses within the specimen before it reaches the surface, $z = \rho z = 0$. For the purpose of our interests, I represents the X-ray intensity which leaves the surface of the sample and I_0 represents the X-ray intensity generated at some position within the X-ray generation volume. Since the X-ray spectrometer is usually placed at an acute angle from the specimen surface, the so-called take-off angle, ψ, the path length from a given depth z is given by PL = z csc ψ, as shown in ◘ Fig. 19.13. When this correction for absorption is applied to each of the many layers $\Delta(\rho z)$ in

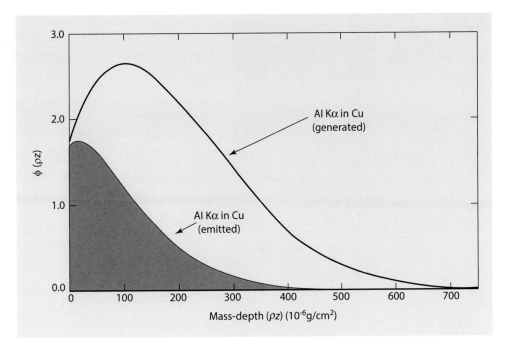

Fig. 19.14 Calculated generated and emitted $\varphi(\rho z)$ curves for Al K-L$_3$ in a Cu matrix at $E_0 = 20$ keV

the $\varphi(\rho z)$ curve, a new curve results, which gives the depth distribution of emitted X-rays. An example of the generated and emitted depth distribution curves for Al K-L$_3$ at an initial electron beam energy of 15 keV (calculated using the PROZA program (Bastin and Heijligers 1990, 1991)) is shown in ◻ Fig. 19.14 for a trace amount (0.1 wt%) of Al in a pure copper matrix. The area under the $\varphi(\rho z)$ curve represents the X-ray intensity. The difference in the integrated area between the generated and emitted $\varphi(\rho z)$ curves represents the total X-ray loss due to absorption. The absorption correction factor in quantitative matrix corrections is calculated on the basis of the $\varphi(\rho z)$ distribution. ◻ Figure 19.5, for example, illustrates the large amount of Ni K-L$_3$ absorbed in the Fe-Ni alloy series as a function of composition.

X-ray absorption is usually the largest correction factor that must be considered in the measurement of elemental composition by electron-excited X-ray microanalysis. For a given X-ray path length, the mass absorption coefficient, (μ/ρ), for each measured characteristic X-ray peak controls the amount of absorption. The value of (μ/ρ) varies greatly from one X-ray to another and is dependent on the matrix elements of the specimen (see ▶ Chapter 4, "X-rays"). For example, the mass absorption coefficient for Fe K-L$_3$ radiation in Ni is 90.0 cm^2/g, while the mass absorption coefficient for Al K-L$_3$ radiation in Ni is 4837 cm^2/g. Using Eq. (19.18) and a nominal path length of 1 μm in a Ni sample containing small amounts of Fe and Al, the ratio of X-rays emitted at the sample surface to the X-rays generated in the sample, I/I_0, is 0.923 for Fe K-L$_3$ radiation but only 0.0135 for Al K-L$_3$ radiation. In this example, Al K-L$_3$ radiation is very heavily absorbed with respect to Fe K-L$_3$ radiation in the Ni sample. Such a large amount of absorption must be taken account of in any quantitative X-ray analysis scheme. Even more serious effects of absorption occur when considering

the measurement of the light elements, for example, Be, B, C, N, O, and so on. For example, the mass absorption coefficient for C K-L radiation in Ni is 17,270 cm^2/g, so large that in most practical analyses, no C K-L radiation can be measured if the absorption path length is 1 μm. Significant amounts of C K-L radiation can only be measured in a Ni sample within 0.1 μm of the surface. In such an analysis situation, the initial electron beam energy should be held below 10 keV so that the C K-L X-ray source is produced close to the sample surface.

As shown in ◻ Fig. 19.12, X-rays are generated up to several micrometers into the specimen. Therefore the X-ray path length (PL = t) and the relative amount of X-rays available to the X-ray detection system after absorption (I/I_0) vary with the depth at which each X-ray is generated in the specimen. In addition to the position, ρz or z, at which a given X-ray is generated within the specimen, the relation of that depth to the X-ray detector is also important since a combination of both factors determine the X-ray path length for absorption. ◻ Figure 19.15 shows the geometrical relationship between the position at which an X-ray is generated and the position of the collimator which allows X-rays into the EDS detector. If the specimen is normal to the electron beam (◻ Fig. 19.15), the angle between the specimen surface and the direction of the X-rays into the detector is the take-off angle ψ. The path length, $t = $ PL, over which X-rays can be absorbed in the sample is calculated by multiplying the depth in the specimen, z, where the X-ray is generated, by the cosecant (the reciprocal of the sine), of the take-off angle, ψ. A larger take-off angle will yield a shorter path length in the specimen and will minimize absorption. The path length can be further minimized by decreasing the depth of X-ray generation, R_x, that is by using the minimum electron beam energy, E_0, consistent with the excitation of

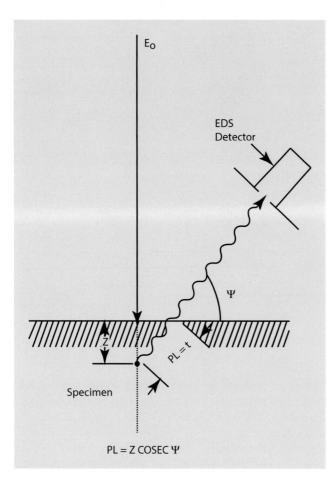

■ **Fig. 19.15** Schematic diagram showing the X-ray absorption path length in a thick, flat-polished sample: PL = absorption path length; ψ = X-ray take-off angle (detector elevation angle above surface)

PL = Z COSEC Ψ

■ **Table 19.4**   Path Length, PL, for Al K-L₃ X-rays in Al

E_0	Take-Off Angle, ψ	R_x (µm)	Path Length, PL, (µm)
10	15	0.3	1.16
10	60	0.3	0.35
30	15	2.0	7.7
30	60	2.0	2.3

the X-ray lines used for analysis. ■ Table 19.4 shows the variation of the path length that can occur if one varies the initial electron beam energy for Al K-L₃ X-rays in Al from 10 to 30 keV and the take-off angle, ψ = 15° and ψ = 60°.

The variation in PL is larger than a factor of 20, from 0.35 µm at the lowest keV and highest take-off angle to 7.7 µm at the highest keV and lowest take-off angle. Clearly the analyst's choices of the initial electron beam energy and the X-ray take-off angle have a major effect on the path length and therefore the amount of absorption that occurs.

In summary, using appropriate formulations for X-ray generation with depth or $\varphi(\rho z)$ curves, the effect of absorption can be obtained by considering absorption of X-rays from element i as they leave the sample. The absorption correction, A_i, can be calculated by taking the ratio of the effect of absorption for the standard, $A_{i,std}$, to X-ray absorption for the unknown, $A_{i,unk}$, for each element, i, in the sample. The effect of absorption can be minimized by decreasing the path length of the X-rays in the specimen through careful choice of the initial beam energy and by selecting, when possible, a high take-off angle.

X-ray Fluorescence, F

Photoelectric absorption results in the ionization of inner atomic shells, and those ionizations can also cause the emission of characteristic X-rays. For fluorescence to occur, an atom species must be present in the target which has a critical excitation energy less than the energy of the characteristic X-rays being absorbed. In such a case, the measured X-ray intensity from this second element will include both the direct electron-excited intensity as well as the additional intensity generated by the fluorescence effect. Generally, the fluorescence effect can be ignored unless the photon energy is less than 5 keV greater than the critical excitation energy, E_c.

The significance of the fluorescence correction, F_i, can be illustrated by considering the binary system Fe-Ni. In this system, the Ni K-L₃ characteristic energy at 7.478 keV is greater than the energy for excitation of Fe K radiation, E_c = 7.11 keV. Therefore, an additional amount of Fe K-L₃ radiation is produced beyond that due to the direct beam on Fe. ■ Figure 19.5 shows the effect of fluorescence in the Fe-Ni system at an initial electron beam energy of 30 keV and a take-off angle, ψ, of 52.5°. Under these conditions, the atomic number effect, Z_{Fe}, and the absorption effect, A_{Fe}, for Fe K-L₃ are very close to 1.0. The measured k_{Fe} ratio lies well above the first approximation straight line relationship. The additional intensity is given by the effect of fluorescence. As an example, for a 10 wt% Fe – 90 wt% Ni alloy, the amount of iron fluorescence is about 25%.

The quantitative calculation of the fluorescence effect requires a knowledge of the depth distribution over which the characteristic X-rays are absorbed. The $\varphi(\rho z)$ curve of electron-generated X-rays is the starting point for the fluorescence calculation, and a new $\varphi(\rho z)$ curve for X-ray-generated X-rays is determined. The electron-generated X-rays are emitted isotropically. From the X-ray intensity generated in each of the layers $\Delta(\rho z)$ of the $\varphi(\rho z)$ distribution, the calculation next considers the propagation of that radiation over a spherical volume centered on the depth ρz of that layer, calculating the absorption based on the radial distance from the starting layer and determining the contributions of absorption to each layer (ρz) in the X-ray-induced $\varphi(\rho z)$ distribution. Because of the longer range of X-rays than electrons in materials, the X-ray-induced $\varphi(\rho z)$ distribution covers a much greater depth, generally an order of magnitude or more than the electron-induced

$\varphi(\rho z)$ distribution. Once the X-ray-induced $\varphi(\rho z)$ generated distribution is determined, the absorption of the outgoing X-ray-induced fluorescence X-rays must be calculated with the absorption path length calculated as above.

The fluorescence factor, F_i, is usually the least important factor in the calculation of composition by evaluating the [ZAF] term in Eq. (19.17). In most analytical cases secondary fluorescence may not occur or the concentration of the element which causes fluorescence may be small. Of the three effects, Z, A, and F, which control X-ray microanalysis calculations, the fluorescence effect, Fi, can be calculated (Reed 1965), with sufficient accuracy so that it rarely limits the development of an accurate analysis.

References

Bastin GF, Heijligers HJM (1990) Quantitative electron probe microanalysis of ultralight elements (boron – oxygen). Scanning 12:225–236

Bastin GF, Heijligers HJM (1991) Quantitative analysis of homogeneous or stratified microvolumes applying the model "PAP". In: Heinrich KFJ, Newbury DE (eds) Electron probe quantitation. Plenum Press, New York City, p 163

Castaing R (1951) "Application of electron probes to local chemical and crystallographic analysis" Ph.D. thesis, University of Paris. (English translation available from the Microanalysis Society at: ► http://www.microanalysissociety.org/

Heinrich KFJ, Myklebust RL, Rasberry SD, Michaelis RE (1971) Preparation and Evaluation of SRM's 481 and 482 Gold-Silver and Gold-Copper Alloys for Microanalysis. U.S. Government Printing Office, Washington, DC. NBS Spec. Publ., pp 260–28

ISO GUM (2008). Guide to the Expression of Uncertainty in Measurement; Guide 98–3:2008: Geneva, Switzerland, Joint Group for Guides in Metrology, Working Group 1. ISO Central. Secretariat, Vernier, Geneva, Switzerland: International Organization for Standards

Marinenko R, Leigh S (2010) Uncertainties in electron probe microanalysis. Proc Eur Microbeam Anal Soc 7(1):012017

Marinenko R, Blackburn D, Bodkin J (1990) Glasses for Microanalysis: SRMs 1871-1875" National Institute of Standards and Technology (U.S.) Special Publication 260–112 (U.S. Gov. Printing Office, Washington) available at: ► http://www.nist.gov/srm/upload/SP260-112.PDF

Newbury DE, Swyt CR, Myklebust RL (1995) 'Standardless' quantitative electron probe microanalysis with energy-dispersive X-ray spectrometry: is it worth the risk? Anal Chem 67:1866

Pouchou J-L, Pichoir F (1991) Quantitative analysis of homogeneous or stratified microvolumes applying the model "PAP". In: Heinrich KFJ, Newbury DE (eds) Electron probe quantitation. Plenum, New York, p 31

Reed SJB (1965) Characteristic fluorescence corrections in electron-probe microanalysis. Brit J Appl Phys 16:913

Saunders S, Karduck P, Sloof W (2004) Certified reference materials for micro-analysis of carbon and nitrogen. Microchim Acta 145:209

Yakowitz H (1975) Methods of quantitative analysis. In: Goldstein JI, Yakowitz H, Newbury DE, Lifshin E, Colby JW, Coleman JR (eds) Practical scanning electron microscopy. Plenum, New York, p 338

Quantitative Analysis: The SEM/EDS Elemental Microanalysis k-ratio Procedure for Bulk Specimens, Step-by-Step

© Springer Science+Business Media LLC 2018
J. Goldstein et al., *Scanning Electron Microscopy and X-Ray Microanalysis*,
https://doi.org/10.1007/978-1-4939-6676-9_20

This chapter discusses the procedure used to perform a rigorous quantitative elemental microanalysis by SEM/EDS following the k-ratio/matrix correction protocol using the NIST DTSA-II software engine for bulk specimens. Bulk specimens have dimensions that are sufficiently large to contain the full range of the direct electron-excited X-ray production (typically 0.5–10 μm) as well as the range of secondary X-ray fluorescence induced by the propagation of the characteristic and continuum X-rays (typically 10–100 μm).

20.1 Requirements Imposed on the Specimen and Standards

The k-ratio/matrix correction protocol for the analysis of bulk specimens has two basic underlying assumptions:

1. The composition is homogeneous throughout the entire volume of the specimen in which primary characteristic X-rays are directly excited by the incident electron beam and in which secondary X-ray fluorescence is induced during the propagation of the primary characteristic and continuum X-rays. A compositionally heterogeneous specimen which does not satisfy this requirement cannot be analyzed by the conventional k-ratio/matrix correction protocol. Examples of such heterogeneous specimens include a horizontally layered specimen such as a thin film on a substrate or an inclusion with dimensions similar to the interaction volume embedded in a matrix. Such specimens must be analyzed with protocols that account for the effects of the particular specimen geometry.

2. The X-ray intensities measured on the location of interest on the specimen and on the standard(s) differ only because the compositions are different. No other factors modify the measured intensities. In particular, geometric effects that arise from physical surface defects, such as scratches, pits, and so on, can modify the interaction of the electron beam (electron backscattering, beam penetration) with the specimen and can alter the subsequent X-ray absorption path length to the detector compared to an ideal flat bulk specimen. This requirement places strict conditions on the surface condition of the specimen and standards. A highly polished, flat surface must be created following the appropriate metallographic preparation protocol for each particular material. The surface should be finished to a surface roughness below 100 nm root mean square (rms) with a typical final polish performed with 100-nm diamond, alumina, ceria or other polishing compound as appropriate. When the analysis involves measuring low energy photons below 1 keV (e.g., for the elements Be, B, C, N, O, and F), the surface finish should be better than 50 nm rms. The preparation protocol should utilize physical

grinding and polishing. "Chemical polishing" should be avoided since chemical reactions may induce shallow, near-surface compositional changes that affect the very shallow region that is excited and sampled by the electron beam. Ion beam milling can be used to shape and finish the specimen, but it must be recognized that implantation of the primary ion and differential material removal caused by differences in the sputtering rates of the elements can modify the composition of a shallow surface layer.

20.2 Instrumentation Requirements

The basis of the k-ratio/matrix corrections protocol is measurement of the X-ray spectra of the specimen and standard(s) under identical conditions of beam energy, known electron dose (the product of beam current and EDS live-time, with accurate dead-time correction), EDS parameters (detector solid angle, time constant, calibration, and window efficiency), target orientation (tilt angle, ideally 0° tilt, i.e., beam perpendicular to the target surface), and EDS take-off angle (i.e., the detector elevation angle above the flat sample surface).

20.2.1 Choosing the EDS Parameters

Consistency in the choice of the EDS parameters is critical for establishing a robust analytical measurement environment, and this is especially important when archived standard spectra are used.

EDS Spectrum Channel Energy Width and Spectrum Energy Span

As shown in ☐ Fig. 20.1, when the energy axis is expanded sufficiently, the EDS spectrum is seen to be a histogram of energy channels of a specific width (e.g., 5 eV, 10 eV, 20 eV) and number (e.g., 1024, 2048, 4096). For accurate peak-fitting purposes, it is desirable to have an adequate number of channels spanning the characteristic X-ray peaks. Because the EDS resolution is a function of photon energy, low photon energy peaks below 1 keV are substantially narrower than higher energy peaks. A choice of 5 eV for the channel energy width will provide a sufficient number of channels to adequately span all of the peaks of analytical interest, including the peaks that occur below 1 keV. C K-L$_{2,3}$ is broadened in EDS to approximately 50 eV full width at half-maximum (FWHM), so a choice of 5-eV/channel will provide at least 10 channels to span the low photon energy peaks, which is important for accurate peak fitting. It is also desirable for the measured spectrum to span the full range of the excited X-ray energy, from an effective threshold of approximately 100 eV to the Duane–Hunt limit, which corresponds to the incident beam

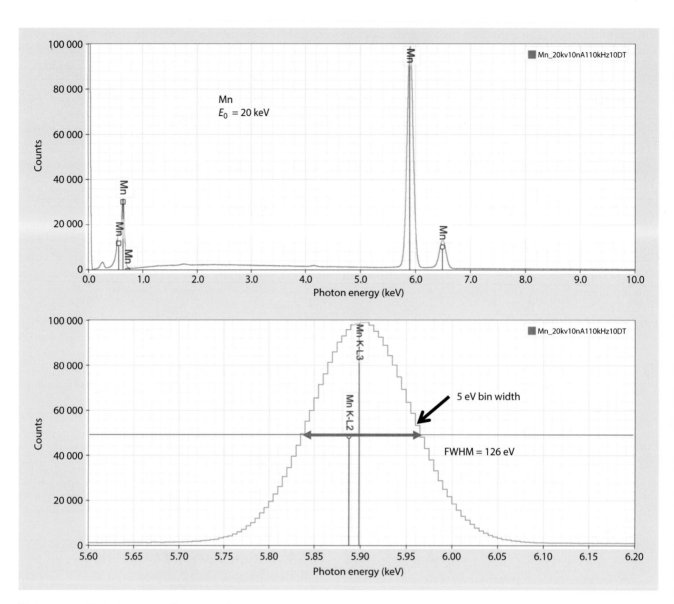

◻ **Fig. 20.1** SDD-EDS spectrum of Mn ($E_0 = 20$ keV)

energy, E_0. The energy span is given by the channel width multiplied by the number of channels. With a 5-eV channel width, a choice of 4096 channels will provide access to photon energies as high as 20.48 keV. For beam energy above 20 keV, the number of channels should be increased to retain the 5-eV channel width, or alternatively the channel width can be increased to 10 eV, but with the consequence that fewer channels will describe each characteristic peak.

EDS Time Constant (Resolution and Throughput)

The EDS is only capable of processing one photon at a time. The basic measurement cycle is the photoelectric absorption of the photon in the detector active volume, measurement of the charge deposited by scattering of the

photoelectron to determine the photon energy, and incrementing the appropriate energy bin in the EDS histogram by one count. The EDS time constant (also known as the shaping time, the processing time, the maximum throughput, or other terms in different vendor EDS systems) effectively determines the amount of time spent on the measurement cycle. A short time constant enables more photons to be processed per unit of real (clock) time, but the trade-off of faster processing is poorer accuracy in assigning the photon energy. While the characteristic X-ray peak has a sharply defined energy, with a natural peak width of a few eV or less, the EDS measurement process inevitably substantially broadens the measured peak. For example, Mn K-$L_{2,3}$ has a natural width of approximately 7 eV (determined as the FWHM) but as displayed in the EDS histogram, the Mn K-$L_{2,3}$ peak is broadened to 122–150 eV

FWHM or more, depending on the choice of time constant. For the particular silicon drift detector (SDD)-EDS and time constant shown in ◘ Fig. 20.1, the broadened EDS peak has a FWHM = 126 eV. The peak width increases (i.e., resolution becomes poorer) as the time constant decreases. The shortest time constant, which gives the highest throughput but the broadest peaks (poorest resolution), is typically chosen for analysis situations where it is important to maximize the total number of X-ray counts per unit of clock (real) time, such as elemental X-ray mapping. For quantitative analysis, better peak resolution is desirable, and thus a longer time constant should be chosen. Whichever time constant strategy is selected, it is important for standards-based quantitative analysis that this same time constant be used for all measurements of unknowns and standards, especially if archived standards are used.

EDS Calibration

Assigning the proper energy bin for a photon measurement depends on the EDS being calibrated. The vendor for a particular EDS system will have a recommended calibration procedure that should be followed on a regular basis as part of establishing a quality measurement environment, with full documentation of the measurements to establish the on-going calibration record. A typical calibration strategy is to choose a material such as Cu that provides (with $E_0 \geq 15$ keV) strongly excited peaks in the low photon energy range (Cu $L_3M_5 = 0.93$ keV) and the high photon energy range (Cu K-$L_{2,3} = 8.04$ keV). Alternatively, some EDS systems that provide a "zero energy reference" signal will use this value with a single high photon energy peak such as Cu K-$L_{2,3}$ or Mn K-$L_{2,3}$ to perform calibration. A good quality assurance practice is to begin each measurement campaign by measuring a spectrum of Cu (or another element, e.g., Mn, Ni, etc., or a compound, e.g., CuS, FeS$_2$, etc.) under the user-defined conditions. This Cu spectrum can be compared to the Cu spectrum that is stored in the archive of standards to confirm that the current measurement conditions are identical to those used to create the archive. This starting Cu spectrum should always be saved as part of the quality assurance plan.

EDS Solid Angle

The solid angle of collection, Ω, is given by

$$\Omega = A / r^2 \qquad (20.1)$$

where A is the active area of the detector and r is the distance from the X-ray source on the specimen to the detector. Some EDS systems are mounted on a retractable arm that enables the analyst to choose the value of r. A consistent and reproducible choice must be made for r since this value has such a strong impact on Ω and thus on the number of photons detected per unit of dose.

20.2.2 Choosing the Beam Energy, E_0

The choice of beam energy depends on the particular aspects of the analysis that the analyst wishes to optimize. As a starting point, a useful general analysis strategy is to optimize the excitation of photon energies up to 12 keV by choosing an incident beam energy of 20 keV, which provides sufficient overvoltage ($E_0/E_c > 1.5$) for K-shell (to Br) and L-shell (elements to Bi) for reasonable excitation. The characteristic peaks of X-ray families that occur in the photon energy range from 4 keV to 12 keV are generally sufficiently separated in energy to be resolved by EDS. When it is important to measure those elements whose characteristic peaks occur below 4 keV, and especially for the low atomic number elements Be, B, C, N, O and F, for which the characteristic peaks occur below 1 keV and suffer high absorption, then analysis with lower beam energy, 10 keV or lower, will be necessary to optimize the results.

20.2.3 Measuring the Beam Current

The SEM should be equipped for beam current measurement, ideally with an in-column Faraday cup which can be selected periodically during the analysis procedure to determine the beam current. As an alternative, a picoammeter can be installed between the electrically isolated specimen stage and the electrical ground to measure the absorbed (specimen) current that must flow to ground to avoid specimen charging. The specimen current is the difference between the beam current and the loss of charge due to BSE and SE emission, both of which vary with composition. To measure the true beam current, BSE and SE emission must be recaptured, which is accomplished by placing the beam within a Faraday cup, which is constructed as a blind hole in a conducting material (e.g., metal or carbon) covered with a small entrance aperture (e.g., an electron microscope aperture of 50 μm diameter or less). This Faraday cup is then placed at a suitable location on the electrically isolated specimen stage. By locating the beam in the center of the Faraday cup aperture opening, the primary beam electrons as well as all BSEs and SEs generated at the inner surfaces are collected with very little loss through the small aperture, so that the current flowing to the electrical ground is the total incident beam current.

20.2.4 **Choosing the Beam Current**

After the analyst has chosen the EDS time constant, the detector solid angle (for a retractable detector), and the beam energy, the beam current should be chosen so as to give a reasonable detector throughput, as expressed by the system dead-time. ◘ Figure 20.2 shows the relationship between the input count rate (ICR) of X-rays that arrive at the detector and the output count rate (OCR) of photons that are actually stored in the measured spectrum. The OCR initially rises linearly with the ICR, but as photons arrive at a progressively greater rate at the detector, photon coincidence begins to occur and the anti-coincidence function begins to reject these coincidence events, reducing the OCR. Eventually a maximum OCR value is reached beyond which the OCR decreases with increasing ICR, eventually falling to zero ("paralyzable dead-time"). A useful measure of the activity state of the EDS detector is the system "dead-time" which is defined as

$$\text{Dead-time}(\%) = \left[(ICR - OCR)/ICR\right] * 100 \quad (20.2)$$

A classic strategy with the low throughput Si(Li)-EDS is to select a beam current on a highly excited pure element such as Al or Si that produces a dead-time of 30 % or less. With SDD-EDS, a more conservative counting strategy is suggested, such that the beam current is chosen so that the dead-time on the most highly excited standard of interest, for example, Al or Si, is less than 10 %. Despite the operation of the anti-coincidence function, SDD-EDS systems typically show evidence of coincidence peaks above a dead-time of 10 % from highly excited parent peaks, as illustrated in ◘ Fig. 20.3, which shows the in-growth of an extensive set of coincidence peaks from several parent peaks. If it is important to measure low intensity X-ray peaks that correspond to minor or trace constituents that occur in spectral regions affected by coincidence peaks, then choosing the low dead-time to minimize coincidence will be an important issue in selecting the general analytical conditions. If there is no interest in measuring X-ray peaks of possible constituents that occur in the region of coincidence peaks, then these regions can be ignored and a counting strategy that involves higher dead-time operation can be used.

Once the analytical conditions (EDS time constant, solid angle, beam energy, and beam current appropriate to the complete suite of standards) have been chosen, these conditions should be used for all standards and unknowns to achieve the basic measurement consistency required for the k-ratio/matrix corrections protocol.

◘ **Fig. 20.2**   Output count rate (OCR) vs. input count rate for an SDD-EDS array of four 10-mm² detectors

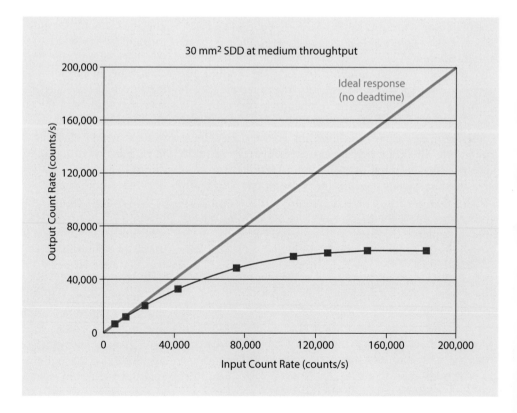

20

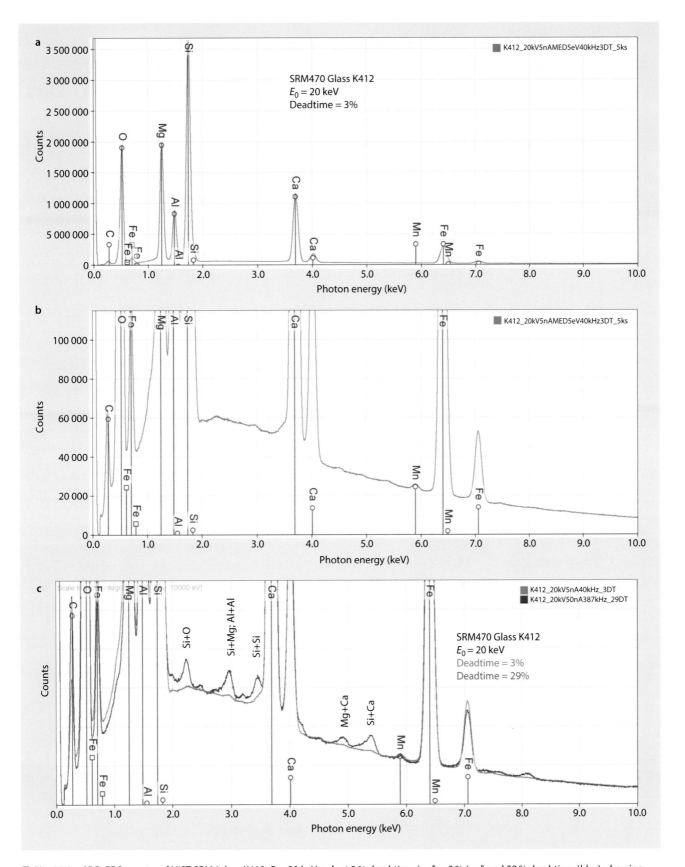

☐ **Fig. 20.3** SDD-EDS spectra of NIST SRM (glass K412, E_0 = 20 keV: **a, b** at 3 % dead-time (*red*); **c** 3 % (*red*) and 29 % dead-time (*blue*), showing in-growth of coincidence peaks

20.3 Examples of the k-ratio/Matrix Correction Protocol with DTSA II (Newbury and Ritchie 2015b)

20.3.1 Analysis of Major Constituents (C > 0.1 Mass Fraction) with Well-Resolved Peaks

The EDS spectra of the minerals pyrite (FeS_2) and troilite (FeS) measured at $E_0 = 20$ keV with a dead-time of ~ 10 % are shown in ◻ Fig. 20.4 and feature well separated peaks for the Fe K- and L- families and the S K-family. These spectra were analyzed with Fe and CuS serving both as peak-fitting references and as standards. CuS is chosen for the S reference and standard rather than elemental S since CuS is stable under electron bombardment while elemental S is not stable. The spectrum for FeS and the residual spectrum after peak-fitting are also shown in ◻ Fig. 20.4. The results for seven replicate analyses are listed in ◻ Table 20.1 (FeS) and ◻ Table 20.2 (FeS_2) along with the ZAF correction factors and the components of

the error budget. In this analysis and the analyses reported below, the relative deviation from the expected value (RDEV) (also referred to as "relative error") is calculated with the "expected" value taken as the stoichiometric formula value or the value obtained from an "absolute" analytical method, just as in gravimetric analysis:

$$RDEV = \left[\frac{(Analyzed\ value - expected\ value)}{expected\ value} \right] \times 100\% \quad (20.3)$$

Optimizing Analysis Strategy

The DTSA II analysis report includes for each analyzed element the ZAF factors and the estimated uncertainties in these factors as well as uncertainties due to the counting statistics associated with the measurements of the unknown and of the standard (Ritchie and Newbury 2012). Careful examination of these factors can be used to refine the analytical strategy to optimize the measurement. Reducing the uncertainty due to the counting statistics requires increasing the dose. The absorption factor A is strongly influenced by the

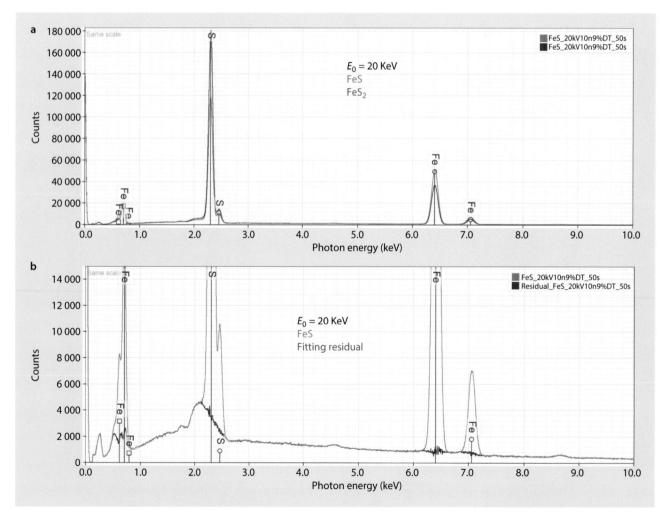

◻ **Fig. 20.4**   **a** SDD-EDS spectra of Pyrite (FeS_2) (*blue*) and meteoritic Troilite (FeS (*red*)) at $E_0 = 20$ keV. **b** NIST DTSA-II analysis of FeS using Fe and CuS as peak-fitting references and as standards. The original spectrum (*red*) and the residual spectrum after peak-fitting (*blue*) are shown

Table 20.1 Analysis of FeS (meteoritic troilite) at $E_0 = 20$ keV with CuS and Fe as fitting references and standards Integrated spectrum count, 0.1–20 keV = 7,048,000; uncertainties expressed in mass fraction. Analysis performed with Fe K-L$_{2,3}$ and S K-L$_{2,3}$

	S	Fe
C_{av} (atom frac)	0.5052	0.4948
Z-correction	0.977	0.95
A-correction	1.118	0.983
F-correction	1.003	1
σ (7 replicates)	0.00075	0.00075
σ$_{Rel}$ (%)	0.15%	0.15%
RDEV (%)	1.00%	−1.00%
C (mass frac, single analysis)	0.3699	0.6305
Counting error, std	0.00020	0.0003
Counting error, unk	0.00020	0.0007
A-factor error	0.0017	0.0002
Z-factor error	2.20×10^5	4.10×10^{-6}
Combined errors	0.0017	0.0008

Table 20.3 Analysis of FeS at $E_0 = 10$ keV with CuS and Fe as fitting references and standards Integrated spectrum count, 0.1–10 keV = 5,630,000; uncertainties expressed in mass fraction. Analysis performed with Fe K-L$_{2,3}$ and S K-L$_{2,3}$

	S	Fe
C_{av} (atom frac)	0.503	0.497
Z-correction	0.973	0.937
A-correction	1.041	0.997
F-correction	1.001	1
σ (7 replicates)	0.00056	0.00056
σ$_{Rel}$ (%)	0.11%	0.11%
RDEV (%)	0.59%	−0.59%
C (mass frac, single analysis)	0.3627	0.6257
Counting error, std	0.0002	0.0008
Counting error, unk	0.0003	0.0018
A-factor error	0.0006	4.10E-05
Z-factor error	2.20×10^{-5}	1.30×10^{-6}
Combined errors	0.0007	0.0019

Table 20.2 Analysis of FeS$_2$ (pyrite) at $E_0 = 20$ keV with CuS and Fe as fitting references and standards Integrated spectrum count, 0.1.–20 keV = 7,765,000; uncertainties expressed in mass fraction. Analysis performed with Fe K-L$_{2,3}$ and S K-L$_{2,3}$

	S	Fe
C_{av} (atom frac)	0.6726	0.3274
Z-correction	0.957	0.928
A-correction	1.181	0.975
F-correction	1.003	1
σ (7 replicates)	0.000314	0.000314
σ$_{Rel}$ (%)	0.05%	0.10%
RDEV (%)	0.88%	−1.80%
C (mass frac, single analysis)	0.5485	0.4657
Counting error, std	0.0003	0.0002
Counting error, unk	0.0003	0.0006
A-factor error	0.0023	0.0002
Z-factor error	3.30×10^{-5}	2.90×10^{-6}
Combined errors	0.0023	0.0007

Table 20.4 Analysis of FeS$_2$ at $E_0 = 10$ keV with CuS and Fe as fitting references and standards Integrated spectrum count, 0.1–10 keV = 6,253,000); uncertainties expressed in mass fraction. Analysis performed with Fe K-L$_{2,3}$ and S K-L$_{2,3}$

	S	Fe
C_{av} (atom frac)	0.671	0.329
Z-correction	0.95	0.91
A-correction	1.061	0.995
F-correction	1.001	1
σ (7 replicates)	0.0007	0.0007
σ$_{Rel}$ (%)	0.11%	0.21%
RDEV (%)	0.65%	−1.30%
C (mass frac, single analysis)	0.537	0.4618
Counting error, std	0.0003	0.0006
Counting error, unk	0.0003	0.0016
A-factor error	0.0008	4.30E-05
Z-factor error	3.20×10^{-5}	9.60×10^{-7}
Combined errors	0.0009	0.0017

choice of beam energy. If the beam energy can be decreased, considering also the constraints imposed by having sufficient overvoltage for all elements to be analyzed, the absorption correction factor and its uncertainty can also be reduced. For the Fe-S examples, lowering the beam energy from 20 to 10 keV gives the results shown in ◻ Tables 20.3 and 20.4. The absorption factor A from is reduced from 1.118 to 1.04 for S in FeS and from 1.18 to 1.06 for S in FeS$_2$, and the relative errors are also reduced slightly, from 1 to 0.59% for S in FeS and from 0.88 to 0.65% for S in FeS$_2$.

20.3.2 Analysis of Major Constituents (C > 0.1 Mass Fraction) with Severely Overlapping Peaks

PbS

The throughput and the peak stability (calibration and resolution) of SDD-EDS spectrometry enable collection of high count, high quality spectra (>5 million counts) within modest measurement time, 100 s or less. High count spectra enable measurements of minor and trace constituents with high precision. High counts and stable peak structures are critical for successful peak intensity measurements by peak-fitting methods, which is especially important for situations where two or more peaks are so close in photon energy that the EDS resolution function convolves the peaks into mutual interference. Despite extreme peak interference, quantitative X-ray microanalysis can be achieved with RDEV values of 5 % relative or less (Newbury and Ritchie 2015a).

PbS (galena) represents a challenging analysis situation for EDS because of the severe interference between the S K-L$_2$ (2.307 keV) and Pb M$_5$-N$_{6,7}$ (2.343 keV), which are

separated by 36 eV, as shown in ◘ Fig. 20.5. Analysis of PbS with DTSA II using CuS and PbSe as peak-fitting references and as standards yields the results in ◘ Table 20.5. Despite the severe peak interference, the relative error based on the formula stoichiometry is only ±1.2 % for S and Pb.

Note that an alternative analytical approach would be to select the beam energy such that $E_0 \geq 20$ keV so that the Pb L-family is excited (L$_{III}$ = 13.04 keV). With this choice of excitation, the Pb L$_3$-M$_{4,5}$ peak at 10.55 keV, which does not suffer interference, could be chosen to measure Pb. Of course, the S K still must be deconvoluted from the interference from the Pb M-family since there is no alternate peak to measure for S.

MoS$_2$

MoS$_2$ represents an even greater analytical challenge because the peaks that must be used for analysis, S K-L$_2$ (2.307 keV) and Mo L$_3$-M$_{4,5}$ (2.293 keV), are separated by only 14 eV, as shown in ◘ Fig. 20.6. Analysis of MoS$_2$ with DTSA II using CuS and Mo as peak-fitting references and as standards yields the results in ◘ Table 20.6. Despite the severe peak interference, the relative error based on the formula stoichiometry is only −0.34 % for S and 0.7 % for Mo.

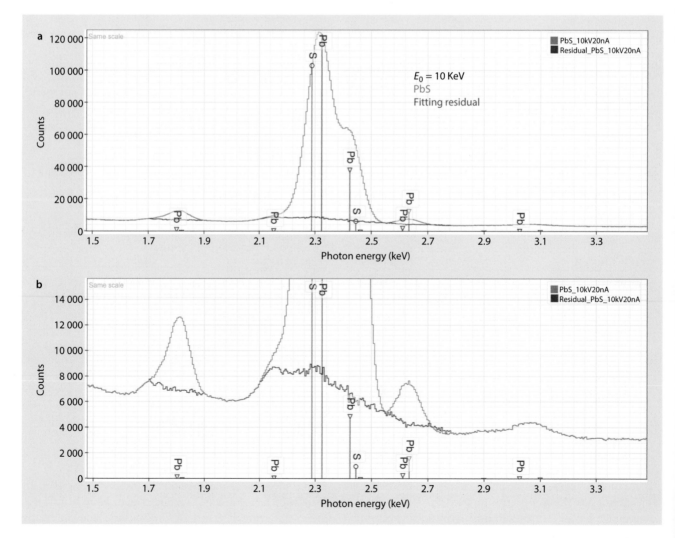

◘ **Fig. 20.5   a** SDD-EDS spectrum of PbS (*red*) and residual (*blue*) after DTSA II analysis using CuS and PbSe as fitting references and standards. **b** Expanded view

Table 20.5 Analysis of PbS at $E_0 = 10$ keV with CuS and PbSe as fitting references and standards; Integrated spectrum count, 0.1–10 keV = 5,482,000; uncertainties expressed in mass fraction. Analysis performed with Pb M_5-$N_{6,7}$ and S K-$L_{2,3}$

	S	Pb
C_{av} (atom frac)	0.4938	0.5062
Z-correction	1.31	0.983
A-correction	1.028	1.056
F-correction	1	1
σ (7 replicates)	0.000953	0.000953
$σ_{Rel}$ (%)	0.19 %	0.19 %
RDEV (%)	−1.20 %	1.2
C (mass frac, single analysis)	0.1306	0.8651
Counting error, std	0.0001	0.0009
Counting error, unk	0.0003	0.001
A-factor error	0.0002	0.0017
Z-factor error	$1.50×10^{-5}$	0.0001
Combined errors	0.0004	0.0022

20.3.3 Analysis of a Minor Constituent with Peak Overlap From a Major Constituent

The problem of accurately recovering peak intensities when overlaps occur is exacerbated when the concentration ratio of the elements producing the overlapping peaks is large, for example, a major constituent ($C > 0.1$ mass fraction) interfering with a minor ($0.01 \leq C \leq 0.1$) constituent. The high throughput (>100 kHz output count rate) of SDD-EDS enables collection of high count EDS spectra in modest collection time (e.g., 10 million counts in 100 s). Moreover, the high throughput of SDD-EDS is achieved with stability in both the peak position (i.e., calibration) and the peak shape (i.e., resolution) across the entire input count rate range. In simultaneous WDS-EDS measurements, this SDD-EDS performance been demonstrated to the spectrum measurement capabilities necessary for robust MLLS peak-fitting to achieve accurate measurement of the interfering peak intensities equal to that of WDS on the spectroscopically resolved peaks (Ritchie et al. 2012).

20.3.4 Ba-Ti Interference in BaTiSi$_3$O$_9$

BaTiSi$_3$O$_9$ (benitoite) provides an example of severe interference between two constituents of identical atomic concentration but with a mass concentration ratio of Ba/Ti = 2.9—Ti K-$L_{2,3}$ (4.510 keV) and Ba L3-$M_{4,5}$ (4.466 keV)—which are separated by 44 eV, as shown in ■ Fig. 20.7. DTSA II analysis of benitoite with Ti and sanbornite (BaSi$_2$O$_5$) as fitting references and standards is given in ■ Table 20.7. Note that in this

analysis, O has been directly analyzed with the k-ratio/matrix corrections protocol and not by the method of assumed stoichiometry. The analytical results are seen to closely match the stoichiometry of the ideal mineral formula.

20.3.5 Ba-Ti Interference: Major/Minor Constituent Interference in K2496 Microanalysis Glass

NIST microanalysis research material K2496 glass contains these same elements, but with Ba as a major constituent ($C = 0.4299$ mass fraction) and Ti as a minor constituent ($C = 0.01799$ mass fraction), giving an elemental ratio of Ba/Ti = 23.9. ■ Figure 20.8a shows the SDD-EDS spectrum and residual after peak fitting, and ■ Table 20.8 contains the results of the analysis. Despite the severe overlap and the large elemental ratio, the concentration for Ti is measured with reasonable accuracy. A reasonable question that the analyst might ask is, If it was not known that the Ti was present, could it be detected? ■ Figure 20.8b shows the fitting residual for an analysis protocol in which Ti was not fit. The peaks for Ti K-$L_{2,3}$ and Ti K-M_3 are revealed in the residual spectrum.

20.4 The Need for an Iterative Qualitative and Quantitative Analysis Strategy

The analysis of NIST glass K2496 demonstrates that rigorous analysis requires an iterative qualitative analysis–quantitative analysis approach. When analyzing an unknown material, it is likely that some constituents at the minor and trace level will not be obvious when the first qualitative analysis is performed due to peak interference from constituents at higher concentrations. An alternating qualitative–quantitative analytical strategy is required to discover possibly hidden minor and trace constituents. In the initial qualitative analysis, the EDS spectrum is evaluated to identify the major and minor elemental constituents whose peaks are readily identifiable. The k-ratio/matrix correction protocol is then applied with appropriate choices for elemental peak-fitting references and for standards, and the "residual" spectrum is constructed that contains the intensity remaining after the fitted peaks have been subtracted. If all constituents have been accounted for, this residual spectrum should only consist of the continuum background and possibly also artifact peaks such as escape and coincidence peaks. However, because of the relative poor energy resolution of EDS, the analyst must perform a second qualitative analysis of the residual spectrum for the presence of previously unrecognized peaks that are associated with constituents that suffer interference from the higher intensity peaks. If such peaks are discovered and assigned to an element(s) not previously recognized, the quantitative analysis must then be repeated with this element(s) included in the peak-fitting and

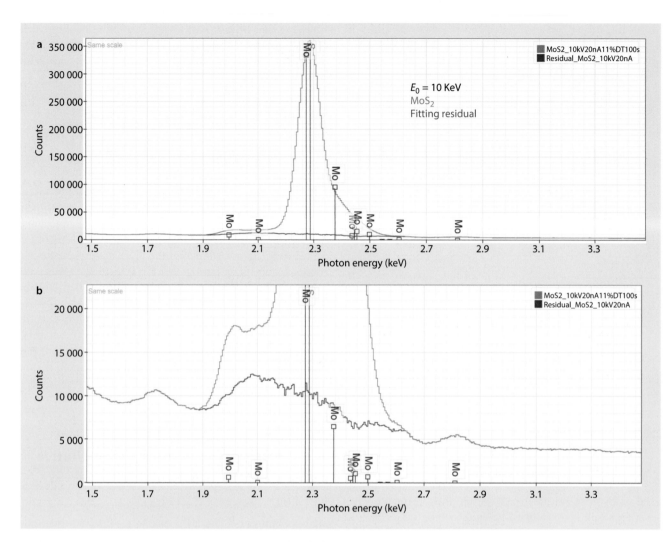

☐ **Fig. 20.6**   **a** SDD-EDS spectrum of MoS$_2$ (*red*) at E_0 = 10 keV (7,326,000 counts) and residual (*blue*) after DTSA II analysis using CuS and Mo as fitting references and standards. **b** Expanded view

☐ **Table 20.6**   Analysis of MoS$_2$ at E_0 = 10 keV with CuS and Mo as fitting references and standards; integrated spectrum count, 0.1–10 keV = 7,326,000; uncertainties expressed in mass fraction. Analysis performed with Mo L$_{2,3}$-M$_{4,5}$ and S K-L$_{2,3}$

	S	Mo
C$_{av}$ (atom frac)	0.6644	0.3356
Z-correction	1.039	0.884
A-correction	1.083	1.024
F-correction	1	1
σ (7 replicates)	0.0022	0.0022
σ$_{Rel}$ (%)	0.33%	0.66%
RDEV (%)	−0.34%	0.70%
C (mass frac, single analysis)	0.3972	0.6046
Counting error, std	0.0003	0.0003
Counting error, unk	0.0006	0.0014
A-factor error	0.0006	0.0006
Z-factor error	2.80×10^{-5}	4.40×10^{-5}
Combined errors	0.0008	0.0015

quantification suite of elements. A third iteration may be necessary to recover constituents present at the trace level near the limits of detection.

20.4.1   Analysis of a Complex Metal Alloy, IN100

IN100 is a nickel-based superalloy which produces the EDS spectrum shown in ☐ Fig. 20.9. In the first qualitative analysis, characteristic X-ray peaks were identified for Al K; the Ti K-family; the Cr, Co, and Ni K- and L- families; and Mo L-family. Analysis with the k-ratio/matrix correction protocol using pure elements as peak-fitting references and as standards gave the results shown in ☐ Table 20.9, with the analytical total slightly below unity. Close inspection of the residual spectrum in ☐ Fig. 20.9 showed an anomaly at the energy of Ti K-M$_{4,5}$ (4.931 keV)) which closely corresponds to the energy of V K-L$_{2,3}$ (4.952 keV) with a separation of 21 eV. When V was included in the suite of fitted elements, the anomaly in the residual spectrum was eliminated, as shown in ☐ Fig. 20.10, and a minor V constituent was recovered in the

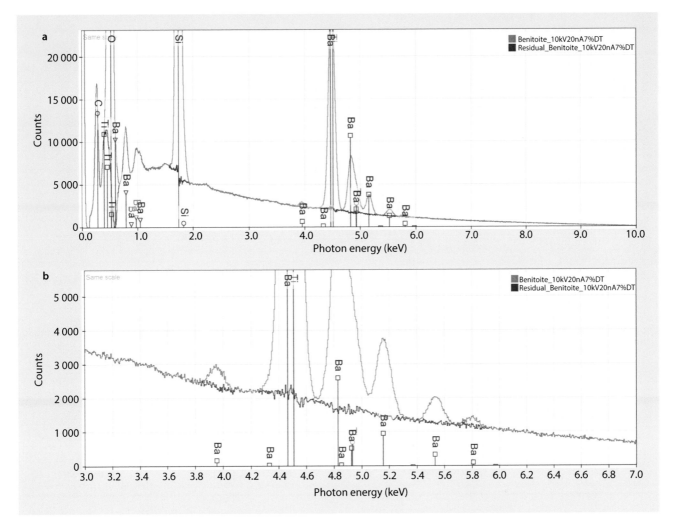

☐ **Fig. 20.7** **a** SDD-EDS spectrum of BaTiSi$_3$O$_9$ (benitoite) (*red*) at $E_0 = 10$ keV (11,137,000 counts) and residual (*blue*) after DTSA II analysis using BaS$_{i2}$O$_5$ (sanbornite) and Ti as fitting references and standards. **b** Expanded view

☐ **Table 20.7** Analysis of BaTiSi$_3$O$_9$ (benitoite) at $E_0 = 10$ keV with Ti and sanbornite (BaSi$_2$O$_5$) as fitting references and standards; integrated spectrum count = 11,366,000. Analysis performed with O K- L$_{2,3}$, Si K-L$_{2,3}$, Ti K-L$_{2,3}$ and Ba L$_3$-M$_{4,5}$

	O	Si	Ti	Ba
C$_{av}$ (atom frac)	0.6416	0.2149	0.07096	0.07256
Z-correction	0.955	0.953	0.947	0.943
A-correction	0.804	1.041	0.989	1.004
F-correction	1	1	1.007	1
σ (7 replicates)	0.000269	0.00016	0.000176	0.000176
σ$_{Rel}$ (%)	0.04 %	0.07 %	0.25 %	0.24 %
RDEV (%)	−0.20 %	0.28 %	−0.66 %	1.60 %
C (mass frac, single analysis)	0.3462	0.2032	0.1143	0.3356
Counting error, std	0.0002	0.0001	7.10×10^{-5}	0.0006
Counting error, unk	0.0002	0.0001	0.0004	0.0009
A-factor error	0.0142	0.0003	2.10×10^{-5}	3.50×10^{-5}
Z-factor error	0.0003	2.40×10^{-5}	1.10×10^{-6}	2.80×10^{-6}
Combined errors	0.0142	0.0003	0.0004	0.0011

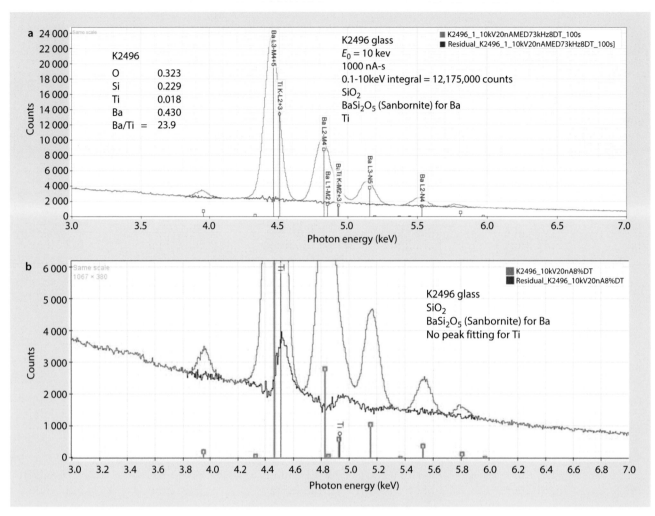

□ **Fig. 20.8** **a** SDD-EDS spectrum of NIST microanalysis glass K2496 (*red*) at $E_0 = 10$ keV (12,175,000 counts) and residual (*blue*) after DTSA II analysis using $BaS_{i2}O_5$ (sanbornite) and Ti as fitting references and stan- dards. **b** Same analysis protocol, but not including Ti in the peak-fitting. Note low level peaks for Ti K-$L_{2,3}$ and Ti K-M_3 (Ba L-family peaks marked as *green lines*)

□ **Table 20.8**   Analysis of NIST microanalysis glass K2496 at $E_0 = 10$ keV with Ti and sanbornite ($BaSi_2O_5$) as fitting references and standards; integrated spectrum count = 12,175,000. Analysis performed with O K-$L_{2,3}$, Si K-$L_{2,3}$, Ti K-$L_{2,3}$ and Ba L_3-$M_{4,5}$

	O	Si	Ti	Ba
C_{av} (atom frac)	0.6228	0.2585	0.01171	0.1069
Z-correction	0.984	0.983	0.983	0.98
A-correction	0.966	1.017	0.986	1.001
F-correction	1	1	1.01	1
σ (7 replicates)	0.000158	0.000277	0.000217	0.000226
$σ_{Rel}$ (%)	0.03 %	0.11 %	1.80 %	0.21 %
RDEV (%)	−1.70 %	0.99 %	−0.64 %	8.70 %
C (mass frac)	0.3066	0.223	0.0177	0.4527
Counting error, std	0.0002	0.0002	1.10×10^{-5}	0.0008
Counting error, unk	0.0002	0.0001	0.0004	0.0007
A-factor error	0.0021	8.80×10^{-5}	3.90×10^{-6}	1.20×10^{-5}
Z-factor error	0.0003	2.70×10^{-5}	1.80×10^{-7}	4.00×10^{-6}
Combined errors	0.0021	0.0002	0.0004	0.0011

20

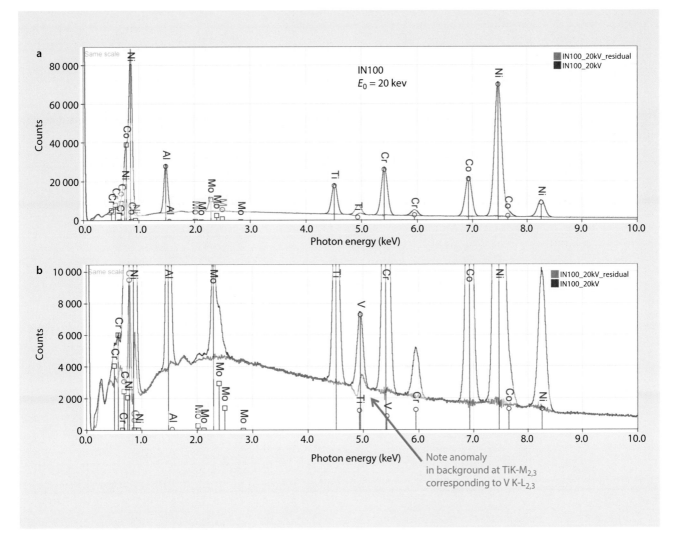

□ **Fig. 20.9** Analysis of IN100 alloy fitting for Al, Ti, Cr, Co, Ni and Mo: **a** full spectrum (*blue*) and residual spectrum after peak-fitting (*red*); **b** expanded view, note anomaly in background at the energy of Ti

K-M4,5 (4.931 keV), which closely corresponds to the energy of V K-L2,3 (4.952 keV)

□ **Table 20.9** Analysis of IN100

	1st quantitative analysis	2nd quantitative analysis
Raw sum	0.9944 ± 0.0011	1.0032 ± 0.0013
Al	0.0559 ± 0.0007	0.0562 ± 0.0007
Ti	0.0473 ± 0.0001	0.0474 ± 0.0001
V		0.0110 ± 0.0002
Cr	0.0981 ± 0.0002	0.0949 ± 0.0006
Co	0.1551 ± 0.0003	0.1553 ± 0.0003
Ni	0.6065 ± 0.0008	0.6069 ± 0.0008
Mo	0.0315 ± 0.0002	0.0315 ± 0.0002

analysis (□ Table 20.9), despite the severe interference from the Ti constituent which has a concentration more than four times higher.

20.4.2 Analysis of a Stainless Steel

When is a Standard Not Suitable as a Peak-Fitting Reference?

One of the great strengths of the k-ratio/matrix correction protocol is simplicity of the required standards. Pure elements can be used for most of the periodic table, and for those elements that are not in a suitable solid form at ambient temperature and at the low chamber pressure, stoichiometric binary compounds that are stable under the beam can be used. This is an excellent situation for the analyst, since it is generally not possible to have a multi-element standard that is homogeneous on the microscopic scale and similar in

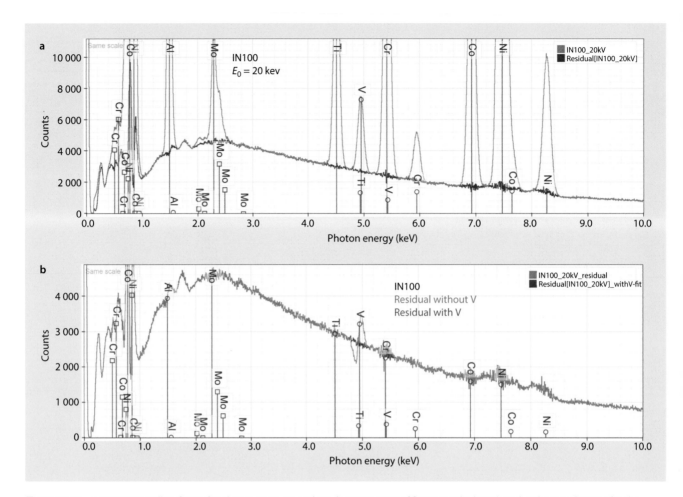

☐ **Fig. 20.10**   **a** IN100 superalloy, fitting for Al, Ti, V, Cr, Co, Ni, and Mo. **b** Comparison of fitting residuals with and without inclusion of V; the background anomaly is eliminated

20

composition to a particular unknown specimen. For many simple binary as well as more complex mixtures of elements, Nature favors heterogeneity on the microscale, and many combinations of elements tend to phase separate to produce chemically heterogeneous microstructures. However, there are important cases where microscopically homogeneous, multi-element compositions are available, such as minerals, glasses, and a few metal alloys. An example is NIST Standard Reference Material 479, an Fe-Cr-Ni alloy which is certified to be homogeneous on a microscopic scale. SRM 479 can serve as a standard for the analysis of another more complex stainless steel. ☐ Figure 20.11 shows the spectrum of a type 316 stainless steel which the initial qualitative analysis shows that in addition to Cr, Fe, and Ni also contains peaks for Si and Mo. While SRM 479 is an ideal standard for this analysis of Cr, Fe, and Ni, it is not suitable to provide peak-fitting references for Cr, Fe, and Ni because of the mutual interference of these peaks. Thus, pure elements for Cr, Fe, and Ni are used for the peak-fitting references, while SRM 479 is used as the standard, reducing the magnitude of the matrix corrections because of the close similarity of the unknown and standard compositions. When the analysis is performed, including elemental Si and Mo as references and standards,

the results given in ☐ Table 20.10 (column 2) are obtained. Close examination of the residual spectrum reveals the peaks of the Mn K-family. When the analysis is repeated including Mn in the suite of fitted elements, the results given in ☐ Table 20.10 (column 3) are obtained with a concentration of Mn = 0.0154, and the residual spectrum no longer contains anomalous peaks, as shown in ☐ Fig. 20.11c, d.

20.4.3 Progressive Discovery: Repeated Qualitative–Quantitative Analysis Sequences

Complex unknowns may require several iterations of qualitative and quantitative analysis to discover all of the constituents. For such situations, the analytical total as well as the residual spectrum serve as powerful guides to reach a successful result. As an example, consider the spectrum of a monazite (a lanthanum-cerium phosphate mineral) shown in ☐ Fig. 20.12a, b. The elements recognized in the first qualitative analysis stage are O, P, La, Ce (major) and Al, Si, Ca, and Th (minor). The first quantitative analysis round for these elements, with O calculated by stoichiometry, yielded

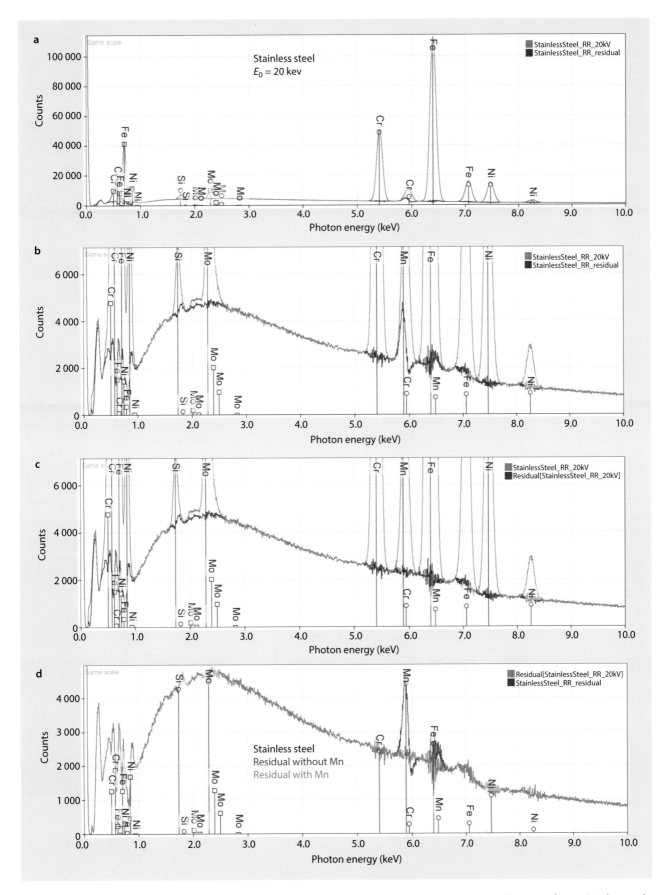

■ **Fig. 20.11** **a** Stainless steel, fitting for Si, Cr, Fe, Ni, and Mo; residual in *blue*. **b** Expanded vertical scale, note detection of Mn. **c** Stainless steel, fitting for Si, Cr, Mn, Fe, Ni, and Mo; residual in *blue*. **d** Comparison of residuals with fitting for Mn (*red*) and without (*blue*)

◻ **Table 20.10**   Analysis of a type 316 stainless steel (mass concentrations)

	1st quantitative analysis	2nd quantitative analysis
Raw sum	0.9861 ± 0.0009	1.0031 ± 0.0009
Si	0.0053 ± 0.0001	0.0053 ± 0.0001
Cr	0.1705 ± 0.0003	0.1711 ± 0.0003
Mn		0.0154 ± 0.0002
Fe	0.6539 ± 0.0006	0.6545 ± 0.0006
Ni	0.1328 ± 0.0005	0.1330 ± 0.0005
Mo	0.0237 ± 0.0002	0.0238 ± 0.0002

the results listed in ◻ Table 20.11 (first analysis). The analytical total is anomalously low at 0.7635. The second round of qualitative analysis of the fitting residuals from the first quantitative analysis in ◻ Fig. 20.12c–e shows several new peaks that are identified: Sr, Y, and Zr in the P region and Ti, Pr, and Nd in the La-Ce region. When these elements are included in the second round of quantitative analysis, as listed in ◻ Table 20.11, the analytical total increases to 0.9629. The third round of qualitative analysis of the fitting residuals from the second quantitative analysis in ◻ Fig. 20.12f,g shows Nb in the P region and Sm and Fe in the La-Ce region. After the third round of quantitative analysis, the analytical total increases to 0.9960, and the third round residuals are shown in ◻ Fig. 20.12h, i superimposed on the residuals from rounds one and two, indicating only minor changes between rounds two and three. Should this analysis be repeated for a fourth round? The level of Nb that has been measured is only 0.0006 (600 ppm), and the confidence is this level is low. There remain some low level structures in the third analysis residuals, but to take this analysis further, the spectrum should be measured for additional time to increase the total count at least by a factor of four.

20.5   Is the Specimen Homogeneous?

For the most part, Nature seems to prefer heterogeneity on a microscopic scale. That is, many combinations of two or more elements spontaneously form two or more phases, where a phase is defined as matter that is distinct in chemical composition and physical state, thus creating a chemical microstructure. Indeed, the great value of electron excited X-ray microanalysis is its capability to measure elemental composition on the spatial scale of a micrometer and finer to characterize this chemical microstructure.

As part of an effective analysis strategy, it is generally wise to make multiple measurements of each distinct region of interest of a specimen rather than just a single measurement. When a material is sampled at multiple locations under carefully controlled, reproducible analytical conditions, there will inevitably be variations in the results due to the natural statistical fluctuations in the numbers of measured X-rays, both for the characteristic and continuum background of the specimen and the standard. The question often arises when examining the variations in such replicate results if the material can be regarded as homogeneous or if the degree of variation in the results is indicative of actual specimen heterogeneity.

In electron-excited X-ray microanalysis, what is of ultimate importance is the precision of the composition rather than just that of an individual intensity measurement for an element. This point has been discussed in detail by Ziebold (1967) and Lifshin et al. (1999). Note first that a k ratio consists actually of the averages of four measurements: N_{sam}, the mean intensity measured on the sample; $N_{sam}(B)$, the corresponding mean background at the same energy; N_{stan}, the intensity measured on the standard; and $N_{stan}(B)$, the corresponding background for the standard:

$$k = \left[N_{sam} - N_{sam}(B) \right] / \left[N_{stan} - N_{stan}(B) \right] \qquad (20.4)$$

In reality a single measurement of each of the four terms in Eq. (20.4) results in only a single estimate of k and many sets of measurements are required to approach the true mean value of k.

For multiple determinations of the k-ratio, Ziebold (1967) showed that the precision in the k-ratio (σ_k) is given by

$$\sigma_k^2 = k^2 \left\{ \begin{array}{l} \left[\left(N_{sam} + N_{sam}(B) \right) / \left[n_{sam} \left(N_{sam} - N_{sam}(B) \right)^2 \right] \right] \\ + \left[\left(N_{stan} + N_{stan}(B) \right) / n_{stan} \left(N_{stan} - N_{stan}(B) \right)^2 \right] \end{array} \right\}$$

$$(20.5)$$

where N represents the mean of the set of measurements for each parameter, for example:

$$N_{sam} = \sum_{i}^{n} N_i / n_{sam} \qquad (20.6)$$

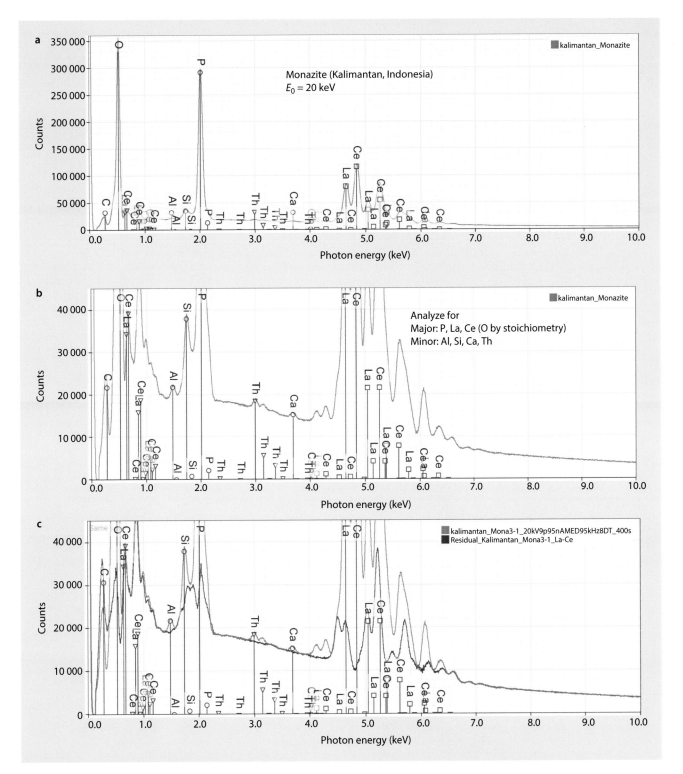

■ **Fig. 20.12** Monazite (lanthanum-cerium phosphate mineral) at $E_0 = 20$ keV; 0.1 keV to 30 keV = 48.7 million counts: **a** original spectrum; **b** vertical expansion; **c** Round 1: full spectrum residuals after first quantitative analysis; **d** P-region, first residuals; **e** La-Ce-region, first residuals; Round 2: **f** P-region, second residuals; **g** La-Ce-region, second residuals; Round 3; **h** P-region, all residuals; **i** La-Ce-region, all residuals

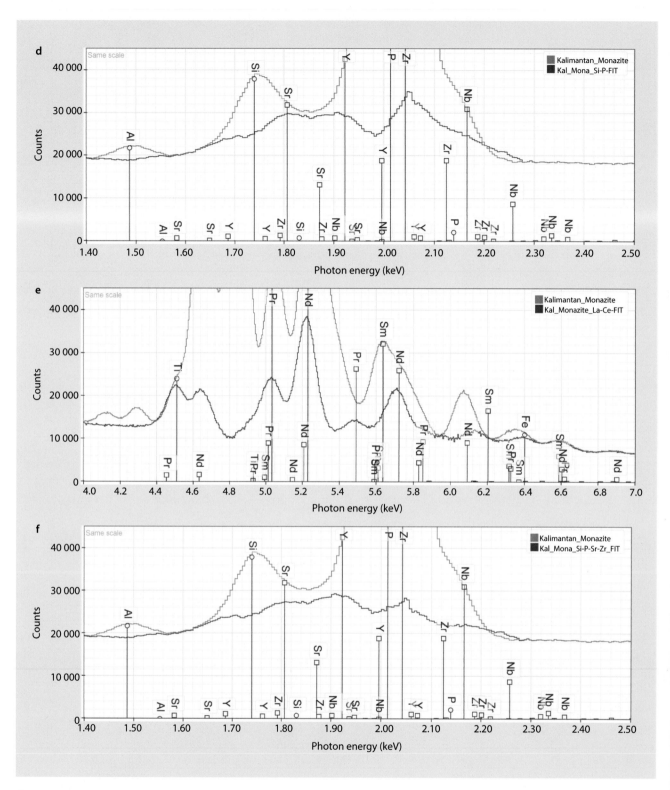

■ **Fig. 20.12**   (continued)

20.5 · Is the Specimen Homogeneous?

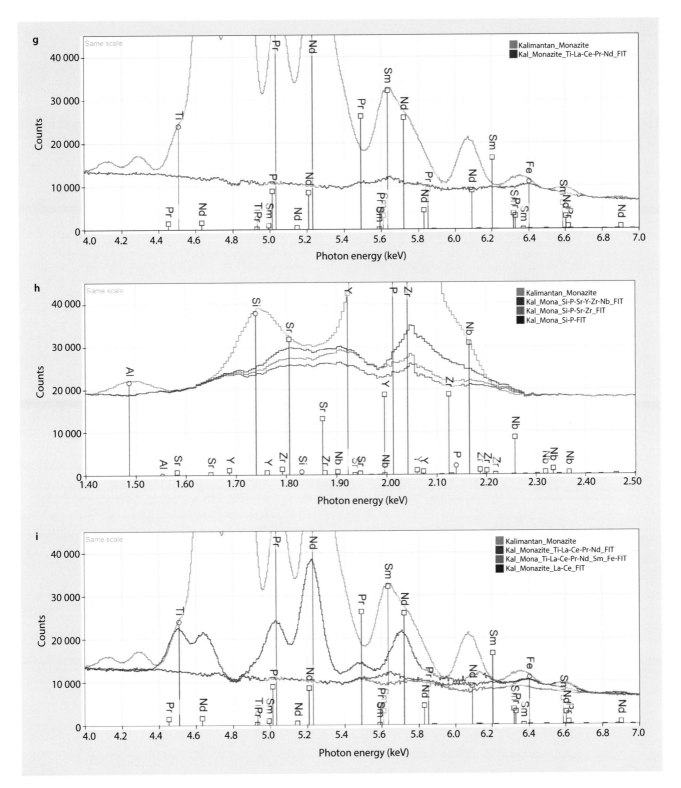

Fig. 20.12 (continued)

◻ **Table 20.11** Analysis of a monazite

Round	O (by assumed stoichiometry)	Al	Si	P	Ca	Ti
First analysis	0.2363 ± 0.0008	0.0013 ± 0.0000	0.0049 ± 0.0000	0.1114 ± 0.0006	0.0007 ± 0.0000	
Second analysis	0.2828 ± 0.0009	0.0016 ± 0.0000	0.0059 ± 0.0001	0.1240 ± 0.0007	0.0007 ± 0.0000	0.0071 ± 0.0001
Third analysis	0.2908 ± 0.0010	0.0015 ± 0.0000	0.0061 ± 0.0001	0.1263 ± 0.0007	0.0007 ± 0.0000	0.0072 ± 0.0001
Element	Fe	Sr	Y	Zr	Nb	La
First analysis						0.1359 ± 0.0002
Second analysis		0.0016 ± 0.0001	0.0098 ± 0.0002	0.0113 ± 0.0002		0.1585 ± 0.0003
Third analysis	0.0022 ± 0.0001	0.0028 ± 0.0001	0.0030 ± 0.0002	0.0117 ± 0.0002	0.0006 ± 0.0001	0.1591 ± 0.0003
Element	Ce	Pr	Nd	Sm	Th	Raw Sum
First analysis	0.2692 ± 0.0004				0.0038 ± 0.0001	0.7635 ± 0.0011
Second analysis	0.2699 ± 0.0004	0.0221 ± 0.0003	0.0750 ± 0.0003		0.0040 ± 0.0001	0.9629 ± 0.0013
Third analysis	0.2709 ± 0.0004	0.0221 ± 0.0003	0.0751 ± 0.0003	0.0073 ± 0.0005	0.0040 ± 0.0001	0.9960 ± 0.0030

and n_{sam} and n_{stan} are the numbers of measurements of the sample and standard. The corresponding precision in the measurement of the concentration is given by

$$\sigma_C^2 = C^2 \frac{\left\{ \begin{array}{l} \left[(N_{sam} + N_{sam}(B))/n_{sam}(N_{sam} - N_{sam}(B))^2 \right] + \\ \left[(N_{stan} + N_{stan}(B))/n_{stan}(N_{stan} - N_{stan}(B))^2 \right] \end{array} \right\}}{\left\{ 1 - \left[(a-1)C/a \right] \right\}^2}$$

(20.7)

where the parameter "a" is the constant in the hyperbolic relation (Ziebold and Ogilvie 1964):

$$(1-k)/k = a\left[(1-C)/C \right]$$ **(20.8)**

The parameter "a" can be calculated using Eq. (20.8) with the measured value of k and the calculated value of C from the quantitative analysis software results.

Equation 20.4 makes it possible to assess statistical uncertainty in an estimate of composition. For example, it can be used to construct a confidence interval (e.g., $\pm 1.96\sigma_C$ gives the 95 % confidence interval) for the difference of two samples or to plan how many counts must be collected to be able to estimate differences between two samples at the desired level of precision. The calculation of the confidence interval is based on the normal distribution of the estimate of C for large samples. This confidence interval is only based on the statistical uncertainty inherent in the X-ray counts. The full error budget requires also estimating the uncertainty in the principal matrix corrections for absorption (A) and scattering/energy loss (Z) (Ritchie and Newbury 2012). NIST DTSA-II provides these error estimates in addition to the error in the measurement of the k-ratio.

The use of Eq. (20.7) to calculate σ_c for an alloy with a composition of 0.215-Mo_0.785-W (21.5 wt % Mo and 78.5 wt % W) and the spectrum shown in ◻ Fig. 20.13 is as follows:

First determine the number of Mo L_3-M_5 and W L_3-M_5 counts measured on the sample and standard as well as the corresponding background counts for each:

At $E_0 = 20$ keV and $i_B = 10$ nA for an SDD-EDS of $\Omega = 0.0077$ sr, the spectrum of the alloy and the residual after peak-fitting, as shown in ◻ Fig. 20.13, gives the following intensities for a single measurement:

Mo L_3-M_5	bkg	W L_3-M_5	bkg
884416	195092	868516	111279

The pure element standards gave the following values for a single measurement:

| 7016889 | 211262 | 1147787 | 134382 |

These intensities yield the following mean k-values:

| 0.1260 | 0.7567 |

From the NIST DTSA-II results and Eq. (20.6):

Mo $k = 0.1235$ $C = 0.2132$ (normalized $C = 0.2148$) $a = 1.92$
W $k = 0.7540$ $C = 0.7792$ (normalized $C = 0.7852$) $a = 1.15$

Substituting these values in Eq. (20.4) gives

Mo	W
$\sigma_C = 0.0003$	$\sigma_C = 0.0012$

Thus, from the statistics of the X-ray counts measured for the alloy and the pure element standards, the 95 % confidence limit for *reproducibility* is given by $\pm 1.96\sigma_C$

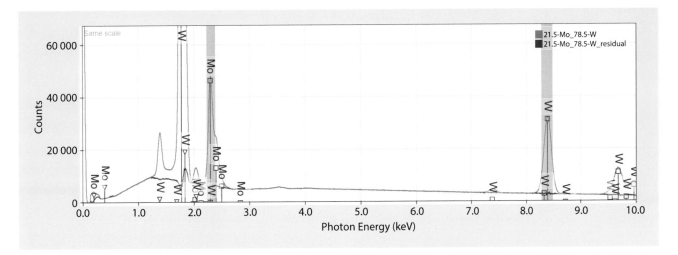

◻ Fig. 20.13   EDS spectrum of 0.215-Mo, 0.785-W alloy showing the residuals after peak fitting; $E_0 = 20$ keV

◻ Table 20.12   Analysis of 0.215-Mo_0.785-W alloy (normalized mass fractions) (analysis in bold used for the calculations above)

Mo	W
0.2166	0.7834
0.2148	0.7852
0.2141	0.7859
0.2177	0.7823
0.2185	0.7815
0.2180	0.7820
0.2175	0.7825
0.2203	0.7797
0.2275	0.7725
0.2312	0.7688
0.2346	0.7654
0.2363	0.7637
0.2358	0.7642
0.2344	0.7656
0.2389	0.7611

Mo $C = 0.2148 \pm 0.0006$ or $0.215 \pm 0.28\%$
W $C = 0.7852 \pm 0.0024$ or $0.785 \pm 0.31\%$

If multiple locations are measured under consistent measurement conditions—e.g., constant beam energy, beam current, and EDS performance—then values that fall outside the ranges given for Mo and W are indicative of heterogeneity, that is, real deviations in the composition of the alloy. ◻ Table 20.12 lists 15 measurements on this alloy made at randomly selected locations, which reveal significant heterogeneity with the most extreme excursion approximately 11 % in the

Mo constituent from the ideal values. This deviation is well outside that expected from natural variations due to statistical fluctuations in the measured counts as calculated above.

The full uncertainty budget reported by DTSA-II, including the estimates for the uncertainties in the A and Z matrix corrections as well as the X-ray statistics is

$$\begin{array}{cc} \text{Mo} & \text{W} \\ \sigma_C = 0.0039 & \sigma_C = 0.0017 \end{array}$$

The large increase in σ_C for Mo beyond the contribution of the X-ray statistics is due to the contribution of the matrix correction factor for absorption, $A = 0.528$.

20.6   Beam-Sensitive Specimens

In some cases, the interaction of the electron beam can damage the specimen and locally alter the composition, often with the effects showing a strong dependence on the total dose, the dose per unit volume, and the dose rate.

20.6.1   Alkali Element Migration

In some insulating materials, especially non-crystalline materials such as glasses, alkali family elements can migrate in response to the local charge injected below the surface by the beam, even when a thin conducting surface layer such as carbon has been applied to discharge the specimen. Migration typically leads to diminishing alkali concentration with time in the excited volume. The phenomenon can be detected by measuring a time series of spectra and carefully comparing the intensity of alkali element peaks to stable matrix peaks such as that of Si, as shown in ◻ Fig. 20.14 for "Corning glass A" which has a high alkali composition with approximately 10 weight percent Na and 2.4 weight percent potassium (listed in ◻ Table 20.11) (Vicenzi et al. 2002). Each spectrum shown in ◻ Fig. 20.14 was recorded for 10 s with a fixed, focused beam, which creates the maximum possible dose per unit volume.

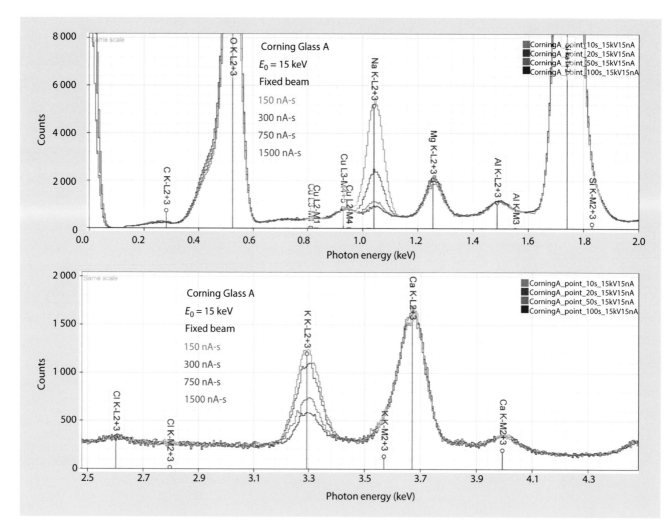

◼ **Fig. 20.14**   Corning glass A, showing Na and K migration as a function of dose for a fixed beam (15 keV, 15 nA)

Comparing the first (150-nA-s dose) and the second spectra (300-nA-s dose), the Na intensity is seen to fall by more than a factor of two as the dose increases, while the K intensity diminishes by approximately 20%. After a dose of 1500 nA-s, the Na peak is reduced to approximately 10% of its intensity after 10 s, while the K peak decreases to approximately 25% of its original value, whereas other non-alkali elements—e.g., Mg, Al, Ca, Si, etc.—remain nearly constant with dose. Even this time series is somewhat misleading. If the initial dose is reduced by a factor of 10, the Na intensity observed is higher by approximately 30%, as shown in ◼ Fig. 20.15, while the K intensity is higher by approximately 5%. At the extremely high volumetric dose created by the fixed point beam in these experiments, significant alkali migration occurs even with the initial short beam dwell (e.g., 1 s, 15 nA). The effects of the dose on the results obtained by quantitative analysis with DTSA-II are given in ◼ Table 20.13. Even in the first analysis (150 nA-s dose), the measured Na concentration is a factor of 2 lower than the synthesized glass composition, and after the maximum dose utilized for this series (1500 nA-s dose), the Na concentration has decreased by a factor of 11.

Methods to reduce alkali element migration are based on modifying the total dose, the dose per unit area (and volume), and/or the dose rate. Reducing the dose per unit area is often one of the most effective ways to control migration. By defocusing the fixed beam or by scanning the focused beam rapidly over a large area, the dose per unit area can be greatly reduced, often by several orders of magnitude, compared to a fixed, focused beam. Because of the basic assumption of the k-ratio/matrix correction protocol that the material being analyzed must have the same composition over the entire volume excited by the electron beam, this increased-area strategy is only valid providing the region of analytical interest is homogeneous over a sufficiently large to accommodate the defocused or rapidly scanned beam. The effect of increasing the scanned area is shown in ◼ Fig. 20.16 for Corning glass A, where the measured Na intensity increases rapidly as the scanned area is increased. ◼ Table 20.14 compares DTSA-II quantitative analyses of spectra with the same dose (15 keV, 1500 nA-s) obtained with a point beam and with that beam rapidly scanning over an area 100 μm square. The scanned area results correspond very closely to the as-synthesized

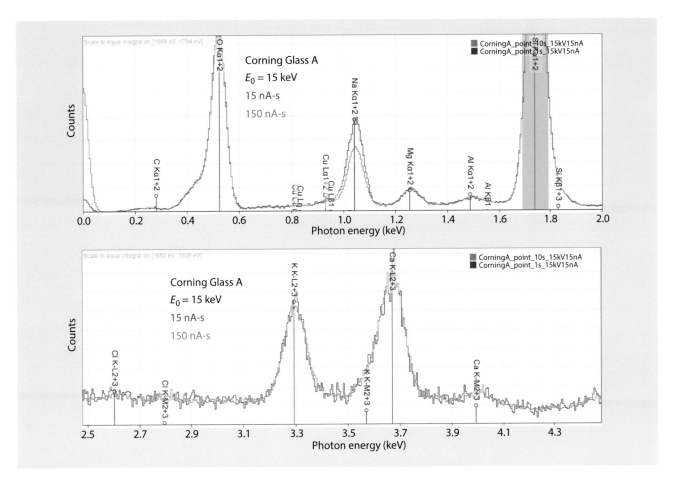

Fig. 20.15 Corning glass A, showing Na and K migration compared as a function of dose with a reduction of a factor of 10 difference for a fixed beam (15 keV, 15 nA); spectra normalized to the Si peak

Table 20.13 DTSA-II analysis of Corning Glass A ($E_0 = 15$ keV), oxygen by assumed stoichiometry, fixed beam

Element	As-synthesized mass conc	150 nA-s raw mass conc	300 nA-s raw mass conc	750 nA-s raw mass conc	1500 nA-s raw mass conc
O	0.4421	0.4316 ± 0.0009	0.4313 ± 0.0009	0.4496 ± 0.0009	0.4644 ± 0.0009
Na	0.1061	0.0519 ± 0.0004	0.0364 ± 0.0004	0.0172 ± 0.0003	0.0098 ± 0.0003
Mg	0.0160	0.0161 ± 0.0002	0.0161 ± 0.0002	0.0178 ± 0.0002	0.0186 ± 0.0002
Al	0.0529	0.0050 ± 0.0001	0.0051 ± 0.0001	0.0052 ± 0.0001	0.0058 ± 0.0001
Si	0.3111	0.3192 ± 0.0006	0.3230 ± 0.0007	0.3438 ± 0.0007	0.3574 ± 0.0007
K	0.0238	0.0251 ± 0.0003	0.0230 ± 0.0003	0.0195 ± 0.0003	0.0166 ± 0.0003
Ca	0.0359	0.0363 ± 0.0003	0.0364 ± 0.0003	0.0381 ± 0.0003	0.0386 ± 0.0003
Ti	0.00474	0.0053 ± 0.0002	0.0059 ± 0.0002	0.0057 ± 0.0002	0.0057 ± 0.0002
Mn	0.00775	0.0086 ± 0.0003	0.0084 ± 0.0003	0.0091 ± 0.0003	0.0101 ± 0.0003
Fe	0.00762	0.0083 ± 0.0003	0.0082 ± 0.0003	0.0074 ± 0.0003	0.0090 ± 0.0003
Cu	0.00935	0.0098 ± 0.0005	0.0104 ± 0.0005	0.0112 ± 0.0005	0.0108 ± 0.0005
Sn	0.00150	0.0030 ± 0.0008	0.0034 ± 0.0007	0.0036 ± 0.0007	0.0054 ± 0.0007
Sb	0.0146	0.0124 ± 0.0007	0.0140 ± 0.0007	0.0139 ± 0.0007	0.0140 ± 0.0007
Ba	0.0050	0.0045 ± 0.0005	0.0036 ± 0.0005	0.0056 ± 0.0005	0.0042 ± 0.0005
Raw total		0.938	0.926	0.9485	0.9718

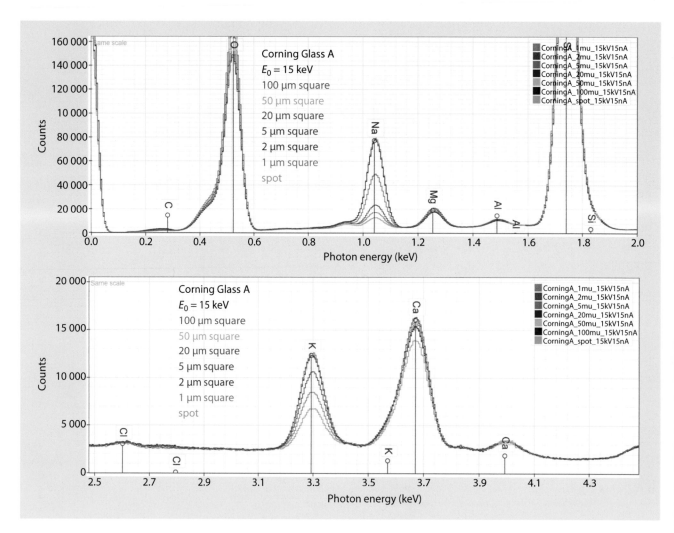

☐ **Fig. 20.16**   Corning glass A, showing Na and K migration as a function of dose for scanning beams covering various areas (20 keV, 10 nA)

values for the glass, including the alkali elements Na and K, whereas the point beam results show reductions in the Na and K concentrations. ☐ Figure 20.17 shows that the measured sodium and potassium concentrations increase to reach the synthesized values as the scanned area dimensions are increased to cover areas above 20 x 20-μm (nominal magnification 5 kX) for the particular dose utilized (15 keV, 1500 nA-s). Thus, while scanning a large homogeneous area obviously concedes the spatial resolution capability of electron-excited X-ray microanalysis, this approach may be the most expedient technique to control and minimize alkali element migration.

Materials that can serve as useful standards for sodium include certain crystalline minerals such as albite (NaAlSi$_3$O$_8$) in which the sodium is much more stable under electron bombardment. However, even for albite the use of a stationary high intensity point beam may produce significant migration effects, as shown in ☐ Fig. 20.18 for spectra collected with a stationary point beam as a function of dose (upper)

and at the same dose with a fixed beam and two different sizes of scanned areas (lower). Thus, the use of a scanned area rather than a fixed beam may be necessary when collecting a standard spectrum, even on a crystalline material.

20.6.2   Materials Subject to Mass Loss During Electron Bombardment—the Marshall-Hall Method

Thin Section Analysis

The X-ray microanalysis of biological and polymeric specimens is made difficult, and sometimes impossible, by several forms of radiation damage that are directly caused by the electron beam. At the beam energies used in the SEM (0.1–30 keV), it is possible for the kinetic energy of individual beam electrons to break and/or rearrange chemical bonds. The radiation damage can release smaller molecules such as

■ Table 20.14 DTSA-II quantitative analysis of Corning glass A: Comparison of results with a fixed beam and scanned beam (100 μm square) (15 keV/15 nA); oxygen by assumed stoichiometry

Element	As-synthesized mass conc	1500 nA-s (fixed beam) raw mass conc	1500 nA-s (100-μm² scan) raw mass conc
O	0.4421	0.4644 ± 0.0009	0.4577 ± 0.0006
Na	0.1061	0.0098 ± 0.0003	0.1076 ± 0.0004
Mg	0.0160	0.0186 ± 0.0002	0.0164 ± 0.0001
Al	0.0529	0.0058 ± 0.0001	0.0056 ± 0.0000
Si	0.3111	0.3574 ± 0.0007	0.3239 ± 0.0005
K	0.0238	0.0166 ± 0.0003	0.0257 ± 0.0002
Ca	0.0359	0.0386 ± 0.0003	0.0350 ± 0.0001
Ti	0.00474	0.0057 ± 0.0002	0.0051 ± 0.0001
Mn	0.00775	0.0101 ± 0.0003	0.0082 ± 0.0001
Fe	0.00762	0.0090 ± 0.0003	0.0077 ± 0.0001
Cu	0.00935	0.0108 ± 0.0005	0.0096 ± 0.0003
Sn	0.00150	0.0054 ± 0.0007	0.0045 ± 0.0003
Sb	0.0146	0.0140 ± 0.0007	0.0125 ± 0.0002
Ba	0.0050	0.0042 ± 0.0005	0.0042 ± 0.0002
Raw total		0.9718	1.0254

■ Fig. 20.17 Results of quantitative analysis of Corning glass A as a function of the size of the area scanned. Nominal magnifications indicated

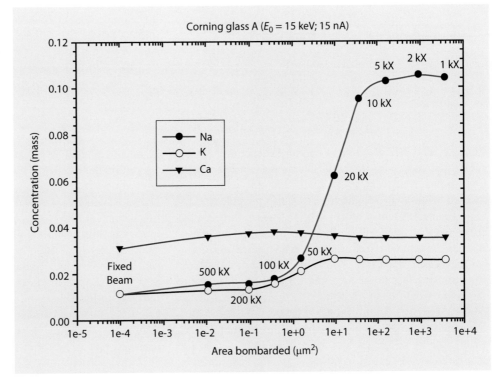

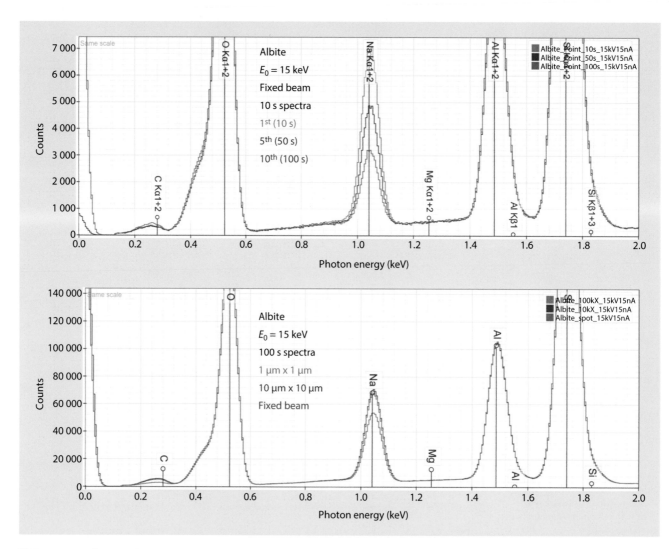

■ **Fig. 20.18** Albite (NaAlSi$_3$O$_8$); $E_0 = 15$ keV, 15 nA: (*upper*) effect of increasing dose on the Na peak; (*lower*) effect of fixed beam versus scanned beam on the Na peak

CO, CO$_2$, and H$_2$O that evaporate into the vacuum, causing substantial mass loss from the interaction volume. At the highest beam currents, typically 10–100 nA, used with a focused beam at a static location, it is also possible to cause highly damaging temperature elevations, which further exacerbate mass loss. Indeed, when analyzing this specimen class it should be assumed that significant mass loss will occur during the measurement at each point of the specimen. If all constituents were lost at the same rate, then simply normalizing the result would compensate for the mass loss that occurs during the accumulation of the X-ray spectrum. Unfortunately, the matrix constituents (principally carbon compounds and water) can be selectively lost, while the heavy elements of interest in biological microanalysis (e. g., Mg, P, S, K, Ca, Fe, etc.) remain in the bombarded region of the specimen and appear to be present at effectively higher concentration than existed in the original specimen. What is then required of any analytical procedure for biological and polymeric specimens is a mechanism to provide a meaningful analysis under these conditions of a specimen that undergoes continuous change.

Marshall and Hall (1966) and Hall (1968) made the original suggestion that the X-ray continuum could serve as an internal standard to monitor specimen changes. This assumption permitted development of the key procedure for beam-sensitive specimens that is used extensively in the biological community and that is also applicable in many types of polymer analysis. This application marks the earliest use of the X-ray continuum as a tool (rather than simply a hindrance) for analysis, and that work forms the basis for the development of the peak-to-local background method applied to challenging geometric forms such as particles and rough surfaces. The technique was initially developed for applications in the high beam current EPMA, but the procedure works well in the SEM environment.

The Marshall–Hall method (Marshall and Hall 1966) requires that several key conditions:

1. The specimen must be in the form of a thin section, where the condition of "thin" is satisfied when the incident beam penetrates with negligible energy loss. For an analytical beam energy of 10–30 keV, the energy loss

passing through a section consisting of carbon approximately 100–200 nm in thickness will be less than 500 eV. This condition permits the beam energy to be treated as a constant, which is critical for the development of the correction formula. Biological specimens are thus usually analyzed in the form of thin sections cut to approximately 100-nm thickness by microtome. Polymers may also be analyzed when similarly prepared as thin sections by microtoming or by ion beam milling. Such a specimen configuration also has a distinct advantage for improving the spatial resolution of the analysis compared to a bulk specimen. The analytical volume in such thin specimens is approximately the cylinder defined by the incident beam diameter and the section thickness, which is at least a factor of 10–100 smaller in linear dimensions than the equivalent bulk specimen case at the same energy, as shown in the polymer etching experiment in the Interaction Volume module.

2. The matrix composition must be dominated by light elements, for example, C, H, N, O, whose contributions will form nearly all of the X-ray continuum and whose concentrations are reasonably well known for the specimen. Elements of analytical interest such as Mg, P, S, Cl, K, Ca, and so on, the concentrations of which are unknown in the specimen, must only be present preferably as trace constituents (<0.01 mass fraction) so that their effect on the X-ray continuum can be neglected. When the concentration rises above the low end of the minor constituent range (e.g., 0.01 to 0.05 mass fraction or more), the analyte contribution to the continuum can no longer be ignored.

3. A standard must be available with a known concentration of the trace/minor analyte of interest and for which the complete composition of low-atomic-number elements is also known and which is stable under electron beam bombardment. Glasses synthesized with low atomic number oxides such as boron oxide are suitable for this role. The closer the low–atomic-number element composition of the standard is to that of the unknown, the more accurate will be the results.

The detailed derivation yields the following general expression for the Marshall–Hall method:

$$\frac{I_{ch}}{I_{cm}} = c\,\frac{\dfrac{C_A}{A_A}}{\sum_i \left[C_i \left(\dfrac{Z_i^2}{A_i} \right) \log_e \left(1.166\,\dfrac{E_0}{J_i} \right) \right]} \qquad (20.9)$$

In this equation, I_{ch} is the characteristic intensity of the peak of interest, for example, S K-L$_{2,3}$ or Ca K-L$_{2,3}$, and I_{cm} is the continuum intensity of a continuum window of width ΔE placed somewhere in the high energy portion of the spectrum, typically above 8 keV, so that absorption effects are negligible and only mass effects are important. C_i is the mass

concentration, Z_i is the atomic number, and A_i is the atomic weight. The subscript "A" identifies a specific trace or minor analyte of interest (e.g., Mg, P, S, Cl, Ca, Fe, etc.) in the organic matrix, while the subscript "i" represents all elements in the electron-excited region. E_0 is the incident beam energy and J is the mean ionization energy, a function only of atomic number as used in the Bethe continuous energy loss equation

Assumption 2 provides that the quantity $\sum(C_i \cdot Z_i^2 / A_i)$ in Eq. (20.9) for the biological or polymeric specimen to be analyzed is dominated by the low-Z constituents of the matrix. (Some representative values of $\sum(C_i \cdot Z_i^2 / A_i)$ are 3.67 (water), 3.01 (nylon), 3.08 (polycarbonate) and 3.28 (protein with S). Typically the range is between 2.8 and 3.8 for most biological and many polymeric materials.) The unknown contribution of the analyte, C_A, to the sum may be neglected when considering the specimen because C_A is low when the analytes are trace constituents.

To perform a quantitative analysis, Eq. (20.9) is used in the following manner: A standard for which all elemental concentrations are known and which contains the analyte(s) of interest "A" is prepared as a thin cross section (satisfying assumption 3). This standard is measured under defined beam and spectrometer parameters to yield a characteristic-to-continuum ratio, I_A/I_{cm}. This measured ratio I_A/I_{cm} is set equal to the right side of Eq. (20.9). Since the target being irradiated is a reference standard, the atomic numbers Z_i, atomic weights A_i and weight fractions C_i are known for all constituents, and the J_i values can be calculated as needed. The only unknown term is then the constant "c" in Eq. (20.9), which can now be determined by dividing the measured intensity ratio, I_A/I_{cm}, by the calculated term. Next, under the same measurement conditions, the characteristic "A" intensity and the continuum intensity at the chosen energy are determined for the specimen location(s). Providing that the low-Z elements that form the matrix of the specimen are similar to the standard, or in the optimum case these concentrations are actually known for the specimen (or can be estimated from other information about the actual, localized, material being irradiated by the electrons, and not some bulk property), then this value of "c" determined from the standard can be used to calculate the weight fraction of the analyte, C_A, for the specimen.

This basic theme can be extended and several analytes—"A," "B," "C," etc.—can be analyzed simultaneously if a suitable standard or suite of standards containing the analytes is available. The method can be extended to higher concentrations, but the details of this extension are beyond the scope of this book; a full description and derivation can be found in Kitazawa et al. (1983). Commercial computer X-ray analyzer systems may have the Marshall–Hall procedure included in their suite of analysis tools. The Marshall–Hall procedure works well for thin specimens in the "conventional" analytical energy regime ($E_0 \geq 10$ keV) of the SEM. The method will not work for specimens where the average atomic number is expected to vary significantly from one analysis point to another, or relative to that of the standard. A bulk specimen where the beam-damaged region is not constrained by the dimensions of the thin section, so that the region continues to change during electron bombardment also violates the fundamental assumptions. Consequently, many materials science applications for "soft" materials cannot be accommodated by the classic Marshall–Hall procedure.

Bulk Biological and Organic Specimens

The quantitative procedures devised by Statham and Pawley (1978) and Small et al. (1979) for the analysis of particles and rough specimens have been adapted to the analysis of bulk biological and organic samples (Roomans 1981, 1988; Echlin 1998). The method is based on the use of the ratio between the intensity of the characteristic and background X-rays defined as P/B, where P and B are measured over the range of energies that defines an EDS peak. The rationale behind the development of the method is that since the characteristic and background X-rays are generated within nearly the same depth distribution, they are subject to the same compositional related absorption and atomic number effects. It is assumed that the percentage of characteristic X-rays absorbed by the sample is the same as the percentage of continuum X-rays of the same energy which are absorbed. In the ratio P/B, the absorption factor (A) is no longer relevant as it has the same value in the numerator as the denominator and thus cancels. Since backscattered electrons are being lost due to changes in atomic number (Z), there is a similar decrease in the efficiency of production of both peak and background. Because the reduced X-ray production affects both peak and background in a similar (although not identical way), this factor is also cancelled out to a first order when the ratio P/B is measured. Additionally, because nearly all biological and/or organic materials consist of low atomic number matrix elements ($Z_{max} = 10$) the secondary fluorescence effect (F) is low and can be treated as a secondary order correction.

Strictly, these assumptions only hold true for homogeneous samples as the characteristic and background X-rays will vary with changes in the average atomic number of the sample. However, this is not considered to have any significant effect in cases where the P/B ratio method is applied to fully hydrated specimens which contain 85–90 % water or to dried organic material containing a small amount of light element salts. The ratio of peak area to the background immediately beneath the peak is relatively insensitive to small changes in surface geometry. However, the sample surface should be as smooth as is practicable because uneven fracture faces give unreliable X-ray data because of preferential masking and absorption.

Spectra are processed by the following procedure. The peaks in the spectra of the unknown and a standard of similar composition are fit by an appropriate procedure, such as multiple linear least squares, to determine the peak area for element i, P_i. The spectrum after peak-fitting and subtraction is then examined again to determine the background intensity remaining in the peak region of interest, giving the corresponding B_i at the same photon energy. Once accurate P/B

ratios are obtained, they can be used for quantitation in a number of ways. The P/B value for one element can be compared with the P/B value for another element in the same sample and to a first order:

$$\left(C_i / C_j\right) = h_{ij}\left[\left(P / B\right)_i / \left(P / B\right)_j\right] \qquad (20.10)$$

where C_i and C_j are the percentage concentrations of elements i and j and h_{ij} is a correction factor which can be obtained from measurements on a standard(s) of known composition very similar in composition to the unknown. Once h_{ij} has been empirically for the element(s) of interest, measurements of the P/B ratio(s) from the unknown can be immediately converted into concentration ratios. An advantage of taking the double ratio of (P/B) in Eq. (20.10) is the suppression of matrix effects, to a first order.

Alternatively, the P/B value for an element in the sample can be compared with the P/B value for the same element in a standard provided there is no significant difference in the matrix composition between sample and standard. If the mean atomic number of a given sample is always the same, the Kramers' relationship shows that the background radiation is proportional to atomic number. If the mean atomic number of the sample is always the same, then

$$C_i = h_i\left(P / B\right)_i \qquad (20.11)$$

where h_i is a constant for each element. If it is possible to analyze all the elements and calculate the concentration of elements such as C, H, O, and N by stoichiometry, then relative concentrations can be readily converted to absolute concentrations.

If there is a significant change in the composition between the unknown and standard(s), then a correction must be applied based upon the dependence of the continuum upon atomic number, following the original Marshall–Hall thin section method:

$$\left(C_i / C_j\right) = h_{ij}\left[\left(P / B\right)_i / \left(P / B\right)_j\right]\left[\left(Z_i^2 / A_i\right) / \left(Z_j^2 / A_j\right)\right] \qquad (20.12)$$

where Z and A are the atomic number and weight.

The peak-to-background ratio method has been found to be as efficient and accurate for biological materials as the more commonly used ZAF algorithms, which have been designed primarily for analyzing non-biological bulk samples. Echlin (1998) gives details of the accuracy and precision of the method as applied to hydrated and organic samples. For the analysis of a frozen hydrated tea leaf standard where independent analysis by atomic absorption spectroscopy was available for comparison, peak-to-background corrections generally gave results within ±10 % relative for trace Mg, Al, Si, and Ca over a range of beam energies from 5 to 20 keV.

References

Echlin P (1998) Low-voltage energy-dispersive X-ray microanalysis of bulk biological specimens. Microsc Microanal 4:577

Hall T (1968) "Some aspects of the microprobe analysis of biological specimens" in Quantitative Electron Probe Microanalysis, ed. Heinrich K. (U.S. Government Printing Office, Washington, DC) 269

Kitazawa T, Shuman H, Somlyo A (1983) Quantitative electron probe analysis: problems and solutions. Ultramicroscopy 11:251

Lifshin E, Doganaksoy N, Sirois J, Gauvin R (1999) Statistical considerations in microanalysis by energy-dispersive spectrometry. Microsc Microanal 4:598

Marshall D, Hall T (1966) "A method for the microanalysis of thin films", in X-ray Optics and Microanalysis, eds. Castaing R, Deschamps P, Philibert J. (Hermann, Paris) 374

Newbury D, Ritchie N (2015a) Review: performing elemental microanalysis with high accuracy and high precision by Scanning Electron Microscopy/Silicon Drift Detector Energy Dispersive X-ray Spectrometry (SEM/SDD-EDS). J Mats Sci 50:493

Newbury D, Ritchie N (2015b) Quantitative electron-excited X-ray microanalysis of Borides, Carbides, Nitrides, Oxides, and Fluorides with Scanning Electron Microscopy/Silicon Drift Detector Energy-Dispersive Spectrometry (SEM/SDD-EDS) and NIST DTSA-II. Microsc Microanal 21:1327

Ritchie N, Newbury D, Davis J (2012) EDS Measurements at WDS precision and accuracy using a silicon drift detector. Microsc Microanal 18:892

Ritchie N, Newbury D (2012) Uncertainty estimates for electron probe X-ray microanalysis measurements. Anal Chem 84:9956

Roomans GM (1981) Quantitative electron probe X-ray microanalysis of biological bulk specimens. SEM/1981/II (AMF O'Hare, IL, SEM, Inc.,) p 345

Roomans GM (1988) Quantitative X-ray microanalysis of biological specimens. J Electron Microsc Tech 9:19

Small J, Heinrich K, Newbury D, Myklebust R (1979) Progress in the development of the peak-to-background method for the quantitative analysis of single particles with the electron probe. SEM/1979/II, p 807

Statham P, Pawley J (1978) A new method for particle X-ray microanalysis based on peak-to-background measurements. SEM/1978/I, p 469

Vicenzi E, Eggins S, Logan A, Wysoczanski R (2002) Microbeam characterization of Corning archeological reference glasses: new additions to the Smithsonian microbeam standard collection. J Res National Inst Stand Technol 107:719

Ziebold T (1967) Precision and sensitivity in electron microprobe analysis. Anal Chem 39:858

Ziebold T, Ogilvie R (1964) An empirical method for electron microanalysis. Anal Chem 36:322

Trace Analysis by SEM/EDS

© Springer Science+Business Media LLC 2018
J. Goldstein et al., Scanning Electron Microscopy and X-Ray Microanalysis,
https://doi.org/10.1007/978-1-4939-6676-9_21

"Trace analysis" refers to the measurement of constituents presents at low fractional levels. For SEM/EDS the following arbitrary but practical definitions have been chosen to designate various constituent classes according to these mass concentration (C) ranges:

Major: $C > 0.1$ mass fraction (greater than 10 wt%)
Minor: $0.01 \leq C \leq 0.1$ (1 wt% to 10 wt%)
Trace: $C < 0.01$ (below 1 wt%)

Note that by these definitions, while "major" and "minor" constituents have defined ranges, "trace" has no minimum. Strictly, the presence of a single atom of the element of interest within the electron-excited mass that is analyzed by X-ray spectrometry represents the ultimate trace level that might be measured for that species, but such detection is far below the practical limit for electron-excited energy dispersive X-ray spectrometry of bulk specimens. In this section trace analysis down to levels approaching a mass fraction of $C = 0.0001$ (100 ppm (ppm) will be demonstrated. While such trace measurements are possible with high count EDS spectra, achieving reliable trace measurements by electron-excited energy dispersive X-ray spectrometry requires careful attention to identifying and eliminating, if possible, pathological contributions to the measured spectrum from unexpected remote radiation sources such as secondary fluorescence and backscattered electrons (Newbury and Ritchie 2016).

21.1   Limits of Detection for SEM/EDS Microanalysis

How does the measured EDS X-ray intensity (counts in the energy window) for an element behave for constituents in the trace regime? As described in the modules on X-ray physics and on quantitative X-ray microanalysis, the X-ray spectrum that is measured is a result of the complex physics of electron-excited X-ray generation and of subsequent propagation of X-rays through the specimen to reach the EDS spectrometer. The intensity measured for the characteristic X-rays of a particular element in the excited volume is affected by all other elements present, creating the so-called matrix effects: (1) the atomic number effect that depends on the rate of electron energy loss and on the number of backscattered electrons and the BSE energy distribution; (2) the absorption effect, where the mass absorption coefficient depends on all elements present; and (3) the secondary fluorescence effect, where the absorption of characteristic and continuum X-rays with photon energies above the ionization energy of the element of interest leads to additional emission for that element. Consider the number of X-rays, N_A, including characteristic plus continuum, measured in the energy window that spans the peak for element "A" as a function of the concentration C_A in a specific mixture of other elements "B," "C," "D," and so on. A simple example would be a binary alloy system where there is complete solid solubility from pure "A" to pure "B," for

example, Au-Cu or Au-Ag. When element "A" is present as a major constituent, as the concentration of "A" is reduced and replaced by "B," the matrix effects are likely to change significantly as the composition is changed. The impact of "B" on electron scattering and X-ray absorption of "A" is likely to produce a complex and non-linear response for N_A as a function of C_A. However, when the element of interest "A" is present in the trace concentration range, the matrix composition is very nearly constant as the concentration of "A" is lowered and replaced by "B," resulting in a monotonic dependence for N_A as a function of C_A, as shown in ◘ Fig. 21.1, known as a "working curve," which in the case of a dilute constituent will be linear. Consider that we have a known point on this linear plot with the values (N_s, C_s) in the trace concentration range that corresponds to the measurement of a known standard or is the result of a quantitative analysis of an unknown. The slope m of this linear function can be calculated from this known point and the 0 concentration point, for which $N_A = 0 + N_{cm}$, since there will still be counts, N_{cm}, in the "A" energy window due to the X-ray continuum produced by the other element(s) that comprise the specimen:

$$\text{slope} = (N_s - N_{cm})/(C_s - 0) \quad (21.1)$$

The slope-intercept form $(y = mx + b)$ for the linear expression for the X-ray counts in the "A" energy window, N_A, as a function of concentration, C_A, can then be constructed as:

$$N_A = \left[(N_s - N_{cm})/(C_s - 0)\right]C_A + N_{cm} \quad (21.2)$$

The y-axis intercept, b, is equal to N_{cm}, as shown in ◘ Fig. 21.1. Because of the presence of the continuum background at all

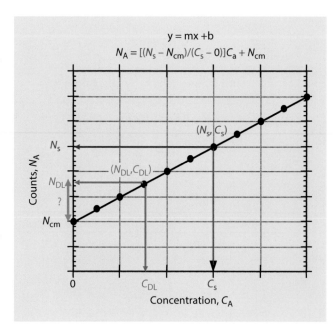

◘ **Fig. 21.1**   Linear working curve for a constituent at low concentration in an effectively constant matrix

photon energies, the concentration limit of detection, C_{DL}, must have a finite, non-zero value that will be found at some point (N_{DL}, C_{DL}) along the linear response between $(N_{cm}, 0)$ and (N_s, C_s). The work of (Currie 1968) can be used to define the condition at which the counts from the characteristic X-ray emission can be distinguished with a high degree of confidence above the natural statistical fluctuations in the background counts: the characteristic counts must exceed three times the standard deviation of the background:

$$N_A > 3 N_{cm}^{\frac{1}{2}} \qquad (21.3)$$

Thus, at the concentration limit of detection, C_{DL}:

$$N_{DL} = \text{continuum} + \text{characteristic} = N_{cm} + 3 N_{cm}^{\frac{1}{2}} \qquad (21.4a)$$

$$N_{DL} - N_{cm} = 3 N_{cm}^{\frac{1}{2}} \qquad (21.4b)$$

Substituting these conditions for (N_{DL}, C_{DL}) in Eq. (21.2):

$$N_{DL} = \left[\left(N_s - N_{cm} \right) / \left(C_s - 0 \right) \right] C_{DL} + N_{cm} \qquad (21.5a)$$

$$N_{DL} - N_{cm} = 3 N_{cm}^{\frac{1}{2}} = \left[\left(N_s - N_{cm} \right) / \left(C_s - 0 \right) \right] C_{DL} \qquad (21.5b)$$

$$C_{DL} = \left[3 N_{cm}^{\frac{1}{2}} / \left(N_s - N_{cm} \right) \right] C_s \qquad (21.5c)$$

Equation (21.5c) enables estimation of C_{DL} from the results of a single analysis of an unknown or from a single measurement of a known standard to provide a value for C_s. The corresponding measured EDS spectrum is used to determine N_s and N_{cm}. If n repeated measurements are made, N_s and N_{cm} are then taken as averages, $\dot{N}_s$ and $\dot{N}_{cm}$, over the n measurements, and Eq. (21.5c) becomes

$$C_{DL} = \left[3 \dot{N}_{cm}^{\frac{1}{2}} / \left(\dot{N}_s - \dot{N}_{cm} \right) n^{1/2} \right] C_s \qquad (21.6)$$

C_{DL} is an estimate of the concentration level of a constituent that can just be detected with a high degree of confidence. Quantification at C_{DL} is not reasonable because the error budget is dominated by the variance of the continuum. To achieve meaningful quantitation of trace constituents, (Currie 1968) further defines a minimum quantifiable concentration, C_{MQ}, which requires that the characteristic intensity exceed $10 N_{cm}^{\frac{1}{2}}$: Inserting this criterion in Eq. 21.5c gives

$$C_{MQ} = \left[10 N_{cm}^{\frac{1}{2}} / \left(N_s - N_{cm} \right) \right] C_s \qquad (21.7)$$

21.2 Estimating the Concentration Limit of Detection, C_{DL}

Equation (21.5c) or (21.6), as appropriate to single or multiple repeated measurements, can be used to estimate the concentration limit of detection for various situations.

21.2.1 Estimating C_{DL} from a Trace or Minor Constituent from Measuring a Known Standard

◻ Figure 21.2 shows a high count silicon drift detector (SDD)-EDS spectrum of K493 (and the residual spectrum after fitting for O, Si, and Pb), in NIST Research Material glass with the composition (as-synthesized) listed in ◻ Table 21.1. ◻ Table 21.1 also lists the measured peak intensity N_s and the background N_{cm} determined for this spectrum with the EDS spectrum measurement tools in DTSA-II. For a single measurement, the values for C_s, N_s, and N_{cm} inserted in Eq. (21.5c) gives the estimate of C_{DL} for each trace element, as also listed in ◻ Table 21.1. If $n = 4$ repeated measurements were made (or a single measurement was performed at four times the dose), C_{DL} would be lowered by a factor of 2. Examination of the values for C_{DL} in ◻ Table 21.1 reveals more than an order-of-magnitude variation depending on atomic number, for example, $C_{DL} = 52$ ppm for Al while $C_{DL} = 754$ ppm for Ta. This strong variation arises from differences in the relative excitation (overvoltage) and fluorescence yield for the various elements, differences in the continuum intensity, and the partitioning of the characteristic X-ray intensity among widely separated peaks for the L-family of the higher atomic number elements, for example, Ce and Ta.

21.2.2 Estimating C_{DL} After Determination of a Minor or Trace Constituent with Severe Peak Interference from a Major Constituent

Because of the relatively poor energy resolution of EDS, peak interference situations are frequently encountered. Multiple linear least squares peak fitting can separate the contributions from two or more peaks within an energy window. This effect is illustrated in ◻ Fig. 21.3 for Corning Glass A, the composition of which is listed in ◻ Table 21.2 along with the DTSA II analysis. There is a significant interference for K and Ca upon Sb and Sn. The initial qualitative analysis identified K and Ca, as shown in ◻ Fig. 21.3(a). When MLLS peak fitting is applied for K and Ca, the Sn and Sb L-family peaks are revealed in the residual spectrum, ◻ Fig. 21.3(b). The limit of detection calculated from the peak for Sn L determined from peak and background intensities determined from this residual spectrum is $C_{DL} = 0.00002$ (200 ppm).

21.2.3 Estimating C_{DL} When a Reference Value for Trace or Minor Element Is Not Available

Another type of problem that may be encountered is the situation where the analyst wishes to estimate C_{DL} for hypothetical

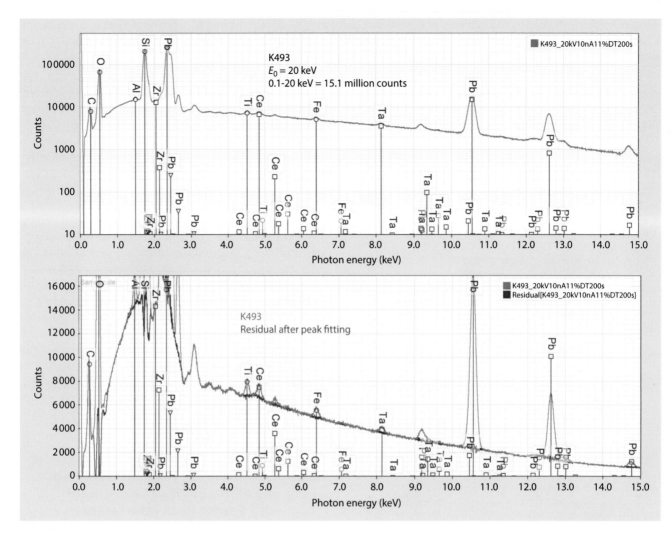

■ **Fig. 21.2**   SDD-EDS spectrum (0.1–20 keV = 29 million counts) of NIST Research Material Glass K493 at E_0 = 20 keV and residual after MLLS peak fitting for O, Si, and Pb

■ **Table 21.1**   Limits of detection estimated from a known standard K493 (E_0 = 20 keV 0.1–20 keV = 29 million counts)

Element	Mass conc	N_s (counts)	N_{cm} (counts)	C_{DL} (mass conc)	C_{DL} (ppm)
O	0.2063				
Al	0.00106	434,999	396,617	0.000052	52 ppm
Si	0.1304				
Ti	0.00192	280,709	255,991	0.000118	118 ppm
Fe	0.00224	229,356	210,799	0.000166	166 ppm
Zr	0.00363	313,728	303,760	0.000602	602 ppm
Ce (L)	0.00554	291,575	269,042	0.000383	383 ppm
Ta (L)	0.00721	156,283	145,340	0.000754	754 ppm
Pb	0.6413				

21

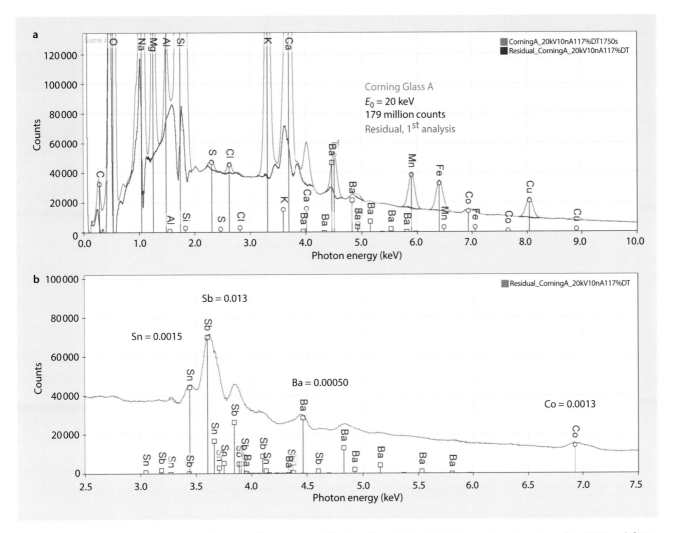

☐ **Fig. 21.3** **a** SDD-EDS spectrum (0.1–20 keV = 7.8 million counts) of Corning Glass A at E_0 = 20 keV and residual spectrum after MLLS peak fitting for K and Ca. **b** Expansion of K and Ca region showing detection of Sn and Sb L-family X-rays

trace constituents that might be present in regions of the spectrum that consist only of the X-ray continuum background. An example is shown in ☐ Fig. 21.4 for high purity Si. The spectrum consists of the Si K-shell X-rays, the associated coincidence peak, and the X-ray continuum background. Consider that the task is to estimate C_{DL} for several elements, for example, Al, Cr, and Cu. In the absence of a specimen of Si with known trace or minor levels of these elements, a reasonable estimate of C_{DL} can be made by determining the threshold k-ratio relative to a pure element, as illustrated for Cr and Cu with the spectra superimposed in ☐ Fig. 21.4(b). Using the energy window for Cr K-$L_{2,3}$ (CrKα), the continuum intensity in the Si spectrum at Cr K-$L_{2,3}$ is measured, $N_{cm\_Si\text{-}at\text{-}Cr}$, and is divided by the Cr K-$L_{2,3}$ intensity from the Cr spectrum at the equivalent dose, giving the k-ratio k_{DL} for detection:

$$k_{DL} = 3 \, N_{cm\_Si\text{-}at\text{-}Cr}^{\frac{1}{2}} / N_{Cr} \qquad (21.8)$$

Values of for k_{DL} for Al, Cr, and Cu as measured for this Si spectrum are listed in ☐ Table 21.3. These k-ratios can be converted into CDL values by calculating the ZAF matrix

correction factors for these constituents at trace levels in Si with DTSA-II, although this is generally a small correction.

21.3 Measurements of Trace Constituents by Electron-Excited Energy Dispersive X-ray Spectrometry

21.3.1 Is a Given Trace Level Measurement Actually Valid?

Trace analysis with high count EDS spectra can be performed to concentrations levels down to 0.0001 mass fraction (100 ppm) in the absence of interferences and 0.0005 (500 ppm) when peak interference occurs. The careful analyst will always ask the question, Is a given trace measurement valid? That is, does the measured trace constituent actually originate within the interaction volume of the specimen that is excited by primary electron beam, or is it the result of remote excitation of another part of the specimen or from components of the SEM itself?

◻ Table 21.2 Corning Glass A, as synthesized and as analyzed with DTSA-II ($E_0 = 20$ keV) [O by stoichiometry; Na (albite); Ca, P (fluoroapatite); S (pyrite); K, Cl (KCl); Sr (SrTiO$_3$); Ba aSi$_2$O$_5$); Pb (PbTe); Mg, Al, Si, Ti, Mn, Fe, Co, Cu, Zn, Sn, Sb (pure elements)]

Element	As-synthesized (mass conc)	DTSA-II analysis (mass conc)
O	0.4407	0.4474 (stoich.)
Na	0.106	0.106 ± 0.0002
Mg	0.0160	0.0163 ± 0.0001
Al	0.0053	0.0051 ± 0.00005
Si	0.3112	0.3171 ± 0.0002
P	0.00057	0.0003 ± 0.0001
S	0.0004	0.00085 ± 0.0001
Cl	0.00069	0.00072 ± 0.00005
K	0.0238	0.0237 ± 0.0001
Ca	0.0360	0.0341 ± 0.0001
Ti	0.00474	0.00485 ± 0.0001
Mn	0.00774	0.00768 ± 0.0001
Fe	0.00762	0.00717 ± 0.0001
Co	0.00134	0.00140 ± 0.0001
Cu	0.00935	0.00933 ± 0.0001
Zn	0.00035	0.00047 ± 0.0001
Sr	0.00085	0.0156 ± 0.0006
Sn	0.0015	0.0011 ± 0.0003
Sb	0.0132	0.0128 ± 0.0002
Ba	0.00502	0.00405 ± 0.0002
Pb	0.00111	0.050 ± 0.0001

The Inevitable Physics of Remote Excitation Within the Specimen: Secondary Fluorescence Beyond the Electron Interaction Volume

The electron interaction volume contains the region within which characteristic and continuum X-rays are directly excited by the beam electrons. Electron excitation effectively creates a volume source of generated X-rays (characteristic and continuum with energies up to the incident beam energy, E_0) embedded in the specimen that propagates out from the interaction volume in all directions. A photon propagating into the specimen will eventually undergo photoelectric absorption which ionizes the absorbing atom, and the subsequent de-excitation of this atom will result in emission of its characteristic X-rays, a process referred to as "secondary fluorescence" to distinguish this source from the "primary

fluorescence" induced directly by the beam electrons. Because the range of X-rays is generally one to two orders of magnitude greater than the range of electrons, depending on the photon energy and the specimen composition, the volume of secondary characteristic generation is much larger than the volume of primary characteristic generation. The range of fluorescence of Fe K-L$_{2,3}$ by Ni K-L$_{2,3}$ in a 75wt% Ni-25wt% Fe alloy at $E_0 = 25$ keV is shown in ◻ Fig. 21.5. The electron range is fully contained with a hemisphere of radius 2.5 μm, but a hemisphere of 80 -μm radius is required to capture 99 % of the secondary fluorescence of Fe K-L$_{2,3}$ by Ni K-L$_{2,3}$. For quantification with the ZAF matrix correction protocol, the secondary fluorescence correction factor, F, corrects the calculated composition for the additional radiation created by secondary fluorescence due to characteristic X-rays. An additional correction, c, is necessary for the continuum-induced secondary fluorescence. The F matrix correction factor is generally small compared to the absorption, A, and atomic number, Z, corrections. For a major constituent, the additional radiation due to secondary fluorescence represents a small perturbation in the apparent concentration, often negligible. However, when a constituent is at the trace level in the electron interaction volume, propagation of the primary characteristic and continuum X-rays into a nearby region of the specimen that is richer in this element will create additional X-rays of the trace element by secondary fluorescence. Because of the wide acceptance area of the EDS, this additional remote source of radiation will still be considered to be part of the spectrum produced at the beam position, possibly severely perturbing the accuracy of the analysis of the trace constituent by elevating the measured concentration above the true concentration. Compensation for this artifact requires careful modeling of the electron and X-ray interactions.

Simulation of Long-Range Secondary X-ray Fluorescence

The Monte Carlo electron trajectory simulation embedded in DTSA-II models the primary electron trajectories and primary X-ray generation, as well as the subsequent propagation of the primary characteristic and continuum X-rays through the target and the generation (and subsequent propagation) of secondary characteristic X-rays. The Monte Carlo menu provides "set-pieces" of analytical interest to predict the significance of secondary fluorescence at the trace level:

NIST DTSA II Simulation: Vertical Interface Between Two Regions of Different Composition in a Flat Bulk Target

◻ Figure 21.6 shows the simulation of an interface between Cu and NIST SRM470 (K-412 glass) for a beam with an incident energy of 25 keV placed 10 μm from the interface in the Cu region. The map of the distribution of excitation reveals the propagation of X-rays from the original electron

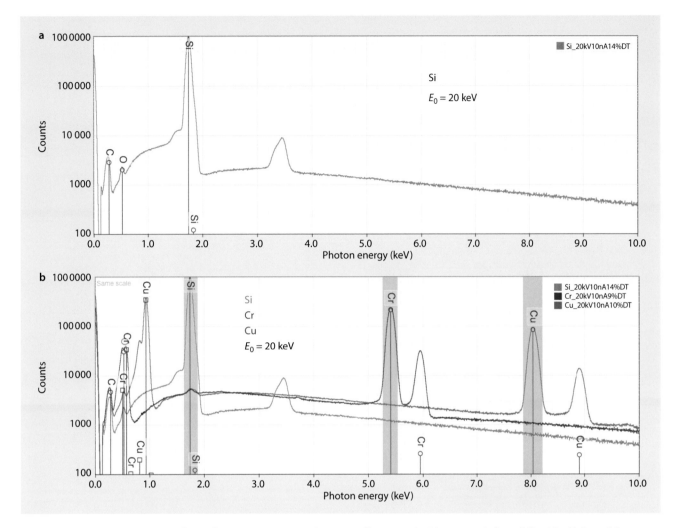

Fig. 21.4 **a** SDD-EDS spectrum of Si (20 keV; 1000 nA-s; 0.1–20 keV = 22 million counts) with energy windows defined for Al, Cr, and Cu. **b** Spectrum for Si with the spectra of Cr and Cu superimposed

Table 21.3 Estimated limits of detection k_{DL} for a Si spectrum with 110 million counts (0.1–20 keV)

Element	k_{DL}	ZAF	C_{DL} (mass conc)	C_{DL} (ppm)
Al	0.000115	1.12	0.000129	129
Cr	0.000133	1.01	0.000134	134
Cu	0.000113	1.00	0.000113	113

interaction volume into the K412 glass to excite secondary fluorescence. The calculated spectrum shown in **Fig. 21.6** shows the presence of an apparent trace level of Fe (and to a lesser extent, Ca, Si, Al, and Mg) in the Cu, corresponding to $k = 0.0028$ relative to a pure Fe standard. The Fe k-ratio as a function of beam position in the Cu is also shown in **Fig. 21.6**. Even with the beam placed in the Cu at a distance of 40 μm from the K-412, there is an apparent Fe trace level in the Cu of k = 0.0004, or about 400 ppm.

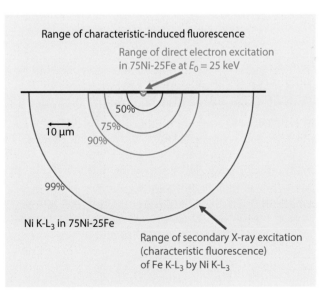

Fig. 21.5 Range of secondary fluorescence of FeKα by NiKα in a 75Ni-25Fe alloy at $E_0 = 25$ keV

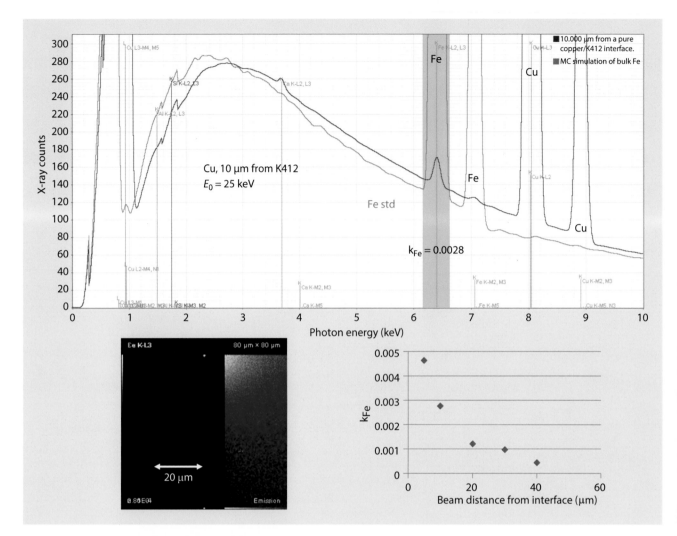

▣ Fig. 21.6 DTSA-II Monte Carlo calculation of fluorescence across a planar boundary between copper and SRM 470 (K-412 glass). The beam is placed in the copper at various distances from the interface. The spectrum calculated for a beam at 10 μm from the interface shows a small Fe peak, which is ratioed to the intensity calculated for pure iron, giving $k_{Fe} = 0.0028$. The inset map of the distribution of secondary FeKα X-ray production shows the extent of penetration of characteristic Cu Kα and Cu Kβ and continuum X-rays into the K-412 glass to fluoresce FeKα. Simulations at other distances give the response plotted in the graph

NIST DTSA II Simulation: Cubic Particle Embedded in a Bulk Matrix

▣ Figure 21.7(a) shows the results of a simulation of a 1-μm cube of K-411 glass embedded in a titanium matrix and excited with a beam energy of 20 keV. For this size and beam energy, the primary electron trajectories penetrate through the sides and bottom of the cube leading to direct electron excitation of the titanium matrix, which is seen as a major peak in the calculated spectrum. When the cube dimension is increased to 20 μm, the beam trajectories at $E_0 = 20$ keV are contained entirely within the K-411 cube. DTSA-II allows calculations with and without implementing the secondary fluorescence calculation. When secondary fluorescence is not implemented, the calculated spectrum ▣ Fig. 21.7(b) shows no Ti characteristic X-rays. When secondary fluorescence is included in the simulation, a small Ti peak is observed, corresponding to an artifact trace level $k = 0.0007$ (700 ppm), demonstrating the long range of the primary X-rays and the creation of a trace level artifact.

When variable pressure SEM operation is considered, the large fraction of gas-scattered electrons creates X-rays from regions up to many millimeters from the beam impact point. Depending on the surroundings, this gas scattering can greatly modify the EDS spectrum from what would originate from the region actually excited by the focused beam. ▣ Figure 21.7(c) shows this effect as simulated with DTSA II, resulting in a large peak for Ti, which is not present in the specimen but which is located in the surrounding region.

21.3 · Measurements of Trace Constituents by Electron-Excited Energy Dispersive X-ray Spectrometry

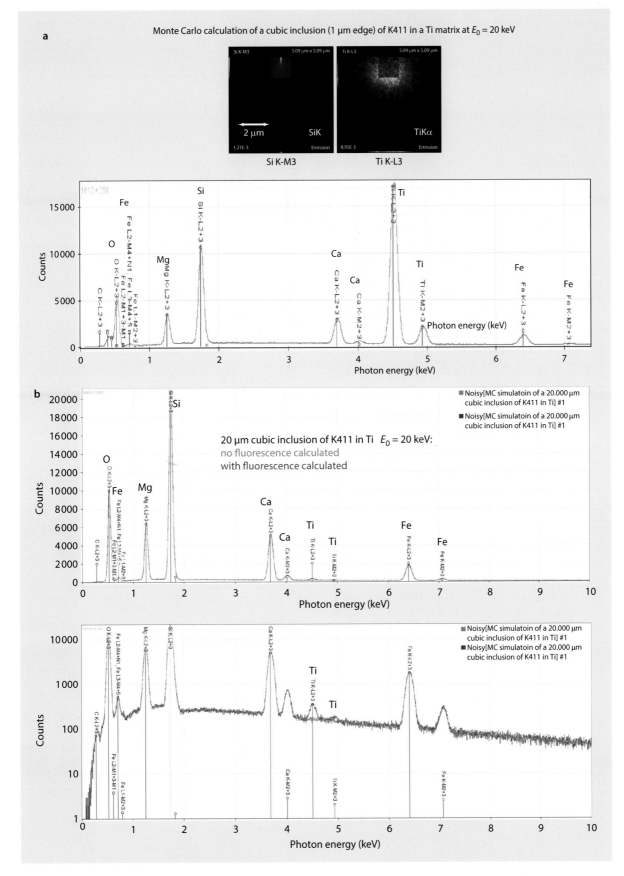

□ **Fig. 21.7   a** DTSA-II Monte Carlo calculation of a 1-μm cubical particle of K411 glass embedded in a Ti matrix with a beam energy of 20 keV, including maps of the distribution of SiK (particle) and TiKα (surrounding matrix). **b** 20-μm cubical particle of K411 glass embedded in a Ti with and without calculation of secondary fluorescence. **c** 20-μm cubical particle of K411 glass embedded in a Ti with calculation of secondary fluorescence and with calculation of gas scattering in VPSEM operation—water vapor; 133 Pa (1 Torr); 10-mm gas path length

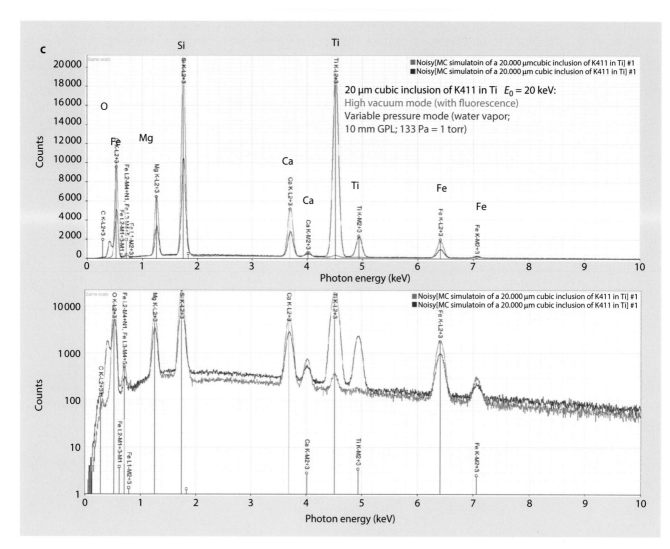

Fig. 21.7   (continued)

21.4   Pathological Electron Scattering Can Produce "Trace" Contributions to EDS Spectra

21.4.1   Instrumental Sources of Trace Analysis Artifacts

While secondary fluorescence that leads to generation of X-rays at a considerable distance from the beam impact is a physical effect which cannot be avoided, there are additional pathological scattering effects that can be minimized or even eliminated. ◘ Figure 21.8 depicts the idealized view of the emission of X-rays generated by the electron beam in the SEM. In this idealized view, the only X-rays that are collected are those emitted into the solid angle of acceptance of the detector, which is defined by a cone whose apex is centered on the specimen interaction volume, whose altitude is the specimen-to-detector distance, and whose base is the active area of the detector that is not shielded by the

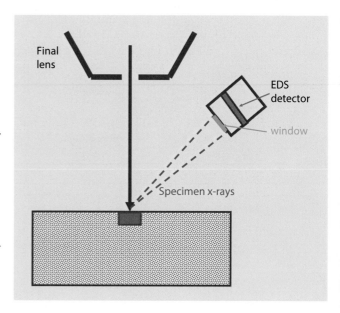

Fig. 21.8   Ideal view of the collection angle of an EDS system

■ **Fig. 21.9** Effect of backscattering to produce remote X-ray sources on SEM components (objective lens, chamber walls, stage, etc.) and use of collimator to block these contributions from reaching the EDS

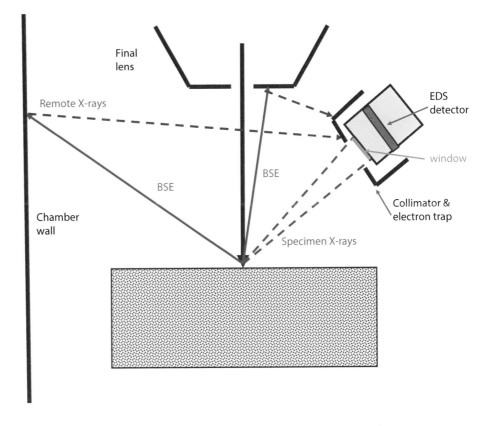

entrance window or other hardware. However, the reality of the EDS measurement is likely to be quite different from this ideal case, at least at the trace constituent level, as a consequence of electron backscattering, shown schematically in ■ Fig. 21.9. For targets of intermediate and high atomic number, a significant fraction of the incident beam is emitted as backscattered electrons, and the majority of these BSEs retain more than half of the incident beam energy. After leaving the specimen, these BSEs are likely to strike the objective lens and the walls of the specimen chamber as well as other hardware, where they generate the characteristic (and continuum) X-rays of those materials. The EDS detector collects X-rays from any source with a line-of-sight to the detector, so to minimize remote BSE-induced contributions to the measured spectrum, the EDS is equipped with a collimator whose function is to restrict the view of the EDS, as illustrated schematically in ■ Fig. 21.9. The solid angle of acceptance of the EDS is substantially reduced by the collimator, minimizing remote contributions from the lens and chamber walls. While the collimator provides a critical improvement to the measured spectrum, it is important for the analyst to understand its inevitable limitations. The actual acceptance solid angle must be constructed by looking out from the detector through the aperture of the collimator, as shown in ■ Fig. 21.10. The typical collimator accepts X-rays

generated in the specimen plane within a circular area with a diameter of several millimeters, a feature that is important for X-ray mapping applications, where the beam is scanned over large lateral areas and X-rays must be accepted from any beam position within the scanned area. Moreover, the acceptance region is three dimensional with a vertical dimension of several millimeters along the beam axis. To determine the true acceptance volume of the EDS collimator, low magnification (maximum scanning area) X-ray mapping of a target such as a blank aluminum sample stub provides a direct view of the transmission of the EDS collimator as a function of x-y position and as a function of the z-position, as shown in ■ Fig. 21.11. For this example, any X-ray generated in a large volume (at least $2.5 \times 3 \times 10$ mm) can potentially be collected by this EDS system despite the otherwise effective collimator. Three important sources of uncontrolled remote excitation within this collimator acceptance volume are shown in ■ Fig. 21.12: (1) beam electrons scattering off the edge of the final aperture (magenta trajectory); (2) beam electrons being stopped by the final aperture and generating the characteristic and continuum X-rays of the aperture material (e.g., Pt; blue dashed trajectory); and (3) re-scattering of BSEs that have struck the final lens and return to the specimen (red trajectory). Both of these sources can create X-rays several millimeters or more from the beam impact location.

■ Fig. 21.10   True extent of acceptance area of EDS constructed by looking out through the collimator

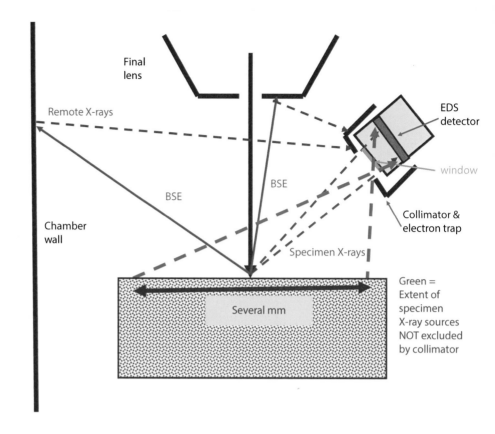

■ Fig. 21.11   X-ray mapping to determine the acceptance volume of the collimator. A series of Al X-ray maps of an aluminum SEM stub at different working distances is shown; the inset graph shows the intensity at the center of each map as a function of working distance

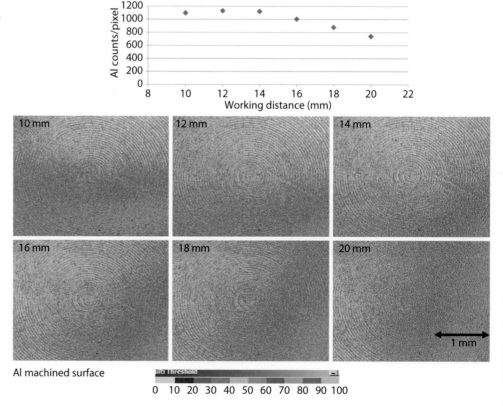

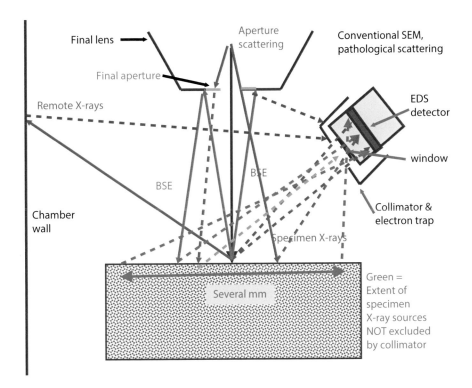

Fig. 21.12 Possible sources of remote excitation: beam electrons scattering off edge of final aperture, beams stopped by aperture generating characteristic and continuum X-rays, and re-scattering of backscattered electrons from the lens

21.4.2 Assessing Remote Excitation Sources in an SEM-EDS System

Remote excitation of X-rays can be assessed by measuring various structures. As shown in ◘ Fig. 21.13, a multi-material Faraday cup can be constructed by placing an SEM aperture (typically 2.5 mm in diameter and made of platinum or another heavy metal) over a blind hole drilled in a block of a different metal, such as a 1-cm diameter aluminum SEM stub, which is then inserted in a hole drilled in a 2.5-cm-diameter brass (Cu-Zn) block. This structure can be used to measure the "in-hole" spectrum to assess sources and magnitude of remote excitation (Williams and Goldstein 1981). The following sequence of measurements is made, as shown in ◘ Fig. 21.14. The beam is successively placed for the same dose on the brass, the Al-stub, the Pt-aperture, and finally in the center of the hole (e.g., 200-µm diameter) of the aperture. Ideally, if there are no electrons scattered outside the beam by interacting with the final aperture or other electron column surfaces, the "in-hole" specimen will have no counts. As shown in ◘ Fig. 21.14 with a logarithmic intensity display, a small number of counts is detected for Pt M, equivalent to $k = 0.00008$ of the intensity measured for the same dose with the beam placed on the Pt aperture. No detectable counts are found for Al from the stub or for the Cu and Zn from the brass block. Thus, for this particular instrument, a small but detectable pathological scattering occurs within approximately 1.5 mm of the central beam axis. While this is a very small effect, the analyst must nonetheless be aware that this unfocused electron or aperture X-ray source might contribute an artifact at the trace level if the element of interest at the beam location is abundant in a nearby region.

While a useful measurement and the place to start in assessing remote excitation, the "in-hole" measurement only detects electrons scattered outside of the beam. Typically, a more serious source of remote excitation is the backscattered electrons (BSEs), which are absent from the "in-hole" measurement. ◘ Figure 21.15 shows a modification of the "in-hole" multi-material target in which the central hole is replaced by a flat, polished scattering target. ◘ Figure 21.16 shows an example of a spectrum in which the central target is high purity carbon, which has a low BSE coefficient of 0.06, surrounded by a 3-mm-diameter region filled with Ag-epoxy, which is surrounded by a Ti block. No detectable counts for characteristic peaks of Ag (conducting epoxy) or Ti (specimen holder) are found.

◘ Figure 21.17 shows a similar measurement for high purity tantalum, which has a high BSE coefficient of 0.45. Both Ag and Ti are detectable at very low relative intensity compared to the intensity measured with the beam placed on pure element targets.

When a three-dimensional target is used for scattering, as illustrated in ◘ Fig. 21.18, additional BSEs are scattered from the tilted surfaces into the regions of the specimen near the beam impact point as well as more distant regions surrounding the specimen. ◘ Figure 21.19 shows such a measurement for a pyramidal fragment of SrF_2 placed on a brass substrate. Low level signals are observed for CuKα and ZnKα, and also for NiKα, which arises from Ni-plating on nearby stage components. This extreme case most closely resembles the challenge posed by a rough, topographic specimen. The uncontrollable scattering renders most trace constituent determinations questionable.

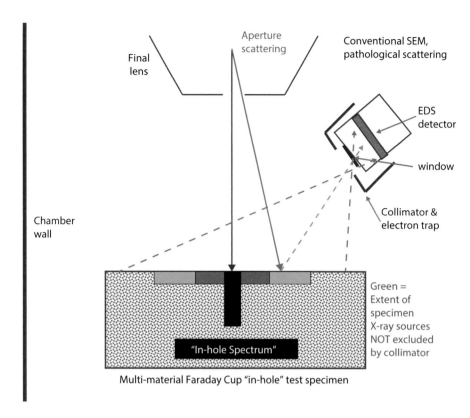

Fig. 21.13 Measurement of "in-hole" spectrum with a multi-material Faraday cup consisting of three materials arranged concentrically

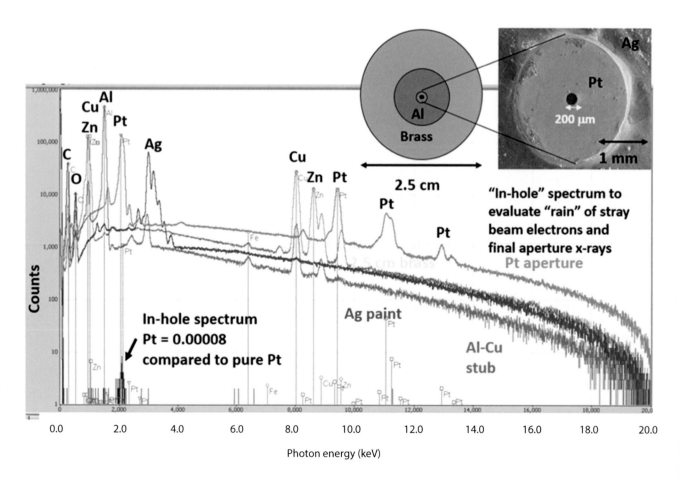

Fig. 21.14 Sequence of measurements for determining pathological scattering effects by the "in-hole" method. For the same dose, EDS spectra are measured on the brass block, aluminum stub, platinum aperture, and at the center of the 200 μm diameter hole of the platinum aperture. A logarithmic display reveals the very low counts in the "in-hole" spectrum compared to the spectra measured on the various materials

Fig. 21.15 Schematic diagram of the "in-hole" configuration with a pure element target placed at the center of a multi-material target

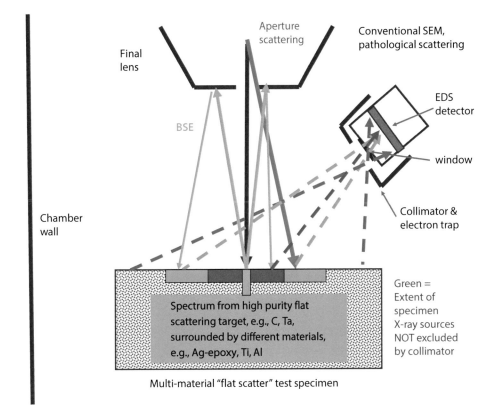

Fig. 21.16 Measurement of high purity C surrounded by Ag (doped epoxy) and titanium; no significant signals for Ag or Ti are observed

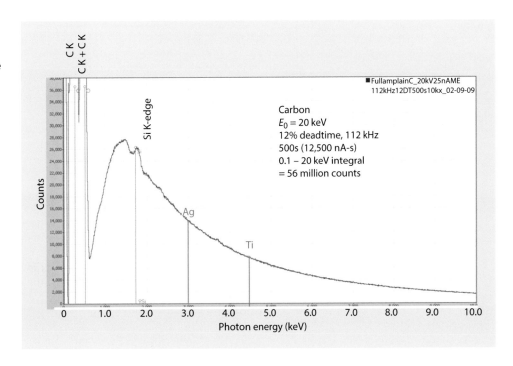

Fig. 21.17 Measurement of high purity Ta surrounded by Ag (doped epoxy) and titanium; Ag or Ti are both observed at very low levels: Ag, k = 0.000003; Ti, k = 0.000005

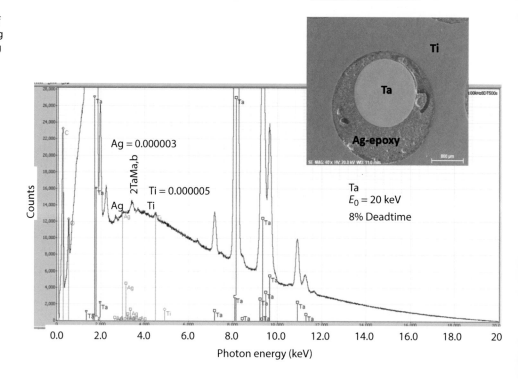

Fig. 21.18 Modification of the "in-hole" configuration with a three-dimensional target to produce the effects of backscattering from inclined surfaces

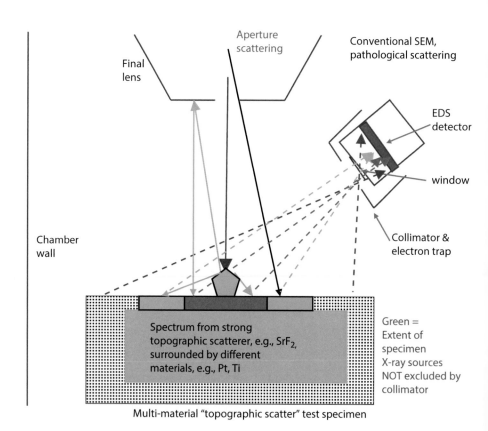

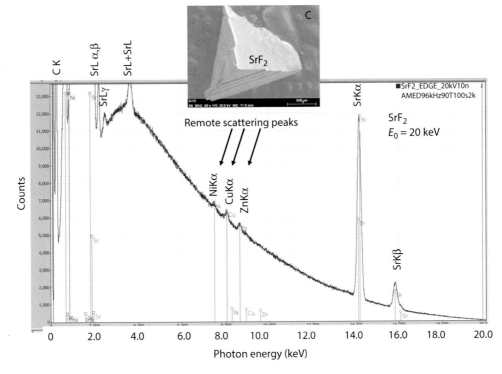

Fig. 21.19 Example with a pyramid of high purity SrF_2 as the scattering target surrounded by a carbon tab (1 cm diameter) on a 2.5 cm diameter brass (Cu and Zn) disk. The Ni signal that is observed likely arises from the Ni-coatings of the specimen stage components

21.5 Summary

High count EDS spectra can be used to achieve limits of detection approaching a mass concentration of $C_{DL} = 0.0001$ (100 ppm) when there are no peak interferences from higher concentration constituents, and $C_{DL} = 0.0005$ (500 ppm) when significant peak interference does occur. However, the analyst must carefully test the SEM/EDS measurement environment to ensure that the trace measurement is meaningful and not a consequence of pathological remote scattering effects.

References

Currie LA (1968) Limits for qualitative detection and quantitative determination. Anal Chem 40:586–593

Newbury DE, Ritchie NWM (2016) Measurement of trace constituents by electron-excited X-ray microanalysis with energy dispersive spectrometry. Micros Microanal 22(3):520–535

Williams DB, Goldstein JI (1981) Artifacts encountered in energy dispersive X-ray spectrometry in the analytical electron microscope. In: Heinrich KFJ, Newbury DE, Myklebust RL, Fiori CE (eds) Energy dispersive X-ray spectrometry, National Bureau of Standards Special Publication 604 (U.S. Department of Commerce, Washington, DC), pp 341–349

Low Beam Energy X-Ray Microanalysis

© Springer Science+Business Media LLC 2018
J. Goldstein et al., *Scanning Electron Microscopy and X-Ray Microanalysis*,
https://doi.org/10.1007/978-1-4939-6676-9_22

22.1 What Constitutes "Low" Beam Energy X-Ray Microanalysis?

The incident beam energy, E_0, is the parameter that determines which characteristic X-rays can be excited: the beam energy must exceed the critical excitation energy, E_c, for an atomic shell to initiate ionization and subsequent emission of characteristic X-rays. This dependence is parameterized with the "overvoltage" U_0, defined as

$$U_0 = E_0 / E_c \qquad (22.1)$$

U_0 must exceed unity for X-ray emission. The intensity, I_{ch}, of characteristic X-ray generation follows an exponential relation:

$$I_{ch} = i_B a (U_0 - 1)^n \qquad (22.2)$$

where i_B is the beam current, a and n are constants, with $1.5 \le n \le 2$.

The intensity of the X-ray continuum (*bremsstrahlung*), I_{cm}, also depends on the incident beam energy:

$$I_{cm} = i_B b Z (U_0 - 1) \qquad (22.3)$$

where b is a constant and Z is the mass-concentration-averaged atomic number of the specimen.

The peak-to-background is then obtained as the ratio of Eqs. (22.2) and (22.3):

$$P / B = I_{ch} / I_{cm} \approx (1 / Z)(U_0 - 1)^{n-1} \qquad (22.4)$$

The value of $n - 1$ in Eq. (22.4) ranges from 0.5 to 1, so that the *P/B* rises slowly as U_0 increases above unity. ◘ Figure 22.1 shows experimental measurements of Si K-L_2 + Si K-M_3 (Si Kα,β) characteristic X-ray intensity as a function of overvoltage. Near the threshold of $U_0 = 1$, the intensity drops sharply, and the Si K-L_2 + Si K-M_3 peak becomes progressively lower relative to the X-ray continuum background, as shown in ◘ Fig. 22.2 for Si measured over a range of beam energies. The peak-to-background strongly influences the limit-of-detection. While X-ray measurements can certainly be made with $1 < U_0 < 1.25$ and the limit-of-detection can be improved by increasing the integrated spectrum intensity by extending the counting time, the detectability within a practical measuring time of a constituent excited in this overvoltage range diminishes. While major constituents may be detected, minor and trace constituents are likely to be below the limit of detection. Thus the situation for $1 < U_0 < 1.25$ must generally be considered "marginally detectable" and is so marked in ◘ Figs. 22.3, 22.4, 22.5, 22.6, 22.7, and 22.8.

◘ **Fig. 22.1**   Production of silicon K-shell X-rays with overvoltage

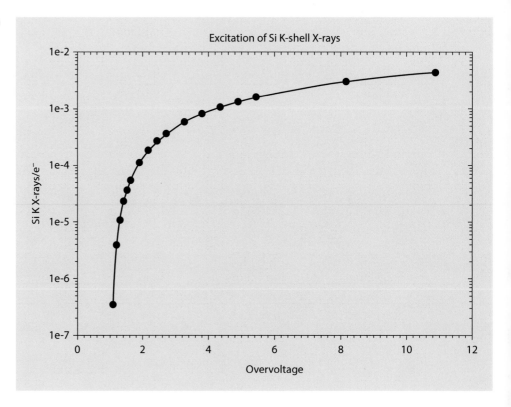

361

22

22.1 · What Constitutes "Low" Beam Energy X-Ray Microanalysis?

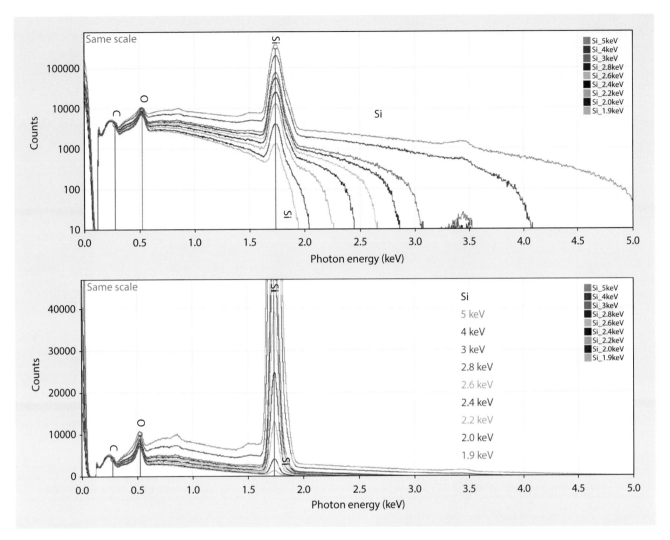

Fig. 22.2 Silicon at various incident beam energies from 5 keV to 1.9 keV showing the decrease in the peak-to-background with decreasing overvoltage

Fig. 22.3 Periodic table illustrating X-ray shell choices for developing analysis strategy within the conventional beam energy range, $E_0 = 20$ keV (Newbury and Ritchie, 2016)

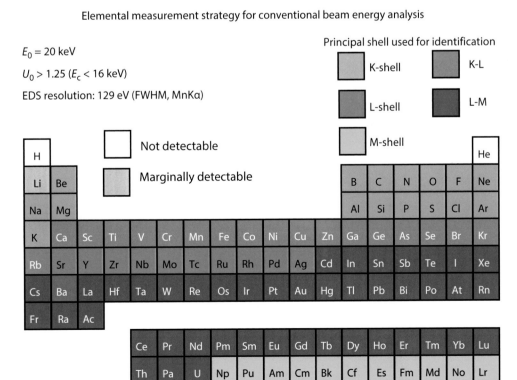

Elemental measurement strategy for conventional beam energy analysis

$E_0 = 20$ keV

$U_0 > 1.25$ ($E_c < 16$ keV)

EDS resolution: 129 eV (FWHM, MnKα)

Principal shell used for identification

- K-shell
- K-L
- L-shell
- L-M
- M-shell
- Not detectable
- Marginally detectable

■ **Fig. 22.4** Periodic table illustrating X-ray shell choices for developing analysis strategy at the lower end of the conventional beam energy range, $E_0 = 10$ keV (Newbury and Ritchie, 2016)

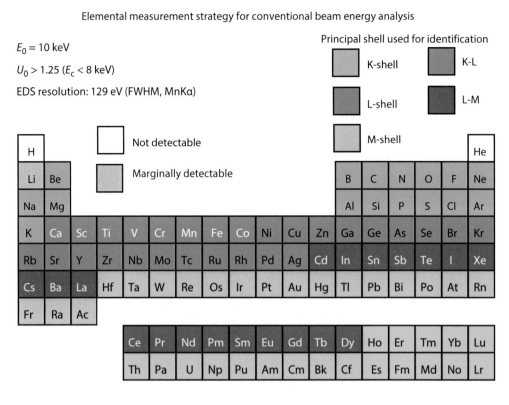

■ **Fig. 22.5** Periodic table illustrating X-ray shell choices for developing analysis strategy for the upper end of the low beam energy range, $E_0 = 5$ keV (Newbury and Ritchie, 2016)

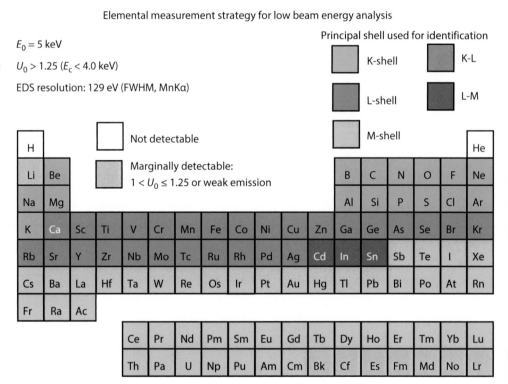

22

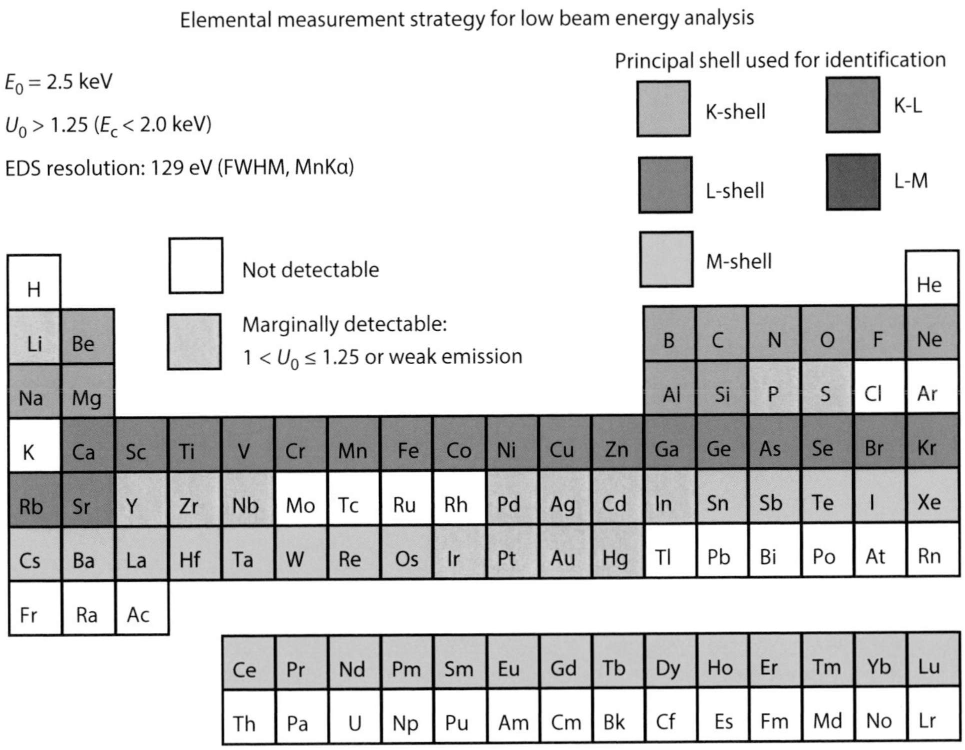

363 **22**

22.1 · What Constitutes "Low" Beam Energy X-Ray Microanalysis?

Fig. 22.6 Periodic table illustrating X-ray shell choices for developing analysis strategy within the low beam energy range, $E_0 = 2.5$ keV (Newbury and Ritchie, 2016)

Elemental measurement strategy for low beam energy analysis

$E_0 = 2.5$ keV

$U_0 > 1.25$ ($E_c < 2.0$ keV)

EDS resolution: 129 eV (FWHM, MnKα)

Principal shell used for identification

K-shell · K-L · L-shell · L-M · M-shell

Not detectable

Marginally detectable: $1 < U_0 \leq 1.25$ or weak emission

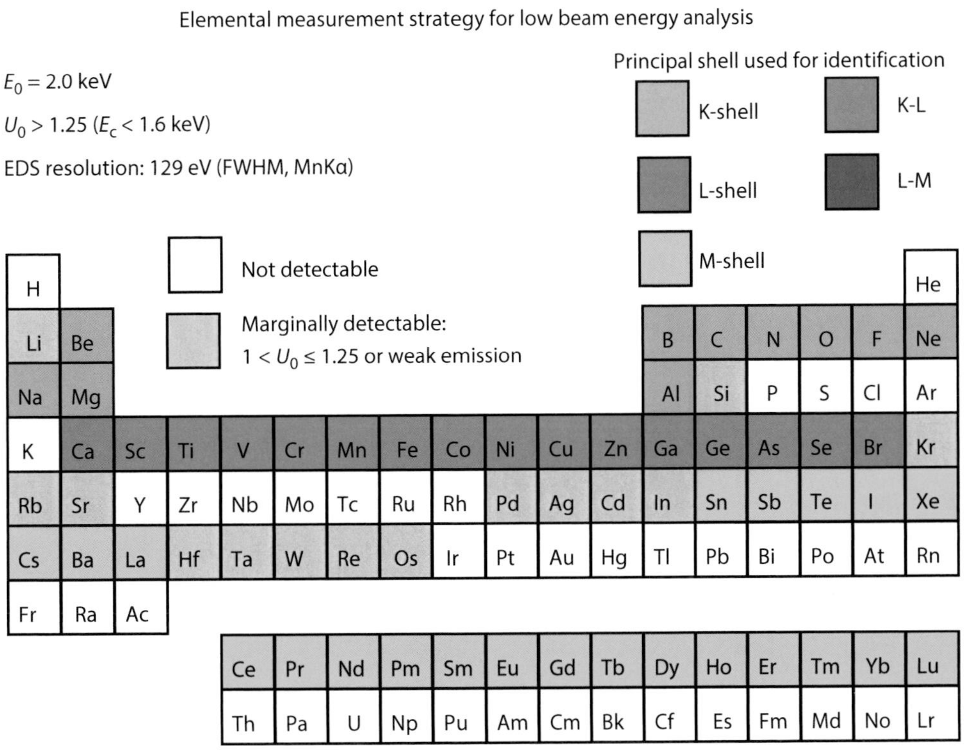

Fig. 22.7 Periodic table illustrating X-ray shell choices for developing analysis strategy within the low beam energy range, $E_0 = 2.0$ keV (Newbury and Ritchie, 2016)

Elemental measurement strategy for low beam energy analysis

$E_0 = 2.0$ keV

$U_0 > 1.25$ ($E_c < 1.6$ keV)

EDS resolution: 129 eV (FWHM, MnKα)

Principal shell used for identification

K-shell · K-L · L-shell · L-M · M-shell

Not detectable

Marginally detectable: $1 < U_0 \leq 1.25$ or weak emission

Fig. 22.8 Periodic table illustrating X-ray shell choices for developing analysis strategy within the low beam energy range, $E_0 = 1.0$ keV (Newbury and Ritchie, 2016)

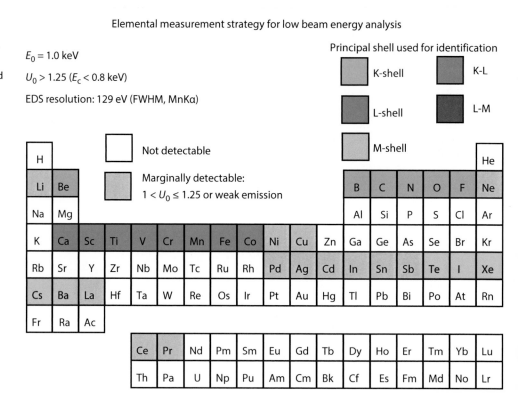

Elemental measurement strategy for low beam energy analysis

$E_0 = 1.0$ keV

$U_0 > 1.25$ ($E_c < 0.8$ keV)

EDS resolution: 129 eV (FWHM, MnKα)

22.1.1 Characteristic X-ray Peak Selection Strategy for Analysis

"Conventional" electron-excited X-ray microanalysis is typically performed with an incident beam energy selected between 10 keV and 30 keV. A beam energy is this range is capable of exciting X-rays from one or more atomic shells for all elements of the periodic table, except for hydrogen and helium, which do not produce characteristic X-ray emission. Li can produce X-ray emissions, but the energy of 0.052 keV is below the practical detection limit of most EDS systems, which typically have a threshold of approximately 0.1 keV. Recent progress in silicon drift detector (SDD)-EDS detector technology and isolation windows is rapidly improving the EDS performance in the photon energy range 50 eV – 250 eV, raising the measurement situation for Li from "undetectable" to the level of "marginally detectable." The choices available for the characteristic X-ray peaks to analyze various elements are illustrated in the periodic table shown in ◻ Fig. 22.3 for $E_0 = 20$ keV. In constructing this diagram, the assumption has been made with the requirement that $U_0 > 1.25$ ($E_c < 16$ keV) to provide for robust detection of major and minor constituents. Note that in constructing ◻ Fig. 22.3, only the excitation of characteristic X-rays has been considered and not the subsequent absorption of X-rays during propagation through the specimen to reach the detector. Absorption has a strong effect on low energy X-rays below 2-keV photon energy and strongly depends on composition and beam energy. Absorption can be minimized by operating at low beam energy, as discussed below, an important factor that must also be considered when developing practical X-ray measurement strategy.

As seen in ◻ Fig. 22.3, when the beam energy is selected at the high end of the conventional range, X-rays from two different atomic shells can be excited for many elements, and the additional information provided by having two X-ray families with two or more peaks to identify greatly increases the confidence that can be placed in an elemental identification. This is especially valuable when peak interference occurs between two elements. For example, a severe interference occurs between S K-$L_{2,3}$ (2.307 keV) and Mo L_3-$M_{4,5}$ (2.293 keV), which are separated by 14 eV. To confirm the presence of Mo when S may also be present, operation with $E_0 > 25$ keV ($U_0 = 1.25$) will also excite Mo K-$L_{2,3}$ (17.48 keV) for unambiguous identification of Mo.

When the beam energy is lowered to the bottom of the conventional analysis range, $E_0 = 10$ keV, the available X-ray shells for measurement are reduced as shown in ◻ Fig. 22.4. Many more elements can only be analyzed with X-rays from one shell, for example, the Ni to Rb L-shells and the Hf to U M-shells.

22.1.2 Low Beam Energy Analysis Range

When the incident beam energy is reduced to $E_0 = 5$ keV, further reduction in the atomic shells that can be excited

22

creates the situation shown in ◼ Fig. 22.5. At 5 keV, only one shell is available for all elements except Ca, Cd, In, and Sn. When the incident energy is reduced below 5 keV, some elements are effectively rendered analytically inaccessible by the restrictions imposed by the X-ray physics. The progressive loss of access to elements in the periodic table is illustrated for $E_0 = 2.5$ keV (◼ Fig. 22.6), $E_0 = 2$ keV (◼ Fig. 22.7), and $E_0 = 1$ keV (◼ Fig. 22.8). Indeed, even with $E_0 = 5$ keV, several elements must be measured with X-rays from shells with low fluorescence yield, such as the Ti L-family and the Ba M-family, resulting in poor peak-to-background.

Based upon the restrictions imposed by the physics of X-ray generation, $E_0 = 5$ keV is the lowest energy which still gives access to the full periodic table, except for H and He, and therefore this value will be considered as the upper bound of the beam energy range for low beam energy microanalysis. The beam energy range from 5 keV to 10 keV represents the transition region between low beam energy microanalysis and conventional X-ray microanalysis.

22.2 Advantage of Low Beam Energy X-Ray Microanalysis

22.2.1 Improved Spatial Resolution

The spatial resolution of X-ray microanalysis is controlled by the range of the electrons for the excitation of characteristic X-rays, as described by the Kanaya–Okayama (1974) range equation modified for the threshold of X-ray production set by the critical excitation energy:

$$R_{K-O}\,(\mu m) = \left(0.0276\, A\, /\, Z^{0.89}\, \rho\right)\left(E_0^{1.67} - E_c^{1.67}\right) \quad (22.5)$$

where A (g/mol) is the atomic weight, Z is the atomic number, ρ (g/cm³) is the density, E_0 (keV) is the beam energy and E_c (keV) is the shell ionization energy. ◼ Figure 22.9 shows the range for the production of Na K-shell X-rays in various matrices: C, Al, Ti, Fe, Ag, and Au. For low beam energy analysis conditions, the range at $E_0 = 5$ keV varies from 0.46 μm (460 nm) for Na in a C matrix to 0.08 μm (80 nm) in

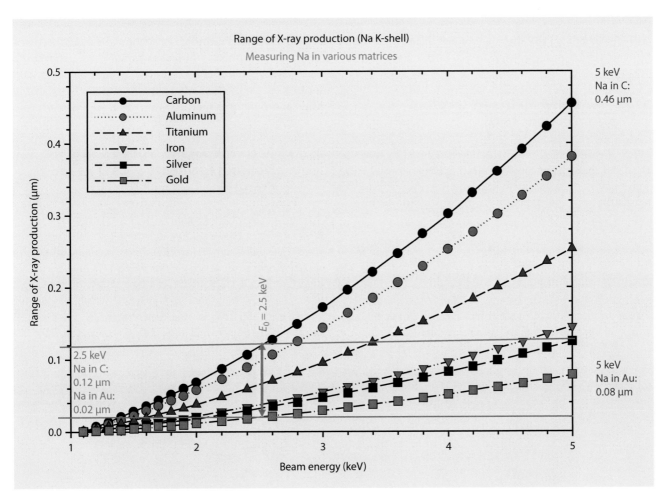

◼ **Fig. 22.9** Range of production of Na K-shell X-rays in various matrices, as calculated with the Kanaya–Okayama range equation

□ **Fig. 22.10**   Absorption correction factor for O K-L$_{2,3}$ (relative to MgO) and Cu L$_3$-M$_{4,5}$ (relative to Cu) as a function of beam energy

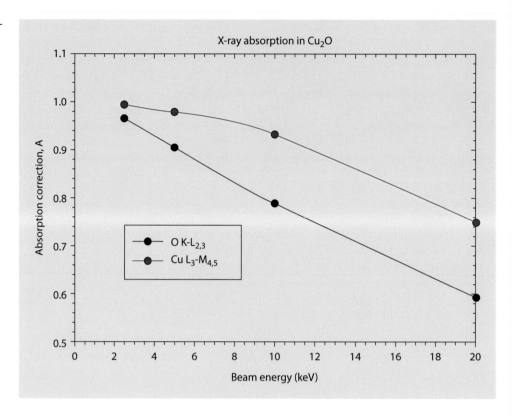

Au, while at $E_0 = 2.5$ keV, the range for Na collapses to 0.12 μm (120 nm) in a C matrix to 0.02 μm (20 nm) in Au.

22.2.2   Reduced Matrix Absorption Correction

When the range of X-ray production is reduced by lowering the beam energy, the generated X-rays undergo lower absorption because of the reduced path length to the surface. This can be a strong effect, because X-ray absorption follows an exponential relationship:

$$I / I_0 = exp\left[-\left(\mu / \rho\right)\rho s\right] \qquad (22.6)$$

where I_0 is the original intensity and I is the intensity remaining after passing through a distance s (cm) of a material of density ρ (g/cm^3) and of mass absorption coefficient, μ/ρ (cm^2/g). For strongly absorbed photons, which is

typically the case for low energy photons, the matrix correction for absorption diminishes rapidly (i.e., approaches unity) as the beam energy is reduced, as shown in □ Fig. 22.10 for O K-L$_{2,3}$ and Cu L$_3$-M$_{4,5}$ in Cu$_2$O (measured relative to MgO and Cu).

22.2.3   Accurate Analysis of Low Atomic Number Elements at Low Beam Energy

Low atomic number elements with $Z \leq 10$ have characteristic X-ray energies below 1 keV, and these low energy photons suffer especially high absorption. By minimizing absorption through operation at low beam energy, Li and Be can be detected; and B, C, N, and F, can be quantitatively analyzed with accuracy such that the analyzed value is generally within ± 5 % relative to the true value, as presented in □ Table 22.1 (borides), □ Table 22.2 (carbides), □ Table 22.3

22

□ **Table 22.1**   Analysis of metal borides at $E_0 = 5$ keV (5 replicates); atomic concentrations

Compound	Metal, C_{av}	Relative accuracy,%	σ_{rel}, %	Boron, C_{av}	Relative accuracy,%	σ_{rel}, %
CrB$_2$	0.3482	4.5	0.32	0.6518	−2.2	0.17
CrB	0.5149	3.0	0.17	0.4851	−3.0	0.19
Cr$_2$B	0.6769	1.5	0.65	0.3231	−3.1	1.4
TiB$_2$	0.3373	1.2	3.6	0.6627	−0.6	1.8

Table 22.2 Analysis of metal carbides at $E_0 = 5$ keV (5 replicates); atomic concentrations

Compound	Metal, C_{av}	Relative accuracy,%	σ_{rel}, %	Carbon, C_{av}	Relative accuracy,%	σ_{rel}, %
SiC	0.4935	−1.3	0.25	0.5065	1.3	0.25
Cr_3C_2	0.6002	0.03	1.4	0.3998	−0.05	2
Fe_3C	0.7479	−0.28	0.23	0.2521	0.84	0.67
ZrC	0.5025	0.49	1.2	0.4975	−0.49	1.2

Table 22.3 Analysis of metal nitrides at $E_0 = 5$ keV (5 replicates); atomic concentrations

Compound	Metal, C_{av}	Relative accuracy,%	σ_{rel}, %	Nitrogen, C_{av}	Relative accuracy,%	σ_{rel}, %
TiN	0.5168	3.4	0.30	0.4832	−3.4	0.32
Cr_2N	0.6606	−0.91	0.55	0.3394	1.8	1.1
Fe_3N	0.7413	−1.1	2	0.2587	3.5	5.7
HfN	0.5050	1.0	1.4	0.495	−1.0	1.4

Table 22.4 Analysis of metal oxides at $E_0 = 5$ keV (5 replicates); atomic concentrations

Compound	Metal, C_{av}	Relative accuracy,%	σ_{rel}, %	Oxygen, C_{av}	Relative accuracy,%	σ_{rel}, %
TiO_2	0.3299	−1.0	0.34	0.6701	0.5	0.17
NiO	0.5110	2.2	0.30	0.4890	−2.2	0.34
CuO	0.5105	2.1	0.10	0.4895	−2.1	0.11
Cu_2O	0.6815	2.2	0.13	0.3185	−4.4	0.28

(nitrides), and ◘ Table 22.4 (oxides) (Newbury and Ritchie 2015). Examples of EDS spectra and the residual spectrum after fitting are shown in ◘ Fig. 22.11 (Cr borides), ◘ Fig. 22.12 (Cr_3C_2), ◘ Fig. 22.13 (Fe_3N), and ◘ Fig. 22.14 (Cu oxides). These examples of analyses for low atomic number elements in compounds with NIST DTSA II used pure elements and stoichiometric compounds (MgO, GaN) as peak-fitting references and standards. Note that with the exception of the Si K-family, the L-shell and M-shell characteristic X-rays of the metallic elements were used as the analytical peaks because the low beam energy was not adequate to ionize the K-shells of these elements (Ti, Cr, Fe, Ni, Cu, Zr).

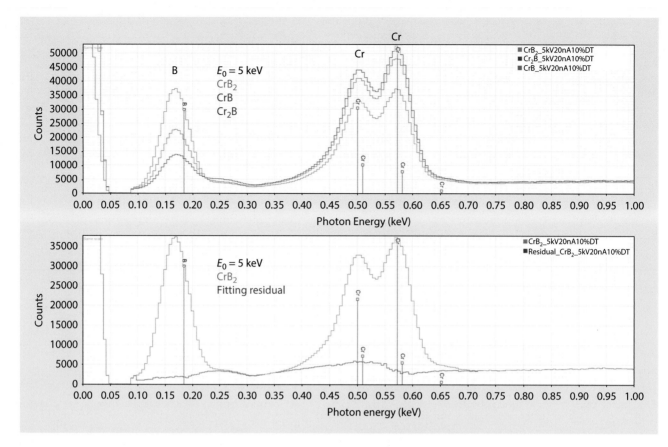

■ **Fig. 22.11**   EDS spectra of chromium borides: CrB2, CrB and Cr$_2$B (upper) and residual after peak fitting for B and Cr in CrB$_2$ (lower); $E_0 = 5$ keV

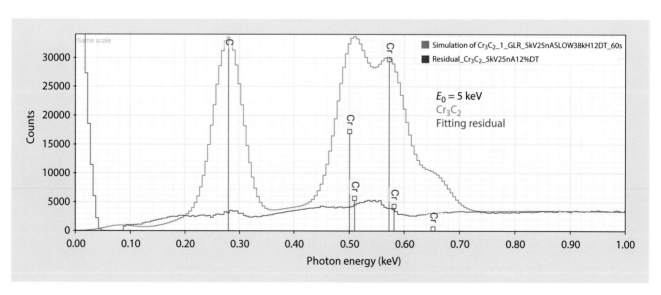

■ **Fig. 22.12**   EDS spectrum of chromium carbide, Cr$_3$C$_2$ and residual after peak fitting for C and Cr; $E_0 = 5$ keV

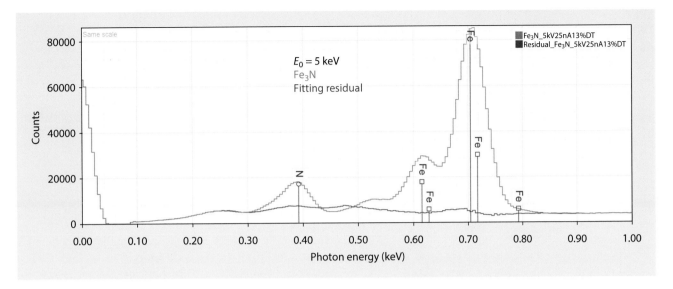

◨ **Fig. 22.13** EDS spectrum of iron nitride, Fe_3N and residual after peak fitting for N and Fe; $E_0 = 5$ keV

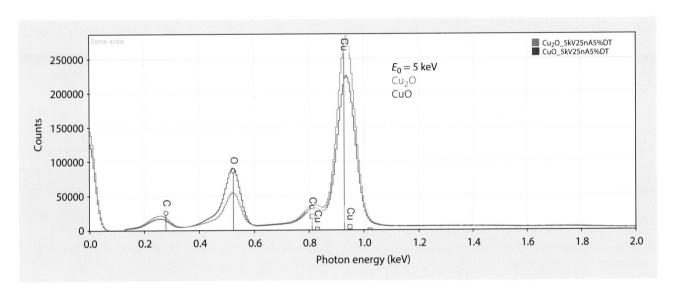

◨ **Fig. 22.14** EDS spectra of copper oxides, Cu_2O and CuO; $E_0 = 5$ keV

22.3 Challenges and Limitations of Low Beam Energy X-Ray Microanalysis

22.3.1 Reduced Access to Elements

High performance SEMs can routinely operate with the beam energy as low as 500 eV; and with special electron optics and/or stage biasing, the landing kinetic energy of the beam can be reduced to 10 eV. Because the beam penetration depth decreases rapidly as the incident energy is reduced, as shown in ◨ Fig. 22.9, which plots the Kanaya–Okayama range for 0 – 5 keV, low kinetic energy provides extreme sensitivity to the surface of the specimen, which

can improve the contrast from surface features of interest. Since the lateral ranges over which the backscattered electron (BSE) and closely related SE_2 signals are emitted are also greatly restricted at low beam energies, these signals closely approach the beam footprint of SE_1 emission and thus contribute to high spatial resolution imaging rather than degrading resolution as they do at high beam energy. Thus, low beam energy operation has strong advantages for SEM imaging down to beam landing energies of tens of eV.

While low beam energy SEM imaging can exploit the full range of landing kinetic energies to seek to maximize contrast from surface features of interest, the situation for low beam energy X-ray microanalysis is much more constrained. As discussed above, as the beam energy is reduced, the atomic shells that can be ionized become more restricted. A beam energy of 5 keV is the lowest energy that provides access to measureable X-rays for elements of the periodic table from $Z = 3$ (Li) to $Z = 94$ (Pu), as shown in ◘ Fig. 22.5. If the beam energy is reduced to $E_0 = 2.5$ keV, EDS X-ray microanalysis of large portions of the periodic table is no longer possible because no atomic shell with useful X-ray yield can be excited or effectively measured for these elements, creating the situation shown in ◘ Fig. 22.6. Further decreases in the beam energy results in losing access to even more elements, with only about half of the elements measureable at $E_0 = 1$ keV, and many of those only marginally so.

Even to achieve the elemental coverage depicted for $E_0 = 5$ keV in ◘ Fig. 22.5, low beam energy EDS X-ray microanalysis requires measurement of characteristic X-rays that are not normally utilized in conventional beam energy analysis for certain elements. Thus Ti must be measured with the Ti L-family when $E_0 \leq 5$ keV, as shown in ◘ Fig. 22.15. Similarly, for Ba, the Ba L-family around 4.5 keV is the usual choice for microanalysis, but the Ba L_3 excitation energy is 5.25 keV, and thus the Ba L-family not excited with $E_0 = 5$ keV, forcing the analyst to utilize the Ba M-family. The EDS

spectrum of $BaCl_2$ with $E_0 = 5$ keV is shown in ◘ Fig. 22.16. Due to the low fluorescence yield of ionizations in the Ba M-shell, the Ba M-family peaks are seen to have a relatively low peak-to-background, despite Ba being present in this case as a major constituent (mass concentration $C = 0.696$), making the measurement of Ba when present as a minor to trace constituent even more problematic. A practical problem that arises when analyzing with the Ba M-family peaks is the difficulty in obtaining suitable Ba M-family peak references that are free of interferences from other elements. While $BaCl_2$ is interference-free in the Ba M-family region, BaF_2 and $BaCO_3$ are not, as shown in ◘ Fig. 22.16. However, $BaCl_2$ shows evidence of degradation under the electron beam, possibly changing the local compositions and thus disqualifying it as a standard. Despite degradation under the beam, $BaCl_2$ can serve as a peak reference, while BaF_2 or another Ba-containing compound or glass that is stable under electron bombardment can serve as a standard. Despite these challenges, successful analysis of the high transition temperature superconducting material $YBa_2Cu_3O_{7-X}$ at $E_0 = 2.5$ keV with CuO, Y_2O_3, and BaF_2 as the standard and $BaCl_2$ as the peak reference is demonstrated in ◘ Fig. 22.17 and ◘ Table 22.5, where analyses with oxygen done directly against a standard (ZnO) and by the method of assumed oxygen stoichiometry of the cations are presented.

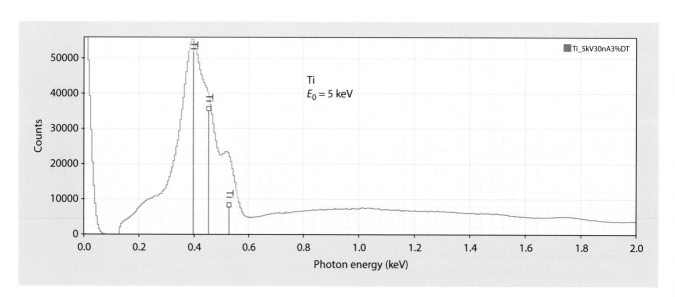

◘ **Fig. 22.15**   EDS spectrum of titanium; $E_0 = 5$ keV

22

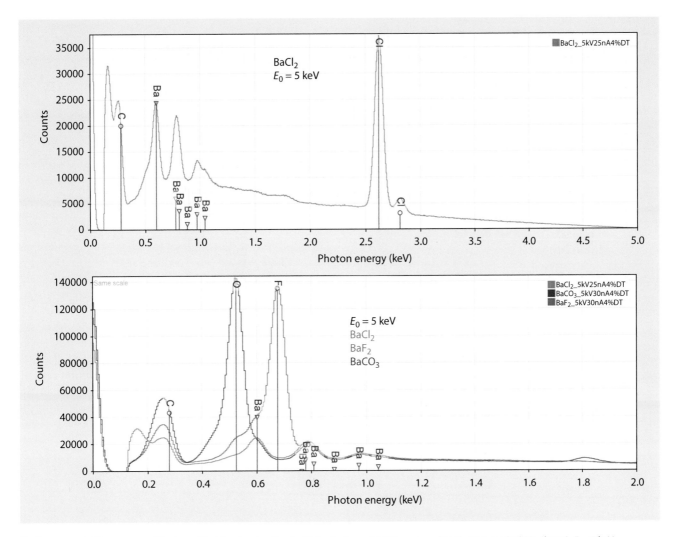

■ **Fig. 22.16** EDS spectrum of barium chloride, showing the Ba M-family (*upper*); EDS spectra of $BaCl_2$, BaF_2, and $BaCO_3$ (*lower*); $E_0 = 5$ keV

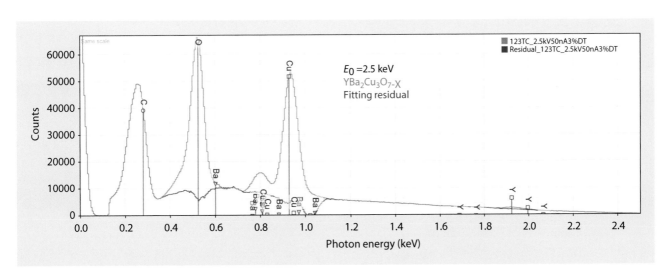

■ **Fig. 22.17** EDS spectrum of $YBa_2Cu_3O_{7-X}$, and residual after peak fitting for O K-L2, the Ba M-family and Cu L-family; $E_0 = 2.5$ keV

Table 22.5   Analysis of $YBa_2Cu_3O_{7-x}$ at $E_0 = 2.5$ keV

Element	C_{av} mass conc	RDEV %	σ_{rel}, %	C_{av} mass conc	RDEV %	σ_{rel}, %
O	0.1574 (stoich)	−6.4	1.1	0.1787 (ZnO)	6.3	1.3
Cu	0.2910	1.7	3.4	0.3024	5.7	1.4
Y	0.1296	−2.9	3.1	0.1322	−0.90	2.4
Ba	0.4220	2.4	3.6	0.3867	−6.2	2.3

22.3.2   Relative Depth of X-Ray Generation: Susceptibility to Vertical Heterogeneity

Another challenge in low beam energy X-ray microanalysis is that the difference in the depth of generation and sampling of characteristic X-rays from different elements imposes strong requirements on the homogeneity of the specimen along the beam axis. While the physics of characteristic X-ray generation is such that relative differences in the generation and emission of X-rays occur at all beam energies, including the conventional beam energy range, in the low beam energy analysis region the effect is exacerbated due to the rapidly changing range as described by Eq. (22.5). It is useful to consider that the photon energy axis of an EDS spectrum can also be thought of as a range axis that describes the depth to which a given photon energy can be generated. Such a range scale is shown parallel to the photon energy axis in ◘ Fig. 22.18 for a ZnS target with $E_0 = 5$ keV. Points on the Kanaya–Okayama range scale corresponding to exciting X-rays with ionization energies of 4 keV, 3 keV, 2 keV and 1 keV are noted. The range scale is non-linear when compared to the energy scale due to the $E_0^{1.67}$ term in the range equation. In ZnS, S K ($E_c = 2.47$ keV) can be excited to a depth of approximately 0.21 μm, while Zn ($E_c = 1.02$ keV) continues to a depth of 0.28 μm. If the ZnS contained Ca as a trace or minor constituent, it would only be generated to a depth of 0.09 μm. Thus, if quantitative analysis is to be successful by means of the k-ratio/matrix corrections protocol performed at a single beam energy in the low beam energy regime, the material must be homogeneous from the surface to the full range of the excited volume.

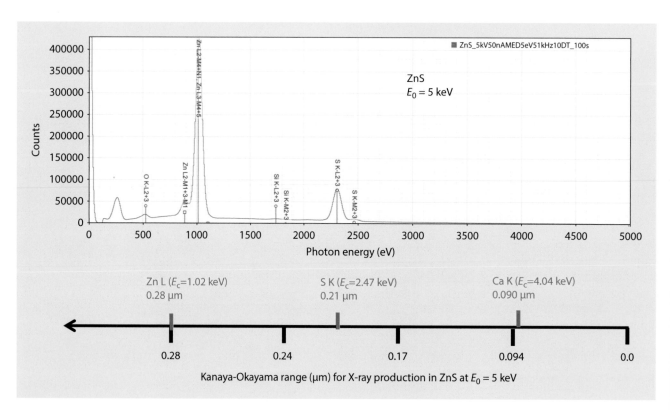

◘ Fig. 22.18   EDS spectrum of ZnS illustrating concept of the energy axis of the spectrum and the corresponding depth of X-ray generation; $E_0 = 5$ keV

22.3.3 At Low Beam Energy, Almost Everything Is Found To Be Layered

Most "pure" elements have surface layers such as native oxide, hydration layers, and others that compromise the requirement for uniform composition throughout the electron-excited volume of both the unknown and the standard(s). For example, when "pure" silicon is used as a standard, the intensity of the O K-$L_{2,3}$ peak, which arises from the SiO_2 layer on Si, increases relative to the Si K-$L_{2,3}$ peak as the beam energy is lowered, as seen in ◻ Fig. 22.19. In conventional analysis with $E_0 \geq 10$ keV, the deviation from "pure" silicon that this surface oxide represents does not constitute a significant source of error since the range is so much greater than the native oxide thickness. However, for low beam energy analysis, the surface oxide constitutes an increasingly significant fraction of the beam excitation volume as the beam energy is reduced, introducing an increasingly larger error because of the uncertainty in the standard composition.

The presence of the O K-$L_{2,3}$ peak from a surface oxide is especially problematic when it interferes with the characteristic peak of interest, such as the Ti L-family, as shown in ◻ Fig. 22.20. O K-$L_{2,3}$ (0.525 keV) is separated from Ti L_1-M_2 (0.529 keV) by

4 eV. Note the large increase in intensity in this region as the beam energy is lowered from 10 keV to 2.5 keV due to the increased contribution from O K-$L_{2,3}$ as the fraction of the interaction volume represented by the surface oxide increases. Obtaining an adequate standard and peak reference for Ti for low beam energy analysis is thus problematic. Even when a compound expected to be oxygen-free such as $TiSi_2$ is selected, there still appears to be excess intensity due to O K-$L_{2,3}$, as shown in ◻ Fig. 22.20. Thus, it may be necessary to use advanced preparation, such as in situ ion milling to clean the surface of Ti to reduce the oxygen contribution to the spectrum.

The conductive coating that is applied to eliminate surface charging in insulating specimens becomes more significant as the beam energy is decreased. This effect is illustrated in ◻ Fig. 22.21 for spectra of the mineral benitoite ($BaTiSi_3O_9$) recorded over a wide range of incident beam energies, where the peak for C K-$L_{2,3}$ is barely detectable at $E_0 = 20$ keV but becomes one of the most prominent peaks in the spectrum at $E_0 = 2.5$ keV. The analyst should try to minimize the carbon contribution to the spectrum by using the thinnest acceptable carbon layer, less than 10 nm thick, and it may be necessary to explore the use of ultrathin (~1 nm) heavy metal coatings as an alternative if it is desired to analyze for carbon.

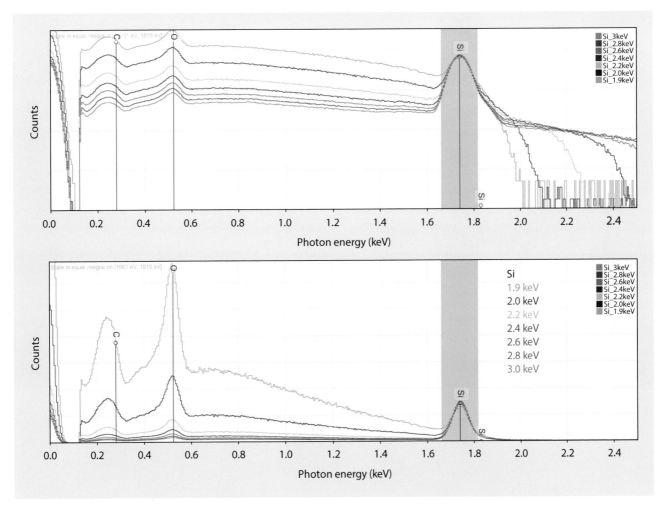

◻ **Fig. 22.19** EDS spectra of Si over a range of beam energies, showing increase in the O K-L_2 peak relative to Si K-L_2; all spectra scaled to Si K-L_2

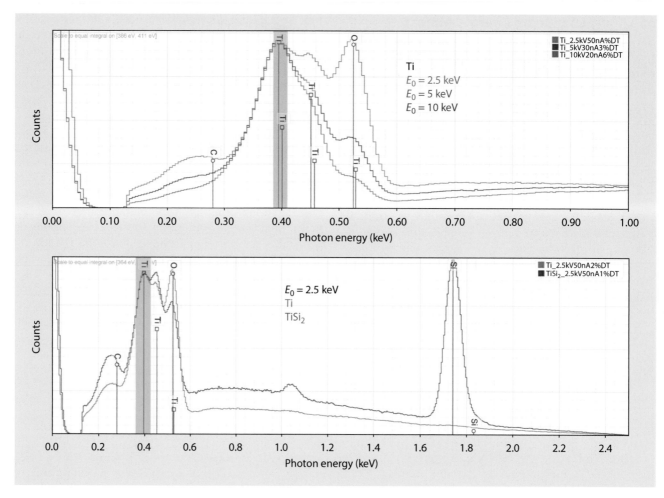

Fig. 22.20 (*Upper*) EDS spectra of Ti at various beam energies showing increase in the O K-L$_2$ peak relative to Ti L-family peaks; (*lower*) EDS spectra of Ti and TiSi$_2$ at $E_0 = 2.5$ keV

Finding an unexpected composition due to surface modification is a common experience when performing low beam energy analysis of materials that must be analyzed in the as-received condition. Without special surface preparation to expose the interior of the material, such as grinding and polishing or ion beam milling, the modified surface region dominates the analysis. ◘ Figure 22.22 (upper spectrum) shows an example of TiB$_2$, where inspection of the fitting residual after analyzing for B and Ti shows significant peaks for C and O. When these elements are included in the analysis, the fitting residual shown in ◘ Fig. 22.22 (lower spectrum) is

obtained, showing no further undiscovered constituents. The analysis results presented in ◘ Table 22.6 reveal significant concentrations of C and O in the TiB$_2$. Note the greater variance in the C and O contaminants compared to the B and Ti host elements.

Analysis of Surface Contamination

Low beam energy analysis samples such a shallow near-surface region that unexpected contamination layers can dominate an analysis. This can lead to the confounding situation where the analysis can be correct, but what is being

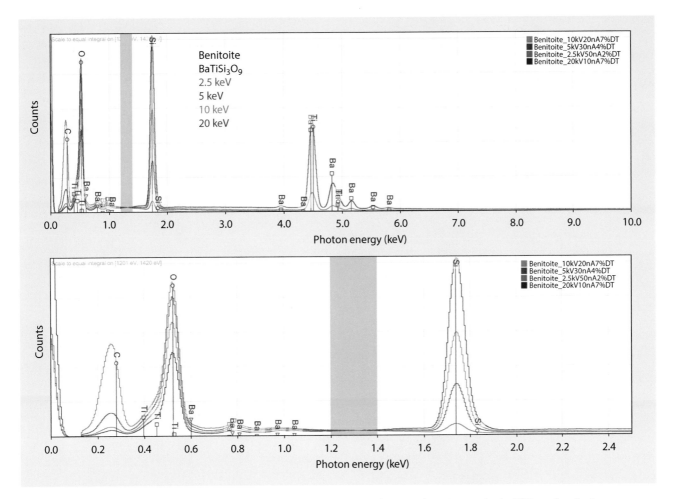

▣ Fig. 22.21   EDS Spectra of benitoite (BaTiSi$_3$O$_9$) over arrange of beam energies showing relative increase in the C K-L$_2$ peak as the beam energy decreases

measured is unanticipated. An example is shown in ▣ Fig. 22.23, which shows a low beam energy SDD-EDS spectrum of NIST SRM 481 (alloy 20Au-80Ag) where the surface was prepared metallographically more than 30 years earlier. The spectrum shows distinct peaks due to S and Cl from the formation of a surface tarnish layer. Quantitative X-ray microanalysis with DTSA-II confirms the high concentrations of S and Cl and very large RDEV values for Ag and Au, as shown in ▣ Table 22.7a. The specimen mount was re-polished with 0.25-μm diamond abrasive, which eliminated the S- and Cl- rich layer, as seen in the spectrum in ▣ Fig. 22.23.

The results of the quantitative analysis after this first repolishing, which are presented in ▣ Table 22.7b, show analytical totals near unity but large RDEV values for Ag and Au, especially for the 20Au-80Ag alloy. This large deviation from the SRM values is likely to be a consequence of the tarnish formation process selectively removing Ag from the alloy. After two additional repolishing steps with 1 μm and 0.25 μm diamond abrasives (▣ Tables 22.7c and 22.7d), this perturbed surface layer was finally removed, exposing the SRM alloy, with the DTSA-II analysis values closely matching the SRM certificate values.

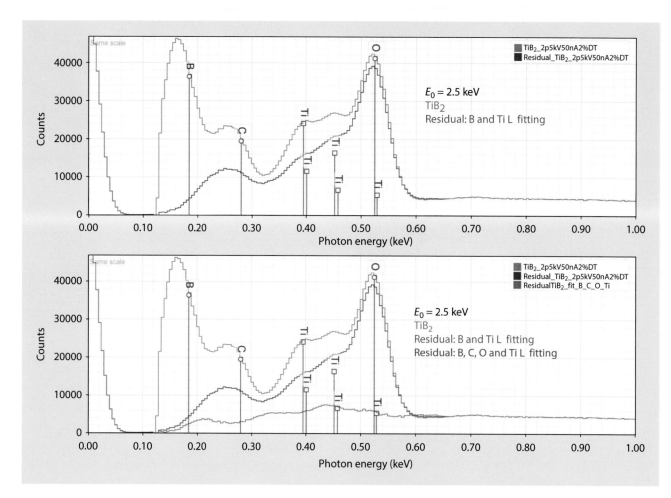

☐ **Fig. 22.22** (*Upper*) EDS spectrum of TiB$_2$ and residual spectrum after fitting for B K-L$_2$ and Ti L-family revealing peaks for C K-L$_2$and O K-L$_2$; (*lower*) after fitting for C K-L$_2$and O K-L$_2$; $E_0 = 2.5$ keV

☐ **Table 22.6**   Analysis of TiB$_2$ at $E_0 = 2.5$ keV

	B (atomic concentration)	C (atomic concentration)	O (atomic concentration)	Ti (atomic concentration)
Mean (5 analyses)	0.5110	0.0708	0.1011	0.3171
σ_{rel}, %	2.7	27	11	2.7

22

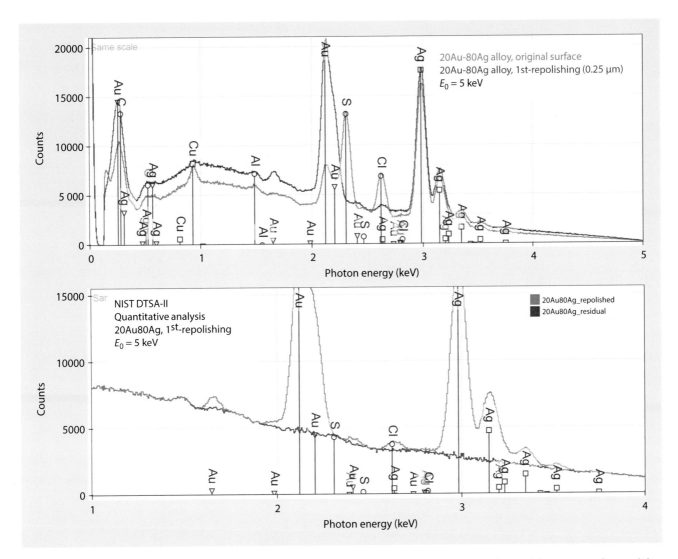

■ **Fig. 22.23** NIST SRM 481 (Au-Ag alloys). Analysis of an old (>30 years) metallographic preparation at $E_0 = 5$ keV, and the spectrum after repolishing with 0.25 μm diamond abrasive

Table 22.7a Analysis of SRM 481 (Au-Ag alloys); 1970s metallographic preparation, original surface; $E_0 = 5$ keV; standards: S (FeS_2); Cl (KCl); Ag, Au

Alloy	Raw analytical total	S (norm mass conc)	σ(%)5 loc	Cl (norm mass conc)	σ(%) 5 loc	Au (norm mass conc)	σ(%)5 loc	RDEV(%)	Ag (norm mass conc)	σ(%) 5 loc	RDEV (%)
20Au–80Ag	0.9855	0.0934	4.4	0.1061	12	0.0796	12	−64	0.7210	0.44	−7
40Au–60Ag	0.9959	0.0311	5.7	0.0965	10	0.3440	6.1	−14	0.5284	2.2	−12
60Au–40Ag	0.9951	0.0094	5	0.045	22	0.5754	4.7	−4.2	0.3706	4.6	−7.2
80Au–20Ag	1.005	0.0051	8.3	0.0365	4.5	0.7340	0.67	−8.3	0.2244	1.6	+12

Table 22.7b Analysis of SRM 481 (Au-Ag alloys); surface after first repolishing with 0.25-μm diamond; $E_0 = 5$ keV; standards: Ag, Au

Alloy	Raw analytical total	Au (norm)	σ(%) 5 loc	RDEV (%)	DTSA-II error budget (%)	Ag (norm)	σ(%) 5 loc	RDEV (%)	DTSA-II error budget (%)
20Au–80Ag	0.9731	0.3050	7.1	+36	0.34	0.6950	3.1	−10.4	0.75
40Au–60Ag	0.9920	0.4460	0.96	+11.4	0.25	0.5540	0.77	−7.6	1.1
60Au–40Ag	0.9907	0.6322	1.0	+5.3	0.21	0.3678	1.8	−7.9	1.5
80Au–20Ag	0.9930	0.8264	0.24	3.2	0.17	0.1736	1.1	−13	2.1

22

◻ Table 22.7c Analysis of SRM 481 (Au-Ag alloys); surface after second repolishing with 1- and 0.25-μm diamond; E_0 = 5 keV; standards: Ag, Au

Alloy	Raw analytical total	Au (norm)	σ(%) 5 loc	RDEV (%)	DTSA-II error budget (%)	Ag (norm)	σ(%) 5 loc	RDEV (%)	DTSA-II error budget (%)
20Au–80Ag	1.005	0.2398	0.87 %	+6.9 %	0.38 %	0.7602	0.28 %	−2.0 %	0.64 %
40Au–60Ag	0.9983	0.4045	0.23	+1.1	0.27	0.5955	0.16	−0.64	0.98
60Au–40Ag	0.9897	0.6084	0.13	+1.3	0.21	0.3916	0.21	−1.9	1.4
80Au–20Ag	0.9998	0.8055	0.26	+0.62	0.19	0.1945	1.1	−2.5	1.9
20Au–80Ag	1.005	0.2398	0.87	+6.9	0.38	0.7602	0.28	−2.0	0.64

◻ Table 22.7d Analysis of SRM 481 (Au-Ag alloys); surface after third repolishing with 1- and 0.25-μm diamond; E_0 = 5 keV; standards: Ag, Au

Alloy	Raw analytical total	Au (norm)	σ(%) 5 loc	RDEV (%)	DTSA-II error budget (%)	Ag (norm)	σ(%) 5 loc	RDEV (%)	DTSA-II error budget (%)
20Au–80Ag	1.017	0.2251	0.13	−0.11	0.39	0.7749	0.13	+0.35	0.61
40Au–60Ag	0.9988	0.3909	0.25	−2.3	0.28	0.6091	0.16	+1.6	0.93
60Au–40Ag	0.9931	0.5931	0.12	−1.2	0.22	0.4069	0.18	+1.9	1.4
80Au–20Ag	0.9957	0.7979	0.17	−0.32	0.18	0.2021	0.68	+1.2	1.9

References

Newbury D, Ritchie N (2015) Quantitative electron-excited X-ray micro-analysis of borides, carbides, nitrides, oxides, and fluorides with scanning electron microscopy/silicon drift detector energy-dispersive spectrometry (SEM/SDD-EDS) and NIST DTSA-II. Micros Microanal 21:1327

Newbury D, Ritchie N (2016) Electron-excited X-ray microanalysis at low beam energy: almost always an adventure! Micros Microanal 22:735–753

Analysis of Specimens with Special Geometry: Irregular Bulk Objects and Particles

© Springer Science+Business Media LLC 2018
J. Goldstein et al., *Scanning Electron Microscopy and X-Ray Microanalysis*,
https://doi.org/10.1007/978-1-4939-6676-9_23

There are two "zero-th level" assumptions that underpin the basis for quantitative electron-excited X-ray microanalysis:

1. The only reason that the measured X-ray intensity differs between the unknown and the standard(s) is that the composition is different. No other factors such as the specimen shape, orientation, or size influence the measured X-ray spectrum.

2. The specimen is homogeneous in composition over the volume excited by the electron beam from which the characteristic and continuum X-rays are emitted, including the secondary radiation induced by absorption of the primary electron-excited radiation.

When either of these conditions is not met, a significant increase in the overall uncertainty budget of the analysis can occur beyond the ideal situation in which the uncertainties arise from counting statistics and from uncertainties in the calculated matrix correction factors.

Considering "zeroth-level" assumption 1, sample geometry can significantly modify the measured X-ray intensity. The ideal specimen is flat and placed at known angles to the incident beam and the X-ray detector(s). Topographic features on bulk specimens (defined as those for which the thickness is much greater than the electron range) or unusual geometric shapes, such as particles with dimensions similar to the electron range, can strongly affect the measured X-rays by modifying X-ray generation and by affecting the loss of X-rays due to absorption. In severe cases, the impact of "geometric factors" on the final concentrations becomes so large as to render the compositional results, as calculated with the conventional standards-based/matrix corrections protocol or the standardless protocol, nearly worthless.

23.1 The Origins of "Geometric Effects": Bulk Specimens

The ideal sample is compositionally homogeneous on a microscopic scale, has a flat surface, and is set at known angles to the incident electron beam and the X-ray spectrometer. Compared to the X-ray spectrum measured from this ideal spectrum, geometric effects occur when the size and shape of the specimen (1) modify the interaction of the electrons with the specimen so as to affect the generated X-ray intensity and (2) alter the length of the absorption path along which the generated X-rays travel to escape the specimen and reach the detector so as to affect the measured X-ray intensity.

Because electron backscattering depends on the local surface inclination to the incident beam, tilted samples generate fewer X-rays compared to a sample at normal beam incidence (0° tilt). An illustration of this effect for bulk copper, as calculated with the Monte Carlo simulation embedded in NIST DTSA-II, is shown in ◧ Fig. 23.1. Even at normal beam incidence where the backscattered electron (BSE) coefficient

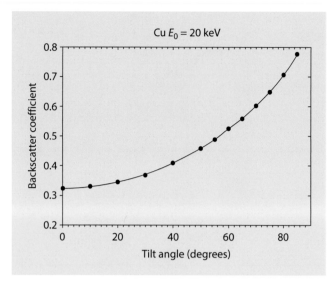

◧ **Fig. 23.1**   Backscatter coefficient η vs. surface tilt θ (inclination) for Cu at $E_0 = 20$ keV as calculated with the Monte Carlo electron trajectory simulation embedded in NIST DTSA-II

is at a minimum, BSEs carry off energy which would have gone to cause additional inner shell ionization events followed by subsequent X-ray emission had those electrons remained in the specimen. As the local surface inclination (tilt) increases, backscattering increases and more X-ray generation is lost compared to normal beam incidence situation. ◧ Figure 23.2 shows Monte Carlo calculations of the Cu K-L$_{2,3}$ X-ray intensity emitted from a flat, bulk copper specimen, expressed as a "k-ratio," where the denominator is the intensity emitted from copper at zero tilt (normal beam incidence). As the local surface inclination (tilt angle) increases above zero degrees, the X-ray production decreases with

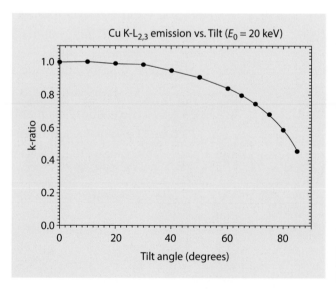

◧ **Fig. 23.2**   Emitted Cu K-L$_{2,3}$ X-ray intensity calculated as the k-ratio relative to the intensity at a tilt of 0°, vs. surface tilt θ (inclination), for Cu at $E_0 = 20$ keV as calculated with the Monte Carlo electron trajectory simulation embedded in NIST DTSA-II

◻ Fig. 23.3 Schematic illustration of the effects of surface topography on the X-ray absorption path length within the specimen

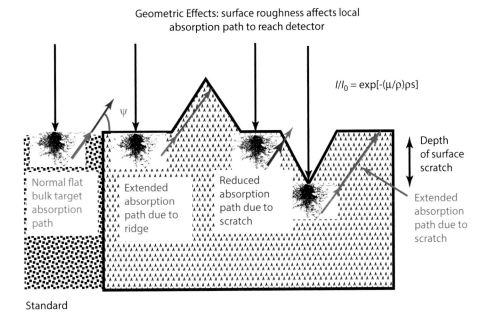

increasing tilt due to the increased backscattering seen in ◻ Fig. 23.1. There is a relatively small decrease in the k-ratio at low tilt angles, but the k-ratio decreases rapidly for tilt angles above approximately 40°. Because of the high photon energy of Cu K-L$_{2,3}$ (8.04 keV) and the relative transparency of any material to its own X-rays, there is no significant absorption so that the behavior shown in ◻ Fig. 23.2 is almost entirely due to the modification of the production of X-rays due to backscatter loss.

Surface topography modifies the X-ray absorption path length to the detector compared to a flat specimen at normal beam incidence. As shown schematically in ◻ Fig. 23.3, topographic features such as scratches and ridges can increase or decrease the absorption path length in the direction of the X-ray detector. X-ray absorption follows an exponential dependence on this path length:

$$I/I_0 = \exp\left[-(\mu/\rho)\,\rho s\right] \quad\quad (23.1)$$

where I_0 is the initial intensity, I is the intensity that remains after passing through a path length s (cm), (μ/ρ) is the mass absorption coefficient (cm^2/g) for the photon energy of interest that depends on the absorption contributions of all elements present, and ρ is the density (g/cm^3). Considering the entire energy range of the generated X-ray spectrum, which extends from a practical minimum threshold of 100 eV to the incident beam energy, E_0 (the Duane–Hunt limit), as the X-ray photon energy decreases, absorption generally increases. Absorption is especially strong if the photon energy is less than 1 keV above the critical ionization energy for any elemental constituent in the specimen. An example of this effect is shown in ◻ Fig. 23.4

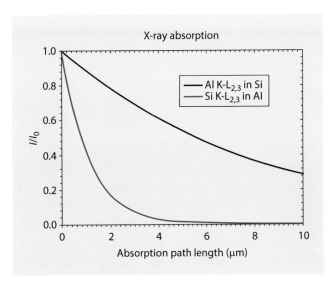

◻ Fig. 23.4 Absorption as a function of path length for Al K-L$_{2,3}$ (1.487 keV) passing through Si (K_{crit} = 1.838 keV), and Si K-L$_{2,3}$ (1.740 keV) passing through Al (K_{crit} = 1.559 keV)

for absorption as a function of path length for two contrasting cases: Al K-L$_{2,3}$ (1.487 keV) passing through Si (K_{crit} = 1.838 keV), and Si K-L$_{2,3}$ (1.740 keV) passing through Al (K_{crit} = 1.559 keV). Because Si K-L$_{2,3}$ is 0.181 keV above the critical ionization energy for Al, Si is very strongly absorbed by Al (μ/ρ = 3282 cm^2/g) such that there is no penetration beyond approximately 6 μm. By comparison, Al K-L$_{2,3}$ is below the critical ionization energy for Si, so it much less strongly absorbed (μ/ρ = 535 cm^2/g), with approximately 50% of Al K-L$_{2,3}$ intensity still remaining after 6-μm penetration through Si.

23.2 What Degree of Surface Finish Is Required for Electron-Excited X-ray Microanalysis To Minimize Geometric Effects?

Early in the history of microanalysis by electron-excited X-ray spectrometry, it was recognized that controlling the surface condition of a specimen was critical to achieving high-accuracy results by reducing geometric effects to a negligible level. Yakowitz and Heinrich (1968) performed a series of experiments in which the metallographic preparation sequence of grinding and polishing was interrupted at various stages. Materials examined included pure elements with two widely different characteristic peak energies, for example, Au M_5-$N_{6,7}$ at 2.123 keV and Au L_3-$M_{4,5}$ at 9.711 keV, and homogeneous binary metal alloys with widely differing characteristic X-ray energies. For each surface condition, the characteristic X-ray intensity was then measured at random locations and along line traverses on the specimen surface to examine the variation in characteristic X-ray intensity that could be ascribed to surface roughness. Results for selected surface conditions for a gold target are listed in ◼ Table 23.1. For the Au L_3-$M_{4,5}$ measurements, a final polish of the surface with 0.5-µm alumina was necessary to reduce the coefficient of

variation for 20 random measurements to a level similar to the expected variation from the random counting statistics, expressed as $3\,n^{1/2}/n$. For the lower photon energy Au M_5-$N_{6,7}$ which suffers stronger absorption, it was necessary to improve the surface polish to 0.1 µm alumina to achieve similar results.

For even lower photon energy peaks, such as those associated with low atomic number elements with $Z \leq 9$ (fluorine) for which $E < 1$ keV, even better surface finish is required to control the geometric effects. Newbury and Ritchie (2013a) simulated X-ray emission from crenelated surfaces with the Monte Carlo simulation embedded in NIST DTSA-II to examine the influence of surface topography on low photon energy peaks. As shown in ◼ Fig. 23.5 for FeO at an incident beam energy $E_0 = 10$ keV, the depth of scratches had to be reduced below 50 nm to reduce the geometric effects on O K, Fe L_3-$M_{4,5}$, and Fe K-$L_{2,3}$ to a negligible level.

23.2.1 No Chemical Etching

In addition to achieving a high degree of surface finish to minimize geometric effects, it is also important to avoid chemical or electrochemical etching of the final surface. For effective optical microscopy of microstructures, chemical

◼ **Table 23.1** Characteristic X-ray intensity measured on gold after various stages of grinding and polishing (Yakowitz and Heinrich 1968)

Surface condition	AuMα (2.123 keV) Coeff. variation,%	AuMα (2.123 keV) $3\,n^{1/2}/n$,%	AuLα (9.711 keV) Coeff. variation,%	AuLα (9.711 keV) $n^{1/2}/n$,%
600 grit SiC	8.6	0.39	1.8	0.93
0.5 µm Al_2O_3	0.7	0.39	1.1	0.93
0.1 µm Al_2O_3	0.46	0.39	0.42	0.93

◼ **Fig. 23.5** Plots of O K, Fe L_3-$M_{4,5}$, and Fe K-$L_{2,3}$ as a function of scratch depth for a crenelated surface as calculated with the Monte Carlo simulation embedded in NIST DTSA-II (Newbury and Ritchie 2013a,b)

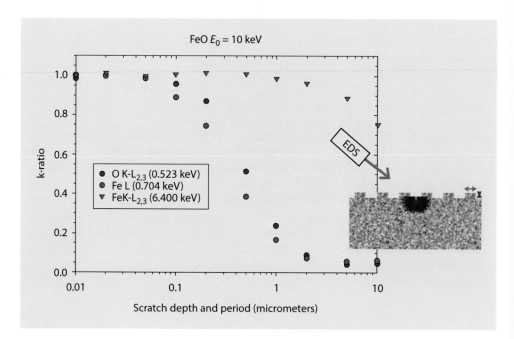

etching is usually required to produce contrast from grains at different crystallographic orientations and from compositionally distinct phases by creating surface relief through differential chemical attack and dissolution or by staining through chemical reactions. Such microscopic physical relief creates unwanted topography similar to mechanically produced scratches that can affect SEM/EDS analysis. Additionally, in some cases chemical etching can actually modify the chemical composition of the surface region, so that it is no longer representative of the bulk of the material.

23.3 Consequences of Attempting Analysis of Bulk Materials With Rough Surfaces

To illustrate the impact of surface topography on microanalysis results, a microscopically homogenous glass (NIST SRM 470 K411) containing several elements (O, Mg, Si, Ca, and Fe) that provide a wide range of range of photon energies, as listed in ◘ Table 23.2, was analyzed with a range of surface topography (Newbury and Ritchie 2013b). Analysis was performed with NIST DTSA-II at $E_0 = 20$ keV using elemental (Mg, Si, Fe) and multi-element (SRM 470 glass K412 for Ca) standards, with oxygen calculated by assumed stoichiometry, followed by normalization of the raw result. When analyzed in the ideal, highly polished (100-nm alumina final polish) flat form, the analyzed concentrations for the Mg and Fe constituents, selected because of their wide difference in photon energies, measured at 20 randomly selected locations show the distribution of results plotted in ◘ Fig. 23.6. The mean of the 20 analyses falls within +1.8 % relative for Fe and −1.0 % relative for Mg (SRM certificate values). Of the 20 analyzed locations, 19 fall within a symmetric cluster that spans

approximately 1% relative along the Mg and Fe concentration axes, while the results for one location fall significantly outside this cluster. This anomalous value was found to be associated with a shallow scratch that remained on the polished surface (location noted on the inset SEM image).

When this highly polished surface was degraded by directional grinding with 1-μm diamond grit, 20 analyses at randomly selected locations produce a much wider scatter in the normalized Mg and Fe concentrations, as shown in ◘ Fig. 23.7, a direct consequence of the effect of surface geometry.

Creating an even more severe topographic feature by gouging the polished surface of K411 with a diamond scribe created the crater seen in the SEM(ET+) SE + BSE image shown in ◘ Fig. 23.8a. Many locations in this gouge crater were analyzed, and the results are plotted in ◘ Fig. 23.8b, showing a very wide range of Mg-Fe results. For comparison, note that the 20 Mg-Fe results from the highly polished surface, including the outlier seen In ◘ Fig. 23.6, are contained within the small red box noted on the plot in ◘ Fig. 23.8b.

◘ Table 23.2   NIST SRM 470 (Glass K411)

Element	Mass concentration	Characteristic X-ray energy (keV)
O	0.4236	0.523
Mg	0.0885	1.254
Si	0.2538	1.740
Ca	0.1106	3.690
Fe	0.1121	6.400

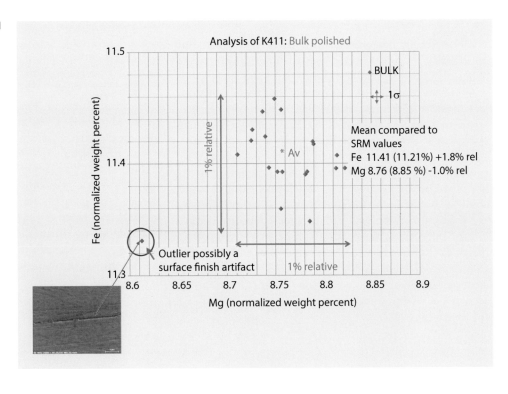

◘ Fig. 23.6 Analysis of polished (0.1-μm alumina final polish) NIST SRM 470 (K411 glass) at $E_0 = 20$ keV with NIST DTSA-II and standards: elemental (Mg, Si, Fe) and multi-element (SRM 470 K412 glass for Ca), with oxygen calculated by assumed stoichiometry. Normalized results. Note cluster of results and one outlier

◘ **Fig. 23.7**   Analysis of SRM470 (K411 glass) with surface roughness produced by abrading with 1-μm diamond grit

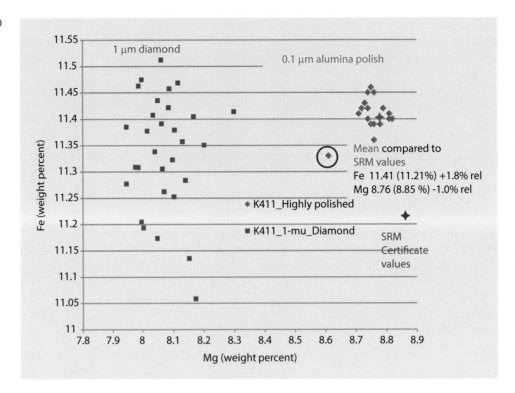

◘ Figure 23.9a shows examples of additional geometric shapes created with K411 glass: deep, narrow pits from diamond scribe impacts, microscopic particles (major dimensions < 100 μm), and macroscopic particles (major dimensions > 500 μm). When random locations are analyzed on these targets, the range of measured Mg and Fe concentration values is shown in ◘ Fig. 23.9b, covering an order of magnitude in both constituents. This huge range occurs with EDS spectra that were readily measurable despite the severe departure from the ideal flat specimen geometry.

These results demonstrate that the SEM microanalyst must realize that just because an EDS spectrum can be obtained when the stationary beam is placed on a topographic feature of interest, the resulting analysis may be subject to such egregious errors so as to be of little use. Sometimes analysis locations that are surprisingly close on a microscopic scale can produce very different results. ◘ Figure 23.10 shows a fractured fragment of pyrite (stoichiometric FeS_2) which has been analyzed at various locations (conditions: $E_0 = 20$ keV; DTSA-II calculations with Fe and CuS as standards, followed by normalization). Despite the proximity of the analyzed locations, the results vary greatly. Thus, analysis at location 3 produces a nearly perfect match to the stoichiometric values with relative deviation from expected value (RDEV) within ±0.15 %, while analysis at nearby location 9 (about 25 μm away) suffers relative accuracy of ±36 %, while at location 7 (about 50 μm away) the RDEV is ±100 %.

■ ■ **The Takeaway**

Just because a feature can be observed in an SEM image and an EDS spectrum can be recorded does not mean that a successful and useful quantitative analysis can be performed!

23.4   Useful Indicators of Geometric Factors Impact on Analysis

There are strong diagnostic indicators that reveal the impact of geometric factors on analysis:

23.4.1   The Raw Analytical Total

The raw analytical total is the sum of all the constituents measured (including any constituents such as oxygen calculated on the basis of assumed stoichiometry of the cations). For an ideal flat sample measured with the beam energy selected in the "conventional range" ($E_0 = 10$ keV to 25 keV) and following a standards-based–matrix correction factor protocol, the analytical total typically will fall between 0.98 and 1.02 mass fraction (98–102 wt %), a consequence of the uncertainties inherent in the measurement process (counting statistics) and in the calculated matrix correction factors. If the raw analytical total exceeds this range, it is usually an indication of a deviation in the measurement conditions (e.g., beam current drift). If the raw analytical total is below this range, this may again indicate a deviation in the

23.4 · Useful Indicators of Geometric Factors Impact on Analysis

☐ **Fig. 23.8   a** SEM (ET+)
SE + BSE image of a crater produced in polished K411 glass
after gouging with a diamond
scribe. **b** Plot of the normalized
Mg and Fe concentrations calculated for measurements at various locations in this crater

Analysis with a compromised sample shape:
surface gouge crater left by tool impact

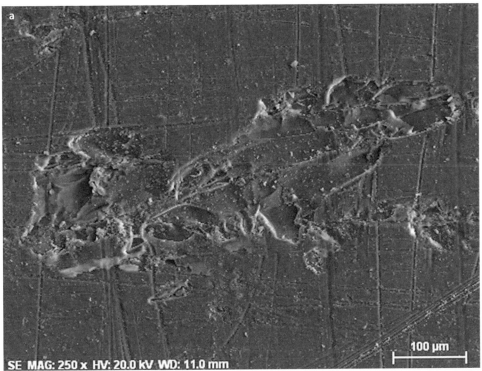

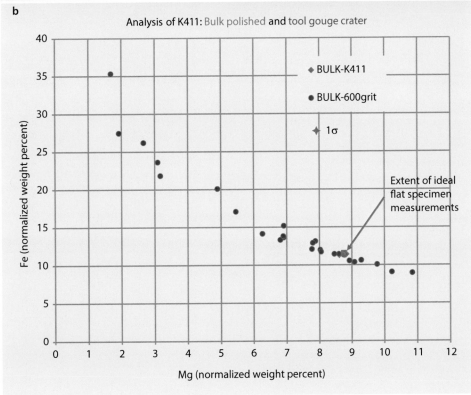

Analysis with a compromised sample shape

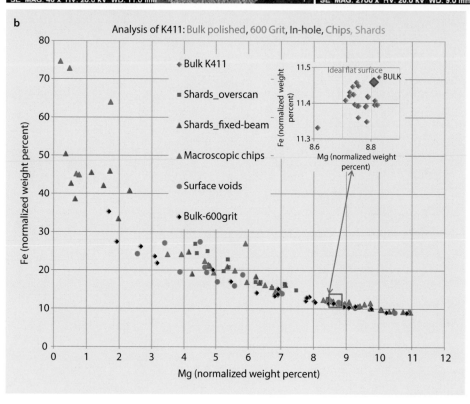

◘ Fig. 23.9   a Examples of deep narrow pits produced in K411 glass by impacts of a diamond scribe; microscopic particles (major dimensions <50 μm); and macroscopic particles (major dimensions >500 μm). **b** Plot of the normalized Mg and Fe concentrations calculated for measurements at various locations on these objects combined with the measurements previously plotted

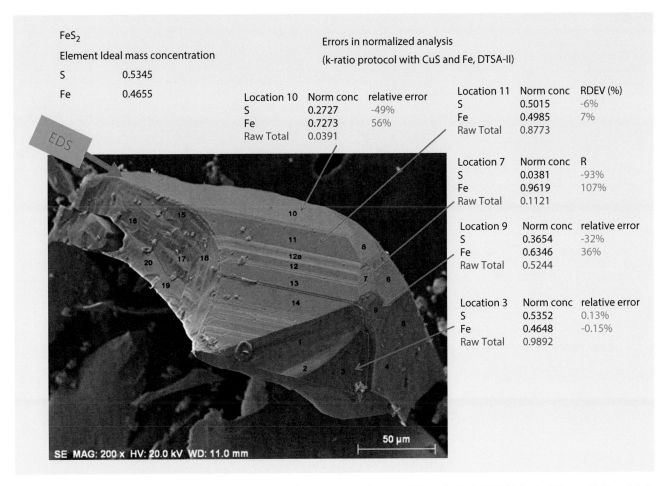

Fig. 23.10 Fragment of pyrite (stoichiometric FeS₂) analyzed at various locations; conditions: E_0 = 20 keV; DTSA-II calculations with Fe and CuS as standards, followed by normalization

measurement conditions, but a low total can also reveal the presence on an unexpected elemental constituent that is not in the list of elements analyzed. For example, an oxidized inclusion in a metallic alloy will typically have an oxygen mass fraction of approximately 0.3, leading to a sharp decrease in the analytical total to ~ 0.7 (70%) if oxygen is not recognized and included in a standards-based analysis of that location compared to the surrounding metallic region. Thus, even for conventional analysis of ideal specimens, the raw analytical total conveys useful information and should always be inspected.

When a specimen deviates from the ideal flat condition and geometric factors affect the analysis, the raw analytical total gives a direct indication, providing a standards-based–matrix correction factor protocol is being used. Note that "standardless analysis" does not provide this critical information on the raw analytical total if this protocol reqwuires a forced normalization to unity mass fraction (100 wt %) since the electron dose information is not considered. "Standardless analysis" schemes that use a locally measured elemental spectrum to establish the dose relationship to the vendor's standard intensity database or which use the peak-to-background method (see below) can provide a meaningful analytical total. Figures 23.11a, b shows the calculated normalized concentrations as a function of the raw analytical total for Mg and Fe in K411 from the suite of spectra obtained from the various

geometric shapes. Note that for this data set, the raw analytical total varies from 0.03 to 1.30 mass fraction (3–130 weight wt %). For this particular composition (K411 glass), the RDEV for Mg and Fe is within a range of 10% relative when the analytical total is in the range 0.8–1.2 mass fraction (80–120 wt %). Different compositions are likely to have different sensitivities to deviations in accuracy, but the general experience is that when the raw analytical total ranges from 0.9 to 1.1 mass fraction (90–110 wt %), the impact of the geometric factors on the analysis will be minimized.

23.4.2 The Shape of the EDS Spectrum

A second powerful indicator that can alert the careful analyst to the possible impact of geometric factors on an analysis is the shape of the EDS spectrum. The shape of the X-ray continuum (*bremsstrahlung*) background from an ideal flat specimen has distinctive properties. Consider the spectrum of pure boron, selected because of the absence of significant characteristic peaks above the energy of boron (0.185 keV), as shown in Fig. 23.12. A small peak of oxygen that arises from the inevitable surface oxide on the boron can be seen in this spectrum, as well as the artifact silicon peak from the absorption and internal fluorescence of the silicon window support grid and the silicon dead layer of the detector. Otherwise, the spectrum consists

■ **Fig. 23.11** Calculated normalized concentrations as a function of the raw analytical total from the suite of spectra obtained from the various geometric shapes: **a** Fe, **b** Mg

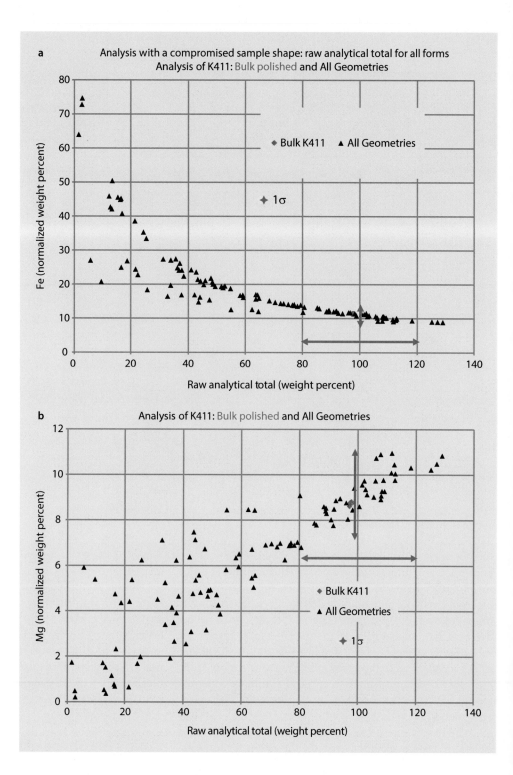

only of the X-ray *bremsstrahlung* that occurs at all photon energies, E_ν, up to the incident beam energy, E_0 (Duane–Hunt limit). Inspecting this spectrum starting at high photon energy, as the photon energy E_ν decreases, the intensity of the continuum, I_{cm}, increases, a consequence of the physics of the generation of the *bremsstrahlung*, which has the following form:

$$I_{cm} \approx i_B\ \dot{Z}\left(E_0\,/\,E_\nu - 1\right) \qquad (23.2)$$

where i_B is the beam current, and $\dot{Z}$ is the average atomic number of the target. For a given material, the mass absorption coefficient increases as the photon energy decreases. Absorption is an exponential effect, so that eventually the increased absorption overwhelms the increase in the continuum intensity so that the intensity reaches a maximum. For boron with $E_0 = 20$ keV this maximum occurs at approximately 1.3 keV. Because of the effect on the electron range and the subsequent X-ray absorption, the exact location of the maximum in *bremsstrahlung* intensity depends on beam energy as well as the specific element(s) acting as the absorber. Moreover, for complex compositions the numerous characteristic peaks are superimposed on the

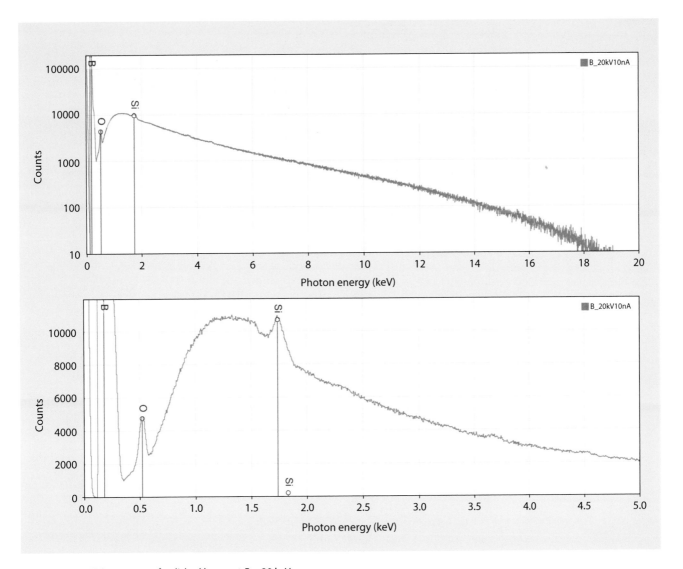

Fig. 23.12   EDS spectrum of polished boron at $E_0 = 20$ keV

background, making it more difficult to discern the position of the *bremsstrahlung* maximum. Nevertheless, the general shape of the EDS *bremsstrahlung* continuum is a powerful indicator of geometric effects that modify X-ray production, both *bremsstrahlung* and characteristic X-rays. ◻ Figure 23.13 shows this effect for ideal flat polished K411 glass and several shards. The deviation in spectral shape from the ideal case is readily apparent, and analysis of spectra with such severe deviations in shape results in large deviations in the relative accuracy of the normalized quantitative results.

23.5 Best Practices for Analysis of Rough Bulk Samples

The optimum approach to the analysis of a rough specimen is obviously to prepare a polished flat surface, but the analyst may be confronted with a situation where no physical modification of the as-received specimen is permitted. That is, the rough surface itself is the object of interest, so that grinding

and polishing would modify or destroy the material that is actually necessary for the final result. How should the analyst proceed in such a case?

The analysis of rough surfaces is inevitably going to be compromised compared to analysis of the ideal flat polished specimen. The analyst must seek to obtain the best possible result under the circumstances, so the analytical strategy must be carefully considered. Electrons that backscatter off rough surfaces are likely to produce remote excitation of X-rays from material(s) that are likely to differ from the location where the beam is striking, as shown schematically in ◻ Fig. 23.14. It may be thought that the collimator on the EDS will restrict the view of the EDS to just the region directly excited by the incident beam. This is not the case. The collimator typically permits acceptance of X-rays with at least 50 % efficiency from an area that is 5 mm in diameter or larger. The exact transmission response depends on the particular EDS detector and its collimator, but the region of transmission can be readily determined by mapping a uniform target, for example, an aluminum SEM mounting stub, at the lowest magnification setting (maximum sized scan

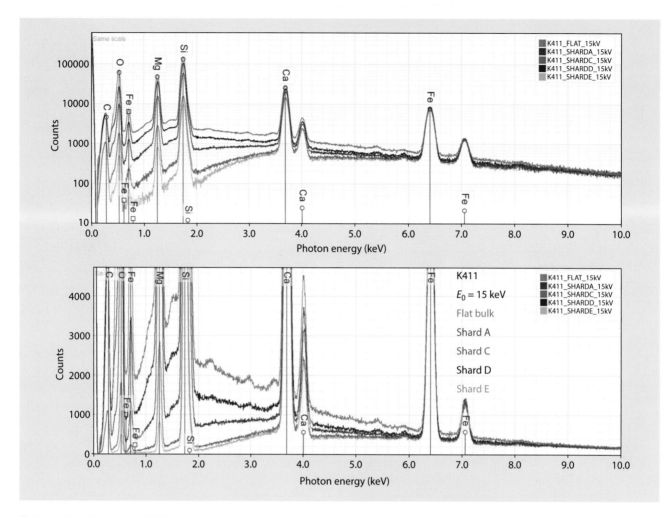

◘ **Fig. 23.13**   EDS spectra of K411 glass in the flat polished condition, and from four shards, showing the deviation in the spectral shape from the ideal; $E_0 = 20$ keV

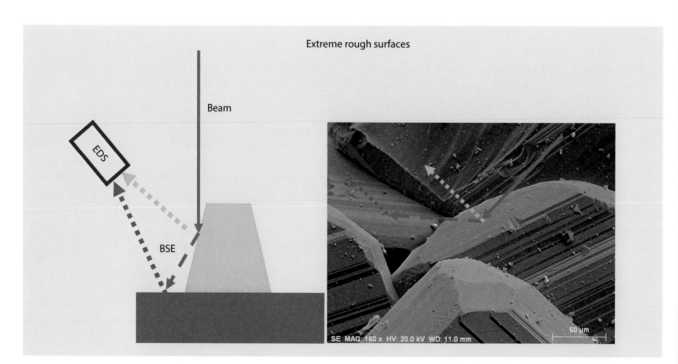

◘ **Fig. 23.14**   SEM (ET+) SE + BSE image of an irregular surface and schematic illustration of electron backscattering from a tilted surface causing remote X-ray excitation

◘ **Fig. 23.15** X-ray mapping experiment to determine extent of collimator acceptance. Large scale low magnification map (3 × 2.5 mm) of an aluminum stub. The graph shows the intensity measured at the center of a series of such maps recorded at different working distances

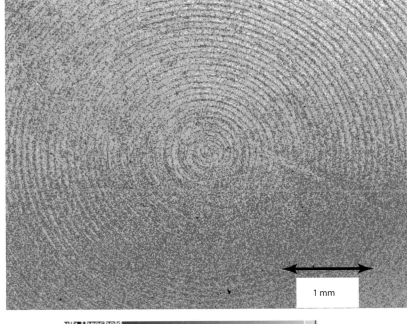

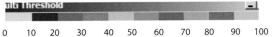

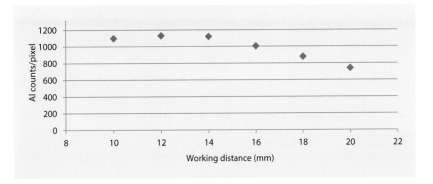

field) of the SEM, as shown in ◘ Fig. 23.15. The aluminum stub was mapped with the specimen plane located at the manufacturer's specified ideal working distance for this SEM/EDS. The Al intensity map encoded with a pseudo-color scale shows that the transmission varies from maximum (100 %) along the top of the image to values in the range 60–70 % at the bottom. The top-to-bottom asymmetry in this map reveals that the collimator on this EDS system is actually misaligned, since with proper orientation the maximum of the collimator transmission should be at the image center (coincident with the optic axis). The transmission function of the collimator as a function of vertical distance along the SEM optical axis can be determined by repeating the mapping at different working distances. The graph in ◘ Fig. 23.15 shows the intensity measured at the center of each map, revealing a decrease of approximately 40 % as the working distance was increased from 10 to 20 mm. This collimator thus allows high transmission from a large volume of space, with dimensions of at least 3 × 2.5 × 10 mm, so that any X-rays produced in this volume with a line-of-sight to the EDS detector will contribute to the measured spectrum.

Optimizing the EDS spectrum measured from a rough, irregularly shaped surface requires careful consideration of the selection of the location on the specimen to be measured. The analyst must be aware of the location of the EDS relative to the measured location to avoid the situation illustrated in ◘ Fig. 23.16a, where the beam location leads to an X-ray path that must pass through the bulk of the specimen to reach the EDS, leading to extremely high absorption. Ideally, using a specimen stage with several rotation axes, a rotation about a vertical axis will bring the feature of interest to directly face the EDS, thus minimizing the absorption, as shown in ◘ Fig. 23.16b. A further rotation about a horizontal axis places the feature perpendicular to beam to minimize backscattering and remote X-ray excitation (◘ Fig. 23.16c). Note that although backscattering is minimized by establishing normal beam incidence (effectively a zero tilt angle), the backscattered electrons are broadly emitted with a cosine distribution so that while the majority are emitted at high angles there still remains a small but significant fraction emitted at low angles to the surface that may strike nearby features and excite the surrounding materials, contributing to the spectrum measured at the beam impact position.

Fig. 23.16 Schematic illustration of orientation movements to optimize EDS collection from a feature of a rough, irregular surface: **a** initial position gives high absorption due to X-ray path through bulk of specimen; **b** rotation about a vertical axis brings feature to directly face EDS; **c** rotation about a horizontal axis places feature perpendicular to beam to minimize backscattering and remote X-ray excitation

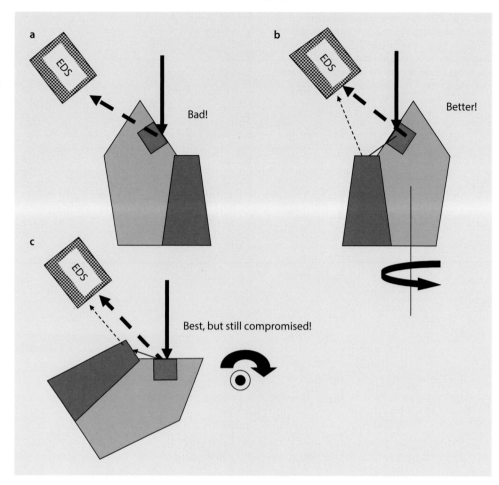

23.6 Particle Analysis

23.6.1 How Do X-ray Measurements of Particles Differ From Bulk Measurements?

The analysis of microscopic particles whose dimensions approach or are smaller than the interaction volume in a bulk target of the same composition is subject to effects similar to analyzing rough, bulk surfaces but with additional challenges. Like the rough bulk case, the curved or locally tilted surface of a particle acts to modify electron backscattering, which alters X-ray production, while deviations from the ideal flat surface alter the X-ray path to the detector, which modifies X-ray absorption compared to a flat bulk target. The geometry of a particle leads to additional loss of beam electrons due to penetration through the sides and bottom of the particle, as shown in the DTSA-II Monte Carlo simulations in **Fig. 23.17**, which depict trajectories in a 1 μm-diameter aluminum particle at different beam energies. At the highest energy simulated, $E_0 = 30$ keV, most trajectories pass through the particle with some lateral

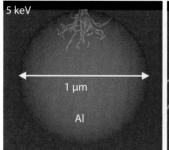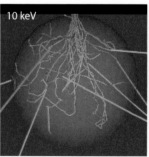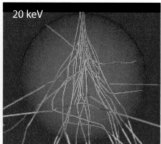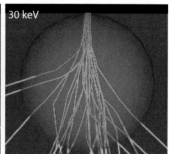

Fig. 23.17 DTSA-II Monte Carlo simulation of electron trajectories in a 1-μm-diameter Al sphere on a bulk C substrate at various beam energies; incident beam diameter = 50 nm. Trajectories inside the Al particle are *blue*. *Green* shows trajectories that emerge from the Al particle which change to orange when they enter the C substrate

23

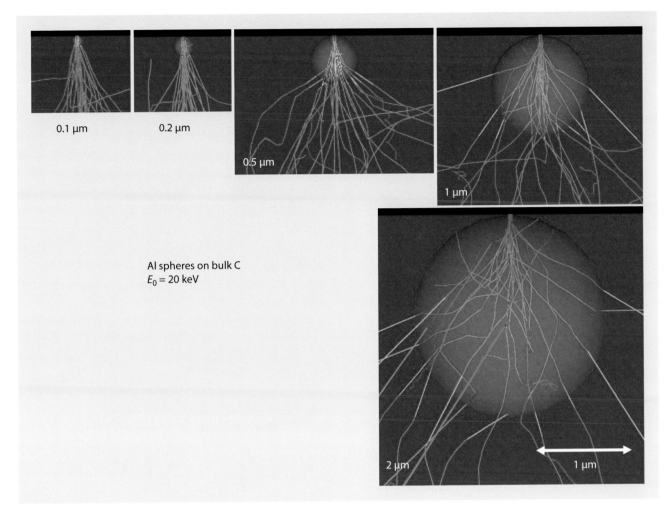

0.1 μm

0.2 μm

0.5 μm

1 μm

Al spheres on bulk C
$E_0 = 20$ keV

2 μm

1 μm

Fig. 23.18 DTSA-II Monte Carlo simulation of electron trajectories in Al spheres of various diameters at $E_0 = 20$ keV on a bulk C substrate. Trajectories inside the Al particle are *blue*. *Green* shows trajectories that emerge from the Al particle which change to orange when they enter the C substrate

scattering. As the beam energy is lowered and the electron range decreases, penetration through the bottom and sides diminishes, and at $E_0 = 5$ keV the interaction volume is completely contained within the 1-μm-diameter aluminum particle. The beam penetration effect also depends on the particle size, as shown in **Fig. 23.18** for particles of various sizes at $E_0 = 20$ keV, and on composition, as shown for particles with a range of atomic numbers in **Fig. 23.19**. Moreover, as opposed to the backscattered electrons in the high angle portion of the cosine distribution which are likely to leave the vicinity of the particle without further interaction, the beam electrons that penetrate through the sides and bottom of the particle are likely to reach the supporting substrate where they will create X-rays of the substrate material that contribute to the overall spectrum measured. This effect can be seen in **Fig. 23.20**, which shows the Al and C peaks calculated by the Monte Carlo simulation for 1-μm diameter Al spherical particles. At a beam energy of $E_0 = 5$ keV, the electron trajectories are contained within the particle which effectively acts like a bulk target. No electrons penetrate the sides or bottom to reach the substrate so there is no C contribution to the EDS spectrum. With increasing incident

energy, electron penetration through the particle into the substrate occurs, and the C peak of the substrate increases relative to the Al-peak from the particle.

23.6.2 Collecting Optimum Spectra From Particles

Before meaningful qualitative and quantitative analysis of particles can be attempted, it is important to optimize the EDS spectrum that is collected. As illustrated in the trajectory plots in the Monte Carlo simulations shown in **Figs. 23.17, 23.18, and 23.19**, and the calculated spectra shown in 23.20, because of the impact of particle size and shape (geometry) on electron interactions, the EDS spectrum of a particle will always be compromised compared to the spectrum of a material of identical composition in the ideal flat, bulk form. The analyst must be aware of the major factors that modify particle spectra and seek to minimize these effects. The particle spectrum can be optimized through careful sample preparation and by understanding of the factors that affect the strategy for beam placement.

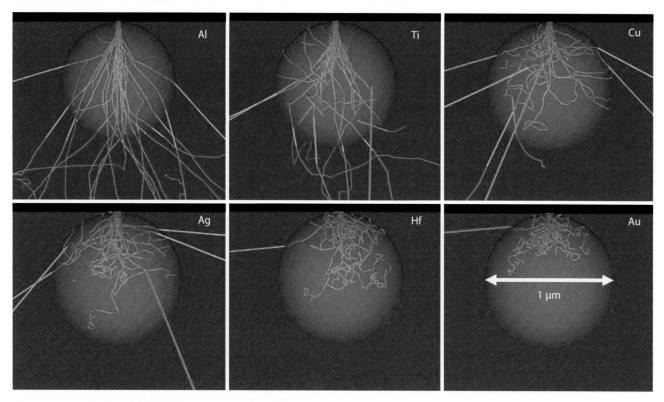

❑ **Fig. 23.19**   DTSA-II Monte Carlo simulation of electron trajectories at $E_0 = 20$ keV in 1 μm-diameter spheres of various elements (Al, Ti, Cu, Ag, Hf, Au) on a bulk C substrate. Trajectories inside the particles are *blue. Green* shows trajectories that emerge from the particle which change to orange when they enter the C substrate

Particle Sample Preparation: Bulk Substrate

Because beam electrons can penetrate through the sides and bottom of a particle and reach the underlying substrate, the measured spectrum will always be a composite of contributions from the particle and the substrate, as shown in ❑ Fig. 23.21 for a K411 glass particle (~1 μm in diameter) on a bulk carbon substrate. To obtain a spectrum that is representative of the particle material alone, it is important whenever possible to choose a substrate consisting of an element(s) not contained in the particle of interest. As part of a quality measurement system, the EDS spectrum of the blank substrate material should be measured to determine what elements are present at the major, minor, and trace level. Additionally, it is desirable that the characteristic peak(s) of the substrate element(s) should not interfere with characteristic peaks from the constituents of the particle.

Carbon is an excellent choice for a bulk substrate since it is available in high purity, is mechanically robust, and supplied by various vendors as planchets with different surface finishes, including a highly polished, glassy surface that is nearly featureless as a background for SEM imaging of small particles. The low energy of the C K X-ray (0.285 keV) is unlikely to cause interference with most elements of interest. In addition to bulk substrates, carbon is often used in the form of carbon-infused tape with a sticky surface to which particles will readily adhere. If a carbon tape preparation is used, the analyst must measure a blank spectrum of the tape since the polymer base of the tape frequently contains elements such as oxygen in addition to carbon. If carbon is of analytical interest, other low atomic number substrates are available, including high purity boron and beryllium (but beware of the health hazard of handling beryllium, especially in the form of beryllium oxide which may be released by surface abrasion). Other pure element substrates such as aluminum and silicon can be used if that element(s) is not of interest, but because of the high degree of excitation of the Al K-$L_{2,3}$ and Si K-$L_{2,3}$ characteristic X-rays under typical operating conditions, a significant fraction of the EDS deadtime will be taken up by the substrate X-rays, diminishing the analytical information collected per unit time on the particle of interest, as well as contributing coincidence peaks that might be misinterpreted as minor or trace constituents, for example, 2 Al K-$L_{2,3}$ (2.974 keV) is close to the energy of Ag L_3-$M_{4,5}$ (2.981 keV) and 2 Si K-$L_{2,3}$ (3.480 keV) is close to the energy of Sn L_3-$M_{4,5}$ (3.440 keV)

When depositing particles on a substrate, the area density should be minimized to avoid situations where electrons scattered off the particle being analyzed can strike nearby particles and excite X-rays, which will then contribute an artifact ("cross talk"). While this remote excitation is likely to be at the equivalent of a minor or trace level constituent, it is critical to understand such contributions when low level constituents are of interest.

23.6 · Particle Analysis

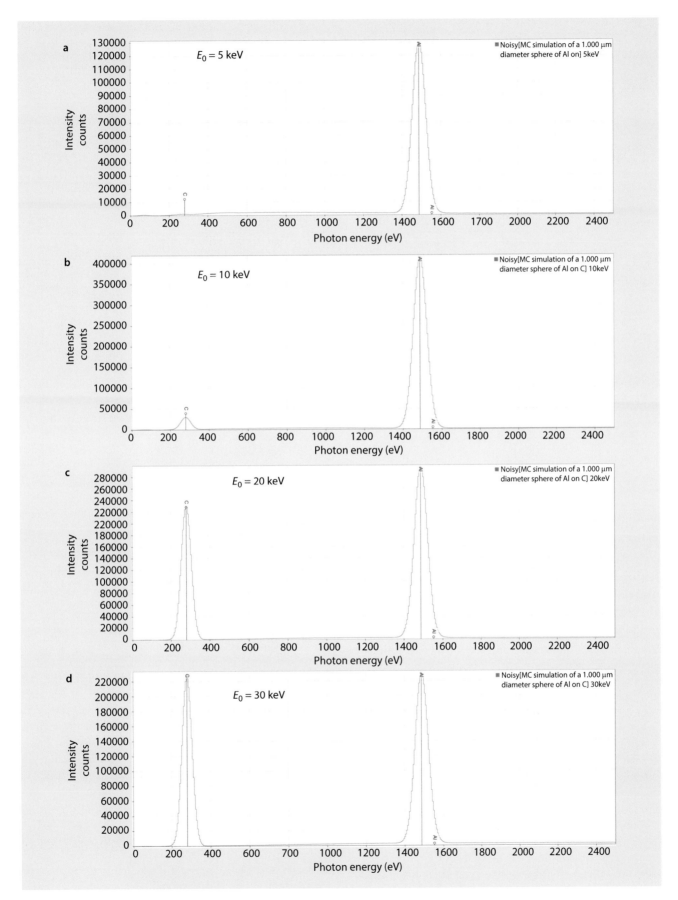

☐ **Fig. 23.20** EDS spectra calculated with the DTSA-II Monte Carlo simulation for a 1-μm-diameter Al particle on a C substrate at various beam energies: **a** 5 keV, **b** 10 keV, **c** 20 keV, and **d** 30 keV. Note the absence of the C peak from the substrate at $E_0 = 5$ keV, its appearance at $E_0 = 10$ keV and the increase relative to the Al peak as the beam energy increases

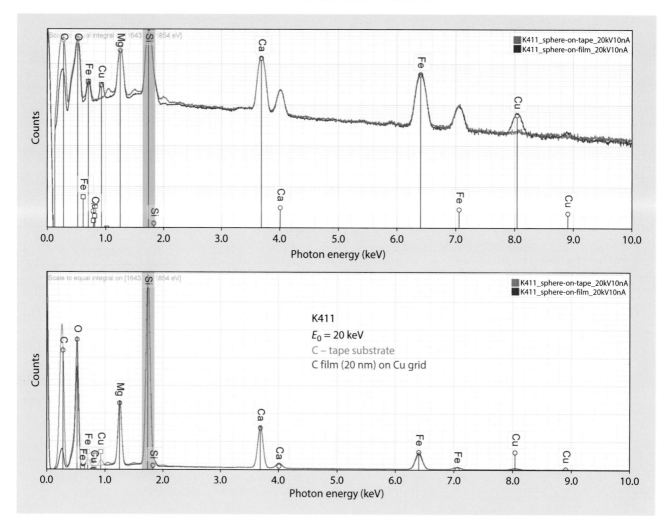

Fig. 23.21 Comparison of EDS spectra similar sized K411 spheres mounted on a bulk carbon substrate (*red*) and on a 20-nm-thick carbon film on a copper support grid *blue*). Note artifact Cu peaks arising from lateral electron scattering from particle to excite the support grid

Particle Sample Preparation: Minimizing Substrate Contributions With a Thin Foil Substrate

As the particle dimensions decrease below 1 μm, electron penetration through the particle increases dramatically, raising the substrate contribution to the X-ray spectrum. Even if the characteristic peak(s) of the substrate material is not of interest and can be ignored, the deadtime due to the substrate spectrum will eventually dominate the overall spectrum measurement, even if the substrate consists of carbon which is relatively poorly excited. Moreover, the contribution of the substrate to the composite spectrum occurs not only from the characteristic peak(s), for example, C K, but also from the continuum background (*bremsstrahlung*) which affects all photon energies. Increased background from the substrate has the effect of lowering the peak-to-background for all characteristic peaks from the elements of the particle, which degrades all aspects of quantitative analysis but especially impacts the limit of detection, raising the minimum concentration that can be reliably measured.

The substrate contribution to the composite spectrum can be minimized by reducing the mass of the substrate. Fine particles, especially those with nanometer dimensions, can be dispersed by various methods, including air-jetting or deposition from fluid drops, onto thin carbon films, typically 20 nm in thickness, which are supported on a grid (copper, nickel, carbon, etc.), as shown in the sequence of images in ▪ Fig. 23.22. To further stabilize the particle deposit, it is typical practice to apply a thin (<10 nm) carbon coating to provide conductivity and also to provide some mechanical constraint. The contribution to the spectrum from the thin carbon support film plus the final coating is much reduced compared to the situation on a bulk or thick tape carbon substrate, as shown in the comparison of spectra from K411 particles of similar size (~1 μm in diameter) shown in ▪ Fig. 23.21. Beam electrons that are scattered laterally from particles can excite the material of the grid, as shown by the presence of copper in the K411 particle spectrum of ▪ Fig. 23.21, but if this system radiation is problematic, this unwanted spectral contribution can be controlled by choosing alternative grid materials, such as carbon.

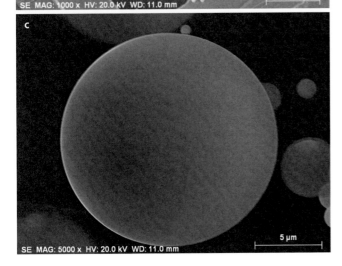

□ **Fig. 23.22** Particles deposited on a thin (~20-nm) carbon film supported on a copper grid shown at various magnifications. **a** Nominal 200 X, field width 351 by 263 micrometers; **b** nominal 1 kX, field width 70 by 53 micrometers; **c** nominal 5 kX, field width 14 by 11 micrometers

The Importance of Beam Placement

The placement of the beam on the particle relative to the EDS detector position can have a strong effect on the measured spectrum. For particles of intermediate size where the interaction volume is contained within the particle, placing the

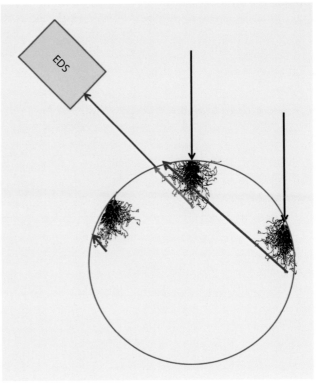

□ **Fig. 23.23** Schematic illustration of the effect of beam placement on a particle on the length of the absorption path to the EDS detector

beam on the side of the particle away from the EDS results in an extended absorption path. The generated X-rays must pass through a large mass of the particle to reach the detector. This effect is illustrated schematically in □ Fig. 23.23, where absorption paths for X-rays generated near the maximum penetration of beam electrons in the particle are compared for three beam positions: top center, and at positions directly facing and away from the EDS. Because absorption follows an exponential dependence on path length and becomes increasingly significant for low energy photons below approximately 4 keV, the low energy portion of the spectrum will show the strongest absorption effects. Spectra recorded at the top of a 5-μm-diameter particle and on the side away from the EDS are compared in □ Fig. 23.24 demonstrating a factor of two difference in the intensity measured at the energy of Mg K-L$_{2,3}$ (1.254 keV). For a beam placed on the curved side of the particle facing the EDS, the absorption path length is actually reduced relative to the top center, leading to extra emission from the low energy photons, creating about 30 % excess for Mg K-L$_{2,3}$, as shown in the comparison of spectra in □ Fig. 23.25. □ Figure 23.26 shows the impact of beam placement on the accuracy of quantitative analysis, as discussed below. It is thus critical that the analyst is always aware of the relative position of the EDS in the SEM image when choosing locations to analyze. A reliable way to locate the EDS is to record an X-ray spectrum image and examine selected X-ray intensity maps, as shown in □ Fig. 23.27. The general rule that the source of the apparent illumination

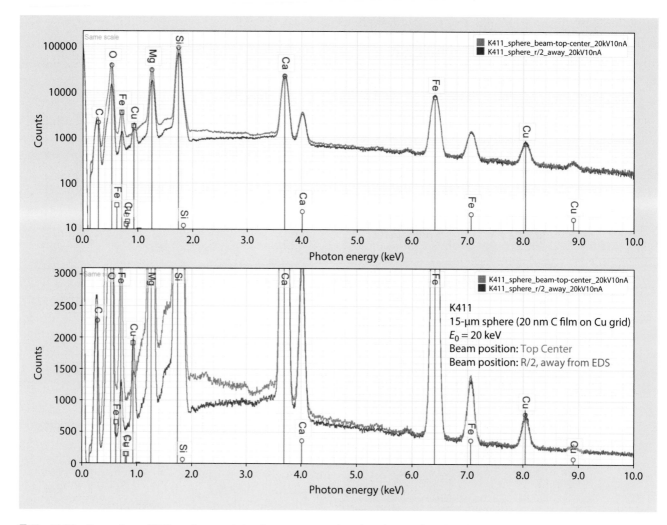

◘ Fig. 23.24   Comparison of EDS spectra recorded at the top center and on the side away from the EDS

appears to come from the position of the detector immediately reveals the true EDS position. For this particular particle, the eclipsing effect of the extended absorption path on the backside of the particle is readily apparent, even in the relatively energetic Co K-L$_{2,3}$ (6.930 keV) intensity map.

Overscanning

"Overscanning" is a strategy by which the beam is continuously scanned over the particle (or rough surface, heterogeneous material, etc.) while the EDS spectrum is being collected. If the magnification control of the SEM is continuously variable, then the size of the scan raster can be adjusted to bracket the particle, minimizing the fraction of the time that the beam spends on the surrounding substrate, as shown in ◘ Fig. 23.28 (yellow box). Alternatively, the scan raster can be adjusted to fill as much of the particle image as possible while remaining within the bounds of the particle, thus avoiding direct beam placement on the substrate, as shown in ◘ Fig. 23.28 (green box).

While useful for gaining qualitative information on the constituents of a particle, overscanning only has utility if the particle (or other target object) is homogeneous. Overscanning discards valuable information on any possible inhomogeneous

structure within the particle. As described below, overscanning is NOT a means by which an "average" composition can be found by quantitative microanalysis.

23.6.3   X-ray Spectrum Imaging: Understanding Heterogeneous Materials

X-ray spectrum imaging (XSI) involves collecting a complete EDS spectrum at every pixel location visited by the scanned beam. When applied to particle analysis, the XSI provides the analyst with an abundance of information which can be recovered by post-collection processing of the XSI datacube. Composition heterogeneity down to the single pixel level can be detected and interpreted. The particle X-ray intensity images shown in ◘ Fig. 23.27a, b were extracted from an XSI, and the localization of Ti and Mo in inclusions is immediately obvious. Solidification dendrites are outlined in the Ti K-L$_{2,3}$ map, a feature that is not obvious in any of the other elemental intensity maps. Software tools enable the analyst to extract spectra that are representative of individual components of the microstructure, such

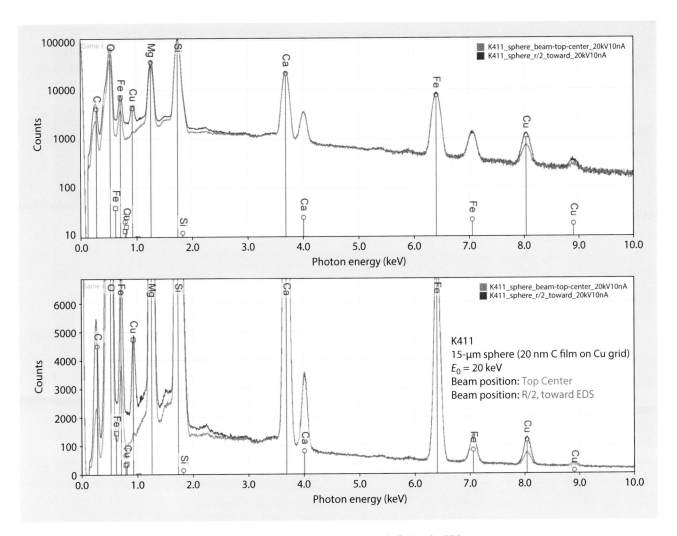

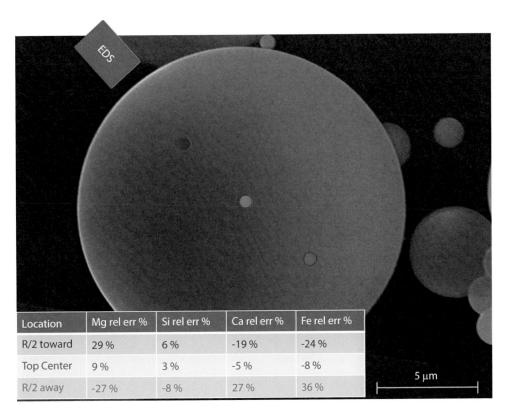

Fig. 23.25 Comparison of EDS spectra recorded at the top center and on the side facing the EDS

Fig. 23.26 Quantitative analysis performed at the static beam locations indicated; relative errors observed for the normalized concentrations (oxygen calculated by stoichiometry)

Location	Mg rel err %	Si rel err %	Ca rel err %	Fe rel err %
R/2 toward	29 %	6 %	-19 %	-24 %
Top Center	9 %	3 %	-5 %	-8 %
R/2 away	-27 %	-8 %	27 %	36 %

◘ **Fig. 23.27** **a** SEM BSE image and X-ray intensity maps for Al K-L$_{2,3}$, Cr K-L$_{2,3}$, and Co K-L$_{2,3}$. Note the shadowing effect of the particle thickness causing attenuation of X-ray intensity. **b** X-ray intensity maps for Mo L$_3$-M$_{4,5}$, and Ni K-L$_{2,3}$, and a color overlay with Mo L$_3$-M$_{4,5}$ = *red*, Ti K-L$_{2,3}$ = *green*, and Ni K-L$_{2,3}$ = *blue*. Note the localization of Ti and Mo in inclusions, and the solidification dendrites outlined by Ti

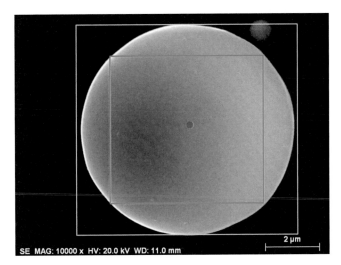

detector (SDD)-EDS inevitably shows the shadowing created by specimen geometry, such shadowing can be greatly diminished by collecting the XSI with an array of SDD-EDS detectors placed symmetrically around the specimen, as shown in ◘ Fig. 23.29.

23.6.4 Particle Geometry Factors Influencing Quantitative Analysis of Particles

There are two principal "particle geometry" effects: the "particle mass effect" and the "particle absorption effect" (Small et al. 1978, 1979).

"Particle Mass Effect"

The penetration of beam electrons through the side and bottom of a particle reduces the X-ray production compared to a flat bulk target, creating the so called "particle mass effect." ◘ Figure 23.30 shows the intensity of Fe K-$L_{2,3}$ produced as a function of sphere diameter for spherical particles of NIST SRM (K411), the composition of which is listed in ◘ Table 23.3, as calculated with the Monte Carlo simulation

◘ **Fig. 23.28** Beam placement strategies for particle analysis: static point beam placed at particle center, overscanning with a scan field that brackets the particle (*yellow box*), overscanning within the particle (*green box*)

as the inclusions, for subsequent analysis. All of this information would be lost if these particles were simply overscanned. While the XSI collected with a single silicon drift

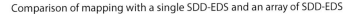

Comparison of mapping with a single SDD-EDS and an array of SDD-EDS

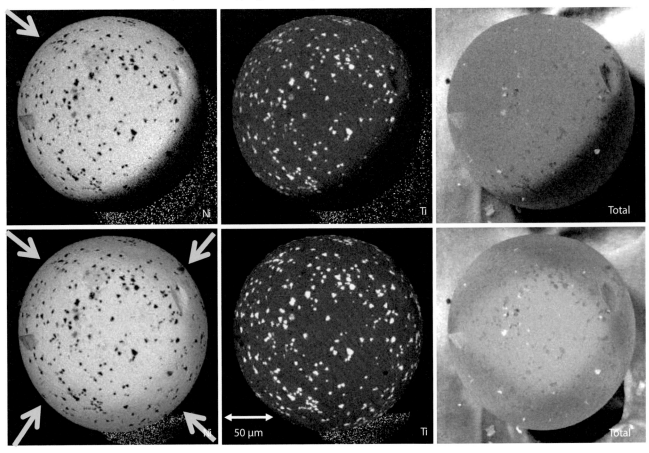

◘ **Fig. 23.29** Comparison of X-ray spectrum imaging with a single EDS and with an array of four EDS detectors; note reduction in shadowing

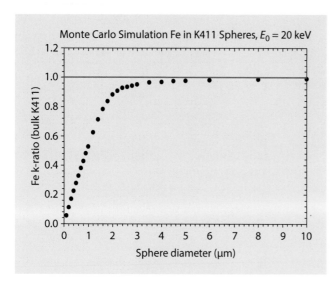

Monte Carlo Simulation Fe in K411 Spheres, $E_0 = 20$ keV

(y-axis: Fe k-ratio (bulk K411), x-axis: Sphere diameter (μm))

Table 23.3 Composition of K411 glass and analysis of flat, bulk target at $E_0 = 20$ keV (Standards: Mg, Si, CaF_2, Fe; oxygen by stoichiometry)

NIST SRM 470 certificate values			
Element	Mass concentration	Analysis	Relative error (%)
O	0.4236	0.4192	−1.0
Mg	0.0885	0.0870	−1.7
Si	0.2538	0.2512	−1.0
Ca	0.1106	0.1082	−2.2
Fe	0.1121	0.1134	1.1

Fig. 23.30 Monte Carlo calculation of the emission of Fe K-L$_{2,3}$ from K411 spheres of various diameters at $E_0 = 20$ keV. The intensity is normalized by the Fe K-L$_{2,3}$ intensity calculated for a flat bulk target of K411

for an incident beam energy of 20 keV. For this plot, the Fe K-L$_{2,3}$ intensity emitted from the particle has been divided by the intensity calculated for a flat bulk K411 target to determine the k-ratio. Note that as the particle diameter increases from zero, the Fe K-L$_{2,3}$ k-ratio increases and asymptotically approaches unity (i.e., equivalent to the flat bulk target) for a diameter of approximately 10 μm. Thus, for the case K411 at $E_0 = 20$ keV, bulk behavior for Fe K-L$_{2,3}$ (6.400 keV) is

observed for spherical particles with diameters of 10 μm and greater.

"Particle Absorption Effect"

The surface curvature of a particle can reduce the absorption path to the detector compared to the interaction volume in a flat bulk target, as shown schematically in **Fig. 23.31**. Since X-ray absorption depends exponentially on the absorption path length, surface curvature can significantly modify the measured intensity, creating the "particle absorption effect." The magnitude of the particle absorption effect depends strongly on the energy of the characteristic photons involved. For the example shown in **Fig. 23.32**, the Fe K-L$_{2,3}$

Fig. 23.31 Schematic illustration of the difference in the X-ray absorption path length in a particle and in a flat bulk target of the same material

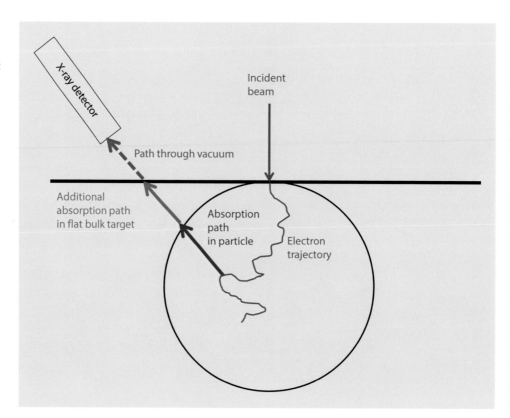

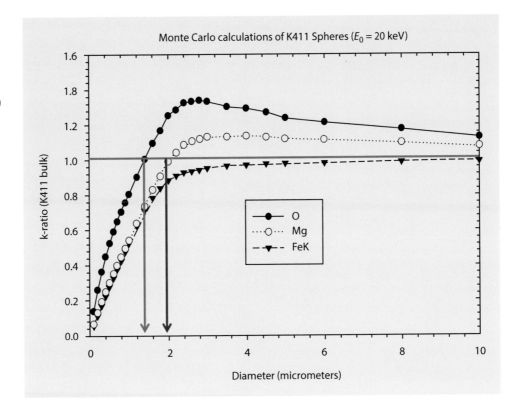

Fig. 23.32 Emission of Fe K-L$_{2,3}$, O K-L$_2$ and Mg K-L$_{2,3}$ (normalized to O K-L$_2$ and Mg K-L$_{2,3}$ from bulk K411); note excursion in the O K-L$_{2,3}$ and Mg K-L$_{2,3}$ above unity (i.e., exceeding the emission from a flat, bulk target)

radiation at 6.40 keV is sufficiently energetic that it undergoes relatively little absorption in the bulk case so that the modification of the absorption path length by particle surface curvature produces a negligible effect. However, the lower energy characteristic photons such as O K-L$_2$ (0.525 keV) and Mg K-L$_{2,3}$ (1.254 keV) in K411 suffer significant absorption in a flat bulk target, so that when these photons are generated in a spherical particle, the reduced absorption path in the direction toward the EDS leads to an increase in X-ray emission compared to a flat bulk target. Figure 23.32 shows that as the particle diameter increases from zero, the k-ratios for O K-L$_2$ and Mg K-L$_{2,3}$ initially increase similarly to FeKα as a result of the particle mass effect. However, for larger particles the reduced absorption path of the curved particle surface causes the emitted O K-L$_2$ k-ratio to actually exceed unity (i.e., higher emission than bulk behavior) for a particle diameter of 1.6 μm, reaching a maximum of 1.35 relative to bulk at a particle diameter of 2.8 μm. For Mg K-L$_{2,3}$, the emission exceeds unity for a particle diameter of 2 μm and reaches a maximum of 1.17 at a diameter of 3.0 μm. For particle diameters beyond the intensity maxima, the O K-L$_2$ and Mg K-L$_{2,3}$ k-ratios gradually decrease with increasing particle diameter, asymptotically approaching the equivalent of bulk behavior at 25 μm diameter for O K-L$_2$, and 18 μm diameter for Mg K-L$_{2,3}$. Thus, for spherical particles of the K-411 composition measured with $E_0 = 20$ keV, effectively bulk behavior is observed for all characteristic X-ray energies for particles with diameters greater than 25 μm for a beam position at the top center of the particle (detector take-off angle 40°). As demonstrated in Fig. 23.23, deviations in the beam placement either toward the EDS or away have significant effects due to the

modification of the absorption path. Particle geometry and its impact on X-ray absorption must be considered when selecting beam locations on a particle for analysis. It is critical that the analyst always be aware of the position of the EDS detector relative to the X-ray source, as demonstrated in Fig. 23.24, to minimize the effects of particle geometry.

23.6.5 Uncertainty in Quantitative Analysis of Particles

Quantitative analysis of particles is performed by following the same k-ratio/matrix correction protocol used for flat, bulk specimens. However, it must be recognized that particle geometry modification of the interaction of beam electrons and the subsequent propagation of X-rays introduce factors which violate the fundamental assumption of the bulk quantification method, namely that the only reason the X-ray intensities measured in the target being analyzed are different from the standards is that the composition(s) is different. Thus, with the impact of the geometric factors, the analytical accuracy of the conventional k-ratio/matrix correction protocol is inevitably compromised. The critical question to consider is the degree to which the uncertainty budget is increased by the systematic error contribution of the particle effects.

The Analytical Total Reveals the Impact of Particle Effects

The analytical total is the sum of the calculated concentrations, including oxygen by stoichiometry if calculated. Table 23.3 shows the results of the analysis of K411 glass in the form of a flat, bulk target. The beam energy was

◻ **Table 23.4** Analysis of a 1.3-μm-diameter spherical particle of K411 glass with fixed beam located at particle center (standards: Mg, Si, CaF_2, Fe; oxygen by stoichiometry)

Element	SRM value	Analysis	Rel error (%)	Normalized	Rel error (%)
O	0.4236	0.1470	−65	0.4224	−0.3
Mg	0.0885	0.0265	−70	0.0761	−14
Si	0.2538	0.0884	−65	0.2541	+0.1
Ca	0.1106	0.0370	−67	0.1064	−3.8
Fe	0.1121	0.0491	−56	0.1410	+26
Raw analytical total	0.3480				

◻ **Table 23.5** Analysis of a 6.1-μm-diameter spherical particle of K411 glass with fixed beam located at particle center (standards: Mg, Si, Ca [K412 glass], Fe; oxygen by stoichiometry)

Element	SRM value	Analysis	Rel error (%)	Normalized	Rel error (%)
O	0.4236	0.4748	+12	0.4353	+2.8
Mg	0.0885	0.1110	+25	0.1018	+15
Si	0.2538	0.2874	+13	0.2636	+3.9
Ca	0.1106	0.1062	−4.0	0.0974	−12
Fe	0.1121	0.1112	−0.8	0.1019	−9.1
Raw analytical total	1.091				

$E_0 = 20$ keV, and the standards used pure elements Mg, Si, and Fe, while SRM 470 K412 glass was used as the Ca standard. The analytical total was 0.9789 (the sum of the SRM certificate values is 0.9886), and the relative errors were all well within the ±5 % error envelope, with the largest error at −2.2 % relative for Ca.

◻ Table 23.4 gives the results of applying the k-ratio/matrix correction factor protocol to the analysis of a 1.3-μm-diameter spherical particle of K411 glass, with the EDS spectrum collected with the beam placed at the center of the particle image. The intensity of each elemental constituent has been measured relative to the same suite of standards: pure elements Mg, Si, and Fe, and CaF_2 as the Ca standard. Oxygen has been calculated by the method of assumed stoichiometry. The analytical total is 0.3480, and the relative errors for the raw calculated concentrations are large and negative, for example, −70 % relative for Mg and −56 % relative for Fe. These large negative relative errors (a negative relative error indicates that the calculated concentration underestimates the true concentration) for both the low photon energy peaks (e.g., Mg K-$L_{2,3}$ and Si K-$L_{2,3}$) and the high photon energy peaks (e.g., Ca Kα and Fe Kα) are a result of the particle mass effect reducing all X-ray intensities compared to bulk behavior. Clearly, these raw concentrations

have such large systematic errors as to offer no realistic meaning. To compensate for the mass effect and thus place the concentrations on a meaningful basis, internal normalization can be applied:

$$C_{n(i)} = C_i / \Sigma C_i \qquad (23.3)$$

Note that normalization is only useful if all constituents present in the analyzed volume are included in the total, including any such as oxygen that are calculated by assumed stoichiometry rather than measured directly. After normalization, the relative errors are reduced in magnitude, as given in ◻ Table 23.4, but the values for Mg (−14 %) and Fe (+26 %) remain well outside the bulk analysis error histogram.

◻ Table 23.5 presents similar measurements and calculations for a 6.1-μm-diameter K411 particle for which the analytical total is 1.091. This particle diameter is sufficiently large so that the X-ray production for the higher energy photons, Ca K-$L_{2,3}$ and Fe K-$L_{2,3}$, has nearly reached equivalence to the flat, bulk target, resulting in relative errors of −5 % or less. For this particle size, the lower energy photon peaks, Mg K-$L_{2,3}$ and Si K-$L_{2,3}$, are still strongly influenced by the particle absorption effect, causing relative errors that are large and positive, since more of these low energy photons escape than

23

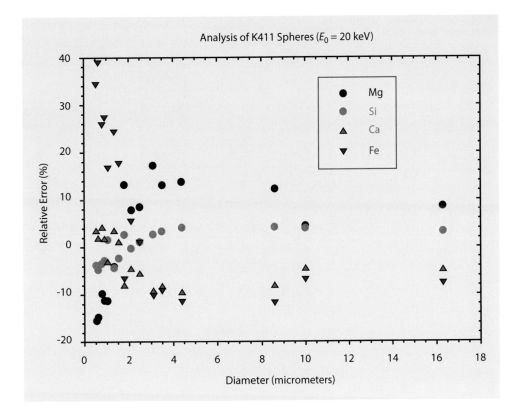

Fig. 23.33 Relative errors observed for various sizes of spherical particles of K411 glass measured with a fixed beam placed at the center of the particle image

is the case for a flat, bulk target. When these raw concentrations are normalized, the relative errors for Mg and Si are reduced, but the relative errors for Ca and Fe are increased after normalization.

These examples demonstrate the complex interplay of X-ray generation and propagation as influenced by particle geometry. While normalization of the raw calculated concentrations is necessary to put particle analyses on a realistic concentration basis, the uncertainty budget for particle analysis is substantially increased compared to that for the ideal flat target. Figure 23.33 plots the relative error envelope for normalized concentrations as a function of particle diameter for spherical particles of K411 glass. For particles whose dimensions are substantially smaller than the bulk interaction volume, the relative errors in the normalized concentrations are large and increase as the particle size decreases. The relative errors decrease as the particle diameter increases, eventually converging with the flat bulk case for particles above approximately 25-µm diameter.

Normalization is most successful when applied to compositions where the measured characteristic X-rays have similar energies, for example, Mg K-$L_{2,3}$, Al K-$L_{2,3}$ and Si K-$L_{2,3}$. Although low atomic number elements such as oxygen can be measured directly, the high absorption of the low energy photons of the characteristic X-ray significantly increases the effect of the particle absorption effect, so that normalization introduces large errors and effectively transfers increased error to the other higher atomic number constituents. In the case of oxygen, the method of assumed stoichiometry is generally much more effective.

Does Overscanning Help?

Because of the difficulty in analyzing particles, overscanning the particle during EDS collection is thought to obtain a spectrum that averages particle effects. In reality, even for homogeneous particles, overscanning does not decrease the relative uncertainties but can actually cause an increase. Figure 23.34a plots the relative errors in the normalized concentrations for the analysis of Mg and Fe in K411 spheres of various sizes, comparing point beam analyses centered on the particle image with continuous overscanning during EDS collection. Mg and Fe are chosen because the large separation in characteristic X-ray energy provides sensitivity to the action of the particle mass effect, which is the only significant influence on energetic Fe K-$L_{2,3}$, while both the mass effect and the absorption effect influence Mg K-$L_{2,3}$. While the error range for point beam analysis is substantially larger than the ideal error histogram for flat bulk target analysis, the effect of overscanning is actually to shift the distribution of results to even more severe relative errors. This is a result of the non-linear nature of X-ray absorption, which can be seen in the beam placement measurements shown in Figs. 23.24 and 23.25. A similar effect is seen for irregular shards in Fig. 23.34b.

◻ Fig. 23.34 Relative errors in normalized analyses for Mg and Fe in K411 glass with point beam centered on the image center of mass and with overscanning: **a** spherical particles, **b** fractured shards

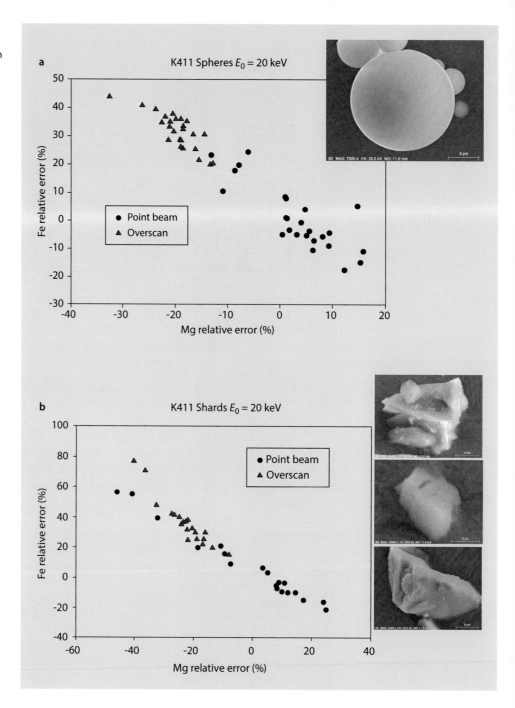

23.6.6 **Peak-to-Background (P/B) Method**

Specimen Geometry Severely Affects the k-ratio, but Not the P/B

Another approach to establishing quantitative X-ray microanalysis for objects of an irregular shape, such as particles and rough surfaces, is the peak-to-background (P/B) method, which is an extension of the Marshall–Hall method for the correction of mass loss in beam-sensitive materials (see ▶ Chap. 20) (Marshall and Hall, 1966; Hall, 1968). The P/B method (Small et al., 1978, 1979a,b; Statham and Pawley, 1978; Statham, 1979; Wendt and Schmidt, 1978; August and

Wernisch, 1991a, b, c) is based on the observation that although the characteristic X-ray intensity emitted from an irregularly shaped object is highly dependent on local geometric effects, the P/B ratio measured between the characteristic X-rays and the continuum X-rays *of the same energy* is much less sensitive to specimen geometry. ◻ Table 23.6 contains measurements of the k-ratio (measured relative to bulk K411) and the P/B from the spectra of the SRM-470 (K-411 glass) shards in Fig. 23.13. The shard spectra show significant deviations from the spectrum of the polished bulk K411, especially at low photon energies below 4 keV. Although the k-ratio for Mg measured for these shards varies by a factor of

23

Table 23.6 K411 shards

Sample	Element	P/B	k-ratio (relative to bulk K411)
Shard A	Mg	4.52	0.545
Shard C	Mg	4.49	0.339
Shard D	Mg	4.52	1.132
Shard E	Mg	4.73	0.389
Bulk	Mg	4.57	1.00
Shard A	Si	16.35	0.617
Shard C	Si	17.32	0.548
Shard D	Si	14.95	1.06
Shard E	Si	17.33	0.447
Bulk	Si	15.80	1.00
Shard A	Ca	6.78	0.835
Shard C	Ca	6.58	0.866
Shard D	Ca	6.43	1.006
Shard E	Ca	7.14	0.710
Bulk	Ca	6.37	1.00
Shard A	Fe	6.48	0.911
Shard C	Fe	6.29	0.941
Shard D	Fe	6.50	0.986
Shard E	Fe	6.82	0.886
Bulk	Fe	6.61	1.00
Range (shard/bulk)	Mg	1.03	2.95
Range (shard/bulk)	Si	1.10	2.24
Range (shard/bulk)	Ca	1.13	1.41
Range (shard/bulk)	Fe	1.03	1.13

P/B peak-to-background

and continuum generation processes, at least to a first order for photons of the same energy. Both types of radiation have a similar, although not identical, depth distribution; thus, the absorption paths to the detector are alike. As the same photon energy is chosen for characteristic and continuum X-rays, the geometric absorption effect is thus comparable for both. When making corrections for an irregularly shaped object, the exact absorption path is very difficult to determine. Because the continuum radiation of the same photon energy is following the same path to the detector that the characteristic radiation follows, regardless of local object shape, this continuum intensity I_B can be used as an automatic internal normalization factor to compensate for the major geometric effects. Furthermore, the P/B ratio is independent of probe current (and thus need not be measured); yet, the quantification results need not be normalized. Because of this, both standards-based and standardless P/B algorithms have been implemented that provide an estimate of the analytical total. More sophisticated models have been developed that account for the second-order (i.e., subtler) differences between the distributions of characteristic and continuum radiation generation (August and Wernisch, 1991a, b, c).

Using the P/B Correspondence

Consider the k-ratio for an object measured relative to a flat, bulk standard of the same composition, $k_{object} = I_{object}/I_{bulk}$. I is the characteristic peak intensity corrected for continuum background at the same energy, I = P − B. The measured k_{object} is a strong function of the object's size and shape, but the ratio $(I_{object}/I_{B,object})/(I_{bulk}/I_{B,\,bulk})$ involving the background at the same photon energy is nearly independent of object size, except for very small particles where the anisotropy of the continuum emission becomes significant (Small et al., 1980). This experimental observation, which has been confirmed by theoretical calculations and Monte Carlo simulations, can be employed in several ways (Small et al., 1978; Statham, and Pawley, 1978). One useful approach is to incorporate the following correction scheme into a conventional ZAF method (Small et al., 1978, 1979b). Given that:

$$\frac{I_{object}}{I_{B,object}} = \frac{I_{bulk}}{I_{B,bulk}} \tag{23.4}$$

a modified particle intensity that compensates for the geometric effects, I^*_{object}, can be calculated that is equivalent to the intensity that would be measured from a flat bulk target of the same composition as the particle:

$$I^*_{object} \approx I_{bulk} = I_{object} * \frac{I_{B,bulk}}{I_{B,object}} \tag{23.5}$$

To apply Eq. 23.5 for the analysis of an irregularly shaped object of unknown composition, the quantities I_{object} and $I_{B,object}$ are determined from the measured X-ray spectrum. Because the composition of the object is unknown, the term $I_{B,bulk}$ in Eq. 23.5 is not known, as a bulk multi-element

2.95 for the most extreme case, the corresponding P/B ratio for Mg only differs from that of the bulk K411 by a factor of 1.03. For the particular combination of elements in K-411 at this beam energy (20 keV), the most extreme variation in the P/B observed for these shards is 1.13 for Ca.

Although the characteristic and continuum X-rays are produced by different physical processes (inner shell ionization versus deceleration in a Coulombic field) that have different behaviors as a function of the exciting electron energy; especially near the ionization threshold for an element, both characteristic and continuum X-rays are generated in nearly the same volume. Both forms of radiation thus scale similar to the geometric mass effect, because the loss of beam electrons due to backscattering and penetration also robs both characteristic

◘ Table 23.7 Relative Deviation from Expected Value (RDEV) observed with peak-to-background corrections compared to the raw concentrations and normalized concentrations after conventional k-ratio/matrix corrections

	C_{bulk}	C_{raw}	RDEV(%)	C_N	RDEV(%)	$C_{P/B}$	RDEV(%)
Al	0.0603	0.0201	−67%	0.0241	−60%	0.0552	−8%
Mo	0.0353	0.0194	−45%	0.0233	−34%	0.0437	+24%
Ti	0.0519	0.0406	−22%	0.0487	−6%	0.0480	−7%
Cr	0.0965	0.0788	−18%	0.0945	−2%	0.0996	+3%
Co	0.155	0.139	−11%	0.166	+7%	0.156	+1%
Ni	0.601	0.536	−11%	0.643	+7%	0.598	−0.5%

Spherical particle: IN-100 alloy, 88 µm diameter, with beam placed at 22 µm from the top center on the backside of particle

standard identical in composition to the unknown object is generally not available. However, an estimate of the concentrations of elements in the unknown object is always available in the ZAF procedure, including the first step, where $C_i = k_i / \Sigma k$. The value of $I_{B,bulk}$ can therefore be estimated from the background measured on pure element standards:

$$I_{B,bulk} = \sum_j C_j I_{j,B,std} \tag{23.6}$$

where $I_{j,B,std}$ is the pure element bremsstrahlung at the energy of interest and C_j is the concentration of element j. An example of an analysis of a complex IN-100 particle with conventional k-ratio/matrix corrections (including normalization of the raw values) and using the k-ratio/matrix corrections augmented with the P/B method is given in ◘ Table 23.7. The relative deviation from the expected value (RDEV) for each element is reduced compared with the simple normalization procedure, especially for Al, which is highly absorbed when measured on the backside of the particle.

The special advantage of the P/B method is that it can be applied to spectra obtained with a focused probe directed at a specific location on a particle. Thus, particles that have a chemically heterogeneous sub-structure can be directly studied. To be effective, the P/B method requires spectra with high counts. Because the ratio of background intensities is used to scale the particle peak intensities, the statistical uncertainty in the background ratio propagates into the error in each concentration value in addition to the statistics of the characteristic peak. Even more importantly, the P/B method depends on the background radiation originating in the excited volume of the specimen only, and not in the surrounding substrate. When an irregularly shaped object such as a particle becomes small relative to the bulk interaction volume, the penetration of the beam into the substrate means that the continuum continues to be produced, even if the substrate is a low atomic number element such as carbon. As noted above, the energy-dispersive X-ray spectroscopy

collimator has a large acceptance area at the specimen. To minimize the extraneous background contributions, the small particles should be mounted on a thin (approximately 10–20 nm) carbon film supported on a metal grid (typically copper, as used in the transmission electron microscope) and mounted over a blind hole drilled into a carbon block. The continuum contribution from such a thin film is negligible relative to particles as small as approximately 250 nm in diameter.

23.7   Summary

1. Particle analysis is inevitably compromised compared to analysis of ideal flat samples, leading to an increased error budget.
2. Careful attention must be paid to optimizing particle sample preparation to minimize substrate contributions to the spectrum and to reduce contributions from nearby particles.
3. Quantitative analysis of particles follows the k-ratio/matrix correction protocol. The analytical total that results from this procedure is an indication of the magnitude of particle geometry effects (mass effect and absorption effect).
4. Normalization of the raw concentrations (including oxygen by stoichiometry, if appropriate) is necessary to place the calculated composition on a realistic basis.
5. Large relative errors, exceeding 10%, are encountered after normalization. The analytical errors are exacerbated when low and high photon peaks must be used for analysis.
6. The analytical errors generally increase as the particle size decreases.
7. Overscanning does not decrease the analytical errors, and may well increase the errors depending on the particular combination of elements being analyzed.

References

August H-J, Wernisch J (1991a) Calculation of the depth distribution function for continuous radiation, Scanning 13:207–215

August H-J, Wernisch J (1991b) Calculation of depth distribution functions for characteristic x-radiation using an electron scattering model. I—theory, X-Ray Spectrom, 20:131–140

August H-J, Wernisch J (1991c) Calculation of depth distribution functions for characteristic x-radiation using an electron scattering model. II—results, X-Ray Spectrom, 20:141–148

Hall T (1968) Some aspects of the microprobe analysis of biological specimens. In: Heinrich K (ed) Quantitative electron probe microanalysis. U.S. Government Printing Office, Washington, DC, p 269

Marshall D, Hall T (1966) A method for the microanalysis of thin films. In: Castaing R, Deschamps P, Philibert J (eds) X-ray optics and microanalysis. Hermann, Paris, p 374

Newbury D, Ritchie W (2013a) Quantitative SEM/EDS, Step 1: What Constitutes a Sufficiently Flat Specimen? Microsc Microanal 19(Suppl 2):1244

Newbury D, Ritchie W (2013b) Is Scanning Electron Microscopy/Energy Dispersive Spectrometry (SEM/EDS) Quantitative? Scanning 35:141

Small JA, Heinrich KFJ, Fiori CE, Myklebust RL, Newbury DE, Dilmore MF (1978) The production and characterization of glass fibers and spheres for microanalysis, In: Johari, O (ed) Scanning electron microscopy/1978/I, IITRI, Chicago, p 445

Small JA, Heinrich KFJ, Newbury DE, Myklebust RL (1979) SEM/1979/II, SEM, Inc., AMF O'Hare, Illinois, p 807

Small JA, Heinrich KFJ, Fiori CE, Myklebust RL, Newbury DE, Dillmore MF (1978) The production and characterization of glass fibers and spheres for microanalysis. In: Johari O (ed) Scanning electron microscopy/1978/I. IITRI, Chicago, p 445

Small JA, Heinrich KFJ, Newbury DE, Myklebust RL (1979a) Progress in the development of the peak-to-background method for the quantitative analysis of single particles with the electron probe in Scanning Electron Microscopy/1979/II, ed. Johari, O. (IITRI, Chicago), p 807

Small JA, Newbury DE, Myklebust RL (1979b) Analysis of particles and rough samples by FRAME P, a ZAF method incorporating peak-to-background measurements, in Microbeam analysis, ed. Newbury, D. (San Francisco Press, San Francisco), p 243

Small JA, Heinrich KFJ, Newbury DE, Myklebust RL, Fiori CE (1980) Procedure for the Quantitative Analysis of Single Particles with the Electron Probe. In: Heinrich KFJ (ed) Characterization of Particles. National Bureau of Standards Special Publication 533, Washington, pp 29–38

Statham PJ, Pawley JB (1978) A new method for particle X-ray microanalysis based on peak to background measurement. In: Johari O (ed) Scanning electron microscopy/1978/I. IITRI, Chicago, p 469

Statham P (1979) A ZAF procedure for microprobe analysis based on measurement of peak-to-background ratios. In: Newbury D (ed) Microbeam Analysis – 1979. San Francisco Press, San Francisco, pp 247–253

Wendt M, Schmidt A (1978) Improved reproducibility of energy-dispersive X-ray microanalysis by normalization to the background, Phys Status Solidi (a) 46:179

Yakowitz H, Heinrich KFJ (1968) Quantitative electron probe microanalysis: fluorescence correction uncertainty. Mikrochim Acta 5:182

Compositional Mapping

© Springer Science+Business Media LLC 2018
J. Goldstein et al., *Scanning Electron Microscopy and X-Ray Microanalysis*,
https://doi.org/10.1007/978-1-4939-6676-9_24

SEM images that show the spatial distribution of the elemental constituents of a specimen ("elemental maps") can be created by using the characteristic X-ray intensity measured for each element with the energy dispersive X-ray spectrometer (EDS) to define the gray level (or color value) at each picture element (pixel) of the scan. Elemental maps based on X-ray intensity provide qualitative information on spatial distributions of elements. Compositional mapping, in which a full EDS spectrum is recorded at each pixel ("X-ray Spectrum Imaging" or XSI) and processed with peak fitting, k-ratio standardization, and matrix corrections, provides a quantitative basis for comparing maps of different elements in the same region, or for the same element from different regions.

24.1   Total Intensity Region-of-Interest Mapping

In the simplest implementation of elemental mapping, energy regions are defined in the spectrum that span the characteristic X-ray peak(s) of interest, as shown in ◘ Fig. 24.1. The total X-ray intensity (counts) within each energy region, I_{Xi}, consisting of both the characteristic peak intensity, including any overlapping peaks, and the continuum background intensity, is digitally recorded for each pixel, creating a set of x-y-I_{Xi} image arrays for the defined suite of elements. Depending on the local concentration of an element, the overvoltage $U_0 = E_0/E_c$ for the measured characteristic peak, the beam current, the solid angle of the detector, and the dwell time per pixel, the number of counts

per elemental window can vary widely from a few counts per pixel to several thousand or more. A typical strategy to avoid saturation is to collect 2-byte deep X-ray intensity data that permits up to 65,536 counts per energy region per pixel. In common with the practice for BSE and SE images, the final elemental map will be displayed with a 1-byte intensity range (0–255 gray levels). To maximize the contrast within each map it is necessary to nearly fill this display range so that it is common practice to automatically scale ("autoscale") the measured intensity in a linear fashion to span slightly less than 1-byte. The displayed gray levels are scaled to range from near black, but avoiding black (gray level zero) to avoid clipping, to near white, but avoiding full white (gray level 255), to prevent saturation. The counts are expanded for elements that span less than 1-byte in the original data collection, while the counts are compressed for elements that extend into the 2-byte range. An example of such total intensity mapping is shown in ◘ Fig. 24.2, which presents a set of maps for Si, Fe, and Mn in a cross section of a deep-sea manganese nodule with a complex microstructure. The EDS spectrum shown in ◘ Fig. 24.1 also reveals a typical problem encountered in simple intensity window mapping. Manganese is one of the most abundant elements in this specimen, and the Mn K-$M_{2,3}$ (6.490 keV) interferes with Fe K-$L_{2,3}$ (6.400 keV), which is especially significant since iron is a minor/trace constituent. To avoid the potential artifacts in this situation, the analyst can instead choose the Fe K-$M_{2,3}$ (7.057 keV) which does not suffer the interference but which is approximately a factor of ten lower intensity than Fe K-$L_{2,3}$. While sacrificing sensitivity, the

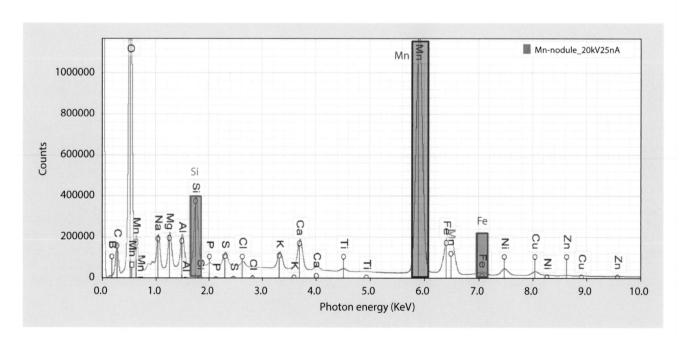

◘ Fig. 24.1   EDS spectrum measured on a cross section of a deep-sea manganese nodule showing peak selection (Si K-L_2, Mn K-$L_{2,3}$, and Fe K-$M_{2,3}$) for total intensity elemental mapping

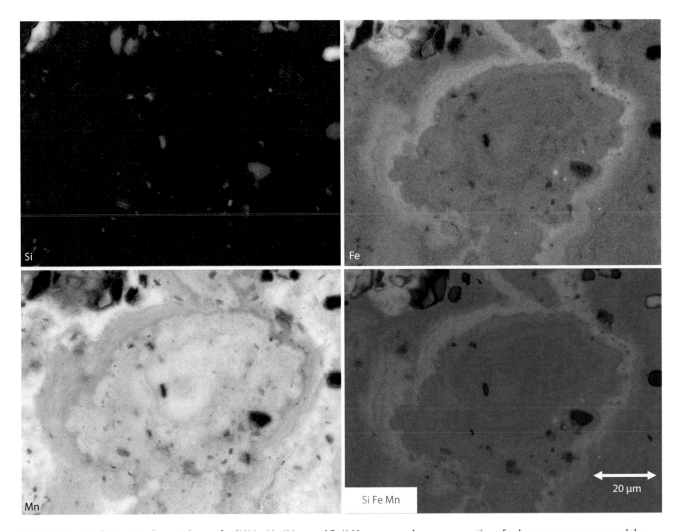

◻ Fig. 24.2 Total intensity elemental maps for Si K-L$_2$, Mn K-L$_{2,3}$, and Fe K-M$_{2,3}$ measured on a cross section of a deep-sea manganese nodule, and the color overlay of the gray-scale maps

success of this strategy is revealed in ◻ Fig. 24.2, where the major Mn and minor Fe are seen to be anti-correlated.

◻ Figure 24.2 also contains a primary color overlay of these three elements, encoded with Si in red, Fe in green, and Mn in blue, a commonly used image display tool which provides an immediate visual comparison of the relative spatial relationships of the three constituents. The appearance of secondary colors shows areas of coincidence of any two elements, for example, the cyan colored region is a combination of green and blue and thus shows the coincidence of Fe and Mn. The other possible binary combinations are yellow (red plus green, Si + Fe) and magenta (red plus blue, Si + Mn), which are not present in this example. If all three elements were present at the same location, white would result.

24.1.1 Limitations of Total Intensity Mapping

While total intensity elemental maps such as those in ◻ Fig. 24.2 are useful for developing a basic understanding of the spatial distributions of the elements that make up the specimen, total intensity mapping is subject to several significant limitations:

1. By selecting only the spectral regions-of-interest, the amount of mass storage is minimized. However, while all spectral regions-of-interest are collected simultaneously, if the analyst needs to evaluate another element not originally selected when the data was collected, the entire image scan must be repeated to recollect the data with that new element included.

2. Total intensity maps convey qualitative information only. The elemental spatial distributions are meaningful only in qualitative terms of interpreting which elements are present at a particular pixel location(s) by comparing different elemental maps of the same area, for example, using the color overlay method. Since the images are recorded simultaneously, the pixel registration is without error even if overall drift or other distortion occurs. However, the intensity information is not quantitative and can only convey relative abundance information within an individual elemental map (e.g., "this location has more of element A than this location because the intensity of A is higher"). The gray levels in maps for different elements cannot be readily compared because the X-ray intensity for each element that defines the gray level range of the map is determined by the local concentration and the complex physics of X-ray generation, propagation, and detection efficiency, all of which vary with the elemental species. The element-to-element differences in the efficiency of X-ray production, propagation, and detection are embedded in the raw measured X-ray intensities, which are then subjected to the autoscaling operation. Unless the autoscaling factor is recorded (typically not), it is not possible after the fact to recover the information that would enable the analyst to standardize and establish a proper basis for inter-comparison of maps of different elements, or even of maps of the same element from different areas. Thus, the sequence of gray levels only has interpretable meaning within an individual elemental map. Gray levels cannot be sensibly compared between total intensity maps of different elements, for example, the near-white level in the autoscaled maps of ■ Fig. 24.2 for Si, Fe, and Mn does not correspond to the same X-ray intensity or concentration for three elements. Because of autoscaling, it is not possible to compare maps for the same element "A" from two different regions, even if recorded with the same dose conditions, since the autoscaling factor will be controlled by the maximum concentration of "A," which may not be the same in two arbitrarily chosen regions of the sample.

3. This lack of quantitative information in elemental total intensity maps extends to the color overlay presentation of elemental maps seen in ■ Fig. 24.2. The color overlay is useful to compare the spatial relationships among the three elements, but the specific color observed at any pixel only depicts elemental coincidence not absolute or relative concentration. The particular color that occurs at a given pixel depends on the complex physics of X-ray generation, propagation, and detection as well as concentration, and the autoscaling of the separate maps that precedes the color overlay, which distorts the apparent relationships among the elemental constituents, also influences the observed colors.

4. When peak interference occurs, the raw intensity in a given energy window may contain contributions from another element, as shown in ■ Fig. 24.1 where the region that includes Fe K-L$_{2,3}$ also contains intensity from Mn K-M$_{2,3}$. While choosing the non-interfered peak Fe K-M$_{2,3}$ gives a useful result in the case of the manganese nodule, if the specimen also contained cobalt at a significant level, Co K-L$_{2,3}$ (6.930 keV) would interfere with Fe K-M$_{2,3}$ (7.057 keV) and invalidate this strategy. The peak interference artifact can be corrected by peak fitting, or by methods in which the measured Mn K-L$_{2,3}$ intensity, which does not suffer interference in this particular case, is used to correct the intensity of the Fe K-L$_{2,3}$ + Mn K-M$_{2,3}$ window using the known Mn K-M$_{2,3}$/K-L$_{2,3}$ ratio.

5. The total intensity window contains both the characteristic peak intensity that is specific to an element and the continuum (background) intensity, which scales with the average atomic number of all of the elements within the excited interaction volume but is not exclusively related to the element that is generating the peak. For the map of an element that constitutes a major constituent (mass concentration $C > 0.1$ or 10 wt %), the non-specific background intensity contribution usually does not constitute a serious mapping artifact. However, for a minor constituent ($0.01 \leq C \leq 0.1$, 1 wt % to 10 wt %), the average atomic number dependence of the continuum background can lead to serious artifacts. ■ Figure 24.3 shows an example of this phenomenon for a Raney nickel alloy containing major Al and Ni with minor Fe. The complex microstructure has four distinct phases, the compositions of which are listed in ■ Table 24.1, one of which contains Fe as a minor constituent at a concentration of approximately $C = 0.04$ (4 wt %). This Fe-rich phase can be readily discerned in the Fe gray-scale map, where the intensity of this phase, being the highest iron-containing region in the image, has been autoscaled to near white. In addition to this Fe-containing phase, there appears to be segregation of lower concentration levels of Fe to the Ni-rich phase relative to the Al-rich phase. However, this effect is at least partially due to the increase in the continuum background in the Ni-rich region relative to the Al-rich region because of the sharp difference in the average atomic number. For trace constituents ($C < 0.01$, 1 wt %), the atomic number dependence of the continuum background can dominate the observed contrast, creating artifacts in the images that render most trace constituent maps nearly useless.

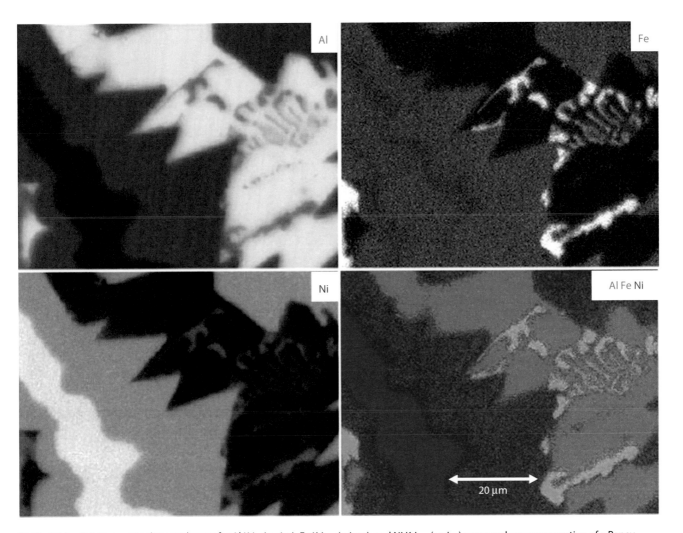

□ **Fig. 24.3** Total intensity elemental maps for Al K-L$_2$ (major), Fe K-L$_{2,3}$ (minor), and Ni K-L$_{2,3}$ (major) measured on a cross section of a Raney nickel alloy, and the color overlay of the gray-scale maps

□ **Table 24.1** Phases in Raney nickel; as measured by electron-excited X-ray microanalysis (mass concentrations)

	Al	Fe	Ni
High Al	0.995	0	0.005
Fe-rich	0.712	0.042	0.246
Intermediate Ni	0.600	0	0.400
High Ni	0.465	0	0.535

24.2 X-Ray Spectrum Imaging

X-ray spectrum imaging (XSI) involves collecting the entire EDS spectrum, I(E), at each pixel location, producing a large data structure [x, y, I(E)] typically referred to as a "datacube"

(Gorlen et al. 1984; Newbury and Ritchie, 2013). Alternatively, the data may be recorded as "position-tagged" photons, whereby as each photon with energy E_p is detected, it is tagged with the current beam location (x, y), giving a database of values (x, y, E_p) which can be subsequently sampled with rules defining the range of Δx and Δy over which to construct a spectrum I(E) at a single pixel or over a defined range of pixels. Depending on the number of pixels and the intensity range of the X-ray count data, the recorded XSI can be very large, ranging from hundreds of megabytes to several gigabytes. Vendor software usually compresses the datacube to save mass storage space, but the resulting compressed datacube can only be decompressed and viewed with the vendor's proprietary software. As an important alternative, if the datacube can be saved in the uncompressed RAW format (a simple block of bytes with a header or an associated file that carries the metadata needed to read the file), the RAW file can be read by publically available, open source software such as NIH ImageJ-Fiji or NIST Lispix (Bright, 2017).

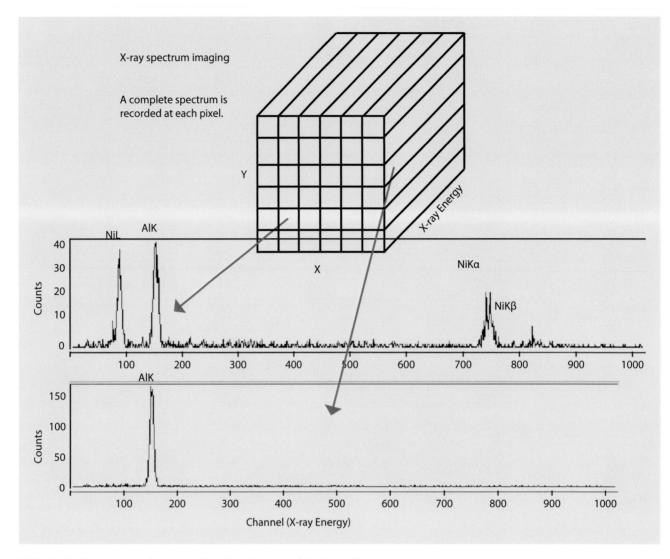

○ **Fig. 24.4**   X-ray spectrum image considered as a datacube of pixels x, y, I(E)

Regardless of how the measured photons are cataloged, the XSI captures all possible elemental information about the region of the specimen being mapped within the limitations set by the excitation energy (E_0), the dose (beam current multiplied by pixel dwell time, $i_B * \tau$), and the EDS spectrometer performance (solid angle and efficiency). ○ Figure 24.4 shows the conceptual nature of the data. The [x, y, I(E)] datacube can be thought of as an x-y array of spectra I(E). The individual pixel spectra can be selected by the analyst for inspection by specifying the x-y location. Depending on the electron beam current, the pixel dwell time, and the solid angle of the EDS detector, the counts in an individual pixel spectrum may be low. For the example XSI of Raney nickel shown in ○ Fig. 24.4, the upper spectrum taken from an Ni-rich area has a maximum of 40 counts per energy channel, while the lower spectrum from a different pixel in an Al-rich region has a maximum of 150 counts. An alternative view of the datacube is shown in ○ Fig. 24.5, where the datacube can be thought of as a stack or deck of x-y image cards, where each x-y image corresponds to a different energy with an energy range ("card thickness") equal to the energy channel width, for example, typically 5 eV or 10 eV. If a card is selected from the stack with an energy that corresponds to a characteristic peak, then the card shows the elemental map for that peak, as shown in ○ Fig. 24.5. A card with an energy that corresponds to spectral background will have fewer counts than a peak card, but such a background card may still have discernible microstructural information because of the atomic number dependence of the X-ray continuum, as shown in ○ Fig. 24.5.

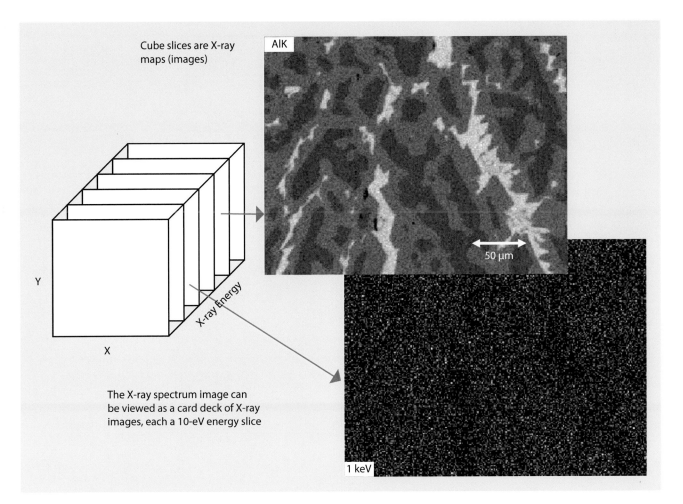

Cube slices are X-ray maps (images)

AlK

50 μm

Y

X-ray Energy

X

The X-ray spectrum image can be viewed as a card deck of X-ray images, each a 10-eV energy slice

1 keV

Fig. 24.5 X-ray spectrum image considered as a stack of x, y images, each corresponding to a specific photon energy E_p

24.2.1 Utilizing XSI Datacubes

In the simplest case, by capturing all possible X-ray information about an imaged region, the XSI permits the analyst to define regions-of-interest for total intensity mapping at any time after collecting the map datacube, creating elemental maps such as those in ▫ Fig. 24.2. If additional information is required about other elements not in the initial set-up, there is no need to relocate the specimen area and repeat the map since the full spectrum has been captured in the XSI.

The challenge to the analyst is to make efficient use of the XSI datacube by discovering ("mining") the useful information that it contains. Software for aiding the analyst in the interpretation of XSI datacubes ranges from simple tools in open source software to highly sophisticated, proprietary vendor software that utilizes statistical comparisons to automatically recognize spatial correlations among elements present in the specimen region that was mapped (Kotula et al. 2003).

24.2.2 Derived Spectra

SUM Spectrum

Simple but highly effective software tools that are present in nearly all vendor XSI platforms as well as the open source software (NIH ImageJ-Fiji and NIST Lispix) include those that calculate "derived spectra." Derived spectra are constructed by systematically applying an algorithm to the datacube to extract carefully defined information. The most basic derived spectrum is the "SUM" spectrum, illustrated schematically in ▫ Fig. 24.6. Conceptually, each energy "card" in the XSI datacube is selected and the counts in all pixels on that card are added together, as indicated by the systematic route through all pixels shown in ▫ Fig. 24.6. This summed count value is then placed in the corresponding energy bin of the SUM spectrum under construction, and the process is then repeated for the next energy card until all energies have been considered. The resulting SUM spectrum has the familiar features of a conventional spectrum: characteristic X-ray peaks and the

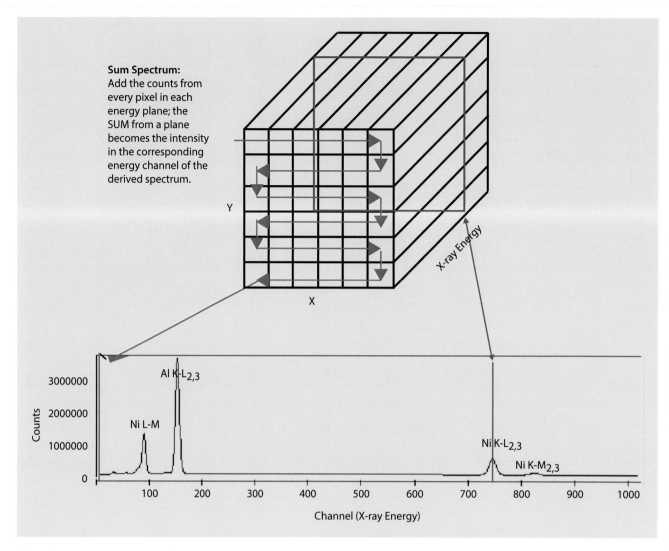

Sum Spectrum:
Add the counts from every pixel in each energy plane; the SUM from a plane becomes the intensity in the corresponding energy channel of the derived spectrum.

☐ **Fig. 24.6** Concept of the SUM spectrum derived from an X-ray spectrum image by adding the counts from all pixels on an individual image at a specific photon energy, E_p

X-ray continuum background, as well as artifacts such as coincidence peaks and Si-escape peaks. The SUM spectrum in ☐ Fig. 24.6 was calculated from the same Raney nickel XSI datacube used for ☐ Fig. 24.4. Note that the count axis now extends to more than 3 million counts, much higher than the count axes of the individual pixel spectra in ☐ Fig. 24.4 because of the large number of pixels that have been added together to construct the SUM spectrum. The characteristic peaks that can be recognized in the SUM spectrum (scaled to the highest intensity) represent the dominant, most abundant elemental features contained in the XSI. The analyst can select a peak

channel and view the corresponding energy card to reveal the elemental map for that peak, as shown in ☐ Fig. 24.7. By selecting a band of adjacent channels that spans the peak and averaging the counts for each x-y pixel in the set of energy cards, an elemental image with reduced noise is obtained, as also shown in ☐ Fig. 24.7. Systematically selecting each of the prominent characteristic X-ray peaks, total intensity maps for all of the major elemental constituents can be obtained. The SUM spectrum can be treated just like a normally recorded spectrum. By expanding the vertical scale or changing from a linear display to a logarithmic display, lower relative intensity

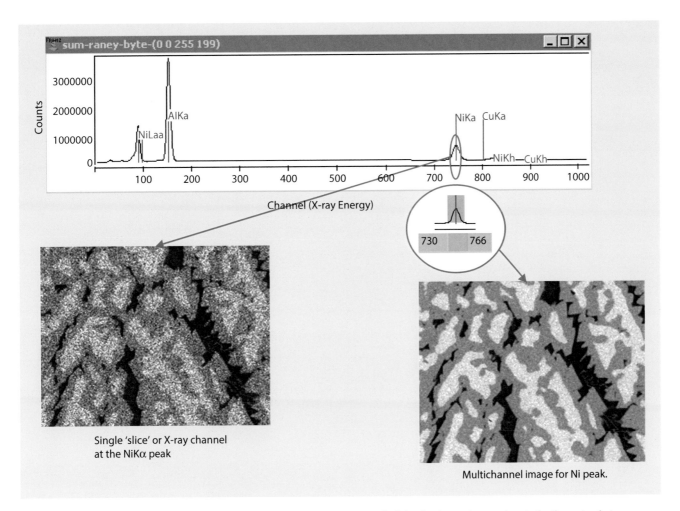

◻ Fig. 24.7 Using the SUM spectrum to interrogate an X-ray spectrum image to find the dominant elemental peaks for the region being mapped

peaks corresponding to minor and trace constituents can be recognized, and the corresponding total-intensity region-of-interest images can be constructed, subject to the increasing influence of the continuum background incorporated in the window. For these lower abundance constituents, it is necessary to use a band of energy channels to reduce the noise in the resulting elemental image, and such images are subject to the artifacts that arise from the atomic number dependence of the X-ray continuum noted above.

MAXIMUM PIXEL Spectrum

The SUM spectrum reveals dominant elemental features in the mapped region. Rare, unexpected elemental features, which in the extreme case may occur at only a single pixel (i.e., looking for a "needle-in-a-haystack" when you don't even know that it

is a needle you are looking for!) can be recognized with the "MAXIMUM PIXEL" derived spectrum (Bright and Newbury 2004). As shown in ◻ Fig. 24.8, the MAXIMUM PIXEL derived spectrum is calculated by making the same tour through all of the pixels on each energy card as is done for the SUM spectrum, but rather than adding the pixel contents, the algorithm now locates the maximum intensity within a card regardless of what pixel it comes from to represent that energy value in the constructed MAXIMUM PIXEL spectrum. An example of the application of the MAXIMUM PIXEL spectrum to an XSI is shown in ◻ Fig. 24.9, where an unanticipated Cr peak is recognized. When the Cr image is selected from the XSI, the Cr is seen to be localized in a small cluster of pixels. Note that in the plot of the SUM, LOG_{10}SUM, and MAXIMUM PIXEL spectra in ◻ Fig. 24.9, the Cr peak is only visible in the MAXIMUM

Fig. 24.8 Concept of the MAXIMUM PIXEL spectrum derived from an X-ray spectrum image by finding the highest count among all the pixels on an individual image at a specific photon energy, E_p

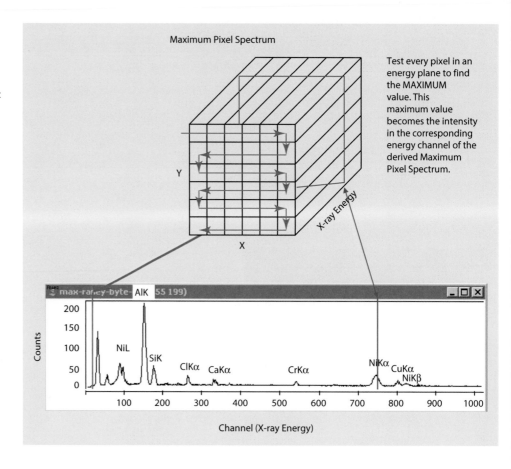

Maximum Pixel Spectrum

Test every pixel in an energy plane to find the MAXIMUM value. This maximum value becomes the intensity in the corresponding energy channel of the derived Maximum Pixel Spectrum.

PIXEL spectrum. Because Cr is represented at only a few pixels, in the SUM spectrum the Cr intensity is overwhelmed by the continuum intensity at all of the other pixels, so that the Cr peak is not visible even in the logarithmic expansion shown in the LOG_{10}SUM plot.

After the elemental maps have been created from the XSI, additional image processing tools can be brought to bear. Spatial regions-of-interest can be defined in the elemental maps or in corresponding BSE or SE images recorded simultaneously with the XSI. These spatial regions can be selected by simple image processing tools that define a group of contiguous pixels that fall within a particular shape (square, rectangle, circle, ellipsoid, or closed free

form are typical choices). Another approach is the creation of a pixel mask by selecting all pixels within a particular elemental map (or BSE or SE image) whose intensities fall within a defined range above a specified gray level threshold. Pixels that satisfy this intensity criterion can occur anywhere in the x–y plane of the XSI and do not have to be contiguous. Selections can range down to an individual pixel. An example of this process is shown in ◻ Figure 24.10, where the Fe-rich phase has been selected from the Fe total intensity map (shown in color overlay with Al and Ni) to create a mask (shown as a binary image in the inset) from which the SUM spectrum representing the Fe-rich phase is constructed.

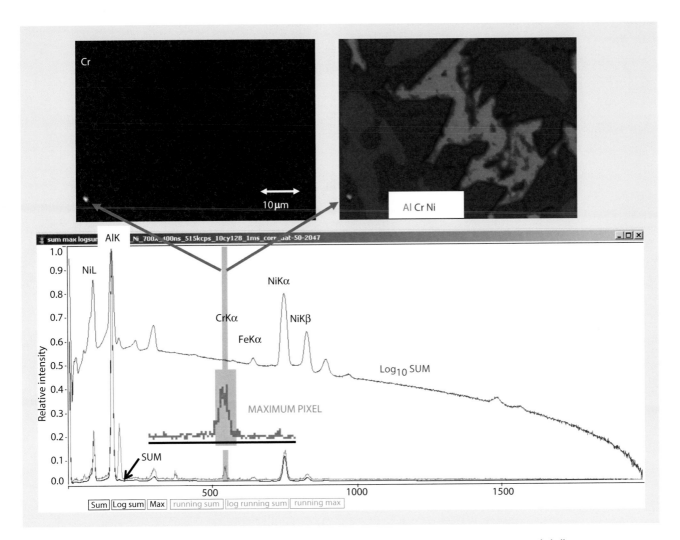

Fig. 24.9 Use of the MAXIMUM PIXEL spectrum to identify and locate an unexpected Cr-rich inclusion in Raney nickel alloy

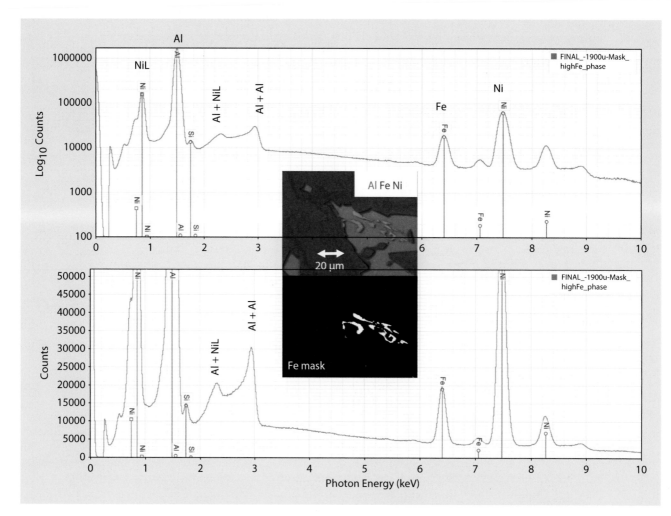

Fig. 24.10 Creation of a pixel mask (inset) using the Fe elemental map for Raney nickel to select pixels from which a SUM spectrum is constructed for the Fe-rich phase

24.3 Quantitative Compositional Mapping

The XSI contains the complete EDS spectrum at each pixel (or the position-tagged photon database that can be used to reconstruct the individual pixel spectra). Quantitative compositional mapping implements the normal fixed measurement location quantitative analysis procedure at every pixel in a map. Qualitative analysis is first performed to identify the peaks in the SUM and MAXIMUM PIXEL spectra of the XSI to determine the suite of elements present within the mapped area, noting that all elements are unlikely to be present at every pixel. The pixel-level EDS spectra can then be individually processed following the same quantitative analysis protocol used for individually measured spectra, ultimately replacing the gray or color scale in total intensity elemental maps, which are based on raw X-ray intensities and which are nearly impossible to compare, with a gray or color scale based on the calculated concentrations, which can be sensibly compared. Most vendor compositional mapping software extracts characteristic peak intensities by applying multiple linear least squares (MLLS) peak fitting or alternatively fits a background model under the peaks. These extracted pixel-level elemental intensities are then quantified by a "standardless analysis" procedure to calculate the individual pixel concentration values. Alternatively, rigorous standards-based quantitative analysis can be performed on the pixel-level intensity data with NIST DTSA II by utilizing the scripting language to sequentially calculate the pixel spectra in a datacube. MLLS peak fitting to extract the characteristic intensities, k-ratio calculation relative to a library of measured standards, and matrix corrections to yield the local concentrations of these elements for each pixel. An example of the NIST DTSA-II procedure applied to an XSI measured on the manganese nodule example of ◘ Fig. 24.2 is presented in ◘ Fig. 24.11, where the Fe $K-L_{2,3}$ total intensity map, which suffers significant interference from Mn $K-M_{2,3}$, shows a change in the apparent level of Fe in the center of the image after quantitative correction, as well as changes in several finer-scale details.

Quantitative compositional maps for major constituents displayed as gray-scale images are virtually identical to raw intensity elemental maps, as shown for the major constituents Al and Ni in ◘ Figs. 24.12 and 24.13. Significant differences between raw intensity maps and compositional maps are found for minor and trace constituents, where correction

◻ Fig. 24.11 Quantitative compositional mapping with DTSA-II on the XSI of a deep-sea manganese nodule from ◻ Fig. 24.2: direct comparison of the total intensity map for Fe and the quantitative compositional map of Fe; note the decrease in the apparent Fe in the high Mn portion in the center of the image after quantification and local changes indicated by *arrows* (*yellow* shows extension of high Fe region; magenta shows elimination of *dark* features)

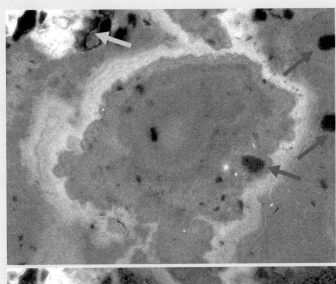

Raw Fe K-$L_{2,3}$ + Mn K-$M_{2,3}$

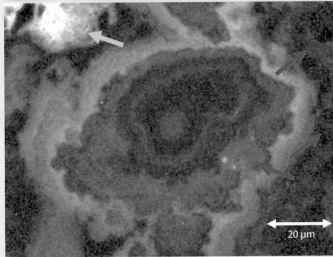

Fe K-$L_{2,3}$ after DTSA-II quantification

20 μm

for the continuum background significantly changes the gray-scale image. As shown for the minor Fe constituent in ◻ Figs. 24.12 and 24.13, the false Fe contrast between the Ni-rich and Al-rich phases that arises due to the atomic number dependence of the X-ray continuum is eliminated in the Fe compositional map.

While rendering the underlying pixel data of quantitative compositional maps in gray scale is a useful starting point, the problem of achieving a quantitatively meaningful display remains. Because of autoscaling, the gray scales of the Al, Ni, and Fe maps in ◻ Fig. 24.12 do not have the same numerical meaning. The Al and Ni concentrations locally reach high enough levels to correspond to brighter gray levels and have sufficient range to create strong contrast with direct gray-scale encoding, as seen in ◻ Figs. 24.12 and 24.13. The Fe constituent which is present at a concentration with a maximum of approximately 0.04 mass fraction (4 wt %) never exceeds the minor constituent range, so the Fe map appears dark with little contrast in the quantitative map of ◻ Fig. 24.13. Autoscaling of the Fe-map in ◻ Fig. 24.14 improves the contrast, but the same limitations

of autoscaling noted above still apply. Various pseudo-color scales, in which bands of contrasting colors are applied to the underlying data, are typically available in image processing software. Such pseudo-color scale can partially overcome the display limitations of gray-scale presentation, but the resulting images are often difficult to interpret. An example of a five band pseudo-color scale applied to the compositional maps using NIST Lispix is shown in ◻ Fig. 24.15. An effective display scheme for quantitative elemental maps that enables a viewer to readily compare concentrations of different elements spanning major, minor and trace ranges can be achieved with the Logarithmic Three-Band Encoding (Newbury and Bright 1999). A band of colors is assigned to each decade of the concentration range with the following characteristics:

Major: $C > 0.1$ to 1 (mass fraction) deep red to red pastel
Minor: $0.01 \leq C \leq 0.1$ deep green to green pastel
Trace: $0.001 \leq C < 0.01$ deep blue to blue pastel

The quantitative elemental maps are displayed with Logarithmic Three-Band Encoding in ◻ Fig. 24.16, and the

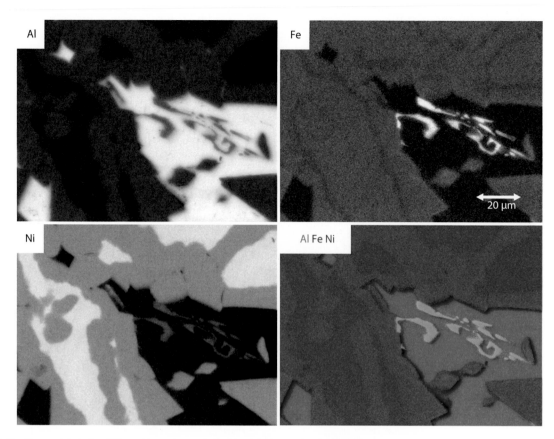

■ **Fig. 24.12**   Raney nickel alloy XSI: total intensity maps for Al, Fe, and Ni with color overlay

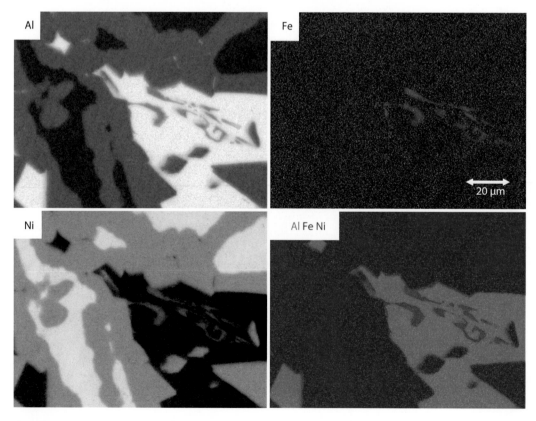

■ **Fig. 24.13**   Quantitative compositional maps of the Raney nickel alloy XSI in 24.13; note low contrast in the Fe image presented without autoscaling

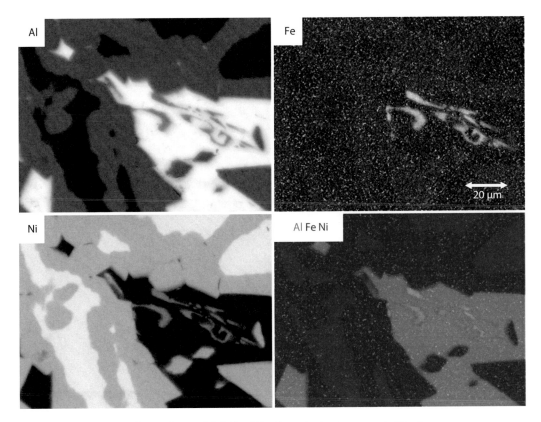

◩ **Fig. 24.14** Quantitative compositional maps Raney nickel alloy XSI with contrast enhancement of the Fe map

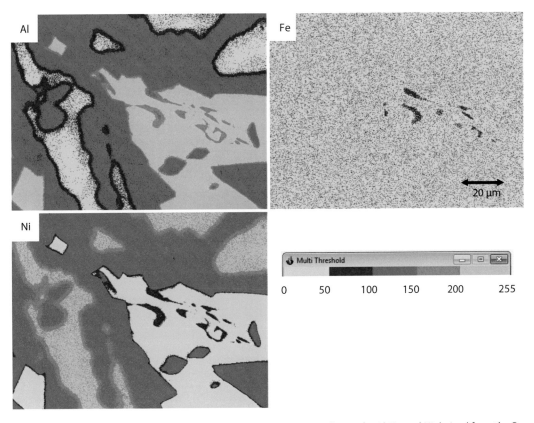

◩ **Fig. 24.15** Five band pseudo-color presentation of the quantitative compositional maps for Al, Fe, and Ni derived from the Raney nickel alloy XSI

major-minor-trace spatial relationships among the elemental constituents are readily discernible.

The Logarithmic Three-Band Encoding of the Fe compositional map in ▢ Fig. 24.16 reveals apparent Fe contrast in the Al-rich and Fe-rich phases, and this contrast differs markedly from the atomic-number dependence of the X-ray continuum seen in the raw Fe intensity image, as shown in ▢ Fig. 24.17. The Logarithmic Three-Band Encoding shows that this apparent contrast occurs near the lower limit of the trace range. Is this trace Fe contrast meaningful or merely an uncorrected

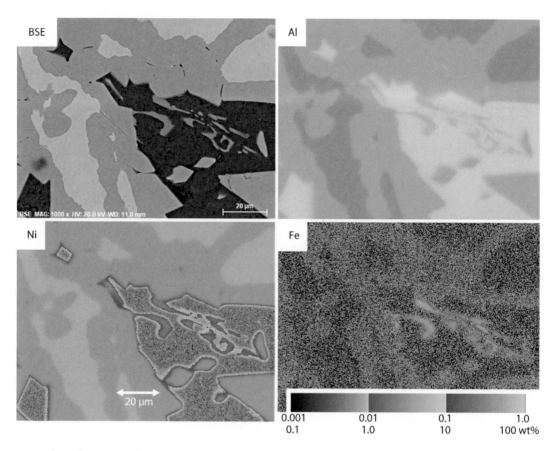

▢ **Fig. 24.16** Logarithmic Three-Band Color encoding of the quantitative compositional maps for Al, Fe, and Ni derived from the Raney nickel alloy XSI

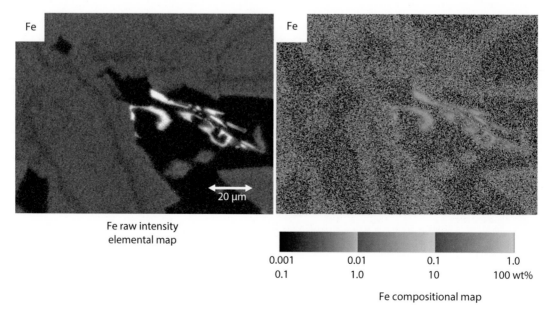

▢ **Fig. 24.17** Direct comparison of the Fe total intensity map and the Logarithmic Three-Band Encoding of the Fe quantitative compositional map. Note the distinct changes in the contrast within the trace concentration ($C < 0.01$) regions of the image

24

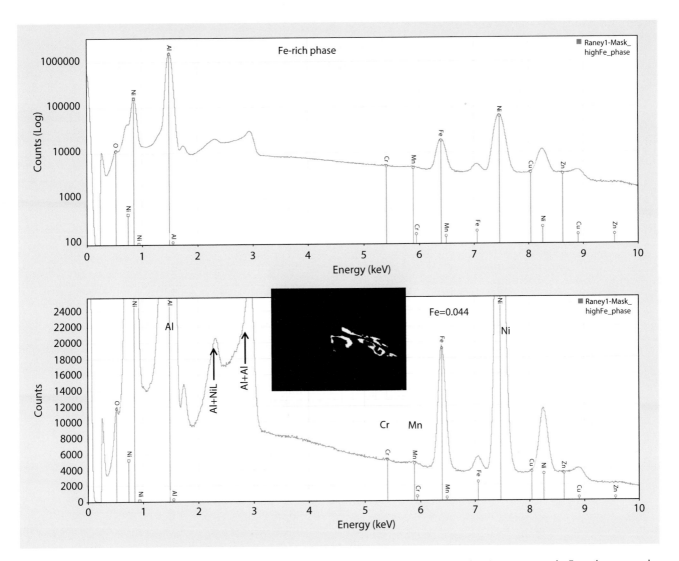

◘ **Fig. 24.18**   Raney nickel alloy XSI: mask of pixels corresponding to the Fe-rich phase and the corresponding SUM spectrum; the Fe peak corresponds to $C = 0.044$ mass fraction

artifact? To examine this question, the analyst can use pixel masks of each phase selected from the Ni compositional map (or the BSE image) to obtain the SUM spectrum, as shown in ◘ Figs. 24.18 (Fe-rich phase), 24.19 (Al-rich phase), 24.20 (Ni-intermediate phase), and 24.21 (Ni-rich phase). These SUM spectra confirm that Fe is indeed present as a trace constituent, with its highest level in the intermediate-Ni phase where $C_{Fe} = 0.0027$ (2700 ppm), falling to $C_{Fe} = 0.00038$ (380 ppm) in the high-Ni phase and to $C_{Fe} = 0.00030$ (300 ppm) in the high-Al phase. Additionally, trace Cr at a similar concentration is found in the Al-rich phase along with the trace Fe.

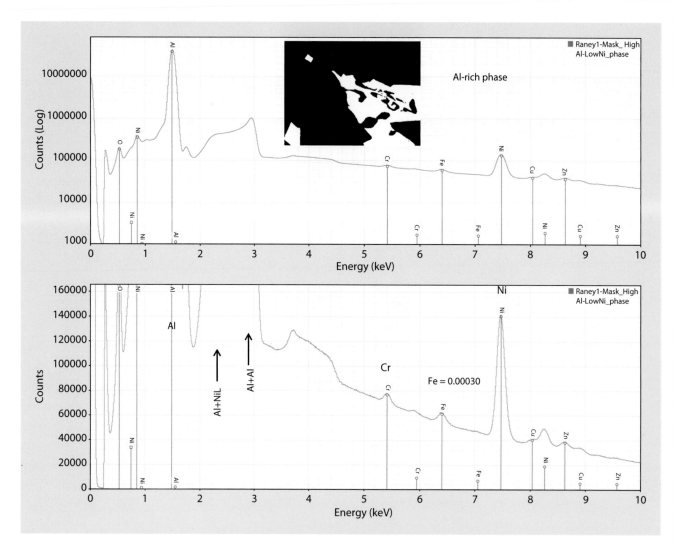

Fig. 24.19 Raney nickel alloy XSI: mask of pixels corresponding to the Al-rich phase and the corresponding SUM spectrum; note the low level peaks for Fe and Cr; the Fe-peak corresponds to $C = 0.00030 = 300$ parts per million

24.4 Strategy for XSI Elemental Mapping Data Collection

24.4.1 Choosing the EDS Dead-Time

The analyst has a choice of time constants in the EDS software, a parameter variously known as shaping time, processing time, etc., and typically expressed as time value (e.g., 200 ns, 400 ns, 1 μs), as a count rate value (e.g., the throughput at the peak of the input–output response), or as a simple integer. The shorter the time constant, the higher the peak throughput, expressed as the output count rate (OCR) versus the input count rate (ICR), but the poorer the resolution. The performance of a silicon drift detector (SDD)-EDS with three time constant choices is illustrated in Fig. 24.22, where the

peak of the OCR vs. ICR plot varies dramatically with the time constant selected. All forms of EDS microanalysis are improved by increasing the number of X-ray counts measured, but elemental mapping is especially dependent on accumulating large numbers of X-rays since mapping divides the total count among a large number of pixels. To obtain adequate counts per pixel for meaningful analytical information at the individual pixel level, it is common practice to accept the resolution penalty to operate on the highest throughput curve in Fig. 24.22. Of course, to produce the X-ray flux necessary to make use of this throughput capability, the EDS detector solid angle should first be maximized by operating at the shortest specimen-to-EDS distance (for a movable EDS) and the beam current should then be adjusted accordingly to produce an acceptable dead-time. An

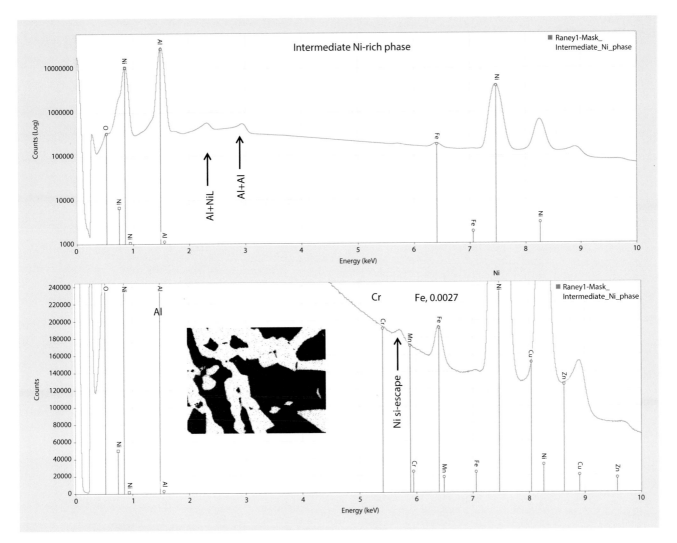

● Fig. 24.20   Raney nickel alloy XSI: mask of pixels corresponding to the intermediate Ni-rich phase and the corresponding SUM spectrum; note the low level peak for Fe; the Fe-peak corresponds to $C = 0.0027 = 2700$ parts per million

acceptable dead-time will depend on the level of spectral arti-facts that the analyst is willing to accept: (1) If the mapping software collects the XSI at constant pixel dwell time without dead-time correction, the resulting map can be subject to severe artifacts if the dead-time is so high that the peak of the OCR versus ICR response is exceeded at some pixels during mapping. As shown in ● Fig. 24.23, operating at very high dead-time can result in the same OCR being produced by two widely different ICR values, which may effectively correspond to two different concentrations. Artifacts produced by this effect are shown in the Al elemental maps recorded at high dead-time shown in ● Fig. 24.24. Dead-time-corrected data collection in elemental mapping avoids this artifact. (2) If the analyst is interested in minor and/or trace level constituents,

coincidence artifacts, which scale with dead-time, must be considered. Coincidence peaks for Al + NiL and Al + Al, as well as the coincidence continuum between these peaks, are illustrated in ● Figs. 24.18, 24.19, 24.20, and 24.21. This por-tion of the EDS spectrum contains K-shell peaks for S and Cl, L-shell peaks for Mo, Tc, Ru, Rh, Pd, and Ag, and M-shell peaks for Hg, Tl, Pb, and Bi. If none of these elements is of interest at the minor or trace level, then the Al + NiL and Al + Al coincidence can be ignored and the advantages of high throughput realized for the other elements of interest. However, if this spectral region is required for one or more of these elements, a lower dead-time should be selected by reducing the beam current to reduce coincidence, which depends strongly on the input count rate.

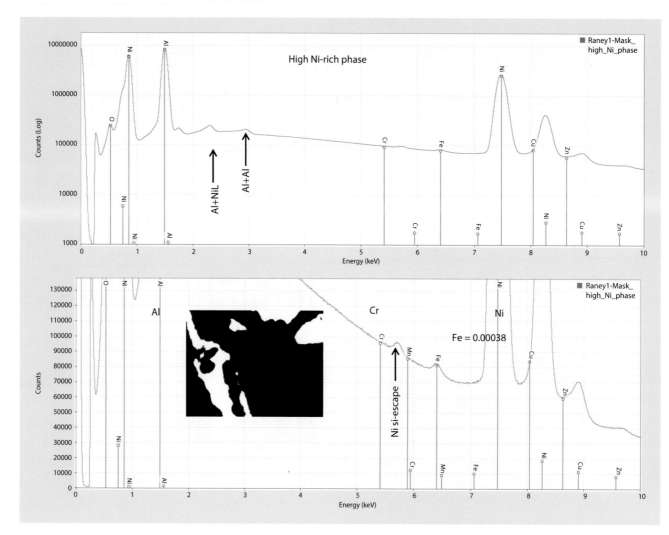

Fig. 24.21   Raney nickel alloy XSI: mask of pixels corresponding to the high Ni-rich phase and the corresponding SUM spectrum; note the low level peak for Fe; the Fe-peak corresponds to $C = 0.00038 = 380$ parts per million

24.4.2   Choosing the Pixel Density

The size of the scanned area, the number of image pixels (n_X, n_Y) and the pixel dwell time, τ, are critical parameters that the analyst must choose when defining an XSI data collection. An estimate of the size of the lateral extent of the X-ray interaction volume, obtained either from the Kanaya–Okayama X-ray range or from Monte Carlo simulation, is also useful, especially for small-area, high-magnification mapping. Choosing the size of the scanned area (magnification) depends on the lateral extent of the specimen features that are the objective of the mapping measurement. As is the case with SEM imaging using BSEs and/or SEs, when large areas are being scanned (low magnification operation), the pixel size may be greater than the lateral extent of the X-ray source size, which is a convolution of the incident beam size and the interaction volume for X-ray production, so that much of the pixel area is effectively unsampled. In principle,

the empty area of the pixel could be "filled in" by increasing the number of pixels to reduce the pixel size, but this would lead to extremely large XSI data structures that would require very long accumulation times. A practical upper limit for XSI mapping is typically 1024×1024 pixels, which for a spectrum of 4096 channels of 2 bytes intensity depth would produce an XSI of 8 Gbytes in the uncompressed RAW format. To reduce the mass storage as well as the subsequent processing time for quantitative compositional mapping calculations, the analyst may choose 512×512 or 256×256 pixel scan fields, especially when a small area is scanned (high magnification operation) leading to overlapping pixels. A pixel overlap of approximately 25% serves to fill all space in the square pixels, but further oversampling provides no additional information, so that the analyst would be better served by lowering the magnification to cover more specimen area with the chosen pixel density, or alternatively, choose a lower pixel density.

Fig. 24.22 Throughput of an SDD-EDS system consisting of four 10-mm² detectors with the outputs summed at three different operating time constants

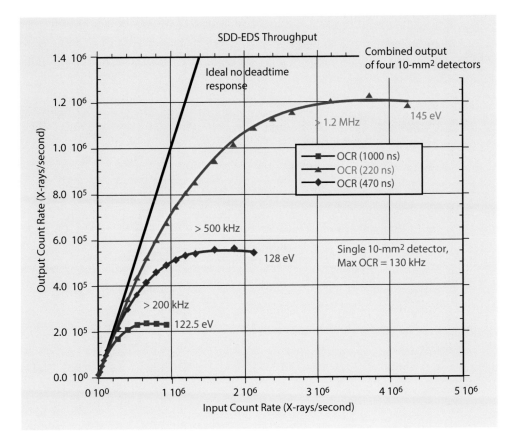

Fig. 24.23 Example of XSI mapping at such high throughput level that count rate based artifacts appear: measured OCR vs. ICR response for an SDD-EDS, showing the same OCR for two different ICR values, which could represent different concentrations of a highly excited element

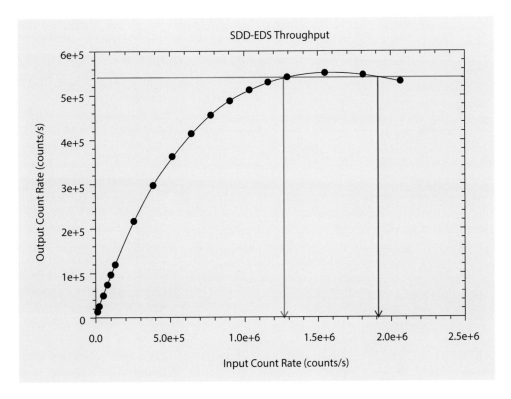

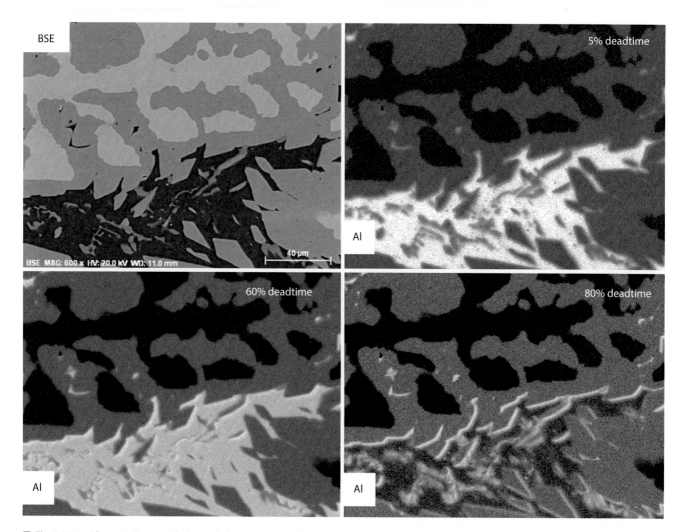

■ **Fig. 24.24** Al map in Raney nickel recorded at 5 %, 60 %, and 80 % dead-time; note changes in the high Al region at 80 % dead-time compared to 5 % dead-time

24.4.3 **Choosing the Pixel Dwell Time**

Once the analyst has chosen the pixel density, the product of the number of pixels in a map and the pixel dwell time gives the total mapping time. The OCR of the EDS and the pixel dwell time determine the number of counts in the individual pixel spectra, which sets the ultimate limit on the compositional information that can be subsequently recovered from the XSI. The high throughput of SDD-EDS, especially when clusters of detectors are used, provides OCR of 10^5/s to 10^6/s, enabling various XSI imaging strategies determined by the number of counts in the individual pixel spectra.

"Flash Mapping"

By operating with a high OCR, major constituents can be mapped in less than 60 s, a mode of operation that can be termed "flash" mapping, which is useful for surveying unknowns. An example of a flash mapping survey of a leaded-brass particle is shown in ■ Fig. 24.25, where the SEM-BSE image in ■ Fig. 24.25a reveals a high-atomic-number inclusion.

An XSI was recorded with 640 by 480 pixels with a 64-μs pixel dwell at on OCR of 750 kHz for a total mapping time of 20 s. The raw intensity maps for Cu, Zn, and Pb and their color overlay are shown in ■ Fig. 24.25b. While very noisy on the single pixel level, these maps nevertheless reveal the localization of the lead corresponding as expected to the bright region in the SEM-BSE image. The color overlay, however, shows numerous dark areas within the particle which do not correspond to Cu, Zn, or Pb. With this short pixel dwell, the pixel level EDS spectra contain only about 50 counts total, the effect of which can be seen in the noisy derived MAXIMUM PIXEL spectrum in ■ Fig. 24.26. (An additional derived spectrum, the RUNNING MAXIMUM PIXEL, which is averaged over three adjacent energy "cards" to reduce the effect of the low count, is also shown.) The SUM spectrum shown in ■ Fig. 24.26, consisting of all counts recorded in 20 s, approximately 15 million, contains abundant information. In addition to the peaks for Cu, Zn, and Pb, a major peak for Ni is observed. When the raw elemental map for Ni is constructed from the XSI, the missing regions in ■ Fig. 24.25b are filled in, as shown

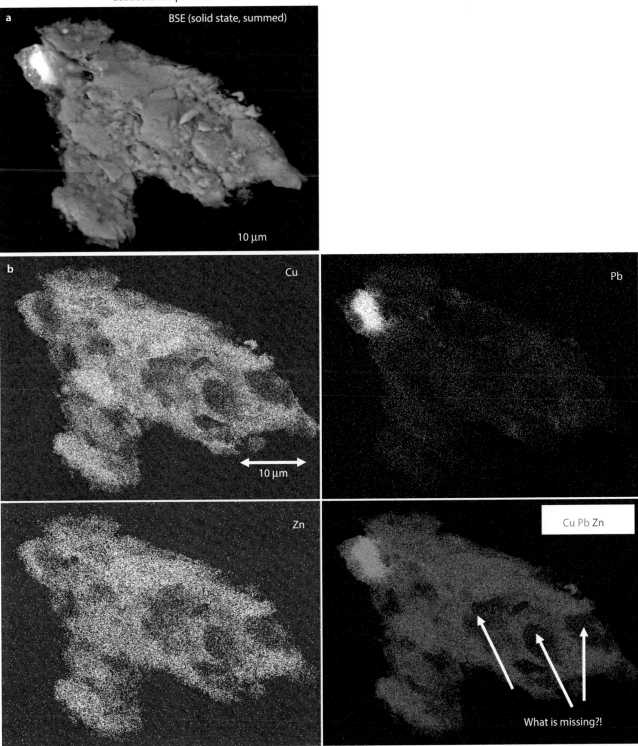

640x480 pixels; 64 μs/pixel = ~ 20 seconds total scan

◻ **Fig. 24.25**   Leaded brass particle XSI: **a** SEM-BSE image; note bright inclusion; **b** XSI recorded with 640 × 480 pixels; 64 μs/pixel = ~ 20 s total scan and total intensity images for Pb, Cu and Zn with color overlay; note apparently missing regions of particle (*dark*)

Fig. 24.26 Leaded brass particle XSI: SUM spectrum, logarithm of the SUM spectrum, MAXIMUM PIXEL spectrum, and RUNNING MAXIMUM PIXEL spectrum (averaged over three consecutive energy "cards"); note unexpected Ni peak

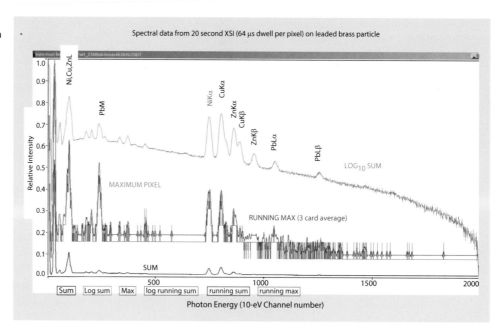

in ■ Fig. 24.27, which also shows the color overlap of Cu, Zn, and Ni. This example demonstrates the value of XSI imaging to augment the information obtained from the atomic number contrast of the SEM-BSE image. In ■ Fig. 24.27b, the SEM-BSE image easily distinguishes the Pb-rich inclusion from the brass matrix but shows no distinct contrast from the Ni-rich regions relative to the Cu-Zn brass matrix. Ni, Cu, and Zn are only separated by one unit of atomic number, so that the BSE atomic number contrast between these phases is very weak and dominated by the contrast produced by the Pb-rich region relative to the brass matrix. The SEM-BSE contrast situation is further complicated by the topographic contrast of the complex surface of the particle. Element-specific compositional imaging reveals the details of the complex microstructure of this particle.

High Count Mapping

The strategy for elemental mapping data collection depends on the nature of the problem to be solved: the most critical question is typically, "What concentration levels are of interest?" If minor and trace level constituents are not important, then short duration (60 s or less), low pixel density (256 × 256 or fewer) XSI maps with a high OCR will usually contain adequate counts, a minimum of approximately 50–500 counts per pixel spectrum, depending on the particular elements and overvoltage, to discern concentration-based contrast for major constituents, as shown in

the examples in ■ Figs. 24.25, 24.26, and 24.27. Of course, by accumulating more counts above this threshold, progressively lower concentration contrast details can be revealed for the major constituents. For problems involving minor and/or trace constituents, longer pixel dwell times are necessary to accumulate at least 500–5000 counts per pixel spectrum, and the beam current should be reduced to keep the dead-time generally below 10% to minimize coincidence peaks. This dead-time condition can be relaxed if the coincidence peaks, which are only produced by high count rate parent peaks associated with major constituents, do not interfere with the minor/trace constituent peaks of interest.

An example of the compositional details that can be observed at the level of approximately 5000 counts per pixel spectrum is shown for a complex Zr-Ni-V alloy with minor Ti, Cr, Mn and Co in ■ Fig. 24.28. Excellent gray-scale (after autoscaling) contrast is obtained between the phases which have relatively small changes in composition for the individual elements. Although the single pixel spectra do not have adequate counts for robust quantification, the analyst can use the images to form pixel masks that contain much higher total counts. The compositional values noted in ■ Fig. 24.28 are based on quantifying SUM spectra taken from the two phases that are specified by the arrows in the elemental maps, and the very small statistical error reported reflects the very high count SUM spectra.

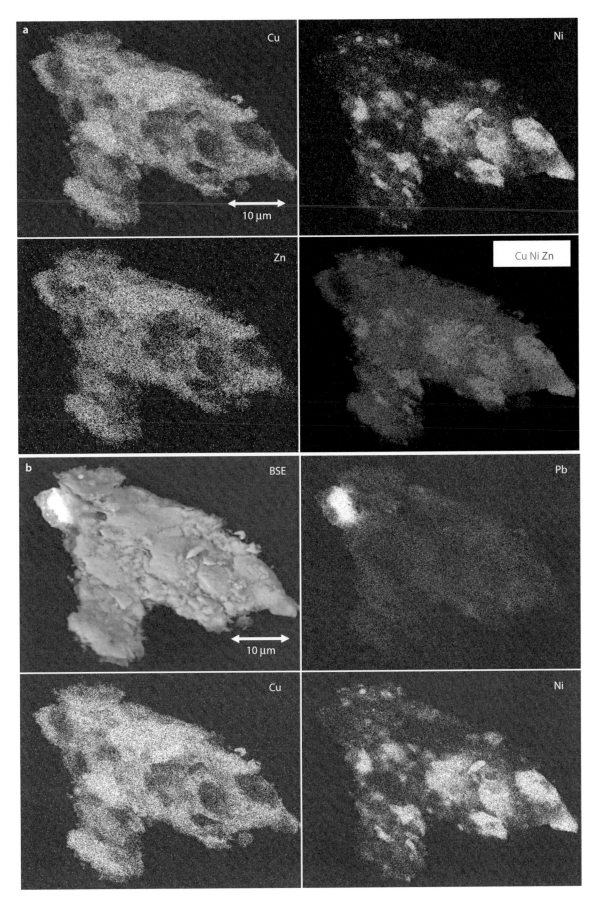

◘ **Fig. 24.27** Leaded brass particle XSI: **a** total intensity images for Cu, Zn, and Ni with color overlay showing Ni filling in empty areas seen in ◘ Fig. 24.25b; **b** direct comparison of SEM-BSE image and Cu, Ni, and Zn total intensity maps; the SEM-BSE image is insensitive to the compositional contrast between the Cu-Zn regions and the Ni-rich regions

Fig. 24.28 Deep gray level mapping of Ni-Zr-V (with minor Ti, Cr, Mn, and Co) hydrogen-storage alloy (XSI conditions: 512 × 384 pixels, 8192 µs = 27 min; OCR 637 kHz = ~ 5200 counts/pixel spec): quantitative compositional maps with autoscaled gray level presentation for **a** major constituents and **b** minor constituents. Compositional values noted are derived from SUM spectra formed from pixel masks of the phases marked by *arrows*

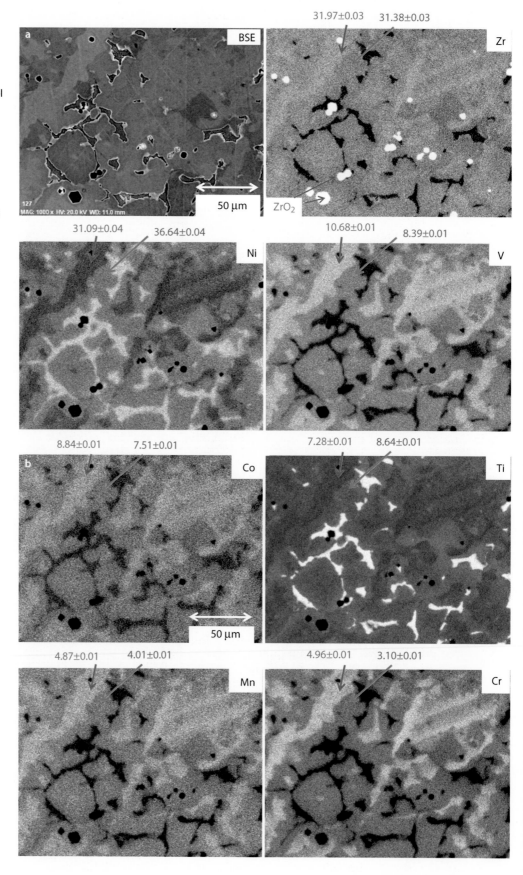

References

Bright D (2017) NIST Lispix, a software engine for image processing, available free at: ► http://www.nist.gov/lispix/doc/contents.htm

Bright D, Newbury D (2004) Maximum pixel spectrum: a new tool for detecting and recovering rare, unanticipated features from spectrum image data cubes. J Microsc 216:186

Gorlen KE, Barden LK, Del Priore JS, Fiori CE, Gibson CC, Leapman RD (1984) Computerized analytical electron microscope for elemental mapping. Rev Sci Instrum 55:912

Kotula P, Keenan MR, Joseph R, Michael JR (2003) Automated analysis of SEM X-ray spectral images: a powerful new microanalysis tool. Microsc Microanal 9:1

Newbury D, Bright D (1999) Logarithmic 3-band color encoding: Robust method for display and comparison of compositional maps in electron probe X-ray microanalysis. Microsc Microanal 5:333

Newbury D, Ritchie N (2013) Elemental mapping of microstructures by scanning electron microscopy –energy dispersive X-ray spectrometry (SEM-EDS): extraordinary advances with the Silicon Drift Detector (SDD). J Anal At Spectrom 28:973

Attempting Electron-Excited X-Ray Microanalysis in the Variable Pressure Scanning Electron Microscope (VPSEM)

© Springer Science+Business Media LLC 2018
J. Goldstein et al., *Scanning Electron Microscopy and X-Ray Microanalysis*,
https://doi.org/10.1007/978-1-4939-6676-9_25

While X-ray analysis can be performed in the Variable Pressure Scanning Electron Microscope (VPSEM), it is not possible to perform uncompromised electron-excited X-ray *microanalysis*. The measured EDS spectrum is inevitably degraded by the effects of electron scattering with the atoms of the environmental gas in the specimen chamber before the beam reaches the specimen. The spectrum is *always* a composite of X-rays generated by the unscattered electrons that remain in the focused beam and strike the intended target mixed with X-rays generated by the gas-scattered electrons that land elsewhere, micrometers to millimeters from the microscopic target of interest.

It is critical to understand how severely the measured spectrum is compromised, what strategies can be followed to minimize these effects, and what "workarounds" can be applied in special circumstances to solve practical problems. The impact of gas-scattered electrons on the legitimacy of the analysis depends on the exact circumstances of the VPSEM conditions (beam energy, gas species, path length through the gas) and the characteristics of the specimen and its surroundings. Gas scattering effects increase in significance as the constituent(s) of interest range in concentration from major (concentration $C > 0.1$ mass fraction) to minor ($0.01 \leq C \leq 0.1$) to trace ($C < 0.01$).

25.1 Gas Scattering Effects in the VPSEM

The VPSEM allows operation with elevated gas pressure in the specimen chamber, typically 10 Pa to 1000 Pa but even higher in the "environmental SEM" (ESEM), where

pressures of 2500 Pa are possible, permitting liquid water to be maintained in equilibrium when the specimen is cooled to ~ 3 °C. Such specimen chamber pressures are several orders of magnitude higher than that of a conventional high-vacuum SEM, which typically operates at 10^{-2} Pa to 10^{-4} Pa or lower. As the beam emerges from the high vacuum of the electron column through the final aperture into the elevated pressure of the specimen chamber, elastic scattering events with the gas atoms begin to occur. Although the volume density of the gas atoms in the chamber is very low compared to the density of a solid material, the path length that the beam electrons must travel typically ranges from 1 mm to 10 mm or more before reaching the specimen. As illustrated in ◼ Fig. 25.1, when elastic scattering occurs along this path, the angular deviation causes beam electrons to substantially deviate out of the focused beam creating a "skirt." The unscattered beam electrons follow the expected path defined by the objective lens field and land in the focused beam footprint identical to the situation at high vacuum but with reduced intensity due to the gas scattering events that rob the beam of electrons. The electrons that remain in the beam behave exactly as they would in a high vacuum SEM, creating the same interaction volume and generating X-rays with exactly the same spatial distribution to produce identically the same spectrum. This "ideal" high vacuum equivalent spectrum represents the true microanalysis condition. However, this ideal spectrum is degraded by the remotely scattered electrons in the skirt which generate characteristic and continuum (*bremsstrahlung*) X-rays from whatever material(s) they strike.

◼ **Fig. 25.1** Schematic diagram of gas scattering in a VPSEM

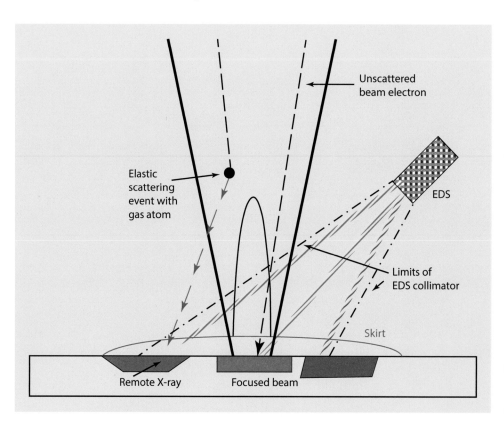

Unscattered beam electron

EDS

Limits of EDS collimator

Elastic scattering event with gas atom

Skirt

Remote X-ray

Focused beam

The extent of the beam skirt can be estimated from the ideal gas law (the density of particles at a pressure p is given by $n/V = p/RT$, where n is the number of moles, V is the volume, R is the gas constant, and T is the temperature) and by assuming single-event elastic scattering (Danilatos 1988):

$$R_s = (0.364 Z / E)(p / T)^{1/2} L^{3/2} x \qquad (25.1)$$

R_s = skirt radius (m)
Z = atomic number of the gas
E = beam energy (keV)
p = pressure (Pa)
T = temperature (K)
L = path length in gas (m) (GPL)

 Figure 25.2 plots the skirt radius for a beam energy of 20 keV as a function of the gas path length through oxygen at several different chamber pressures. For a pressure of 100 Pa and a gas path length of 5 mm, the skirt radius is calculated to be 30 μm. Consider the change in scale due to gas scattering. The high vacuum microanalysis footprint can be estimated with the Kanaya-Okayama range equation. For a copper specimen and $E_0 = 20$ keV, the full range $R_{K-O} = 1.5$ μm, which is a good estimate of the diameter of the interaction volume projected on the entrance surface, the "microanalysis

footprint." The gas scattering skirt is thus a factor of 40 larger in diameter, but a factor of 1600 larger in area.

While Eq. (25.1) is useful to estimate the extent of the gas scattering on the spatial resolution of X-ray microanalysis under VPSEM conditions, it provides no information on the relative fractions of the spectrum that arise from the unscattered beam electrons (exactly equivalent to the high vacuum microanalysis footprint) and the skirt. The Monte Carlo simulation embedded in NIST DTSA-II enables explicit treatment of gas scattering to provide detailed information on the unscattered beam electrons and the distribution of electrons scattered into the skirt. The VPSEM menu of DTSA-II is shown in Fig. 25.3 and allows selection of the gas path length, the gas pressure, and the gas species (He, N_2, O_2, H_2O, or Ar). An example of a portion of the electron scattering information produced by the Monte Carlo simulation is listed in Table 25.1 for a gas path length of 5 mm through 100 Pa of oxygen with a 20-keV beam energy; the full table extends to 1000 μm. This data set is plotted as the cumulative electron intensity as a function of radial distance out to 50 μm from the beam center in Fig. 25.4. For these conditions the unscattered beam retains 0.70 of the beam intensity that enters the specimen chamber. The skirt out to a radius of 30 μm contains a cumulative intensity of 0.84 of the incident beam current. To capture 0.95 of the total current for a 5-mm gas path length in

 Fig. 25.2 Radial dimension of gas scattering skirt as a function of gas path length at various pressures for O_2 and $E_0 = 20$ keV

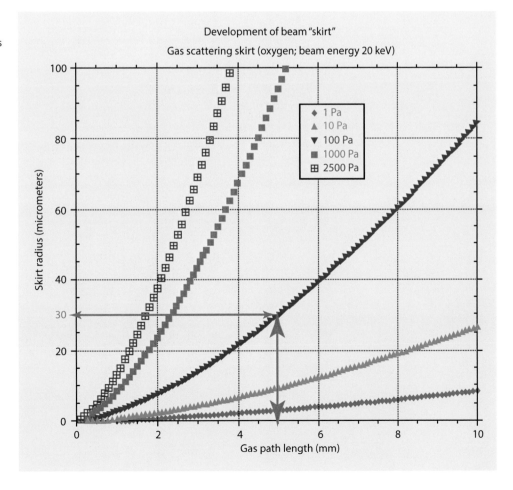

■ **Fig. 25.3** Selection of VPSEM gas parameters in DTSA-II for simulation

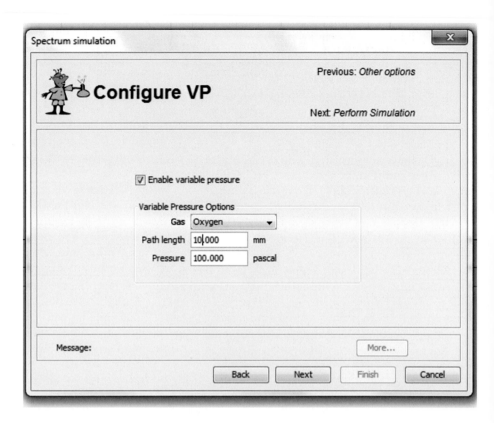

■ **Table 25.1** Portion of the electron scattering table for VPSEM simulation. Note the unscattered fraction of the 20 keV beam, 0.701 (100 Pa, 5-mm gas path length, oxygen). The full table extends to 1000 μm

Ring	Inner radius, μm	Outer radius, μm	Ring area, μm²	Electron count	Electron fraction	Cumulative, %
Undeflected	–	–	–	701	0.701	–
1	0.0	2.5	19.6	755	0.755	75.5
2	2.5	5.0	58.9	23	0.023	77.8
3	5.0	7.5	98.2	11	0.011	78.9
4	7.5	10.0	137.4	10	0.010	79. 9
5	10.0	12.5	176.7	7	0.007	80.6
6	12.5	15.0	216.0	6	0.006	81.2
7	15.0	17.5	255.3	9	0.009	82.1
8	17.5	20.0	294.5	10	0.010	83.1
9	20.0	22.5	333.8	4	0.004	83.5
10	22.5	25.0	373.1	3i	0.003	83.8
11	25.0	27.5	412.3	6	0.006	84.4
12	27.5	30.0	451.6	4	0.004	84.8
13	30.0	32.5	490.9	6	0.006	85.4
14	32.5	35.0	530.1	2	0.002	85.6
15	35.0	37.5	569.4	2	0.002	85.8
16	37.5	40.0	608.7	3	0.003	86.1
17	40.0	42.5	648.0	4	0.004	86.5

◻ Fig. 25.4 DTSA-II Monte Carlo calculation of gas scattering in a VPSEM: $E_0 = 20$ keV; oxygen; 100 Pa; 3-, 5-, and 10-mm gas path lengths (GPL) to a radial distance of 50 μm

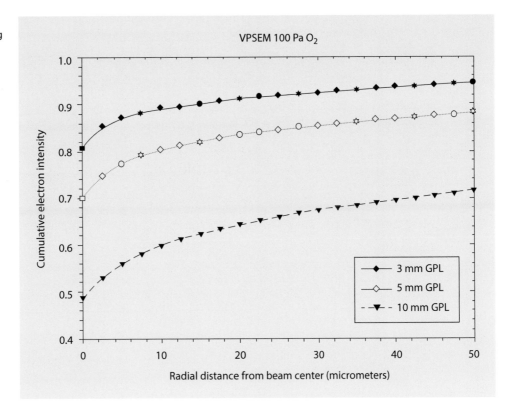

◻ Fig. 25.5 DTSA-II Monte Carlo calculation of gas scattering in a VPSEM: $E_0 = 20$ keV; oxygen; 100 Pa; 3-, 5-, and 10-mm gas path lengths to a radial distance of 1000 μm

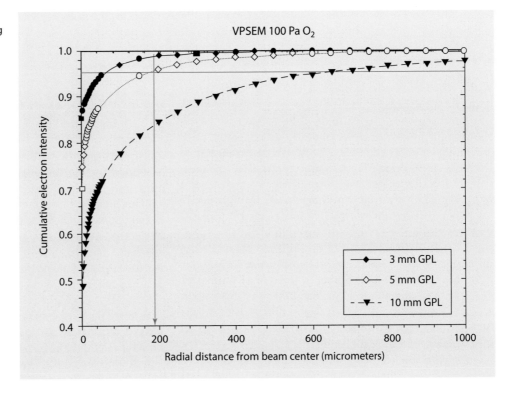

100 Pa of oxygen requires a radial distance of approximately 190 μm, as shown in ◻ Fig. 25.5, which plots the skirt distribution out to 1000 μm (1 mm). The strong effect of the gas path length on the skirt radius, which follows a 3/2 exponent in the scattering Eq. 25.1, can be seen in ◻ Fig. 25.5 in the plots for 3 mm, 5 mm, and 10 mm gas path lengths.

The extent of the degradation of the measured EDS spectrum by gas scattering is illustrated in the experiment shown in ◻ Fig. 25.6. The incident beam is placed at the center of a polished cross section of a 40 wt % Cu – 60 wt % Au alloy wire 500 μm in diameter surrounded by a 2.5-cm-diameter Al disk. For a beam energy of 20 keV and a gas path length of

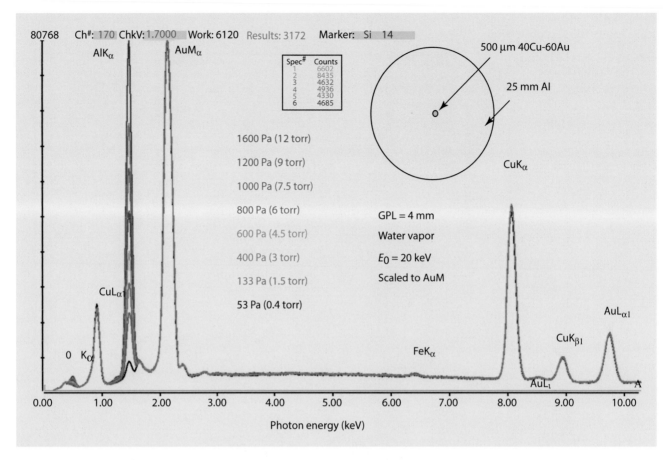

□ **Fig. 25.6**   EDS spectra measured with the beam placed in the center of a 500 μm diameter wire of 40 wt % Cu–60 wt %Au surrounded by a 2.5-cm-diameter Al disk; $E_0 = 20$ keV; gas path length = 4 mm; oxygen at various pressures

4 mm through water vapor, the EDS spectra measured over a pressure range from 53 Pa to 1600 Pa are superimposed, showing the in-growth of the Al peak with increasing pressure. Even at 53 Pa, a detectable Al peak is observed, despite the beam center being 250 μm away from the Al. As the pressure is increased, the Al peak ranges from an apparent trace to minor and finally major constituent peak.

DTSA-II also simulates the composite spectrum created by these two classes of electrons as they strike the specimen. Various configurations of two different materials can be specified, one that the unscattered beam strikes, for example, a particle, and the other by the skirt electrons, for example, the surrounding matrix. □ Figure 25.7 shows spectra simulated for the example of □ Fig. 25.6, the 500-μm-diameter 40 wt %Cu–60 wt %Au wire in the Al disk with a 4-mm-gas path length through water vapor. The simulation of the lowest VPSEM gas pressure of 53 Pa produces a low level Al peak similar to the experimental measurement. Thus, even at this low pressure and short gas path length for which 89 % of the electrons remain in the focused beam, there are still gas-scattered electrons falling at least 250 μm from the beam impact. As the pressure is progressively increased, the in-growth of the Al peak due to the skirt electrons is well modeled by the Monte Carlo simulation.

25.1.1   Why Doesn't the EDS Collimator Exclude the Remote Skirt X-Rays?

Gas scattering in the VPSEM mode *always* degrades the incident beam, transferring a significant fraction of the beam electrons into the skirt. The radius of the skirt can reach a millimeter or more from the focused beam impact. It might be thought that the EDS collimator would restrict the acceptance area of the EDS to exclude most of the skirt. As shown in the schematic diagram in □ Fig. 25.8, while a simple collimator acts to successfully shield the EDS from accepting X-rays produced by backscattered electrons striking the lens and chamber walls, the acceptance volume near the column optic axis is quite large. The EDS acceptance is not defined by looking back at the detector from the specimen space as the cone of rays whose apex is at the beam impact on the specimen and whose base is the detector active area (the dashed red lines in □ Fig. 25.8). While the red lines define the solid angle of the detector for emission from the beam impact point, the acceptance region is actually defined by looking from the detector through the collimator at the specimen space (the dashed green lines in □ Fig. 25.8). The true area of acceptance can be readily determined by conducting X-ray mapping measurements. □ Fig. 25.9 shows a series of measurements of X-ray maps of a machined Al disk taken at the lowest available

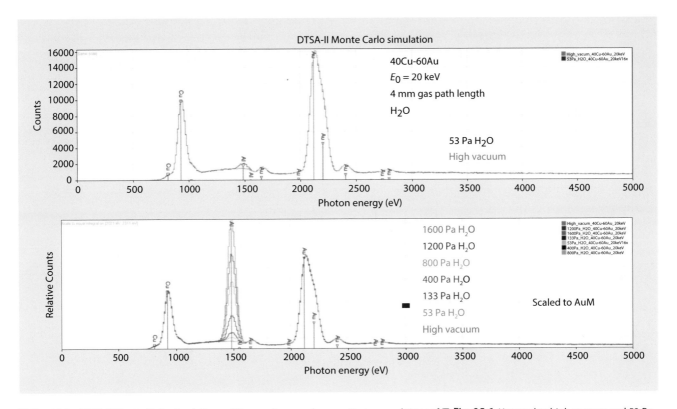

■ **Fig. 25.7**   DTSA-II Monte Carlo simulations of the specimen and gas scattering conditions of ■ Fig. 25.6. Upper plot: high vacuum and 53 Pa (4 mm GPL, H_2O). Lower plot: high vacuum and various pressures from 53 Pa to 1600 Pa

■ **Fig. 25.8**   Schematic diagram showing the acceptance area of the EDS collimator

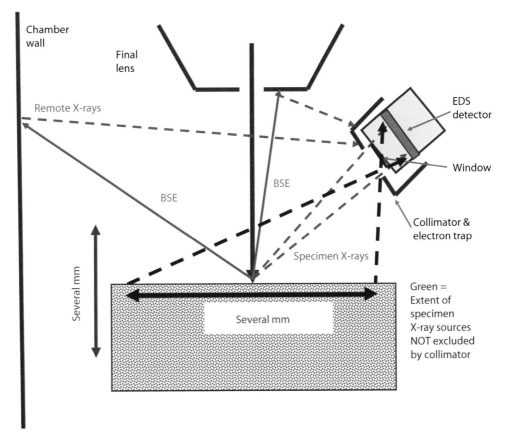

25

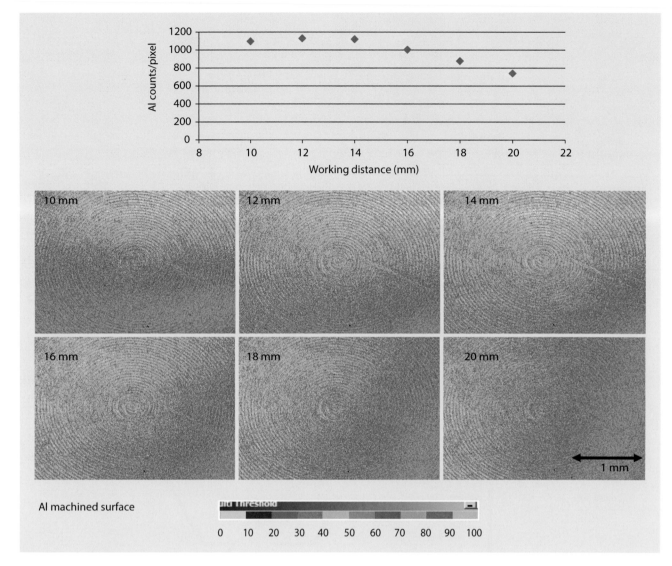

Al machined surface

0 10 20 30 40 50 60 70 80 90 100

◻ **Fig. 25.9**   Collimator acceptance volume as determined by mapping an Al disk at various working distances (10–20 mm) at the widest scanned field (i.e., lowest magnification); the plot shows the intensity measured at the center of the scan field as a function of working distance

magnification (largest scan field) over a series of working distances. The false color scale shows the while the intensity is not uniform within a map, it generally varies by less than 30 % over the full map, a distance of millimeters, and moreover, as the maps are measured at different working distances, there is only about 30 % variation over the vertical range, which is confirmed by the plot of the intensity measured at the center of each map. Thus, X-rays generated throughout a large volume are accepted through the collimator by the EDS so that collimation provides virtually no relief from the effects of remote X-ray generation caused by gas scattering in the VPSEM.

25.1.2   Other Artifacts Observed in VPSEM X-Ray Spectrometry

Inelastic scattering of the beam electrons and backscattered electrons with the atoms of the environmental gas causes

inner shell ionization leading to X-ray emission that contributes to the measured spectrum. The density of gas atoms (number/unit volume) is orders of magnitude lower than the density in the solid specimen, but the distance that the beam travels in the gas is orders of magnitude longer than it travels in the solid. The contribution of the environmental gas is illustrated for a simple experiment in ◻ Fig. 25.10, where the beam strikes a carbon target at a series of different pressures. The in-growth of the oxygen peak from ionization of the water vapor used as the environmental gas can be seen. A detectable oxygen peak is seen for pressures of 133 Pa (1 torr) and higher for this particular gas path length (6 mm) and beam energy (20 keV). ◻ Figure 25.11 shows an example of the contribution of the environmental gas to the spectrum measured from a 50 μm diameter glass shard with the composition listed in the figure placed on a carbon substrate. At the lower pressure (266 pA = 2 torr) for the gas path length used (3 mm), the footprint of the focused beam and skirt

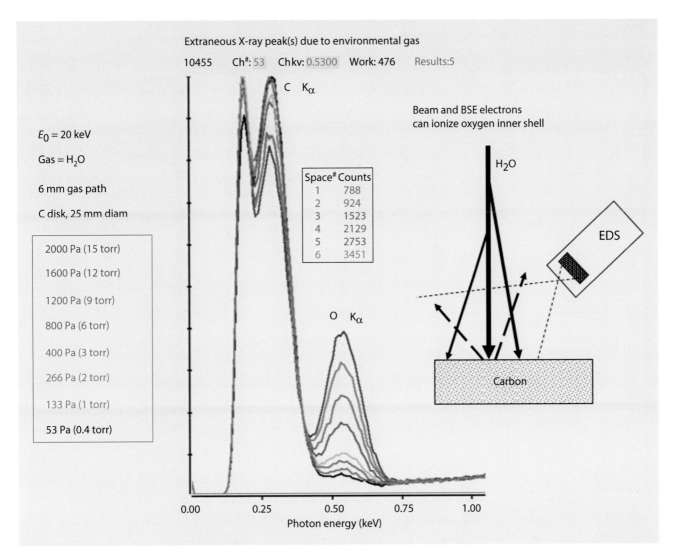

Fig. 25.10 Generation of O K X-rays from the environmental gas as a function of chamber pressure

remain within the 50 μm diameter glass shard, as evidenced by the negligible C intensity in the spectrum. An O peak is also observed, at least some of which is actually from the specimen. When the pressure is increased by a factor of five to 1330 Pa (10 torr), the C intensity rises significantly because the skirt now extends beyond the boundary of the particle, and the O peak intensity increases by a factor of four, all of which is due to the environmental gas. It is also worth noting that in addition to characteristic X-rays from the environmental gas, there is increased *bremsstrahlung* generation as well from the inelastic scattering of beam and backscattered electrons with the gas atoms. In ◘ Fig. 25.11, the background is substantially higher for the spectrum measured at the elevated pressure, leading to a reduced peak-to-background, which is easily seen for the Zn L-family and Al K-L$_2$ peaks, an effect which makes for poorer limits of detection.

If the environmental gas can contribute to the spectrum, can the gas also absorb X-rays from the specimen? Because of the low gas density, this effect might be expected to be negligible, and as listed in ◘ Table 25.2, which is calculated for a 40–mm-path through the gas from the X-ray source at the beam impact on the specimen to the EDS, for the lowest pressure considered, 10 Pa, over 99 % of the X-rays of all energies leaving the specimen in the direction of the detector arrive there, even for F K, the energy of which, 0.677 keV, is just above the O K-shell absorption energy, 0.535 keV, which results in a large mass absorption coefficient. When the pressure is increased to 100 Pa, the loss of F K due to absorption increases to ~ 6 %, and at a pressure of 2500 Pa (18.8 torr), ~ 80 % of the F K radiation is lost to gas absorption, and even Al K-L$_2$ suffers a 20 % loss in intensity compared to the conventional high vacuum SEM situation.

25

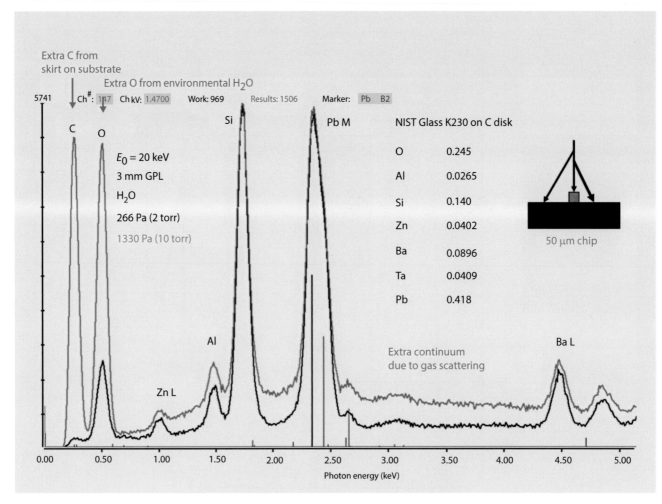

○ **Fig. 25.11**   Modification of the measured X-ray spectrum by gas scattering, including in-growth of the O K peak from contributions of the environmental gas as well as increased background due to increased *bremsstrahlung* created by the gas scattering

○ **Table 25.2**   Absorption of X-rays by the environmental gas (O_2) (40-mm source to EDS)

Element/X-ray	I/I_0 (2500 Pa)	I/I_0 (100 Pa)	I/I_0 (10 Pa)
F K (0.677 keV)	0.194	0.940	0.994
NaK (1.041 keV)	0.572	0.979	0.998
AlK (1.487 keV)	0.805	0.992	0.9992
SiK (1.740 keV)	0.868	0.995	0.9995
S K (2.307 keV)	0.939	0.998	0.9998
ClK (2.622 keV)	0.957	0.998	0.9998
K K (3.312 keV)	0.986	0.999	0.9999
CaK (3.690 keV)	0.990	0.9996	0.9999

25.2   What Can Be Done To Minimize gas Scattering in VPSEM?

Manipulating the parameters in Eq. (25.1) provides the basis for minimizing, but not eliminating, the effects of gas scattering:

1. Z, atomic number of the scattering gas: By lowering the atomic number of the gas, the skirt radius is reduced. This effect is illustrated in ○ Fig. 25.12, which is derived from DTSA-II Monte Carlo simulations comparing the skirt radius for He and O_2 for a 10-mm gas path length and 100-Pa gas pressure with a beam energy $E_0 = 20$ keV.
2. E, beam energy (keV). Operating at the highest possible beam energy reduces the gas scattering skirt.
3. p, the gas pressure (Pa): Operating with the lowest possible gas pressure minimizes the gas scattering skirt

25.2 · What Can Be Done To Minimize gas Scattering in VPSEM?

451

25

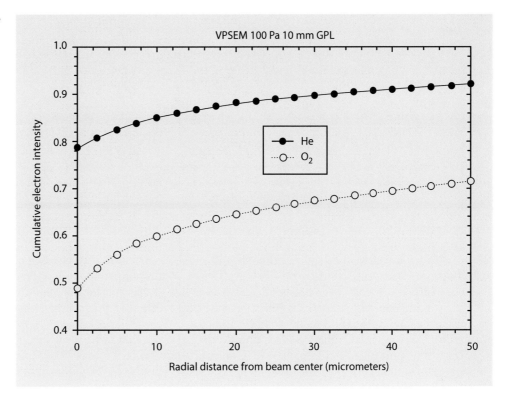

◘ Fig. 25.12 Comparison of the scattering skirt for He and O_2

4. T, the sample chamber temperature (K): The scattering skirt is reduced by operating at the lowest possible temperature.
5. L, the gas path length (m). The shorter the gas path length, the smaller the gas scattering skirt, as shown in ◘ Fig. 25.4, where the skirt is compared for gas path lengths of 3 mm, 5 mm, and 10 mm. Note that the gas path length appears in Eq. (25.1) with a 3/2 power, so that the skirt radius is more sensitive to this parameter than the other parameters in Eq. (25.1). A modification to the vacuum system of the VPSEM that minimizes the gas path length consists of using a small diameter tube to extend the high vacuum of the electron column into the sample chamber.

25.2.1 Workarounds To Solve Practical Problems

Gas scattering effects can be minimized but not avoided, and for many VPSEM measurements and *in situ* experiments, the microscopist/microanalyst may be significantly constrained in the extent to which any of the parameters in Eq. (25.1) can actually be changed to reduce the gas scattering skirt without losing the advantages of VPSEM operation. The measured EDS spectrum is always compromised, but by carefully choosing the problems to study, successful X-ray analysis can still be performed. A useful way to consider the impact of gas scattering is the general concentration level at which the remote scattering corrupts the measured spectrum:

Major constituent: concentration $C > 0.1$ mass fraction
Minor constituent: $0.01 \leq C \leq 0.1$
Trace constituent: $C < 0.01$

Depending on the exact nature of the specimen, gas scattering will almost always introduce spectral artifacts at the trace and minor constituent levels, and in severe cases artifacts will appear at the level of apparent major constituents.

25.2.2 Favorable Sample Characteristics

Given that the LVSEM operating conditions have been selected to minimize the gas scattering skirt, what specimen types are most likely to yield useful microanalysis results? If most of the gas scattering skirt falls on background material that contains an element or elements that are different from the elements of the target and of no interest, then by following a measurement protocol to identify the extraneous elements, the measured spectrum can still have value for identifying the elements within the target area, always recognizing that the target is being excited by the focused beam and the skirt.

Particle Analysis

Particle samples comprise a broad class of problems related to the environment, technology, forensics, failure analysis, and other areas. Particles are very often insulating in nature so that a conductive coating is required for examination in the conventional high vacuum SEM, and the complex morphologies

452 **Chapter 25** · Attempting Electron-Excited X-Ray Microanalysis in the Variable Pressure Scanning Electron Microscope (VPSEM)

25

of particles often make it difficult to apply a suitable coating. The VPSEM with its charge dissipation through gas ionization is an attractive alternative to achieve successful particle imaging. When VPSEM X-ray analysis measurements are needed to characterize particles, specimen preparation is critical to achieve a useful result. There are many methods available for particle preparation, but the general goal for successful VPSEM X-ray analysis is to broadly disperse the particles on a suitable substrate so that the unscattered beam and the inner-most intense portion of the skirt immediately surrounding the beam can be placed on individual particles without exciting nearby particles. The more distant portions of the beam skirt may still excite other particles in the dispersion, but the relative fraction of the electrons that falls on these particles is likely to be sufficiently small that the artifacts introduced in the measured spectrum will be equivalent to trace ($C < 0.01$ mass fraction) constituents.

1. When particles are collected on a smooth (i.e., not tortuous path) medium such as a porous polycarbonate filter, the loaded filter can be studied directly in the VPSEM with no preparation other than to attach a portion of the filter to a support stub. Prior to attempting X-ray measurements of individual particles, the X-ray spectrum of the filter material should be measured under the VPSEM operating conditions as the first stage of determining the analytical blank (that is, the spectral contributions of all the materials involved in the preparation *except* the specimen itself). In addition to revealing the elemental constituents of the filter, this blank spectrum will also reveal the contribution of the environmental gas to the spectrum.

2. When particles are to be transferred from the collection medium, such as a tortuous path filter, or simply obtained from a loose mass in a container, the choice of the sample substrate is the first question to resolve. Conceding that VPSEM operation will lead to significant remote scattering that will excite the substrate, the sample substrate should be chosen to consist of an element that is not of interest in the analysis of individual particles. Carbon is a typical choice for the substrate material, but if the analysis of carbon in the particles is important, then an alternative material such as beryllium (but beware of the health hazards of beryllium and its oxide) or boron can be selected. If certain higher atomic number elements can be safely ignored in the analysis, then additional materials may be suitable for substrates, such as aluminum, silicon,

germanium, or gold (often as a thick film evaporated on silicon). Again, whatever the choice of substrate, the X-ray spectrum of the bare substrate should be measured to establish the analytical blank prior to analyzing particles on that substrate.

■ Figure 25.13a shows a VPSEM image of a particle cluster prepared on carbon tape (which has a blank spectrum consisting of major C and minor O from the polymer base) and the EDS X-ray spectrum obtained with the beam placed on particle "A". The spectrum is seen to contain a high intensity peak for Si, lower intensity peaks for O, Al, and K, and peaks just at the threshold of detection for Ca, Ti, and Fe. With the other particles nearby, how much of this spectrum can be reliably assigned to particle "A"? A local "operational blank" can be measured by placing the beam on several nearby substrate locations so that focused beam only excites the substrate while the skirt continues to excite the specimen over its extended reach. Examples of the "working blanks" for this particular specimen are shown in ■ Fig. 25.13b, c and are revealed to be surprisingly similar, considering the separation of location "BL3" from "BL1" and the particle array. Inspection of these working blanks show Al and Si at the minor level, and Ca and Fe at the trace level, as estimated from the peak-to-background ratio. Thus, the interpretation of the spectrum of particle "A" can make use of the local analytical blank, as shown in ■ Fig. 25.13d. The major Al and Si peaks are not significantly perturbed by the low blank contributions for these elements, and the minor K is not present in the working blank and therefore it can be considered valid. However, the low levels of Ca and Fe observed in the blank are similar to those in the spectrum on particle "A", and thus they should be removed these from consideration as legitimate trace constituents. Note that despite the size of particle "A," which is approximately 50 μm in its longest dimension, it must be remembered that measurement sensitivity to any possible heterogeneity within particle "A" is likely to be lost in the VPSEM mode because of the large fraction of the incident current that is transferred within the 25-μm radius, as demonstrated in ■ Table 25.1. Another example of the use of the working blank is shown in ■ Fig. 25.13e for particle "D". Here the Ca is much higher than in the working blank, and so it can be considered a legitimate particle constituent, as can the Mg, which is not in the working blank. The low Fe peak is similar in both the particle spectrum and the working blank, so it must not be considered legitimate.

■ **Fig. 25.13** **a** VPSEM image of a cluster of particles and an EDS X-ray spectrum measured with the beam placed on one of them, particle "A." **b** EDS X-ray spectrum measured with the beam placed on the substrate at location BL1 so that only the beam skirt excites the particles. **c** EDS X-ray spectrum measured with the beam placed on the substrate at location BL3 so that only the beam skirt excites the particles. **d** Use of analytical blank to aid interpretation of the EDS spectrum of Particle "A." **e** Use of analytical blank to aid interpretation of the EDS spectrum of Particle "A"

453

25

25.2 · What Can Be Done To Minimize gas Scattering in VPSEM?

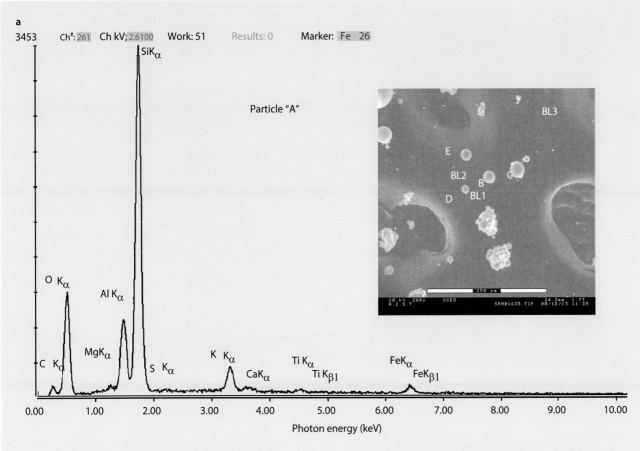

a

3453   Ch#: 261   Ch kV; 2.6100   Work: 51   Results: 0   Marker: Fe 26

SiK$_\alpha$

Particle "A"

O K$_\alpha$

Al K$_\alpha$

C K$_\alpha$ MgK$_\alpha$ S K$_\alpha$ K K$_\alpha$ CaK$_\alpha$ Ti K$_\alpha$ Ti K$_{\beta 1}$ FeK$_\alpha$ FeK$_{\beta 1}$

0.00   1.00   2.00   3.00   4.00   5.00   6.00   7.00   8.00   9.00   10.00

Photon energy (keV)

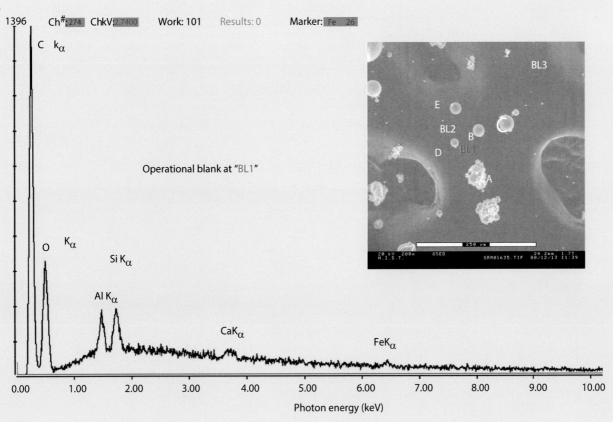

b

1396   Ch#: 274   ChkV: 2.7400   Work: 101   Results: 0   Marker: Fe 26

C k$_\alpha$

Operational blank at "BL1"

O K$_\alpha$

Si K$_\alpha$

Al K$_\alpha$

CaK$_\alpha$

FeK$_\alpha$

0.00   1.00   2.00   3.00   4.00   5.00   6.00   7.00   8.00   9.00   10.00

Photon energy (keV)

25

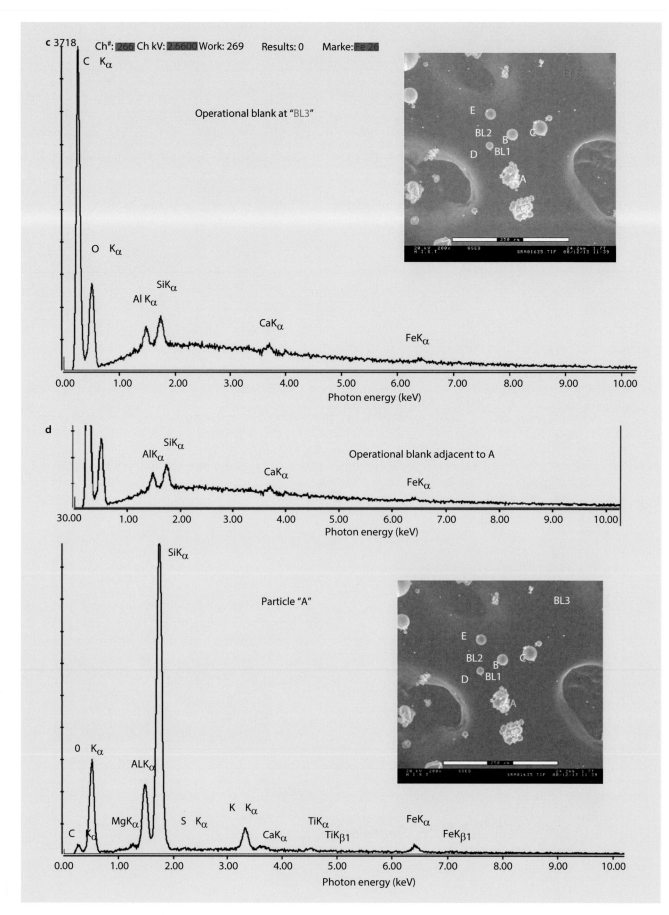

■ Fig. 25.13 (continued)

455 25

25.2 · What Can Be Done To Minimize gas Scattering in VPSEM?

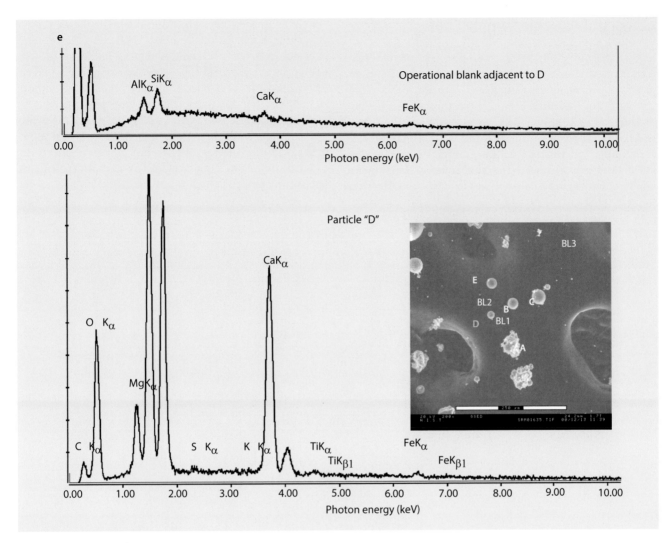

■ Fig. 25.13 (continued)

25.2.3   Unfavorable Sample Characteristics

EDS analysis in VPSEM is most problematic for specimens consisting of densely packed microscopic features, for example, most solid materials, natural and synthetic, with a microstructure. The measured EDS X-ray spectrum in such a case depends very strongly on the VPSEM conditions, the dimensions of the target area of interest, and the exact nature of the surrounding microstructure. An example is presented in ◻ Fig. 25.14, which shows the microstructure of Raney nickel, an aluminum-nickel catalyst. Based on the contrast in this SEM-BSE image, there are three different phases present denoted D, I, and B, with different Al/Ni ratios. The spectral intensities for Al and Ni show differences in the low pressure EDS spectra shown in ◻ Fig. 25.15a that are sufficient to readily distinguish the phases despite the gas scattering ($E_0 = 20$ keV; 50 Pa; water vapor; 6-mm gas path length). When the pressure is increased to 665 Pa, two of the phases can no longer be distinguished in the EDS spectrum shown in ◻ Fig. 25.15b. This loss of phase recognition is also seen as a loss of contrast in X-ray mapping, as seen in the elemental intensity maps shown in ◻ Fig. 25.16a, b.

◻ **Fig. 25.14**   SEM-BSE image of Raney nickel in a VPSEM

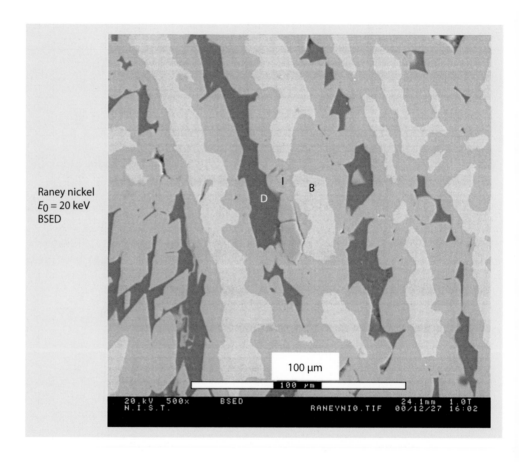

457   **25**

25.2 · What Can Be Done To Minimize gas Scattering in VPSEM?

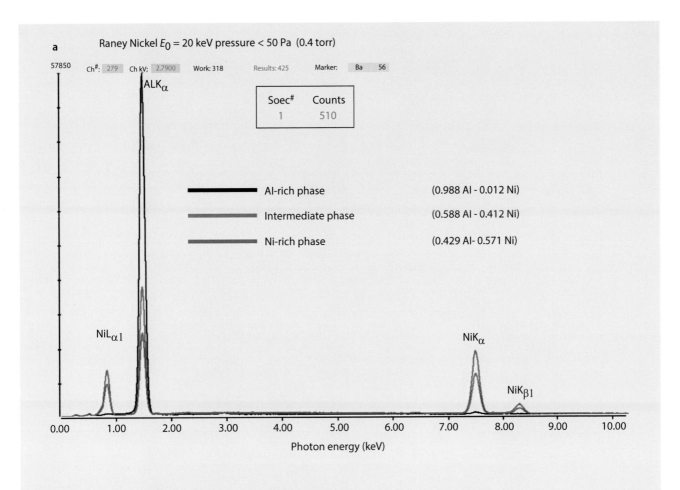

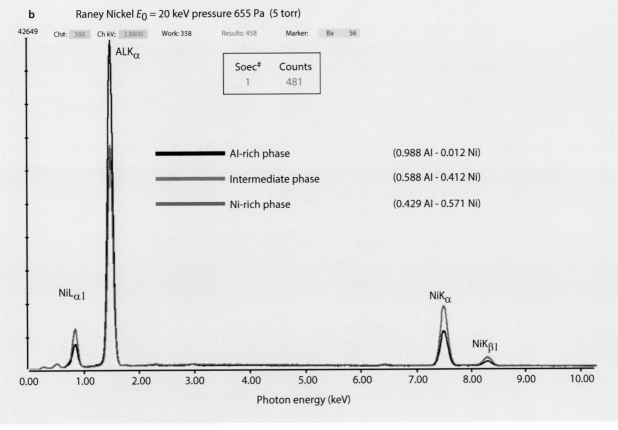

☐ **Fig. 25.15**   **a** EDS spectra obtained at locations "D", "I," and "B." $E_0 = 20$ keV; water vapor; 50 Pa, 6-mm gas path length. **b** EDS spectra obtained at locations "D," "I," and "B." $E_0 = 20$ keV; water vapor; 665 Pa, 6-mm gas path length

25

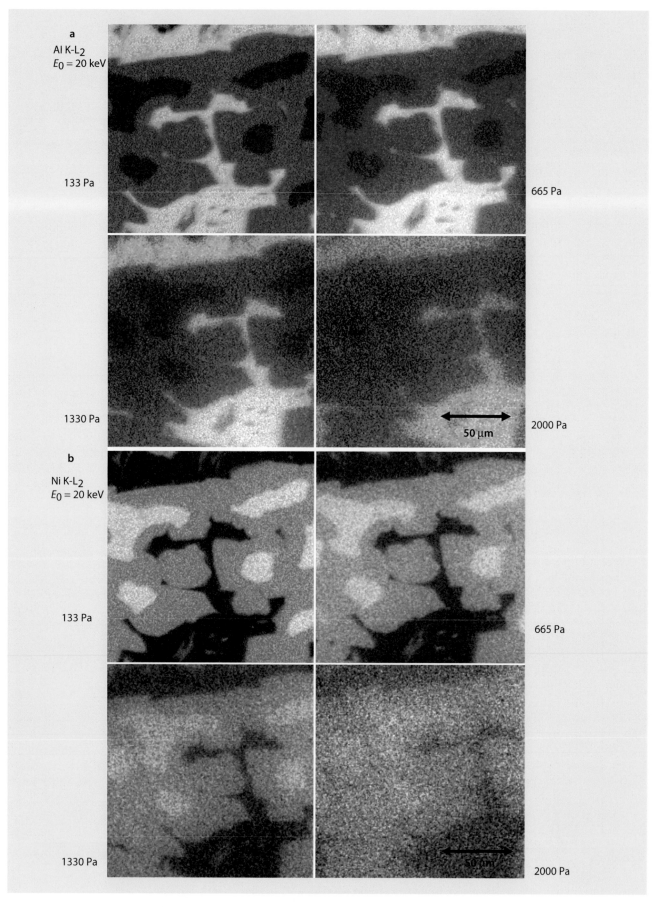

□ **Fig. 25.16**   **a** Elemental intensity map for Al K-L$_2$ at various pressures ($E_0 = 20$ keV, water vapor and 6-mm gas path length). **b** Elemental intensity map for NiKα at various pressures ($E_0 = 20$ keV, water vapor and 6-mm gas path length)

References

Danilatos GD (1988) Foundations of environmental scanning electron. Microscopy Adv Electronics Electron Phys 71:109

Danilatos GD (1993) Introduction to the ESEM instrument. Micros Res Tech 25:354

Newbury D (2002) X-Ray microanalysis in the variable pressure (environmental) scanning electron microscope. J Res Natl Inst Stand Technol 107:567

Energy Dispersive X-Ray Microanalysis Checklist

© Springer Science+Business Media LLC 2018
J. Goldstein et al., *Scanning Electron Microscopy and X-Ray Microanalysis*,
https://doi.org/10.1007/978-1-4939-6676-9_26

26

26.1   Instrumentation

What you will need to prepare.

26.1.1   SEM

The starting point is a scanning electron microscope (SEM) with an energy dispersive X-ray spectrometer. A good and useful measurement doesn't necessarily require a fancy SEM. Almost any functional SEM will do. However, there are a handful of basic requirements.

- The SEM should be capable of beam energies of at least 15 keV (20–30 keV maximum is better). While it is possible to make many measurements at lower voltages, 15 keV is a useful practical minimum to begin an analytical measurement campaign.
- The probe current must remain stable to within a percent or better for minutes or hours. Most thermal sources (tungsten filament, LaB_6 or Schottky field-emitters) will work fine. Cold field emitters can be problematic due to current drift issues. The more stable the probe current the easier and the more accurate standards-based measurements will be. Note that "standardless" analysis methods do not require a stable beam current or knowledge of the value of the beam current. While convenient, this is bought at the price of losing the analytical total, which has considerable value.

26.1.2   EDS Detector

Any functional EDS detector/pulse processor system is likely to be adequate. Older Si(Li)-EDS detectors with poorer resolution and less throughput may require patience and may produce slightly less good results. The one caveat is that some older detectors utilize thick beryllium windows that absorb virtually all X-rays below 1-keV photon energy. These detectors will not be able to measure X-rays from C, O, F, and other light elements. If you need to measure these elements you will need a detector with a thin polymer ("Moxtek") or other high-transparency window. In the end, patience and care is more important than owning the latest equipment.

- Resolution at Mn K-$L_{2,3}$ of 150 eV or better
- Throughput of 1,000 cps second or better on bulk Cu
- A "light element" window capable of seeing C
- Software to acquire spectra and export the spectra in the msa (ISO-22029, 2012) format

26.1.3   Probe Current Measurement Device

You will need a mechanism to measure the probe current—a measure of the number of electrons striking the sample. There are two ways to measure the probe current. You can measure the probe current directly using a Faraday cup and picoammeter. Alternatively, you can monitor the probe current indirectly by measuring an X-ray spectrum of a specific element, for example, Cu, obtain the integrated spectrum count, e.g., from 0.1 keV to E_0, and then use the proportionality between probe current and X-ray emission to calibrate the probe current over the course of a measurement campaign.

Direct Measurement: Using a Faraday Cup and Picoammeter

The classic procedure to measure the probe current is with a Faraday cup and a picoammeter. Some instruments integrate a Faraday cup into the stage. A few instruments implement a Faraday cup as a retractable device within the optics column. Most SEMs require a user-provided Faraday cup.

To perform a beam current measurement you will need the following:

A Faraday Cup

A Faraday cup is essentially a hole which electrons enter but never leave. It consists of a metal aperture with a 10–100-μm diameter orifice mounted over a millimeter-diameter void, for example, a blind hole drilled in a block. Carbon is an excellent material for the block because its low backscattered electron (BSE) coefficient minimizes re-backscattering from the walls of the hole. For a metal block, the interior of the void can be coated in a material with a low backscatter coefficient (such as carbon dag) to minimize possible loss through multiple backscatter events. The Faraday cup should be accessible on the SEM stage simultaneously with the specimen so it just a matter of driving the stage to the appropriate locations, which is usually an automatable function. It should NOT be necessary to break the vacuum and interrupt the specimen measurements to insert the Faraday cup.

Electrically Isolated Stage

The stage should be electrically isolated from the instrument ground. The electrically isolated stage is connected by a single wire to a vacuum electrical feed-through so that the only path to ground passes through the connected picoammeter.

- A picoammeter—A current meter capable of measuring currents between picoamps and hundreds of nanoamps.
- A cable to connect the vacuum feed-through to the picoammeter.

Procedure:

- Drive the SEM stage to the location of the Faraday cup, and center the aperture in the SEM image. Zoom up in magnification until the aperture fills the screen, and then further increase the magnification so the scanned field is well within the aperture. Choose the appropriate current range on the picoammeter and read the current.
- Note: When the beam is not within the Faraday cup, the current flowing to the picoammeter will represent the "specimen current," which is the beam current minus the backscattered electron and secondary electron emission. Because η and δ change with composition, specimen current is NOT a useful measure of absolute beam current or of stability.

Indirect Measurement: Using a Calibration Spectrum

A carefully collected spectrum can be used as an alternative to a Faraday cup and picoammeter. Ultimately, the probe current always appears in formulas for standards-based quantitative analysis as a ratio. We can therefore replace the probe current with any quantity which is proportional to the probe current, even if we don't know the proportionality constant. A carefully collected spectrum can provide such a metric.

The calibration spectrum must be collected—

1. At a consistent beam energy
2. At a consistent working distance (The same working distance that the unknown and standard spectra are collected)
3. On a consistent material (flat, polished)
4. For a consistent live-time (acquisition duration)

Typically, a pure metal such as copper, nickel, aluminum, silicon, among others, is used and the number of counts in the range of energies around the K characteristic lines is used as the proxy for the probe current, or the entire spectrum can be integrated from a threshold, for example, 0.1 keV to E_0.

Precise calibration of the probe current using a spectrum takes patience. For 0.1 % precision, you will need to integrate at least 1,000,000 counts in the range of energies used for the calibration.

Disadvantages:

1. It takes longer to acquire a calibration spectrum of equivalent precision than it takes to measure the probe current with a picoammeter.

Advantages:

1. The polished standard used for calibration can be a member of the set being used for the analysis so that no extra material or preparation needed.
2. A calibration spectrum can compensate for slight differences in measurement geometry.
3. A Faraday cup and picoammeter aren't necessary.
4. The calibration spectrum can also be used for quality control purposes.

26.1.4 Conductive Coating

Many samples and standard materials are non-conductive and need to be coated with a nanoscale coating of a conductive material. Typical equipment used includes—

1. Carbon coater (usually the best choice for X-ray microanalysis unless C, N, or O must be optimized in the analysis)
2. AuPd or other heavy metal/metal alloy coater (usually chosen when secondary electron image performance is important)

The coater should be capable of laying a controlled thickness of a conductive film over the sample. Most coaters rotate to ensure that all sides of the sample are coated. Use the thinnest coating that discharges the specimen, for example, 5–10 nm of carbon should be sufficient, or 1–3 nm of a high Z coating such as Au or Au-Pd.

CRITICAL: The applied coating must actually be connected to electrical ground. Do not assume that an electrical path is automatically established by the coating. The sides of a tall insulating specimen may not actually become adequately coated. Use a strip of conducting adhesive tape from the coated surface to the conducting support stub to ensure the electrical path.

26.2 Sample Preparation

Regardless of whether you are performing standards-based or standardless quantitative analysis, you will need to ensure that your unknowns and standards, if used, are prepared in a suitable fashion for analysis. How you prepare your samples will depend very much on the type of samples that you analyze. For the most accurate analysis of bulk materials, including that utilizing the standardless protocol, the specimen must be prepared with a flat surface, with the degree of flatness becoming increasingly critical for measurements with X-rays having energies below 1 keV—e.g., Be, B, C, N, O, F—where the surface roughness should be reduced below 50 nm root mean square. Some samples require very little preparation (e.g., a silicon wafer) and others (anything that isn't flat) require a lot. You may need to embed the samples and standards in epoxy mounting medium (preferably conductive) and use suitable equipment to grind and polish the samples. Appropriate preparation protocols are so specialized that rather than provide an exhaustive list of possible procedures, the analyst is referred to the rich literature on sample preparation:

— Echlin, P., Handbook of Sample Preparation for Scanning Electron Microscopy and X-ray Microanalysis (Springer, New York, 2009)
— Geels, K., Metallographic and Materialographic Specimen Preparation, Light Microscopy, Image Analysis, and Hardness Testing (ASTM, West Conshohocken, PA, 2006)
— McCall, J., Metallographic Specimen Preparation: Optical and Electron Microscopy (Springer, New York, 2012).
— Vander Voort, G., Metallography, Principles and Practice (ASM International, Materials Park, OH, 1999)

26.2.1 Standard Materials

Standards are specially prepared materials that serve to provide X-ray data for the elements present in the unknown. You will need a standard for each element in the unknown material. Note that some standards can serve for two or more elements, for example, FeS2.

To be useful, a standard must be—
1. Suitably sized—Practically speaking, most standards range in size from tens of micrometers to a tens of millimeters
2. Polished to a smooth, flat surface with surface texture of less than 100 nm (50 nm for low energy X-rays like oxygen)
3. Mounted in a manner that facilitates placement on the stage perpendicular to the beam
4. Conductive or coated with a conductive material
5. There are several different classes of standards, in order of the ease of use:
   (i) Pure element standards:
        The easiest standards to use are pure elements and this is where a novice should start whenever possible.
   (ii) Simple compound standards:
        Some elements are not stable as pure elements—N, O, F, Cl, Br, S—and must be provided as compounds. The easiest to use are typically stable, stoichiometric compounds like alumina (Al_2O_3) or calcium fluoride (CaF_2). With stable stoichiometric compounds, the true composition of the standard is unambiguous. Ideally, the compound is chosen so that there are no interferences between the element's characteristic peaks.

(iii) Complex standards:
      In many cases, complex standards similar to the unknown sample will produce the most accurate results. However, complex standards often are more difficult to work with and it is often hard to find reliable compositional data. Use of complex standards is an advanced topic and rarely justified unless suitable pure element and simple compound standards are not available.

26.2.2 Peak Reference Materials

Peak reference spectra serve as examples of the shape of an element's characteristic peaks. A peak reference can serve as a reference for one or more families. Standards materials can be used as references if there are no interferences for that element. Paradoxically, some materials suitable to serve as peak references are not suitable for use as standards, generally due to instability under the electron beam; for example, $BaCl_2$ provides an excellent peak reference for the Ba M-family, but it is unstable under electron bombardment and thus cannot serve as a standard.

26.3 Initial Set-Up

26.3.1 Calibrating the EDS Detector

Most EDS detectors allow you to configure two different characteristics of the detector—the pulse process time and the energy calibration. Some detectors provide other more advanced options. You will need to consult the manufacturer's documentation to determine the optimal setting for these parameters. In general, the goal should be to ensure that the detector is configured the same way day-in/day-out even if it means making small compromises to the ultimate performance.

Selecting a Pulse Process Time Constant
1. Most EDS detectors will allow you to make a trade-off between X-ray throughput and detector resolution. Higher throughput comes at the price of poorer detector resolution. This setting is called different things by various vendors—throughput, pulse processor setting, shaping time, or time-constant.
2. Typically, a pulse processor time constant selected in the middle of the allowed range represents the best trade-off. Despite common belief, the highest resolution setting is rarely the optimal choice as this setting is usually accompanied by severe throughput limitations. Usually, it is better to select a moderate resolution and obtain a moderate throughput.
3. On a modern silicon drift detector (SDD), throughput is typically not limited by the pulse-process time but rather through pulse pile-up (coincidence) events. Selecting a very fast process time won't actually improve usable throughput.

4. Be sure to turn off "adaptive shaping" or other mechanisms that adapt the process time dynamically depending upon X-ray flux. Adaptive shaping changes the shape of the characteristic peaks as the count rate varies due to variations in local composition and makes linear peak fitting less accurate.

Energy Calibration

1. Select one material that you will always use to calibrate your detector. Mount a piece of it near your Faraday cup. Copper is a good choice but there are other materials that have both low energy and higher energy X-ray peaks that can be used for calibration.

2. Select a channel width and channel count that fully covers the range of beam energies that you may use. A width of 10 eV/channel and 2,048 channels is a good choice for a 20-keV beam energy. A width of 10 eV/channel and 4,096 channels is suitable for higher-energy work. Since EDS spectra comprise relatively small size files and computer storage is inexpensive, consider 5 eV per channel, especially for low photon energy work to provide adequate channels to describe the peak. Whatever choice is made, it is important to consistently use that choice.

3. Each day before you start collecting data, collect and store an initial spectrum from your selected calibration material, for example, Cu. Use this spectrum and your vendor's software to calibrate the energy axis. Usually this involves adjusting the electronics to ensure that the measured peaks are centered around the correct channels. Most modern detectors automatically and dynamically adjust the zero offset and the calibration is just a matter of the software automatically selecting a gain that produces the desired number of eV/channel. Older detectors may require a manual gain and zero offset adjustment using external potentiometers. Regardless of the mechanism, at 10 eV/channel a 0.1 % mis-calibration will correspond to a single channel error in the position of a peak at 10 keV, so the detector calibration should ideally be maintained to better than 0.1 %. Fortunately, modern detectors are able to hold this precision of calibration for days or weeks.

4. Once the detector is calibrated, set the beam energy and probe current to established nominal values and collect a spectrum for a established live time. Save this spectrum as a demonstration that your detector is performing correctly. DTSA-II provides tools to extract and track performance metrics and then plot the results over time as a control chart.

Quality Control

Sooner or later, you will be asked by a client to demonstrate that your instrument was performing correctly when their data was collected. The easiest way to satisfy this requirement and to impress the client is to keep a long term record and present the data as a control chart. DTSA-II provides functionality which permits you to archive the spectra you just used to calibrate your detector as a record of the detectors performance. The program extracts efficiency, calibration and other pieces of instrumental data and records them in a database. This database can then be exported as control charts or tabulations. It is a nice way to make the most of the calibration data you have already spent the time to collect.

Maintaining the Working Distance/ Specimen-to-EDS Distance

Maintaining consistent experimental conditions is critical for achieving rigorous quantitative microanalysis. A critical parameter is the specimen-to-EDS distance, S_{EDS}, since the solid angle of the EDS varies as $1/S_{EDS}^2$. The SEM and/or EDS manufacturer(s) will have specified the ideal SEM working distance (WD) at which the EDS central axis intersects the SEM optic axis. An electron probe microanalyzer uses a fixed-focus optical microscope with a very shallow depth-of-focus to bring the specimen to this ideal WD position (to which the wavelength dispersive spectrometers are also focused) on a consistent basis. While very useful, such an optical microscope is rarely available on an analytical SEM, so that another method must be used to select the working distance properly and consistently. The following procedure assumes the use of a flat polished specimen.

1. Load the specimen so that its Z-height is approximately correct for the ideal working distance specified by the manufacturer. Most SEMs provide a mechanical mounting reference frame to approximately set this initial specimen altitude.

2. Using the manufacturer's specified value of the ideal SEM working distance for EDS (e.g., 10 mm), set the SEM objective lens strength to focus at this optimal working distance, making use of whatever objective lens meter reading is available to monitor this adjustment.

3. Select a low magnification to begin (e.g., 100×). Despite care in polishing, there are almost always a few fine scale scratches or other irregularities to be found. Locate one, and use the stage Z-motion (i.e., motion along the optic axis of the SEM) to bring this scratch into approximate focus, without adjusting the objective lens strength. Increase the magnification in stages to 5000× and refine the focus with the Z-axis motion, not by changing the objective lens strength. If the SEM stage automation system permits, save this Z-parameter.

4. This procedure is likely to be more reproducible at lower probe currents where the convergence angle is larger and depth-of-focus is smaller.

5. Consistency is critical. Always collect your spectra at this ideal working distance. Before collecting each spectrum ensure that the sample surface is at the optimal working distance by setting the objective lens/working distance and bringing the sample into focus using the vertical stage axis.

26

Sample Orientation

Sample orientation is also a critical parameter to hold constant.

1. The ideal sample orientation has the sample perpendicular to the electron column's optical axis. The electrons strike the sample normal to the surface and their behavior in the sample is best understood.
2. Tilts of a few degrees from the ideal perpendicular orientation can significantly degrade the accuracy of quantitative measurements for highly absorbed X-rays, such as low energy photons below 1 keV.
3. The best way to ensure that the samples are oriented correctly is by checking the orientation of the stage using a spirit level and then ensuring that the sample's surface is parallel to the stage datum. Check both orientations as a stage may be perpendicular to the optic axis in one direction (x or y) and tilted on the other (y or x).
4. Sometimes it is not possible to use a level to ensure orientation, in these cases you may be able to use a flat portion of the sample and the image focus to ensure that the sample maintains focus as the stage moves. A 1° tilt corresponds to a change of working distance of 17 μm over a travel of 1 mm.

Detector Position

Maintaining the position of the detector relative to the sample is critical. The sample position aspect of this has been discussed above in terms of setting the proper working distance. On some instruments, it is also possible to position the X-ray detector along a slide mount with a screw drive (manual or motor driven) that moves the detector along its central axis out of the chamber and away from the sample.

1. Usually, the optimal location for the detector is as close to the sample as possible without obstruction or collision.
2. The detector snout should not touch anything inside the chamber. The snout should be electrically isolated from the chamber.
3. The position of the detector should be maintained by setting a stop that ensures that the detector is returned to precisely the same position each time it is removed and returned.
4. The solid angle and therefore also the throughput is a function of the distance of the detector to the sample squared. Thus small variations in this distance can contribute to large differences in measured X-ray intensities. A 1 % error in distance leads to a 2 % error in intensity.
5. On a few instruments, the take-off angle can be varied. A single take-off angle should be selected and maintained. All else remaining the same, higher take-off angles are better than lower ones.

Probe Current

1. It is not necessary to maintain exactly the same probe current throughout the entire measurement process but it is beneficial to try to maintain the probe current to a narrow range because this minimizes errors resulting from non-linearities in the picoammeter.
2. Typically, the probe current is selected to maximize the X-ray throughput on a selected material (e.g., Cu), while simultaneously maintaining a low rate of coincidence events (pulse pile-up).
3. Coincidence events occur at all throughputs but become much more common at higher throughputs (scaling roughly as the square of the throughput). The acceptable coincidence rate is composition dependent. Lower probe currents and lower coincidence rates are required for trace element analysis and when a coincidence event occurs at the same energy as a minor elemental peak.
4. Older Si(Li) detectors with lower throughputs typically had less of an issue with coincidence events and maintaining a dead time of 30 % was close to optimal for all vendors and most samples.
5. Silicon drift detectors are capable of higher throughput but are also more susceptible to coincidence events so an acceptable coincidence rate, rather than a specific dead time, should be used as a metric to select the probe current. Note that coincidence depends strongly on how many high-count-rate peaks are present in the spectrum, so the degree of coincidence will vary with composition.
6. A good starting point is to observe the coincidence events that occur with alumina (Al_2O_3). Adjust the probe current to maintain the amplitude of the largest coincidence peak on this challenging sample as less than 1 % of the amplitude of the parent peak.

26.4 Collecting Data

26.4.1 Exploratory Spectrum

1. Before proceeding to collect final data on a sample, it is generally a good idea to collect an exploratory spectrum. This exploratory spectrum should be collected for sufficiently long to ensure that all the elements (major, minor, and trace) present in specimen can be identified. This exploratory spectrum can also be used to determine how long an acquisition will be necessary to get the precision you desire for each element in the spectrum.
2. Subject the exploratory spectrum to qualitative analysis using the vendor-supplied automatic peak identification tool but always follow with manual inspection of the suggested elemental identifications to determine validity. The elements identified will determine the standards that will be necessary for the analysis.
3. If you suspect that an element may also be present but is obscured by another element, a standard should be collected for that element too.
4. It may be beneficial to perform a standardless analysis on the exploratory spectrum to get a rough idea of the composition of the unknown (Be sure that the manually confirmed element list, not the raw automatic peak identification list, is used for the standardless analysis.)

26.4.2 Experiment Optimization

Determining the peak fitting reference requirements and the optimal acquisition times can be a challenge. DTSA-II provides a tool to help. The user provides an estimate of the composition of the unknown (crude estimates are fine), the standards you are going to use and the desired measurement precision and the tool will tell you when references are required and suggest approximately how long to acquire each standard, reference and unknown spectrum.

26.4.3 Selecting Standards

1. Select a standard for each element that you believe is present in the unknown.
2. In certain problem domains, you may be able to omit a standard if you calculate oxygen by assumed stoichiometry or assume an element by difference from an analytical total of one.
3. It's generally best to assume that you will measure every element and collect a standard for each. You can always change your mind later.
4. Initially, it is probably best to select a simple standard (one for which no peak shape references are required)

26.4.4 Reference Spectra

References serve two purposes:
1. To resolve interferences in multi-element standards
   - Example: In BaF_2, the Ba M lines interfere with the F K lines. To use BaF_2 as a standard for Ba, two references (e.g., $BaCl_2$ for Ba and CaF_2 for F) are required to provide clear, interference-free views of the Ba M-family peaks and the F K-family peaks in the range of energies in which the Ba M lines interfere with the F K lines.
2. To strip elements which are known to contribute to the unknown spectrum but are not really present in the material
   - Example: A reference to strip a thin conductive coating of Au, Pt, or C.

26.4.5 Collecting Standards

1. Since standards will typically contribute to many measurements, it is generally a good idea to spend extra time to ensure that standards are of high quality.
2. The total acquisition time necessary depends upon the measurement goals and can be determined by examining the intensity in an element's characteristic peaks. In particular, you should examine the characteristic peaks that will be used to perform the quantification. The background corrected peak integral should contain at least 10,000 counts for approximately 1 % precision or 160,000 counts for 0.25 % precision.

3. It is generally better to collect multiple shorter acquisition spectra from multiple points on the standard and build a single high-quality standard spectrum from the best of these. Collecting multiple spectra allows you to discern problems with individual spectra that may otherwise go unnoticed.
4. Collect N where $N > 2$ spectra from various different points on the standard.
   - Measure and record the probe current before collecting each spectrum and after the last.
   - Note any significant changes in probe current where "significant" is determined by the desired measurement precision.
5. Compare the N spectra but plotting them simultaneously using the same vertical scale for all. Examine carefully the intensity in the characteristic peak of choice. Discard any spectra which differ systematically by more than counting statistics. Sum the remaining spectra together to form a single spectrum.
6. Ensure the following properties are assigned to the standard spectrum.
   (i) Beam energy
   (ii) Probe current
   (iii) Live-time
   (iv) Composition of the standard
7. Save the standard to disk.
8. Repeat for each required standard.

26.4.6 Collecting Peak-Fitting References

References serve as examples of an element's characteristic peak shapes. The factors that make for a good reference are different from those that make a good standard.
1. A good material for a reference need not be as carefully mounted and prepared as a standard or unknown.
2. Particles or rougher surfaces can serve as adequate reference materials.
3. You don't need to know the probe current or live-time.
4. While it is generally a good idea to collect the reference at the same beam energy (particularly for lower overvoltages on the L and M lines), it is not always necessary.
5. References should be high count spectra but are less susceptible to count statistics than the standards. In general, a reference should have significantly more than 10,000 counts in the element's useful characteristic peak to provide an adequate perspective of the peak shape.
6. Simple elemental standards can always be reused as references.
7. Save the references to disk.

26.4.7 Collecting Spectra From the Unknown

1. Examine the exploratory spectrum to determine how long an acquisition time to use for the unknown

26

2. Examine the intensity in the measured characteristic peak for each element in the unknown to make realistic precision goals for each element:
   - 1 % or better precision is realistic for major elements (C > 0.1 mass fraction)
     - 1 % precision requires at least 10,000 counts in the unknown's quantified characteristic peak
   - 3–1 % is realistic for minor elements ($0.01 \leq C \leq 0.1$ mass fraction)
     - 3 % precision requires at least 1,000 counts in the unknown's quantified characteristic peak
3. Each element is likely to require a different acquisition time. Select the longest.
4. Collect multiple spectra from different positions on the unknown as you did with the standards, compare the unknown spectra looking for differences. If one or more spectra are different, try to identify the source of the differences. The differences may reflect real inhomogeneities or they may represent measurement artifacts like surface contamination, roughness, voids, or probe instability.
5. Collect an image with the spectrum so that if there is a problem with a particular spectrum, you can assess whether there may be a sample-related problem.
6. You may reasonably eliminate (and potentially recollect) spectra that differ due to recognized measurement artifacts. However, unexpected differences may be important clues that the specimen is locally different in some unexpected manner that this difference comprises real information that you do not want to ignore. Reality on the micrometer-scale is often more complex than we expect.
7. Identify the subset of the acquired unknown spectra that you are going to quantify. Ensure each spectrum from the unknown contains the following data items:
   (i)   Beam energy
   (ii) Probe current
   (iii) Live-time

26.5   Data Analysis

26.5.1   Organizing the Data

By this point, you should have collected all the pieces of data you need to perform the quantification:
1. Standard spectra—One high-quality standard for each element in the unknown
2. Reference spectra—For each standard with an interference and for each element to strip
3. Unknown spectra

26.5.2   Quantification

DTSA-II quantification proceeds as follows:
1. Select the unknown spectrum or group of spectra to quantify.
2. Select the standards. If a standard has two or more elements, you will be asked which of the elements are to be considered for this analysis: for example, if FeS_2 is selected, you will can select Fe and/or S. Only one standard can be selected per element. If an element present in the standard has already been selected, it will be grayed out.
3. Select the references. DTSA-II will inform you if a standard cannot serve as a reference due to a conflict from interfering peaks; for example, for BaF_2, the Ba M-family and F K-peak mutually interfere. If a reference is needed that is different from the standard, then select the element in dispute (which appears in red) from the list and select an appropriate spectrum to serve as a reference, for example, CaF_2 for F and $BaCl_2$ for Ba M-family.
4. DTSA-II will then execute and return a report with the k-ratios measured, the elemental concentrations calculated, the residual spectrum after peak fitting, and the uncertainty budget consisting of the uncertainties due to counting statistics of the unknown and standard, the atomic number correction, and the absorption correction.

26.6   Quality Check

26.6.1   Check the Residual Spectrum After Peak Fitting

You aren't done until you've checked for blunders, mistakes, and surprises. Two of the most common mistakes are missed elements and misidentified elements. Both of these mistakes can be identified using the residual spectrum. The residual spectrum is a derived spectrum computed from the unknown spectrum in which all the quantified characteristic peaks are removed.
1. Missed element: One of the most common surprises is a missed element. Sometimes, the element is hiding under the characteristic lines for another element. You won't know about the other element until you've performed peak fitting for the intensity contributions from the elements that you already know about. A missed element will show up as a peak in the residual. If you have missed an element, you will need to add an appropriate standard for that element (and possibly a reference) and re-quantify the data.

2. Misidentified element: Many characteristic peaks can be mistaken for another element. A peak may be at the correct energy for another element but the residual will reveal the mistake. The shape of peaks will be different and the residual will appear irregular and non-physical. Alternatively, there may be other peaks that remain in the residual unaccounted for by your initial choice of standards. To fix this problem you should replace the standard for the misidentified element with one for the correct element and re-quantify the data.

26.6.2 Check the Analytic Total

The analytic total is the sum of the mass fractions measured for each element in the unknown. The analytic total should be close to unity if all elements have been recognized and included in quantitative calculations. When a typical material is analyzed under typical conditions, the analytical total may reasonably vary 1 % or 2 % from unity due to measurement uncertainties. Simple measurements based on energetic K peak transitions can be reliably measured to better than a percent. More complex measurements involving low energy X-rays (like carbides and oxides) are likely to have larger deviations. A deviation of more than a percent or two should inspire you to start asking questions.

1. Have I missed an element?
   (i) Check the residual. Is there a peak that hasn't been quantified?
   (ii) Is it possible that there is an unmeasurable element like H, He, or Li in the unknown?
2. Is there a problem with the measurement process?
   (i) Is the sample preparation adequate?
   (ii) Is the sample tilted?
   (iii) Is the sample at the correct working distance?
   (iv) Is the probe current being measured accurately?
   (v) Did the probe current drift?
   (vi) Did the specimen charge?

26.6.3 Intercompare the Measurements

Whenever feasible you should make multiple measurements of each material. As part of the quality control process, you should compare these measurements.
1. How do the measurements vary among themselves?
2. How does the variance calculated from the measured values compare with the estimated uncertainties (particularly the uncertainties due to precision)?
3. Do the measured variances suggest that the material is homogeneous or inhomogeneous?

4. Is there an outlier? Can you explain the outlier? Examine the SEM image of the region. Is there any obvious difference in the image compared to other areas? If there is no obvious reason for the outlier, does it suggest something about the sample or does it says something about the measurement process? Can you reproduce the outlier by re-measuring the spectrum at the same location?

Report the Results
• What to Report
The analytical report should be a concise description of what request was made to the analyst, what analytical strategy was developed, how the measurement was performed and the results.

• Analytical Procedure
The analytical procedure should provide sufficient detail that you or someone else with the correct instrumentation could replicate the measurement.
1. Scanning electron microscope
   – Manufacturer and model
   – Beam energy
   – Nominal probe current
2. X-ray detector
   – Manufacturer and model
   – Configuration settings
3. Other aspects
   – Picoammeter
   – Software
4. Standards
   – Identity, composition, source, live time, probe current, sample preparation, coating
5. References
   – Identity, composition, characteristic line assignment, sample preparation, coating
6. Unknown
   – Identity, sample preparation, coating
7. How the locations for analysis were selected. Was it based on the customer's directions or if the analyst selected the locations, what criteria were used?

• Results
Each spectrum should be reported independently. For multiple spectra, tabular form works well.
1. Reporting the elemental data
   – It is generally best to report the non-normalized mass-fraction unless there is precedent for using an alternative format. It is only acceptable to report the normalized mass-fraction if the analytical total is also reported.
2. If you report in oxide fraction or atom fraction be sure to include the analytical total since the act of converting

from mass concentration (weight fraction) to atomic fraction or oxide fraction includes normalization.

3. DTSA-II provides an estimate of the measurement uncertainty for each element. The estimate should make clear which factors were considered and whether the estimate should be considered an estimate of accuracy or just an estimate of precision. A measurement without an uncertainty estimate is open to misuse. The client may assume that the result is more accurate than it really is and draw conclusions that cannot be justified by the data. Alternatively, the client may not trust the data or may assume that it is less accurate than it is and fail to draw conclusions that are justified. Either way, data presented without uncertainties is of limited utility.

4. If the spectra represent a nominally homogeneous region (or one you suspect to be), add descriptive statistics (mean, standard deviation) summarizing the variation between locations for each element. Compare this value with the uncertainty estimate for a single measurement to detect heterogeneity.

- Conclusions

Conclusions should be pithy. You should be very careful only to report that which is directly supported by the measurement results. In other words, stick to the facts and avoid conjecture. Don't answer questions that go beyond the data and your personal expertise.

Reference

International Organization for Standardization (2012) Standard file format for spectral data exchange ISO 22029:2012. ▶ https://www.iso.org/standard/56211.html

X-Ray Microanalysis Case Studies

© Springer Science+Business Media LLC 2018
J. Goldstein et al., *Scanning Electron Microscopy and X-Ray Microanalysis*,
https://doi.org/10.1007/978-1-4939-6676-9_27

27.1   Case Study: Characterization of a Hard-Facing Alloy Bearing Surface

Background: As part of a study into the in-service failure of the bearing surface of a large water pump, characterization was requested of the hard-facing alloy, which was observed to have separated from the stainless steel substrate, causing the failure.

This problem illustrates the critical importance of careful specimen preparation of a macroscopic object with centimeter dimensions to locate regions of microscopic interest with micrometer dimensions. Metallographic preparation produced a polished cross section of an intact portion of the hard-facing alloy layer as deposited onto the stainless steel base, as shown in ◘ Fig. 27.1. SEM-EDS analysis with DTSA-II gave the results shown in ◘ Fig. 27.2 for the stainless steel base and at one location in the hard-facing alloy. EDS elemental mapping produced the images shown in ◘ Fig. 27.3. The SEM-BSE image (◘ Fig. 27.1) revealed the presence of numerous voids that were predominantly located (white arrows) at the interface between the hard-facing alloy and the stainless steel substrate, with a smaller population of voids located within the hard-facing alloy (yellow arrows). The elemental maps (◘ Fig. 27.3) revealed that the hard-facing alloy layer had a very complex microstructure with distinct heterogeneity. The voids were found to be closely associated with regions with that had elevated Cr, both near the hard-facing alloy-stainless steel interface and within the hard-facing alloy. These voids severely compromised the performance of the bearing surface.

Complex fine-grained regions were also observed within the hard-facing alloy, as shown in the elemental maps in

◘ Fig. 27.4 and the SEM-BSE image in ◘ Fig. 27.5, which shows strong atomic number contrast. SEM-EDS analysis with DTSA-II gave the results presented in ◘ Fig. 27.5, which confirm the gray scale sequence as a function of composition, with the sharp rise in W (see table in ◘ Fig. 27.5) dominating backscattering.

The information provided by SEM-EDS enabled metallurgists to modify the hard-facing alloy composition and deposition conditions to eliminate the void formation during deposition, producing satisfactory service behavior.

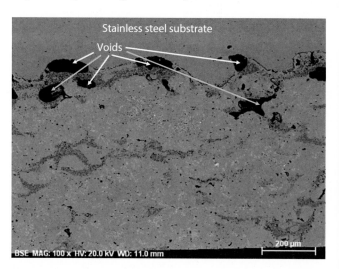

◘ **Fig. 27.1** SEM-BSE image of the cross-section of a hard-facing alloy deposited on a stainless steel substrate. Note the voids at the interface between the hard-facing alloy and the stainless substrate (*white arrows*), as well as a smaller population of voids entirely within the hard-facing alloy (*green arrows*)

◘ **Fig. 27.2** SEM-BSE image showing locations of SEM-EDS analyses with NIST DTSA-II

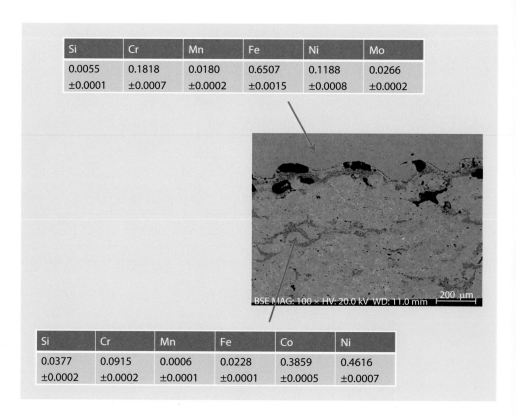

Si	Cr	Mn	Fe	Ni	Mo
0.0055	0.1818	0.0180	0.6507	0.1188	0.0266
±0.0001	±0.0007	±0.0002	±0.0015	±0.0008	±0.0002

Si	Cr	Mn	Fe	Co	Ni
0.0377	0.0915	0.0006	0.0228	0.3859	0.4616
±0.0002	±0.0002	±0.0001	±0.0001	±0.0005	±0.0007

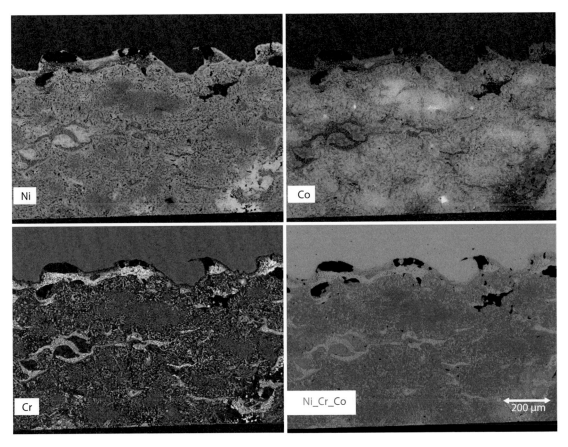

Fig. 27.3 Elemental mapping, with color overlay: Ni = red; Cr = green; Co = blue

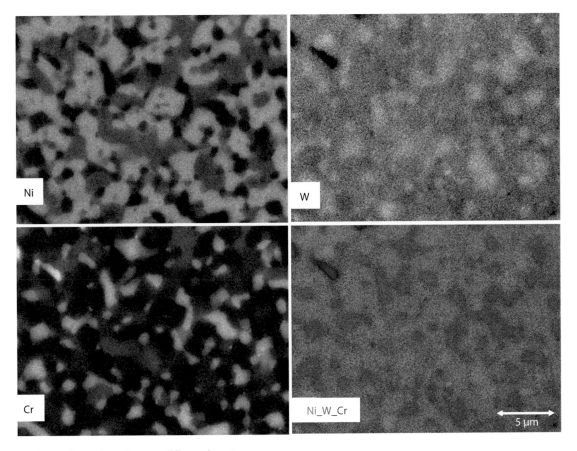

Fig. 27.4 Elemental mapping of an area of fine-scale grains

27

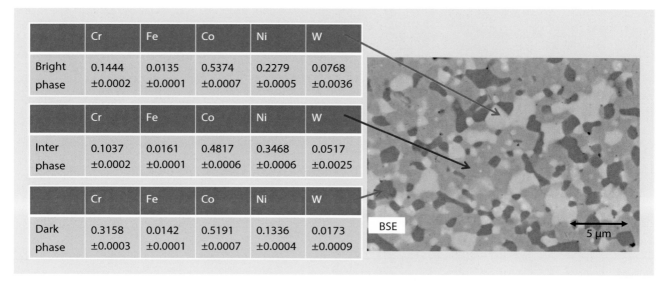

	Cr	Fe	Co	Ni	W
Bright phase	0.1444 ±0.0002	0.0135 ±0.0001	0.5374 ±0.0007	0.2279 ±0.0005	0.0768 ±0.0036

	Cr	Fe	Co	Ni	W
Inter phase	0.1037 ±0.0002	0.0161 ±0.0001	0.4817 ±0.0006	0.3468 ±0.0006	0.0517 ±0.0025

	Cr	Fe	Co	Ni	W
Dark phase	0.3158 ±0.0003	0.0142 ±0.0001	0.5191 ±0.0007	0.1336 ±0.0004	0.0173 ±0.0009

◻ **Fig. 27.5**   SEM-BSE image and DTSA-II analyses of selected grains in the fine-scale region

27.2   Case Study: Aluminum Wire Failures in Residential Wiring

Background: In the early 1970s, aluminum wire was used extensively as a substitute for more expensive copper wire in residential and commercial wiring, specifically for 110 V electrical outlets that used steel screw compressive clamping of the wire against a brass or steel plate. The aluminum wire–steel screw junctions were observed to fail catastrophically through a process of overheating, leading in extreme cases to initiation of structural fires (Meese and Beausoliel 1977; Rabinow 1978). ◻ Figure 27.6 shows an example of the damage to the wire-screw junction and the surrounding plastic housing and wire insulation caused during an overheating event observed in a laboratory test. This failure was a puzzling occurrence, since aluminum is an excellent electrical conductor and was long used successfully in high voltage electrical transmission lines. Moreover, the vast majority of Al wire–screw connections provided proper service without overheating. However, those connections that did fail in service often produced such catastrophic effects that the critical evidence of the initiation of the failure was destroyed. Capturing an event like that shown in ◻ Fig. 27.6 required intensive laboratory studies in which thousands of junction boxes were tested and continuously monitored with thermal sensors until a failure initiated, which was then automatically interrupted to prevent complete destruction of the evidence.

This problem illustrates the "macro to micro" sampling problem. The failure mechanism was eventually discovered by SEM/EDS characterization to have a microscopic point of origin, but this microscopic failure origin with micrometer dimensions was hidden within a complex macroscopic structure with centimeter dimensions. Solving the problem required a careful sample preparation strategy to locate the unknown feature(s) of interest. The metallographer mounted the entire Al-wire/steel screw/brass plate assembly in epoxy, as shown in ◻ Fig. 27.7, and sequentially ground and polished

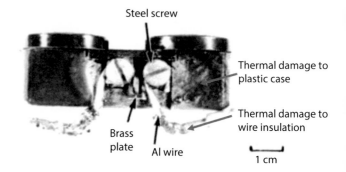

◻ **Fig. 27.6**   Residential electrical outlet wired with aluminum. The laboratory test was interrupted after the thermal event initiated and was automatically detected, but significant thermal damage to the plastic casing and wire insulation still occurred

◻ **Fig. 27.7**   Metallographic mount (2.5-cm diameter) showing the cross section of the steel screw, aluminum wire, and brass plate

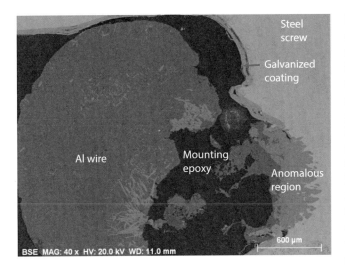

Fig. 27.8 SEM-BSE image of an anomalous zone of contact between the Al wire and the Fe screw

through the structure until an anomalous region was revealed (**Fig. 27.8**). As shown with SEM/BSE imaging and elemental mapping in **Fig. 27.9**, in this anomalous region the aluminum and iron had reacted to form two distinct Al-Fe zones (Newbury and Greenwald 1980; Newbury 1982). Fixed

beam quantitative X-ray microanalysis with NIST DTSA II and pure element standards (analyses performed during a revisiting of the 1980 specimens) produced the results shown in **Fig. 27.10**, where zone 1 is found to correspond closely to the intermetallic compound $FeAl_3$, while zone 2 corresponds to Fe_2Al_5. The presence of these intermetallic compounds is significant because of their resistivity. $FeAl_3$ and Fe_2Al_5 have electrical resistivities of approximately 1 μΩ–m, similar to that of the alloy nichrome (1.1 μΩ–m), which is used for resistive heating elements and which is a factor of 38 higher than pure Al and 10.3 higher than pure Fe. The formation of these intermetallic compounds at the screw-wire contact was initiated when electrical arcing occurred because the connection became mechanically loose due to creep of the Al wire and the poor compliance (springiness) of the wire–screw clamp. Once the local formation of the intermetallic compounds had been initiated by arcing followed by local welding of the Al wire and the steel screw, the increased resistivity caused localized resistive heating that stimulated the interdiffusion of Al and Fe, leading to the further intermetallic compound growth in a runaway positive feedback. Eventually this intermetallic compound zone expanded to dimensions of several hundred micrometers, as seen in **Fig. 27.7**, creating a resistive heating element that caused

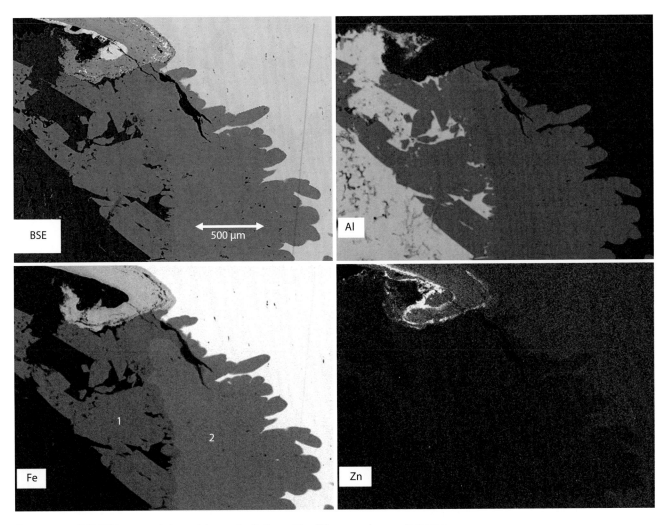

Fig. 27.9 SEM-BSE image and elemental maps for Al, Fe, and Zn of the anomalous contact zone

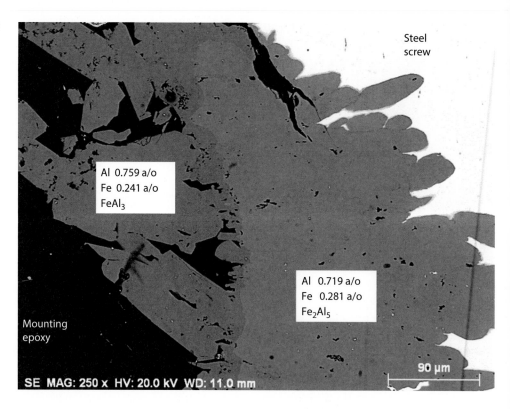

Fig. 27.10 SEM-BSE image of the anomalous zone of contact with quantitative X-ray microanalysis results from fixed-beam analysis in the two distinct Al-Fe regions (note the contrast in the BSE image)

Al 0.759 a/o
Fe 0.241 a/o
FeAl$_3$

Al 0.719 a/o
Fe 0.281 a/o
Fe$_2$Al$_5$

Steel screw

Mounting epoxy

SE MAG: 250 x HV: 20.0 kV WD: 11.0 mm

90 µm

the damage seen in ◘ Fig. 27.6. (Note that the practical solution to this problem was to modify the wire connections to provide much greater springiness to eliminate the opening of gaps that allowed arcing to occur.)

27.3 Case Study: Characterizing the Microstructure of a Manganese Nodule

"Manganese nodules" are rock concretions that form on the deep sea floor through the action of microorganisms that precipitate solid chemical forms from metals dissolved in the water, often in close association with hydrothermal vents.

The elemental composition of a polished cross section of a manganese nodule, shown in ◘ Fig. 27.11, was examined by SEM/BSE and by SEM/EDS X-ray spectrum imaging elemental mapping. The SEM/BSE image in ◘ Fig. 27.12 reveals a complex layered microstructure that suggests non-uniform deposition of the precipitated minerals over time. This non-uniform deposition is confirmed by the elemental maps for O, Mn, and Ni and color overlay shown in ◘ Fig. 27.13 and for the Mn, Fe, and Ni maps shown in ◘ Fig. 27.14. Note the oxygen-rich areas (green) in ◘ Fig. 27.13. These regions correspond to silica and aluminosilicate grains within the manganese nodule, as revealed in ◘ Fig. 27.15. The composition measured with a fixed beam placed at the center of the field of view is listed in ◘ Table 27.1, showing the high abundance of Mn as a major constituent and the presence of other transition elements (e.g., Fe, Ni, and Cu) as minor constituents. ◘ Figure 27.16 shows the results of quan-

Fig. 27.11 Manganese nodule

titative processing of the XSI with the k-ratio/matrix correction protocol using DTSA-II. The resulting concentration maps have been encoded with the logarithmic three-band color scheme shown in ◘ Fig. 27.16, enabling quantitative comparison of the constituents, using NIST Lispix.

Note that some features in the elemental maps are a result of artifacts. Thus, the cracks noted in the SEM/BSE of ◘ Fig. 27.11 are also seen in the O elemental map, but not in the Mn or Ni maps. The origin of this artifact is the difference in the photon energies of these elements. The O K-shell X-ray

27.3 · Case Study: Characterizing the Microstructure of a Manganese Nodule

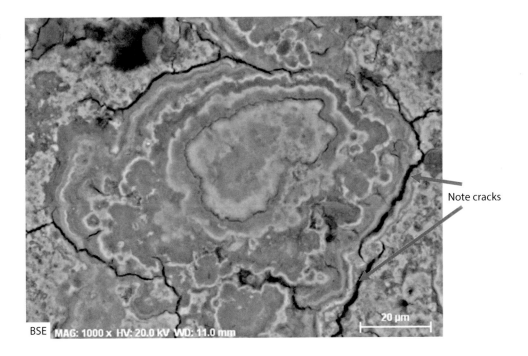

▣ Fig. 27.12 SEM/BSE image of a polished cross section; note the cracks

Note cracks

BSE MAG: 1000 x HV: 20.0 kV WD: 11.0 mm

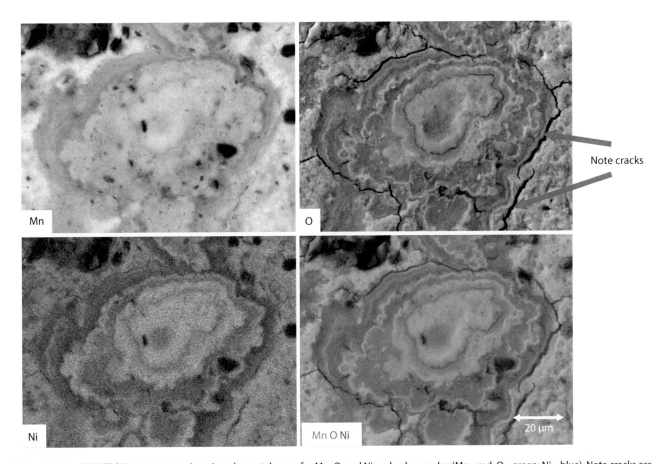

Mn

O

Note cracks

Ni

Mn O Ni

20 µm

▣ Fig. 27.13 SEM/EDS X-ray spectrum imaging elemental maps for Mn, O, and Ni and color overlay (Mn = red; O = green; Ni = blue). Note cracks are observed in the O map but are much less visible in Mn and Ni

27

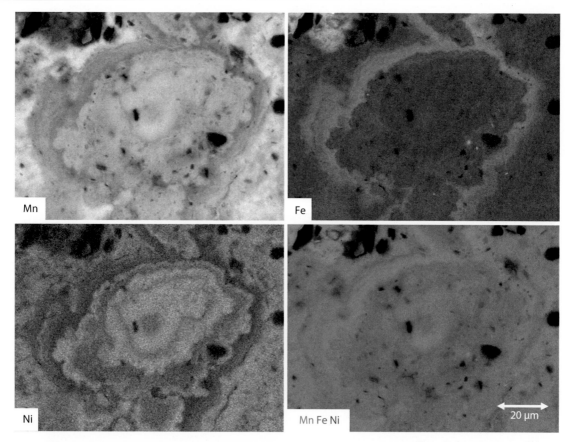

☐ **Fig. 27.14**   SEM/EDS X-ray spectrum imaging elemental maps for Mn, Fe, and Ni and color overlay (Mn = red; Fe = green; Ni = blue)

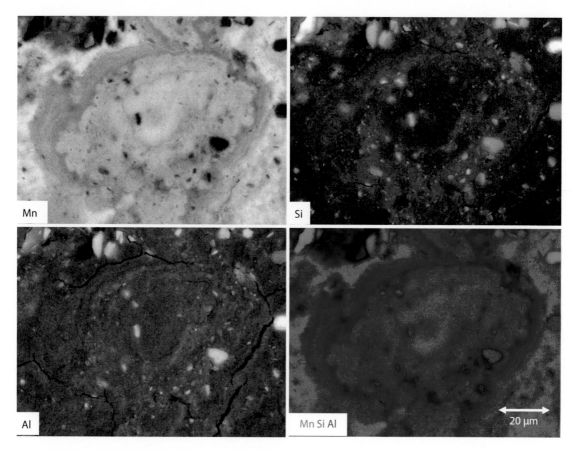

☐ **Fig. 27.15**   SEM/EDS X-ray spectrum imaging elemental maps for Mn, Si, and Al and color overlay (Mn = red; Si = green; Al = blue)

Table 27.1   Fixed-beam analysis at the center of ☐ Fig. 27.12; NIST DTSA II analysis with pure element and microanalysis glass standards

O	Na	Mg	Al	Si	K	Ca	Ti	Mn	Fe	Ni	Cu
0.2742 ± 0.0003	0.0420 ± 0.0005	0.0196 ± 0.0002	0.0210 ± 0.0001	0.0458 ± 0.0002	0.0100 ± 0.0000	0.0267 ± 0.0001	0.0031 ± 0.0000	0.4412 ± 0.0003	0.0834 ± 0.0002	0.0196 ± 0.0001	0.0134 ± 0.0002

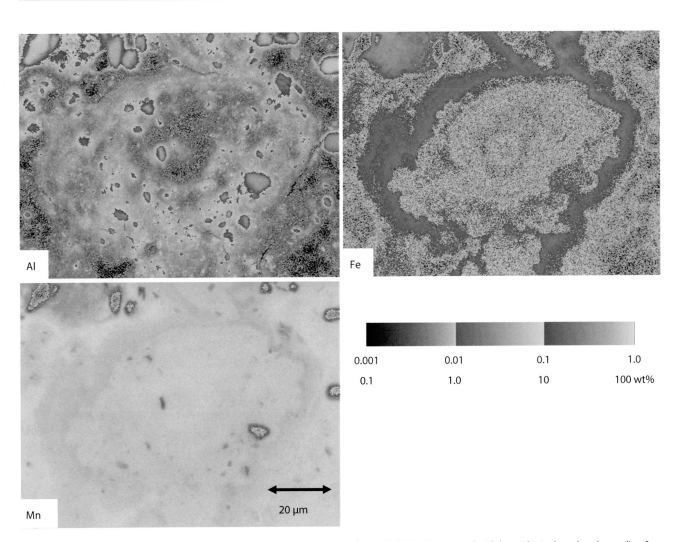

Fig. 27.16   SEM/EDS X-ray spectrum imaging maps after quantitative analysis with DTSA-II presented with logarithmic three-band encoding for Al, Fe, and Mn. Note Fe-enrichment band

at an energy of 0.523 keV suffers strong absorption when the electron beam is located in the crack, so that the intensity is greatly reduced, producing an accurate representation of the crack. The MnK-$L_{2,3}$ (5.898 keV) and NiK-$L_{2,3}$ (7.477 keV) photons have higher energy and suffer much less absorption, so that most of those photons generated when the beam is in the crack still escape despite having to pass through more material to reach the EDS, greatly reducing the contrast of the cracks relative to the surrounding matrix.

References

Meese W, Beausoliel R (1977) Exploratory study of glowing electrical connections". Nat Bur Stand (US) Build Sci Ser 103:29

Newbury D (1982) What is causing failures of aluminum wire connections in residential circuits? Anal Chem 54:1059A–1064A

Newbury D, Greenwald S (1980) Observations on the mechanisms of high resistance junction formation in aluminum wire connections. J Res Nat Bur Stand US 85:429–440

Rabinow J (1978) Some thoughts on electrical connections. Nat Bur Stand US NBSIR:78–1507

Cathodoluminescence

© Springer Science+Business Media LLC 2018
J. Goldstein et al., *Scanning Electron Microscopy and X-Ray Microanalysis*,
https://doi.org/10.1007/978-1-4939-6676-9_28

28

28.1   Origin

Cathodoluminescence (CL) is the emission of low energy photons in the range from approximately 1 eV to 5 eV (infrared, visible, and ultraviolet light) as a result of inelastic scattering of the high energy beam electrons (■ Fig. 28.1). Materials that can emit such photons are insulators or semiconductors which have an electronic structure with a filled valence band of allowed energy states that is separated by a gap of disallowed energy states from the empty conduction band, as shown schematically in ■ Fig. 28.2a. Inelastic scattering of the beam electron can transfer energy to a weakly bound valence electron promoting it to the empty conduction band, leaving a positively charged "hole" in the conduction band. When a free electron and a positive hole are attracted and recombine, the energy difference is expressed as a photon, as illustrated in ■ Fig. 28.2b. Because the possible energy transitions and the resulting photon emission are defined by the intrinsic properties of a high purity material, such as the band-gap energy but also including energy levels that result from physical defects such as lattice vacancies, rather than by the influence of impurity atoms, this type of CL is referred to as "intrinsic CL emission." Since the valence electron promoted to the conduction band can receive a range of possible kinetic energies depending on the details of the initial scattering, the photons emitted during free electron–hole recombination can have a range of energies, resulting in broad band CL photon emission. Because of the great mismatch in the velocity of the high energy (keV) beam electron and the low energy (eV) valence electron, this is not an efficient process and in general CL emission is very weak. The ionization cross section is maximized for electrons with three to five times the binding energy of the valence electrons, so that most efficient energy transfer to initiate CL emission occurs from the more energetic slow SE (>10 eV) and the fast SE (hundreds of eV) also created by inelastic scattering of the primary electron.

In more complex materials that are modified by impurities, the presence of impurity atoms in the host crystal lattice can create sharply defined energy levels within the band gap to which valence electrons can be scattered, as illustrated in ■ Fig. 28.3. The subsequent electron–hole transitions that involve these well-defined energy states create a photon or series of photons with a sharply defined energy or series of energies ("extrinsic CL emission"). This sharp line spectrum may be superimposed on a broad range intrinsic spectrum which can still occur.

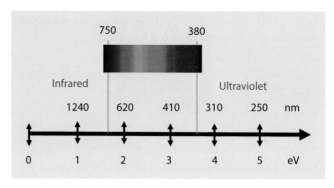

■ **Fig. 28.1**   Range of photon energies and wavelengths for cathodoluminescence

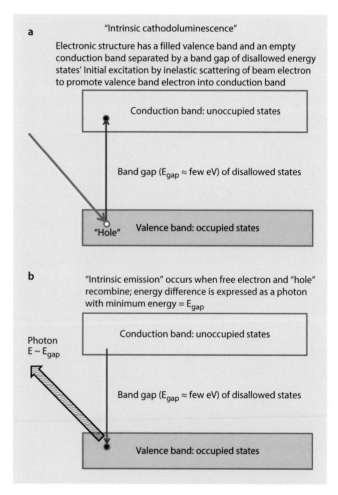

■ **Fig. 28.2**   **a** Origin of intrinsic CL: the material's electron energy states fill the valence band which is separated by a band-gap of several eV from an empty conduction band. Inelastic scattering of the beam electron promotes a valence band electron to the conduction band, leaving a positively charged hole in the valence band. **b** Origin of intrinsic CL: The free electron and hole are mutually attracted and recombine, with the energy released as an electromagnetic photon with a minimum energy equal to the band-gap energy

Fig. 28.3 Origin of extrinsic CL: the presence of impurity atoms in the host lattice creates narrow energy levels within the band-gap. Inelastic scattering of the beam electron can promote valence band electrons to the conduction band as well as to the impurity levels in the band-gap. Electron transitions can occur between the various energy levels, creating both broadband CL emission and sharply defined CL emission

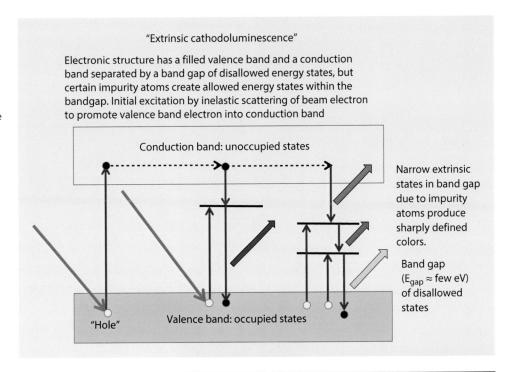

"Extrinsic cathodoluminescence"

Electronic structure has a filled valence band and a conduction band separated by a band gap of disallowed energy states, but certain impurity atoms create allowed energy states within the bandgap. Initial excitation by inelastic scattering of beam electron to promote valence band electron into conduction band

Conduction band: unoccupied states

Narrow extrinsic states in band gap due to impurity atoms produce sharply defined colors.

Band gap ($E_{gap} \approx$ few eV) of disallowed states

"Hole"

Valence band: occupied states

28.2 Measuring Cathodoluminescence

Some materials such as the mineral benitoite and the compound ZnS produce such strong CL that the emitted light can be readily observed by the unaided eye, as shown in ☐ Figs. 28.4 and 28.5. However, in general the CL phenomenon is weak, requiring the use of an efficient optical collection scheme for successful measurement.

28.2.1 Collection of CL

To maximize photon collection, the specimen can be surrounded by a mirror configured to collect a solid angle of nearly 2π sr. An example shown in ☐ Fig. 28.6 of an ellipsoidal mirror in which the specimen is placed at one mirror focus, where an aperture permits the electron beam to reach the specimen, and the collector, typically a fiber optic, is placed at the other focus. For the examples of direct observation of CL shown in ☐ Figs. 28.4 and 28.5, the CL emission was collected through the high collection angle optical microscope integrated into an electron optical column (specifically, an electron probe microanalyzer configuration with wavelength dispersive spectrometry, EPMA/WDS, in which the optical microscope serves as the critical positioning aid for the WDS optimization).

28.2.2 Detection of CL

The photomultiplier (PM) is a useful detector for CL because of its sensitivity to photons in the energy range of interest, its fast response (hundreds of picoseconds to nanoseconds), and

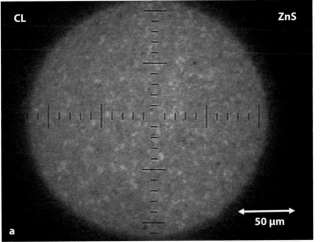

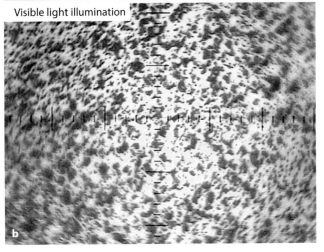

Fig. 28.4 **a** CL emission from ZnS observed with a defocused 220-μm-diameter beam with 500 nA of beam current at $E_0 = 20$ keV. **b** Corresponding white light illumination image with the electron beam blanked into a Faraday cup

28

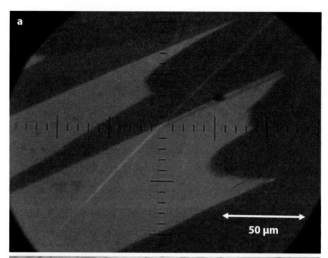

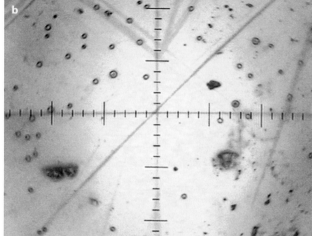

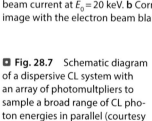

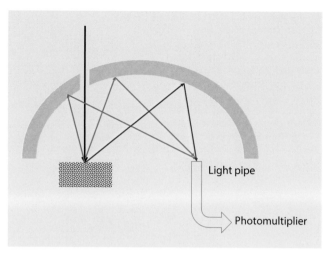

■ **Fig. 28.6** Schematic diagram of a high efficiency CL collection optic based upon an ellipsoidal mirror

its operation as a high gain (gain ~10^6), low noise amplifier. The choice of the photocathode material, which converts the photons to low energy electrons at the first stage of the electron cascade, determines the spectral response, quantum efficiency, and sensitivity as a function of photon energy, typically showing a strong dependence as a function of wavelength. By choosing different photocathodes, PMs can be optimized for efficiently detecting different photon energies. ■ Figure 28.7 illustrates a dispersive system in which the CL is scattered by a grating onto an array of PMs optimized to detect the different photon energies. Such a parallel detection strategy is critical to optimize measurements from weakly emitting systems, as well as to study systems with complex emission.

Another detection scheme makes use of a single wide energy response photomultiplier to detect CL photons across the energy range. To separate the different color

■ **Fig. 28.5** **a** CL emission from the mineral Benitoite (BaTiSi$_3$O$_9$) observed with a defocused 220-μm-diameter beam with 500 nA of beam current at $E_0 = 20$ keV. **b** Corresponding white light illumination image with the electron beam blanked into a Faraday cup

■ **Fig. 28.7** Schematic diagram of a dispersive CL system with an array of photomultpliers to sample a broad range of CL photon energies in parallel (courtesy of Gatan, Inc.)

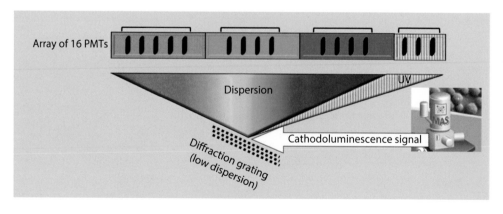

components of the CL emission, repeated scans of the area of interest are made with color filters (red, green, and blue) in the optical path to the PM, and the R-, G-, and B-images are combined.

Systems in which the photons are passed into a scanning optical spectrometer make possible detailed study of the CL emission in a single channel mode of operation.

28.3 Applications of CL

28.3.1 Geology

Carbonado Diamond
Carbonado is a rare variety of black diamond found in porous aggregates of fine polycrystals.

◻ Figure 28.8 shows CL imaging (using RGB filters) of a carbonado sample where the polycrystalline diamonds interacted with trace uranium during the Precambrian. The resulting MeV radiation damage is evident as metamict halos leaving the defects in the diamond structure (Magee et al. 2016). These haloes were not evident in SEM SE or BSE images or in elemental maps, indicating the sensitivity of CL emission to the presence of lattice defects that do not otherwise significantly affect the electron beam specimen interaction.

Ancient Impact Zircons
◻ Figure 28.9 shows CL imaging (using RGB filters) of an impact zircon collected using $E_0 = 20$ keV and 5-keV beam energies. The details of zircon disproportionation to ZrO_2 and silicate glass, as well as the zircon interior, are more evident at using a low energy beam (Zanetti et al. 2015).

28.3.2 Materials Science

Semiconductors
As an example of the detailed information that can be obtained with CL spectrometry, ◻ Fig. 28.10 shows a study of hexagonal GaN structures grown on a silicon substrate with a patterned mask to define the shape, as shown in the SEM image ($E_0 = 2$ keV) (◻ Fig. 28.10a). The panchromatic CL image formed by the integrated intensity of all measurable wavelengths (◻ Fig. 28.10b), reveals crystal defects which are observed as dark regions in the CL image including: (i) planar defects which can be seen as dark lines where they intersect the face(s) of the pyramid and (ii) regions with a high concentration of point defects or impurities. CL spectrum imaging was used to map the near band edge emission of GaN as a 3D datacube where a full CL spectrum (1D) was acquired at every pixel in the region of interest (2D). Changes in the near band edge

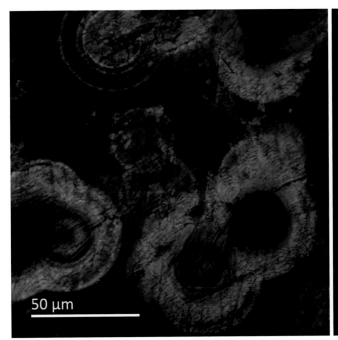

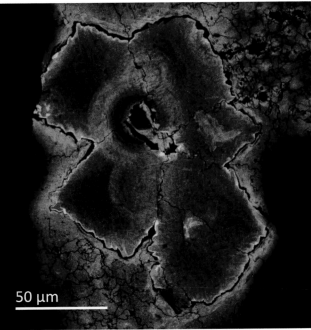

◻ **Fig. 28.8** CL imaging (RGB) of a carbonado, a rare variety of diamond. The polycrystalline material interacted with uranium during the Precambrian and MeV radiation damage is evident as metamict halos leaving the defects in the diamond structure (Magee et al. 2016) (Images courtesy of E. Vicenzi (Smithsonian Institution))

28

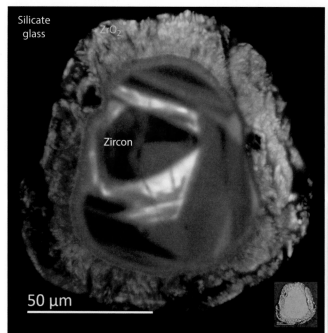

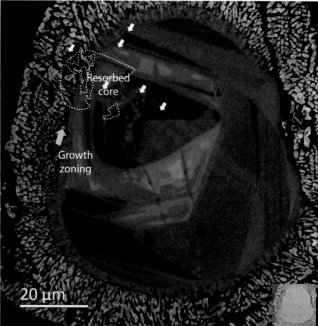

■ **Fig. 28.9** CL imaging (RGB) of an impact zircon collected using $E_0 = 20$ keV (*left*) and 5 keV (*right*). The details of zircon disproportionation to ZrO_2 and silicate glass, as well as the zircon interior, are more evident using a low energy beam (Zanetti et al. 2015) (Images courtesy of E. Vicenzi (Smithsonian Institution))

emission were investigated by spectral peak fitting. The derived central wavelength map reveals the strain level of the material with noticeable shifts appearing along the pyramid ridges. The RGB composite image is created from: high strain state GaN (yellow), low strain state (purple), defect (red).

Lead-Acid Battery Plate Reactions

It is possible to detect relatively strong CL signals with modified BSE detectors, as most of these solid state detectors are very sensitive to light. The challenge is to exclude BSE from the detected signal. The easiest way to do this is to coat a glass coverslip with a conductive transparent material like indium tin oxide (ITO). The glass cover slip prevents the electrons from interacting with the solid state detector and the conductive transparent coating prevents charging while allowing the CL signal to reach the detector.

It is interesting to note that some of the low vacuum or variable pressure SEMs that are commercially available use the CL light generated by the interaction of the secondary electrons with the gas in the chamber to produce a signal during variable pressure operation. One can use the same detector system, which consists of a glass light guide located near the sample that is coupled to a photomultiplier, for direct detection of CL emission. This type of detector is sensitive to low levels of light and thus when used in high vacuum mode can be a very simple but effective CL detector.

During the charging and discharging of lead acid battery plates, a variety of lead containing phases can form on the surface of the lead plate. Two important phases are lead sulfide (PbS) and lead sulfate ($PbSO_4$). Note that these compounds are similar in backscattering making it difficult to determine PbS from $PbSO_4$ in images. ■ Figure 28.11a shows a secondary electron image of the surface of a lead plate after it has been exposed to conditions that may occur in a lead-acid battery. Numerous euhedral crystals were observed on the surface of the lead plate. The EDS spectra indicated Pb and S as the major constituents and possibly O. Oxygen can be detected in the EDS spectrum of $PbSO_4$ from an ideal flat specimen, as shown in the DTSA-II simulation in ■ Fig. 28.11c, but because of the high absorption from Pb, the time requirement for mapping O to locate $PbSO_4$ becomes prohibitive. Moreover, given the complex topography of the sample shown in ■ Fig. 28.11a, mapping oxygen is likely to be badly compromised by strong X-ray absorption artifacts from the topography. CL can be of great use in this system as PbS does not exhibit CL while $PbSO_4$ strongly exhibits CL, enabling these compounds to be rapidly distinguished. In order to determine the likely compound, a simple CL system consisting of a light guide attached to a photomultiplier tube was used. ■ Figure 28.11b is an image obtained using this simple CL detector. Note that the euhedral crystals strongly exhibit CL, and this clearly indicates that these crystals are most likely $PbSO_4$ and not PbS.

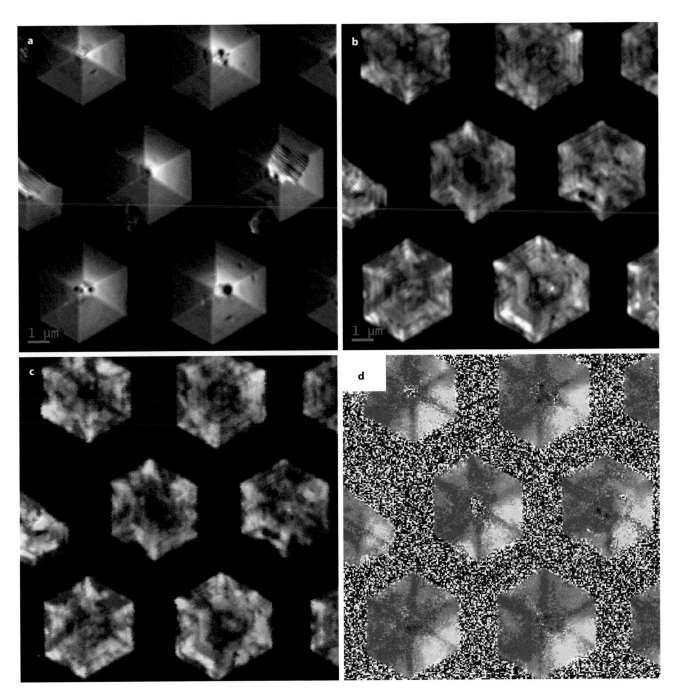

☐ **Fig. 28.10** CL study of GaN structures grown on Si: **a** SEM SE image ($E_0 = 2$ keV); **b** panchromatic CL image; **c** CL spectrum image data analyzed to derive the central wavelength value; **d** RGB composite (Example courtesy of D. Stowe, Gatan, Inc.)

28

◻ **Fig. 28.11   a** SEM SE image of deposits on lead-acid battery. **b** CL image of the same area; **c** DTSA-II simulation of the EDS spectra of PbS and PbSO$_4$ from an ideal flat surface at $E_0 = 15$ keV (courtesy of Sandia National Laboratory)

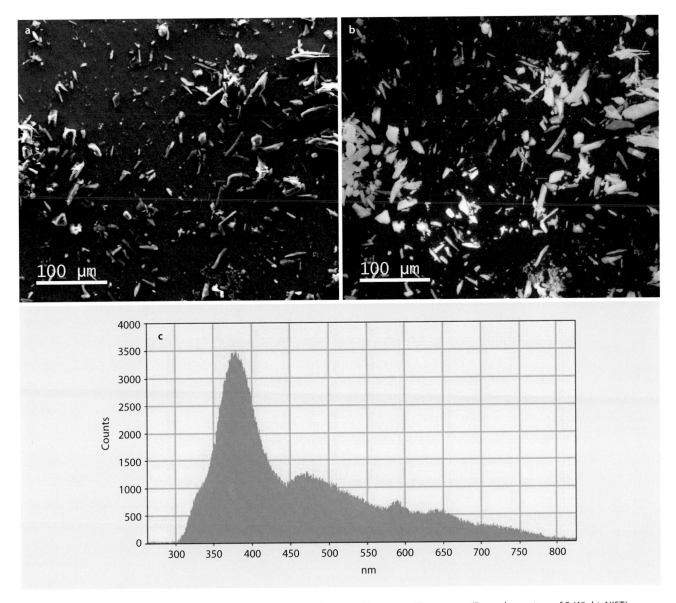

Fig. 28.12 Acetaminophen, $E_0 = 20$ keV: **a** SEM SE image; **b** panchromatic CL image; **c** CL spectrum (Example courtesy of S. Wight, NIST)

28.3.3 Organic Compounds

Despite the general vulnerability of organic compounds to radiation damage under electron bombardment, some organic compounds can be examined with CL spectrometry. ■ Figure 28.12 shows an SEM SE image and a panchromatic CL image of acetaminophen (paracetamol, *N*-(4-hydroxyphenyl) ethanamide *N*-(4-hydroxyphenyl)acetamide) along with a CL spectrum showing broad CL bands.

References

Magee C, Teles G, Vicenzi E, Taylor W, Heaney P (2016) Uranium irradiation history of carbonado diamond; implications for Paleoarchean oxidation in the São Francisco craton (South America). Geology 44. doi: 10.1130/G37749.1

Zanetti M, Wittman A, Carpenter PK, Joliff BL, Vicenzi E, Nemchin A, Timms N (2015) Using EPMA, Raman LS, Hyperspetcral CL, SIMS, and EBSD to study impact melt- induced decomposition of zircon. Microsc Microanal 21:S3 1447–1448

Characterizing Crystalline Materials in the SEM

© Springer Science+Business Media LLC 2018
J. Goldstein et al., *Scanning Electron Microscopy and X-Ray Microanalysis*,
https://doi.org/10.1007/978-1-4939-6676-9_29

29

While amorphous substances such as glass are encountered both in natural and artificial materials, most inorganic materials are found to be crystalline on some scale, ranging from sub-nanometer to centimeter or larger. A crystal consists of a regular arrangement of atoms, the so-called "unit cell," which is repeated in a two- or three-dimensional pattern. In the previous discussion of electron beam–specimen interactions, the crystal structure of the target was not considered as a variable in the electron range equation or in the Monte Carlo electron trajectory simulation. To a first order, the crystal structure does not have a strong effect on the electron–specimen interactions. However, through the phenomenon of channeling of charged particles through the crystal lattice, crystal orientation can cause small perturbations in the total electron backscattering coefficient that can be utilized to image crystallographic microstructure through the mechanism designated "electron channeling contrast," also referred to as "orientation contrast" (Newbury et al. 1986). The characteristics of a crystal (e.g., interplanar angles and spacings) and its relative orientation can be determined through diffraction of the high-energy backscattered electrons (BSE) to form "electron backscatter diffraction patterns (EBSD)."

29.1 Imaging Crystalline Materials with Electron Channeling Contrast

29.1.1 Single Crystals

The regular arrangement of atoms in crystalline solids can influence the backscattering of electrons because of the regular three-dimensional variations in atomic density in the crystal compared to those same atoms placed in the near random three-dimensional distribution of an amorphous solid. If a well-collimated (i.e., highly parallel) electron beam is directed at a crystal array of atoms along a series of different directions, the density of atoms that the beam encounters will vary with the crystal orientation, as shown in the simple schematic of ◘ Fig. 29.1. A sense of this effect can be obtained by manual manipulation of a macroscopic, three-dimensional ball-and-stick model of a crystal. For certain orientations of the model, the observer can see through the "atoms" along the open gaps between the planes in the model. In a real solid, the atoms are tightly packed, limited in their approach by the repulsive interaction of their atomic shells. The "channels" in reality are regions of the crystal where the atomic packing creates lower charge density with which the beam electrons will interact more weakly. When the beam is aligned with the channels, a small fraction of the beam electrons penetrate more deeply into the crystal before beginning to scatter. For beam electrons that start scattering deeper in the crystal, the probability that they will return to the surface as backscattered electrons is reduced compared to the amorphous target case, and so the measured backscatter coefficient is lowered compared to the average value from the amorphous target.

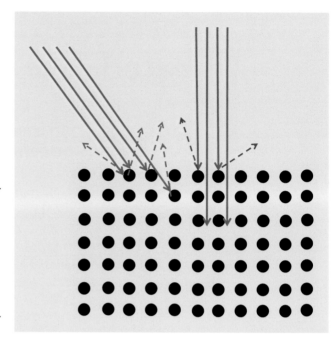

◘ **Fig. 29.1** Schematic illustration of the channeling effect: the atomic area density that the beam encounters depends on its orientation relative to the crystal

For other crystal orientations where denser atom packing is found, the beam electrons begin to scatter immediately at the surface, increasing the backscatter coefficient relative to the amorphous target case. As seen in the Monte Carlo simulation of an amorphous target, the elastic scattering of beam electrons rapidly randomizes their trajectories out of their initially well-collimated condition, reducing and eventually eliminating sensitivity to channeling in the crystal. For a bulk target, the modulation of the backscatter coefficient between the maximum and minimum channeling case is small, typically only about a 2–5 % difference. Nevertheless, this crystallographic or electron channeling contrast can be used to form SEM images that contain information about the important class of crystalline materials.

To determine the likelihood of electron channeling, the properties of the electron beam (i.e., its energy, E, or equivalently, wavelength, λ) are related to the critical crystal property, namely the spacing of the atomic planes, d, through the Bragg diffraction relation:

$$n\lambda = 2d\sin\theta_B \tag{29.1}$$

where n is the integer order of diffraction ($n = 1, 2, 3$, etc.). Equation 29.1 defines a special beam incidence angle relative to a particular set of the crystal planes with spacing d (referred to as the "Bragg angle," θ_B) at which the channeling condition changes sharply from weak to strong as the beam incidence angle increases relative to that particular set of crystal planes. Since a real crystal contains many different possible sets of atom planes, the degree of channeling encountered for an arbitrary orientation of the beam to the crystal is actually the sum of the contributions of many planes. Generally, these contributions tend to cancel giving the amorphous average backscattering except when the Bragg condition for a certain set of planes dominates the mix, giving rise to a sharp change in the channeling condition. What is the magnitude of Bragg angles for typical SEM beam and target conditions? Consider an incident beam of $E_0 = 15$ keV, where the electron wavelength, λ_e is given by the de Broglie equation:

$$\lambda_e = h/p \qquad (29.2a)$$

where h is Planck's constant and p is the momentum ($m_0 v$). In terms of beam energy, the wavelength is given by

$$\lambda_e (\text{nm}) = 1.226 / \left[E^{0.5} \left(1 + 0.979E - 6\,E \right)^{0.5} \right] \qquad (29.2b)$$

where E is expressed in eV (Hirsch et al. 1965).

At 15 keV, $\lambda_e = 0.00994$ nm. If this 15 keV-beam is directed at a crystal of silicon, which has a diamond-cubic crystal lattice with a fundamental cube dimension of $a = 0.543$ nm, the series of possible "allowed" Bragg angles (expressed in the "Miller indices" [hkl] which designate the crystal planes) is given in ◻ Table 29.1.

We can directly image the effects of crystallographic channeling on the total backscattered electron intensity for the special case of a large single crystal viewed with a large scanned area (i.e., low magnification). As shown schematically in ◻ Fig. 29.2, the act of image scanning not only moves the beam laterally in x- and y-directions, but at each beam position the angle of the beam relative to the surface normal changes. For a 10-mm working distance from the scan rocking point, a scan excursion of 2 mm in width causes the beam incidence angle to change by $\pm 12°$ across the field. As seen in ◻ Table 29.1, the allowed Bragg angles for a 15-keV electron beam striking Si have values of a fraction of a degree to a few degrees, so that a scan angle of $\pm 12°$ will certainly cause the beam to pass through the Bragg condition for at least some of

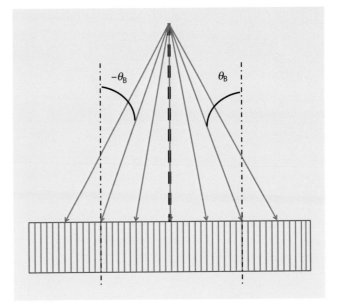

◻ **Fig. 29.2** Wide field ("low magnification") image scanning produces sufficiently large changes in the scan incidence angle to pass through the Bragg conditions for the particular set of crystal planes. Note that the Bragg condition is satisfied in positive and negative going angles, giving rise to two sharp changes in contrast

the crystal planes. A large area image created at $E_0 = 15$ keV by scanning a flat, topographically featureless silicon crystal—prepared with the surface nearly parallel to the (001) plane—is shown in ◻ Fig. 29.3. The image consists of a pattern of bands, each running parallel to a particular crystal plane, creating a so-called electron channeling pattern (Coates 1967). The channeling effect appears as this series of prominent bands that span the crystal because the intersection of the crystal planes with the surface defines a linear trace, and the Bragg condition is satisfied along lines parallel to this trace where the scan angle relative to the planes equals the Bragg angle, $\pm \theta_B$. "Higher order" Bragg angles ($n = 2, 3, 4$, etc., with the specific integers that appear depending on "allowed reflections") are satisfied as a series of progressively fainter lines parallel to the central bands, which can be seen in ◻ Fig. 29.3 (b) for the family of {220} type bands. Note that it is appropriate to apply both a dimensional marker and an angular marker to this image. When the scanned area is decreased, that is, the magnification is increased, the range of the angular scan is reduced, and consequently, the bands on the display appear to widen and the angular extent of the electron channeling pattern decreases, as shown in ◻ Fig. 29.4a, b. If the magnification is made high enough and the angular range is sufficiently reduced, the beam will scan through such a small angle that the observer will see the center of the pattern swell to eventually fill the image with a single gray level. Effectively, the crystal presents a single atomic density that restricts the channeling effect to a single intensity value.

◻ **Table 29.1** Bragg angles for Si with $E_0 = 15$ keV

Planes (hkl)	Spacing (nm)	θ_B (degrees)
111	0.314	0.908
220	0.192	1.48
311	0.164	1.74
400	0.136	2.10
422	0.111	2.57
511	0.105	2.72

29

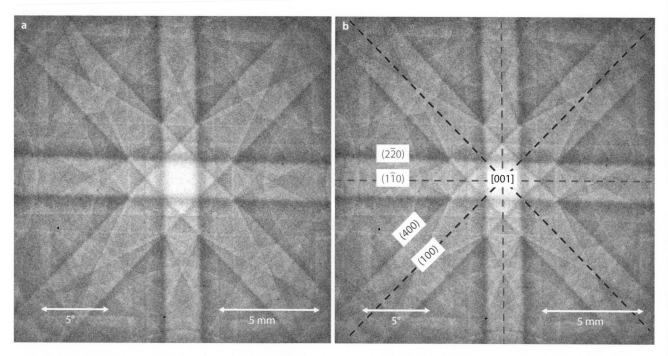

■ **Fig. 29.3   a** Wide field scanning (BSE image) of a Si single crystal wafer whose surface is parallel to the (001) plane, thus looking along the [001] pole; $E_0 = 15$ keV. **b** The traces of two different sets of crystal planes are marked, as well as the parallel channeling bands defined when the scan angle to the planes equals $\pm\theta_B$. Images processed with ImageJ-Fiji CLAHE function

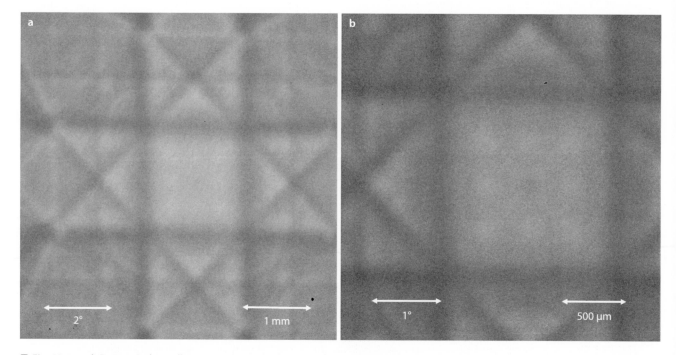

■ **Fig. 29.4   a, b** Progressively smaller scanning areas (higher magnifications) restrict the angular range and thus the portion of the electron channeling pattern that is observed. Images processed with ImageJ-Fiji CLAHE function

29.1.2   Polycrystalline Materials

Most crystalline materials are not single crystals. During solidification from a molten state, numerous crystals nucleate randomly in the liquid and grow in size until they encounter each other, producing a three-dimensional microstructure of randomly oriented crystalline units called "grains." Materials fabrication processes such as

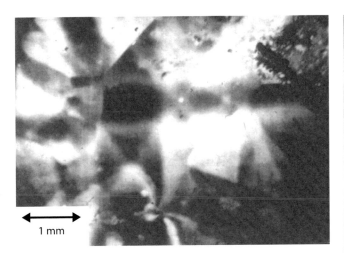

☐ **Fig. 29.5** Electron channeling contrast from a coarse grain size in polycrystalline Fe-3.2Si

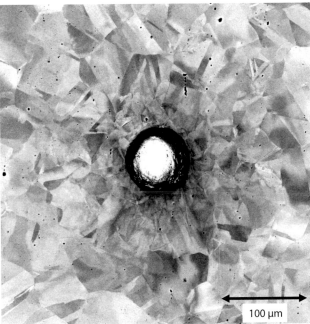

☐ **Fig. 29.7** Electron channeling contrast from grains in polycrystalline 75Fe-25Ni deformed by a diamond scribe impact; $E_0 = 15$ keV; BSE detector

☐ **Fig. 29.6** Electron channeling contrast from grains in polycrystalline 75Fe-25Ni; $E_0 = 15$ keV; BSE detector

rolling, extrusion, forging, etc., and subsequent thermal treatment result further modify the crystalline microstructure, resulting in grain sizes ranging from centimeters to nanometers depending on composition and thermomechanical history and often inducing "preferred orientations" where certain crystal directions will align among a subset of the grains. What happens to the channeling contrast image as the grain size is reduced from the large single crystal case? ☐ Fig. 29.5 shows a coarse polycrystal with grain dimensions in the millimeter range, again viewed with a large scan field at low magnification. The scan angle is nearly the same as for ☐ Fig. 29.3, but the presence of boundaries between the crystals ("grain boundaries") interrupts the channeling pattern so that we can no longer observe enough of the pattern to determine the orientation. We can, however, detect the grains themselves because of the pattern discontinuities. In ☐ Fig. 29.6, the grain size has been further reduced, so that the magnification must be increased so that the angular change across each grain is now very small. The grains in this 75Fe-25Ni alloy are seen with nearly uniform shades of gray because each grain orientation effectively provides an easy, intermediate, or hard orientation for channeling resulting in low, intermediate, or high BSE emission. How far down in scale can crystalline features be observed? With special high resolution SEMs and energy selecting BSE detectors, the angular distortions introduced by the strain fields of individual crystal dislocations have been seen (Morin et al. 1979; Kamaladasa and Picard 2010).

Finally, the long-range effects of plastic deformation, which introduces defects and residual stress patterns into ductile metals, can be directly observed in channeling contrast images. ☐ Figure 29.7 shows the channeling contrast from the damaged area around a large diamond stylus hardness indent in the same 75Fe-25Ni alloy specimen shown in ☐ Fig. 29.6. A similar effect around a hardness indentation in polycrystalline nickel is shown in ☐ Fig. 29.8, revealing the extent of plastic deformation around the indent.

29

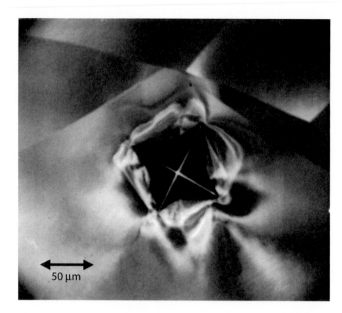

50 µm

◻ **Fig. 29.8** Electron channeling contrast from grains in polycrystalline Ni deformed by a diamond indentation placed in a single grain; $E_0 = 20$ keV; BSE detector

29.1.3 Conditions for Detecting Electron Channeling Contrast

Specimen Preparation

Channeling contrast effects are created in a shallow near-surface layer that is 10–100 nm in depth, depending on the incident beam energy and the material, where the incident beam has not yet undergone sufficient elastic scatter to destroy the initial beam collimation. Thus, the condition of the sample surface is of extreme importance for a successful channeling contrast measurement. The low level of channeling contrast, 2–5 %, also requires eliminating other competing sources of contrast, especially surface topography. Thus, a typical sample preparation will involve grinding and polishing to achieve a flat, topography-free surface. However, the inevitable surface damage layer produced in many materials, especially metals, by mechanical abrasion, including the "Beilby layer" that remains after the final polishing with the finest scale abrasives, must be removed by chemical or electrochemical polishing or low energy ion beam sputtering. (Note that high energy ion sputtering can produce subsurface damage that can destroy electron channeling contrast.)

Instrument Conditions

Detecting channeling effects that produce weak contrast in the range 2–5 % in an SEM image requires careful choice of operating conditions. The low contrast creates a requirement for a high threshold current, so that typically a beam current of 10 nA or more is needed when scanning at rapid frame rates to find features of interest. After such features have been located, a smaller beam diameter can be used to improve resolution at the inevitable cost of a lower beam current, which then requires a longer frame time to maintain contrast

visibility. Because a small convergence angle is also desirable to maximize channeling contrast, a high electron source brightness is important to maximize the beam current into a focused beam of useful size. A high beam energy in the range from 10 to 30 keV is desirable both to increase source brightness and to create high energy BSEs for efficient detection. Channeling contrast is carried by the high energy fraction of the backscattered electrons, and to maximize that signal, a large solid angle BSE detector (solid state or passive scintillator) is the best choice with the specimen set normal to the beam. The positively biased Everhart–Thornley (E–T) detector, with its high efficiency for BSEs through capture of the SE_2 and SE_3 signals, is also satisfactory. Because of the weak contrast, the live image display must be enhanced by careful adjustment of the "black level" and amplifier settings to spread the channeling contrast over the full black-to-white range of the final display to render channeling contrast visible. Post-collection image processing by various tools, e.g., CLAHE in ImageJ-Fiji, can be very effective at recovering fine scale details.

Crystallographic contrast by electron channeling provides images of the crystallographic microstructure of materials. For many applications in materials science, it is also important to measure the actual orientation of the microstructure on a local basis with high spatial resolution of 1 micrometer laterally or even finer. The technique of electron backscatter diffraction (EBSD) patterns provides the ideal complement to channeling contrast microscopy.

29.2 Electron Backscatter Diffraction in the Scanning Electron Microscope

An understanding of the crystallography of a material is required to fully describe the structure property relationships that control a material's physical properties. By linking the microstructure to the crystallography of the sample, a full picture of the sample can be developed. The development of electron backscattering diffraction (EBSD) equipment and its placement in commercial tools for phase identification and orientation determination has provided new insights into microstructure, crystallography and materials physical properties.

EBSD in the SEM has been developed for two different purposes. The oldest application of EBSD is for the measurement of texture on a grain by grain basis. Texture determined in this way is called the microtexture of the sample (Randle 2013). An example of this is ◻ Fig. 29.9, where the point-by-point orientation of an assembly of ZnO crystals is shown where the color coding indicates the orientation of the crystal. The other use of EBSD is for the identification or discrimination of micrometer or sub-micrometer crystalline phases (Michael 2000; Dingley and Wright 2009). ◻ Figure 29.10 is an example of this, showing a dual-phase steel with both ferrite (body-centered cubic) and austenite (face-centered cubic) present. EBSD can easily discriminate

☐ **Fig. 29.9** Inverse pole figure map of ZnO crystals

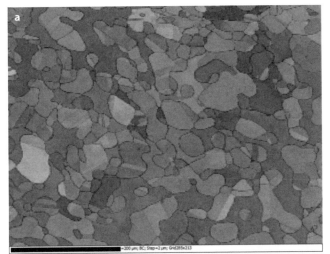

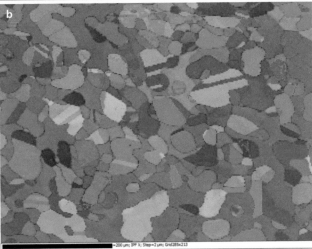

☐ **Fig. 29.10** EBSD of a dual phase steel that contains both austenite and ferrite. EBSD of multiphase samples can discriminate phases with different crystal structures. **a** This is a band contrast map, basically a measure of the pattern sharpness, which accurately reflects the grain structure of the sample. **b** Orientation map of the austenite phase while the ferrite is shown in the underlying band contrast image (Bar = 200 μm)

chemistry through energy dispersive spectrometry and the crystallography of the sample by electron channeling contrast imaging and EBSD. EBSD techniques have recently been developed that allow both the elastic and plastic strains present in a microstructure to be determined and this has been called high resolution EBSD or HREBSD. A more useful description would be high angular resolution EBSD. This technique is beyond the scope of this chapter (Wilkinson et al. 2006).

Microtexture is a term that means a population of individual orientations that are usually related to some feature of the sample microstructure. A simple example of this is the relationship of the individual grain size to grain orientation. The concept of microtexture may also be extended to include the misorientation between grains, often termed the mesotexture. It is now possible using EBSD to collect and analyze thousands of orientations per minute, thus allowing excellent statistics in various distributions. The ability to link texture and microstructure has enabled significant progress in the understanding of recrystallization, grain boundary structure and properties, grain growth and many other important physical phenomena.

The identification of phases in the SEM has usually been obtained by determining the composition of the phase and then inferring the identity of the phase. This technique is subject to the inherent inaccuracies in quantitative analysis in the SEM using EDS or WDS. In addition, this technique is not useful when one is attempting to identify a phase that has multiple crystal structures, but only one composition. A good example of this concern is TiO_2, with three different crystal structures. Another important technological problem is the identification of austenite in ferrite in engineering steels. Austenite has a face centered cubic crystal structure and ferrite is body centered cubic and both phases may have very similar chemistries.

Improved resolution for EBSD has been achieved by utilizing thin samples that allow the transmission and diffraction of electrons using accelerating voltages (20–30 kV) achievable in the SEM. The patterns generated in this way have very similar characteristics to EBSD but are formed in transmission mode and thus, due to the thin sample, have much improved spatial resolution as compared to EBSD performed on bulk samples. Transmission Kikuchi diffraction (TKD) allows the microtexture of ultrafine grained crystalline materials to be studied (Keller and Geiss 2012; Trimby 2012).

In order to best use EBSD, it is helpful if the reader has some knowledge of crystallography and how crystallography is represented. There are a number of excellent books on this subject (McKie and McKie 1986; Rousseau 1998). This module will first describe the origin of the EBSD pattern in the SEM. The detectors or cameras used to detect EBSD patterns will be described. As sample preparation is critical to the success of an EBSD investigation, it is important to understand the methods needed to produce samples of appropriate quality. Finally, details of the actual experiment and the resulting output will be discussed.

between these two crystal structures. Both applications add an important new tool to the SEM. The SEM now has the capability to study the morphology of a sample through either secondary or backscattered electron imaging, the

29

29.2.1 Origin of EBSD Patterns

EBSD patterns are obtained in the SEM by illuminating a highly tilted specimen with a stationary electron beam. The beam electrons interact with the sample and are initially inelastically scattered. There are two current methods that are employed. The first method (conventional EBSD) uses the sample in a mode where backscattered electrons are detected (i.e., from a bulk sample). The second method (often termed TKD or transmission-EBSD although the use of the latter implies the impossibility of simultaneous transmission and backscattering of the electrons) relies on a very thin sample and the patterns are then formed by the transmitted electrons. In the case of the bulk sample and backscattered electron detection, the sample is held at a steep angle with respect to the electron beam. When a thin sample is used for TKD, the sample does not need to be tilted at such a high angle, and in fact the electron beam may be used to illuminate a sample which is not tilted. In either case, backscatter or transmission, some of these inelastically scattered electrons that have lost little of their original energy satisfy the diffraction condition with the crystalline planes within the sample. When this interaction happens near the surface of the sample these electrons escape and form the EBSD pattern that is observed. These backscattered electrons appear to originate from a virtual point below the surface of the specimen. These types of patterns were first described by Kikuchi and are often referred to as Kikuchi patterns (Randle 2013). Some of the backscattered electrons satisfy the Bragg conditions for diffraction ($+\theta$ and $-\theta$) and are diffracted into cones of intensity with a semi-angle of ($90\text{-}\theta$), with the cone axis normal to the diffracting plane. As shown by the Bragg Eq. (29.1), and ◨ Table 29.1, the short wavelength of the electron (at typical accelerating voltages of 10–30 kV used in the SEM) results in a small Bragg angle of less than 2°. Each plane yields two cones of intensity. The cones are quite flat and when they intercept the imaging plane they are imaged as two nearly straight lines separated by an angle of twice the Bragg angle. An alternative but equivalent description is the single event model. In this model, it is argued that the inelastic and elastic scattering events are intimately related and may be thought of as one event. In this case the electron channels out of the sample and forms the EBSD pattern (Winkelmann 2009; Randle 2013). ◨ Figure 29.11 are two examples of an EBSD patterns collected from the mineral hematite at 5 kV and 40 kV. Note that the Kikuchi bands appear as nearly straight lines. The effect of the accelerating voltage is clearly seen. As the accelerating voltage is increased the Bragg angle decreases resulting in more narrowly spaced Kikuchi bands in the patterns. It is also important to note that the patterns are fixed with respect to the crystal orientation and so the positions of the line traces in the patterns have not moved. Fortunately we need not understand the exact physics of EBSD pattern formation in order to use these patterns for crystallographic analysis.

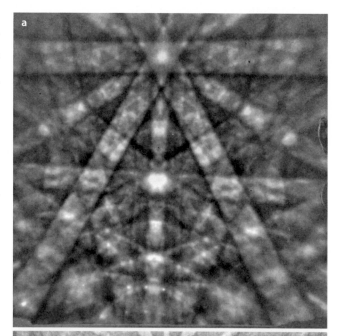

◨ **Fig. 29.11** EBSD patterns acquired from the mineral hematite (*trigonal*) that demonstrate the pattern changes that result from the acceleration voltage change. Note that as expected from the Bragg equation the band widths decrease with increasing accelerating voltage. **a** 5 kV, **b** 40 kV

EBSD patterns consist of what appear to be nearly straight bands (they are actually conic sections) which may have bright or dark centers with respect to the rest of the pattern. These straight bands are the Kikuchi lines discussed previously. The Kikuchi bands intersect in many locations and these are called zone axes which are actual crystallographic directions within the unit cell of the crystal. The angles between the Kikuchi bands and the angles between the zone axes are specific for a given crystal structure. These features can be seen in ◨ Fig. 29.11b where the Kikuchi bands and the zone axes are clearly visible. In high quality patterns as shown in ◘ Fig. 29.11, there are many other features that are visible and may be used to understand the unit cell from which the pattern was collected.

29.2.2 Cameras for EBSD Pattern Detection

It is relatively easy to collect EBSD patterns in the SEM. The earliest EBSD images were collected by directly exposing photographic emulsions inside the specimen chamber of the SEM. As discussed previously, EBSD patterns are obtained from bulk samples by illuminating a highly tilted surface with the electron beam and then collecting the patterns on a position sensitive detector, sometimes referred to as a camera, that is 1–2 cm from the tilted sample surface. Generally the detector surface is located normal to the sample tilt axis so that tilting of the sample is easily achieved but the detector may also be tilted a few tens of degrees away from the horizontal position. The first electronic capture of EBSD patterns utilized low-light video rate cameras that produced useable but somewhat noisy images requiring the use of very high beam currents in the SEM. Current EBSD cameras consist of a fluorescent screen, either circular or rectangular, that is about 2 cm in diameter if circular or 2 cm on a side if square or rectangular formats are utilized. The screen is coated with a thin Al coating to avoid charging and to exclude light. The fluorescent screen is kept thin so that it can be imaged from the opposite side using either a fiber optic bundle or a lens to transfer the image on the phosphor to a charge coupled device (CCD) or a CMOS type solid state imager. Camera designs strive to minimize the loss of light from the fluorescent screen to increase the speed at which patterns can be collected or to increase the quality of the detected patterns. The transfer optics must also be designed to minimize any optical distortions to the collected patterns (Schwarzer et al. 2009).

The imager should have a sufficient number of pixel elements so that the angular resolution is adequate for detection. In addition, it is extremely useful to be able to bin the pixels in the detector. Binning pixels simply means that adjacent pixels are summed together (normally in specific patterns like 2 × 2 or 4 × 4) to increase the signal but with a decrease in pattern resolution. Currently, detectors are capable of collecting more than 1000 patterns per second when heavily binned and with sufficient electron beam current.

The entire detector must be mounted on a precision retractable stage. The precision retractable stage allows the detector assembly to be moved into the same position each time the detector is inserted to maintain the geometrical calibration of the system. It is important to position the detector close to the sample, generally the sample to detector distance should be on the order of or even slightly less than the fluorescent screen diameter so that a large portion of the EBSD pattern may be collected. The insertion mechanism is motorized to allow the camera to be moved into position in a matter of few seconds. A typical arrangement of the sample and

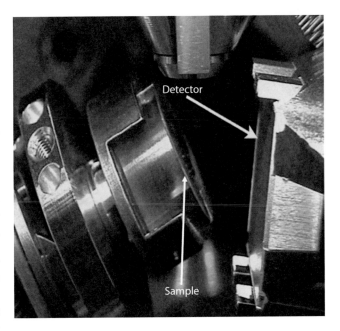

◘ **Fig. 29.12**   An image from the inside of an SEM sample region with the sample tilted for EBSD and the EBSD camera or detector inserted. Also note at the bottom of the detector screen there are small rectangular solid state electron detectors for imaging the sample while it is highly tilted. These detectors result in remarkable crystallographic contrast

the EBSD camera or detector is shown in ◘ Fig. 29.12. Also visible at the bottom of the EBSD detector are the forescattered electron detectors. These solid state detectors provide excellent crystallographic contrast when the sample is highly tilted and standard backscattered or secondary electron imaging are not optimal.

29.2.3 EBSD Spatial Resolution

It is important to understand the volume from which the diffraction pattern is generated in the SEM. In many applications the highest resolution is not required to utilize EBSD for the mapping of the texture of a microstructure. However, there are applications where high spatial resolution is required. Examples are the mapping of deformed microstructures or the study of fine grained materials.

The formation of EBSD patterns depends on the scattering and subsequent diffraction of electrons. Bragg's Law describes the diffraction of specific wavelengths of electrons from the planes in a crystalline solid. If there is a large spread in the energy of the electrons that are exiting the sample surface we would not expect to see the lines that we see in our EBSD patterns due to diffraction. These lines are in nearly the exactly correct position as described by the Bragg diffraction of the electrons based on the beam voltage used to produce the patterns. From this we must conclude that the electrons leaving the surface are of nearly the same energy. Electrons that have lost larger amounts of energy contribute to the background intensity and do not contribute to the diffraction features that we want to observe (Deal et al. 2008).

29

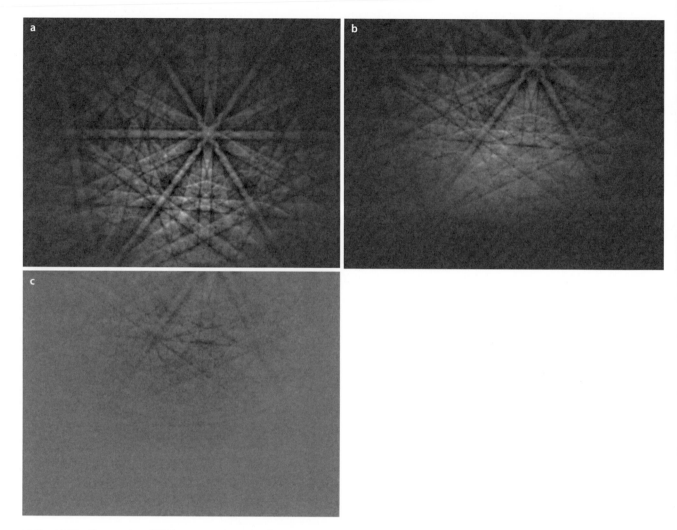

Fig. 29.13   These series of EBSD patterns show the effect of sample tilt on the EBSD pattern quality. **a** 60°, **b** 50°, and **c** 40°

For EBSD, we tilt the sample at a high angle with respect to the electron beam. This tilt is important to obtaining the high quality patterns that are needed for EBSD. As shown in ☐ Fig. 29.13, the tilt angle has a large effect on the pattern quality. The best pattern qualities are obtained at higher tilt angles with some optimal condition between the pattern quality and the ease of imaging of the highly tilted sample. ☐ Figure 29.14a shows the backscattered electron energy distribution that is developed from a highly tilted sample (70° tilt) compared to 0° tilt. Note that the energy is highly peaked toward the operating voltage of the microscope at 70° tilt when compared to the same distribution from an electron beam that is normal to the sample surface. Further, since the EBSD pattern is clearly visible, they must form from only the electrons that have lost either no or a small amount of their energy. The electrons that have lost only a small amount of their energy are emitted from the sample in regions that are very close to the original beam foot print on the surface of the sample. This is shown in ☐ Fig. 29.14b, which is a Monte Carlo simulation of the distance from the initial beam impact point that the electrons emerge from the sample surface for all the backscattered electrons, those that lost up to 2 kV and

those that lost only 0.2 kV. The pattern is formed from these low loss electrons and thus the resolution is quite good. Thus, although EBSD involves backscattered electrons, it is a technique with much higher spatial resolution than standard backscattered electron imaging. Typically, resolutions of better than 0.1 μm can be achieved for a beam voltage of 20 kV in the transition metals. The electrons that have lost larger amounts of energy will contribute the background intensity in the EBSD pattern. These electrons are the main reason why there is limited contrast in EBSD patterns without some form of background removal either through division or subtraction or some other computational method of removal (Michael and Goehner 1996).

The high tilt angle limits the spatial resolution attainable due to the elongation of the electron beam foot print on the sample surface. The resolution parallel to the tilt axis is much better than the resolution perpendicular to the tilt axis due to the high sample tilt angles used to acquire EBSD patterns. The resolution perpendicular to the tilt axis is related to the resolution parallel to the tilt axis by

$$L_{\mathrm{perp}} = L_{\mathrm{para}}\left(1/\cos\theta\right)$$ (29.3)

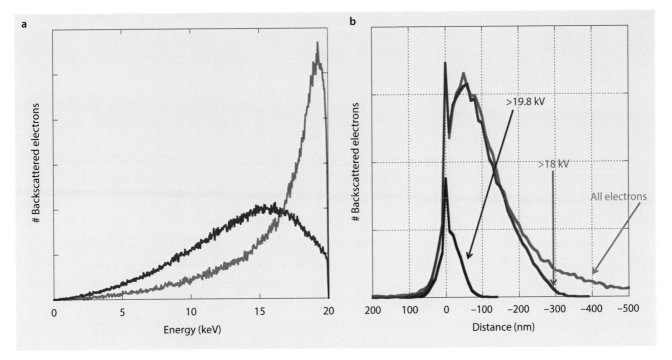

a

b

Fig. 29.14 **a** Monte Carlo electron trajectory simulations of the back-scattered electron distributions for Ni at 20 kV at a tilt of 70° (*red*) and for a sample at 0° tilt (*blue*) that is normal to the electron beam. **b** Backscattered electron distributions from Ni at 20 kV and a sample tilt of 70° for different levels of energy loss as a function of distance from the beam impact point

where θ is the sample tilt with respect to the horizontal, L_{perp} is the resolution perpendicular to the tilt axis, and L_{para} is the resolution parallel to the tilt axis. The resolution perpendicular to the tilt axis is roughly three times the resolution parallel to the tilt axis for a tilt angle of 70° and increases to 5.75 times for a tilt angle of 80°. Thus, it is best to work at the lowest sample tilt angles possible consistent with obtaining good EBSD patterns.

29.2.4 How Does a Modern EBSD System Index Patterns

Modern EBSD systems all use the same basic steps to go from a collected EBSD pattern to indexing and to the determination of the crystal orientation with respect to a reference frame. Once the pattern is collected, the necessary information for indexing of the pattern (indexing refers to assigning a consistent set of crystallographic directions to the pattern with respect to a given or a set of given candidate structures) must be derived from the diffraction pattern. Once the pattern has been indexed correctly (various vendors use different measures of what constitutes correct indexing), it is a simple matter to determine the crystallographic orientation represented by the EBSD pattern.

The most important factor in obtaining a quality orientation is that the system have good quality patterns for indexing. The quality of the detected patterns needed for a given experimental measurement can be influenced by choices made over how the camera is operated. EBSD patterns are of low inherent contrast mainly due to electrons that have lost significant amounts of energy contributing to the overall intensity of the pattern background. To compensate for this, methods of removing the background are utilized. Some systems use a software-generated background to compensate for the background and increase the pattern contrast. A more established approach uses a background image obtained by scanning over a large number of grains that produce the background signal. This background is then used to normalize the raw EBSD pattern to produce a pattern with high contrast (Michael and Goehner 1996).

EBSD cameras are generally able to collect patterns at higher angular resolution per pixel than is needed for most orientation mapping. However, if high accuracy is needed, it is possible to use the full resolution of the camera to produce EBSD patterns. The disadvantage of using the full resolution of the EBSD camera is that longer exposure times are needed to produce usable EBSD patterns. This longer exposure can really slow down map acquisition. It is possible to bin the camera resolution, usually by factors of 2, to produce lower resolution patterns but with the ability to collect the patterns at a much higher collection rate. Thus, if the high angular resolution is not required, the camera should be used at a reduced resolution, (2×2, 4×4, or 8×8 binning) to speed up the acquisition. The tradeoff is between orientation accuracy and speed of the measurement.

Once the camera is set correctly for the given experiment, the patterns should be of high quality with good contrast. The next step is for the software to extract the line positions from the collected pattern. In all modern systems this is accomplished with an algorithm called the Hough Transform, which takes straight lines and transfers them into points

29

which are easier for image analysis software to find. It is possible to adjust the resolution parameters of the Hough transform to give higher accuracy results, but again at the expense of time.

The next step is to compare the angles between the bands in the patterns with a look up table populated with the calculated values of interplanar angles from a candidate phase crystal structure. Once a consistent set of indices are found the pattern is considered to be indexed. No indexing will ever be perfect so the deviation between the calculated pattern and the experimentally measured lines is used as a quality metric that is stored with the data. The orientation of the crystal at each pixel can be calculated with respect to the axes of the microscope and then stored.

It is important to store as much data as possible for each pixel. Sometimes it may be advantageous to store the complete EBSD pattern at each pixel, particularly when one may wish to reanalyze the data using a different match unit to help figure out missing pixels that did not index. The data that should be stored with each pixel at a minimum should include the pixel coordinates, the phase match, representation of the pixel orientation, some sort of pattern quality metric, and the indexing confidence or error. This procedure of pattern acquisition, line position identification, pattern indexing, and data storage is repeated for every pixel in the map area.

29.2.5 Steps in Typical EBSD Measurements

The conduct of an EBSD experiment in order to collect orientation maps or perform phase identification from a crystalline sample requires that the user complete a number of steps, as shown in ◘ Fig. 29.15. These steps consist of the following:

1. Prepare Sample—samples must be free of polishing-induced damage
2. Align sample in the SEM—select the optimal sample tilt and mounting
3. Adjust SEM and set EBSD parameters—select voltage and current and detector settings
4. Check for patterns—make sure patterns are the highest quality consistent with the needed speed
5. Run automated map—start the EBSD map and monitor quality of map in terms of the number of pixels indexed.

Each step in this sequence is necessary to obtaining high-quality EBSD results and will be considered separately below.

Sample Preparation for EBSD

As in all materials characterization using the SEM, the preparation of a sample that is representative of the starting material is critical to the success of an EBSD measurement. Although some samples may not require any sample preparation at all, such as grown thin films on substrates, naturally occurring crystals, and others, most EBSD samples will require preparation. The main requirement of sample preparation for EBSD is that the sample analysis surface is free from

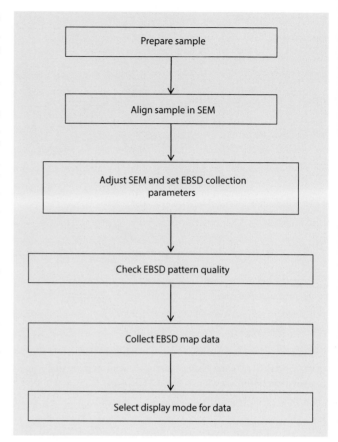

◘ Fig. 29.15 Steps to accomplishing a successful EBSD map. One of the most important steps is sample preparation; without good sample preparation, quality EBSD is difficult to achieve

damage associated with the sectioning and polishing steps. In most cases, this simply implies standard metallographic polishing techniques applied carefully to the sample. Standard preparation techniques include sectioning, mounting grinding and final polishing are the starting point for preparing samples suitable for EBSD. It is the final polishing step that must be conducted in a manner that leaves the surface free of any sectioning damage. As noted previously, like electron channeling the EBSD pattern originates from a very thin surface layer of the sample. Final polishing with colloidal suspensions, electro-polishing, and ion polishing are all methods that have been proven effective for EBSD. Chemical etching is also sometimes useful but should be applied with caution as the surface topography and any oxide films should be avoided for the best EBSD results. It is of course quite useful to assess the microstructure with optical microscopy prior to attempting any EBSD measurements.

Once the sample is prepared and placed in the SEM chamber it is often possible to assess the odds of success before any EBSD measurements are made. One of the most helpful methods is channeling contrast or grain contrast imaging using short working distances and a solid state back scattered electron detector. In this method the sample is moved relatively close to the backscatter detector (4–10-mm working distance depending on the sample and the SEM

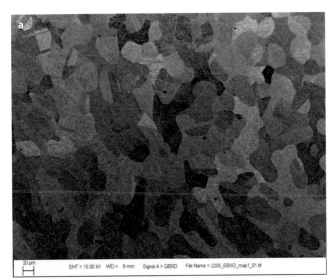

Fig. 29.16 Backscattered electron images of metallographically polished dual phase steel for EBSD analysis. Short working distance backscattered imaging results in excellent crystallographic contrast. **a** Standard metallographic practice shows numerous fine scratches that are visible in channeling contrast. EBSD mapping is still possible but not optimal. **b** Standard practice plus 4 h of vibratory polishing on colloidal silica results in a higher-quality surface finish and optimal EBSD mapping (Bar = 20 μm)

instrumentation particulars) and a backscattered electron image obtained. These images are sensitive to the crystallography of the sample and the grain orientation will determine the backscatter coefficient leading to beautiful images of the sample grain structure. In many cases, subsurface scratches and other sample preparation artifacts will be visible in this type of image that is not seen in secondary electron images or optical microscopy. ◘ Figure 29.16 is an example of a high-quality metallographically polished sample image of a dual phase stainless steel. Note that the sample as polished with no colloidal silica polishing exhibits a large number of subsurface scratches that limit the EBSD results. After colloidal polishing for a few hours the difference in surface condition is readily apparent. Generally, results like these are a good predictor of a successful EBSD analysis.

Align Sample in the SEM

After the sample is imaged and the sample preparation found to be of good quality, the sample orientation within the SEM must be considered. EBSD is often used to determine the texture or distribution of crystallographic orientations with respect to some external frame of reference related to the sample or how the sample was processed. A good example of this is when metal plates are produced by mechanical rolling. It is often desired to determine the texture of the sample with respect to the rolling direction, the transverse direction, or the plate normal direction. Or for the case where only one direction of the sample is expected to show a preferred texture, as in a wire where the preferred crystallographic direction lies along the wire axis, this axis should be mounted parallel to one of the primary directions in the SEM. Other times it is an assessment of the crystallographic growth direction and thus the sample should be carefully aligned so that the growth direction corresponds to one of the primary axes of the scanned image. Thus, the sample must be carefully aligned in the SEM, and one must be aware of the tilt axis of the SEM stage and the directions of the x- and y-stage movements (Britton et al. 2016).

Once the external reference directions have been set parallel to the microscope reference direction, then the sample still must be tilted for EBSD examination. For the best EBSD the sample must be tilted to about 70° with respect to the electron beam. This tilt is best performed with a view of the inside of the chamber with a video image so that the high tilt does not result in the sample contacting the microscope pole piece or other hard expensive components of the microscope. Once the sample is tilted to the appropriate angle and the area of interest brought under the electron beam then the EBSD camera must be moved into position. The best camera position is close to the sample so that the camera captures about 60–90° of the pattern. Also the brightest part of the pattern should be located about one third of the total camera area down from the top of the active area. This position will give the best possibility of good patterns being obtained.

Check for EBSD Patterns

Once the sample has been prepared appropriately and positioned correctly within the SEM, it is sensible to have a quick collection of a few randomly located points to ensure that good patterns can be obtained. This is the opportunity to optimize the SEM parameters for the acquisition. This is also an excellent opportunity to determine that the phases selected for indexing are chosen correctly. Correct completion of these steps will go a long way to ensure that a quality orientation map can be collected.

The first step is to select correct microscope parameters for EBSD acquisition. The accelerating voltage is not extremely critical for success but should be set somewhere between 10 and 20 kV depending on the sample that is to be analyzed. The other microscope-critical setting is the current in the electron beam. In general, the best choice is the highest beam current that the microscope can generate consistent with the beam's size not being larger than the features of interest within the sample. Modern EBSD cameras have been developed to be more sensitive and capable of dealing with lower beam currents. This in practice means that one should try settings that generate 1–2 nA as a starting point. Higher beam currents will be desirable for faster EBSD map acquisition. Once these conditions are set, a few patterns should be collected. If patterns are of poor quality it may be necessary to revisit the sample preparation steps used and modify them in order to reduce the surface damage. Although some fine tuning of the microscope operating conditions may improve the pattern quality, microscope operating conditions cannot make up for poor quality patterns due to inadequate sample preparation.

If good quality patterns are obtained, it now an excellent time to determine that appropriate match units have been selected from a database of crystallographic structures. For the highest-quality EBSD work, one should try and calibrate the EBSD camera geometry for each experiment. Modern systems will use one of the known phases from the sample (if you have a known phase) and the crystallographic data file selected to try and index one or multiple diffraction patterns. Once a match has been found, the software will attempt to vary the calibration parameters to select the parameters that give the best fit to the observed EBSD pattern. Once this is done one can assume the system is adequately calibrated. It is recommended though that multiple points be tried to make sure calibration is optimal.

As discussed, EBSD is not a method that is routinely used to determine the crystal structure of the sample, although there has been work in this area, but it is a method that requires suitable match units to successfully index the EBSD patterns. Excellent quality EBSD patterns will not be indexed if an incorrect candidate unit cell has been selected. One should attempt to index a few of the patterns obtained from the sample with the selected candidate unit cells. If indexing is not possible, then it may be necessary to change the candidate unit cell and attempt indexing. In the case of a multiphase sample it is important to collect patterns from all of the phases and insure that they can be indexed from the list of candidate unit cells.

Adjust SEM and Select EBSD Map Parameters

The analysis is now at the point where the quality of the EBSD patterns and therefore the sample preparation has been assessed and found to be adequate. The sample has been correctly positioned within the microscope with respect to collecting data that is referenced to the sample and the microscope operating conditions have been optimized. The EBSD collection is nearly ready to proceed with a few other details.

First, due to the high tilt of the sample it will be apparent at lower magnifications that the areas of the sample away from the focus point will be out of focus. The best way to

deal with this issue is the use of dynamic focus as discussed elsewhere in this book. In brief, the dynamic focus adjusts the focus so that the electron beam focus is changed with respect to the position of the electron beam on the sample. This is of most importance for low magnification detailed scans but may still provide some advantages at higher magnifications as well. Once the dynamic focus has been properly set up, one can collect the desired SEM images of the sample.

EBSD is useful for a range of samples that include electrical conductors, like metals through electrically insulating oxide. For insulating samples or metal samples mounted for metallography, methods must be employed to reduce charging of the sample. Sample charging is an important issue as charging in the extreme can result in pattern distortions to the point that the patterns are no longer useful and if only a minor effect sample drift may result causing the resulting images and maps to be distorted. It is sometimes adequate to sputter coat samples to ensure conductivity. However, this coating should be kept as thin as possible and it is best to use metal films rather than carbon coating as the conductivity of metal coating are usually much higher. If the sample cannot be coated there is another approach that has been found to be quite useful and that is the use of variable pressure in the sample chamber. The introduction of a low pressure, generally only a few 10–20 Pa, is required to reduce charging effects for EBSD. If a pressure that is too high is utilized there will be a noticeable degradation of the pattern quality due to scattering of the electrons between the sample and the phosphor screen. Thus, some trial and error may be required to set the conditions correctly (El-Dasher and Torres 2009).

There is no absolutely correct way to set up EBSD map parameters due to the variety of sample types and the information that may be required. There are a few simple things that can be done to ensure that quality orientation data are collected. One, the most important is to determine the step size that will be used for the EBSD map. If too large of a step size is selected, important details of the microstructure may be missed; and conversely if too fine of a step size is used, the resulting maps may be of high quality but will have taken much longer to acquire than if a larger step size had been selected. One strategy for an unknown sample is to start with a coarse step size so that a rather quick map can be obtained and allow better sampling strategies to be developed. The large step size map will allow the user to determine if the candidate match phase or phases are appropriate, get an estimate of the grain size and to begin to visualize the microstructure of the sample. The large-step-size image will then allow a better selection of step size for more detailed studies. One rule of thumb is that at least 10 points across the diameter of a grain are required for a reasonable assessment of the grain size.

Run the Automated Map

Once all of the previous steps have been carried out it is now time to collect the crystallographic orientations. Once the run is started the software will collect an EBSD pattern at each pixel, detect the bands in the EBSD pattern, calculate the best fit to the band positions using the candidate crystal structures, calculate the unit cell orientation, save the information and move to the next pixel and repeat this series. Even though modern EBSD systems are capable of running 100–1000 patterns each second, larger maps may consist of more than a million pixels and thus require hours or days to collect. A map of 2000 × 1000 pixels taken at a setting that allows 100 patterns to be collected and analyzed each second will require nearly 6 h to complete. Faster acquisition rates are available but at the expense of orientation accuracy or noise. Quite often it is most efficient to run these longer acquisitions overnight when the SEM is not being utilized anyway. These long acquisition times put extra emphasis in the microscope's environment in order that sample drift due to temperature changes or other disturbances are minimized. It is also quite useful to post a note on the operating panel of the SEM so that the next user does not disturb an ongoing acquisition or assume that the microscope has been left in a safe condition with respect to the inserted EBSD cameras.

29.2.6 Display of the Acquired Data

EBSD is somewhat unique in analytical techniques as there are so many ways to present the collected data in meaningful ways. Of course, everyone likes to produce beautiful color maps of microstructures, but in reality it is not just the images that are important but the crystallographic data that is contained in every point in an image that is important. In order to get that data, one must begin to use and understand crystallographic representations of the sample that are not simple images. Although it is beyond the scope of this chapter to describe every possible way that EBSD data can be displayed it is important to at least introduce these and how they might be utilized (Randle and Engler 2000).

Quite often, the first map that EBSD practitioners display is called a band contrast or an image quality map or other names depending on the particular vendor involved. These images are most commonly shown as gray-scale images where the gray level is scaled by some measure of the quality of the pixel by pixel EBSD pattern. The sharper or more defined the pattern is the higher the gray level in the image. These images can be striking representations of the microstructure of the sample and can reveal microstructural details not clearly visible in either light-optical microscope images or secondary or backscattered electron images in the SEM.

One of the most common ways to display EBSD orientation data is with orientation maps. More accurately these are called inverse pole figure maps with respect to a specific

physical direction. In order to understand inverse pole figure maps it is important to understand what are pole figures and inverse pole figures and how to interpret them.

Pole figures are used to answer the question, Where does a particular crystallographic direction or plane fall in space relative to some arbitrary physical sample direction or plane? Pole figures have been used for many years and are very common in the preferred orientation or texture literature. In many cases a single pole figure does not provide sufficient information and additional pole figures of other crystallographic directions are required. A pole figure is simply a stereogram with the axes defined by the external reference frame. It is common for evaporated or deposited thin films to have a specific crystallographic direction parallel to the film growth direction. ◘ Figure 29.17 is a series of pole figures from an evaporated Au thin film. In this example we show the [111] and the [110] pole figures. For the <111> pole figure we note that there is a large number of poles plotted in the center of the pole figure. This shows that many of the pixels in the data set have the <111> direction aligned with the sample normal. There are additional rings also in the <111> pole figure. These are a result of there being more than one <111> type direction in a cubic crystal (in fact there are actually four of these present). In the <110> pole figure we see that there is a number of poles in the center of the pole figure indicating that there are a number of measured pixels with <110> directions parallel to the sample normal direction.

Inverse pole figures are used to answer the question, What crystallographic poles or planes are preferentially parallel or perpendicular to a specific sample direction? We usually again display these with respect to the physical axes of the microscope or the sample as described for the pole figures. For inverse pole figures we plot all of the directions that are pointing in a specific direction of the sample. Inverse pole figures are extremely useful for samples where there are specific axes of the sample that have a preferred crystallographic direction. ◘ Figure 29.18 is an example of inverse pole figures plotted for the same Au-thin film shown in ◘ Fig. 29.17. It is often helpful to show all three orthogonal directions so that the preferred texture can be visualized. Note that the inverse pole figures of ◘ Fig. 29.18 show exactly the same information that is shown in ◘ Fig. 29.17. It is sometimes helpful to first plot the inverse pole figures as they will give an indication of the samples texture without plotting pole figures of many different directions. The Z or surface normal direction of the inverse pole figures of ◘ Fig. 29.18 is the one that carries the most information about the sample. Here we see the high density of pixel orientations that are clustered around the <111> and less so around the <110> directions.

Once we are familiar with the concepts of pole figures and inverse pole figures, we can then go ahead and understand inverse pole figure maps which are one of the more common ways that EBSD orientation data is shown. An inverse pole figure map extends the idea of an inverse pole

29

◘ **Fig. 29.17** Pole figures for an evaporated Au thin film constructed from EBSD data. In the <110> pole figure all of the <110> directions are shown with respect to the growth direction. Note that there is some intensity in the center of the pole figure as a result of some of the grains having a <110> pole parallel to the sample normal direction. In the <111> pole figure note that there is large number of <111> pole plotted in the center of the pole figure. This demonstrates that the majority of the grains have a <111> direction parallel to the sample surface normal

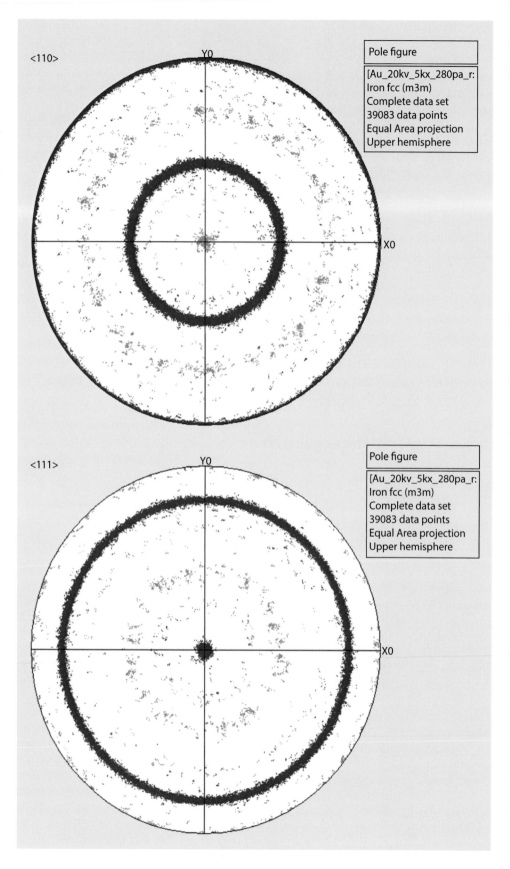

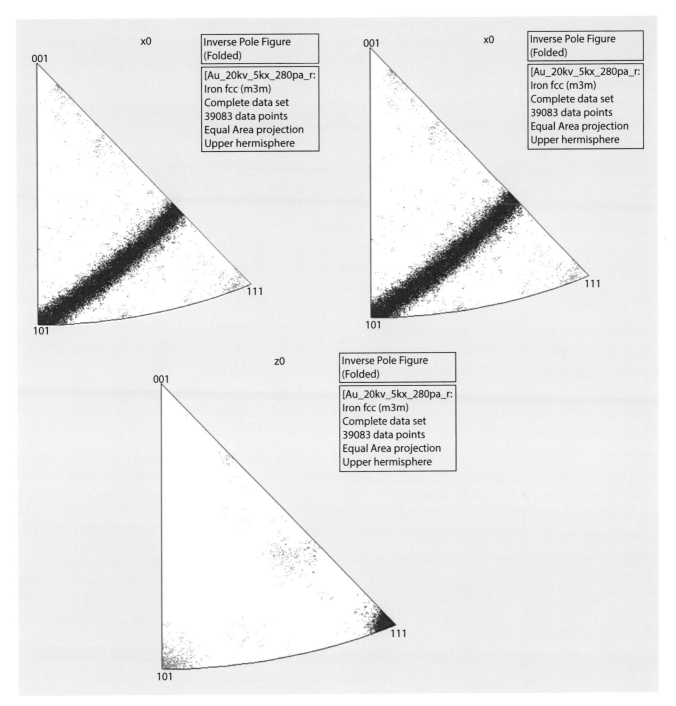

Fig. 29.18 Inverse pole figures that show the same data as the pole figures in ■ Fig. 29.9. The three directions X, Y, and Z (parallel to the samples surface normal) are shown. The high density of poles plotted near the <111> apex of the stereographic triangle indicates that many of the measured pixels had <111> parallel to the sample surface normal direction. The X and Y pole figures show where the other <111> poles plot. The smaller density of poles near the <110> apex of the triangle indicates that there are a few pixels with a <110> direction parallel to the sample surface normal

figure to an image. In order to do this we must first assign a color to each direction in the inverse pole figure stereographic triangle. A common way this is done is shown in ■ Fig. 29.19. Inverse pole figure maps are then plotted by mapping the orientation of each pixel onto the color key shown in ■ Fig. 29.19. ■ Figure 29.20 is an inverse pole figure map produced from the same data that is shown as pole figures (■ Fig. 29.17) or inverse pole figures (■ Fig. 29.18). When inverse pole figure maps are displayed it is important to always include the reference direction, otherwise it is difficult or impossible for the viewer to make sense of the information shown. ■ Figure 29.20 is an inverse pole figure map with respect to the surface normal direction or the Z direction. Thus, the map is mostly blue, indicating that most of the

29

grains have an <111> direction normal to the sample surface. There are also some areas that are green in the map and these areas have an <110> direction parallel to the surface normal. The inverse pole figure map with respect to the X direction looks totally different to the Z inverse pole figure map. Both of these are shown in ◘ Fig. 29.20.

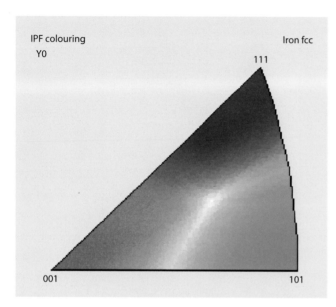

◘ **Fig. 29.19**   Typical color key used for inverse pole figure maps. This color key is used to color each pixel in an image to produce an inverse pole figure map or image. Thus, if a pixel has a <001> direction parallel to a specific direction then we use this inverse pole figure key to plot that pixel as red. Or conversely, if we observe a red pixel in an orientation map we know the orientation of that pixel is close to <100> parallel to the plotted direction

29.2.7   Other Map Components

Once the orientations of the pixels in an array are known, it is now possible to add additional information to the maps. There are many possible components the can be plotted based on EBSD data. One of the easiest components to add is that of grain boundaries. Grain boundaries are the planes (which intersect the planar surface as lines) that separate two regions of different crystallographic orientations. It is fairly trivial to calculate the change in orientations between two pixels. If we do this for an entire map we can then plot lines where the difference in orientations between two adjacent pixels exceeds a predefined limit. It is typically assumed that a grain boundary is represented by a change in orientation that exceeds 10°. ◘ Figure 29.20 has black lines plotted where the change in orientation exceeds this limit and thus the map shows the size of individual grains even though they are of consistent color due to the grain orientations with respect to the sample normal.

29.2.8   Dangers and Practice of "Cleaning" EBSD Data

Many of the currently available software platforms for EBSD allow users to modify the inverse pole figure maps in order to improve their appearance only (Brewer and Michael 2010). There are always pixels in a map that are either not indexed due to pattern quality or many other reasons or that are mis-indexed due to some sort of symmetry issues or multiple possible solutions to the bands that are found. These cleaning routines normally perform two separate operations. The first step is to remove the mis-indexed pixels. These pixels usually

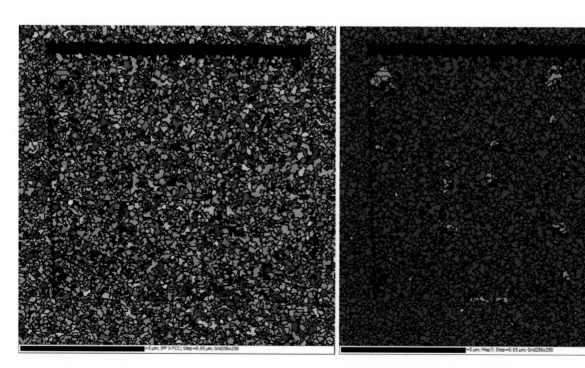

◘ **Fig. 29.20**   Inverse pole figure maps from the X-direction (*left*) and the Z-direction (*right*). This is the same data that is shown in ◘ Figs. 29.9 and 29.10. Inverse pole figure maps show the pixel-by-pixel or spatial arrangements of the crystal orientations (Bar = 5 μm)

show up as single pixels surrounded by correctly indexed pixels. The software searches through the data and where there is a single pixel of a different orientation surrounded by pixels of a different but similar orientation, the single mis-indexed pixel is replaced by the average of the surrounding orientations. The next step is to deal with the pixels that are not indexed. The same procedure is applied as with the single mis-indexed pixels in that kernel math is used. Each individual pixel has six nearest neighbors. If a cleaning procedure that requires six nearest neighbors to agree then the single un-indexed pixel is assigned the average orientation of it's nearest neighbors. These procedures allow the user to select the number of nearest neighbors that must agree and using less than six nearest neighbors allows larger regions of pixels with no correct indexing to be replaced. Use of these procedures can be very dangerous and must be disclosed when inverse pole figure maps are published.

29.2.9 Transmission Kikuchi Diffraction in the SEM

One of the limitations of EBSD is that the resolution is compromised by the fact that the patterns are formed by backscattered electrons and that the sample is highly tilted leading to the previously mentioned fact that the resolution perpendicular to the tilt axis is much worse than that parallel to the tilt axis. The resolution can be greatly improved if the backscattered volume is reduced and the geometrical factors are reduced. One could image EBSD at reduced voltages to reduce the interaction volume, but this process has practical limitations related to the need for increased sample preparation quality. Also, improved EBSD cameras would be needed to take advantage of lower voltage operation. Lower voltage operation does nothing to reduce the geometrical effects.

Transmission Kikuchi diffraction (TKD, although some in the literature have referred to this method as t-EBSD, which is

the acronym for transmission electron backscattered diffraction, which is a rather non-physical description due to the inclusion of both transmission and backscattered in the name) is the transmission analog to EBSD is a way to achieve extremely high spatial resolution for crystallographic orientation or phase mapping (Keller and Geiss 2012; Trimby 2012). TKD is realized by using an electron-transparent sample, as in the transmission electron microscope, that is held at normal or near-normal incidence with respect to the electron beam while a standard EBSD camera is placed at the exit surface of the sample, as shown in ◘ Fig. 29.21. Here the electron beam accelerating voltage must be high and the sample must be thin in order for transmission of the electrons to occur. The maximum beam energy typically available in modern SEMs is 30 kV, which requires that the sample must be quite thin. The use of a thin sample limits the size of the beam interaction volume within the sample and immediately improves the spatial resolution of TKD. In addition, the fact that the sample may be oriented normal to the incident electron beam further improves the resolution to the point that 2-nm spatial resolution has been achieved in TKD. It is incredibly fortuitous that the TKD patterns are extremely similar in appearance to EBSD patterns, as shown in ◘ Fig. 29.22, and can be collected with the same cameras and the same analysis software as is used for EBSD. There are two main disadvantages to TKD. First, it may be difficult to produce appropriate thin samples. However, most laboratories will have access to a dual-platform FIB/SEM and thin samples prepared with FIB are perfectly adequate for TKD. The second disadvantage is that when a small pixel size is needed, it is difficult to map larger regions. For example, if a map step size of 4 nm is selected, a 1000×1000 pixel map will only cover $4 \times 4\ \mu m$. However, if orientation mapping of extremely fine grained material is needed, TKD may be the only way to achieve the resolution needed.

A typical TKD map acquired at 30 kV of a thinned sample of polycrystalline Si layers in a semiconductor device is shown in ◘ Fig. 29.23. This map was acquired with a step size

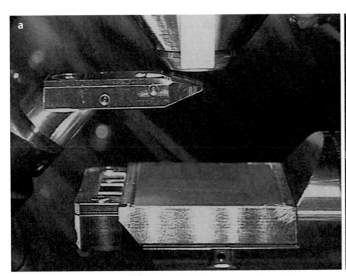
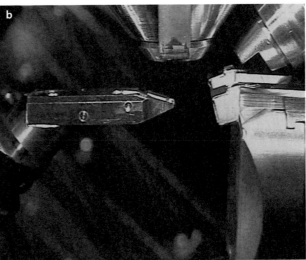

◘ **Fig. 29.21** Detector and sample positioning for transmission Kikuchi diffraction. **a** The detector shown is an on-axis detector. Sample and detector positioned and inserted for TKD with an on-axis detector. Note row of solid state detectors located on the detector for collecting transmission images of the sample. **b** Sample and detector arrangement for TKD with a conventional style EBSD detector

29

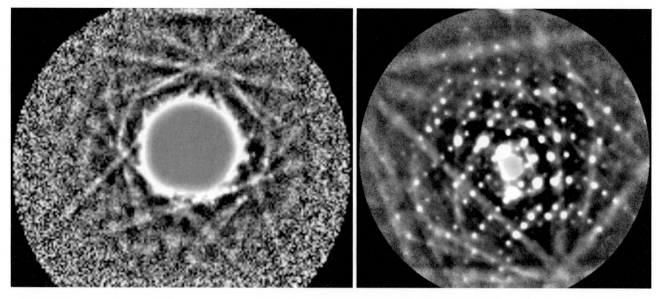

□ **Fig. 29.22** TKD patterns collected at 30 kV from a thin sample of austenite. The imaging conditions and sample thickness result in either typical-appearing Kikuchi patterns, as shown in the left image; or if the sample is very thin, spot patterns can be collected (*right*)

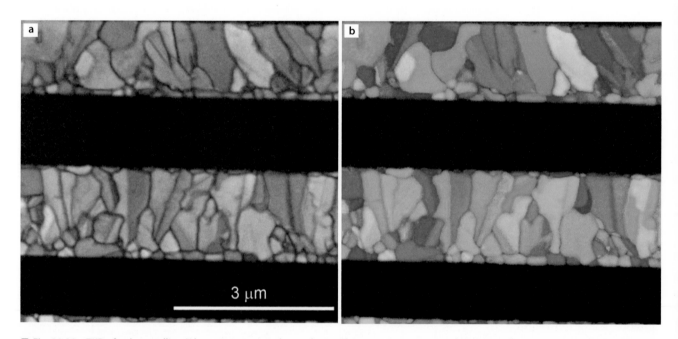

3 μm

□ **Fig. 29.23** TKD of polycrystalline Si layers in a semiconductor device. These maps were acquired at 30 kV with a 6-nm step size. **a** Band quality image of the Si layers. **b** Orientation map with respect to the growth direction of the polycrystalline Si layers

of 6 nm and demonstrates the superior resolution that can be attained with think samples and the TKD method.

29.2.10 Application Example

Application of EBSD To Understand Meteorite Formation

EBSD has found application in many materials studies from ceramics, to semiconductors to metals and alloys. It has also has been applied to understanding metallic meteorites and

their thermal history. One example of this will be illustrated with the Gibeon meteorite. There has been interest in understanding the beautiful Widmanstatten pattern that is seen in these meteorites and how this two phase structure of ferrite (body-centered cubic crystal structure) and austenite (face-centered cubic crystal structure). Previous work had studied the formation of this structure and most of those studies had assumed that at high temperatures in the parent asteroid the meteorite consisted of very large grains of austenite. During the cooling of this meteorite in space over many millions of years the austenite was assumed

Fig. 29.24 **a** EBSD inverse pole figure map with respect to the sample normal direction that is constructed by tiling 220 separate 1 × 1-mm maps. The interesting Widmanstatten pattern can be seen in the large ferrite plates that are running nearly the length of the image. This map contains both indexed austenite and ferrite; although at this scale only the larger ferrite is visible. **b** This is the same area as shown in **Fig. 29.23a**, but now we present the ferrite as a band contrast or a measure of the pattern sharpness and the austenite in the colors of the inverse pole figure map with respect to the sample normal direction. The amazing observation from this EBSD data is that the austenite has the same orientation throughout the large 22 × 10-mm area, which leads to the interpretation that the austenite is retained from the original parent body

10 mm

to have fully transformed to ferrite. On further cooling, the ferrite, super-saturated with Ni, then was thought to decompose to the two phase ferrite plus austenite structure. However, EBSD has shown that this is not be the correct path for the observed microstructural evolution (Goldstein and Michael 2006).

Figure 29.24a is a large-area EBSD map acquired by mapping smaller areas of about 1 × 1 mm and tiling together 220 of these tiles into a large area map that covers 22 × 10 mm with a 3-μm step size. The sample was mechanically polished using standard metallographic practice followed by a few hours of vibratory polishing on colloidal silica. The entire map shown consists of a more than 25 million individual pixels. The general microstructure at a low magnification is clearly visible in **Fig. 29.24a**. **Figure 29.24b** shows a band contrast image of the ferrite and the austenite as an inverse pole figure map with respect to the sample surface normal. Note that all of the austenite has the same or very close to the same orientation, as shown by the austenite all of the same color in the inverse pole figure map. This is an important observation as the austenite could not have formed from precipitation from the ferrite but must be remnants of the original large austenite grains found in the parent meteorite body at elevated temperatures early in the meteorites life. This is further demonstrated by the pole figures shown in **Fig. 29.25**. The austenite pole figures show that there is only a single orientation of austenite in the 22 mm × 10 mm area.

The ferrite pole figures are much more complicated and are a result of the many variants of ferrite that form from a single orientation of austenite.

There are also regions in **Fig. 29.24a** that are very fine grained and difficult to resolve with the 3-μm step size used. Further examination of the microstructure showed that these regions were extremely fine grained and required higher resolution than can be achieved with using bulk EBSD. Due to the small feature size in these areas, TKD is an excellent method to utilize. **Figure 29.26** is a secondary electron image of a focused ion beam produced thin sample. Also shown is a scanning transmission electron image acquired at 30 kV which demonstrates that the sample is sufficiently thin for the transmission of 30 kV electrons. **Figure 29.27** is the resulting TKD map and phase information obtained from the thin sample using an on-axis TKD detector. The step size for this image was 4 nm. It is now clear from these images that the fine-grained regions in **Fig. 29.24a** consist of regions of single crystal austenite that can be seen in **Fig. 29.24b** but also regions of ferrite that have begun to decompose during cooling to the equilibrium austenite plus ferrite that would be expected. The presence of twinned austenite precipitates is somewhat surprising, but may be explained by some of the stress in the sample during transformation.

This example shows how EBSD and TKD may be applied to complex microstructures and how the use of TKD is extremely complementary to EBSD. The visualization of the

29

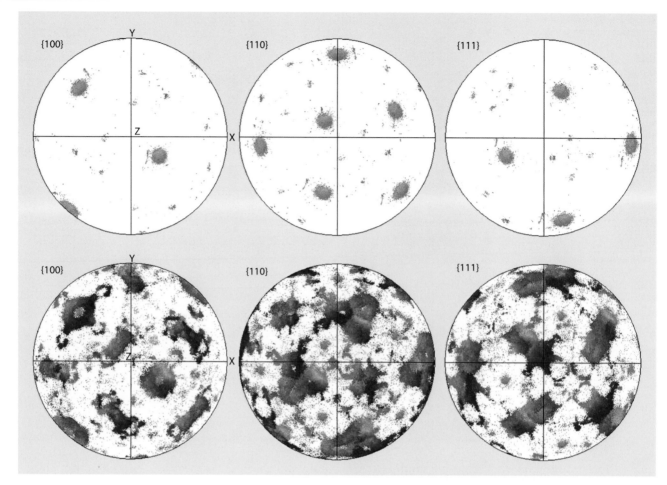

◘ **Fig. 29.25** Pole figures from the austenite (*top*) and the ferrite (*bottom*). In the austenite pole figures the single orientation of the austenite is shown by the arrangements of the poles that are shown. This was also observed from ◘ Fig. 29.23b. The complexity of the ferrite pole figures is due to the many crystallographic variants of ferrite that form from the single orientation of austenite

◘ **Fig. 29.26** A thin sample made for TKD of the fine two phase regions in the Gibeon meteorite. The sample was prepared with conventional FIB followed by low voltage ion FIB milling at 5 kV. **a** Secondary electron SEM image of the thinned sample. **b** Scanning transmission electron image of the sample at 30 kV

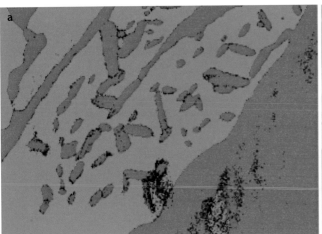
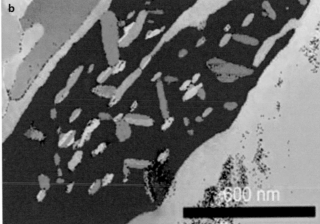

🔲 **Fig. 29.27** TKD phase and orientation maps acquired from the thin sample shown in 🔲 Fig. 29.18. The data was acquired at a beam voltage of 30 kV with a step size of 4 nm. This data was collected using an on-axis TKD detector, as shown in 🔲 Fig. 29.13a. **a** Map with austenite colored red and ferrite green. **b** Orientation map of both the austenite and ferrite phases. Note the fine-scale twinning that occurs in the austenite precipitates

texture of the samples and an understanding of the fine details present in the microstructure were extremely useful in determining the cooling history and the microstructural development in this meteorite. It should also not be lost on the reader that EBSD and the related TKD span a huge range of length scales from collecting large mm sized regions to using extremely small step sizes to elucidate the nm scaledetails of the microstructure.

29.2.11   Summary

EBSD and the related technique of TKD are an important part of crystalline materials characterization in the SEM. With EBSD the SEM can now be considered a complete materials characterization tool that can not only take excellent quality images of samples but can determine the elemental constituents of the sample as well as the detailed crystallography.

29.2.12   Electron Backscatter Diffraction Checklist

Specimen Considerations

EBSD requires a properly prepared sample that is securely attached to an appropriate support as the sample will be tilted to high angles to facilitate EBSD pattern acquisition. The sample surface should be free of artifacts due to sample preparation. Due to the high sample tilt and the generally long acquisition times required for EBSD, carbon tape is not a good choice as it tends to creep allowing the sample to move causing drift related image issues. Samples mounted in epoxy metallographic mounting materials are subject to drift caused

by charging of the polymer material. It is satisfactory to lightly conductively coat the specimen for EBSD analysis provided the coating is kept as thin as possible while remaining adequate for charge removal.

Non-conductive samples may also be lightly coated for analysis as discussed above. It is also possible to utilize the variable pressure mode of operation to reduce charging related artifacts. One must carefully choose the correct pressure to be used as too high of a pressure will result in blurred patterns due to scattering in the gas and too low of a pressure may not control the charging.

Proper positioning of the specimen within the SEM is critical. One must remember that the sample will be tilted to a high angle for analysis. The high tilt required for EBSD will limit how tall the sample is and how it must be mounted on the SEM sample stage. A sample that is too short may also present difficulties in positioning the sample in the proper location.

EBSD Detector

Due to the limited space in most SEM sample chambers, it is best to position the sample so that the sample is tilted appropriately and the area to be analyzed is in the field of view. Once the sample position has been established the EBSD detector should be introduced. The exact location of the detector is not extremely important, but the sample to detector distance should be sufficiently short so that a large solid angle can be obtained and the brightest part of the pattern located in the upper half of the detector image. It is very important to keep in mind the position of the sample and the detector as there are two kinds of people who use EBSD: Those who have hit the EBSD detector and those who will. So be careful when moving the sample and the detector. A chamberscope is absolutely required for safe EBSD operation.

Selection of Candidate Crystallographic Phases

EBSD requires the possible phases that will be analyzed to be selected before the analysis is started. Generally, EBSD is conducted on samples that are already well characterized with respect to the phases that are present. Modern systems are capable of sorting through a large number of phases to match with the experimental patterns, but the operator should try to keep this list to a minimum to allow maximum speed of acquisition. There are many databases that provide crystallographic data and it is also possible for the operator to input specific descriptions of unit cells.

Microscope Operating Conditions and Pattern Optimization

It is difficult to recommend specific operating conditions for EBSD of all samples, but there are starting conditions that should allow the system to be set up efficiently. It is suggested that 20 kV and a beam current of a few nanoamperes is a good starting point for EBSD analysis. The quality of EBSD patterns can be rapidly assessed under these conditions. For faster acquisition higher beam current is always better, as long as the resolution is consistent with the microstructural length scales that are to be studied. Higher operating voltages are also sometimes useful with coated samples, and lower voltages may be used to provide an improved spatial resolution at the expense of acquisition speed. The operator should strive for clear patterns. In most commercial systems, the operator has a choice of the pattern resolution that is to be collected. For many orientation studies, the largest number of pixels is almost never needed and the EBSD detector is binned to produce larger pixels. For example, a typical EBSD detector may have a maximum pixel resolution of 1600×1200, but one would not use the full resolution and would select to bin the result; so, for example, 4×4 binning would result in an EBSD pattern with 400×300 pixels. Binning helps with pattern quality as larger pixels collect more signal and thus increasing the S/N of the pattern. Binning of the detector allows higher speed acquisitions to be achieved. Additional increases in pattern quality can be achieved at the expense of collection speed.

Once the detector settings have been determined it is also necessary to select the background removal method. Modern EBSD systems have many methods for background removal while older systems will be limited. Correct background correction is important to maximize the signal content of the patterns while suppressing the high background contribution that is always present. For polycrystalline samples, it is easy to scan a large representative region of the sample which collects the average background levels without the sharp diffraction features. Other methods that utilize a software blurring algorithm may also be utilized and can be better than the collected back ground method.

Now that the sample and the detector position are set and beam conditions that provide useful EBSD patterns are established, it is now time to calibrate the system. Calibration on modern systems is entirely automatic provided a suitable match unit has been specified. It is important that once a calibration has been established that the sample to detector geometry not be altered or a new calibration will need to be determined.

Selection of EBSD Acquisition Parameters

Successful orientation mapping will depend on careful selection of the mapping parameters and the most important of these is specifying the step size or the spacing between individual measurement points. A selection of a spacing that is too large risks missing the important microstructural features and a spacing that is too small will require longer acquisition times with little gain in information. A good starting point is to plan on between four and ten pixels or measurements across the smallest features to be studied. This sampling will provide quality images and data while not wasting time acquiring redundant information.

At this time it is useful to acquire electron images of the region of interest. Secondary electron imaging may show some surface features but imaging of highly polished samples may not provide useful information. The use of forescattered detectors is recommended as there is a good signal level and forescattered images often show surprising high grain contrast.

Collect the Orientation Map

Now that the experimental conditions for the EBSD acquisition have been selected it is often best to collect a small map to determine if the parameters selected are capable of producing a quality result. One of the most common ways to judge a quality result is to look at the number or the fraction of pixels in the map that have been successfully indexed to a certain level of confidence. Modern systems on well-polished samples can be capable of indexing 95 % or more of the pixels. Of course second phase fractions and grain size can influence the number of indexed patterns. It is not always necessary to have 95 % of the pixels indexed to obtain a useful result. If a high fraction of the pixels are not indexed it is important to understand the reasons. If the correct phase or phases have been selected then it may be that the system was not calibrated adequately. If the selected phases are correct and the calibration is correct then it is possible that sample preparation was not optimal or the sample is heavily deformed, leading to the low fraction of indexed pixels. Once a satisfactory indexing rate is achieved in the test map it is now reasonable to select a larger area for orientation mapping and proceed with mapping.

References

Brewer L, Michael J (2010) Risks of 'cleaning' electron backscatter data. Microsc Today 18:10

Britton T, Jiang J, Guo Y, Vilalta-Clemente A, Wallis D, Hansen L, Winkelmann A, Wilkinson A (2016) Tutorial: crystal orientations and EBSD – or which way is up. Mater Charact 117:113

Coates D (1967) Kikuchi-like reflection patterns observed in the scanning electron microscope. Philos Mag 16:1179

Deal A, Hooghan T, Eades A (2008) Energy-filtered electron backscatter diffraction. Ultramicroscopy 108:116

Dingley D, Wright S (2009) Phase identification through symmetry determination in EBSD patterns. In: Schwarz AJ, Kumar M, Adams BL, Field DP (eds) Electron backscatter diffraction in materials science, 2nd edn. Kluwer Academic/Plenum Publishers, New York, p 97

El-Dasher BS, Torres SG (2009) Electron backscatter diffraction in low vacuum conditions. In: Schwarz AJ, Kumar M, Adams BL, Field DP (eds) Electron backscatter diffraction in materials science, 2nd edn. Kluwer Academic/Plenum Publishers, New York

Goldstein J, Michael J (2006) The formation of plessite in meteoritic metal. Meteorit Planet Sci 41:553

Hirsch P, Howie A, Nicholson R, Pashley D, Whelan M (1965) Electron microscopy of thin crystals. Butterworths, London, p 85

Kamaladasa R, Picard Y (2010) Basic principles and application of electron channeling in a scanning electron microscope for dislocation analysis. In: Mendez-Villas A, Diez J (eds) Microscopy: Science, Technology, applications and education. Formatex: Spain

Keller R, Geiss R (2012) Transmission EBSD from 10 nm domains in a scanning electron microscope. J Microsc 245:245

McKie D, McKie C (1986) Essentials of crystallography. Blackwell Scientific Publications, Boston

Michael J (2000) Phase identification using EBSD in the SEM. In: Schwarz AJ, Kumar M, Adams BL (eds) Electron backscatter diffraction in materials science. Kluwer Academic/Plenum Publishers, New York, p 75

Michael J, Goehner R (1996) Phase identification in a scanning electron microscope using backscattered electron Kikuchi patterns. J Res Natl Inst Stand Technol 101:301

Morin P, Pitaval M, Besnard D, Fontaine G (1979) Electron channeling imaging a scanning electron microscopy. Philos Mag 40:511–524

Newbury D, Joy D, Echlin P, Fiori C, Goldstein J (1986) Electron channeling contrast in the SEM. In: Advanced scanning electron microscopy and X-ray microanalysis. Plenum Press, New York, p 87

Randle V (2013) Microtexture determination and its applications, 2nd edn. Maney, London

Randle V, Engler O (2000) Introduction to texture analysis: macrotexture, microtexture and orientation mapping. Gordon and Breach Science Publications, Amsterdam

Rousseau JJ (1998) Basic crystallography. Wiley, New York

Schwarzer R, Field D, Adams B, Kumar M, Schwartz A (2009) Present state of electron backscatter diffraction and prospective developments. In: Schwarz AJ, Kumar M, Adams BL, Field DP (eds) Electron backscatter diffraction in materials science, 2nd edn. Kluwer Academic/Plenum Publishers, New York, p 1

Trimby P (2012) Orientation mapping of nanostructured materials using transmission Kikuchi diffraction in the scanning electron microscope. Ultramicroscopy 120:16

Wilkinson A, Meaden G, Dingley D (2006) High-resolution elastic strain measurement from electron backscatter diffraction patterns: new levels of sensitivity. Mater Sci Technol 22:1271

Winkelmann A (2009) Dynamical simulation of electron backscatter diffraction patterns. In: Electron backscatter diffraction in materials science. Springer US, pp 21–33

Focused Ion Beam Applications in the SEM Laboratory

© Springer Science+Business Media LLC 2018
J. Goldstein et al., *Scanning Electron Microscopy and X-Ray Microanalysis*,
https://doi.org/10.1007/978-1-4939-6676-9_30

30

30.1   Introduction

The use of focused ion beams (FIB) in the field of electron microscopy for the preparation of site specific samples and for imaging has become very common. Site specific sample preparation of cross-section samples is probably the most common use of the focused ion beam tools, although there are uses for imaging with secondary electrons produced by the ion beam. These tools are generally referred to as FIB tools, but this name covers a large range of actual tools. There are single beam FIB tools which consist of the FIB column on a chamber and also the FIB/SEM platforms that include both a FIB column for sample preparation and an SEM column for observing the sample during preparation and for analyzing the sample post-preparation using all of the imaging modalities and analytical tools available on a standard SEM column. A vast majority of the FIB tools presently in use are equipped with liquid metal ion sources (LMIS) and the most common ion species used is Ga. Recent developments have produced plasma sources for high current ion beams. The gas field ion source (GFIS) is discussed in module 31 on helium ion microscopy in this book.

This chapter will first review ion/solid interactions that are important to our use of FIB tools to produce samples that are representative of the original material. This discussion will then be followed by how FIB tools are used for specialized imaging of samples and how they are used to prepare samples for a variety of SEM techniques.

30.2   Ion–Solid Interactions

It is important to understand some of the ion-solid interactions that occur so that the user can appreciate why certain methods and procedures are followed during sample preparation. There are many events that occur when an energetic ion interacts with the atoms in a solid, but for the case of SEM sample preparation and ion imaging we are mainly interested in sputtering, secondary electron production and damage to the sample in terms of ion implantation and loss of crystalline structure. Sputtering is the process that removes atoms from the target. Secondary electron production is important as images formed with secondary electrons induced by ions have some important advantages over electron-induced secondary electron imaging. Finally, it is important to realize that it is impossible to have an ion beam interact with a sample without some form of damage occurring that leaves the sample different than before the ion irradiation.

A schematic diagram of the interactions is shown in ◻ Fig. 30.1. Here an energetic ion is injected into a crystalline sample. The ion enters the sample at position 1. The ion is then deflected by interactions with the atomic nuclei and the electron charges. As the ion moves through the sample it has sufficient energy to knock other atoms off their respective lattice positions as shown at position 2. The target atoms that are knocked off their atomic positions can have enough energy to knock other target atoms off their atomic positions as shown at position 3. Some of the atoms that have been knocked from their atomic positions may reoccupy a lattice position or may end up in interstitial sites. There can also be lattice sites that are not reoccupied by target atoms and are left as vacancies. Both interstitials and vacancies are considered damage to the crystalline structure of the sample as shown in position 4. Most of the time, the original beam ion will end up coming to rest within the sample. This is termed *ion implantation* and is shown at position 5. Ion implantation results in the detection of the ion beam species in the sample. Many of the collision cascades will eventually reach the surface of the sample. Sufficient energy may be imparted to knock an atom from the surface into the vacuum. This process is called sputtering and results in a net loss of material from the sample as shown in position 6. At the same time when the ion is either entering or leaving the sample, secondary electrons are generated that are useful for producing images of the sample surface scanned by the ion beam. It is important to remember that scanning an energetic ion beam over the surface of the sample will always result in some damage to the sample. Understanding the interaction of ions

◻ **Fig. 30.1**   Schematic of some of the important ion–solid interactions that can occur

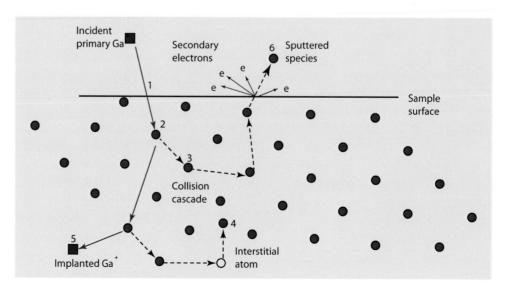

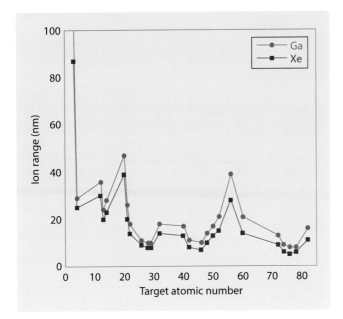

Fig. 30.2 Calculated range for 30-kV Xe and Ga ions as a function of atomic number

with the target material is helpful in reducing the amount of damage to acceptable or tolerable levels through appropriate sample preparation techniques. For a complete review of ion–solid interactions see Nastasi et al. (1996).

In module 1 on electron–beam specimen interactions, the range that an electron travels in a sample is discussed. Generally for medium-energy electrons (15–20 keV) the electron range in the transition elements is on the order of 1 μm. The ranges of heavy ions like Ga and Xe are extremely short when compared to electron ranges for similar energies. ◘ Figure 30.2 is a plot of the ion ranges for Xe and Ga as a function of the atomic number of the target material. Note that the range for either Ga or Xe ions with an energy of 30 keV is generally much less than 50 nm. Thus, ions travel very short distances in solid targets and as a result the near surface region of the sample is where the ion interactions take place and where we expect to see the crystalline sample damage discussed in ◘ Fig. 30.1 (Nastasi et al. 1996; Ziegler and Biersack 1985).

30.3 Focused Ion Beam Systems

Modern focused ion beam tools are almost always two-column systems with a FIB column and an SEM column mounted on one chamber and both columns focused precisely on the same region of the sample. This allows one to use the SEM column to monitor the progress of the FIB milling that is being performed and to image the sample immediately after preparation. This arrangement also enables the use of sequential FIB milling and SEM imaging leading to the capability to produce 3D data sets. There are still some highly specialized uses for single beam FIB tools; for example, integrated circuit modification is done generally with single beam FIB tools. Modern FIB systems utilize either a LMIS

source or a plasma source to produce ion beams with variable current to allow both large volume removal and fine scale polishing of the sample. Both the LMIS and the plasma source have advantages and disadvantages. The most common LMIS produces Ga ions, while it is common for the plasma sources to utilize inert gases like Ne, Ar, or Xe. Ga ion sources have higher brightness as compared to the plasma sources and thus have higher current densities in the focused ion spots. Ga is a fairly reactive element, however, and may result in artifacts when implanted into some materials. Plasma sources have lower brightness but higher overall current than the LMIS sources resulting lower current densities in the focused spot but much higher total current that allows much faster material removal rates when compared to LMIS ion sources. Both the LMIS and the plasma ion source utilize similar optics to produce focused scanning beams of ions. ◘ Figure 30.3 shows a schematic of a typical ion column. This column resembles a simple SEM column with the exception that the magnetic lenses that are used to focus electrons are

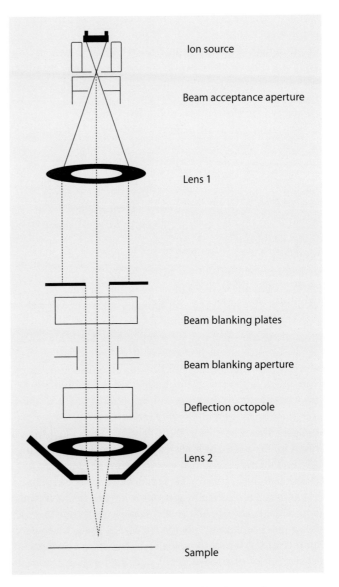

Fig. 30.3 Schematic of a typical focused ion beam column

30

not capable of focusing the heavier ions. For ions, the lenses are electrostatic and require high voltages to focus energetic ions due to their relatively large mass to charge ratio.

Sample stages must be accurate and reproducible for easy and efficient FIB processing of samples. There can be multiple stage moves during the preparation of samples in a FIB tool and each move should be reproducible so that the operator can easily return to the region of interest. Stage accuracy and reproducibility has become more important as automated routines have become common place during the production of samples or during sequential milling and imaging operations (Giannuzzi 2006; Orloff et al. 2003).

30.4 Imaging with Ions

Although FIB is usually used to remove material via sputtering, ions produce a large yield of secondary electrons. The secondary electrons signal can be collected and imaged just like secondary electrons produced with an electron beam, which means we can use all of the detectors that we are very familiar with from electron beam imaging. This leads to the use of the FIB as an imaging tool sometimes referred to as scanning ion microscopy. Secondary electrons produced with ions are referred to as ion induced secondary electrons (iSE). FIB columns are all equipped with one or more iSE detectors with the most common one being the ET detector as used in the SEM. iSE imaging has some advantages over SE imaging in the SEM. First, the ion beam produces many more SEs per incident particle than does a similar current electron beam resulting in a high signal, low noise image. Also, it is interesting to note that the iSE signal collected in a FIB is free of the backscattered electron component that reduces contrast in electron beam induced SE images. iSE imaging of surfaces shows topographical contrast that is familiar to anyone who has operated an SEM. iSE imaging of crystalline samples can produce very striking high contrast grain images due to the higher propensity for ions to channel along specific crystal planes resulting in a varying iSE yield as a function of grain orientation.

During ion imaging of a sample there are other signals produced that are of limited use. The interaction with the sample causes secondary ions to be ejected from the sample. These can be used to form images. However there are few secondary ions produced as compared to the iSE and therefore the signals tend to be noisy requiring either longer scan times or higher beam currents to be used, both of which will result in increased levels of sample damage during imaging.

Resolution of scanning ion microscope images is not just a function of the beam size that is generated by the ion column. The ultimate resolution of a scanning ion image is a convolution of the beam size and some measure of the rate at which the sample is milled. Resolution is worse for materials that mill quickly and better for those that have a slower sputter rate (Orloff et al. 1996, 2003). Just like the SEM, as the current in the probe is increased, the beam size increases. ◘ Figure 30.4 shows spot burns from an LMIS (Ga) column

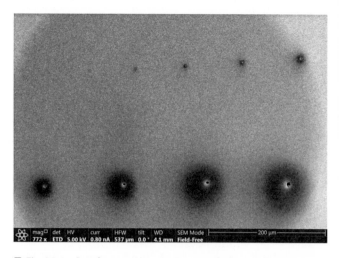

◘ **Fig. 30.4** Spot burns on a tungsten coated silicon wafer. The beam was held stationary at each point for 10 s. *Top row* from *left to right*: 24 pA, 80 pA, 0.23 nA, 0.43 nA, 0.79 nA, and 2.5 nA. *Bottom row* from *left to right*: 9 nA, 23 nA, 47 nA and 65 nA (Bar = 200 µm)

where the ion beam is put in spot mode and left stationary so that the substrate is milled away. The beam size is then some measure of the size of the spot including the halos around the milled area as ion beams suffer significant aberrations as occur with lenses in the SEM. In both the LMIS and the plasma cases, the ion columns are capable of producing symmetric ion beams with the exception that at the larger currents various optical aberrations become dominating and result in a less well defined ion beams.

The collection of secondary ions requires an additional detector that is sensitive to secondary ions and can reject the signal produced by the secondary electrons. These secondary ion detectors are often now optional on modern instruments. ◘ Figure 30.5 is a comparison between secondary electron imaging induced by electrons and secondary ion imaging induced by ion scanning. The sample consists of a tungsten wire and a human hair. The electron image shows good surface detail, but the uncoated hair is charging and the surface information is somewhat obscured. The secondary ion image shows no charging artifacts on the human hair but also shows an increased level of surface detail due to the small escape length of the secondary ions. One must always remember that imaging with ions will always cause some level of damage within the sample (Giannuzzi 2006; Mayer 2007; Michael 2011).

◘ Figure 30.6 is an electron-induced secondary electron (SE) image of free-machining brass coated with layers of Cu, Ni, and Au. Note that free-machining brass contains particles of Pb. ◘ Figure 30.6 shows the contrast that we have come to expect from electron beam induced SE imaging. The higher atomic number regions appear brighter as a result of higher secondary electron yield that results from the SE_2 contributions from the unavoidable backscattered electrons (BSEs). ◘ Figure 30.7 is an iSE image (30 kV Ga) of the identical area of the sample. The various layers are immediately obvious because the ion channeling contrast is quite strong and the grain structure of each layer is revealed. It can also be observed that the Pb region and the Au layer are no longer

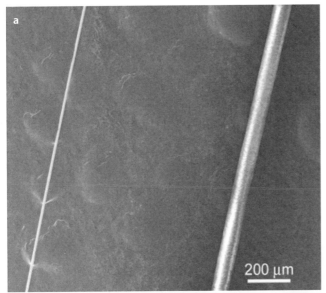

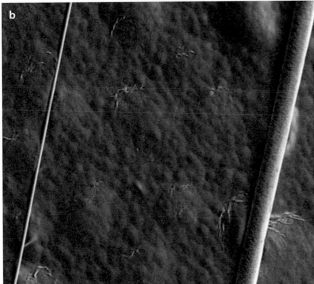

Fig. 30.5 Images of a tungsten wire (*left*) and a human hair (*right*). **a** Secondary electron image induced by 5-kV electrons. **b** Ion induced secondary ion images with 30-kV Ga ions. Note the reduced charging in the ISE image and the increased surface detail visible in the secondary ion image

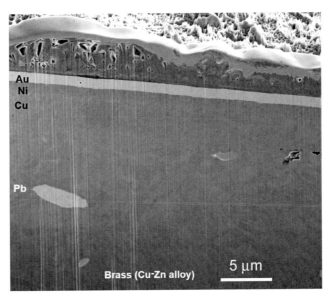

Fig. 30.6 Electron beam (5-kV) induced secondary electron imaging of a free-machining brass (Cu-Zn alloy) that has been coated with layers of Cu, Ni and Au. Note that the various layers are faintly visible and that the contrast is as expected with the highest atomic number region (Pb) brighter than the brass or the Cu and Ni. Au also appears bright

Fig. 30.7 Ion beam (30-kV) induced secondary electron imaging of a free-machining brass (Cu-Zn alloy) that has been coated with layers of Cu, Ni and Au. Note that the various layers are much more easily visualized due to the high contrast crystallographic contrast. Also, note that the brightness of the phases can no longer be interpreted strictly by atomic number. Here the Pb and the Au regions appear with lower signal levels than does the brass or the Cu and Ni layers

bright relative to the brass as in ☐ Fig. 30.6. ☐ Figure 30.7 demonstrates that the iSE yield is not a simple function of the target atomic number (Joy and Michael 2014). Also, ☐ Fig. 30.7 demonstrates the very strong grain contrast that can be observed in many crystalline materials. This grain contrast is due to the way that the crystallography of the sample impacts the penetration depth of the primary beam ions and therefore the iSE yield. Small changes in the ion beam incident angle can change the grain contrast that is observed. Thus, if you are trying to determine the nature of the contrast that is observed in a crystalline sample it is a simple matter to tilt the sample 2–4° and observe how the contrast changes. If the contrast is due to ion channeling then the contrast between grains should change (Giannuzzi and Michael 2013).

30.5 Preparation of Samples for SEM

FIB–based sample preparation for SEM is a large field due to the versatility of modern SEMs and the many techniques that are utilized. One of the most common uses of FIB is for subtractive processing of the sample in that the FIB is used to

30

sputter site specific regions of the sample to produce cross sections. It is also common to produce samples that are manipulated from the bulk for study. One example is the extraction and production of thin samples for transmission Kikuchi diffraction as discussed in module 29 on electron backscatter diffraction. FIB has become an indispensable tool in the production of electron transparent samples for STEM/TEM and now the FIB can produce samples that are sufficiently thin for transmission imaging in the SEM at 30 keV using the STEM-in-SEM technique. FIB sample preparation is applicable to many materials classes that are imaged in the SEM including polymers, metal, semiconductors, and ceramics.

30.5.1   Cross-Section Preparation

In materials characterization it is often of interest to image a section of the sample that is not visible from a planar section. In this case, the FIB is used to mill away material to expose a cross sectional view of the sample, as shown schematically in ◘ Fig. 30.8. There are many ways to accomplish this and in many modern FIB tools this process is fully automated and the user must simply indicate where the cross section is to be made. The following example will show the steps that are needed for a cross section to be produced.

As was discussed earlier, any time the sample is exposed to the ion beam damage will occur. It is necessary to image the sample during preparation so the easiest way to eliminate the damage to the sample from the beam is to place a protective layer over the area to be cross sectioned. FIB tools come

equipped with gas injectors that can be used during sample preparation. Common precursor gasses used can deposit tungsten, platinum, or carbon. Each of these materials works quite well to protect the region of interest from the ion beam. The precursor gasses are delivered to the sample surface through a small needle that is placed in very close proximity to the area of interest. Either the ion beam or the electron beam can be scanned over the area of interest. Some of the gas molecules that are delivered through the needle absorb on to the sample surface where combined action of the primary beam (electron or ion) and the secondary electrons produced by the interaction of the beam with the sample decomposes the absorbed gas. This leaves behind a deposit that contains the desired material but also includes the ion beam species (if ions were used) and some residual organics from the precursor. Due to the short range of ions in materials, the protective layer need not be very thick; but typically most applications use about 1 μm to provide protection and for ease of subsequent milling of the sample. ◘ Figure 30.9 shows a cross section produced in a sulfide copper test coupon where the objective of the experiment was to measure the rate of sulfide growth in accelerated aging conditions. The first step (◘ Fig. 30.9a) is to deposit the protective platinum layer on the area to be sectioned. The goal is to have the completed cross section positioned under the platinum protective layer. A coarse first cut is made with a large current ion beam, the result of which is shown in ◘ Fig. 30.9b. Although the cross section is relatively clean at this point it is not adequate for quality SEM imaging. Further polishing of the cross section is completed with lower current ion beams in this case ◘ Fig. 30.9c was completed with a 1-nA ion beam and ◘ Fig. 30.9d is the cross section after final polishing with a 300-pA ion beam. The total time to produce this cross section is about 20 min. The completed cross section imaged at higher resolution is shown in ◘ Fig. 30.10. Note that all of the important microstructural features are easily observed in this very smooth polished cross section.

FIB prepared surfaces can be very smooth and nearly featureless. The relative brightness between different materials can often provide sufficient image contrast. Materials contrast by itself may be insufficient and a method to enhance the contrast may be needed. One way to do this is to introduce gasses near the sample that react with the ion beam and the sample surface to etch the FIB polished surface. One chemical that does this for semiconductor devices is trifluoracetic acid (TFA) that etches oxides and nitrides preferentially to silicon and metals. ◘ Figure 30.11 demonstrates the use of TFA for enhancing the contrast in FIB milled surfaces. The sample was prepared using standard cross sectioning procedures in the FIB. After cross sectioning the sample was rotated and tilted so that the milled surface was normal to the ion beam. A low beam current is selected and then the gas is introduced through a fine needle while the ion beam is scanned over the milled surface. While milling, the sample image provides an etching progress monitor so that the etching progress can be terminated when the correct degree of etching has been achieved.

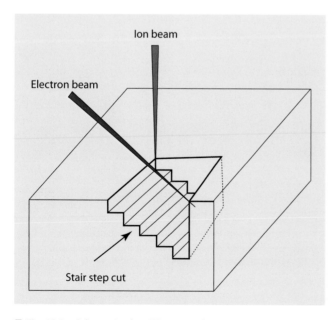

◘ **Fig. 30.8**   Schematic of an FIB-prepared cross section. A stair-step is milled using the ion beam and then the exposed section that is perpendicular to the original sample surface is polished with a series of lower current ion beams. The cross section can be immediately imaged with the SEM beam that is at some inclined angle with respect to the ion beam

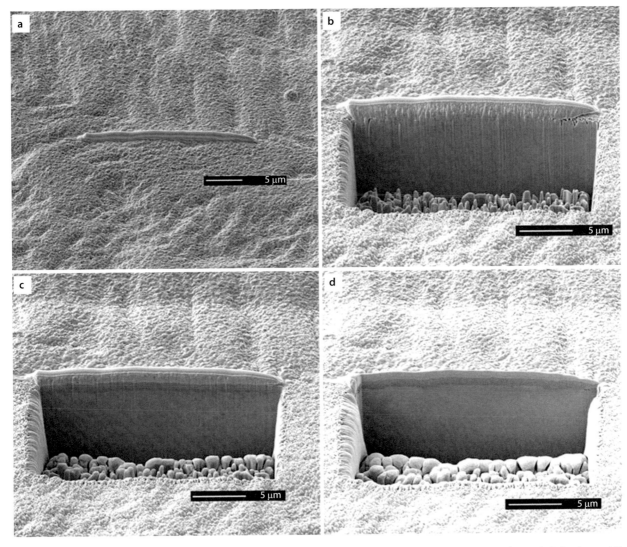

🔲 **Fig. 30.9** Required steps to making a quality cross section of the surface of a corroded copper test coupon in the FIB. **a** Deposition of platinum protective layer. **b** Coarse milling of the cross section with a 7-nA Ga ion beam. **c** Further polishing with a 1-nA Ga ion beam **d** Final polishing with a 300-pA Ga ion beam to produce the completed cross section

🔲 **Fig. 30.10** High resolution image of the sulfide layer on the top of the test coupon sectioned in 🔲 Fig. 30.9. Note the surface smoothness and the feature sizes that can be observed on the ion milled section

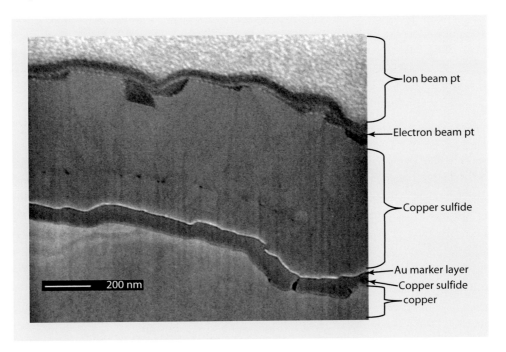

30

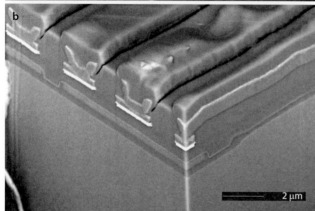

☐ **Fig. 30.11**   This figure demonstrates the enhancement of image contrast using a TFA as a beam assisted etchant. **a** As-milled cross section of a semiconductor device. Note the surface smoothness and the low contrast. **b** Same surface after beam assisted etching with TFA. Note that process adds a small amount of topography to the milled surface allowing the different materials to be more easily imaged

SEM applications of FIB sample preparation can also utilize samples that have been removed from the bulk material. This is generally done when the research requires EDS, EBSD, or STEM-in-SEM to be conducted as each of these techniques will not work optimally with standard cross sections (Prasad et al. 2003). Lift-out samples are made by milling trenches on both sides of the area of interest after a protective layer was deposited. The FIB beam is used to cut the sample free from the bulk material and then it can either be polished in the trench or lifted out to a support grid for subsequent polishing and if needed thinning to an acceptable thickness. Figure 30.12 shows some of the steps required to lift-out a 150-μm-wide sample. This was accomplished with a Xe plasma FIB but the steps are the same for a Ga FIB. *Ex situ* lift-out was used to remove the sample from the bulk followed by attachment to the Cu support structure to allow safe handling of the sample.

Once the sample was attached to the support structure, final sample polishing was performed and the sample was ready for analysis. EBSD results obtained from the sample shown in ☐ Fig. 30.12 are shown in ☐ Fig. 30.13. Note that the sample surface is nearly ideal for EBSD as the number of mis-indexed or not indexed pixels is quite low.

This technique of sample lift-out is applicable to all imaging and analysis modes in the SEM. Once the sample is mounted flat on a surface or a support structure it is in an ideal sample orientation for imaging with secondary electrons of backscattered electrons. Lift-out samples also provide ideal sample orientations for EDS, WDS, or EBSD analysis. This is not true of cross sections that are milled into the bulk and not lifted out as access to the sample for some imaging and analysis techniques is not close to ideal depending on the particular FIB/SEM platform chosen (Giannuzzi 2006).

30.5.2   FIB Sample Preparation for 3D Techniques and Imaging

One of the truly important advances in FIB applications is the ability to automate the FIB operation and coordinate it with the SEM imaging or analysis using EDS or EBSD. This coordination allows the FIB column to be used to mill specified volumes from a sample face and then allow that same slice to be imaged or an analytical technique applied and then the process can be repeated. This is often referred to as serial sectioning. In this way a direct reconstruction (direct tomograph) of a 3D volume can be developed. One must always remember though that in this sort of work there is no ability to go back and start over unless a second suitable region is available. It is also important to remember the length scales that are practically reached with FIB methods. The typical area that can be accessed with LMIS-based FIB columns is 50 μm wide by 10 μm deep. The number of slices is limited by the operator's patience and the stability of the FIB/SEM being utilized.

The first step is to deposit a protective layer over the entire region that is to be milled. This step is necessary, as discussed for single sections, to protect the sample region from damage by the ion beam. The larger the area to be sectioned, the larger the protective layer has to be and this deposition may be quite time consuming. Once the protective layer has been deposited, there are two ways to proceed with serial sectioning and imaging. The easiest method is to simply produce a cross section of the sample, as described previously. The polished face is then used as the starting point of the sectioning. One must be very careful when doing this to ensure that the

◻ Fig. 30.12 Gibeon meteorite sample produced by the lift-out method to produce large area samples suitable for EBSD, TKD, EDS, or STEM imaging. **a** Milled sample ready for ex situ lift-out. **b** Lift-out sample on the end of the micromanipulator needle. **c** Cu support for the sample. It was placed across the large gap in the middle and epoxied in place. **d** Sample after final milling of the surface. **e** Sample arranged on a pre-tilted sample holder for EBSD

30

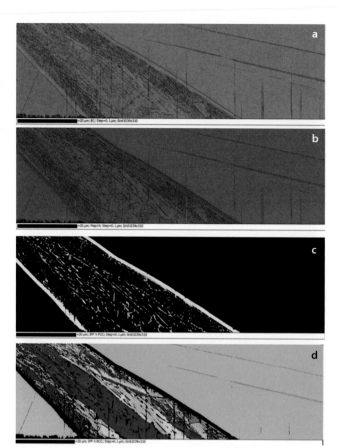

Fig. 30.13   EBSD results of the FIB prepared sample shown in Fig. 30.12. **a** Pattern contrast image demonstrates that the sample has little curtaining. **b** Phase map with ferrite (*BCC*) in red and Austenite (*FCC*) in blue. **c** Orientation map for the Austenite phase **d** Orientation map of the Ferrite phase (Bar = 20 μm)

aligned and stacked followed by the user selecting the planes of interest for image reconstruction.

A much faster method requires the volume of interest to be milled using any means into a cantilever-like beam that is then sliced starting at the free end. This method has numerous advantages over the bulk sample method as there is much easier access to the sample for imaging and analysis. This can also be accomplished by milling a chunk that contains the region of interest from the sample and then mounting the chunk onto a suitable support structure. The chunk then represents the cantilevered beam sample and is sequentially milled from the free side of the sample. This method is faster as much less material needs to be removed for each slice and there is no issue with re-deposition of the sputtered material. Figure 30.15 shows an example of the cantilever beam method for serial sectioning through a tin whisker on a copper substrate. In this case it was important to first protect the whisker with electron beam deposited platinum followed by ion beam deposited platinum. Once the feature of interest is protected from the ion beam, the material around the whisker is removed so that actual sectioning time during the serial sectioning will be minimized. EBSD orientation maps were collected at every slice during serial sectioning. Some commercially available FIB/SEMs require the sample to be repositioned for EBSD and then FIB slicing, while others possess a geometry where the sample does not have to be moved between sectioning and analytical acquisitions. For systems requiring movement between sectioning and EBSD, accurate alignment using fiducial marks is mandatory. Figure 30.16 is a reconstruction of the EBSD maps obtained from the tin whisker shown in Fig. 30.15. This data was acquired with a 200-nm slice thickness and an EBSD step size of 200 nm, leading to a voxel dimension of 200 × 200 × 200 nm. The acquisition required 75 sections that required a total time of 48 h to section and collect the EBSD data. Once this data is obtained and aligned and reconstructed then further examination of the spatial relationships between grains and the whisker are possible leading to an improved understanding of whisker growth.

30.6 Summary

The combination of FIB and SEM is now an established and important technique for materials and biological sample preparation and has enable precise site specific samples to be produced. LMIS sources (mostly Ga) and plasma sources (mostly Xe) have been developed. The LMIS-equipped FIB tools produce much smaller probes that allow more precise sectioning due to a smaller probe size with higher current densities while the plasma FIB tools are finding application where large amounts of material need to be removed efficiently. The applications of FIB include sample preparation for imaging with electrons and ions and for a variety of analytical techniques including EBSD and EDS.

trench that is cut is sufficiently wide to prevent the accumulation of redeposited material (material sputtered from the sample will often fill in the sides of the trench) from obscuring the region of interest. Once the initial trench has been prepared, the FIB/SEM can be set in automatic mode to proceed with the milling and imaging operations. This method is best used for imaging modes of operation (backscatter or secondary electron imaging) only as the access to the milled sample surface is limited. The resulting take-off angle for EDS in this mode is often sub-optimal, although good EDS spectrum imaging results have been obtained in this manner (Kotula et al. 2006).

 Figure 30.14 is an example of the first method of serial sectioning where a volume of interest is imaged in the center of a sample. The sample is an electroplated coating on a substrate. The serial sectioning was accomplished by sequentially milling the exposed cross section followed by imaging with secondary electrons with the SEM column. Figure 30.14a contains examples of the "real" images obtained from the slicing and imaging process. The remaining images shown in Fig. 30.14b, c are obtained after the individual slices are

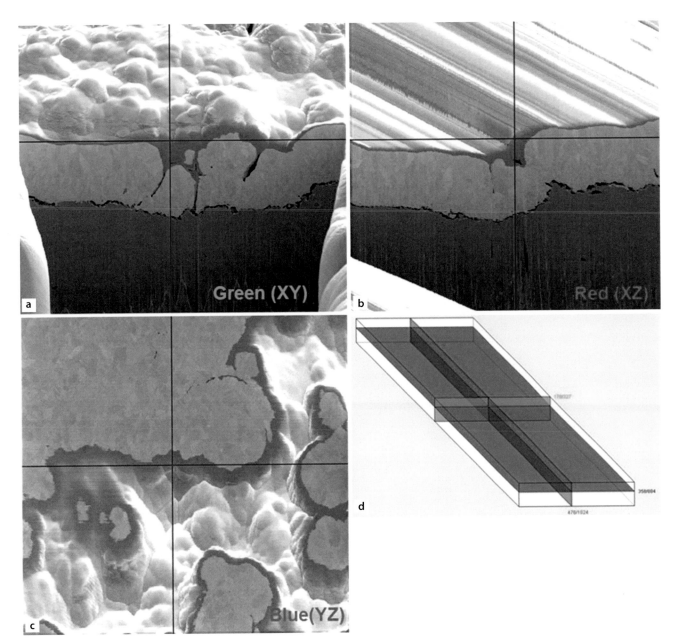

□ **Fig. 30.14** Image reconstruction of a plated stainless steel test coupon. Each section was milled perpendicular to the sample surface. These reconstructions were made from a series of 360 milled slices and required approximately 3 h to collect. The width of the milled area is 20 µm. **a** A secondary electron image of one milled cross section that is the green orientation shown in **d**. This image does not need to be reconstructed as it is the collected data. **b** Reconstructed slice along the red plane shown in **d**. This image is reconstructed once the slice thickness is known. The resolution in this direction is limited by the FIB milled slice thickness. **c** This is a reconstructed image of a slice parallel to the sample surface shown in blue in **d**. **d** Schematic of milled volume

30

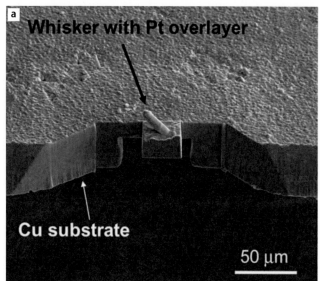

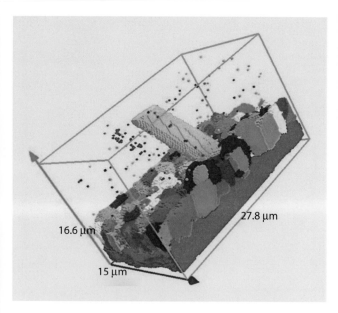

Fig. 30.15 Preparation of a cantilever beam style sample for serial sectioning. **a** The sample before sectioning consist of the tin whisker coated extensively with platinum using first the electron beam and then the ion beam. The cantilever beam was shaped with the FIB and thinned to maximize the speed of cutting. **b** The same beam after serial sectioning. EBSD was performed at every slice. Note the large cross used as a fiducial to align images

Fig. 30.16 EBSD 3D reconstruction of a tin whisker from serial sectioning data in the FIB. The acquisition required 75 200-nm-thick sections and took nearly 48 h to complete sectioning and data acquisition

References

Giannuzzi L, Michael J (2013) Comparison of channeling contrast between ion and electron images. Microsc Microanal 19:344

Giannuzzi L et al (2006) Introduction to focused ion beams: instrumentation, theory, techniques and practice. Springer Science & Business Media, New York

Joy D, Michael J (2014) Modeling ion-solid interactions for imaging applications. MRS Bulletin 39:342

Kotula P, Keenan M, Michael J (2006) Tomographic spectral imaging with multivariate statistical analysis: comprehensive 3D microanalysis. Microsc Microanal 12:36

Mayer J, Giannuzz LA, Kamino T, Michael J (2007) TEM sample preparation and FIB-induced damage. MRS bulletin 32(05):400–407

Michael J (2011) Focused ion beam-induced microstructural alterations: texture development, grain growth, and intermetallic formation. Microsc Microanal 17:386

Nastasi M, Mayer J, Hirvonen J (1996) Ion-solid interactions: fundamentals and applications. Cambridge University Press, Cambridge

Orloff J, Swanson L, Utlaut M (1996) Fundamental limits to imaging resolution for focused ion beams. J Vac Sci Technol B 14:3759

Orloff J, Swanson L, Utlaut M (2003) High resolution focused ion beams: FIB and its applications: Fib and its applications: the physics of liquid metal ion sources and ion optics and their application to focused ion beam technology. Springer Science & Business Media, New York

Prasad S, Michael J, Christenson T (2003) EBSD studies on wear-induced subsurface regions in LIGA nickel. Scr Mater 48:255

Ziegler J, Biersack J (1985) The stopping and range of ions in matter. In: Treatise on heavy-ion science. Springer US, New York, pp 93–129

Ion Beam Microscopy

© Springer Science+Business Media LLC 2018
J. Goldstein et al., *Scanning Electron Microscopy and X-Ray Microanalysis*,
https://doi.org/10.1007/978-1-4939-6676-9_31

31

Electron beams have made possible the development of the versatile, high performance electron microscopes described in the earlier chapters of this book. Techniques for the generation and application of electron beams are now well documented and understood, and a wide variety of images and data can be produced using readily available instruments. While the scanning electron microscope (SEM) is the most widely used tool for high performance imaging and microanalysis, it is not the only option and may not even always be the best instrument to choose to solve a particular problem. In this chapter we will discuss how, by replacing the beam of electrons with a beam of ions, it is possible to produce a high performance microscope which resembles an SEM in many respects and shares some of its capabilities but which also offers additional and important modes of operation.

31.1 What Is So Useful About Ions?

The smallest object or feature that can be imaged by any optical instrument is determined by the diameter "δ" of the smallest spot of illumination that can be directed on to a specimen. As first shown by Fresnel (1817, 1826), this parameter "δ" is found to be

$$\delta = (k\lambda)/\alpha \qquad (31.1)$$

where λ is the wavelength of the emitted beam, k is a constant of the order of unity, and α is the convergence angle of the beam and so cannot exceed $\pi/2$ radians (i.e., 90°). In practice α must usually be chosen to be much smaller than $\pi/2$ in magnitude to minimize the effects of aberrations in the imaging lenses of the microscope. Consequently, an SEM operating in the energy range 10 keV to 30 keV will generally only offer a limiting resolution of the order of 1 nm, even under the very best conditions. Although as noted elsewhere in this volume, further improvements in high resolution SEM can be achieved by advanced electron optical engineering, this requires complex and expensive aberration correction schemes. The constraints imposed on the minimum convergence angle that can be used result in a high resolution SEM image that has inevitable limitations on the achievable depth of field, which is typically only tens of nanometers or worse. Fortunately this performance limitation can now be overcome by using a beam of ions rather than of electrons.

Ions are much more massive than electrons, so at any given energy their wavelength λ is significantly shorter, as shown in ◻ Fig. 31.1. For example, at an energy of 10 keV the wavelength of an electron is 0.012 nm, but for the same energy a hydrogen (H⁺) ion has a wavelength that is smaller than the electron wavelength by a factor of 43. A helium ion

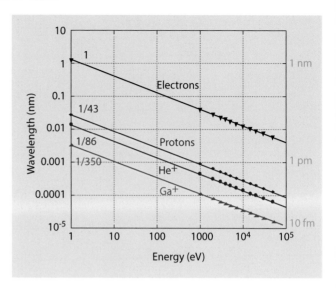

◻ **Fig. 31.1**   Comparison of wavelengths of various particles as a function of beam energy

has a wavelength that is 86× smaller, so potentially offering a beam spot of a few picometers (pm), and making correspondingly enhanced image resolution possible. ◻ Figure 31.2a shows an example of high resolution electron and He⁺ imaging of gold islands on carbon with conditions optimized for SEM (i.e., low beam energy) and HIM (high beam energy). The extraordinary fine-scale detail that can be obtained with HIM imaging of Au on C is shown in ◻ Fig. 31.2b. ◻ Figure 31.3 shows a more challenging imaging problem, that of soft material, as viewed by the SEM in the "conventional" beam energy range and by HIM, where much finer details are visible in the ion beam image.

An additional benefit of ion beam imaging is that because the ion beam wavelengths λ are so short it is possible to significantly reduce the convergence angle α of the ion beam, thus extending the imaging depth of field by a factor of $1/\alpha$, while still offering a much superior resolution. ◻ Figure 31.4a shows an example of the large depth of field that can be achieved simultaneously with wide field-of-view. Unlike high resolution SEM images, which are essentially two-dimensional because of the poor depth of field, high resolution ion beam images can also offer three-dimensional information, as shown in ◻ Fig. 31.4b, where the depth of field is at least 1.5 μm for 1-μm image width. ◻ Figure 31.5 illustrates the value of this combination of high resolution and high depth of field in ion beam imaging by revealing details throughout an image of a complex three-dimensional specimen of human pancreatic cells.

All electrons are the same—but all ions are not, and so many different ion-based microscopes can be configured and optimized for the various tasks. At present the most widely used ion beam for high resolution imaging is helium (He⁺),

◖ **Fig. 31.2**   **a** Gold islands on carbon as viewed by SEM and HIM under optimized conditions for each method: SEM (20-pA beam current at 1 keV); HIM (1-pA beam current at 30 keV). **b** High magnification HIM image of gold on carbon sample with lacey contamination visible. Field of View is 200 nm (Image acquired by Shawn McVey using the ORION Plus HIM) (Image courtesy of Carl Zeiss)

◖ **Fig. 31.3**   SEM and HIM images of soft material (polymer fibers)

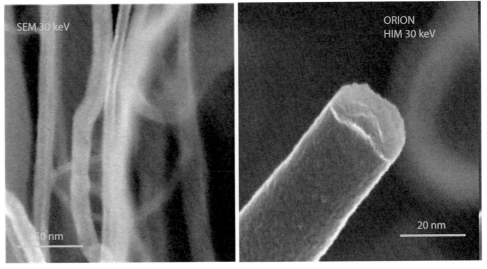

Electrons   He ions-solid surface

31

■ **Fig. 31.4** **a** Image of two tungsten wires wetted by gallium. In this large field of view, the depth of field is over 800 μm, and all features are in focus. (Image acquired by Shawn McVey using the ORION NanoFab) (Image courtesy of Carl Zeiss) (Bar = 100 μm). **b** Illustration of the depth-of-field of helium ion microscopy at high resolution. The depth-of-field is at least 1.5 μm for an image width of 1 μm

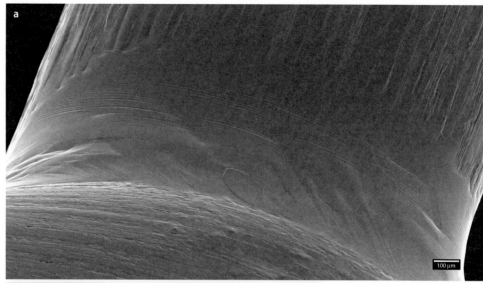

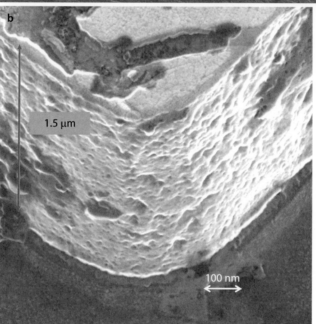

while gallium (Ga⁺) is widely used for ion beam machining of materials. Historically the earliest ion beam microscope (Levi-Setti 1974) used protons (H⁺), while more recently argon (Ar⁺), neon (Ne⁺), and lithium (Li⁺) beams have all also been put into use for nanoscale imaging and fabrication. At present negatively charged ions are much less widely employed than the corresponding positive species, but it can be safely anticipated that they will eventually be employed for special purposes because of the additional versatility that they can offer.

Although in principle any ion could be employed for microscopy, using a heavier ion requires a correspondingly higher accelerating voltage to operate than does a lighter ion, and this heavy ion bombardment may cause significantly more damage to the specimen under examination. Heavy ions also usually have multiple ionization states and these may result in emitted beams whose incident energy is effectively distributed across several values spanning a wide range. This energy spread, in turn, may result in chromatic aberrations in the lenses which may then further degrade the achievable resolution.

In addition, while all energetic ions cause sputtering of the target atoms to some degree and also implant into the

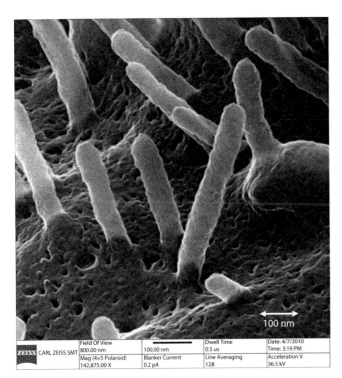

| ZEISS CARL ZEISS SMT | Field Of View 800.00 nm | 100.00 nm | Dwell Time 0.5 us | Date: 4/7/2010 Time: 3:19 PM |
| | Mag (4×5 Polaroid) 142,875.00 X | Blanker Current 0.2 pA | Line Averaging 128 | Acceleration V 36.5 kV |

◻ **Fig. 31.5** High resolution helium ion imaging with high depth-of-field of a soft tissue sample, human pancreatic cells (Sample courtesy of Paul Walther, Univ. of Ulm)

target, which can locally alter the target composition, ions of higher mass, such as Ga⁺ and above, have higher sputtering rates that may substantially alter the target. While acceptable for some tasks such as thinning or machining materials, the level of damage per incident heavy ion is undesirable for imaging purposes because such significant local alteration occurs that the fine spatial details of the specimen are lost before an image representative of the original material can be successfully captured.

31.2 Generating Ion Beams

The approach now most commonly employed for producing a beam of ions for use in microscopy is a development of the method originally employed by Prof. Erwin Muller at Penn State University in the 1940s and 1950s (Muller 1965). Muller's device was a sealed metal cylinder containing helium gas which was cooled to a temperature of just a few degrees Kelvin (K). At one end of the cylinder was a sharply pointed metal needle connected to a power supply capable of supporting a positive voltage of up to a few kilovolts, while at the other end was a fluorescent imaging screen. When neutral helium atoms drifted towards the needle they became

ionized due to the high electric field gradient and acquired a positive potential which then resulted in them being accelerated away from the tip and towards the viewing screen where they formed into an image. On October 11, 1955, Muller and his students were able to demonstrate for the first time ever the direct observation of an atomic structure. Almost 20 years later Professor Riccardo Levi-Setti of the University of Chicago developed a modification of Muller's ion source which allowed it to generate ion beams that could be focused, scanned, and used for imaging in the same way as electrons in a conventional SEM (Levi-Setti 1974). It was this development that became the basis for the source for the present helium ion microscope (HIM).

In present-day ion beam microscopes, the emitter is once again fabricated into the form of a needle, but now the exact size and shape of this tip is very carefully optimized. Using patented, and proprietary, procedures developed by the Zeiss company the emitter tip is shaped and sharpened until it contains just three atoms (Notte et al. 2006). This "trimer" configuration is inherently more stable than any more random arrangement and also ensures that the maximum emitted ion current is directed parallel to the axis of the ion beam. A carefully placed, moveable aperture can then be used to select any one of the three "trimer" ion emission peaks to be used as the beam source for the instrument. The available ion beam current varies with the magnitude of the helium, or other gas, pressure and can reach values as high of several hundred picoamperes.

Low energy ions, i.e., those with less than about 1 MeV of energy, are not significantly affected by magnetic fields so all of the lenses must be electrostatic in type rather than magnetic. The ion beam then travels along the microscope column until it reaches the specimen where its interaction generates ion-induced secondary electrons (iSE) as well as, in some cases, other signals. The ion emitter is adequately stable and provides a bright signal for periods from 5–10 h before the emitter needs to be re-optimized. Once the available beam current becomes too low in intensity, or too unstable to be useful, then the tip must be reformed. This can be done in situ by the operator using an automated procedure which takes some 10–20 min to complete.

Changing from the familiar He⁺ beam to a Ne⁺ beam, or to some other source of emission, requires that any residual gas in the chamber must be first pumped away. The desired new gas of choice can then be injected into the chamber, and the system can be brought back into operation by reforming the "trimer" as described earlier. The usable overall lifetime of these emitters is typically of the order of many months when they are treated with reasonable care and attention.

Although in many ways operating a helium ion beam (HIM) microscope is similar to operating a conventional

31

SEM, to fully optimize the performance of the HIM it is necessary to pay more attention to certain details. The ion source offers just two adjustable controls—the "extractor" which determines the field at the tip and so controls the emission of the source, and the "accelerator" which determines the landing energy of the beam on to the specimen surface. The extractor module "floats" on the top of the potential determined by the accelerator setting, and both the brightness and the stability of the ion source are affected by the extractor module settings. As the extractor potential is increased, the emission current rises before reaching a plateau at the so-called "best imaging voltage" (BIV), which for helium ions occurs at a field strength of about 44 V/nm depending on the geometry of the tip itself. Exceeding the BIV will make the beam less stable, and can result in so much damage to the emitter that it may become necessary to reform the tip again before reliable operation can be restored. The "accelerator" control determines the actual landing energy of the ions on the specimen. For a He$^+$ beam this energy is usually in the range 25–45 keV while for a heavier ion such as Ne$^+$ the corresponding value is more typically in the 20–35-keV range. In either case the yield of secondary electrons increases with the accelerator setting, but continuous operation with the system at energies of 45 keV or higher may cause problems such as insulator breakdowns and discharges. It is therefore good practice to record and save all the experimental parameters likely to be encountered.

31.3   Signal Generation in the HIM

Energetic electrons and ions travel through solid materials while undergoing elastic and inelastic scattering events until they either deposit all of their initial energy and come to a halt, or are re-emitted from a sample surface and subsequently escape while generating secondary electrons and backscattered electrons or ions as they leave. Every such beam particle trajectory is unique and so it is not possible to predict in advance how deep, or how far, any particular incident ion or electron might travel. Examples of Monte Carlo simulations for ion beam trajectories are shown in ◘ Fig. 31.6.

The most probable depth reached by the beam, and the horizontal spread of the beam, can both be estimated using the formula developed by Kanaya and Okayama (1972) which assumes that the range R depends only on the incident energy E of the incident particle and the density ρ of the material through which the beam is traveling. The "K–O" range is then given as

$$R_{\text{K–O}} = \kappa E^p / \rho \qquad (31.2)$$

where $R_{\text{K–O}}$ is the beam range (in nm), ρ is the density of the target material (in g/cm^3), k is a constant depending on the choice of particle, i.e., electrons or ions, and P is a scaling constant. For example, when using a helium ion beam then

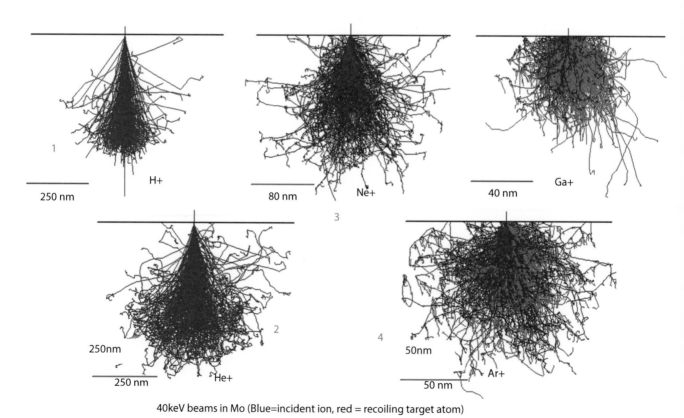

40keV beams in Mo (Blue=incident ion, red = recoiling target atom)

◘ **Fig. 31.6**   Monte Carlo ion beam trajectory simulations for various ion species. $E_0 = 40$ keV

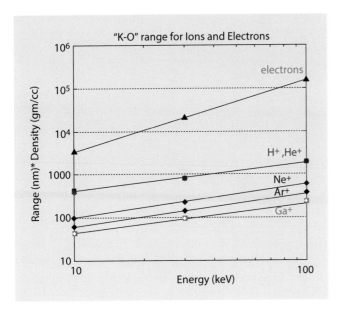

Fig. 31.7 Plot of Kanaya–Okayama range for electrons and various ion species

$\kappa = 80$ nm, and $P = 0.72$, so the range versus density plot has the form shown in ◘ Fig. 31.7 which also shows the corresponding $R_{K\text{-}O}$ for electrons. In the energy range most likely to be of interest, the majority of the emitted secondary signal under electron bombardment from any material of interest most likely comes from the SE_2 component of the signal, i.e., it is generated by backscattered electrons as they exit through the sample surface with reduced resolution compared to the SE_1 component produced by the incident beam. By comparison the beam range of He^+ ions is not only much shorter than the corresponding electron range but also increases more slowly with energy. The iSE yield is typically three to five times larger than the comparable electron-excited SE values and mostly consists of the high resolution SE_1 generated by the incident ion beam with little or no SE_2 component to degrade the resolution. Secondary electrons have often been considered to be of limited value and for modest resolution use only, but recent work (e.g., Zhu et al. 2009) has shown that, to the contrary, the secondary electron signal is a uniquely powerful tool. The SE signal can efficiently capture and display imaging information ranging in scale from millimeters all the way down to single atoms, and even to the subatomic range, providing the incident beam footprint can be made small enough.

As shown by Bethe (1930, 1933) the generation rate $N(SE)$ of secondary electrons by energetic ions or electrons depends on the instantaneous magnitude of the electron stopping power $(-dE/dS)$ of the beam, so

$$N(SE) = -(1/\varepsilon).(dE/ds) \qquad (31.3)$$

where ε is a constant whose value depends on the target material, E is the instantaneous energy of the incident charged particle, and s is the distance travelled along the trajectory. The ion-generated SE yield is always larger than the corresponding SE yield from electrons because ions deposit their energy much more rapidly, and much closer to the surface, than electrons can do. The effect of changing the beam energy on the limiting imaging performance based upon the secondary electron yield is also different in the electron and ion cases. For an SEM operating in the conventional beam energy range above 10 keV, SE image quality does not improve very much with beam energy because the effect of increasing the beam brightness is mostly offset by the fall in the SE yield with increasing energy. For ion beams however, raising the beam energy increases both the yield of ions from the gun and the stopping power of the target which increases the generation rate of the iSE, both of which effects contribute to improved SE imaging performance.

Based on these ideas it is now possible to predict how the emitted yield of secondary electrons for a given material will vary for both ion and electron generation. The SE generation rate at a given depth in the sample varies as the stopping power at that point and which is in turn a function of the velocity of the incoming particle. Secondary electrons which are generated beneath the specimen surface must diffuse back to the surface before they can be detected. The yield of iSE which reach the surface and so could escape is then predicted to be

$$\text{Yield} = 0.5 * \exp(-z / \lambda_{SE}) \qquad (31.4)$$

where λ_{SE} is the appropriate mean free path range for secondary electrons, and z is the distance from the generation point to the nearest exit surface. Detailed predictions of iSE yields can now be made based on this model; see, for example, Ramachandra et al. (2009) and Dapore (2011). The magnitude, and the form, of the iSE yield curves varies with the energy of the incident beam, as shown in ◘ Fig. 31.8 for Si, as well as with the choice of beam and target material. For a helium beam and a carbon target the iSE yield reaches a maximum value of about 4 which is achieved at a He + energy of 750 keV. For a gold target the iSE yield peaks at a value of 6.4 and at an energy of about 1000 keV. The details of these iSE yield curves vary with the choice of both the incident ion and the target material. For example, when using H^+ as the beam of choice the iSE signal reaches its maximum yield of 1.7 at an energy of only about 100 keV, while for an Ar^+ beam the iSE yield reaches its maximum yield, which is in excess of 50, at an energy of 30 MeV. Simulations make it possible to determine how to optimize resolution, maximize contrast, and minimize damage.

31

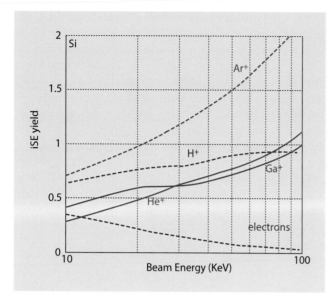

◘ Fig. 31.8 Yield of secondary electrons from Si for electrons and various ions as a function of energy

31.4 Current Generation and Data Collection in the HIM

A typical ion beam microscope operates at energies selected between 10 and 35 keV, and generates an incident beam current of the order of 0.1 pA to >100 pA at the specimen. These ions interact with the specimen of interest generating an iSE secondary electron signal which can be collected by an Everhart–Thornley (ET) detector very similar to that used in the conventional SEM. The signal is then passed through an "analog to digital" (A/D) convertor for real-time display and for storage. Because the incident ion beam currents are quite low, the A/D convertor, which samples the incoming iSE signal at a fixed repetition period of 100 ns, will likely completely miss many of these sparsely generated signal pulses. To overcome this problem, an additional image control labeled as "image intensity" (II) has been added to the familiar SEM brightness and contrast controls. This "image intensity" control allows the operator to choose between (a) the true averaging mode in which N successive A/D conversions are made and the total yield is then divided by N and (b) the true integration mode in which N successive A/D conversions are summed and that value is then reported, or any arbitrary setting between these two extremes. (J. Notte 2015, personal communication). The addition of this illumination control provides considerable additional imaging flexibility while ensuring that the usual "brightness" and "contrast" controls are still able to determine the overall appearance of the image.

In practice, operating the HIM is generally similar to operating an SEM, but the HIM achieves superior image resolution and contrast. The small beam probe diameter and the much enlarged depth of field together produce highly detailed images of even the most complicated three dimensional structures, as shown in ◘ Fig. 31.9 for Ga-ion-beam etched directionally-solidified Al-Cu eutectic alloy. The

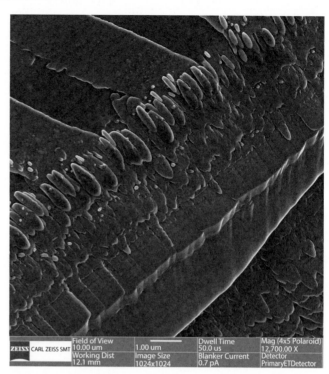

◘ Fig. 31.9 High spatial resolution and high depth of field HIM imaging of Ga-ion-beam etched Al-Cu aligned eutectic alloy. Vertical relief approximately 5 μm. (Bar = 1 μm) (A. Vladar and D. Newbury, NIST)

limited penetration of the ion beam into materials provides highly detailed images of surface, and near-surface features are visible in HIM images that would likely never be evident in a conventional SEM image.

The strategy for operating the HIM is different than that of a conventional SEM because the incident beam energy is generally held fixed at the highest possible energy, typically in the range 30–40 keV, because this simultaneously optimizes both the signal-to-noise ratio and the image resolution. When there is a requirement to examine sub-surface detail this can best be achieved by exploiting the inevitable removal of surface layers by the ion beam as it "rasters" across the sample. Material can then be removed at the rate of a few tens of nanometers per minute, with images stored every few seconds to yield a full three-dimensional reconstruction of the sampled volume.

Charging is an inevitable problem for the HIM. The high SE coefficient of ion beams tends to cause positive charging at the surface, which is further exacerbated by the positive charge injected by the positive He$^+$ ions. The simplest and most reliable technique to control such positive charging is to periodically flood the specimen surface with a very low energy electron beam so as to re-establish charge balance before starting the next scan, although this approach requires care to eliminate both under- and over-compensation. An alternative approach is to inject air into the specimen chamber through a small jet positioned just a short distance away from the desired sample region. This is easy to implement and requires little supervision once an initial charge balance has been achieved.

The HIM operating mode equivalent to the familiar SEM "backscattered electron signal" is "Rutherford backscattered

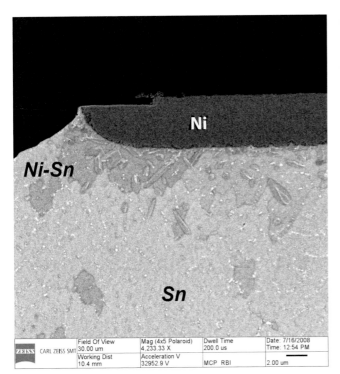

Field Of View 30.00 um | Mag (4x5 Polaroid) 4,233.33 X | Dwell Time 200.0 us | Date: 7/16/2008 Time: 12:54 PM
Working Dist 10.4 mm | Acceleration V 32952.9 V | MCP RBI | 2.00 um
ZEISS CARL ZEISS SMT

Fig. 31.10 Reaction zone at Sn-Ni interface as imaged with ORION Plus HIM from backscattered helium striking the annular microchannel plate. Field of view is 30 µm. Material composition is easily distinguished by atomic number contrast (Bar = 2 µm) (Sample provided by F. Altmann of the Fraunhofer Institute for Mechanics of Materials, Halle, Germany; Image courtesy of Carl Zeiss)

Field Of View 3.00 um | 200.00 nm | Dwell Time 1.0 us | Date: 10/4/2011 Time: 1:38 PM
Working Dist 8.1 mm | Blanker Current 8.5 pA | Frame Averaging 32 | Acceleration V 30.0 kV
ZEISS

Fig. 31.11 Ion channeling contrast observed in deposited copper showing the distinct crystal grains. The field of view is 3 µm, and the incident beam was helium at $E_0 = 30$ keV (Bar = 200 nm) (Imaged with the E–T detector using the ORION Plus, courtesy of Carl Zeiss)

imaging (RBI)" (Rutherford 1911). Some fraction of the incident ion beam is deflected though an angle greater than 90° as a result of high angle elastic scattering events. This elastic scattering generates a signal carrying information about the crystalline geometry and the orientation of the specimen. An example of the Ni-Sn reaction zone at an interface between Ni and Sn to form intermetallic compounds as imaged with backscattered He ions is shown in ◘ Fig. 31.10. The RBI signal increases in magnitude as a function of the atomic number of the target material, but is also affected by a periodic variation of the RBI signal in which the signal falls to a much lower intensity whenever the outer shell of electrons is filled. This behavior occurs at the atomic numbers 2, 8, 18, and son on, and so rather restricts the quantitative utility of this mode as an analytical tool.

When ions travel through a crystalline material they may experience "channeling contrast," which is a variation in the signal level that is dependent on the orientation of any currently placed crystalline target material to the incident beam. When using a helium ion beam, signal variations of 40 % or more in magnitude can be obtained in this way so producing highly visible maps of sample crystallography, as shown in ◘ Fig. 31.11 for a polycrystalline copper sample. A similar effect also occurs for electrons but the corresponding contrast is an order-of-magnitude smaller at approximately 2–5 %. The utility of this mode of operation for analytical purposes is enhanced by the fact that the convergence angle

of the ion beam itself is also one to two orders of magnitude smaller than for an electron beam making it readily possible to retrieve crystal orientation data from regions at the bottom of trenches and holes while still having adequate signal intensity with which to work.

31.5 Patterning with Ion Beams

When ions strike a target they always remove a finite amount of material from the top surface by the process called "sputtering," which involves ion-atom collisions that dislodge atoms and transfer sufficient kinetic energy so that a small fraction escape the sample. While this is sometimes a problem when imaging is the main application of a microscope, the ability of ion beams to remove material in a controlled fashion from a surface is of considerable and increasing value. At its simplest this capability can be applied to generate simple structures such as arrays of holes, or to pattern substrates. The typical beam choice for this process is gallium because of its high sputtering coefficient, but residual gallium becomes embedded in the material being machined. Neon is also a useful candidate for the gas as it can remove material at about 60 % of the speed of gallium but without permanently depositing anything into the target material. For the very softest or most fragile materials a helium ion beam can also pattern at a slow but acceptable rate.

An important current use of this technology relies on "Graphene" as the material of choice. Graphene comes in the form of flexible sheets which may be only by a few atom layers in thickness and yet can readily be cut and shaped as required. Graphene provides the basis to fabricate nanoscale semiconductor devices, and varieties of other active structures.

31.6   Operating the HIM

Using a He$^+$, or some other, ion beam for imaging materials is somewhat different from, although not really much more complex than, conventional imaging with electrons. However some planning prior to turning on the ion beam will help by

---

Step 1

Choose the initial ion beam to be employed. If only one beam type is available, for example, Ga$^+$, then this step is not crucial because it should imply that the required operating choices have already been optimized. However, if several different ion beam species are available then a planned strategy becomes necessary. If there is only a limited amount of the sample then the user should start with the softest ion beam available— helium—so as to minimize damage in the near-surface region of the specimen—and image as required. Only when this set of images is completed should a more aggressive beam choice be tried.

Step 2

Select the desired beam energy. The best choice is always the highest voltage available because this will enhance image resolution, and the ion source will be running at its optimum efficiency. Even at modestly high energies, for example 45 keV, the depth of penetration of ions into most specimens will generally be less than that of a conventional SEM operating at just a few hundred electron volts. If the signal-to-noise of the image appears to be good enough to record useful information then it is worth trying to reduce the incident beam current by using a smaller beam current in the column to see if the damage rate of the sample can be reduced while simultaneously increasing the imaging depth of field.

Step 3

Next choose a working distance (WD) from the source to the sample. Obtaining the shortest possible WD is not as important in an ion microscope as it is in a conventional SEM because the aberrations of the ion beam-optical system are much less severe than those in conventional electron optics. Using a longer working distance will allow safer specimen tilting and manipulation. Before proceeding also ensure that the specimen stage and its specimens are not touching anything else, nor obscuring access to the various detectors, probes, and accessories in the column and chamber.

Step 4

If these are the first observations of the sample which is now in the HIM, then try recording images from some expendable region of the specimen at several different magnifications and currents to confirm whether or not ion-induced beam damage is going to be a problem. In any case always start imaging at the lowest possible magnification and work up from there. Working "high to low" is likely to be unsatisfactory because the damage induced at higher magnification will be obvious as the magnification is reduced. Also check the imaged field of view from time to time for signs of damage, or any indications of drift.

Step 5

Maintain a constant check on the incident ion beam current. If this is falling, or if the images are becoming noisy and unstable, then the emitter tip should be reformed before continuing.

---

minimizing beam damage optimizing conditions, and improving specimen throughput.

31.7   Chemical Microanalysis with Ion Beams

The ability of a microscope to image a material while simultaneously performing a chemical microanalysis of the material is of great value in all areas of science and technology. As a result a high percentage of all SEMs are equipped with X-ray detectors for the identification and analysis of specimen composition. However, X-ray generation is only possible when the velocity of the incoming charged particle (electrons or ions) equals or exceeds the velocity of the orbiting electron within the sample. For an incident beam of electrons these two energies will only be numerically equal in value when their velocities are the same. When the kinetic energy of a beam electron matches the ionization (binding) energy of an atomic shell electron, inelastic scattering of the beam electron becomes possible, causing ejection of that atomic electron. But when the incident beam is made of helium ions, which have a mass 7300 times heavier than an electron, then the energy of the ion also has to be increased by a factor of 7300× by accelerating it before atomic ionization becomes possible. X-ray microanalysis with an ion beam is therefore only possible with ion beam energies in the MeV range, which is far above the operating energy of the HIM.

Possibly the most promising approach for chemical analysis using ion beams at present is Time of Flight- Secondary Ion Mass Spectrometry (TOF-SIMS), which directly analyzes the specimen atoms sputtered atoms from the surface by the primary ion beam. In this method the specimen of interest is contained in the sample chamber in the usual way, but on one side of the chamber is a shutter which can be opened or closed at very high speed as required. The ion beam is then switched on for a short period of time, striking the specimen and sputtering the top two or three atomic layers of the specimen, including neutral, negative, and positive fragments and at the same time the shutter is opened. Which fragment of the sample, positive or negative, is to be analyzed is determined by the polarity of the electric field applied in the chamber. The chosen charged particles from the sample drift toward the shutter at which point they are abruptly accelerated by a strong electric field and travel on until they reach the detector which records the time taken from the original electrical pulse. This time interval will depend on the mass as well as the charge of the particle and so by measuring the transit time required to reach the detector, the mass-to-charge ratio of each individual atom cluster can be deduced. Given accurate "time of transit" measurement capability on the system, this mass-to-charge ratio data can provide rapid, reliable, and unique identification of all the atomic and cluster ions ejected during ion bombardment. After a suitable time interval the shutter is again closed, the system is briefly allowed to stabilize, and the operation is then repeated as often as required to achieve adequate statistics.

TOF-SIMS offers some significant advantages over other possible approaches. The incident ion beam can either be held stationary at a point, or scanned across a chosen area as required to produce a chemical map. Each cycle of this process will then physically remove a few atomic layers from the analyzed area, so repeated measurements on the same area can eventually provide a three-dimensional view of the sampled region. Most important of all, this approach to analysis provides detailed information about the actual chemistry of the area selected, identifying the chemistry of the various compounds, clusters, and elements that are present whereas conventional X-ray microanalysis can only identify which elements are present. This technology is presently only available on a few dual-beam Ga^+ "focused ion beam source" (SEM–FIB) instruments, but many groups are now working to extend this capability to the HIM.

References

Bethe H (1930) Theory of the transmission of corpuscular radiation through matter. Ann Phys Leipzig 5:325

Bethe HA (1933) Charged particle energy loss in solids. In: Handbook of physics, vol 24. Springer, Berlin, p 273

Dapor M (2011) Secondary electron emission yield calculation performed using two different Monte Carlo strategies. Nucl Instr Methods Res Phys 269:B269–1671

Everhart T, Thornley R (1960) Wide-band detector for micro-microampere low-energy electron currents. J Sci Instrum 37:246

Fresnel AJ (1817) Memoire sur les modifications que la reflexion imprime ala lumiere polarisse. FO 1:441

Fresnel A (1826) Diffraction of light. Mem Acad Sci Paris 5:339–487

Kanaya K, Okayama S (1972) Penetration and energy-loss theory of electrons in solid targets. J Appl Phys 5:43

Levi-Setti RL (1974) Proton scanning microscopy: feasibility and promise. Scanning Electron Microscopy Part 1:125

Muller EW (1965) Field Ion Microscopy. Science 149:591

Notte J, Hill R, McVey S, Farkas L, Percival R, Ward B (2006) An introduction to helium ion microscopy. Micros. Microanal. 12:126

Ramachandra R, Griffin B, Joy DC (2009) A model of secondary electron imaging in the helium ion scanning microscope. Ultramicroscopy 109:748

Rutherford ER (1911) Radioactive Transformations. Yale University Press, New Haven

Zhu Y, Inada H, Nakamura M, Wall J (2009) Imaging single atoms using secondary electrons with an aberration-corrected electron microscope. Nat Mater 8:808

Supplementary Information

© Springer Science+Business Media LLC 2018
J. Goldstein et al., *Scanning Electron Microscopy and X-Ray Microanalysis*,
https://doi.org/10.1007/978-1-4939-6676-9

Appendix

A Database of Electron–Solid Interactions

Compiled by
David C Joy
EM Facility, University of Tennessee, and
Oak Ridge National Laboratory
Revision # 12–1

For comments, errata, and suggestions please contact me at djoy@utk.edu

Database can be found in chapter 3 on SpringerLink: http://link.springer.com/chapter/10.1007/978-1-4939-6676-9_3.

A Database of Electron–Solid Interactions

David C Joy
EM Facility, University of Tennessee,
Knoxville, TN 37996–0810
and
Oak Ridge National Laboratory,
Oak Ridge, TN 37831–6064

- **Abstract**

A collection of data comprising secondary and backscattered electron yields, measurements of electron stopping powers, and X-ray ionization cross sections has been assembled from published sources. Values are provided for both elements and many compounds, although the quality and quantity of the available data vary widely from one material to another. These compilations provide the basic framework for understanding and interpreting electron beam images in a quantitative way—as is required for example in semiconductor device metrology—and also form a comprehensive source of experimental data for testing analytical and Monte Carlo models of electron beam interactions.

Introduction

The year 1997 marked the one hundredth anniversary of the discovery of the electron. Within a year of that event Starke (1898) in Germany, and Campbell-Swinton (1899) in England, independently showed that electrons were backscattered from solid specimens and so made the first quantitative measurements of the interaction of electrons with material. Over the century since then many dozens of papers have been published that contain information on various aspects of electron–solid interactions. Unfortunately no systematic collections of the results of such investigations appear to be available for any part of the field of electron microscopy and microanalysis. As a result, anyone requiring a specific piece of data—such as the backscattering yield of molybdenum at 15 keV, or the secondary electron yield from gallium arsenide at 3 keV to take two random examples—has no option but to search the literature in the hope of finding a value which must then, in the absence of any other comparable evidence, be taken as correct. What is required is a source which collects and collates all the values available so as to provide the user with not only a value, but some indication as to its likely reliability.

Structure of the Database

The database presented here is an attempt to present as complete a survey as possible of the published results on backscattering yields, secondary electron yields, stopping powers, X-ray ionization cross sections, and fluorescent yields. Computer-aided literature searches have been conducted to try and find all published references in this general area for the period from 1898 to the present day. Clearly no claim can be made as to the completeness of any such search, and it is perhaps to be hoped that some major body of work has been overlooked because, as discussed below, there are otherwise major omissions in the materials available.

The rules for the data included in this collection are simple:

(a) Only experimental results are included. Values that are not specifically indicated by the author(s) as being experimental, or values that are clearly the result of interpolation, extrapolation, or curve fitting, have been expunged.

(b) No attempt has been made to critically assess the accuracy or precision of the data, nor to remove any results on the basis of their presumed quality.

(c) Values have been tabulated primarily for the energy range up to 30 keV, although data points for incident energies up to 100 keV have been included where they are available.

The decision not to engage in any judgment of the quality of any of the sets of results may seem to be a significant drawback to the utility of the database. However, until so much data has been collated for each element or compound that rogue values can infallibly be distinguished and eliminated, there is no criterion on which to reject any particular result. Further it is conceivable that two tabulated values of a given parameter may differ substantially and yet still both be of value. This is because of an inherent contradiction in the nature of the measurements that are being made. A measurement made in a UHV electron scattering machine with *in situ* sample cleaning and baking facilities will naturally be more "reliable" than a measurement made inside a typical scanning electron microscope. But the values recorded in the microscope are more "representative" of the conditions usually employed on a day-to-day basis in an e-beam tool than those obtained in the environment of a specialist instrument. All types of results are, therefore, reported so that users of the database can make their own judgment as to the suitability, or otherwise, of any given piece of information.

The database currently contains several thousand individual values collected from more than 100 published papers and reports spanning the period from 1898 to the present day. Since this is a "work in progress" the compilation is constantly being extended as additional values become available. As far as possible a consistent style of presentation is used so that data for different elements and compounds may readily be compared. All of the available data sets are grouped by element or compound name for each of the major information groups (SE yields, BSE yields, stopping powers, X-ray ionization cross sections). The data is presented in a simple two-column format with the origin of each of the data sets (numbered #1 to #n) indicated by a number in parenthesis referring back to the bibliography.

Backscattered Electrons

The data on the backscattered electron (BSE) yield as a function of the atomic number of the target and of the incident beam energy is of particular importance in Monte Carlo computations because it provides the best test of the scattering models that are used in the simulation. This data is therefore both the starting point for the construction of a Monte Carlo model, and the source of values against which the simulation can be tested. The backscattered electron section contains data for 40 or more elements spread across the periodic table, as well as for a selection of compounds. If Castaing's rule can be assumed to be correct, then the backscattering yield of a compound can be found if the backscattering coefficients and the atomic fraction of the elements that form it are known. Hence a desirable long-term goal is to obtain a complete set of BSE yield curves for elements. At present the BSE section contains information on more than 45 elements, which is barely half of the solid elements in the periodic table, and of this number perhaps only 25 % of the data sets are of the highest quality, so much experimental work remains to be done especially at the lower energies.

Secondary Yields

With the increasing interest in the simulation of secondary electron (SE) line profiles and images, there is a need to have detailed information on secondary electron yields as a function of atomic number and incident beam energy. Secondary electron emission was the subject of intense experimental study for a period of 20 years or more from the early 1930s, resulting in the publication of no less than six full-length books on the topic. This effort did not, however, produce as much experimental data as would have been expected, because the aim of much of the work that was done was to demonstrate that the SE yield versus energy curve followed a "universal law" (Seiler 1984), and to find the parameters describing this curve. As a result the data actually published was usually given in a normalized format that makes it difficult to derive absolute values. The database currently contains

yields for about 40 or elements, and a collection of inorganic compounds and polymers.

The clear discrepancies that often exist between the comparable sets of original yield figures for the same material may be the result of surface contamination, or the result of a different assumption about the appropriate emitted energy range for secondary electrons (usually now taken to be 0–50 eV, although in some early work 0–70 or even 0–100 eV was used). In addition, since many of the materials documented are poorly conducting the effects of charging must also be considered. For example, in studies of the oxides (e.g., Whetten and Laponsky 1957) maximum SE yields of $\delta > 10$ were measured using pulsed electron-beam techniques. Clearly no non-conducting material can sustain this level of emission for any significant period of time, since it will become positively charged and recollect its own secondaries. Similarly at higher energies, where the SE yield $\delta < 1$ and negative charging occurs, the incident beam energy must be corrected for any negative surface potential acquired by the sample to give a correct result (although there is no little evidence in the original papers that this has been done). Consequently, all SE yield results for insulators must be treated with caution unless the provenance of the original data is well documented.

Since there is no sum rule for secondary yields, data must be acquired for every compound of interest over the energy range required, a task which will be a lengthy one unless suitably automated procedures can be developed and applied. In addition it will be necessary to repeat many of the measurements reported here using better techniques before any level of precision and accuracy can be obtained. In summary the SE data is in an even less satisfactory state than that for the BS electrons, even though a wider range of materials is covered, because the quality of much of the data is poor.

Stopping Powers

The stopping power of an electron in a solid, that is, the rate at which the electron transfers its energy to the material through which it is passing, is a quantity of the highest importance for all studies of electron-solid interactions since it determines, among other parameters the electron range (Bethe 1930), the rate of secondary electron production (Bethe 1941), the lateral distribution and the distribution in depth of X-ray production, and the generation and distribution of electron–hole pairs. Despite its importance there is no body of experimental measurements of stopping power at those energies of interest to electron microscopy and microanalysis. Instead stopping powers, and the quantities which depend on them, have been deduced by analyzing measurements of the transmission energy spectrum of MeV-energy β-particles to yield a value for the mean ionization potential I of the specimen (ICRU 1983), and then Bethe's (1930) analytical expression for the stopping power has been invoked to compute the stopping power at the energy of interest. While this procedure is of acceptable accuracy at high energies (>10 keV) it is not reliable at lower energies because some of the interactions included in the value of I (e.g., inner shell ionizations) no longer contribute.

The database contains experimentally determined stopping power curves for a collection of elements and compounds. The method for obtaining this information from electron energy loss spectra has been described elsewhere (Luo et al. 1991). The data is plotted in units of eV/Å as a function of the incident energy in keV. At the high energy end of the profiles the data corresponds closely to values deduced from Bethe's (1930) law and using the I-values from the ICRU tables. At lower energies, however, significant deviations occur as the Bethe model becomes physically unrealistic although good agreement has been found with values computed from a dielectric model of the solids (Ashley et al. 1979).

The stopping power of a compound is the weighted sum of the stopping power of its constituents; thus a key priority for future work should be to complete the set of stopping power profiles for elements rather than to acquire more data on compounds.

X-ray Ionization Cross Sections

Measured values of the X-ray ionization cross sections for various elements and emission lines as a function of incident beam energy are also of great importance in microanalysis. Unfortunately, as a brief study of the graphs included here will show, the amount of data available is small for K-shells, negligible for the L-shells, and all but non-existent for the M-shells and higher. This is the result of pervasive experimental difficulties, in particular, the fact that any measurement couples together the ionization cross section and the fluorescent yield ω. Since, as can be seen from the plots in section 5 of the database, the value of the fluorescent yield ω is poorly known for the L- and M-shells this causes a significant degree in uncertainty in the cross section deduced from this data. A more practically useful approach is, instead, to quote an "X-ray generation" cross section which is the product of the ionization cross section and the fluorescent yield term. Because the fluorescent term is never required separately in X-ray microanalysis this result looses nothing of its generality but is much more robust. Future updates of this database will include results in this format. For completeness section 5 tabulates all the available fluorescent yield data for K-, L-, and M-shells.

Conclusions

This database is a first step toward the goal of providing a comprehensive collection of the parameters which describe electron–solid interactions. In addition to meeting the needs of those working in Monte Carlo modeling, it is hoped that a systematic collection of data such as this may also be of value in experimental electron microscopy. The quality and quantity of the data that has been amassed varies widely from one material, and from one topic, to another, so that while a few elements can be considered as well characterized, the overall situation is poor, especially for materials used in such areas as integrated circuit device fabrication.

Acknowledgments   The original version of the database was supported by the Semiconductor Research Corporation (SRC) under contract 96-LJ-413.001, contract monitor is Dr. D Herr, and by a grant from International SEMATECH.

References

Ashley JC, Tung CJ, and Ritchie RH (1979) Surf Sci 81:409
Bethe HA (1930) Ann Phys 5:325
Bethe HA (1941) Phys Rev 59:940
Bishop H (1966) Ph.D. Thesis University of Cambridge
Campbell-Swinton AA (1899) Proc Roy Soc 64:377
I.C.R.U (1983) "Stopping powers of electrons and positrons", Report #37 to International Committee on Radiation Units (ICRU:Bethesda, MD)
Luo S, Zhang X, Joy DC (1991) Rad. Effects and Defects in Solids 117:235
Seiler H (1984) J Appl Phys 54:R1
Starke H (1898) Ann Phys 66:49
Whetten NR, Laponsky AB (1957) J Appl Phys 28:515

Index

Reference List

1. Hunger HJ, Kuchler L (1979) Phys Stat Sol (a)56:K45
2. Reimer L, Tolkamp C (1980) Scanning 3:35
3. Moncrieff DA, Barker PR (1976) Scanning 1:195
4. Bongeler R, Golla U, Kussens M, Reimer L, Schendler B, Senkel R, Spranck M (1993) Scanning 15:1
5. Shimizu R (1974) J Appl Phys 45:2107
6. Bishop HE (1963) Ph.D. Thesis, Cambridge
7. Philibert J, Weinryb E (1963) Proc 3rd Conf on X-ray optics and microanalysis p163; also Weinryb E, Philibert J (1964) C R Acad Sci 256:4535
8. Heinrich KFJ (1966) Proc 4th Conf. on X-ray Optics and Microanalysis. In: Castaing R et al. (ed). Hermann, Paris, p 159
9. Kanaya K, Ono S (1984) In: Kyser D (ed) Electron beam interactions with solids. SEM Inc., Chicago, p 69–98
10. Dione GF (1973) J Appl Phys 44:5361
11. Knoll M (1935) Z Tech Phys 16:467

12. Bruining H (1942) Die Sekundt-Elektronen Emission Fester Korper. Springer-Verlag, Berlin
13. Neubert G, Rogaschewski S (1980) Phys Stat Sol (a) 59:35
14. Kanter M (1961) Phys Rev 121:1677
15. Drescher H, Reimer L, Seidel M (1970) Z Angew Physik 29:331
16. Sommerkamp P (1970) Z Angew Physik 4:202
17. Czaja W (1966) J Appl Phys 37:4236
18. Whetten NR, Laponsky AB (1957) J Appl Phys 28:515
19. Whetten NR (1964) J Appl Phys 35:3279
20. Dawson PH (1966) J Appl Phys 37:3644. (Note that this SE data was taken for energy cutoffs varying from 30eV to 100eV)
21. Bruining H, De Boer JM (1938) Physica V:17
22. Cosslett VE, Thomas RN (1965) Brit J Appl Phys J Phys D 16:774
23. Koshikawa T, Shimizu R (1973) J Phy D Appl Phys 6:1369
24. Johnson JB, McKay KG (1954) Phys Rev 93:668
25. Dionne GF (1975) J Appl Phys 46:3347
26. Kanaya K, Kawakatsu M (1972) J Phys D Appl Phys 5:1727
27. Hazelton RC, Yadlowsky EJ, Churchill RJ, Parker LW (1981) IEEE Trans on Nucl Sci NS-28:4541
28. Gross B, von Seggern H, Berraissoui A (1983) IEEE Trans on Electrical Insulation EI-22:23
29. Kishimoto Y, Ohshima T, Hashimoto H, Hayashi T (1990) J Appl Polymer Sci 39:2055
30. Matskevich TL (1959) Fiz Tvend Tela Acad Nauk SSSR 1:227
31. Kazantsev AP, Matskevich TL (1965) Sov Phys Sol State 6:1898
32. Willis RF, Skinner RF (1973) Solid State Comm 13:685
33. Gair S, quoted in Burke EA (1980) IEEE Trans. on Nucl Sci NS-27:1760
34. Sternglass EJ (1954) Phys Rev 95:345
35. Palluel P (1947) Compt Rendu 224:1492
36. Clark E (1935) Phys Rev 48:30
37. Tawara H, Harison KG, De Heer FJ (1973) Physica 63:351
38. Hink W, Pashke H (1971) Phys Rev A4:507; also Hink W, Pashke H (1971) Z Phys 244:140
39. Isaacson M (1972) J Chem Phys 56:1813
40. Colliex PC, Jouffrey B (1972) Phil Mag 25:491
41. Egerton RF (1975) Phil Mag 31:199
42. Glupe G, Melhorne W (1971) J Physique Suppl 32 C4 – 40; also Glupe G, Melhorne W (1967) Phys Lett 25A:274
43. Shima K, Nakagawa T, Umetani K, Fikumo T (1981) Phys Rev A24:72
44. Davis DV, Mistry VD, Quarles CA (1972) Phys Lett 38A:169
45. Platton H, Schiwietz G, Nolte G (1985) Phys Lett 107A:83
46. Fitting HJ (1974) Phys Stat Sol (a)26:525
47. Luo S, Zhang X, Joy DC (1991) Rad Eff and Defects in Solids 117:235
48. Sorenson H (1977) J Appl Phys 48:2244
49. Kurrelmeyer B, Hayner LJ (1937) Phys Rev 52:952
50. Kamiya M, Kuwako A, Ishii K, Morita S, Oyamada M (1980) Phys Rev A22:413
51. McDonald SC, Spicer BM (1988) Phys Rev A37:985
52. Hoffman DHH, Brendel C, Genz H, Low W, Muller S, Richter A (1979) Z Phys A293:187; also -Hoffman DHH, Genz H, Low W, Richter A (1978) Phys Lett 65A:304
53. Hink W, Ziegler A (1969) Z Phys 226:222
54. Luo S "Study of Electron-Solid Interactions," PhD Thesis, University of Tennessee May 1994
55. Meyer F, Vrakking JJ (1973) Phys Lett 44A:511; see also Vrakking JJ, Meyer F (1974) Phys Rev A 9:1932
56. Hippler R, Sneed K, McGregor I, Kleinpoppen H (1982) Z Phys A307:83
57. Quarles C, Semaan M (1982) Phys Rev A26:3147
58. Jessenberger J, Hink W (1975) Z Phys A275:331
59. Shima K (1980) Phys Lett 237
60. Pockman LT, Webster DL, Kirkpatrick P, Harworth K (1947) Phys Rev 71:330
61. Myers HP (1952) Proc Roy Soc 215:329
62. Hubner H, Ilgen K, Hoffman KW (1972) Z Phys255:269
63. Kiss K, Kalman G, Palinkas J, Schlank B (1981) Acta Phys Hung 50:97
64. Salem SI, Moreland LD (1971) Phys Lett 37A:161
65. Green M (1964) Proc Roy Soc 83:435
66. Hansen H, Flammersfeld A (1966) Nucl Phys 79:134
67. Smick AE, Kirkpatrick P (1945) Phys Rev 67:153
68. Wittry DB (1966) Proc. 4th Conf. on X-ray optics and microanalysis. In: Castaing R et al. (ed). Hermann, Paris, p 168
69. Rothwell TE, Russell PE (1988) In: Newbury DE (ed) Microbeam analysis—1988. San Francisco, San Francisco Press, p 149
70. Whetten NR (1962) In: Methods in experimental physics. NY, Academic Press, p IV
71. Green M, Cosslett VE (1968) Brit J Appl Phys (J Phys D) 1:425
72. Sudarshan TS (1976) IEEE Trans. on Electrical Insulation EI-11:32
73. LaVerne JA, Mozumder A (1985) J Phys Chem 89:4219
74. Al-Ahmad KO, Watt DE (1983) J Phys D Appl Phys 16:2257
75. Ishigure N, Mori C, Watanabe T (1978) J Phys Soc Japan 44:973
76. Garber FW, Nakai MY, Harter JA, Birkhoff RD (1971) J Appl Phys 42:1149
77. Luo S (1994), Ph.D Thesis University of Tennessee
78. Kalil F, Stone WG, Hubell HH, Birkhoff RD (1959) ORNL Report 2731
79. Arifov UA, Khadzhimukhamedov Kh. Kh. Makhmudov (1971) In: Secondary emission and structural properties of solids. In: Arifov UA (ed) Consultants Bureau, New York, p 30
80. Arifov UA, Kasymov A Kh (1971) In: Arifov UA (ed) Secondary emission and structural properties of solids. Consultants Bureau, New York, p 78
81. Septier A, Belgarovi M (1985) IEEE Trans. on Electrical Insulators EI-20:725
82. Ahearn AJ (1931) Phys Rev 38:1858
83. Flinn EA, Salehi M (1980) J Appl Phys 51:3441
84. Flinn EA, Salehi M (1979) J Appl Phys 50:3674
85. Dunn B, Ooka K, Mackenzie JD (1973) J Am Cer Soc 56:9
86. Joy DC, Joy CS, SEMATECH Report TT# 96063130A-TR, August 1996
87. Kaneff W (1960) Annalen der Physik 7:84
88. Petzel B (1960) Annalen der Physik 6:55
89. Hovington P, Joy DC, Gauvin R, Evans N (1995) unpublished data. For a summary see Joy DC, Luo S, Gauvin R, Hovington P, Evans N (1996) Scanning Microscopy 10:653–666
90. Buhl R (1959) Zeit F Phys 155; see also Keller M (1961) Zeit F Phys 164:274
91. Miyake S (1938) Proc Phys Math Soc Japan 20:280
92. Lashkarev VE, Chaban AS (1935) Phys Z Soviet 8:240
93. Westbrook GL, Quarles CA (1987) Nucl Instrum and Methods Phys Res B24-25:196
94. Motz JW, Placious RC (1964) Phys Rev 136:662
95. Broyles C, Thomas D, Haynes S (1953) Phys Rev 89:715
96. Lay H (1934) Z Physik 91:533
97. Frey WF, Johnston RE, Hopkins JI (1958) Phys Rev 113:1057
98. Konstantinov AA, Sazonova TE, Perepelkin VV (1960) Izv Akad Nauk SSSR, Ser Fiz 24:1480; also from Konstantinov et al. (1961) Izv Akad Nauk SSSR Ser Fiz 25:219; Konstantinov et al. (1964) Izv Akad Nauk SSSR Ser Fiz 28:107; Konstantinov et al. (1965) Izv Akad Nauk SSSR Ser Fiz 29:302
99. Gray PR (1956) Phys Rev 101:1306
100. Data taken from miscellaneous references in Fink RW, Jopson RC, Mark H, Swift CD (1966) Rev Mod Phys 38:513
101. Data taken from miscellaneous references in Bambynek W, Crasemann B, Fink RW, Freund HU, Mark H, Swift CD, Price RE, Rao PV (1972) Rev Mod Phys 44:716
102. Jopson RC, Mark H, Swift CD (1962) Phys Rev 128:2671
103. Sidhu NPS, Grewal BS, Sahota HS (1988) Xray spectrometry 17:29
104. Fischer B, Hoffmann KW (1967) Z fur Physik 204:122
105. Henrich VE, Fan JCC (1974) J App Phys 45:5484
106. Bronstein IM, Fraiman BS (1969) In: Vtorichnaya Elektronnaya Emissiya. Nauka, Moskva, p 340
107. El Gomati MM, Assad AMD (1997) Proc. 5th European Workshop on Microbeam Analysis (in press)
108. Rossouw CJ, Whelan MJ (1979) J Phys D Appl Phys 12:797
109. Allawadhi KL, Sood BS, Raj Mittal, Singh N, Sharma JK (1996) X-ray Spectr 25:233

110. Horakeri LD, Hanumaiah B, Thontadarya SR (1997) X-ray Spectr 26:69

111. Mearini GT, Krainsky IL, Dayton JA, Wang X, Zorman CA, Angus JC, Hoffman DF, Anderson DF (1995) Appl Phys Lett 66:242

112. Thomas S, Pattinson EB (1969) Brit J Appl Phys II-2:1539

113. He FQ, Long X, Peng X, Luo Z, An Z (1996) Nucl Inst Meth Phys v 114: 213

114. He FQ, Long X, Peng X, Luo Z, An Z (1997) Nucl Inst Meth Phys v 129: 445, v 6: 697

115. Suszcynsky DM, Borovsky JE, Goertz CK (1992) J Geophysical Res E2 97:2611–2619

116. Horakeri LD, Hanumaiah, Thontadarya SR (1998) X-ray Spectrometry 27:344

117. An Z, Li TH, Wang LM, Xia XY, Luo ZM (1996) Phys Rev 54:3067

118. Luo Z, An Z, Li T, Wang LM, Zhu Q, Xia X (1997) J Phys B At Mol Opt Phys 30:2681

119. Luo Z, An Z, He F, Li T, Long X, Peng X (1996) J Phys B At Mol Opt Phys 29:4001

120. He FQ, Peng XF, Long XG, Luo ZM, An Z (1997) Nucl Instrum and Methods in Physical Research B129:445

121. Peng X, He F, Long XG, Luo ZM, An Z (1998) Phys Rev A58:2034

122. Thiel BL, Stokes DJ, Phifer D (1999) Secondary electron yield curve for liquid water, microscopy and microanalysis 5(Suppl 2)(Proc. Microscopy & Microanalysis 99):282–283

123. Llovet X, Merlet C, Salvat F (2000) J Phys B At Mol Opt Phys 33:3761–3772

124. Schneider H, Tobehn I, Ebel F, Hippler R (1993) Phys Rev Lett 71:2707–22709

125. An Z, Tang CH, Zhou CG, Luo ZM (2000) J Phys B At Mol Opt Phys 33:3677–3684

126. Llovet X, Merlet C, Salvat F (2001) J Phys B At Mol Opt Phys 35:973–982

127. Hilleret N, Bojko J, Grobner O, Henrist B, Scheuerlein C, Taborelli M (2000) "The secondary electron yield of technical materials and its Variation with surface treatments" in Proc.7th European Particle Accelerator Conference, Vienna

128. Campos CS, Vasconcellos MAZ, Llovet X, Salvat F (2002) Measurements of L-shell X-ray production cross sections of W, Pt, and Au by 10–30 keV electrons. Phys Rev A, 66

129. Prasad MS (2002) personal communication

130. Ushio Y, Banno T, Matuda N, Baba S, Kinbara A (1988) Thin Solid Films 167:299

131. Lin Y (2003) personal communication

132. Sim K-S, Kamel N (2003) Phys Lett A 317: 378

133. Sorenson H, Schou J (1982) J Appl Phys 53:5230

134. Pinato CA, Hessel R (2000) J Appl Phys 88:478

135. Martsinovskaya EG (1965) Soviet Phys Solid State 7:662

136. Yang KY, Hoffman RW (1987) Surf Interf Anal 10:121

137. Dawson PH (1966) J Appl Phys 36:3644

138. Merlet C, Llovet X, Salvat F (2008) Phys Rev A 78:022704

139. Merlet C, Llovet X, Fernandez-Varea JM (2006) Phys Rev A 73:062719

140. Merlet C, Llovet X, Salvat F (2004) Phys Rev A69:032708

Index

Printed by Books on Demand, Germany